Lecture Notes in Mathematics

Edited by J.-M. Morel, F. Takens and B. Teissier

Editorial Policy for Multi-Author Publications: Summer Schools / Intensive Courses

1. Lecture Notes aim to report new developments in all areas of mathematics and their applications – quickly, informally and at a high level. Mathematical texts analysing new developments in modelling and numerical simulation are welcome. Manuscripts should be reasonably self-contained and rounded off. Thus they may, and often will, present not only results of the author but also related work by other people. They should provide sufficient motivation, examples and applications. There should also be an introduction making the text comprehensible to a wider audience. This clearly distinguishes Lecture Notes from journal articles or technical reports which normally are very concise. Articles intended for a journal but too long to be accepted by most journals, usually do not have this "lecture notes" character.

2. In general SUMMER SCHOOLS and other similar INTENSIVE COURSES are held to present mathematical topics that are close to the frontiers of recent research to an audience at the beginning or intermediate graduate level, who may want to continue with this area of work, for a thesis or later. This makes demands on the didactic aspects of the presentation. Because the subjects of such schools are advanced, there often exists no textbook, and so ideally, the publication resulting from such a school could be a first approximation to such a textbook.
   Usually several authors are involved in the writing, so it is not always simple to obtain a unified approach to the presentation.
   For prospective publication in LNM, the resulting manuscript should not be just a collection of course notes, each of which has been developed by an individual author with little or no co-ordination with the others, and with little or no common concept. The subject matter should dictate the structure of the book, and the authorship of each part or chapter should take secondary importance. Of course the choice of authors is crucial to the quality of the material at the school and in the book, and the intention here is not to belittle their impact, but simply to say that the book should be planned to be written by these authors jointly, and not just assembled as a result of what these authors happen to submit.
   This represents considerable preparatory work (as it is imperative to ensure that the authors know these criteria before they invest work on a manuscript), and also considerable editing work afterwards, to get the book into final shape. Still it is the form that holds the most promise of a successful book that will be used by its intended audience, rather than yet another volume of proceedings for the library shelf.

3. Manuscripts should be submitted (preferably in duplicate) either to Springer's mathematics editorial in Heidelberg, or to one of the series editors (with a copy to Springer). Volume editors are expected to arrange for the refereeing, to the usual scientific standards, of the individual contributions. If the resulting reports can be forwarded to us (series editors or Springer) this is very helpful. If no reports are forwarded or if other questions remain unclear in respect of homogeneity etc, the series editors may wish to consult external referees for an overall evaluation of the volume. A final decision to publish can be made only on the basis of the complete manuscript, however a preliminary decision can be based on a pre-final or incomplete manuscript. The strict minimum amount of material that will be considered should include a detailed outline describing the planned contents of each chapter.
   Volume editors and authors should be aware that incomplete or insufficiently close to final manuscripts almost always result in longer evaluation times. They should also be aware that parallel submission of their manuscript to another publisher while under consideration for LNM will in general lead to immediate rejection.

Continued on inside back-cover

Lecture Notes in Mathematics 1855

Editors:
J.-M. Morel, Cachan
F. Takens, Groningen
B. Teissier, Paris

G. Da Prato   P. C. Kunstmann   I. Lasiecka
A. Lunardi   R. Schnaubelt   L. Weis

Functional Analytic Methods for Evolution Equations

Editors: M. Iannelli
         R. Nagel
         S. Piazzera

 Springer

Authors

Giuseppe Da Prato
Scuola Normale Superiore
Piazza dei Cavalieri 7
56126 Pisa, Italy
e-mail: g.daprato@sns.it

Peer C. Kunstmann
Lutz Weis
Mathematisches Institut I
Universität Karlsruhe
Englerstrasse 2
76128 Karlsruhe, Germany
e-mail: peer.kunstmann@math.uni-karlsruhe.de
lutz.weis@math.uni-karlsruhe.de

Irena Lasiecka
Department of Mathematics
University of Virginia
Kerchof Hall, P.O. Box 400137
Charlottesville
VA 22904-4137, U.S.A.
e-mail: il2v@weyl.math.virginia.edu

Alessandra Lunardi
Department of Mathematics
University of Parma
via D'Azeglio 85/A
43100 Parma, Italy
e-mail: alessandra.lunardi@unipr.it

Roland Schnaubelt
FB Mathematik und Informatik
Martin-Luther Universität
Theodor-Lieser-Str. 5
6099 Halle, Germany
e-mail: schnaubelt@mathematik.uni-halle.de

Editors

Mimmo Iannelli
Department of Mathematics
University of Trento
via Sommarive 14
38050 Trento, Italy
e-mail: iannelli@science.unitn.it

Rainer Nagel
Susanna Piazzera
Mathematisches Institut
Universität Tübingen
Auf der Morgenstelle 10
72076 Tübingen, Germany
e-mail: rana@fa.uni-tuebingen.de
supi@fa.uni-tuebingen.de

Library of Congress Control Number: 200411249

Mathematics Subject Classification (2000):
34Gxx, 34K30, 35K90, 42A45, 47Axx, 47D06, 47D07, 49J20, 60J25, 93B28

ISSN 0075-8434
ISBN 3-540-23030-0 Springer Berlin Heidelberg New York
DOI: 10.1007/b100449

Springer is a part of Springer Science + Business Media
springeronline.com
© Springer-Verlag Berlin Heidelberg 2004
Printed in Germany

Typesetting: Camera-ready TeX output by the authors

41/3142/du - 543210 - Printed on acid-free paper

Preface

Evolution equations describe time dependent processes as they occur in physics, biology, economy or other sciences. Mathematically, they appear in quite different forms, e.g., as parabolic or hyperbolic partial differential equations, as integrodifferential equations, as delay or difference differential equations or more general functional differential equations. While each class of equations has its own well established theory with specific and sophisticated methods, the need for a unifying view becomes more and more urgent.

To this purpose functional analytic methods have been applied in recent years with increasing success. In particular the concepts of *Abstract Cauchy Problems* and of *Operator Semigroups* on Banach spaces allow a systematic treatment of general evolution equations preparing the ground for a better theory even for special equations.

It is the aim of this volume to make this evident. Five contributions by leading experts present recent research on functional analytic aspects of evolution equations.

In the first contribution, **Giuseppe Da Prato** from the Scuola Normale Superiore in Pisa (Italy) gives an introduction to stochastic processes on infinite dimensional spaces. His approach is based on the concept of Markov semigroups and does not require familiarity with probability theory. The main emphasis is on important qualitative properties of these semigroups and the Ornstein-Uhlenbeck semigroup serves as his major example.

In the second contribution, which is by far the longest, **Peer Kunstmann** and **Lutz Weis** (both from the University of Karlsruhe in Germany) discuss the (maximal) regularity of the solutions of inhomogeneous parabolic Cauchy problems.

Regularity properties are fundamental for a theory of nonlinear parabolic equations. Since 1998 this theory has made enormous progress with spectacular breakthroughs based on new Fourier multiplier theorems with operator-valued functions and "square function estimates" for the holomorphic H^∞-functional calculus (some of these results are due to the authors). This contribution is a unifying and accessible presentation of this theory and some of its applications.

Control theoretic aspects of evolution equations in finite dimensions have been studied for a long time. Thanks to functional analytic tools there is now a well established infinite dimensional theory described in some recent monographs. However, this theory does not cover so-called boundary and point control problems. **Irena Lasiecka** from the University of Virginia (USA) developed (mostly with Roberto Triggiani) a systematic approach to these problems using a beautiful combination of abstract semigroups methods and sharp PDE estimates. In her contribution she explains this approach and discusses illustrating examples such as systems of coupled wave, plate and heat equations.

While in the previous contributions the focus is on the dynamics of the state variable, **Alessandra Lunardi** (University of Parma, Italy) studies moving boundary problems. In order to explain these highly nonlinear problems, she concentrates on the heat equation on a moving domain, first in one and then in higher dimensions. Her work is intended as an introduction to an important new field, to very recent results, and to interesting open problems.

Roland Schnaubelt (University of Halle, Germany) shows in the last contribution how nonautonomous linear evolution equations can be studied by a reduction to an autonomous problem to which semigroup methods apply. In particular, the well developed spectral theory for semigroups allows a systematic characterisation of, e.g., exponential dichotomy of the solutions. He then applies these results to obtain qualitative properties of the solutions to nonlinear equations.

These contributions were the basis of lectures given at the Autumn School on "Evolution Equations and Semigroups" at Levico Terme (Trento, Italy) from October 28 to November 2, 2001, within the program of the CIRM (Centro Internazionale per la Ricerca Matematica). We thank Professor Mario Miranda for the support provided to the School. Thanks are also due to the speakers for their cooperation and the permission to collect the expanded notes of their lectures. We hope that this volume will be valuable for beginners as well as for experts in evolution equations.

Trento and Tübingen,   *Mimmo Iannelli*
March, 2004   *Rainer Nagel*
                                                              Susanna Piazzera

Contents

An Introduction to Markov Semigroups

Lectures held at the Autumn School in Levico Terme (Trento, Italy), October 28-November 2, 2001

Giuseppe Da Prato

Scuola Normale Superiore, Piazza dei Cavalieri 7, 56126 Pisa, Italy.
g.daprato@sns.it

Preface

This paper contains the notes of a short course on Markov semigroups. The main aim was to give an introduction to some important properties as: ergodicity, irreducibility, strong Feller property, invariant measures, relevant to some important Markov semigroups arising in infinite dimensional analysis and in stochastic dynamical systems. We have considered in particular the heat semigroup in infinite dimensions, the Ornstein–Uhlenbeck semigroup, the transition semigroup of a one dimensional dynamical system perturbed by noise.

The lectures were designed for an audience having a basic knowledge of functional analysis and measure theory but not familiar with probability. An effort has been done in order to make the lectures as self–contained as possible. In this spirit, the first part was devoted to collect some basic properties of Gaussian measures in Hilbert spaces including the reproducing kernel and the Cameron–Martin formula, a tool that was systematically employed.

Several concepts and results contained in this course are taken from the the notes [3] and the monographs [4], [5], [6].

1 Gaussian Measures in Hilbert Spaces

In all this section H represents a separable Hilbert space, (inner product $\langle \cdot, \cdot \rangle$, norm $|\cdot|$), and $L(H)$ the set of all linear continuous operators from H into H. We denote by $\Sigma(H)$ the subset of $L(H)$ consisting of all symmetric operators and we set

$$L^+(H) = \{T \in \Sigma(H) : \langle Tx, x \rangle \geq 0, \quad x, y \in H\}.$$

An important role will be played by symmetric nonnegative *trace class* operators,

$$L_1^+(H) = \{T \in L^+(H) : \operatorname{Tr} T < +\infty\},$$

where

$$\operatorname{Tr} T = \sum_{k=1}^{\infty} \langle Te_k, e_k \rangle$$

and (e_k) is a complete orthonormal system in H.

We recall that $\operatorname{Tr} T$ does not depend on the choice of the orthonormal system (e_k). Moreover if T is of trace class then it is compact and $\operatorname{Tr} T$ is the sum of its eigenvalues repeated according to their multiplicity, see e. g. [18].

Finally we shall denote by $\mathcal{B}(H)$ the σ-algebra generated by all open (or closed) subsets of H.

1.1 Measures in Hilbert Spaces

Let μ be a probability measure on $(H, \mathcal{B}(H))$. Assume that its first moment is finite,

$$\int_H |x| \mu(dx) < +\infty.$$

Then the linear functional $F : H \to \mathbb{R}$ defined as

$$F(h) = \int_H \langle x, h \rangle \mu(dx), \quad h \in H,$$

is continuous since $|F(h)| \le \int_H |x| \mu(dx) |h|, \quad h \in H$. By the Riesz representation theorem there exists $m_u \in H$ such that

$$\langle m_\mu, h \rangle = \int_H \langle x, h \rangle \mu(dx), \quad h \in H.$$

m_μ is called the *mean* of μ. We shall write $m_\mu = \int_H x \mu(dx)$.

Assume now that the second moment of μ is finite,

$$\int_H |x|^2 \mu(dx) < +\infty.$$

Then we can consider the bilinear form $G : H \times H \to \mathbb{R}$ defined as

$$G(h, k) = \int_H \langle h, x - m_\mu \rangle \langle k, x - m_\mu \rangle \mu(dx), \quad h, k \in H.$$

G is continuous since

$$|G(h, k)| \le \int_H |x - m_\mu|^2 \mu(dx) |h| |k|, \quad h, k \in H.$$

Therefore there is a unique $Q_\mu \in L(H)$ such that

$$\langle Q_\mu h, k \rangle = \int_H \langle h, x - m_\mu \rangle \langle k, x - m_\mu \rangle \mu(dx), \quad h, k \in H.$$

Q_μ is called the *covariance* of μ.

Proposition 1.1. $Q_\mu \in L_1^+(H)$ *that is, is symmetric positive and of trace class.*

Proof. Symmetry and positivity of Q_μ are clear. To prove that Q_μ is of trace class fix a complete orthonormal system (e_k) in H and note that, since

$$\langle Q_\mu e_k, e_k \rangle = \int_H |\langle x - m_\mu, e_k \rangle|^2 \mu(dx), \ k \in \mathbb{N},$$

we have, by the monotone convergence theorem and the Parseval identity, that

$$\text{Tr } Q_\mu = \sum_{k=1}^\infty \int_H |\langle x - m_\mu, e_k \rangle|^2 \mu(dx) = \int_H |x - m_\mu|^2 \mu(dx) < +\infty.$$

$\square$

If μ is a probability measure on $(H, \mathcal{B}(H))$ we define its *Fourier transform* by setting

$$\hat{\mu}(h) = \int_H e^{i\langle h, x \rangle} \mu(dx), \quad h \in H.$$

The following result holds, see e.g. [17].

Proposition 1.2. *Let μ and ν be probability measures on $(H, \mathcal{B}(H))$ such that $\hat{\mu}(h) = \hat{\nu}(h)$ for all $h \in H$. Then $\mu = \nu$.*

Let K be another Hilbert space and let $X : H \to K$ be a Borel mapping ([1]). If μ is a probability measure on $(H, \mathcal{B}(H))$ we denote by $\mathcal{L}(X)$ or μ_X the *law* of X. $\mathcal{L}(X)$ is the probability measure on $(K, \mathcal{B}(K))$ defined as

$$\mathcal{L}(X)(I) = \mu_X(I) = \mu(X^{-1}(I)), \quad I \in \mathcal{B}(K).$$

The following formula of change of variables is basic.

Proposition 1.3. *Let $X : H \to K$ be a Borel mapping and $\varphi : K \to \mathbb{R}$ a bounded Borel mapping. Then we have*

$$\int_H \varphi(X(x))\mu(dx) = \int_K \varphi(y)\mu_X(dy). \tag{1.1}$$

---

[1] that is $F \in \mathcal{B}(K) \Rightarrow X^{-1}(F) \in \mathcal{B}(H)$.

Proof. It is enough to prove (1.1) when φ is a simple function, see e.g. [15]. Let $\varphi = \chi_A$ where $A \in \mathcal{B}(K)$ ([2]). Then we have $\varphi(X) = \chi_{X^{-1}(A)}$, and so

$$\int_H \varphi(X(x))\mu(dx) = \mu(X^{-1}(A)) = \mu_X(A).$$

On the other hand we have

$$\int_K \varphi(y)\mu_X(dy) = \int_K \chi_A(y)\mu_X(dy) = \mu_X(A).$$

So, the conclusion follows. $\square$

1.2 Gaussian Measures

Let us first consider Gaussian measures on $\mathbb{R}$. For any $m \in \mathbb{R}$ and $\lambda \geq 0$, we define the Gaussian measure $N_{m,\lambda}$ on $(\mathbb{R}, \mathcal{B}(\mathbb{R}))$ as

$$N_{m,\lambda}(d\xi) = \begin{cases} \frac{1}{\sqrt{2\pi\lambda}} e^{-\frac{(\xi-m)^2}{2\lambda}} d\xi \ \ \text{if } \lambda > 0, \\[2mm] \delta_m(d\xi) \ \ \text{if } \lambda = 0, \end{cases} \tag{1.2}$$

where δ_m is the Dirac measure concentrated at m:

$$\delta_m(I) = \begin{cases} 1 \ \ \text{if } m \in I \\[2mm] 0 \ \ \text{if } m \notin I, \end{cases} \qquad I \in \mathcal{B}(H).$$

The following identities are easy to check.

$$\int_{-\infty}^{+\infty} \xi N_{m,\lambda}(d\xi) = m, \tag{1.3}$$

$$\int_{-\infty}^{+\infty} (\xi - m)^2 N_{m,\lambda}(d\xi) = \lambda, \tag{1.4}$$

$$\int_{-\infty}^{+\infty} e^{i\alpha\xi} N_{m,\lambda}(d\xi) = e^{i\alpha m - \frac{\lambda}{2}\alpha^2}. \tag{1.5}$$

We now consider a general separable Hilbert space H. For any $a \in H$ and any $Q \in L(H)$ we want to define a Gaussian measure $N_{a,Q}$ on $(H, \mathcal{B}(H))$, having mean a, covariance operator Q, and Fourier transform given by

$$\widehat{N_{a,Q}}(h) = \exp\left\{ i\langle a, h\rangle - \frac{1}{2}\langle Qh, h\rangle \right\}, \ h \in H. \tag{1.6}$$

----

[2] For any Borel set A we denote by χ_A the function that holds 1 on A and 0 on the complement of A.

Notice that, in view of Proposition 1.1, the fact that Q is symmetric, positive and of trace class is a necessary requirement.

Let us consider the natural isomorphism γ between H and the Hilbert space ℓ^2 of all sequences (x_k) of real numbers such that

$$\sum_{k=1}^{\infty} |x_k|^2 < +\infty,$$

defined by $\gamma(x) = (x_k)$, $x \in \ell^2$. In all this section we shall identify H with ℓ^2.

Since Q is of trace class it is compact, so there exist a complete orthonormal basis (e_k) on H and a sequence of nonnegative numbers (λ_k) such that

$$Qe_k = \lambda_k e_k, \quad k \in \mathbb{N}.$$

For any $x \in H$ we set $x_k = \langle x, e_k \rangle$, $k \in \mathbb{N}$.

We are going to define $N_{a,Q}$ as a product measure

$$N_{a,Q} = \underset{k=1}{\overset{\infty}{\times}} N_{a_k, \lambda_k}.$$

For this we first recall, following [10, Section 38], some general results about countable products of measures. Let μ_k be a sequence of probability measures on $(\mathbb{R}, \mathcal{B}(\mathbb{R}))$. We shall define a product measure

$$\mu = \underset{k=1}{\overset{\infty}{\times}} \mu_k,$$

on the space $\mathbb{R}^\infty = \underset{k=1}{\overset{\infty}{\times}} \mathbb{R}$, consisting of all sequences of real numbers (endowed with the product topology).

We first define μ on all *cylindrical subsets* $I_{k_1,...k_n;A}$ of $\mathbb{R}^\infty$, where $n, k_1 < ... < k_n$ are positive integers, and $A \in \mathcal{B}(\mathbb{R}^n)$:

$$I_{k_1,...k_n;A} = \{(x_j) \in \mathbb{R}^\infty : (x_{k_1}, ..., x_{k_n}) \in A\}.$$

It is easy to see that the family of all cylindrical subsets of $\mathbb{R}^\infty$ is an algebra, denoted by $\mathcal{C}$ and that μ is additive on $\mathcal{C}$. Moreover the σ–algebra generated by $\mathcal{C}$ coincides with $\mathcal{B}(\mathbb{R}^\infty)$. See e.g. [7, page 9].

We define

$$\mu(I_{k_1,...k_n;A}) = (\mu_{k_1} \times ... \times \mu_{k_n})(A), \quad I_{k_1,...k_n;A} \in \mathcal{C}$$

and show that μ is σ–additive on $\mathcal{C}$. This will imply, see e.g. [10], that μ can be extended to a probability measure on the product σ–algebra $\underset{k=1}{\overset{\infty}{\times}} \mathcal{B}(\mathbb{R}) = \mathcal{B}(\mathbb{R}^\infty)$.

Proposition 1.4. *μ is σ–additive on $\mathcal{C}$ and has a unique extension to a probability measure on $(\mathbb{R}^\infty, \mathcal{B}(\mathbb{R}^\infty))$.*

Proof. To prove σ-additivity of μ it is enough to show continuity of μ at 0. This is equivalent to prove that if (E_j) is a decreasing sequence on $\mathcal{C}$ such that $\mu(E_j) \geq \varepsilon$, $j \in \mathbb{N}$, for some $\varepsilon > 0$, then we have

$$\bigcap_{j=1}^{\infty} E_j \neq \emptyset.$$

To prove this fact, let us consider the *sections* of E_j defined as

$$E_j(\alpha) = \{x \in \mathbb{R}_1^{\infty} : (\alpha, x) \in E_j\}, \ \alpha \in \mathbb{R},$$

where we have used the notation $\mathbb{R}_n^{\infty} = \underset{k=n+1}{\overset{\infty}{\times}} \mathbb{R}$, $n \in \mathbb{N}$. Set

$$F_j^{(1)} = \left\{\alpha \in \mathbb{R} : \mu^{(1)}(E_j(\alpha)) \geq \frac{\varepsilon}{2}\right\}, \ j \in \mathbb{N},$$

where $\mu^{(n)} = \underset{k=n+1}{\overset{\infty}{\times}} \mu_k$, $n \in \mathbb{N}$. Then by the Fubini theorem we have

$$\mu(E_j) = \int_{\mathbb{R}} \mu^{(1)}(E_j(\alpha))\mu_1(d\alpha)$$

$$= \int_{F_j^{(1)}} \mu^{(1)}(E_j(\alpha))\mu_1(d\alpha) + \int_{[F_j^{(1)}]^c} \mu^{(1)}(E(\alpha))\mu_1(d\alpha)$$

$$\leq \mu_1(F_j^{(1)}) + \frac{\varepsilon}{2}.$$

Therefore $\mu_1(F_j^{(1)}) \geq \frac{\varepsilon}{2}$.

Since μ_1 is a probability measure, it is continuous at 0. Therefore, since $(F_j^{(1)})$ is decreasing, there exists $\overline{\alpha_1} \in \mathbb{R}$ such that

$$\mu^1(E_j(\overline{\alpha_1})) \geq \frac{\varepsilon}{2}, \ j \in \mathbb{N},$$

and consequently we have
$$E_j(\overline{\alpha_1}) \neq \emptyset. \tag{1.7}$$

Now set

$$E_j(\overline{\alpha_1}, \alpha_2) = \{x_2 \in \mathbb{R}_2^{\infty} : (\overline{\alpha_1}, \alpha_2, x) \in E_j\}, \ j \in \mathbb{N}, \ \alpha_2 \in \mathbb{R},$$

and
$$F_j^{(2)} = \left\{\alpha_2 \in \mathbb{R} : \mu^{(2)}(E_j(\alpha)) \geq \frac{\varepsilon}{2}\right\}, \ j \in \mathbb{N}.$$

Then by the Fubini theorem we have

$$\mu^1(E_j(\overline{\alpha_1})) = \int_{\mathbb{R}} \mu^{(2)}(E_j(\overline{\alpha_1}, \alpha_2))\mu_2(d\alpha_2)$$

$$= \int_{F_j^{(2)}} \mu^{(2)}(E_j(\overline{\alpha_1}, \alpha_2))\mu_2(d\alpha_2) + \int_{[F_j^{(2)}]^c} \mu^{(2)}(E_j(\overline{\alpha_1}, \alpha_2))\mu_2(d\alpha_2)$$

$$\leq \mu_2(F_j^{(2)}) + \frac{\varepsilon}{4}.$$

Therefore $\mu_2(F_j^{(2)}) \geq \frac{\varepsilon}{4}$. Since $(F_j^{(2)})$ is decreasing, there exists $\overline{\alpha_2} \in \mathbb{R}$ such that

$$\mu^2(E_j(\overline{\alpha_1}, \overline{\alpha_2})) \geq \frac{\varepsilon}{4}, \quad j \in \mathbb{N},$$

and consequently we have

$$E_j(\overline{\alpha_1}, \overline{\alpha_2}) \neq \emptyset. \tag{1.8}$$

Arguing in a similar way we see that there exists a sequence $(\overline{\alpha_k}) \in \mathbb{R}^\infty$ such that

$$E_j(\overline{\alpha_1}, ..., \overline{\alpha_n}) \neq \emptyset, \tag{1.9}$$

where

$$E_j(\alpha_1, ..., \alpha_n) = \{x \in \mathbb{R}_n^\infty : (\alpha_1, ..., \alpha_n, x) \in E_j\}, \quad n \in \mathbb{N}.$$

This implies, as easily seen, that

$$(\alpha_n) \in \bigcap_{j=1}^\infty E_j.$$

Therefore $\bigcap_{j=1}^\infty E_j$ is not empty as required. Thus we have proved that μ is σ–additive on $\mathcal{C}$. Now the second statement follows since a σ–additive function on an algebra $\mathcal{A}$ can be uniquely extended to a probability measure on the σ–algebra generated by $\mathcal{A}$, see [10]. $\square$

We can now define the Gaussian measure $N_{a,Q}$.

Theorem 1.5. *For any $a \in H$ and any $Q \in L_1^+(H)$ there exists a unique measure μ such that its Fourier transform $\widehat{\mu}$ is given by*

$$\widehat{\mu}(h) = e^{i\langle a,h\rangle}e^{-\frac{1}{2}\langle Qh,h\rangle}, \quad h \in H. \tag{1.10}$$

Moreover,

$$\int_H |x|^2 N_{a,Q}(dx) = \operatorname{Tr} Q + |a|^2, \tag{1.11}$$

$$\int_H \langle x, h\rangle \mu(dx) = a, \quad h \in H, \tag{1.12}$$

$$\int_H \langle x - a, h\rangle\langle x - a, k\rangle\mu(dx) = \langle Qh, k\rangle, \quad h, k \in H. \tag{1.13}$$

We call μ the Gaussian measure in H with mean 0 and covariance operator Q and we set $\mu = N_{a,Q}$.

Proof. Let us consider the product measure on $(\mathbb{R}^\infty, \mathcal{B}(\mathbb{R}^\infty))$,

$$\mu = \underset{k=1}{\overset{\infty}{\times}} N_{a_k, \lambda_k},$$

defined by Proposition 1.4. We claim that the support of μ is included in ℓ^2 that is that $\mu(\ell^2) = 1$ (^{3}). We have in fact, from the monotone convergence theorem

$$\int_{\mathbb{R}^\infty} \sum_{k=1}^\infty x_k^2 \mu(dx) = \sum_{k=1}^\infty \int_\mathbb{R} x_k^2 \mu_k(dx_k) = \sum_{k=1}^\infty (\lambda_k + a_k^2), \qquad (1.14)$$

where $\mu_k = N_{a_k, \lambda_k}$. Therefore

$$\mu\left(\{x \in \mathbb{R}^\infty : |x|_{\ell^2}^2 < \infty\}\right) = 1,$$

as claimed.

Now we define the Gaussian measure $N_{a,Q}$ as the restriction of μ to ℓ^2. To check (1.10) it is useful to introduce a sequence (P_n) of projections in H :

$$P_n x = \sum_{k=1}^n \langle x, e_k \rangle e_k, \quad x \in H, \ n \in \mathbb{N}.$$

Obviously $\lim_{n\to\infty} P_n x = x$, $x \in H$. Consequently, by the dominated convergence theorem we have, recalling (1.5),

$$\int_H e^{i\langle h, x \rangle} \mu(dx) = \lim_{n\to\infty} \int_H e^{i\langle P_n h, P_n x \rangle} \mu(dx) = \lim_{n\to\infty} \prod_{k=1}^n \int_\mathbb{R} e^{ih_n x_n} N_{a_n \lambda_n}(dx)$$

$$= \prod_{k=1}^\infty e^{ia_n h_n} e^{-\frac{1}{2}\lambda_n h_n^2} = e^{i\langle a, h \rangle} e^{-\frac{1}{2}\langle Qh, h \rangle},$$

and so (1.10) is proved. (1.11) follows from (1.14).

Let us prove (1.12). Since $|\langle x, h \rangle| \leq |x| \, |h|$ and $\int_H |x| \mu(dx)$ is finite by (1.11), we have, by the dominated convergence theorem,

$$\int_H \langle x, h \rangle \mu(dx) = \lim_{n\to\infty} \int_H \langle P_n x, h \rangle \mu(dx).$$

But

---

^{3} It is easy to see that ℓ^2 is a Borel subset of $\mathbb{R}^\infty$.

$$\int_H \langle P_n x, h \rangle \mu(dx) = \sum_{k=1}^{n} \int_H x_k h_k \mu(dx)$$

$$= \sum_{k=1}^{n} h_k \int_{\mathbb{R}} x_k N_{a_k, \lambda_k}(dx_k) = \sum_{k=1}^{n} h_k a_k = \langle P_n a, h \rangle \to \langle a, h \rangle,$$

as $n \to \infty$. Thus (1.12) is proved. The proof of (1.13) is similar. $\square$

The last part of this section is devoted to the computation of some Gaussian integrals, which will be often used in what follows. For the sake of simplicity we assume that $\mathrm{Ker}\, Q = \{0\}$ and that (this is not a restriction), $\lambda_1 \geq \lambda_2 \geq \ldots \lambda_n \geq \ldots$

To formulate the next result notice that for any $\varepsilon < \frac{1}{\lambda_1}$, the linear operator $1 - \varepsilon Q$ is invertible and $(1 - \varepsilon Q)^{-1}$ is bounded. We have in fact, as easily checked,

$$(1 - \varepsilon Q)^{-1} x = \sum_{k=1}^{\infty} \frac{1}{1 - \varepsilon \lambda_k} \langle x, e_k \rangle e_k, \quad x \in H.$$

In this case we can define the *determinant* of $(1 - \varepsilon Q)$ by setting

$$\det(1 - \varepsilon Q) = \lim_{n \to \infty} \prod_{k=1}^{n} (1 - \varepsilon \lambda_k) := \prod_{k=1}^{\infty} (1 - \varepsilon \lambda_k).$$

It is easy to see that, in view of the assumption $\sum_{k=1}^{\infty} \lambda_k < +\infty$, the product below is finite and positive.

Proposition 1.6. *Let $\mu = N_{a,Q}$ and $\varepsilon \in \mathbb{R}$. Then we have*

$$\int_H e^{\frac{\varepsilon}{2}|x|^2} \mu(dx) = \begin{cases} [\det(1 - \varepsilon Q)]^{-1/2} e^{-\frac{\varepsilon}{2}\langle(1 - \varepsilon Q)^{-1} a, a \rangle}, & \text{if } \varepsilon < \frac{1}{\lambda_1}, \\[2mm] +\infty, & \text{otherwise.} \end{cases} \tag{1.15}$$

Proof. For any $n \in \mathbb{N}$ we have

$$\int_H e^{\frac{\varepsilon}{2}|P_n x|^2} \mu(dx) = \prod_{k=1}^{n} \int_{\mathbb{R}} e^{\frac{\varepsilon}{2} x_k^2} N_{a_k, \lambda_k}(dx_k).$$

Since $|P_n x|^2 \uparrow |x|^2$ as $n \to \infty$ and, by an elementary computation,

$$\int_{\mathbb{R}} e^{\frac{\varepsilon}{2} x_k^2} N_{a_k, \lambda_k}(dx_k) = \frac{1}{\sqrt{1 - \varepsilon \lambda_k}} \, e^{-\frac{\varepsilon}{2} \frac{a_k^2}{1 - \varepsilon \lambda_k}},$$

the conclusion follows from the monotone converge theorem. $\square$

Exercise 1.7. Compute the integral

$$J_m = \int_H |x|^{2m} \mu(dx), \quad m \in \mathbb{N}.$$

Hint. Notice that $J_m = 2^m F^{(m)}(0)$, where

$$F(\varepsilon) = \int_H e^{\frac{\varepsilon}{2}|x|^2} \mu(dx).$$

Proposition 1.8. *Let* $\mu = N_{a,Q}$. *Then we have*

$$\int_H e^{\langle h,x \rangle} \mu(dx) = e^{\langle a,h \rangle} e^{\frac{1}{2}\langle Qh,h \rangle}, \quad h \in H. \tag{1.16}$$

Proof. For any $\varepsilon > 0$ we have

$$e^{\langle h,x \rangle} \le e^{|x|\,|h|} \le e^{\varepsilon |x|^2}\, e^{\frac{1}{\varepsilon}|x|^2}.$$

Choose $\varepsilon < \frac{1}{\lambda_1}$; then we see by Proposition 1.6 that the function $x \to e^{\langle h,x \rangle}$ is integrable with respect to μ. Consequently, by the dominate convergence theorem, it follows that

$$\int_H e^{\langle h,x \rangle} \mu(dx) = \lim_{n \to \infty} \int_H e^{\langle h, P_n x \rangle} \mu(dx),$$

which yields the conclusion. □

2 Gaussian Random Variables

Let H and K be separable Hilbert spaces and let μ be a probability measure on $(H, \mathcal{B}(H))$. A *random variable* X in H with values in K is a Borel mapping $X : H \to K$, that is

$$I \in \mathcal{B}(K) \Rightarrow X^{-1}(I) \in \mathcal{B}(H).$$

When $K = \mathbb{R}$ we call X a *real random variable*.

X is called *Gaussian* if its law $\mathcal{L}(X) = \mu_X$ is a Gaussian measure on K. We say also that X is a *Gaussian random variable* in H taking values in K.

Proposition 2.1. *Let* $X : H \to K$ *be a random variable on* $(H, \mathcal{B}(H), \mu)$ *such that* $\int_H X(x)\mu(dx) = 0$. *Then the covariance* Q_X *of the law of* X *is given by*

$$\langle Q_X \alpha, \alpha \rangle_K = \int_H \langle X(x), \alpha \rangle_K^2 \mu(dx), \quad \alpha \in K. \tag{2.1}$$

Proof. We have in fact by the change of variables formula (1.1),

$$\langle Q_X \alpha, \alpha \rangle_K = \int_K \langle y, \alpha \rangle_K^2 \mu_X(dy) = \int_H \langle X(x), \alpha \rangle_K^2 \mu(dx).$$

□

2.1 Change of Variables of Gaussian Measures

Proposition 2.2 (Translations). *Let $\mu = N_{a,Q}$ be a Gaussian measure on $(H, \mathcal{B}(H))$ and $b \in H$. Then $X(x) = x + b$ is a random Gaussian variable and $\mu_X = N_{a+b,Q}$.*

Proof. By Proposition 1.2 it is enough to show that

$$\int_H e^{i\langle k,y\rangle} \mu_X(dy) = e^{i\langle a+b,k\rangle} e^{-\frac{1}{2}\langle Qk,k\rangle}, \quad k \in H.$$

We have in fact by the change of variables formula (1.1),

$$\int_H e^{i\langle k,y\rangle} \mu_X(dy) = \int_H e^{i\langle x+b,k\rangle} \mu(dx) = e^{\langle a+b,k\rangle} e^{-\frac{1}{2}\langle Qk,k\rangle}, \quad k \in H.$$

$\square$

Proposition 2.3 (Linear transformations). *Let $\mu = N_{a,Q}$ be a Gaussian measure on $(H, \mathcal{B}(H))$ and $T \in L(H, K)$. Then we have $\mu_T = N_{Ta,TQT^*}$, where T^* is the transpose of T.*

Proof. Again by Proposition 1.2 it is enough to show that

$$\int_K e^{i\langle k,y\rangle_K} \mu_T(dy) = e^{\langle T^*k,a\rangle} e^{-\frac{1}{2}\langle TQT^*k,k\rangle}, \quad k \in K.$$

We have in fact, by the change of variables formula (1.1),

$$\int_K e^{i\langle k,y\rangle_K} \mu_T(dy) = \int_H e^{i\langle k,Tx\rangle_K} \mu(dx)$$

$$= \int_H e^{i\langle T^*k,x\rangle_K} \mu(dx) = e^{\langle T^*k,a\rangle} e^{-\frac{1}{2}\langle TQT^*k,k\rangle}, \quad k \in H.$$

$\square$

Corollary 2.4. *Let $\mu = N_Q$ be a Gaussian measure on $(H, \mathcal{B}(H))$, and $h^1, \ldots, h^k \in H$. Let $X : H \to \mathbb{R}^k$ be defined by*

$$X(x) = \left(\langle x, h^1 \rangle, \ldots, \langle x, h^k \rangle \right), \quad x \in H.$$

Then μ_X is a Gaussian measure $N_{Q'}$ on $\mathbb{R}^k$ with

$$Q'_{i,j} = \langle Qh^i, h^j \rangle, \ i, j = 1, \ldots, k.$$

Proof. By an easy computation we see that the transpose X^* of X is given by

$$X^*(\xi) = \sum_{i=1}^k \xi_i h^i, \quad \xi = (\xi_1, \ldots, \xi_k) \in \mathbb{R}^k.$$

Consequently we have

$$XQX^*(\xi) = \sum_{i=1}^k \xi_i (Qh^i) = \sum_{i=1}^k \xi_i \left(\langle Qh^i, h^1 \rangle, \ldots, \langle Qh^i, h^k \rangle \right),$$

and the conclusion follows from Proposition 2.3. $\square$

2.2 Independent Gaussian Random Variables

Let $(\Omega, \mathcal{F}, \mathbb{P})$ be a probability space, and let $X_1, ..., X_k$ be real random variables. Let us define a mapping $X : H \to \mathbb{R}^k$ by setting

$$X(\omega) = (X_1(\omega), ..., X_k(\omega)), \ \omega \in \Omega.$$

Then X is a random variable with values in $\mathbb{R}^k$.

We say that $X_1, ..., X_k$ are *independent* if

$$\mathbb{P}_X = \underset{j=1}{\overset{k}{\times}} \mathbb{P}_{X_j}.$$

Exercise 2.5. *Let $\varphi_1, ..., \varphi_k$ be real Borel functions. Show that $X_1, ..., X_k$ are independent if and only if*

$$\int_\Omega \varphi_1(X_1) \ldots \varphi_k(X_k) \, d\mathbb{P} = \prod_{j=1}^k \int_\Omega \varphi_j(X_j) \, d\mathbb{P},$$

for any choice of $\varphi_1, \ldots, \varphi_k$.

Exercise 2.6. *Let us consider the probability space $(H, \mathcal{B}(H), \mu)$ where $\mu = N_Q$, and let (e_k) and (λ_k) be such that*

$$Qe_k = \lambda_k e_k, \ k = 1, ..., d.$$

Show that the random variables

$$X_1(x) = \langle x, e_1 \rangle, ..., X_d(x) = \langle x, e_d \rangle,$$

are independent.

Proposition 2.7. *Let $\mu = N_Q$ be a Gaussian measure on $(H, \mathcal{B}(H))$, and $h^1, \ldots, h^k \in H$. Let*

$$X_1(x) = \langle x, h^1 \rangle, ..., X_k(x) = \langle x, h^k \rangle.$$

Then $X_1, ..., X_n$ are independent if and only if

$$\langle Qh^i, h^j \rangle = 0 \quad \text{if } h^i \neq h^j.$$

Proof. By Corollary 2.4 it follows that

$$\mu_{(X_1,...,X_n)} = N_{(Q_{i,j})}, \ Q_{i,j} = \langle Qh^i, h^j \rangle, \ i, j = 1, ..., k,$$

and

$$\mu_{X_j} = N_{\langle Qh^j, h^j \rangle}, \ j = 1, ..., k.$$

Now the conclusion follows easily. $\square$

2.3 The Reproducing Kernel and the White Noise Function

In this section we consider a Gaussian measure $\mu = N_Q$, where $Q \in L_1^+(H)$. We assume for simplicity that Ker $Q = \{0\}$.

We denote by (e_k) a complete orthonormal system on H, such that $Qe_k = \lambda_k e_k$, $k \in H$, where (λ_k) are eigenvalues of Q. We set $x_k = \langle x, e_k \rangle$, $k \in \mathbb{N}$.

Lemma 2.8. *Assume that H is infinite dimensional. Then the range $Q(H)$ of Q is a dense subspace of H different from H, given by*

$$Q(H) = \left\{ y \in H : \sum_{k=1}^{\infty} y_k^2 \lambda_k^{-2} < +\infty \right\}. \tag{2.2}$$

Proof. Let $y \in Q(H)$ and let $x \in H$ such that $Qx = y$. Since $y_k = \lambda_k x_k$, $k \in \mathbb{N}$, we find

$$x = \sum_{k=1}^{\infty} x_k e_k = \sum_{k=1}^{\infty} \frac{y_k e_k}{\lambda_k},$$

and consequently we have $\sum_{k=1}^{\infty} \frac{y_k^2}{\lambda_k^2} < +\infty$. Conversely if $y \in H$ and $\sum_{k=1}^{\infty} \frac{y_k^2}{\lambda_k^2} < +\infty$, setting $x = \sum_{k=1}^{\infty} \frac{y_k e_k}{\lambda_k}$, we have $Qx = y$ so that $y \in Q(H)$. Thus (2.2) is proved. It is clear that $Q(H)$ does not coincide with H. It remains to prove the density of $Q(H)$ in H. Let x_0 be an element of H orthogonal to $Q(H)$. Then we have

$$\langle Qx, x_0 \rangle = \langle x, Qx_0 \rangle = 0, \quad \forall x \in H,$$

that yields $Qx_0 = 0$, and so $x_0 = 0$. This shows that $Q(H)$ is dense in H. □

It is also useful to introduce the operator $Q^{1/2}$ defined as

$$Q^{1/2}x = \sum_{k=1}^{\infty} \sqrt{\lambda_k} \langle x, e_k \rangle e_k, \quad x \in H.$$

Its range $Q^{1/2}(H)$ is called the *reproducing kernel* of the measure N_Q. It will play an important role in the following. Arguing as before we see that $Q^{1/2}(H)$ is a proper subspace of H, given by

$$Q^{1/2}(H) = \left\{ y \in H : \sum_{k=1}^{\infty} y_k^2 \lambda_k^{-1} < +\infty \right\},$$

and that $Q^{1/2}(H)$ is dense in H.

Proposition 2.9. *We have $\mu(Q^{1/2}(H)) = 0$.*

Proof. For any $n, k \in \mathbb{N}$ set

$$U_n = \left\{ y \in H : \sum_{h=1}^{\infty} \frac{y_h^2}{\lambda_h} < n^2 \right\},$$

and

$$U_{n,k} = \left\{ y \in H : \sum_{h=1}^{2k} \frac{y_h^2}{\lambda_h} < n^2 \right\}.$$

Clearly $U_n \uparrow Q^{1/2}(H)$ as $n \to \infty$, and for any $n \in \mathbb{N}$, $U_{n,k} \downarrow U_n$ as $k \to \infty$. So it is enough to show that

$$\mu(U_n) = \lim_{k \to \infty} \mu(U_{n,k}) = 0. \tag{2.3}$$

We have in fact

$$\mu(U_{n,k}) = \int_{\left\{ y \in \mathbb{R}^k : \sum_{h=1}^{2k} \frac{y_h^2}{\lambda_h} < n^2 \right\}} \prod_{h=1}^{2k} N_{\lambda_k}(dy_k),$$

that is equivalent to

$$\mu(U_{n,k}) = \int_{\{y \in \mathbb{R}^{2k} : |y| < n\}} N_{I_{2k}}(dy),$$

where I_{2k} is the identity in $\mathbb{R}^{2k}$. We have $dy = r^{2k-1} d\omega_{2k} dr$, where $d\omega_{2k}$ is the area element of the unitary sphere on $\mathbb{R}^{2k}$. Consequently

$$\mu(U_{n,k}) = \frac{\mu(U_{n,k})}{\mu(H)} = \frac{\int_0^n e^{-\frac{r^2}{2}} r^{2k-1} dr}{\int_0^{+\infty} e^{-\frac{r^2}{2}} r^{2k-1} dr} = \frac{\int_0^{n^2/2} e^{-\rho} \rho^{k-1} d\rho}{\int_0^{+\infty} e^{-\rho} \rho^{k-1} d\rho}.$$

Therefore

$$\mu(U_{n,k}) = \frac{1}{(k-1)!} \int_0^{n^2/2} e^{-\rho} \rho^{k-1} d\rho$$

$$\leq \frac{1}{(k-1)!} \int_0^{n^2/2} \rho^{k-1} d\rho = \frac{1}{k!} \left(\frac{n^2}{2} \right)^k.$$

Now (2.3) follows easily. □

Given $f \in Q^{1/2}(H)$ let us consider the function $W_f \in L^2(H, \mu)$ defined as

$$W_f(x) = \langle Q^{-1/2} f, x \rangle, \ x \in H.$$

It is important to define W_f for any $f \in H$. We could try to set

$$W_f(x) = \langle Q^{-1/2} x, f \rangle, \ x \in Q^{1/2}(H),$$

but this definition would not be useful because, as we have just seen, $\mu(Q^{1/2}(H)) = 0$. However it is possible to define in a natural way W_f as we shall see now. Let us consider the mapping W

$$W : Q^{1/2}(H) \subset H \to L^2(H, \mu), \ f \to W_f,$$

where $W_f(x) = \langle x, Q^{-1/2}f \rangle$, $x \in H$. Then we can extend W to all H since it is an isomorphism. We have in fact by (2.5) for any $f, g \in Q^{1/2}(H)$

$$\int_H W_f(x)W_g(x)\mu(dx) = \langle QQ^{-1/2}f, Q^{-1/2}g \rangle = \langle f, g \rangle. \qquad (2.4)$$

The function W will play an important role in what follows, it is called the *white noise* function.

Often we will write for simplicity

$$W_f(x) = \langle Q^{-1/2}x, f \rangle, \ x, f \in H,$$

but one has to keep in mind the correct meaning of W_f.

The following result shows that e^{W_f} is square integrable for any $f \in H$.

Proposition 2.10. *For any $f \in H$, we have $e^{W_f} \in L^2(H, \mu)$, and*

$$e^{W_f} = \lim_{n \to \infty} e^{W_{f_n}}, \ \text{in } L^2(H, \mu), \qquad (2.5)$$

where $f_n = \sum_{k=1}^n \langle f, e_k \rangle e_k = P_n f$. Moreover

$$\int_H e^{W_f}(x)\mu(dx) = e^{\frac{1}{2}|f|^2}, \ f \in H. \qquad (2.6)$$

Proof. Let us prove that $(e^{W_{f_n}})$ is a Cauchy sequence in $L^2(H, \mu)$. If $n, m \in \mathbb{N}$ we have in fact, taking into account (1.16),

$$\int_H \left[e^{W_{f_n}(x)} - e^{W_{f_m}(x)} \right]^2 \mu(dx)$$

$$= \int_H \left[e^{2W_{f_n}(x)} - 2e^{W_{f_n}(x)+W_{f_m}(x)} + e^{2W_{f_m}(x)} \right] \mu(dx)$$

$$= e^{2|P_n f|^2} - 2e^{\frac{1}{2}|(P_n+P_m)f|^2} + e^{2|P_m f|^2},$$

and the conclusion follows since $P_n f \to f$ as $n \to \infty$. $\square$

Exercise 2.11. *Show that*

$$\int_H e^{iW_f(x)}\mu(dx) = e^{-\frac{1}{2}|f|^2}. \qquad (2.7)$$

We end this section with an important property of the function W.

Proposition 2.12. *Let $n \in \mathbb{N}$, $f_1, ..., f_n \in H$. Then the following statements hold.*

(i) $(W_{f_1}, ..., W_{f_n})$ is a Gaussian random variable in $\mathbb{R}^n$ with mean 0 and covariance operator

$$(Q_{i,j}) = \langle f_i, f_j \rangle, \ i, j = 1, ..., n.$$

(ii) $W_{f_1}, ..., W_{f_n}$ are independent if and only if $(f_1, ..., f_n)$ is an orthogonal system.

Proof. The conclusion follows from the fact that W is an isomorphism, arguing as in the proof of Proposition 2.7. □

2.4 Cameron–Martin Formula

We consider a Gaussian measure $\mu = N_Q$, where $Q \in L_1^+(H)$, such that $\text{Ker } Q = \{0\}$. We denote by (e_k) a complete orthonormal system on H, such that $Qe_k = \lambda_k e_k$, $k \in H$, where (λ_k) are eigenvalues of Q. We set $x_k = \langle x, e_k \rangle$, $k \in \mathbb{N}$. For any $n \in \mathbb{N}$ we consider the application $\Pi_n : H \to \mathbb{R}^n$:

$$\Pi_n(x) = (x_1, ..., x_n), \quad x \in H.$$

Theorem 2.13. *Let $Q \in L_1^+(H)$ and $a \in Q^{1/2}(H)$. Then the measures $N_{a,Q}$ and N_Q are equivalent. Moreover*

$$\frac{dN_{a,Q}}{dN_Q}(x) = d(a, x), \ x \in H, \tag{2.8}$$

where

$$d(a, x) = e^{-\frac{1}{2}|Q^{-1/2}a|^2 + W_{Q^{-1/2}a}(x)} = e^{-\frac{1}{2}|Q^{-1/2}a|^2 + \langle Q^{-1/2}a, Q^{-1/2}x \rangle}, \ x \in H. \tag{2.9}$$

Proof. We have in fact

$$(N_{a,Q})_{\Pi_n} = N_{\Pi_n x_0, \text{diag } (\lambda_1, ..., \lambda_n)}, \quad (N_Q)_{\Pi_n} = N_{0, \text{diag } (\lambda_1, ..., \lambda_n)}.$$

Clearly the measures $(N_{a,Q})_{\Pi_n}$ and $(N_Q)_{\Pi_n}$ are equivalent, because $\lambda_k > 0$ for all $k \in \mathbb{N}$, and we have

$$\frac{d(N_{a,Q})_{\Pi_n}}{d(N_Q)_{\Pi_n}}(\xi) = e^{-\sum_{k=1}^{n} \frac{1}{2} \frac{\alpha_k^2}{2\lambda_k} + \frac{\alpha_k \xi_k}{\lambda_k}} \quad \xi \in \mathbb{R}^n.$$

It follows that

$$\int_{\mathbb{R}^n} \varphi(\xi)(N_{a,Q})_{\Pi_n}(d\xi) = \int_{\mathbb{R}^n} \varphi(\xi) e^{-\sum_{k=1}^{n} \frac{1}{2} \frac{\alpha_k^2}{2\lambda_k} + \frac{\alpha_k \xi_k}{\lambda_k}} (N_Q)_{\Pi_n}(d\xi).$$

So, by the change of variables formula (1.1), we have that

$$\int_H \varphi(\Pi_n x) N_{\alpha,Q}(dx) = \int_H \varphi(x) e^{-\frac{1}{2}|Q_n^{-1/2}\alpha|^2 + \langle Q_n^{-1/2}\alpha, Q_n^{-1/2}x \rangle} N_Q(dx).$$

The conclusion follows from the dominate convergence theorem and Proposition 2.10. □

3 Markov Semigroups

We are given a separable Hilbert space H, (inner product $\langle \cdot, \cdot \rangle$, norm $|\cdot|$). We shall use the following notations:
$C_b(H)$ is the Banach space of all uniformly continuous and bounded mappings $\varphi : H \to \mathbb{R}$ endowed with the norm

$$\|\varphi\|_0 = \sup_{x \in H} |\varphi(x)|.$$

$C_b^*(H)$ is the topological dual of $C_b(H)$.
For any $k \in \mathbb{N}$, $C_b^k(H)$ is the subspace of $C_b(H)$ of all functions $\varphi : H \to \mathbb{R}$ which are k times Fréchet differentiable on H with uniformly continuous and bounded derivatives $D^h\varphi$ with h less than or equal to k. We set

$$[\varphi]_k := \sup_{x \in H} \|D^k\varphi(x)\|, \quad \|\varphi\|_k = \|\varphi\|_0 + [\varphi]_k, \quad \varphi \in C_b^k(H).$$

We set moreover

$$C_b^\infty(H) = \bigcap_{k=1}^{\infty} C_b^k(H).$$

$C_b^+(H)$ is the cone in $C_b(H)$ consisting of all nonnegative functions.
If $\varphi \in C_b(H)$, setting

$$\varphi^+(x) = \frac{1}{2}\left(|\varphi(x)| + \varphi(x)\right), \quad \varphi^-(x) = \frac{1}{2}\left(|\varphi(x)| - \varphi(x)\right), \quad x \in H,$$

we have that $\varphi = \varphi^+ - \varphi^-$ and $\varphi^+, \varphi^- \in C_b^+(H)$.
$\mathbb{1}$ is the function on H identically equal to 1.
$B_b(H)$ is the Banach space of all Borel and bounded mappings $\varphi : H \to \mathbb{R}$ endowed with the norm

$$\|\varphi\|_0 = \sup_{x \in H} |\varphi(x)|.$$

$M(H)$ is the space of all probability measures on $(H, \mathcal{B}(H))$.

There is a natural embedding of $M(H)$ into $C_b^*(H)$. Namely, for any $\mu \in M(H)$ we set

$$F_\mu(\varphi) = \int_H \varphi(x)\mu(dx).$$

In the following we shall often identify μ with F_μ.
A linear operator $T \in L(C_b(H))$ is said to be *positive* if $f \in C_b^+(H) \Rightarrow Tf \in C_b^+(H)$. If in addition for any $x \in H$ there exists $\nu_x \in M(H)$ such that

$$T\varphi(x) = \int_H \varphi(y)\nu_x(dy), \quad \varphi \in C_b(H),$$

we say that T is an *integral operator* and that ν_x is a *probability kernel*.

If T is an integral operator it can be uniquely extended to all Borel and bounded real functions $B_b(H)$; we shall still denote by T this extension.

The following result will be used later.

Proposition 3.1. *Let (T_n) be a sequence of positive integral operators and let $T \in L(C_b(H))$ be positive. Assume that for any $x \in H$, $K > 0$ and $\varepsilon > 0$ there exists $n_{x,K,\varepsilon} \in \mathbb{N}$ such that the following implication holds:*

$$\varphi \in C_b(H), \|\varphi\|_0 \le K, \ n \ge n_{x,K,\varepsilon} \Rightarrow |T\varphi(x) - T_n\varphi(x)| \le \varepsilon.$$

Then T is an integral operator.

Proof. Given $x \in H$ we consider the functional $F_x \in C_b^*(H)$ defined as

$$F_x(\varphi) = T\varphi(x), \quad \varphi \in C_b(H).$$

We have to show that $F_x \in M(H)$. By the Daniell theorem, see [15, Proposition II-7-1], it is enough to show that, given a sequence of nonnegative functions $(\varphi_j) \subset C_b(H)$ such that $\varphi_j(x) \uparrow \varphi(x)$, $x \in H$ with $\varphi \in C_b(H)$ we have $T\varphi_j(x) \uparrow T\varphi(x)$, $x \in H$.

Let $\varepsilon > 0$ and set $K = \|\varphi\|_0$. Then if $n_0 := n_{x,K,\varepsilon}$ we have

$$|T\varphi(x) - T\varphi_j(x)| \le |T\varphi(x) - T_{n_0}\varphi(x)| + |T_{n_0}\varphi(x) - T_{n_0}\varphi_j(x)|$$

$$+|T\varphi_j(x) - T_{n_0}\varphi_j(x)| \le 2\varepsilon + |T_{n_0}\varphi(x) - T_{n_0}\varphi_j(x)|, \quad j \in \mathbb{N}.$$

The conclusion follows now letting j tend to ∞. $\square$

3.1 Definition and First Properties

A *Markov semigroup* P_t on $C_b(H)$ is a mapping

$$[0, +\infty) \to L(C_b(H)), \ t \to P_t,$$

such that

(i) $P_0 = 1$, $P_{t+s} = P_t P_s$, $\quad t, s \ge 0$.
(ii) P_t is a positive integral operator and $P_t 1 = 1$, for any $t \ge 0$.
(iii) For any $\varphi \in C_b(H)$ (resp. $\varphi \in B_b(H)$) and $T > 0$ the mapping

$$[0, T] \times H \to \mathbb{R}, \ (t, x) \to P_t\varphi(x)$$

is continuous (resp. Borel).

We notice that there are more general definitions of Markov processes, but the definition (i)—(iii) is enough for our purposes. The fact that $P_t\varphi \in C_b(H)$ for any $\varphi \in C_b(H)$ is commonly referred as the *Feller* property.

Let P_t be a Markov semigroup. Then, by the assumptions above there exists a mapping

$$[0, +\infty) \times H \to M(H), \ (t, x) \to \lambda_{t,x}, \tag{3.1}$$

such that $\lambda_{0,x} = \delta_x$, and

$$P_t\varphi(x) = \int_H \varphi(y)\lambda_{t,x}(dy), \quad \varphi \in C_b(H). \tag{3.2}$$

$\lambda_{t,x}$ is called a *Markov probability kernel*.

Since, for any $\varphi \in C_b(H)$, we have

$$-\|\varphi\|_0 \le \varphi(x) \le \|\varphi\|_0, \quad x \in H,$$

then, using the positivity of P_t, it follows that

$$|P_t\varphi(x)| \le \|\varphi\|_0, \quad x \in H.$$

Consequently, P_t is a semigroup of contractions on $C_b(H)$. Note that P_t is not strongly continuous in general, see Example 3.4 below. There are in the literature several papers devoted to generalization of strongly continuous semigroups. Among some recent papers related to Markov semigroups we quote [2], [19], [13]. We follow here the approach by [19].

The Infinitesimal Generator of P_t.

For any $t > 0$ we set

$$\Delta_t\varphi = \frac{1}{t}(P_t\varphi - \varphi), \quad \varphi \in C_b(H).$$

Then we define the *infinitesimal generator N of P_t* by setting

$$D(N) = \Big\{ \varphi \in C_b(H) : \ \exists f \in C_b(H), \ \lim_{t\to 0^+} \Delta_t\varphi(x) = f(x), \quad x \in H$$

$$\text{and} \ \sup_{t\in(0,1]} \|\Delta_t\varphi\|_0 < +\infty \Big\}, \tag{3.3}$$

and

$$N\varphi(x) = \lim_{t\to 0^+} \Delta_t\varphi(x) = f(x), \quad x \in H, \ \varphi \in D(N). \tag{3.4}$$

N is called the *infinitesimal generator of P_t in $C_b(H)$*.

We now study some basic properties of the resolvent set $\rho(N)$ and of the resolvent $(\lambda - N)^{-1}$ of N. Results and proofs are straightforward generalizations of the classical Hille–Yosida theorem, see e.g. [9], [20] (the difference being that P_t is not strongly continuous but only pointwise continuous), and so they will be rapidly sketched, for more details see [19].

Proposition 3.2. *The half–right* $(0, +\infty)$ *belongs to* $\rho(N)$ *and for any* $\lambda > 0$ *we have*

$$(\lambda - N)^{-1}\varphi(x) = \int_0^{+\infty} e^{-\lambda t} P_t\varphi(x)dt, \quad \varphi \in C_b(H), \ x \in H. \tag{3.5}$$

Moreover

$$\|(\lambda - N)^{-1}\varphi\|_0 \le \frac{1}{\lambda} \|\varphi\|_0, \ \varphi \in C_b(H). \tag{3.6}$$

Proof. Let $\varphi \in C_b(H)$, and set for any $\lambda > 0$ and any $x \in H$,

$$F(\lambda)\varphi(x) = \int_0^{+\infty} e^{-\lambda t} P_t\varphi(x)dt.$$

It is easy to see that $F(\lambda)\varphi \in C_b(H)$, see [19]. Let us prove that $F(\lambda)\varphi \in D(N)$. In fact for any $h > 0$ and $x \in H$ we have

$$P_h F(\lambda)\varphi(x) = e^{\lambda h} \int_h^{+\infty} e^{-\lambda s} P_s\varphi(x)ds.$$

It follows that

$$D_h P_h F(\lambda)\varphi(x)|_{h=0} = \lambda \int_0^{+\infty} e^{-\lambda s} P_s\varphi(x)ds - \varphi(x)$$
$$= \lambda F(\lambda)\varphi(x) - \varphi(x), \tag{3.7}$$

and

$$|P_h F(\lambda)\varphi(x) - F(\lambda)\varphi(x)| \le (e^{\lambda h} - 1)\left|\int_h^{+\infty} e^{-\lambda s} P_s\varphi(x)ds\right|$$
$$+ \left|\int_0^h e^{-\lambda s} P_s\varphi(x)ds\right| \le \|\varphi\|_0 \left[\frac{e^{\lambda h} - 1}{\lambda} e^{-\lambda h} + \frac{1 - e^{-\lambda h}}{\lambda}\right] \le Ch. \tag{3.8}$$

By (3.7), (3.8) it follows that $F(\lambda)\varphi \in D(N)$ and $(\lambda - L)F(\lambda)\varphi = \varphi$.

Let us show now that if $\varphi \in D(N)$ we have $F(\lambda)(\lambda - N)\varphi = \varphi$. This will yield the statement. We have in fact

$$F(\lambda)(\lambda - N)\varphi(x) = \int_0^{+\infty} e^{-\lambda s} P_s(\lambda\varphi(x) - N\varphi(x))ds$$

$$= \lambda F(\lambda)(x) - \int_0^{+\infty} e^{-\lambda s} D_s F_s\varphi(x)dt.$$

Now the conclusion follows by integrating by parts. $\square$

Extension of P_t to $B_b(H)$

Since P_t is an integral operator it can be extended to all Borel and bounded real functions $B_b(H)$; we shall still denote by P_t this extension. Moreover if $\Gamma \in \mathcal{B}(H)$ and χ_Γ denotes the characteristic function of Γ, we have

$$P_t \chi_\Gamma(x) = \lambda_{t,x}(\Gamma), \quad x \in H, \ t \geq 0. \tag{3.9}$$

The following result will be used later.

Proposition 3.3. *For any $t, s \geq 0$, $x \in H$ and $\Gamma \in \mathcal{B}(H)$ we have*

$$\lambda_{t+s,x}(\Gamma) = \int_H \lambda_{s,y}(\Gamma) \lambda_{t,x}(dy). \tag{3.10}$$

Proof. We have in fact, taking into account (3.9),

$$P_{t+s}\chi_\Gamma(x) = P_t P_s \chi_\Gamma(x) = \int_H P_s \chi_\Gamma(y) \lambda_{t,x}(dy) = \int_H \lambda_{s,y}(\Gamma) \lambda_{t,x}(dy).$$

Since $P_{t+s}\chi_\Gamma(x) = \lambda_{t+s,x}(\Gamma)$, the conclusion follows. $\square$

Example 3.4. Let us consider the problem

$$\begin{cases} Z'(t) = b(Z(t)), \ t \geq 0, \\ \\ Z(0) = x \in \mathbb{R}, \end{cases} \tag{3.11}$$

where $b : \mathbb{R} \to \mathbb{R}$ is Lipschitz continuous. It is well known that problem (3.11) has a unique solution $Z(\cdot) = Z(\cdot, x) \in C^1([0, +\infty); H)$. Moreover

$$Z(t + s, x) = Z(t, Z(s, x)), \ t, s \in [0, +\infty).$$

Now we associate to the dynamic system (3.11) (which is non linear in general) a semigroup of *linear* operators defined on the space $C_b(H)$ by setting,

$$P_t \varphi(x) = \varphi(Z(t, x)), \quad x \in H, \ t \geq 0. \tag{3.12}$$

$P_t \varphi$ describes the evolution of the "observable" φ. As we shall see, P_t is important in studying asymptotic properties of (3.11) as ergodicity, mixing etc...

From (3.12) the semigroup law, $P_{t+s} = P_t P_s$, $t, s \geq 0$, follows immediately. Moreover (3.2) holds with $\lambda_{t,x} = \delta_{Z(t,x)}$. Thus P_t is a Markov semigroup.

In section 6 we shall consider a stochastic perturbation of (3.11) and the corresponding transition semigroup.

Exercise 3.5. Prove that the semigroup P_t defined by (3.12) is strongly continuous in $C_b(H)$ if and only if b is bounded.

3.2 Invariant Measures

In this section P_t represents a Markov semigroup on $C_b(H)$.

A probability measure $\mu \in M(H)$ is said to be *invariant* for P_t if

$$\int_H P_t \varphi d\mu = \int_H \varphi d\mu, \quad \varphi \in C_b(H), \ t \geq 0. \tag{3.13}$$

This condition is clearly equivalent (after identification of μ with F_μ) to

$$P_t^* \mu = \mu, \quad t \geq 0, \tag{3.14}$$

where P_t^* is the transpose operator of P_t,

$$\langle \varphi, P_t^* F \rangle = \langle P_t \varphi, F \rangle, \quad \varphi \in C_b(H), \quad F \in C_b^*(H).$$

Exercise 3.6. Assume that P_t is a Markov semigroup. Prove that μ is invariant if and only if

$$\mu(A) = \int_H \lambda_{t,x}(A) \mu(dx), \ A \in \mathcal{B}(H). \tag{3.15}$$

A basic result is the following.

Theorem 3.7. *Assume that the Markov semigroup P_t has an invariant measure μ. Then for all $t \geq 0$, $p \geq 1$, P_t is uniquely extendible to a linear bounded operator on $L^p(H, \mu)$ that we still denote by P_t. Moreover*

$$\|P_t\|_{L(L^p(H,\mu))} \leq 1, \quad t \geq 0. \tag{3.16}$$

Finally P_t is a strongly continuous semigroup on $L^p(H, \mu)$.

Proof. Let $\varphi \in C_b(H)$. By the Hölder inequality we have that

$$|P_t \varphi(x)|^p \leq \int_H |\varphi(y)|^p \lambda_{t,x}(dy) = P_t(|\varphi|^p)(x), \quad x \in H.$$

Now, integrating with respect to μ gives

$$\int_H |P_t \varphi(x)|^p \mu(dx) \leq \int_H P_t(|\varphi|^p)(x) \mu(dx) = \int_H |\varphi(x)|^p \mu(dx)$$

in view of the invariance of μ. Since $C_b(H)$ is dense in $L^p(H, \mu)$, then P_t is uniquely extendible to $L^p(H, \mu)$ and (3.16) follows.

Let us show finally that P_t is strongly continuous on $L^p(H, \mu)$. Since $C_b(H)$ is dense in $L^p(H, \mu)$, it is enough to show continuity of $P_t \varphi$ for any $\varphi \in C_b(H)$. Since $P_t \varphi(x)$ is continuous in (t, x), in this case the conclusion follows easily using the dominated convergence theorem. $\square$

From now on we shall assume that there exists an invariant measure μ for P_t. Our goal is to study the asymptotic behaviour of P_t.

We denote by Σ the linear subspace of $L^2(H, \mu)$ consisting of all *stationary* points of P_t,

$$\Sigma := \{\varphi \in L^2(H, \mu) : P_t\varphi = \varphi \text{ for all } t \geq 0\}.$$

It is important to notice that Σ is a lattice, as is proved in the next proposition.

Proposition 3.8. *Assume that φ and ψ belong to Σ. Then*

(i) $|\varphi| \in \Sigma$.
(ii) φ^+, φ^-, $\varphi \vee \psi$, $\varphi \wedge \psi \in \Sigma$.
(iii) For any $a \in \mathbb{R}$ the characteristic function of the set

$$\{x \in H : \varphi(x) > a\},$$

belongs to Σ.

Proof. Let us prove (i). Assume that $\varphi \in \Sigma$, so that $\varphi(x) = P_t\varphi(x)$, μ–a.e. Then we have

$$|\varphi(x)| = |P_t\varphi(x)| \leq P_t|\varphi|(x), \quad \mu\text{-a.e..} \tag{3.17}$$

We claim that

$$|\varphi(x)| = P_t|\varphi|(x), \quad \mu \text{ a.e.}$$

Assume by contradiction that there is a Borel subset $I \subset H$ such that $\mu(I) > 0$ and

$$|\varphi(x)| < P_t|\varphi|(x), \quad x \in I.$$

Then we have

$$\int_H |\varphi(x)|\mu(dx) < \int_H P_t|\varphi|(x)\mu(dx).$$

Since, by the invariance of μ,

$$\int_H P_t|\varphi|(x)\mu(dx) = \int_H |\varphi|(x)\mu(dx),$$

we find a contradiction.

Statements (ii) follow from the obvious identities

$$\varphi^+ = \frac{1}{2}(\varphi + |\varphi|), \quad \varphi^- = \frac{1}{2}(\varphi - |\varphi|),$$

$$\varphi \vee \psi = (\varphi - \psi)^+ + \psi, \quad \varphi \wedge \psi = -(\varphi - \psi)^+ + \varphi.$$

Finally let us prove (iii). It is enough to show that $\chi_{\{\varphi>0\}}$ belongs to Σ. We have in fact, as easily checked,

$$\chi_{\{\varphi>0\}} = \lim_{n\to\infty} \varphi_n(x), \quad x \in H,$$

where $\varphi_n = (n\varphi^+) \wedge 1$, $n \in \mathbb{N}$, belongs to Σ by (ii). Therefore $\{\varphi > 0\} \in \Sigma$. $\square$

If $\varphi \notin \Sigma$ the limit

$$\lim_{t \to +\infty} P_t\varphi(x), \ x \in H, \tag{3.18}$$

does not exist in general, as the following example shows.

Example 3.9. Let $H = \mathbb{R}^2$. Consider the differential equation

$$Z_1'(t) = -Z_2(t), \ Z_2'(t) = Z_1(t), \ Z_1(0) = x_1, \ Z_2(0) = x_2,$$

whose solution is given by

$$Z_1(t) = x_1 \cos t - x_2 \sin t, \ \ Z_2(t) = x_1 \sin t + x_2 \cos t, \ t \geq 0,$$

and the Markov semigroup:

$$P_t\varphi(x) = \varphi(Z(t, x)), \ \ x \in \mathbb{R}^2, \ t \geq 0.$$

Then the limit (3.18) exists only for $x = 0$.

We shall see in the next section that the limit of the ergodic means

$$\lim_{T \to +\infty} \frac{1}{T} \int_0^T P_t\varphi(x)dt, \ x \in H,$$

always exists (Von Neumann theorem).

3.3 Ergodicity

We want to study here the asymptotic behaviour as $T \to \infty$ of the following mean

$$M(T)\varphi = \frac{1}{T} \int_0^T P_s\varphi ds, \ \varphi \in L^2(H, \mu), \ T > 0. \tag{3.19}$$

For this we first recall a classic result about the ergodic means of a linear operator T.

Proposition 3.10. *Let T be a bounded operator on a Hilbert space E, (norm $\|\cdot\|$, inner product $\langle\cdot, \cdot\rangle$). Set*

$$M_n = \frac{1}{n} \sum_{k=0}^{n-1} T^k, \ \ n \in \mathbb{N},$$

and assume that there is $K > 0$ such that $\|T^n\| \leq K$, $n \in \mathbb{N}$. Then, for any $x \in E$, there exists the limit

$$\lim_{n \to \infty} M_n x := M_\infty x. \tag{3.20}$$

Moreover $M_\infty \in L(E)$, $M_\infty^2 = M_\infty$, and $M_\infty(E) = \mathrm{Ker}\,(1 - T)$.

Proof. First we notice that the limit in (3.20) obviously exists when x belongs to Ker $(1-T)$ or to $(1-T)(E)$. In fact in the first case we have $\lim_{n\to\infty} M_n x = x$ whereas in the latter we have $\lim_{n\to\infty} M_n x = 0$, because $\|T^n\| \le K$ and

$$M_n(1-T) = (1-T)M_n = \frac{1}{n}(1-T^n), \quad n \in \mathbb{N}. \tag{3.21}$$

Consequently we also have

$$\lim_{n\to\infty} M_n x = 0, \quad x \in \overline{(1-T)(E)}. \tag{3.22}$$

Now let $x \in E$ be fixed. Since $\|M_n x\| \le K\|x\|$, there exist a sub-sequence (n_k) of $\mathbb{N}$, and an element $y \in E$ such that $M_{n_k} x \to y$ weakly as $k \to \infty$. By (3.21) it follows also that $TM_{n_k} x \to Ty = y$, so that $y \in \text{Ker}\,(1-T)$.

Claim. $\lim_{n\to\infty} M_n x = y$.

First note that, since $y \in \text{Ker}\,(1-T)$, we have $M_n y = y$, and so

$$M_n x = M_n y + M_n(x-y) = y + M_n(x-y). \tag{3.23}$$

Moreover $x - y \in \overline{(1-T)(E)}$, since $x - y = \lim_{k\to\infty}(x - M_{n_k}x)$, and $x - M_{n_k}x \in (1-T)(E)$ because

$$x - M_{n_k}x = \frac{1}{n_k}\sum_{h=0}^{n_k-1}(1-T^h)x = \frac{1}{n_k}\sum_{h=0}^{n_k-1}(1+T+\dots+T^{h-1})(1-T)x.$$

Consequently the claim (as well as (3.20)) follows from (3.22).

Finally, since $(1-T)M_n \to 0$, we have $M_\infty = TM_\infty$, so that $T^k M_\infty = M_\infty$, $k \in \mathbb{N}$, and $M_\infty = M_n M_\infty$, that yields as $n \to \infty$, $M_\infty = (M_\infty)^2$, as required. $\square$

We are now ready to prove the following important theorem due to Von Neumann.

Theorem 3.11. *Let P_t be a Markov semigroup and μ an invariant measure for P_t. Set*

$$M(T)\varphi(x) = \frac{1}{T}\int_0^T P_s\varphi(x)ds, \quad \varphi \in L^2(H,\mu), \; x \in H.$$

Then there exists the limit

$$\lim_{T\to\infty} M(T)\varphi =: M_\infty \varphi \quad \text{in } L^2(H,\mu). \tag{3.24}$$

Moreover $M_\infty^2 = M_\infty$, $M_\infty(L^2(H,\mu)) = \Sigma$, (where Σ is the subspace of all stationary points of P_t) and

$$\int_H M_\infty \varphi d\mu = \int_H \varphi d\mu. \tag{3.25}$$

Proof. For all $T > 0$ we set

$$T = n_T + r_T, \quad n_T \in \mathbb{N} \cup \{0\}, \ r_T \in [0, 1).$$

Let $\varphi \in L^2(H, \mu)$, then we have

$$M(T)\varphi = \frac{1}{T} \sum_{k=0}^{n_T - 1} \int_k^{k+1} P_s\varphi ds + \frac{1}{T} \int_{n_T}^T P_s\varphi ds$$

$$= \frac{1}{T} \sum_{k=0}^{n_T - 1} \int_0^1 P_{s+k}\varphi ds + \frac{1}{T} \int_0^{r_T} P_{s+n(T)}\varphi ds \qquad (3.26)$$

$$= \frac{n_T}{T} \frac{1}{n_T} \sum_{k=0}^{n_T - 1} (P_1)^k M(1)\varphi + \frac{r_T}{T} (P_1)^{n_T} M(r_T)\varphi.$$

Since obviously

$$\lim_{T \to \infty} \frac{n_T}{T} = 1, \ \lim_{T \to \infty} \frac{r_T}{T} = 0,$$

letting $n \to \infty$ in (3.26) and recalling Proposition 3.10, we get (3.24).

Let us prove now that

$$M_\infty P_t = P_t M_\infty = M_\infty, \quad t \geq 0. \qquad (3.27)$$

In fact, given $t \geq 0$ we have

$$M_\infty P_t \varphi = \lim_{T \to \infty} \frac{1}{T} \int_0^T P_{t+s}\varphi ds = \lim_{T \to \infty} \frac{1}{T} \int_t^{t+T} P_s\varphi ds$$

$$= \lim_{T \to \infty} \frac{1}{T} \left\{ \int_0^T P_s\varphi ds - \int_0^t P_s\varphi ds + \int_T^{T+t} P_s\varphi ds \right\}$$

$$= M_\infty \varphi,$$

and this yields (3.27).

By (3.27) it follows that $M_\infty \varphi \in \Sigma$ for all $\varphi \in L^2(H, \mu)$ and moreover that

$$P_\infty M(T) = M(T) P_\infty = P_\infty,$$

that yields as $T \to \infty$, $M_\infty^2 = M_\infty$. Finally, (3.25) follows easily from (3.13). $\square$

Remark 3.12. One can show, Birkhoff theorem, that the convergence in the limit (3.24) holds pointwise, see e.g. [16].

Let us now define *ergodicity* and *mixing*. An invariant measure μ of P_t is said to be ergodic if

$$\lim_{T\to+\infty} \frac{1}{T}\int_0^T P_t\varphi dt = M_\infty\varphi = \overline{\varphi}, \quad \varphi \in L^2(H,\mu), \qquad (3.28)$$

where $\overline{\varphi}$ is the mean of φ,

$$\overline{\varphi} = \int_H \varphi(x)\mu(dx).$$

If the following stronger condition holds

$$\lim_{t\to+\infty} P_t\varphi(x) = \overline{\varphi}, \quad \varphi \in L^2(H,\mu), \ x \in H, \ \mu \text{ a.e.}, \qquad (3.29)$$

we say that μ is *strongly mixing*.

Proposition 3.13. *μ is ergodic if and only if the dimension of the subspace Σ of all stationary points of P_t is 1.*

Proof. Let μ be ergodic. Then, recalling that M_∞ is a projector on Σ, we see by (3.28) that the dimension of Σ must be 1.

Assume conversely that the dimension of Σ is 1. Then there is a linear bounded functional F on $L^2(H,\mu)$ such that

$$M_\infty\varphi = F(\varphi)1, \quad \varphi \in L^2(H,\mu).$$

By the Riesz representation theorem there exists an element $\varphi_0 \in L^2(H,\mu)$ such that $F(\varphi) = \langle \varphi, \varphi_0 \rangle$. Integrating this equality on H with respect to μ gives, taking into account the invariance of M_∞ (see (3.25)),

$$\int_H M_\infty\varphi d\mu = \int_H \varphi d\mu = \int_H \varphi\varphi_0 d\mu, \quad \varphi \in L^2(H,\mu).$$

Now, by the arbitrariness of φ, it follows that $\varphi_0 = 1$. $\square$

We want now to give a useful characterization of ergodic measures. Let us first give a definition.

Let μ be an invariant measure for P_t. A Borel set $\Gamma \in \mathcal{B}(H)$ is said to be *invariant* for P_t if its characteristic function χ_Γ belongs to Σ. If $\mu(\Gamma)$ is equal either to 0 or to 1, we say that Γ is *trivial*, otherwise that it is *non trivial*.

Theorem 3.14. *Let μ be an invariant measure for P_t. Then μ is ergodic if and only if each invariant set of μ is trivial.*

Proof. Let Γ be invariant for the ergodic measure μ. Then χ_Γ must be μ–a.e. constant (otherwise dim $\Sigma \geq 2$) and so Γ is trivial.

Assume conversely that all invariant sets for μ are trivial and, in contradiction, that μ is not ergodic. Then there exists a non constant function $\varphi_0 \in \Sigma$. Therefore by Proposition 3.8 there exists some $\lambda \in \mathbb{R}$ such that the invariant set $\{\varphi_0 > \lambda\}$ is not trivial, a contradiction. $\square$

3.4 The Set Λ of All Invariant Measures of P_t

We assume here that P_t is a Markov semigroup on H. We denote by Λ the set of all its invariant measures. Clearly Λ is a convex subset of $C_b^*(H)$ which we assume to be non empty.

Theorem 3.15. *Suppose that Λ consists exactly of one element μ. Then μ is ergodic.*

Proof. Assume in contradiction that μ is not ergodic. Then there is a non trivial invariant set Γ. Let us prove that the measure μ_Γ:

$$\mu_\Gamma(A) = \frac{\mu(A \cap \Gamma)}{\mu(\Gamma)}, \quad A \in \mathcal{B}(H),$$

belongs to Λ. Since $\mu_\Gamma \neq \mu$, this will give a contradiction.

To prove that μ_Γ is invariant, it is enough to show (see (3.15)), that

$$\mu_\Gamma(A) = \int_H \lambda_{t,x}(A) \mu_\Gamma(dx), \quad A \in \mathcal{B}(H),$$

or, equivalently, that

$$\mu(A \cap \Gamma) = \int_\Gamma \lambda_{t,x}(A) \mu(dx), \quad A \in \mathcal{B}(H). \tag{3.30}$$

In fact, since Γ is invariant, we have

$$P_t \chi_\Gamma = \chi_\Gamma, \quad P_t \chi_{\Gamma^c} = \chi_{\Gamma^c}, \ t \geq 0,$$

and so

$$\lambda_{t,x}(\Gamma) = \chi_\Gamma(x), \quad \lambda_{t,x}(\Gamma^c) = \chi_{\Gamma^c}(x), \ t \geq 0.$$

Consequently

$$\lambda_{t,x}(A \cap \Gamma^c) = 0, \text{ for } x - \mu \text{ a.e. in } \Gamma, \quad \lambda_{t,x}(A \cap \Gamma) = 0, \text{ for } x - \mu \text{ a.e. in } \Gamma^c,$$

and so

$$\int_\Gamma \lambda_{t,x}(A) \mu(dx) = \int_\Gamma \lambda_{t,x}(A \cap \Gamma) \mu(dx) + \int_\Gamma \lambda_{t,x}(A \cap \Gamma^c) \mu(dx)$$

$$= \int_\Gamma \lambda_{t,x}(A \cap \Gamma) \mu(dx) = \int_H \lambda_{t,x}(A \cap \Gamma) \mu(dx) = \mu(A \cap \Gamma),$$

and (3.30) holds. $\square$

We want now to show that the set of all *extremal* points of Λ (see e. g. [20, page 362]), is precisely the set of all ergodic measures. For this we need a lemma.

Lemma 3.16. *Let $\mu, \nu \in \Lambda$ with μ ergodic and ν absolutely continuous with respect to μ. Then $\mu = \nu$.*

Proof. Let $\Gamma \in \mathcal{B}(H)$. By the Von Neumann theorem there exists $T_n \uparrow \infty$ such that

$$\lim_{n \to \infty} \frac{1}{T_n} \int_0^{T_n} P_t \chi_\Gamma dt = \mu(\Gamma), \quad \mu \text{ a.e.} \tag{3.31}$$

Since $\nu << \mu$, identity (3.31) holds also ν–a.e. Now integrating (3.31) with respect to ν gives

$$\lim_{n \to \infty} \frac{1}{T_n} \int_0^{T_n} \left(\int_H P_t \chi_\Gamma d\nu \right) dt = \nu(\Gamma), \quad \mu \text{ a.e.}$$

Consequently $\nu(\Gamma) = \mu(\Gamma)$ as required. $\square$

We can now prove the announced characterization.

Theorem 3.17. *The set of all invariant ergodic measures coincides with the set of extremal points of Λ.*

Proof. We first prove that if μ is ergodic then it is an extremal point of Λ. Assume in contradiction that μ is not extremal. Then there exist $\mu_1, \mu_2 \in \Lambda$ with $\mu_1 \neq \mu_2$, and $\alpha \in (0, 1)$ such that

$$\mu = \alpha \mu_1 + (1 - \alpha) \mu_2.$$

Clearly $\mu_1 << \mu$ and $\mu_2 << \mu$. By Lemma 3.16 this implies $\mu_1 = \mu_2 = \mu$, a contradiction.

We finally prove that if μ is an extremal point of Λ, then it is ergodic. Assume in contradiction that μ is not ergodic. Then there exists a non trivial invariant set Γ. Consequently, arguing as in the proof of Theorem 3.15, we have $\mu_\Gamma, \mu_{\Gamma^c} \in \Lambda$. Since

$$\mu = \mu(\Gamma) \mu_\Gamma + (1 - \mu(\Gamma)) \mu_{\Gamma^c},$$

we find that μ is not extremal, a contradiction. $\square$

Theorem 3.18. *If $\mu, \nu \in \Lambda$ with ν different of μ. Then μ and ν are singular.*

Proof. Let $\Gamma \in \mathcal{B}(H)$ be such that

$$\mu(\Gamma) \neq \nu(\Gamma). \tag{3.32}$$

From the Von Neumann Theorem it follows that there exists $T_n \uparrow +\infty$ and two Borel sets M and N such that $\mu(M) = 1, \nu(N) = 1$, and

$$\lim_{n \to \infty} \frac{1}{T_n} \int_0^{T_n} (P_t \chi_\Gamma)(x) dt = \mu(\Gamma), \; \forall \, x \in M,$$

$$\lim_{n \to \infty} \frac{1}{T_n} \int_0^{T_n} (P_t \chi_\Gamma)(x) dt = \nu(\Gamma), \; \forall \, x \in N.$$

Since $\mu(\Gamma) \neq \nu(\Gamma)$ this implies $M \cap N = \emptyset$, and so μ and ν are singular. $\square$

4 Existence and Uniqueness of Invariant Measures

4.1 The Prokhorov Theorem

A sequence $(\mu_k) \subset M(H)$ is said to be *weakly convergent* to $\mu \in M(H)$ if we have

$$\lim_{k \to \infty} \int_H \varphi(x)\mu_k(dx) = \int_H \varphi(x)\mu(dx), \quad \varphi \in C_b(H).$$

In this case we shall write $\mu_k \rightharpoonup \mu$.

A subset $\Lambda \in M(H)$ is said to be *weakly relatively compact* if from any sequence in Λ, we can extract a subsequence weakly convergent to an element of $M(H)$.

A subset $\Lambda \subset M(H)$ is said to be *tight* if there exists an increasing sequence (K_n) of compact sets of H such that

$$\lim_{n \to \infty} \mu(K_n) = 1 \quad \text{uniformly on} \quad \Lambda,$$

or, equivalently, if for any $\varepsilon > 0$ there exists a compact set K_ε such that

$$\mu(K_\varepsilon) \geq 1 - \varepsilon, \quad \mu \in \Lambda.$$

Let us state the *Prokhorov theorem*, for a proof see e. g. [12].

Theorem 4.1. *A subset Λ of $M(H)$ is tight if and only if it is weakly relatively compact.*

4.2 The Krylov–Bogoliubov Theorem

We are given a Markov semigroup P_t in H. For any $T > 0$ and any $x_0 \in H$ we denote by $\mu_{T,x_0} \in M(H)$ the mean

$$\mu_{T,x_0} = \frac{1}{T} \int_0^T \lambda_{t,x_0} dt.$$

Let us prove the following *Krylov–Bogoliubov* theorem.

Theorem 4.2. *Assume that for some $x_0 \in H$ the set $(\mu_{T,x_0})_{T>0}$ is tight. Then there is an invariant measure for P_t.*

Proof. Since $(\mu_{T,x_0})_{T>0}$ is tight, by the Prokhorov theorem there exists a sequence $T_n \uparrow \infty$ and a probability measure $\mu \in M(H)$ such that

$$\lim_{n \to \infty} \int_H \varphi(x)\mu_{T_n,x_0}(dx) = \int_H \varphi(x)\mu(dx), \quad \varphi \in C_b(H).$$

This is equivalent to

$$\lim_{n \to \infty} \frac{1}{T_n} \int_0^{T_n} P_t\varphi(x_0)dt = \int_H \varphi(x)\mu(dx), \quad \varphi \in C_b(H).$$

Setting $\varphi = P_s\psi$ we see that

$$\lim_{n\to\infty} \frac{1}{T_n} \int_0^{T_n} P_{t+s}\psi(x_0)dt = \int_H P_s\psi(x)\mu(dx), \quad \psi \in C_b(H). \qquad (4.1)$$

Let us prove that the left hand side of (4.1) is equal to $\int_H \psi d\mu$. By the arbitrariness of ψ, this will prove that μ is invariant. We have in fact

$$\frac{1}{T_n} \int_0^{T_n} P_{t+s}\psi(x_0)dt = \frac{1}{T_n} \int_s^{T_n+s} P_t\psi(x_0)dt$$

$$= \frac{1}{T_n} \int_0^{T_n} P_t\psi(x_0)dt + \frac{1}{T_n} \int_{T_n}^{T_n+s} P_t\psi(x_0)dt - \frac{1}{T_n} \int_0^s P_t\psi(x_0)dt$$

$$\to \int_H \psi(x)\mu(dx) \text{ as } n \to \infty.$$

$\square$

Remark 4.3. Obviously if the set $(\lambda_{t,x_0})_{t>0}$ (instead of the set of the means), is tight then the assumptions of Theorem 4.2 are fulfilled and so there exists an invariant measure for P_t.

Let us give some sufficient conditions for the tightness of $(\lambda_{t,x_0})_{t>0}$.

Proposition 4.4. *Assume that there exists a Borel nonnegative function* $V : H \to \mathbb{R}$, *such that*

(i) $\lim\limits_{|x|\to+\infty} V(x) = +\infty$.

(ii) *For any* $R > 0$ *the set* $\Gamma_R := \{x \in H : V(x) \leq R\}$ *is compact.*

(iii) *There exists* $x_0 \in H$ *and* $C > 0$ *such that*

$$P_t V(x_0) = \int_H V(x)\lambda_{t,x_0}(dx) \leq C, \quad t \geq 0. \qquad (4.2)$$

Then $(\lambda_{t,x_0})_{t>0}$ *is tight and therefore there exists an invariant measure for* P_t.

Proof. We first notice that by the assumptions it follows that

$$\bigcup_{R>0} \Gamma_R = H.$$

Let us prove that

$$\lim_{R\to+\infty} \lambda_{t,x_0}(\Gamma_R^c) = 0 \text{ uniformly in } t \geq 0. \qquad (4.3)$$

We have in fact

$$\lambda_{t,x_0}(\Gamma_R^c) = \int_{\{x \in H \colon V(x) > R\}} \lambda_{t,x_0}(dx) \le \frac{1}{R} \int_H V(x)\lambda_{t,x_0}(dx) \le \frac{C}{R}.$$

By (4.3) it follows that $\{\lambda_{t,x_0}\}_{t>0}$ is tight, and so the conclusion follows from the Krylov–Bogoliubov theorem. $\square$

Proposition 4.5. *Assume that Λ is non empty and that there exists a Borel nonnegative function $V : H \to \mathbb{R}$, such that*

(i) $\lim_{|x| \to +\infty} V(x) = +\infty$.

(ii) For any $R > 0$ the set $\Gamma_R := \{x \in H : V(x) \le R\}$ is compact.

(iii) For any $t > 0$ there exists $K(t) > 0$ such that

$$P_t V(x) = \int_H V(y)\lambda_{t,x}(dx) \le K(t), \quad t \ge 0, \ x \in H. \qquad (4.4)$$

Then Λ is tight and weakly compact. Moreover there exists an ergodic invariant measure of P_t.

Proof. Let $\nu \in \Lambda$. Then, integrating (4.4) with respect to ν and taking into account the invariance of ν, yields

$$\int_H P_t V(x)\nu(dx) = \int_H V(x)\nu(dx) \le K(t), \quad t \ge 0.$$

It follows that

$$\int_H V(x)\nu(dx) \le \inf_{t \ge 0} K(t) := C_1.$$

This implies that Λ is tight. In fact for any $\nu \in \Lambda$ and any $R > 0$ we have

$$\nu(\Gamma_R^c) = \int_{\{x \in H \colon V(x) > r\}} \nu(dx) \le \frac{1}{R} \int_H V(x)\nu(dx) \le \frac{C_1}{r}.$$

Moreover Λ is weakly compact since if $(\mu_n) \subset \Lambda$ and $\mu_n \rightharpoonup \mu$ we have clearly $\mu \in \Lambda$. Finally the last statement follows from the Krein–Milman theorem, see e.g [20, page 362]. $\square$

4.3 Uniqueness of Invariant Measures

The uniqueness of the invariant measure is related to some properties of the Markov semigroup P_t that we introduce now. We shall use the following notation

$$B(z, \varepsilon) = \{x \in H : |z - x| < \varepsilon\}, \ z \in H, \ \varepsilon > 0.$$

- P_t is said to be *irreducible* if for any ball $B(z, \varepsilon)$ one has $P_t \chi_{B(z,\varepsilon)}(x) > 0$ or, equivalently, $\lambda_{t,x}(B(z, \varepsilon)) > 0$.
  It is obvious that if P_t is irreducible, then it is *positivity improving*; that is if $\varphi \in C_b(H)$ is nonnegative and it is strictly positive on some ball, then $P_t\varphi(x) > 0$ for all $t > 0, x \in H$.

- P_t is said to be *strong Feller* if for any $\varphi \in B_b(H)$ ([4]) and any $t > 0$ we have $P_t\varphi \in C_b(H)$.
- P_t is said to be *regular* if all probabilities $\lambda_{t,x}$, $t > 0$, $x \in H$, are equivalent.

Let us give a sufficient condition in order that P_t is strong Feller.

Proposition 4.6. *Let P_t be a Markov semigroup. Assume that for any $t > 0$ and $\varphi \in C_b(H)$, we have $P_t\varphi \in C_b^1(H)$ and that for any $t > 0$ there exists a positive constant $K(t)$ such that*

$$\|DP_t\varphi\|_0 \leq K(t)\|\varphi\|_0.$$

Then P_t is strong Feller and we have

$$|P_t\varphi(x) - P_t\varphi(y)| \leq K(t)\|\varphi\|_0|x - y|, \quad x, y \in H, \varphi \in B_b(H).$$

Proof. Let us fix $\varphi \in B_b(H)$, $t > 0, x, y \in H$ and define a (signed) measure ζ by setting

$$\zeta = \zeta_{t,x,y} = \lambda_{t,x} - \lambda_{t,y}.$$

Let $(\varphi_n) \subset C_b(H)$ be a sequence such that

(i) $\lim_{n\to\infty} \varphi_n(z) = \varphi(z)$ for $\zeta_{t,x,y}$–almost $z \in H$.
(ii) $\|\varphi_n\|_0 \leq \|\psi\|_0$.

Therefore, by the dominated convergence theorem,

$$P_t\varphi(x) - P_t\varphi(y) = \lim_{n\to\infty} \int_H \varphi_n(z)\zeta(dz) = \lim_{n\to\infty}(P_t\varphi_n(x) - P_t\varphi_n(y))$$

and consequently

$$|P_t\varphi(x) - P_t\varphi(y)| \leq K(t)\|\varphi\|_0|x - y|.$$

This implies that $P_t\varphi$ is Lipschitz continuous. Thus P_t is strong Feller as claimed. $\square$

Proposition 4.7. *Assume that P_t is strong Feller and μ is an invariant measure for P_t. Then for any $t > 0$ and $x \in H$, $\lambda_{t,x}$ is absolutely continuous with respect to μ.*

Proof. Recall first that by (3.15) we have for any $\Gamma \in \mathcal{B}(H)$

$$\mu(\Gamma) = \int_H \lambda_{t,x}(\Gamma)\mu(dx) = \int_H P_t\chi_\Gamma(x)\mu(dx).$$

Now, let $t > 0$, $x \in H$ and assume that $\mu(\Gamma) = 0$. Then, from the identity above it follows that $\lambda_{t,x}(\Gamma) = 0$, since $P_t\chi_\Gamma$ is continuous and nonnegative. $\square$

---

[4] We recall that $B_b(H)$ is the set of all mappings $\varphi : H \to \mathbb{R}$ bounded and Borel.

Corollary 4.8. *Assume that P_t is strong Feller and $\mu(\Gamma) = 0$ for some $\Gamma \in \mathcal{B}(H)$. Then $P_t\chi_\Gamma(x) = 0$ for all $t > 0$ and all $x \in H$.*

Proposition 4.9. *Assume that P_t is strong Feller and that μ is an invariant measure for P_t. Assume moreover that μ is full, that is*

$$\mu(B(x,r)) > 0, \quad x \in H, \, r > 0.$$

Then μ is ergodic.

Proof. Assume by contradiction that there exists a non trivial invariant set $\Gamma \in \mathcal{B}(H)$, that is $0 < \mu(\Gamma) < 1$ and

$$P_t\chi_\Gamma(x) = \chi_\Gamma(x), \quad \mu - \text{a.e.}$$

Fix $t > 0$ and set

$$A_1 = \{x \in H : \ P_t\chi_\Gamma(x) = 1\}, \quad A_0 = \{x \in H : \ P_t\chi_\Gamma(x) = 0\}.$$

A_1 and A_0 are non empty, otherwise the set Γ would be trivial. Moreover, we know that $\mu(A_1 \cup A_0) = 1$. Set finally

$$B = \left\{x \in H : \ \frac{1}{4} < P_t\chi_\Gamma(x) < \frac{3}{4}\right\},$$

so that $\mu(B) = 0$. But B contains a ball because P_t is strong Feller and consequently $P_t\chi_\Gamma$ is continuous; this contradicts the fact that μ is full. $\square$

Proposition 4.10. *Assume that P_t is regular. Then it possesses at most one invariant measure. Moreover if μ is an invariant measure for P_t, then μ is equivalent to $\lambda_{t,x}$ for all $t > 0$ and all $x \in H$ and it is ergodic.*

Proof. Assume that μ is an invariant measure for P_t and let $t_0 > 0$ and $x_0 \in H$. Then by identity (3.15),

$$\mu(A) = \int_H \lambda_{t,x}(A)\mu(dx), \ A \in \mathcal{B}(H), \tag{4.5}$$

we see that μ is equivalent to λ_{t_0,x_0}. In fact if for some $A \in \mathcal{B}(H)$ we have $\lambda_{t_0,x_0}(A) = 0$ then $\lambda_{t,x}(A) = 0$ for any $t > 0$ and any $x \in H$ since all probabilities $\lambda_{t,x}$, $t > 0$, $x \in H$, are equivalent. Therefore $\mu(A) = 0$ by (4.5) and so $\mu << \lambda_{t_0,x_0}$.

Conversely if $\mu(A) = 0$, again by (4.5), we have that $\lambda_{t,x}(A) = 0$ for almost all (and consequently for all) $x \in H, t > 0$. Thus $\lambda_{t_0,x_0} << \mu$.

We prove now that μ is ergodic. Let $\Gamma \in \mathcal{B}(H)$ be an invariant set. Since by the invariance

$$P_t\chi_\Gamma(x) = \lambda_{t,x}(\Gamma) = \chi_\Gamma(x), \ x - \mu \text{ a.s.},$$

we have that $\lambda_{t,x}(\Gamma)$ is either 0 or 1, $x\mu$ a.e. and so for all $x \in H$, by the regularity of P_t. This implies by (4.5) that (Γ) is trivial. Thus μ is ergodic.

Let us finally prove the uniqueness of μ. From what we have just proved, we know that the set Λ of all invariant measures of P_t consists of ergodic measures. Therefore, by Theorem 3.17 Λ must consist of exactly one element. □

To prove regularity an important tool is the following result due to Khas'minskii.

Theorem 4.11. *If P_t is strong Feller and irreducible, then it is regular.*

Proof. It is enough to show that the measures $\lambda_{t,x}$, $t > 0, x \in H$ have the same null sets. This means that if $\Gamma \in \mathcal{B}(H)$ then:

(i) either $\lambda_{t,x}(\Gamma) = 0$ for all $t > 0, x \in H$,
(ii) or $\lambda_{t,x}(\Gamma) > 0$ for all $t > 0, x \in H$.

Let $\Gamma \in \mathcal{B}(H)$ and assume that (i) does not hold. Then there exists $x_0 \in H$ and $t_0 > 0$ such that

$$P_{t_0}\chi_\Gamma(x_0) > 0.$$

Now $P_t\chi_\Gamma$ is continuous by the strong Feller property, and so $P_t\chi_\Gamma(x) > 0$ for x in a ball of H. Since P_t is positivity improving, it follows

$$P_t\chi_\Gamma(x) > 0, \ \forall\, t \geq t_0, \ x \in H.$$

We claim that this holds also for $t \leq t_0$. In fact for any $t_1 < t_0$ there exists $x_1 \in H$ such that

$$P_{t_1}\chi_\Gamma(x_1) > 0,$$

(otherwise $P_t\chi_\Gamma$ would be identically 0). By the same argument as before, we have

$$P_{t_0}\chi_\Gamma(x) > 0, \quad x \in H,$$

and the conclusion follows. □

Corollary 4.12. *If P_t is strong Feller and irreducible, then it has at most one invariant measure.*

5 Examples of Markov Semigroups

5.1 Introduction

In this section we present some examples of Markov semigroups in a separable Hilbert space H. We study, in each case, properties of irreducibility, strong Feller and regularity.

It is useful to introduce the set $\mathcal{E}(H)$ of all *exponential functions*, that is the linear the span in $C_b(H)$ of all functions $\varphi_h, h \in H$, of the form ([5]):

$$\varphi_h(x) = e^{i\langle h,x \rangle}, \ x \in H.$$

The following lemma shows that any function of $C_b(H)$ can be pointwise approximated by elements of $\mathcal{E}(H)$ equibounded in the sup norm.

Lemma 5.1. *For all $\varphi \in C_b(H)$, there exists a sequence $(\varphi_{n,k}) \subset \mathcal{E}(H)$ such that*

$$(i) \quad \lim_{k \to \infty} \lim_{n \to \infty} \varphi_{k,n}(x) = \varphi(x), \quad x \in H,$$

$$(5.1)$$

$$(ii) \ \|\varphi_{k,n}\|_0 \leq \|\varphi\|_0, \quad n, k \in \mathbb{N}.$$

Proof. We divide the proof in two steps.

Step 1. We assume that H is finite–dimensional.

Let $\dim H = N$ and let $\varphi \in C_b(H)$. For any $n \in \mathbb{N}$ there exists a function $\varphi_n \in C_b(H)$ with the following properties

(i) φ_n is n periodic with period n in all its coordinates.
(ii) $\varphi_n(x) = \varphi(x)$ for all $x \in [-n + 1/2, n - 1/2]^N$.
(iii) $\|\varphi_n\|_0 \leq \|\varphi\|_0$.

It is easy to check that the sequence (φ_n) is pointwise convergent to φ. Moreover any function φ_n can be approximated, through a suitable Fourier series, by elements of $\mathcal{E}(H)$.

Step 2 Case when $\dim H = \infty$.

Given $\varphi \in C_b(H)$, set

$$\psi_k(x) = \varphi(P_k x), \ x \in H, \ k \in \mathbb{N}.$$

where

$$P_k x = \sum_{j=1}^{k} \langle x, e_j \rangle e_j, \ x \in H,$$

and (e_j) is a complete orthonormal system in H. By Step 1, for any $k \in \mathbb{N}$, there exists a sequence $(\varphi_{k,n})_{n \in \mathbb{N}} \subset \mathcal{E}(H)$ such that

$$\lim_{n \to \infty} \varphi_{k,n}(x) = \psi_k(x), \ x \in H,$$

and

$$\|\varphi_{k,n}\|_0 \leq \|\psi_k\|_0 \leq \|\varphi\|_0.$$

Now, for any $x \in H$ we have

$$\lim_{k \to \infty} \lim_{n \to \infty} \varphi_{k,n}(x) = \varphi(x), \ \forall \, x \in H,$$

and the conclusion follows. $\square$

---

[5] Since the Hilbert space H is real one has to take the real parts of functions in $\mathcal{E}(H)$.

5.2 The Heat Semigroup

We are given a linear operator $Q \in L_1^+(H)$. Then there exists a complete orthonormal system (e_k) in H, and a sequence (λ_k) of nonnegative numbers such that

$$Qe_k = \lambda_k e_k, \quad k \in \mathbb{N}.$$

Let us define

$$P_t\varphi(x) = \int_H \varphi(y)N_{x,tQ}(dy), \quad \varphi \in C_b(H), \ t \geq 0. \tag{5.2}$$

Other equivalent expressions of P_t are, as easily checked,

$$P_t\varphi(x) = \int_H \varphi(x+y)N_{tQ}(dy) = \int_H \varphi(x+\sqrt{t}y)N_Q(dy), \quad \varphi \in C_b(H), \ t \geq 0. \tag{5.3}$$

Proposition 5.2. P_t is a Markov semigroup with probability kernel $\lambda_{t,x} = N_{x,tQ}$, $t > 0, x \in H$. Moreover P_t is strongly continuous.

Proof. Let first prove that the semigroup law

$$P_{t+s}\varphi = P_t P_s\varphi, \ t, s \geq 0, \tag{5.4}$$

holds. In view of Lemma 5.1, it is enough to prove (5.4) for any function $\varphi_h(x) = e^{i\langle x,h\rangle}$ with $x, h \in H$. But in this case (5.4) follows immediately from the identity

$$P_t\varphi_h(x) = \int_H e^{i\langle y,h\rangle} N_{x,tQ}(dy) = e^{i\langle h,x\rangle}e^{-\frac{1}{2}t\langle Qh,h\rangle}. \tag{5.5}$$

It remains to prove that P_t is strongly continuous. Let $\varphi \in C_b(H)$. Since

$$P_t\varphi(x) = \int_H \varphi(x+\sqrt{t}\,z)N_Q(dz),$$

we have

$$|P_t\varphi(x) - \varphi(x)| = \left|\int_H [\varphi(x+\sqrt{t}\,z) - \varphi(x)]N_Q(dz)\right|$$

$$\leq \int_H \omega_\varphi(\sqrt{t}\,z)N_Q(dz),$$

where ω_φ is the uniform continuity modulus of φ. Therefore the conclusion follows letting t tend to 0. $\square$

P_t is called the *Heat semigroup*. If H has finite dimension d and $Q = I$, it is easy to see that

$$P_t\varphi(x) = (2\pi t)^{-d/2} \int_{\mathbb{R}^d} e^{-\frac{1}{2t}|x-y|^2} \varphi(y)dy, \ \varphi \in C_b(H). \tag{5.6}$$

In this case $u(t,x) = P_t\varphi(x)$ is the solution to the heat equation:

$$\begin{cases} D_t u(t,x) = \dfrac{1}{2} \Delta u(t,x), \ x \in H, \ t > 0, \\[2mm] u(0,x) = \varphi(x), \ x \in H. \end{cases} \tag{5.7}$$

Remark 5.3. If H is infinite dimensional, then by the Feldman–Hajek theorem, see e.g [18] it follows that P_t is not regular because the measures N_{tQ} and N_{sQ} are singular for $s \neq t$.

Assume that $Q \neq 0$. Then it is easy to see that (P_t) has no invariant probability measures. Assume in fact in contradiction that there exists $\mu \in M(H)$ invariant for P_t. Then for any exponential function $\varphi_h(x) = e^{i\langle x,h\rangle}$ we should have, taking into account (5.5),

$$\hat{\mu}(h) = \int_H e^{i\langle x,h\rangle}\mu(dx) = \int_H P_t e^{i\langle x,h\rangle}\mu(dx) = e^{-\frac{t}{2}\langle Qh,h\rangle}\tilde{\mu}(h),$$

for all $t > 0$; a contradiction.

It is well known, that in finite dimensions when Ker $Q = \{0\}$ the heat semigroup (5.6) is regularizing in the sense that for any $\varphi \in C_b(H)$ and any $t > 0$, we have that $P_t\varphi \in C_b^\infty(H)$.

In infinite dimensions this result is wrong, we can only show that P_t regularizes in the directions of the reproducing kernel $Q^{1/2}(H)$.

Proposition 5.4. *Let $\varphi \in C_b(H)$ and $z \in H$. Then there exists the directional derivative*

$$\langle DP_t\varphi(x), Q^{1/2}z\rangle = \lim_{h\to 0} \frac{1}{h}\left(P_t\varphi(x+hQ^{1/2}z) - P_t\varphi(x)\right),$$

given by

$$\langle DP_t\varphi(x), Q^{1/2}z\rangle = \frac{1}{\sqrt{t}} \int_H \langle z, (tQ)^{-1/2}y\rangle\varphi(x+y)N_{tQ}(dy). \tag{5.8}$$

Proof. We have

$$P_t\varphi(x+hQ^{1/2}z) = \int_H \varphi(x+hQ^{1/2}z+y)N_{tQ}(dy)$$

$$= \int_H \varphi(x+y)N_{hQ^{1/2}z,tQ}(dy).$$

Moreover, by the Cameron–Martin formula we have

$$\frac{dN_{hQ^{1/2}z,tQ}(dy)}{dN_{tQ}(dy)}(y) = e^{-\frac{h^2}{2t}|z|^2 + \frac{h}{\sqrt{t}}\langle z,(tQ)^{-1/2}y\rangle}, \ y \in H.$$

It follows

$$P_t\varphi(x + hQ^{1/2}z) = \int_H \varphi(x + hQ^{1/2}z + y)e^{-\frac{h^2}{2t}|z|^2 + \frac{h}{\sqrt{t}}\langle z,(tQ)^{-1/2}y\rangle}N_{tQ}(dy).$$

Taking the derivative with respect to h and setting $h = 0$ gives (5.8). $\square$

We prove now that if $\varphi \in C_b^2(H)$, then $u(t,x) = P_t\varphi(x)$ is the unique solution of the Heat equation:

$$\begin{cases} D_t u(t,x) = \dfrac{1}{2} \operatorname{Tr}[QD^2u(t,x)] = (Lu(t,\cdot))(x), \ t \geq 0, \ x \in H, \\ u(0,x) = \varphi(x). \end{cases} \tag{5.9}$$

Proposition 5.5. *Let $\varphi \in C_b^2(H)$, and set*

$$u(t,x) = P_t\varphi(x), \ t \geq 0, \ x \in H.$$

Then u is the unique solution of problem (5.9)

Proof. Existence. By (5.3) we have

$$D_t u(t,x) = \frac{1}{2\sqrt{t}} \int_H \langle D\varphi(x + \sqrt{t}y), y\rangle N_Q(dy). \tag{5.10}$$

Consequently by (5.8) it follows that

$$\langle D^2u(t,x)e_k, e_k\rangle = \frac{1}{\sqrt{t}\,\lambda_k} \int_H y_k D_k\varphi(x + \sqrt{t}\,y)N_{tQ}(dy).$$

Therefore

$$\frac{1}{2}\operatorname{Tr}[QD^2u(t,x)] = \sum_{k=1}^{\infty} \lambda_k \langle D^2u(t,x)e_k, e_k\rangle$$

$$= \frac{1}{\sqrt{t}} \int_H \langle y, D\varphi(x + \sqrt{t}\,y)\rangle N_{tQ}(dy).$$

Comparing with (5.10) yields the conclusions.

Uniqueness. Let $v : [0,T] \times H \to \mathbb{R}$ be a continuous function such that

$$D_t v(t,\cdot) = L(t,\cdot), \ v(0,\cdot) = \varphi.$$

Then we have

$$D_s P_{t-s}v(s,\cdot) = -LP_{t-s}v(s,\cdot) + P_{t-s}v(s,\cdot), \ s \in [0,t].$$

Since
$$LP_t\psi = P_tL\psi, \ \psi \in C_b^2(H),$$

we obtain
$$D_sP_{t-s}v(s, \cdot) = 0, \ s \in [0, t].$$

This implies that $D_sP_{t-s}v(s, \cdot)$ is constant so that $v(s, \cdot) = P_s\varphi$. This proves uniqueness. $\square$

We show now that if Ker $Q = \{0\}$ then P_t is irreducible. This follows from the next proposition.

Proposition 5.6. *Let* $Q \in L_1^+(H)$ *such that Ker* $Q = \{0\}$ *and let* $a \in H$. *Then we have* $0 < N_{a,Q}(B(x, r)) < 1$, *where* $B(x, r)$ *is the ball of center* x *and radius* r.

Proof. Let us assume for simplicity $x = 0$ and set $B_r = B(0, r)$. For any $n \in \mathbb{N}$ the following inclusion obviously holds

$$B_r \supset \left\{ x \in H : \sum_{k=1}^n x_k^2 \le \frac{r^2}{2}, \ \sum_{k=n+1}^\infty x_k^2 < \frac{r^2}{2} \right\}.$$

Consequently

$$N_{a,Q}(B_r) \ge N_{a,Q}\left(\sum_{k=1}^n x_k^2 \le \frac{r^2}{2}\right) N_{a,Q}\left(\sum_{k=n+1}^\infty x_k^2 < \frac{r^2}{2}\right).$$

Now the first factor is obviously positive, thus it is enough to show that the second is positive as well. We have in fact

$$\mu\left(\sum_{k=n+1}^\infty x_k^2 < \frac{r^2}{2}\right) = 1 - \mu\left(\sum_{k=n+1}^\infty x_k^2 \ge \frac{r^2}{2}\right)$$

$$\ge 1 - \frac{2}{r^2}\sum_{k=n+1}^\infty (\lambda_k + |a_k|^2) > 0,$$

for n sufficiently large. $\square$

If H is finite dimensional it is easy to see that P_t is strong Feller if and only if Ker $Q = \{0\}$. In the infinite dimensional case the strong Feller property never holds. We have in fact the following result, see [6].

Proposition 5.7. *Let Ker* $Q = \{0\}$ *and let* H *be infinite dimensional. Then* P_t *is not strong Feller.*

Proof. Let us choose a subspace $H_0 \subset H$, with Borel embedding, H_0 different from H and such that $Q^{1/2}(H) \subset H_0$ and $N_{tQ}(H_0) = 1$, $t \ge 0$. For this it is enough to take a non decreasing sequence (α_k) of positive numbers such that $\alpha_k \uparrow +\infty$, and

$$\sum_{k=1}^{\infty} \alpha_k \lambda_k < +\infty,$$

and to define

$$H_0 = \left\{ x \in H : \sum_{k=1}^{\infty} \alpha_k |x_k|^2 < +\infty \right\}.$$

Then $Q^{1/2}(H) \subset H_0$ and $N_{tQ}(H_0) = 1$ since

$$\int_H \sum_{k=1}^{\infty} \alpha_k |x_k|^2 N_{tQ}(dx) = t \sum_{k=1}^{\infty} \alpha_k \lambda_k < +\infty.$$

We show now that

$$P_t \chi_{H_0} = \chi_{H_0}, \ t \geq 0, \tag{5.11}$$

this will prove that P_t is not strong Feller.

We have in fact

$$P_t \chi_{H_0}(x) = \int_{H_0} \chi_{H_0}(x+y) N_{tQ}(dy) + \int_{H_0^c} \chi_{H_0}(x+y) N_{tQ}(dy).$$

Now (5.11) follows taking into account that $x \in H_0$, $H_0 + x = H_0$, and for $x \notin H_0$, $(H_0 + x) \cap H_0 = \emptyset$. $\square$

We say that a function φ is *harmonic* if $P_t \varphi = \varphi$, $t \geq 0$.

The following generalization of the Liouville Theorem holds, see [6].

Proposition 5.8. *Assume that* $Ker \ Q = \{0\}$ *and* $\varphi \in C_b(H)$ *is harmonic. Then* φ *is constant.*

Proof. We have in fact from (5.8)

$$|\langle DP_t \varphi(x), Q^{1/2} z \rangle| \leq \frac{1}{\sqrt{t}} \|\varphi\|_0. \tag{5.12}$$

Since $P_t \varphi = \varphi$ this implies that

$$\langle D\varphi(x), Q^{1/2} z \rangle = 0, \ \forall \, z \in H.$$

Consequently φ is constant on the reproducing kernel $Q^{1/2}(H)$. Since $Q^{1/2}(H)$ is dense in H this implies that φ is constant. $\square$

5.3 The Ornstein–Uhlenbeck Semigroup

Let A be the infinitesimal generator of a strongly continuous semigroup e^{tA} in H and let $C \in L^+(H)$. We shall assume in this section that for any $t > 0$ the linear operator:

$$Q_t x := \int_0^t e^{sA} C e^{sA^*} x \, ds, \ x \in H,$$

is of trace–class. Then for all $t \geq 0$, $x \in H$, we set

$$R_t\varphi(x) = \int_H \varphi(y) N_{e^{tA}x, Q_t}(dy) = \int_H \varphi(e^{tA}x + y) N_{Q_t}(dy), \quad \varphi \in C_b(H).$$
(5.13)

For all $t \geq 0$, R_t maps $\mathcal{E}(H)$ into itself. In fact if $\varphi_h(x) = e^{i\langle h, x\rangle}$ we have, recalling formula (1.10) for the Fourier transform of a Gaussian measure,

$$R_t\varphi_h(x) = e^{i\langle h, e^{tA}x\rangle} \int_0^t e^{i\langle h, y\rangle} N_{Q_t}(dy) = e^{-\frac{1}{2}\langle Q_t h, h\rangle} \varphi_{e^{tA^*}h}(x), \quad x \in H.$$
(5.14)

Exercise 5.9. *Show that*

$$R_t(|x|^2) = |e^{tA}x|^2 + \operatorname{Tr} Q_t, \ t \geq 0.$$
(5.15)

Proposition 5.10. *R_t is a Markov semigroup with probability kernel $\lambda_{t,x} = N_{e^{tA}x, Q_t}$.*

Proof. First we prove the semigroup law. In view of Lemma 5.1 it is enough to show that

$$R_{t+s}\varphi_h = R_t R_s \varphi_h, \ t, s \geq 0,$$

for all exponential function φ_h, $h \in H$. In fact by (5.14) it follows that

$$R_s R_t \varphi_h(x) = \exp\{i\langle e^{(t+s)A}x, h\rangle - \frac{1}{2}\langle (Q_t + e^{tA}Q_s e^{tA^*})h, h\rangle\}.$$

Since $Q_t + e^{tA}Q_s e^{tA^*} = Q_{t+s}$, the conclusion follows.

Let us finally prove the continuity of $(t, x) \to P_t\varphi(x)$ in $[0, T] \times H$ for all $\varphi \in C_b(H)$. Since $C_b^1(H)$ is dense in $C_b(H)$, see [6, Theorem 2.2.1] it is enough to consider the case when $\varphi \in C_b^1(H)$. Let $t_0 \in [0, T]$ and $x_0 \in H$. Then for any $t \in [0, T]$ and $x \in H$ we have

$$|R_t\varphi(x) - R_{t_0}\varphi(x_0)| \leq |R_t\varphi(x) - R_t\varphi(x_0)| + |R_t\varphi(x_0) - R_{t_0}\varphi(x_0)|$$

$$\leq \|\varphi\|_1 \|e^{tA}\| |x - x_0| + |R_t\varphi(x_0) - R_{t_0}\varphi(x_0)|.$$

Moreover

$$|R_t\varphi(x_0) - R_{t_0}\varphi(x_0)| \leq \int_H |R_{t_0}\varphi(e^{(t-t_0)A}x_0 + y) - R_{t_0}\varphi(x_0)| N_{Q_{t-t_0}}(dy)$$

$$\leq \|\varphi\|_1 |e^{(t-t_0)A}x_0 - x_0| + \|\varphi\|_1 \int_H |y| N_{Q_{t-t_0}}(dy).$$

Since, by the Hölder inequality, we have

$$\left[\int_H |y| N_{Q_{t-t_0}}(dy)\right]^2 \leq \int_H |y|^2 N_{Q_{t-t_0}}(dy) = \operatorname{Tr}(Q_{t-t_0}),$$

the conclusion follows. □

By Proposition 5.6 it follows that R_t is irreducible if and only if Ker $Q_t = \{0\}$. This happens for instance when Ker $C = \{0\}$. In fact if $x_0 \in H$ is such that $Q_t x_0 = 0$ we have

$$0 = \langle Q_t x_0, x_0 \rangle = \int_0^t |\sqrt{C} e^{sA} x_0|^2 ds = 0,$$

which implies $x_0 = 0$.

The semigroup R_t is not strongly continuous when A is different from 0. In fact in this case, from (5.14) we see that $R_t \varphi_h$ does not converge to 0 in $C_b(H)$ for all $h \neq 0$.

We can define the infinitesimal generator L of R_t by (3.3), (3.4).

Regularizing Property of R_t

In this subsection we present a situation where the following implication holds:

$$\varphi \in C_b(H),\ t \geq 0\ \Rightarrow\ R_t \varphi \in C_b^\infty(H).$$

By Proposition 4.6 this implies that R_t is strong Feller.

We shall need the the following assumption.

$$e^{tA}(H) \subset Q_t^{1/2}(H),\quad t > 0. \tag{5.16}$$

If (5.16) holds we set

$$\Gamma(t) = Q_t^{-1/2} e^{tA},\ t \geq 0, \tag{5.17}$$

where $Q_t^{-1/2}$ is the pseudo–inverse of $Q_t^{1/2}$ ([6]). In view of the closed graph theorem, $\Gamma(t)$ is a bounded operator for any $t > 0$.

Remark 5.11. The condition (5.16) is fulfilled for instance when $C = I$. In this case, see [6], there exists $K > 0$ such that

$$\|\Gamma(t)\| \leq \frac{M}{\sqrt{t}},\quad t > 0. \tag{5.18}$$

Theorem 5.12. *Assume that hypothesis* (5.16) *holds. Then if* $\varphi \in C_b(H)$ *we have* $R_t \varphi \in C_b^\infty(H)$. *In particular*

$$\langle DR_t \varphi(x), h \rangle = \int_H \langle \Gamma(t)h, Q_t^{-1/2} y \rangle \varphi(e^{tA} x + y) N_{Q_t}(dy),\quad h \in H, \tag{5.19}$$

and

$$|DR_t \varphi(x)| \leq \|\Gamma(t)\|\, \|\varphi\|_0,\quad x \in H,\ t > 0. \tag{5.20}$$

---

[6] Let $T \in L^+(H)$, and $y \in T(H)$. We denote by $T^{-1}(y)$ the element of $T^{-1}(\{y\})$ having minimal norm.

Proof. Let $x \in H$ and $t > 0$. Since, by (5.16), $e^{tA}x \in Q_t^{1/2}(H)$, the measures $N_{e^{tA}x, Q_t}$ and N_{Q_t} are equivalent in view of the Cameron–Martin formula, see Theorem 2.13. Moreover

$$\frac{dN_{e^{tA}x, Q_t}}{dN_{Q_t}}(y) =: \rho_t(x, y),$$

where $\rho_t(x, y) = e^{-\frac{1}{2}|\Gamma(t)x|^2 + \langle \Gamma(t)x, Q_t^{-1/2}y \rangle}$. It follows

$$R_t\varphi(x) = \int_H \varphi(y) e^{-\frac{1}{2}|\Gamma(t)x|^2 + \langle \Gamma(t)x, Q_t^{-1/2}y \rangle} N_{Q_t}(dy).$$

From this identity it is not difficult to show $R_t\varphi$ is differentiable for $t > 0$ and that for any $h \in H$ we have

$$\langle DR_t\varphi(x), h \rangle = \int_H \varphi(y) \left[\langle \Gamma(t)h, Q_t^{-1/2}y \rangle - \langle \Gamma(t)h, \Gamma(t)x \rangle \right]$$

$$\times \rho_t(x, y) N_{Q_t}(dy),$$

$$= \int_H \langle \Gamma(t)h, Q_t^{-1/2}y \rangle \varphi(e^{tA}x + y) N_{Q_t}(dy).$$

Thus $R_t\varphi \in C_b^1(H)$. Arguing in a similar way, we can prove that $R_t\varphi \in C_b^\infty(H)$. $\square$

We consider now the particular case when, besides (5.16), the Laplace transform of $\|\Gamma(t)\|$

$$\tilde{\gamma}(\lambda) := \int_0^{+\infty} e^{-\lambda t} \|\Gamma(t)\| dt \tag{5.21}$$

exists for all $\lambda > 0$. This happens for instance when $C = I$, see Remark 5.11.

In this case we can prove the following result.

Proposition 5.13. *Assume that (5.16) and (1.16) hold. Let $f \in C_b(H)$. Then $(\lambda - L)^{-1}f \in C_b^1(H)$ and*

$$|D(\lambda - L)^{-1}f(x)| \leq \tilde{\gamma}(\lambda)\|f\|_0, \quad x \in H. \tag{5.22}$$

Proof. It is enough to take the Laplace transform of (5.20). $\square$

Corollary 5.14. *We have $D(L) \subset C_b^1(H)$, with continuous embedding.*

Proof. Let $\varphi \in D(L)$ and set $f = \varphi - L\varphi$, so that $\varphi = (1 - L)^{-1}f$. Then by Proposition 5.13 we have that $\varphi \in C_b^1(H)$ and

$$|D\varphi(x)| \leq \tilde{\gamma}\|f\|_0 \leq \tilde{\gamma}(\|\varphi\|_0 + \|L\varphi\|_0). \square$$

Invariant Measures

We assume here in addition that A is of negative type, that is

$$\|e^{tA}\| \le Me^{-\omega t}, \ \ t \ge 0,$$

for some $M > 0, \omega > 0$.

Under this assumption the linear operator

$$Qx := \int_0^{+\infty} e^{tA}Ce^{tA^*}x\,dt, \ \ \ x \in H, \tag{5.23}$$

is clearly well defined.

Lemma 5.15. Q *is of trace class.*

Proof. We have in fact

$$Qx = \sum_{k=1}^{\infty} \int_{k-1}^{k} e^{(s+k-1)A}Ce^{(s+k-1)A^*}x\,ds = \sum_{k=1}^{\infty} e^{(k-1)A}Q_1 e^{(k-1)A^*}x\,ds.$$

It follows $\operatorname{Tr} Q \le M \sum_{k=1}^{\infty} e^{-2\omega(k-1)} \operatorname{Tr} Q_1 < +\infty.$ $\ \square$

Theorem 5.16. *A probability measure μ on $(H, \mathcal{B}(H))$ is invariant for R_t if and only if $\mu = N_Q$. In this case for any $\varphi \in L^2(H, \mu)$ we have*

$$\lim_{t \to +\infty} R_t\varphi(x) = \int_H \varphi(y)N_Q(dy) \ \ \ \text{in } L^2(H,\mu), \ \ x \in H, \tag{5.24}$$

thus μ is ergodic and strongly mixing.

Proof. Existence. We prove that $\mu = N_Q$, is invariant. For this it is enough to check that for any exponential function φ_h we have

$$\int_H R_t\varphi_h(x)\mu(dx) = \int_H \varphi_h(x)\mu(dx). \tag{5.25}$$

In fact, taking into account (5.14), we see that (5.25) is equivalent to

$$\langle Qe^{tA^*}h, e^{tA^*}h \rangle + \langle Q_t h, h \rangle = \langle Qh, h \rangle, \ h \in H,$$

which is also equivalent to $e^{tA}Qe^{tA^*} + Q_t = Q$, which clearly holds.

Uniqueness. Assume that μ is invariant. Then we have

$$\hat{\mu}(e^{tA^*}h)\, e^{-\frac{1}{2}\langle Q_t h, h \rangle} = \hat{\mu}(h),$$

where $\hat{\mu}$ is the Fourier transform of μ. As $t \to \infty$ we find

$$\hat{\mu}(h) = e^{-\frac{1}{2}\langle Qh, h \rangle}.$$

This implies, by the uniqueness of the Fourier transform (Proposition 1.2), $\mu = N_Q$. □

Since μ is invariant, R_t can be uniquely extended to a strongly continuous semigroup of contractions R_t^p on $L^p(H, \mu)$ by Theorem 3.7. We shall denote by L_p its infinitesimal generator.

Exercise 5.17. *Let $p \geq 1$, $h \in H$ and $\varphi_h(x) = e^{i\langle x, h \rangle}$. Prove that $\varphi_h \in D(L_p)$ if and only if $h \in D(A^*)$ where A^* is the adjoint of A. If $h \in D(A^*)$ show that*

$$L_p \varphi_h(x) = -\frac{1}{2} \langle Ch, h \rangle + i\langle x, A^* h \rangle. \tag{5.26}$$

Exercise 5.17 suggest to introduce a special class of exponential functions. We denote by $\mathcal{E}_A(H)$ the linear span of all functions $\varphi_h(x) = e^{i\langle h, x \rangle}$ with $h \in D(A^*)$. It is easy to check that $\mathcal{E}_A(H)$ is invariant for R_t and belongs to $D(L_p)$ for all $p \geq 1$. Moreover by (5.26) it follows that

$$L_p \varphi = \frac{1}{2} \operatorname{Tr} [D^2 \varphi] + \langle x, A^* D\varphi \rangle, \quad \varphi \in \mathcal{E}_A(H).$$

Moreover, the following result holds.

Proposition 5.18. *For all $p \geq 1$, $\mathcal{E}_A(H)$ is a core for L_p.*

Proof. In fact $\mathcal{E}_A(H)$ is dense in $L^p(H, \mu)$, and it is invariant for R_t. Now the conclusion follows from a classical result, see e. g. [8]. □

5.4 The Sobolev Space $W^{1,2}(H, \mu)$

We are here concerned with the Gaussian measure $\mu = N_Q$ where Q is defined by (5.23). We denote by (e_k) a complete orthonormal system in H and by (λ_k) a sequence of positive numbers such that

$$Q e_k = \lambda_k e_k, \quad k \in \mathbb{N},$$

and by D_k the partial derivative in the direction e_k, $k \in \mathbb{N}$, defined on $\mathcal{E}(H)$. One can show, see e.g. [6], that D_k is closable on $L^2(H, \mu)$. We shall still denote by D_k its closure.

Now we define $W^{1,2}(H, \mu)$ as the set of all $\varphi \in L^2(H, \mu)$ such that

$$\sum_{k=1}^{\infty} \int_H |D_k \varphi|^2 d\mu < +\infty.$$

For general properties of $W^{1,2}(H, \mu)$ see [6]. We shall use the fact that the embedding of $W^{1,2}(H, \mu)$ into $L^2(H, \mu)$ is compact.

Now we show that $R_t \varphi$ is regularizing for any $t > 0$ and any $\varphi \in L^2(H, \mu)$.

Proposition 5.19. *For all* $\varphi \in L^2(H, \mu)$, *we have* $R_t \varphi \in W^{1,2}(H, \mu)$ *and*

$$\|DR_t\varphi\|_{L^2(H,\mu)} \leq \|\Gamma(t)\| \, \|\varphi\|_{L^2(H,\mu)}, \quad t > 0. \tag{5.27}$$

Moreover, $D(L_2) \subset W^{1,2}(H, \mu)$.

Proof. Since $C_b(H)$ is dense in $L^2(H, \mu)$ it is enough to show (5.27) for $\varphi \in C_b(H)$. By (5.19) and the Hölder inequality it follows that

$$|DR_t\varphi(x)|^2 \leq \|\Gamma(t)\|^2 R_t(\varphi^2), \quad t > 0, x \in H.$$

Integrating with respect to μ, and taking into account the invariance of μ, gives

$$\int_H |DR_t\varphi(x)|^2 \mu(dx) \leq \|\Gamma(t)\|^2 \int_H |\varphi(x)|^2 \mu(dx) \, t > 0.$$

Thus (5.27) is proved. Finally, the last statement can be proved as Corollary 5.14. □

Since the embedding $W^{1,2}(H, \mu) \subset L^2(H, \mu)$ is compact, we have the following result.

Corollary 5.20. *For all* $t > 0$, *the operator* R_t *is compact.*

6 Bounded Perturbations of Ornstein–Uhlenbeck Generators

In this section we shall consider two linear operators $A : D(A) \subset H \to H$, $C \in L^+(H)$ and a mapping $F : H \to H$ Lipschitz continuous and bounded. We shall assume that

(i) $A : D(A) \subset H \to H$ is the infinitesimal generator of a strongly continuous semigroup e^{tA} in H. There are $M > 0$ and $\omega > 0$ such that

$$\|e^{tA}\| \leq Me^{-\omega t}, \ t \geq 0. \tag{6.1}$$

(ii) We have

$$\int_0^{+\infty} \text{Tr}\,[e^{tA}Ce^{tA^*}]dt < +\infty, \tag{6.2}$$

where A^* is the adjoint of A.

(iii) We have

$$e^{tA}(H) \subset Q_t^{1/2}(H), \quad t > 0. \tag{6.3}$$

Moreover there exists $N > 0$ and $\alpha \in (0, 1)$ such that

$$\|\Gamma(t)\| \leq Nt^{-\alpha}e^{-\omega t}, \quad t > 0, \tag{6.4}$$

where $\Gamma(t) = Q_t^{-1/2}e^{tA}$, $t > 0$.

Under these assumptions we know by Theorem 5.12 that the semigroup R_t

$$R_t\varphi(x) = \int_H \varphi(e^{tA}x + y)N_{Q_t}(dy), \quad t \geq 0, \quad \varphi \in C_b(H), \qquad (6.5)$$

is strong Feller and that $D(L) \subset C_b^1(H)$, where L is the infinitesimal generator of R_t. Moreover, from Theorem 5.16 the Gaussian measure $\mu = N_Q$, where

$$Qx = \int_0^{+\infty} e^{sA}Ce^{sA^*}x\,ds, \quad t \geq 0, \ x \in H,$$

is invariant for R_t.

We want here to construct a Markov semigroup P_t whose infinitesimal generator is given by

$$N\varphi(x) = L\varphi(x) + \langle F(x), D\varphi(x)\rangle, \quad \varphi \in D(L). \qquad (6.6)$$

Notice that N maps $D(L)$ into $C_b(H)$ since $D(L) \subset C_b^1(H)$ and F is continuous and bounded.

We will show that P_t is given by $P_t\varphi(x) = u(t,x)$ where $u(t,x)$ is the unique continuous solution of the following integral equation

$$u(t,\cdot) = R_t\varphi + \int_0^t R_{t-s}(\langle F(\cdot), Du(s,\cdot)\rangle)ds, \quad t \geq 0. \qquad (6.7)$$

Notice that formally (6.7) is the integral (or mild) form of the following problem

$$D_t u = Lu + \langle F, Du\rangle, \quad u(0,\cdot) = \varphi.$$

Proposition 6.1. *For any $\varphi \in C_b(H)$ and for any $T > 0$ there is a unique continuous solution u of equation (6.7). Moreover there exists $N_1 > 0$ such that*

$$\|Du(t,\cdot)\|_0 \leq N_1 t^{-\alpha}e^{-\omega t}\|\varphi\|_0, \quad t > 0. \qquad (6.8)$$

Proof. We fix $T > 0$ and write equation (6.7) as $u = f + \gamma(v)$, where

$$f(t,\cdot) = R_t\varphi, \quad \gamma(u)(t,\cdot) = \int_0^t R_{t-s}(\langle F(\cdot), Du(s,\cdot)\rangle)ds, \quad t \in [0,T].$$

Then we shall solve equation (6.7) by a fixed point argument, on the space Z_T consisting of the set of all functions $u : [0,T] \times H \to \mathbb{R}$ such that

(i) u is continuous and bounded.
(ii) for all $t > 0$, $u(t,\cdot) \in C_b^1(H)$,
(iii) $\sup\limits_{t \in (0,T]} t^\alpha \|Du(t,\cdot)\|_0 < +\infty.$

It is easy to see that Z_T, endowed with the norm

$$\|u\|_{Z_T} := \|u\|_0 + \sup_{t \in (0,T]} t^\alpha \|Du(t,\cdot)\|_0,$$

is a Banach space. Notice that f belongs to Z_T in view of assumption (6.4) and Theorem 5.12. Now, let us show that γ is a contraction on Z_T, provided T is sufficiently small. We have in fact

$$|\gamma(u)(t,x)| \le \|F\|_0 \int_0^t \|Du(s,\cdot)\|_0 \, ds \le \frac{t^{1-\alpha}}{1-\alpha} \|u\|_{Z_T}, \qquad (6.9)$$

and, since

$$D\gamma(u)(t,x) = \int_0^t D[R_{t-s}(\langle F(\cdot), Du(s,\cdot)\rangle)]ds,$$

we obtain

$$t^\alpha |D\gamma(u)(t,x)| \le t^\alpha N\|F\|_0 \int_0^t (t-s)^{-\alpha} s^{-\alpha} ds \|u\|_{Z_T}$$

$$= t^{1-\alpha} N\|F\|_0 \int_0^1 (1-s)^{-\alpha} s^{-\alpha} ds \, \|u\|_{Z_T}. \qquad (6.10)$$

Thus γ is a contraction, provided T is sufficiently small, say $T \le T_0$. Consequently, there is a unique solution of equation (6.7) in $[0, T_0]$. Now, by repeating the previous argument in the interval $[T_0, 2T_0]$ and so on, we conclude the proof of existence and uniqueness.

Let us now show (6.8). Since, in view of (6.4), we have

$$\|Du(t,\cdot)\|_0 \le Nt^{-\alpha} e^{-\omega t}\|\varphi\|_0 + N\|F\|_0 \int_0^t (t-s)^{-\alpha} e^{-\omega(t-s)}\|Du(s,\cdot)\|_0 \, ds,$$

the conclusion follows from a well known generalization of the Gronwall lemma, see e.g. [11]. $\square$

It is easy to see that, setting $P_t\varphi = u(t,\cdot)$, P_t is a one–parameter semigroup on $C_b(H)$ and we have $P_t 1 = 1$. We want now to show that P_t is a Markov semigroup. Notice that from (6.7) even the fact that P_t is a positive operator is not apparent. For this it is convenient to introduce a family of approximating problems depending on a parameter $\beta > 0$,

$$\begin{cases} D_t u_\beta(t,\cdot) = L u_\beta(t,\cdot) + \mathcal{F}_\beta(u_\beta(t,\cdot)), & t \ge 0, \\ u_\beta(0,\cdot) = \varphi. \end{cases} \qquad (6.11)$$

$\mathcal{F}_\beta$ is defined as follows

$$\mathcal{F}_\beta(\psi)(x) = \frac{1}{\beta} \left(\psi(\eta(\beta, x)) - \psi(x) \right), \quad \psi \in C_b(H), \ x \in H, \qquad (6.12)$$

where η is the solution of the initial value problem

$$D_t\eta(t,x) = F(\eta(t,x)), \quad \eta(0,x) = x. \tag{6.13}$$

Notice that, as easily checked, for any $\varphi \in C_b^1(H)$ we have

$$\lim_{\beta \to 0} \mathcal{F}_\beta(\varphi) = \langle F, D\varphi \rangle \quad \text{in } C_b(H).$$

Obviously, (6.11) is equivalent to the following problem

$$\begin{cases} D_t u_\beta(t,x) = \left(L - \dfrac{1}{\beta}\right) u_\beta(t,x) + \dfrac{1}{\beta}\, u_\beta(t,\eta(\beta,x)), \quad t \ge 0, \\ u_\beta(0,\cdot) = \varphi. \end{cases} \tag{6.14}$$

Since $\mathcal{F}_\beta$ is Lipschitz continuous,

$$\|\mathcal{F}_\beta(\psi) - \mathcal{F}_\beta(\psi_1)\|_0 \le \frac{2}{\beta}\, \|\psi - \psi_1\|_0, \quad \psi, \psi_1 \in C_b(H),$$

it is clear that problem (6.11) has a unique continuous solution u_β. Moreover, by the variation of constant formula, problem (6.14) is equivalent to the following integral equation,

$$u_\beta(t,x) = e^{-\frac{t}{\beta}} R_t\varphi(x) + \frac{1}{\beta} \int_0^t e^{-\frac{t-s}{\beta}} R_{t-s} u_\beta(t-s, \eta(\beta,x))ds, \tag{6.15}$$

whose solution we denote by $u_\beta(t,x) = u_\beta(t,x;\varphi)$. It is well known that $u_\beta(t,x;\varphi)$ can be obtained as a limit of successive approximations $(u_\beta^n(t,x;\varphi))_{n\in\mathbb{N}}$ of (6.15), defined by

$$u_\beta^0(t,x;\varphi) = e^{-\frac{t}{\beta}} R_t\varphi(x),$$

$$u_\beta^n(t,x;\varphi) = e^{-\frac{t}{\beta}} R_t\varphi(x) + \frac{1}{\beta} \int_0^t e^{-\frac{t-s}{\beta}} u_\beta^{n-1}(t-s, \eta(\beta,x);\varphi)ds, \quad n \in \mathbb{N}.$$

We set also

$$P_t^{n,\beta}\varphi(x) := u_\beta^n(t,x;\varphi), \quad n \in \mathbb{N}, \ x \in H, \ \varphi \in C_b(H).$$

We can prove now the following result.

Proposition 6.2. P_t *is a Markov semigroup.*

Proof. By a recurrence argument on successive approximations of u_β^n we can check that $u_\beta(t,x;\varphi)$ has the following properties

(i) $\|u_\beta(t,\cdot;\varphi)\|_0 \le \|\varphi\|_0$.
(ii) If $\varphi \ge 0$ we have $u_\beta(t,\cdot;\varphi) \ge 0$, $\varphi \in C_b(H)$.
(iii) $P_t^{n,\beta}$ is an integral operator in $C_b(H)$ for any $\beta > 0, t > 0, n \in \mathbb{N}$.

Moreover, always by recurrence, we see that for all $n \in \mathbb{N}$,

$$|P_t^\beta \varphi(x) - P_t^{n,\beta}\varphi(x)|_0 \leq \sum_{k=1}^{\infty} \frac{t^k}{k! \beta^k} \|\varphi\|_0 \quad x \in H, \; n \in \mathbb{N}.$$

Consequently, by Proposition 3.1, P_t^β is an integral operator for all $t > 0$ and all $\beta > 0$.

Now, notice that u_β is also the solution of the following integral equation

$$u_\beta(t, \cdot) = R_t\varphi + \int_0^t R_{t-s}(\mathcal{F}_\beta(u_\beta(s, \cdot)))ds, \quad t \geq 0, \tag{6.16}$$

which we can write as $u^\beta = f + \gamma^\beta(v)$, where

$$f(t, x) \quad = R_t\varphi(x), \quad t \in [0, T], \quad x \in H,$$

$$\gamma_\beta(v)(t, \cdot) = \int_0^t R_{t-s}(\mathcal{F}_\beta(v(s, \cdot)))ds, \quad t \in [0, T], \; v \in C_b(H).$$

Notice that γ_β fulfills estimates quite similar to (6.9) and (6.10). This follows in fact by the inequality

$$|\mathcal{F}_\beta(\psi(x))| \leq \frac{1}{\beta} \|D\varphi\|_0 |\eta(\beta, x) - x| \leq \|F\|_0 \|D\varphi\|_0.$$

Therefore, using a straightforward argument concerning fixed points depending by the parameter β, we see that

$$\lim_{\beta \to 0} u_\beta(t, x; \varphi) = u(t, x; \varphi), \quad t \geq 0, \; x \in H,$$

where $u(t, x; \varphi) = u(t, x)$ is the solution of (6.7). Therefore we obtain that the following properties hold

(i) $\|u(t, \cdot; \varphi)\|_0 \leq \|\varphi\|_0$.
(ii) If $\varphi \geq 0$ we have $u(t, \cdot; \varphi) \geq 0$, $\varphi \in C_b(H)$.

Also, invoking again by Proposition 3.1, we see that P_t is an integral operator for all $t > 0$. The proof is complete. $\square$

Proposition 6.3. *The infinitesimal generator of P_t is N.*

Proof. First we prove that the resolvent set of N contains an half–line $(\lambda_0, +\infty]$ and we have

$$(\lambda - N)^{-1} = (\lambda - L)^{-1}(1 - T_\lambda)^{-1}, \quad \lambda > \lambda_0, \tag{6.17}$$

where

$$T_\lambda\psi = \langle F, D(\lambda - L)^{-1}\psi\rangle, \quad \psi \in C_b(H). \tag{6.18}$$

Let us consider in fact the equation

$$\lambda\varphi - L\varphi - \langle F(x), D\varphi \rangle = f. \tag{6.19}$$

Setting $\psi = \lambda\varphi - L\varphi$ equation (6.19) becomes

$$\psi - T_\lambda\psi = f. \tag{6.20}$$

Taking into account (5.22) we see that, for a suitable constant C_1, we have

$$\|T_\lambda\psi\|_0 \le C_1\lambda^{\alpha-1}\|F\|_0\|\psi\|_0$$

and so T_λ is a contraction for all $\lambda > \lambda_0$, so that (6.17) follows.

Denote by N_1 the infinitesimal generator of P_t. Then, taking the Laplace transform of both sides of equation (6.7) yields

$$(\lambda - N_1)^{-1}\varphi = (\lambda - L)^{-1}\varphi + (\lambda - L)^{-1}\langle F, D(\lambda - N_1)^{-1}\varphi \rangle,$$

which is equivalent to

$$(\lambda - L)(\lambda - N_1)^{-1}\varphi = \varphi + \langle F, D(\lambda - N_1)^{-1}\varphi \rangle.$$

Setting $(\lambda - N_1)^{-1}\varphi = \psi$ we have

$$(\lambda - L)\psi = \varphi + \langle F, D\psi \rangle.$$

But this implies $\psi = (\lambda - N)^{-1}\varphi$ for all $\lambda > \lambda_0$. Therefore $N = N_1$ as claimed.
$\square$

6.1 Existence and Uniqueness of Invariant Measure for P_t

To prove existence of an invariant measure we will first show that the semigroup P_t is extendible to a strongly continuous semigroup P_t^2 on $L^2(H, \mu)$. Note that μ is not in general invariant for P_t, so that the existence of such an extension is not a priori granted.

The existence of an invariant measure for P_t^2 and, consequently for P_t could be proved by using the Perron–Frobenius theorem, see [8], [1]. We present here a direct proof.

Proposition 6.4. *For any $\varphi \in L^2(H, \mu)$ there is a unique function $u \in C([0, +\infty); L^2(H, \mu))$ solution of equation (6.7). Setting $P_t^2\varphi(x) = u(t, x)$, P_t^2 is a strongly continuous semigroup P_t^2 on $L^2(H, \mu)$ which extends P_t. Moreover, for all $t > 0$, $u(t, \cdot) \in W^{1,2}(H, \mu)$ and there exists $K_1 > 0$ such that*

$$\|Du(t, \cdot)\|_{L^2(H,\mu)} \le K_1 t^{\alpha-1}\|\varphi\|_{L^2(H,\mu)}, \quad t > 0. \tag{6.21}$$

Finally, the semigroup P_t^2 is compact for all $t > 0$ and its infinitesimal generator is given by

$$N_2\varphi := L_2\varphi + \langle F, D\varphi \rangle, \quad \varphi \in D(N_2) = D(L_2).$$

Proof. The proof is very similar to those of Propositions 6.1 and 6.3. The main difference is that, instead of Theorem 5.12 we have to use Proposition 5.19.

Finally, compactness of P_t follows from (6.21) and the compactness of the embedding $W^{1,2}(H,\mu) \subset L^2(H,\mu)$. $\square$

We now consider the adjoint semigroup $(P_t^2)^*$; we denote by N_2^* its infinitesimal generator, and by Σ^* the set of all its stationary points:

$$\Sigma^* = \left\{ \varphi \in L^2(H,\mu) : \ (P_t^2)^*\varphi = \varphi, \ t \geq 0 \right\}.$$

Notice that $(P_t^2)^*$ is not a Markov semigroup in general. In particular the equality $(P_t^2)^*1 = 1$ does not hold in general.

The proof of the following result is completely similar to that of Proposition 3.8; so it will be omitted.

Lemma 6.5. $(P_t^2)^*$ *has the following properties:*

(i) For all $\varphi \geq 0$ μ -a.e, one has $(P_t^2)^*\varphi \geq 0$ μ -a.e.
(ii) Σ^* is a lattice, that is if $\varphi \in \Sigma^*$ then $|\varphi| \in \Sigma^*$.

Proposition 6.6. *There exists a unique invariant measure ν of P_t such that $\int_H |x|\nu(dx) < +\infty$. Moreover ν is absolutely continuous with respect to μ and $d\nu/d\mu \in L^2(H,\mu)$.*

Proof. Existence. Let $\lambda > 0$ be fixed. Clearly $1 \in D(N_2)$ and we have $N_2 1 = 0$. Consequently $1/\lambda$ is an eigenvalue of $R(\lambda, N_2)$ since

$$(\lambda - N_2)^{-1}1 = \frac{1}{\lambda}.$$

Moreover $1/\lambda$ is a simple eigenvalue because μ is ergodic. Since the embedding $W^{1,2}(H,\mu) \subset L^2(H,\mu)$ is compact and $D(L_2) \subset W^{1,2}(H,\mu)$ by Proposition 5.19, it follows that $(\lambda - N_2)^{-1}$ is compact too for all $\lambda > 0$. Therefore $(\lambda - N_2^*)^{-1}$ is compact as well and $1/\lambda$ is a simple eigenvalue for $(\lambda - N_2^*)^{-1}$. Consequently there exists $\rho \in L^2(H,\mu)$ such that

$$(\lambda - N_2^*)^{-1}\rho = \frac{1}{\lambda}\,\rho. \tag{6.22}$$

It follows that $\rho \in D(N_2^*)$ and $N_2^*\rho = 0$. Since Σ^* is a lattice, ρ can be chosen to be nonnegative and such that $\int_H \rho d\mu = 1$.

Now set

$$\nu(dx) = \rho(x)\mu(dx), \ x \in H.$$

We claim that ν is an invariant measure for P_t^2. In fact taking the inverse Laplace transform in (6.22) we find

$$(P_t^2)^*\rho = \rho$$

that implies for any $\varphi \in L^2(H,\mu)$

$$\int_H P_t^2 \varphi d\nu = \int_H P_t^2 \varphi \, \rho d\mu = \int_H \varphi (P_t^2)^* \rho d\mu = \int_H \varphi d\nu.$$

Note finally that

$$\int_H |x| \nu(x) = \int_H |x| \rho(x) \mu(dx)$$

$$\leq \left(\int_H |x|^2 \mu(dx) \right)^{1/2} \left(\int_H \rho^2(x) \mu(dx) \right)^{1/2} < +\infty.$$

Uniqueness. Let ζ be another invariant measure for P_t such that $\int_H |x| \zeta(dx) < +\infty$. We have for any $t > 0$ and $x \in H$,

$$\left| P_t \varphi(x) - \int_H \varphi(y) \zeta(dy) \right| \leq \int_H |P_t \varphi(x) - P_t \varphi(y)| \zeta(dy).$$

In view of (6.8) it follows that

$$|P_t \varphi(x) - P_t \varphi(y)| \leq \int_0^1 |D P_t \varphi(\theta x + (1 - \theta) y) \cdot (x - y)| d\theta$$

$$\leq K_1 t^{-\alpha} e^{-\omega t} \|\varphi\|_0 \, |x - y|.$$

Now, integrating in y with respect to ζ, we obtain

$$\left| P(t)\varphi(x) - \int_H \varphi(y) \zeta(dy) \right| \leq K_1 t^{-\alpha} e^{-\omega t} \|\varphi\|_0 \left(|x| + \int_H |y| \zeta(dy) \right).$$

This implies that

$$\lim_{t \to \infty} P_t \varphi(x) = \int_H \varphi(y) \zeta(dy), \quad x \in H.$$

Consequently, letting t tend to ∞ in the identity

$$\int_H P_t \varphi(x) \nu(dx) = \int_H \varphi(x) \zeta(dx),$$

yields

$$\int_H \varphi(y) \nu(dy) = \int_H \varphi(x) \zeta(dx), \quad \varphi \in C_b^1(H).$$

Hence $\zeta = \nu$. $\square$

Remark 6.7. The condition that the first moment of ν is finite can be dropped, see [5].

7 Diffusion Semigroups

7.1 Introduction

Let us consider a dynamical system in $H = \mathbb{R}$ (for simplicity) governed by a differential equation

$$
\begin{cases}
Z'(t) = b(Z(t)), & t \geq s, \\
Z(s) = x \in \mathbb{R},
\end{cases}
\tag{7.1}
$$

where $b : \mathbb{R} \to \mathbb{R}$ is Lipschitz continuous. It is well known that equation (7.1) has a unique solution $Z(\cdot, s, x) \in C^1([s, +\infty); \mathbb{R})$, and that (7.1) is equivalent to the integral equation

$$
Z(t) = x + \int_s^t b(Z(s))ds, \ t \geq 0.
$$

To take into account random perturbations one is led to add to the right hand side a *stochastic* perturbation $B(t) - B(s)$ where B is a real stochastic process, that is a collection $(B(t))_{t \geq 0}$ of real random variables, on a probability space $(\Omega, \mathcal{F}, \mathbb{P})$. So, we consider the family of equations, indexed by $\omega \in \Omega$,

$$
X(t, \omega) = x + \int_s^t b(X(r, \omega))dr + B(t, \omega) - B(s, \omega), \quad t \geq s, \ \omega \in \Omega. \tag{7.2}
$$

We will often omit the variable ω in what follows.

One of the more common choices, is to take for B a Brownian motion. A *Brownian motion* B on the probability space $(\Omega, \mathcal{F}, \mathbb{P})$ is a stochastic process such that

(i) $B(0) = 0$ and if $0 \leq s < t$, $B(t) - B(s)$ is a real Gaussian random variable with law N_{t-s}.
(ii) If $0 < t_1 < ... < t_n$, the random variables

$$
B(t_1), \ B(t_2) - B(t_1), ..., B(t_n) - B(t_{n-1})
$$

are independent.
(iii) For almost all $\omega \in \Omega$, $W(\cdot)(\omega)$ is continuous.

Remark 7.1. It is not difficult to construct a real Brownian motion. Let us consider in fact the probability space $(H, \mathcal{B}(H), \mu)$, where $H = L^2(0, +\infty)$, and $\mu = N_Q$, where Q is any operator in $L_1^+(H)$ such that $\operatorname{Ker} Q = \{0\}$ and set $B(t) = W_{\chi_{[0,t]}}$, $t \geq 0$, where

$$
\chi_{[0,t]}(s) = \begin{cases}
1 \text{ if } s \in [0, t], \\
0 \text{ otherwise,}
\end{cases}
$$

and W is the white noise function introduced in Chapter 1. Then clearly $B(0) = 0$. Moreover, since for $t > s$,

$$B(t) - B(s) = W_{\chi_{[0,t]}} - W_{\chi_{[0,s]}} = W_{\chi_{(s,t]}},$$

we have by Proposition 2.12 that $B(t) - B(s)$ is a real Gaussian random variable with law N_{t-s}, and (i) is holds. Let us consider (ii). Since the system of elements of H,

$$(\chi_{[0,t_1]}, \chi_{(t_1,t_2]}, \cdots, \chi_{(t_{n-1},t_n]}),$$

is orthogonal, we have again by Proposition 2.12 that the random variables $B(t_1)$, $B(t_2) - B(t_1)$, ..., $B(t_n) - B(t_{n-1})$ are independent. Thus (ii) is proved. The proof of continuity of B, is more technical, see e. g. [6, Section 1.4].

In the following $B(t)$ represents a Brownian motion on a probability space $(\Omega, \mathcal{F}, \mathbb{P})$. We shall denote by $\mathbb{E}$ integration with respect to $\mathbb{P}$ and, by $\mathcal{F}_t$ the σ–algebra generated by $\{B(s); \ s \leq t\}$. One can say that $\mathcal{F}_t$ contains all the "story" of the Brownian motion, up to t.

Let us come back to equation (7.2). To solve it, it is enough to solve for any $T > 0$ the deterministic integral equation

$$u(t) = x + \int_s^t b(u(r))ds + f(t) - f(s), \ t \geq s, \tag{7.3}$$

where $x \in \mathbb{R}$, $f \in C([s,T]; \mathbb{R})$ ([7]).

Remark 7.2. When $f \in C^1([s,T]; \mathbb{R})$, the solution u of (7.3) coincides with the solution of the Cauchy problem

$$u'(t) = b(u(t)) + f'(t), \ \ u(s) = x. \tag{7.4}$$

We can assume that $B(\cdot)(\omega)$ is continuous for any $\omega \in \Omega$. However, one can show that the set of those ω such that $B(\cdot)(\omega)$ is differentiable has probability 0, see e. g. [3].

Equation (7.3) can be easily solved by the classical method of successive approximations. Let us set

$$u_0(t) = x, \ u_{n+1}(t) = x + \int_s^t b(u_n(r))dr + f(t), \ n \in \mathbb{N}, \ t \in [s,T]. \tag{7.5}$$

Then the following result holds.

---

[7] $C([s,T]; \mathbb{R})$ is the space of all continuous functions $f : [s,T] \to \mathbb{R}$, endowed with the norm $\|f\|_0 = \sup_{t \in [s,T]} |f(t)|$. $C^1([s,T]; H)$ is the subspace of $C([s,T]; H)$ of all continuously differentiable functions. We set $\|f\|_1 = \|f\|_0 + \|Df\|_0$.

Lemma 7.3. *Let $x \in H$, $0 \leq s \leq T$, $f \in C([s,T]; \mathbb{R})$. Then there exists a unique $u = u(\cdot, s, x) \in C([s,T]; H)$ fulfilling equation (7.3), and such that*

$$u = \lim_{n \to \infty} u_n, \text{ in } C([s,T]; \mathbb{R}). \tag{7.6}$$

Moreover, if $0 \leq s < \sigma < t$, we have

$$u(t, \sigma, u(\sigma, s, x)) = u(t, s, x). \tag{7.7}$$

Finally, if in addition b is of class C^k for some $k \in \mathbb{N}$, then u is of class C^k.

Proof. We only sketch the proof, since it is standard. We have

$$|u_1(t) - u_0(t)| \leq |b(x)|T + \|f\|_0, \ t \in [0, T],$$

and, by recurrence,

$$|u_{n+1}(t) - u_n(t)| \leq (|b(x)|T + \|f\|_0) \frac{M^n T^n}{n!}, \ n \in \mathbb{N}, \ t \in [0, T],$$

and so (7.6) follows easily by a classical argument.

Finally the last statement is standard. □

We now define a mapping γ_s by setting

$$\gamma_s : \mathbb{R} \times C([s,T]; \mathbb{R}) \to C([s,T]; \mathbb{R}), \ (x, f) \to \gamma_s(x, f) = u, \tag{7.8}$$

Now we can prove:

Proposition 7.4. *Let $\eta \in L^2(\Omega, \mathcal{F}, \mathbb{P})$. Then the following statements hold.*

(i) There exists a unique continuous stochastic process $X(\cdot, \eta)$ in $[s, +\infty)$ solution of the integral equation $\mathbb{P}$ almost all $\omega \in \Omega$,

$$X(t, s, \eta) = \eta + \int_s^t b(X(r, s, \eta)) dr + B(t) - B(s); \tag{7.9}$$

$X(\cdot, s, \eta)$ is given by

$$X(\cdot, s, \eta(\omega)) = \gamma(\eta, B(\cdot)(\omega)), \ \omega \in \Omega, \tag{7.10}$$

where γ is defined by (7.8).
(ii) For any $T > 0$ we have

$$X(\cdot, s, \eta) = \lim_{n \to \infty} X_n(\cdot, s, \eta), \text{ in } C([s,T]; L^2(\Omega, \mathcal{F}, \mathbb{P})), \tag{7.11}$$

where X_n, $n \in \mathbb{N}$ is defined by recurrence as

$$X_0(t, s, \eta) = \eta, \ X_{n+1}(t, s, \eta) = \eta + \int_s^t b(X_n(r, s, \eta)) ds + B(t) - B(s).$$

(iii) If $0 \le s < \sigma < t$, we have

$$X(t, \sigma, X(\sigma, s, x)) = X(t, s, x). \tag{7.12}$$

(iv) If $\eta = x$ is constant, the law of $X(t, s, x)$, $t \ge s$, is independent of the choice of the particular Brownian motion B.

Proof. (i), (ii) and (iii) are immediate consequences of Lemma 7.3 and (iv) can be easily proved by recurrence ([8]). □

Remark 7.5. Notice that, since $B(t)(\omega)$ is not differentiable in t for almost all $\omega \in \Omega$, the same happens for $X(t, s, x)(\omega)$.

Proposition 7.6. *For any $h > 0$, $X(t, s, x)$ has the same law as $X(t+h, s+h, x)$.*

Proof. We have in fact

$$X(t+h, s+h, x) = x + \int_{s+h}^{t+h} b(X(\tau, s+h, x))d\tau + B(t+h) - B(s+h)$$

$$= x + \int_{s}^{t} b(X(\tau+h, s+h, x))d\tau + B(t+h) - B(s+h).$$

This shows that $X(t+h, s+h, x)$ fulfills (7.9) with $B(t)$ replaced by $B_1(t) = B(t) - B(h)$. Since $B_1(t)$ is a Brownian motion, as easily checked, the conclusion follows from Proposition 7.4–(iv). □

In the following we set

$$X(t, x) = X(t, 0, x), \quad t \ge 0, \ x \in H.$$

Now we define the so called *transition semigroup*. We set

$$P_{s,t}\varphi(x) = \mathbb{E}[\varphi(X(t, s, x))], \quad t \ge s \ge 0, \ \varphi \in C_b(H),$$

and

$$P_t\varphi(x) = \mathbb{E}[\varphi(X(t, x))], \quad t \ge 0, \ \varphi \in C_b(H). \tag{7.13}$$

Proposition 7.7. *P_t is a Markov semigroup.*

Proof. We have just to show that P_t is a one parameter semigroup on $C_b(H)$ because then the probability kernel will be given by

$$\lambda_{t,x} = \mathcal{L}(X(t, 0, x)), \quad t \ge 0, \ x \in H.$$

---

[8] Notice that the law of $B(\cdot)$ in $C([0, T])$ does not depend on the choice of the Brownian motion, see e. g. [3]

Step 1. Let $\eta \in L^2(\Omega, \mathcal{F}_s, \mathbb{P})$. Then we have

$$\mathbb{E}[P_{s,t}\varphi(\eta)] = \mathbb{E}[\varphi(X(t,s,\eta))], \quad t \geq 0, \ \varphi \in C_b(H). \tag{7.14}$$

It is enough to consider the case when η is simple, that is

$$\eta = \sum_{k=1}^{n} c_k \chi_{A_k},$$

where $n \in \mathbb{N}$, $c_1, ..., c_n \in \mathbb{R}$ and $A_k \in \mathcal{F}_s$ for $k = 1, .., n$. By recurrence we can check that

$$X(t,s,\eta) = \sum_{k=1}^{n} \chi_{A_k} X(t,s,c_k),$$

and consequently

$$\varphi(X(t,s,\eta)) = \sum_{k=1}^{n} \chi_{A_k} \varphi(X(t,s,c_k)).$$

Since clearly χ_{A_k} and $\varphi(X(t,s,c_k))$ are independent, we have

$$\mathbb{E}[\varphi(X(t,s,\eta))] = \sum_{k=1}^{n} \mathbb{P}(A_k)\mathbb{E}[\varphi(X(t,s,c_k))]$$

$$= \sum_{k=1}^{n} \mathbb{P}(A_k)\mathbb{E}[P_{s,t}\varphi(c_k)] = \mathbb{E}[P_{s,t}\varphi(\eta)].$$

Step 2. For $0 < \sigma < s < t$ we have

$$P_{\sigma,s}P_{s,t}\varphi = P_{\sigma,t}\varphi \quad \varphi \in C_b(H). \tag{7.15}$$

We have in fact

$$P_{\sigma,s}P_{s,t}\varphi(x) = \mathbb{E}[P_{s,t}\varphi(X(s,\sigma,x))].$$

Then, applying (7.14) with $\eta = X(s,\sigma,x)$ yields

$$P_{\sigma,s}P_{s,t}\varphi(x) = \mathbb{E}[P_{s,t}\varphi(X(s,\sigma,x))] = \mathbb{E}[\varphi(X(t,s,X(s,\sigma,x)))] = \mathbb{E}[\varphi(X(t,\sigma,x))],$$

by Proposition 7.4–(iii).

Finally the semigroup law for P_t follows obviously from (7.15). $\square$

We want finally give some informations about the infinitesimal generator N of P_t. We will show that

$$N\varphi = \frac{1}{2} D^2\varphi + b(x)D\varphi, \quad \varphi \in C_b^2(H). \tag{7.16}$$

For this we need the following result.

Proposition 7.8 (Itô's formula). *For any* $\varphi \in C_b^2(H)$ *the function* $t \to \mathbb{E}[\varphi(X(t,x))]$ *is differentiable in* t *and we have* $(^9)$

$$\frac{d}{dt}\, \mathbb{E}\left[\varphi(X(t,x))\right] = \mathbb{E}\left[(N\varphi)(X(t,x))\right], \quad t \geq 0, \ x \in H. \qquad (7.17)$$

Proof. Let $h > 0$ and set $X(t+h,x) - X(t,x) = \delta_h$. Using the Taylor formula for φ we find

$$\frac{1}{h}\left(\mathbb{E}[\varphi(X(t+h,x))] - \mathbb{E}[\varphi(X(t,x))]\right)$$

$$= \frac{1}{h}\,\mathbb{E}[D\varphi(X(t,x))\delta_h] + \frac{1}{2h}\,\mathbb{E}[D^2\varphi(X(t,x))\delta_h^2]$$

$$+\frac{1}{h}\,\mathbb{E}\left[\int_0^1 (1-\xi)][D^2\varphi((1-\xi)X(t,x) + \xi X(t+h,x)) - D^2\varphi(X(t,x))\delta_h^2 d\xi\right]$$

$$= I_1 + I_2 + I_3. \tag*{(7.18)}$$

On the other hand we have

$$\delta_h = \int_t^{t+h} b(X(\tau,x))d\tau + (B(t+h) - B(t)).$$

Therefore

$$I_1 = \frac{1}{h}\,\mathbb{E}\left[\int_t^{t+h} b(X(\tau,x))d\tau\, D\varphi(X(t,x))\right]$$

$$+\frac{1}{h}\,\mathbb{E}[D\varphi(X(t,x))\,(B(t+h) - B(t)].$$

Since $B(t+h) - B(t)$ is independent of $D\varphi(X(t,x))$, because $B(t)$ is a process with independent increments, we have that the second term in the identity above vanishes. Thus, letting h tend to 0, we find

$$\lim_{h\to 0} I_1 = \mathbb{E}\left[D\varphi(X(t,x))b(X(t,x))\right]. \tag{7.19}$$

Concerning I_2 we have

---

[9] Notice that the differentiability of $\mathbb{E}[\varphi(X(t,x))]$ is not obvious, since $X(t,x)(\omega)$ is not differentiable in t in general.

$$I_2 = \frac{1}{2h} \, \mathbb{E}[D^2\varphi(X(t,x)) \left[\int_t^{t+h} b(X(\tau,x))d\tau \right]^2$$

$$+ \frac{1}{h} \, \mathbb{E}\left[D^2\varphi(X(t,x)) \int_t^{t+h} b(X(\tau,x))d\tau \, (B(t+h) - B(t)) \right]$$

$$+ \frac{1}{2h} \, \mathbb{E}\left[D^2\varphi(X(t,x)) \cdot (B(t+h) - B(t))^2 \right]$$

$$= I_{2,1} + I_{2,2} + I_{2,3}.$$

Clearly

$$\lim_{h \to 0} I_{2,1} = \lim_{h \to 0} I_{2,2} = 0. \tag{7.20}$$

Concerning $I_{2,3}$ we have, using again the fact that $B(t+h) - B(t)$ is independent of $D\varphi(X(t,x))$,

$$I_{2,3} = \frac{1}{2h} \, \mathbb{E}[D^2\varphi(X(t,x))]\mathbb{E}[(B(t+h) - B(t))^2] = \frac{1}{2} \, \mathbb{E}[D^2\varphi(X(t,x))),$$

because $\mathbb{E}[(B(t+h) - B(t))^2] = h$ since $B(t+h) - B(t)$ is a Gaussian random variable N_h. Therefore

$$\lim_{h \to 0} I_{2,3} = \frac{1}{2} \, \mathbb{E}[(D^2\varphi(X(t,x)))]. \tag{7.21}$$

Let us show finally that

$$\lim_{h \to 0} I_3 = 0 \tag{7.22}$$

To prove this we set

$$I_3 = \frac{1}{h} \, \mathbb{E} \int_0^1 g(\xi, h)\delta_h^2 d\xi,$$

where

$$g(\xi, h) = (1 - \xi) \left[D^2\varphi((1-\xi)X(t,x) + \xi X(t+h,x)) - D^2\varphi(X(t,x)) \right].$$

By the Hölder inequality we have

$$|I_3|^2 \le \frac{1}{h^2} \, \mathbb{E} \int_0^1 \|g(\xi, h)\|^2 d\xi \, \mathbb{E} \, |\delta_h|^4.$$

Thus, to prove (7.22) it is enough to show that

(i) $\lim_{h \to 0} \mathbb{E} \int_0^1 \|g(\xi, h)\|^2 d\xi = 0,$

(ii) $\frac{1}{h^2} \, \mathbb{E} \, |\delta_h|^4 \le c$, for some $c > 0.$

(i) follows from the dominated convergence theorem since $\lim_{h \to 0} g(\xi, h) = 0$, for all $h \in (0, 1]$ and $\mathbb{P}$-a.s, and $\|g(\xi, h)\| \le 2\|\varphi\|_2$. Let us prove (ii). Since $\mathbb{E}\,|B(t + h) - B(t)|^4 = (2n + n^2)h^2$, we have ([10])

$$\frac{1}{h^2}\,\mathbb{E}\,|\delta_h|^4 \le \frac{8}{h^2}\,\mathbb{E}\left|\int_t^{t+h} b(X(s, x))ds\right|^4 + \frac{8}{h^2}\,\mathbb{E}|B(t+h) - B(t)|^4$$

$$\le \frac{8}{h^2}\,\mathbb{E}\left|\int_t^{t+h} b(X(s, x))ds\right|^4 + 2n + n^2,$$

and (ii) is proved.

Now the conclusion follows from (7.19)–(7.20). $\square$

Finally, setting $t = 0$ in (7.17) yields

Proposition 7.9. *We have* $C_b^2(H) \subset D(N)$, *and if* $\varphi \in D(N)$ *then* (7.16) *holds.*

References

1. W. Arendt, A. Grabosch, G. Greiner, U. Groh, H. P. Lotz, U. Moustakas, R. Nagel, F. Neubrander and U. Schlotterbeck, One-parameter semigroups of positive operators. *Lecture Notes in Mathematics*, 1184. Springer-Verlag, Berlin, 1986.
2. S. Cerrai, A Hille-Yosida theorem for weakly continuous semigroups, *Semigroup Forum*, **49**, 349-367, 1994.
3. G. Da Prato, *An introduction to infinite dimensional analysis*, Appunti SNS Pisa, 2001.
4. G. Da Prato and J. Zabczyk, *Stochastic equations in infinite dimensions,* Encyclopedia of Mathematics and its Applications, Cambridge University Press, 1992.
5. G. Da Prato and J. Zabczyk, *Ergodicity for Infinite Dimensional Systems*, London Mathematical Society Lecture Notes, n.229, Cambridge University Press, 1996.
6. G. Da Prato and J. Zabczyk, *Second Order Partial Differential Equations in Hilbert Spaces,* London Mathematical Society Lecture Notes, n.293, Cambridge University Press, 2002.
7. C. Dellacherie and P.A. Meyer, *Probabilité et Potentiel, vol 1*, Hermann, 1975.
8. E. B. Davies, *One parameter semigroups*, Academic Press, 1980.
9. K. Engel and R. Nagel, *One-parameter semigroups for linear evolution equations,* with contributions by S. Brendle, M. Campiti, T. Hahn, G. Metafune, G. Nickel, D. Pallara, C. Perazzoli, A. Rhandi, S. Romanelli and R. Schnaubelt, *Graduate Texts in Mathematics*, **194**, Springer-Verlag, 2000.
10. R. S. Halmos, *Measure Theory*, Van Nostrand, 1950.
11. D. Henry, *Geometric Theory of Semilinear Parabolic Equations*, Springer-Verlag, 1981

---

[10] Notice that $E|B(t + h) - B(t)|^4 = (2\pi h)^{-1/2} \int_{\mathbb{R}} e^{-\frac{\xi^2}{2}} \xi^4 d\xi = 3h^2$.

12. N. Ikeda and S. Watanabe, *Stochastic differential equations and diffusion processes*, North Holland, 1981.

13. F. Kühnemund, Bi-continuous semigroups on spaces with two topologies: Theory and applications, Dissertation der Mathematischen Fakultät der Eberhard-Karls-Universität Tübingen zur Erlangung des Grades eines Doktors der Naturwissenschaften, 2001.

14. A. Lunardi, *Analytic semigroups and optimal regularity in parabolic problems*, Birkhäuser, 1995.

15. J. Neveu, *Bases mathématiques du Calcul des Probabilités*, Masson, 1970.

16. K. Petersen, *Ergodic theory*, Cambridge Studies in Avanced Mathematics, 1983

17. K. P. Parthasarathy, *Probability measures in metric spaces*, Academic Press, New York, 1967.

18. A. Pietsch, *Nuclear locally convex spaces*, Springer-Verlag, Berlin 1972.

19. E. Priola, On a class of Markov type semigroups in spaces of uniformly continuous and bounded functions, *Studia Math.*, **136**, 271-295, 1999.

20. K. Yosida, *Functional Analysis*, Springer–Verlag, 1965.

Maximal L_p-regularity for Parabolic Equations, Fourier Multiplier Theorems and H^∞-functional Calculus

Peer C. Kunstmann and Lutz Weis

Mathematisches Institut I, Universität Karlsruhe, Englerstrasse 2,
D-76128 Karlsruhe, Germany.
{peer.kunstmann,lutz.weis}@math.uni-karlsruhe.de

Abstract: In these lecture notes we report on recent breakthroughs in the functional analytic approach to maximal regularity for parabolic evolution equations, which set off a wave of activity in the last years and allowed to establish maximal L_p-regularity for large classes of classical partial differential operators and systems.

In the first chapter (Sections 2-8) we concentrate on the singular integral approach to maximal regularity. In particular we present effective Mihlin multiplier theorems for operator-valued multiplier functions in UMD-spaces as an interesting blend of ideas from the geometry of Banach spaces and harmonic analysis with R-boundedness at its center. As a corollary of this result we obtain a characterization of maximal regularity in terms of R-boundedness. We also show how the multiplier theorems "bootstrap" to give the R-boundedness of large classes of classical operators. Then we apply the theory to systems of elliptic differential operators on $\mathbb{R}^n$ or with some common boundary conditions and to elliptic operators in divergence form.

In Chapter II (Sections 9-15) we construct the H^∞-calculus, give various characterizations for its boundedness, and explain its connection with the "operator-sum" method and R-boundedness. In particular, we extend McIntosh's square function method form the Hilbert space to the Banach space setting. With this tool we prove, e.g., a theorem on the closedness of sums of operators which is general enough to yield the characterization theorem of maximal L_p-regularity. We also prove perturbation theorems that allow us to show boundedness of the H^∞-calculus for various classes of differential operators we studied before. In an appendix we provide the necessary background on fractional powers of sectorial operators.

0 Introduction

Some History: Consider an elliptic differential operator A on $X = L_q(\Omega)$, $1 < q < \infty$, and the Cauchy problem

$$y'(t) = Ay(t) + f(t), \qquad y(0) = 0 \qquad (0.1)$$

with given $f \in L_q([0, T], X)$. Then A has maximal-L_p-regularity if the solution $y(t)$ of (0.1) satisfies the a priori estimate

$$\int_0^T \|y'(t)\|_{L_q(\Omega)}^p dt + \int_0^T \|Ay(t)\|_{L_q(\Omega)}^p dt \leq C \int_0^T \|f(t)\|_{L_q(\Omega)}^p dt \qquad (0.2)$$

which states that y' and Ay are well defined and have the same (i. e. maximal) regularity as the right hand side f. The interest in maximal regularity stems from the fact that it is an important tool in the theory of non-linear differential equations. For example it allows to reduce quasilinear systems to linear systems via linearization techniques and fixed-point theorems (see e. g. [5], [47], [49], [172]).

The maximal L_p-regularity estimate (0.2) has a long tradition. The classical potential theoretic approach to such estimates is summarized in the book [150] by O. A. Ladyzhenskaya, V. A. Solonnikov, and N. N. Uraltseva.

Around the same time, P. E. Sobolevskii (e. g. in [203]), and then P. Grisvard ([110]) and G. DaPrato ([59]) recast the problem in the framework of analytic semigroups on Banach spaces and introduced two basic approaches to establish the maximal L_p regularity inequality (0.2)

- estimates for singular integral operators which were instrumental in proving maximal L_p-regularity for arbitrary analytic generators on Hilbert spaces (L. De Simon, [64], see also [44]), for dissipative operators on the L_q-scale (D. Lamberton, [151]), and for generators satisfying Gaussian estimates (M. Hieber and J. Prüss, [119] see also [52]).

- the operator-sum method, which gives maximal L_p-regularity for analytic generators A on real interpolation spaces (G. DaPrato, P. Grisvard, [59]), and for A with bounded imaginary powers on UMD-spaces. (G. Dore and A. Venni, [73], [70], [158])

In spite of this progress a central question raised by H. Brézis (presented in [54]) in the eighties remained open for many years: Does every generator of an analytic semigroup on L_q, $1 < q < \infty$, have maximal L_p-regularity? A deeper understanding of this question needed additional tools from the geometry of Banach spaces. In 1999, N. J. Kalton and and G. Lancien [132] used the theory of bases in Banach spaces, to show that (among Banach spaces with an unconditional basis) Hilbert spaces are the only spaces, in which Brézis' question has a positive answer. At the same time L. Weis [220], [221] gave a characterization of maximal L_p-regularity for an individual generator

A on UMD-spaces in terms of the *R-boundedness* of the resolvent of A or the semigroup generated by A. This notion, which originates in the work of J. Bourgain [39] on vector-valued multiplier theorems and was studied in its own right in [46], proved to be very adaptable to the task of checking maximal regularity for elliptic partial differential operators. In short sequence, simplified proofs for known results were found ([146], [135]), but more importantly, large classes of operators with rather general coefficient and boundary conditions ([141], [74], [66], [113]), operators in divergence form ([33], [34]) and Stokes operators ([67], [183], [95], [96], [97]) could be treated.

Again there were two approaches to the basic characterization theorem of maximal regularity. In [220] an operator-valued Fourier multiplier theorem in the style of Mihlin's theorem was used. This theorem has found already many extensions to more general situations ([210], [114], [14], [104], [103], [123], [126]) and applications to elliptic boundary value problems ([66]).

In [134], a second proof was given using a closedness result for sums of closed operators in the tradition of Dore and Venni. This method was applied to boundary value problems in [74]. The most important tool here is the H^∞-functional calculus which was developed over the years by McIntosh and his collaborators ([176], [56]) and which was used in the context of maximal regularity already in [4], [152] and [158].

The Notes: In these notes we give an introduction into these new developments. Section 1 gives a more detailed overview of ideas and techniques, in form of a mathematical essay that leaves the details of proofs to the following sections.

In Chapter I we give first some Banach space background for R-boundedness, including Kahane's inequality and a convexity result. The latter allows us to connect with operator theory by stating some simple but basic criteria for R-boundedness (Section 2). Following the approach of J. Bourgain we then prove operator-valued Fourier multiplier theorems by showing how we can build more general Fourier multipliers from the most basic one, the Hilbert transform. This is how UMD-spaces enter the story on maximal regularity: For a Banach space X we need the boundedness of the Hilbert transform on the Bochner space $L_p(X)$ (Sections 3 and 4). Our vector-valued setting has an additional advantage: it allows us to deduce the R-boundedness of large classes of classical operators in function spaces from the multiplier theorem itself.

In particular this leads to simple proofs for the maximal regularity of semigroups on $L_p(\Omega)$ satisfying Gaussian estimates, or that are contractive on the L_p-scale. The latter class includes most operators generating a stochastic processes (Section 5).

Then we apply the theory to systems of elliptic differential operators on $\mathbb{R}^N$ or with some common boundary condition. Here we did not strive for the greatest generality in order to explain the method without obscuring by excessive technical detail (Sections 6 and 7). We also explain how weighted norm estimates that generalize Gaussian kernel bounds may be used to establish maximal regularity and apply this method to elliptic operators in divergence form (Section 8).

In Chapter II we consider operator-theoretic tools for maximal regularity, in particular we give a careful construction of the holomorphic H^∞-functional calculus (Section 9). We first present the standard examples of sectorial operators with a bounded H^∞-calculus such as generators of contraction semigroups on Hilbert spaces or on the L_p-scale and explain the connection with dilation theorems (Section 10). Then we approach the beautiful characterization of operators with a bounded H^∞-calculus on Hilbert space through the square function method of A. McIntosh, which is again inspired by harmonic analysis (Section 11). Following these clues we develop a Banach space theory for square functions which allows us to construct a functional calculus for operator-valued functions, a joint functional calculus for commuting operators. This leads us back to maximal regularity via the sum theorem. Here R-boundedness plays a crucial role again (Section 12). Then we present comparison and perturbation results that allow to show that elliptic operators have a bounded H^∞-calculus if their highest order coefficients are sufficiently smooth (Section 13). Finally, we take up the weighted norm estimates from Section 8 and show how they may be used to extrapolate a bounded H^∞-calculus from L_2 to other L_p-spaces. Again, this is applied to elliptic operators in divergence form (Section 14). In an appendix (Section 15) we collect the necessary information on fractional powers and extrapolation spaces and give proofs based on the H^∞-calculus.

Prerequisites: These notes are an expanded version of a course the second author gave at the CIRM-autumn school on "Evolution equations and semigroups" in Levico Terme in October 2001. The course was addressed to advanced students and this is also the audience we have in mind writing these lecture notes. We tried to preserve somewhat the informal style of lectures, to spell out explicitly the underlying ideas and reduce technicalities to a minimum. Often this effort lead us to new or substantially simplified proofs (e.g., for some of the main results in Sections 3, 4, 5, 11, and 12, but also for some of the applications to differential operators in Sections 6, 7, and 13).

We assume a course in functional analysis as well as some familiarity with the Fourier transform and semigroups. An introduction such as in [197], Sections IX.1, IX.2, X.8 and [222] V.2 should suffice. At the end of each section we give historical notes, comment on variants of presented results, and indicate references for further reading.

For a presentation of this material more in the style of a research monograph and the most general results on boundary value problems we refer to [66].

Acknowledgement: The authors express their sincere gratitude to their colleagues in Karlsruhe, N. Basmer, J. Dettweiler, A. Fröhlich, M. Girardi,

V. Goersmeyer, B. Haak, C. Kaiser, C. Schmoeger, and J. Zimmerschied, for generous help and most valuable support. We also thank the editors of this volume, M. Iannelli, R. Nagel, and S. Piazzera, for constant encouragement and patience. Without all of them this work would not have been possible.

1 An Overview: Two Approaches to Maximal Regularity

In this introductory section we give an overview of our main results on maximal regularity, Fourier multiplier theorems and the bounded holomorphic functional calculus, and show how they are connected. We try to include enough details to give a flavour of the results, the mathematics involved and a feeling for the interplay of the various techniques. Detailed proofs, for the most part, will have to wait for the following sections.

We use the semigroup approach to evolution equations, which we recall now briefly. Many Cauchy problems for elliptic and parabolic partial differential equations can be written abstractly in the form

$$y' = Ay + f, \qquad\qquad y(0) = x_0, \qquad\qquad (1.1)$$

where A is a closed and densely defined operator on a Banach space X, $f : \mathbb{R}_+ \mapsto X$ is a locally integrable function and $x_0 \in X$ an initial value. In these notes we will concentrate on operators A which have their spectrum outside a sector $\Sigma_\sigma = \{\lambda \in \mathbb{C} : \lambda \neq 0, |\arg \lambda| < \sigma\}$ a little larger than a half plane, i.e. $\sigma > \frac{\pi}{2}$, and their resolvent $R(\lambda, A) = (\lambda - A)^{-1}$ satisfy a "parabolic" estimate

$$\|\lambda R(\lambda, A)\| \leq C, \qquad\qquad \lambda \in \Sigma_\sigma$$

on Σ_σ. For such an A we can define a family of operators by a contour integral

$$T_z = \frac{1}{2\pi i} \int_\Gamma e^{\lambda z} R(\lambda, A) d\lambda, \qquad\qquad z \in \Sigma_{\sigma - \frac{\pi}{2}}, \qquad (1.2)$$

where $\Gamma = \partial(\Sigma_\nu \setminus \{\lambda : |\lambda| \leq \varepsilon\})$ for some $\frac{\pi}{2} < \nu < \sigma$ and $\varepsilon > 0$, with the following properties:

$$\frac{d}{dz} T_z = A T_z, \qquad\qquad z \in \Sigma_{\sigma - \frac{\pi}{2}}, \qquad (1.3)$$

$$T_{z_1} T_{z_2} = T_{z_1 + z_2}, \qquad\qquad z_1, z_2 \in \Sigma_{\sigma - \frac{\pi}{2}}. \qquad (1.4)$$

Thus A generates the *analytic semigroup* (T_z) and we can recover the resolvent from (T_z) by

$$R(\lambda, A) = \int_0^\infty e^{-\lambda t} T_t \, dt, \qquad\qquad \text{Re}\lambda > 0. \qquad (1.5)$$

Hence (1.2) is the inversion of the Laplace transform (1.5). These observations lead to the following characterization of analytic semigroups and their generators.

1.1 Theorem. *For a densely defined closed operator A on a Banach space X, the following are equivalent:*

(i) *$\{\lambda R(\lambda, A) : \lambda \in \Sigma_\sigma\}$ is bounded for some $\sigma > \frac{\pi}{2}$.*
(ii) *There is an analytic semigroup (T_z) on a sector $\Sigma_\delta, \delta > 0$, such that (1.3) holds and $\{T_z : z \in \Sigma(\delta)\}$ is bounded.*
(iii) *There is a C_0-semigroup $T_t, t \in \mathbb{R}_+$, such that (1.3) holds for $t \in \mathbb{R}_+$ and $\{T_t, tAT_t : t > 0\}$ is bounded.*

Remark. The supremum of all $\sigma > 0$ for which (i) holds equals the supremum of all $(\frac{\pi}{2} + \delta)$ with $\delta > 0$ for which (ii) holds.
Condition (1.3) implies in particular that the solution of the Cauchy problem (1.1) can be given formally in terms of (T_t) by the classical "Variation of constants formula":

$$y(t) = T_t x_0 + \int_0^t T_{t-s}(f(s)) ds. \tag{1.6}$$

Here are some important examples of generators of analytic semigroups that will play a prominent role in what follows.

1.2 Examples. Let $X = L_p(\mathbb{R}^n)$ for $1 \le p < \infty$.

a) Everybody's favorite example is certainly the *Laplace operator*

$$A = \Delta = \sum_{j=1}^n \frac{\partial^2}{\partial x_j^2} \qquad D(\Delta) = W_p^2(\mathbb{R}^n).$$

Here $W_p^2(\mathbb{R}^n)$ is the completion of $C_c^\infty(\mathbb{R}^n)$ in the norm

$$\|f\| = \sum_{|\alpha| \le 2} \|D^\alpha f\|_{L_p}.$$

Δ generates the Gaussian semigroup (see Section 5)

$$(G_t f)(u) = (4\pi t)^{-n/2} \int_{\mathbb{R}^n} e^{-|u-v|^2/4t} f(v) dv. \tag{1.7}$$

b) A little bit more fancy are systems of *elliptic partial differential equations* i.e. we consider operators $-A$ of order $2m$ on $L_p(\mathbb{R}^n, \mathbb{C}^n)$ where

$$A = \sum_{|\alpha| \le 2m} a_\alpha D^\alpha$$

with matrix-valued coefficients $a_\alpha \in L_\infty(\mathbb{R}^n, \mathbb{C}^{N \times N})$. Here we use the following multiindex notation: $D^\alpha = D_1^{\alpha_1} \cdot \ldots \cdot D_n^{\alpha_n}$ for $\alpha = (\alpha_1, \ldots, \alpha_n) \in \mathbb{N}_0^n$ where $D_j := -i\frac{\partial}{\partial x_j}$. Denote the principle symbol of A by

$$\mathcal{A}_\pi(x,\xi) = \sum_{|\alpha|=2m} a_\alpha(x)\xi^\alpha, \qquad (x,\xi) \in \mathbb{R}^n \times \mathbb{R}^n.$$

Then A is called elliptic if the eigenvalues of $\mathcal{A}_\pi(x,\xi)$, $\xi \neq 0$, belong to Σ_θ for some $\theta \in [0,\pi)$ and

$$\sup\{|\mathcal{A}_\pi(x,\xi)^{-1}| : x \in \mathbb{R}^n, |\xi| = 1\} < \infty,$$

(see Section 6).

c) In mathematical physics *Schrödinger operators* of the form

$$-A = -\Delta + V$$

are prominent. Here the "potential" V belongs to $L_p(\mathbb{R}^n) + L_\infty(\mathbb{R}^n)$ for $p > \frac{n}{2} \geq 1$, or more generally to the Kato class (see Section 8).

d) The operator $A = (-\Delta)^{1/2}$ generates the *Poisson semigroup*

$$T_t f(u) = \frac{1}{\pi} \int_{\mathbb{R}^N} \frac{t}{t^2 + |u-v|^{\frac{N+1}{2}}} f(v)dt, \qquad (1.8)$$

which plays an important role in harmonic analysis (see Section 5).

e) Most of the examples we just mentioned generate a semigroup (T_t) which satisfies a *Gaussian bound* in the sense that for a constant C and all $f \in L_p(\mathbb{R}^n)$

$$|T_t f(u)| \leq CG_t|f|(u), \qquad u \in \mathbb{R}^n, t \in \mathbb{R}_+$$

where (G_t) is the semigroup in (1.7). There are much larger classes of differential operators, for which such Gaussian bounds are satisfied. This is important since Gaussian bounds are a powerful tool to extend properties of classical partial differential operators, such as the Laplace operator, to larger classes of differential operators. These bounds often allow to transfer Hilbert space results for partial differential operators to the L_p-setting. By now Gaussian bounds have been established for large classes of differential operators (see Sections 5 and 8).

1.3 Maximal Regularity. Let A be the generator of a bounded analytic semigroup on a Banach space X. We say that A has *maximal L_p-regularity* for $1 < p < \infty$ on $[0,T)$, $0 < T \leq \infty$, if for $x_0 = 0$ and all $f \in L_p([0,T), X)$ the solution (1.6) of the Cauchy problem (1.1) is (Fréchet-)differentiable a.e., takes its values in $D(A)$ a.e. and, most importantly, y' and Ay belong to $L_p([0,T), X)$. In this case by the closed graph theorem there is a constant C_p such that

$$\|y'\|_{L_p([0,T),X)} + \|Ay\|_{L_p([0,T),X)} \leq C_p\|f\|_{L_p([0,T),X)}. \qquad (1.9)$$

The word 'maximal' refers to the fact that by (1.1) $f = y' - Ay$ and therefore y' and Ay cannot be in a 'better' function space than f. Maximal L_p-regularity has many important applications to evolution equations, e.g. it is an important tool in the investigation of the following problems:

- existence and uniqueness of solutions to non-autonomous evolution equations;
- existence and uniqueness of solutions to quasi-linear partial differential equations;
- stability theory for evolution equations, i.e. construction of central manifolds, and detection of bifurcations;
- regularity of solutions of elliptic differential equations;
- existence and uniqueness of solutions of Volterra integral equations;
- uniqueness of mild solutions of the Navier Stokes equation.

Usually maximal L_p-regularity is used in these applications to reduce the non-autonomous or nonlinear problem via a fixed point argument to an autonomous or linear problem, respectively. In some cases maximal regularity is needed to apply an implicit function theorem (see the notes). We illustrate the use of fixed point arguments in a simple case.

1.4 Illustration. We consider the non-autonomous Cauchy problem

$$y'(t) = A(t)y(t) + f(t), \qquad y(0) = x, \tag{1.10}$$

where $A(t)$, $t \in [0, T]$, is a family of analytic generators that satisfies $\|R(\lambda, A(t))\| \leq C(1 + |\lambda|)^{-1}$ on the closed right half plane and the maximal regularity estimate (1.9) with a constant C independent of $t \in [0, T]$. Furthermore, we assume that $D(A(t))$ is isomorphic to a fixed Banach space $D \subset X$ for all t and $t \in [0, T] \mapsto A(t) \in B(D, X)$ is continuous.

Such an equation may be given as a differential equation

$$\partial_t u(t, \omega) = \sum_{i,j=1}^{n} a_{ij}(t, \omega) \frac{\partial^2}{\partial x_i \partial x_j} u(t, \omega), \qquad u(0, \omega) = x(\omega),$$

where $A(t) = \sum a_{ij}(t, \cdot) \frac{\partial^2}{\partial x_i \partial x_j}$ are uniformly elliptic differential operators with coefficients a_{ij} continuous in t and ω.

Notice that, in contrast to (0.1) in the introduction, we now have a non-zero initial value x. In this case we also need that the solution for $f = 0$ given by $y(s) = e^{sA(t)}x$ belongs to the space $W_p^1([0, T], X) \cap L_p([0, T], D)$. The space of all $x \in X$ with this property is a certain intermediate space between X and D, which in our case does not depend on t. We denote it by $(Y, \|\cdot\|_Y)$ and refer to the notes for further details. So for $x \in Y$ we have an estimate

$$\|s \mapsto e^{sA(t)}x\|_{W_p^1([0,T],X)} + \|s \mapsto e^{sA(t)}x\|_{L_p([0,T],D)} \leq C'\|x\|_Y$$

for a constant C' uniformly in $t \in [0, T]$.

In order to prove the existence of a solution of (1.10) for $x \in Y$ we write

$$y'(t) = A(0)y(t) + g_y(t), \qquad g_y(t) = [A(t) - A(0)]y(t) + f(t).$$

Denote by $L(y)$ the solution of the autonomous problem

$$u'(t) = A(0)u(t) + g_y(t), \qquad u(0) = x. \tag{1.11}$$

Then, at least formally, the solutions of (1.10) are the fixed points of L. Maximal regularity will now be used to show that L is a contraction in $L_p([0, a], D)$ with $1 < p < \infty$ if $a \leq T$ is small enough. Indeed, if $y_1, y_2 \in L_p([0, a], D)$, then $L(y_1) - L(y_2)$ equals the solution u of

$$u'(t) = A(0)u(t) + g_{y_1}(t) - g_{y_2}(t), \quad u(0) = 0.$$

Since D is isomorphic to $D(A(0))$ with the norm $x \mapsto \|A(0)x\|$ (note that $0 \in \rho(A(0))$) we obtain with the maximal regularity inequality (1.9) for equation (1.11) that

$$
\begin{aligned}
\|L(y_1) - L(y_2)\|_{L_p([0,a],D)} &\leq C\|A(0)u\|_{L_p([0,a],X)} \\
&\leq C\|g_{y_1} - g_{y_2}\|_{L_p([0,a],X)} \\
&= C\|[A(\cdot) - A(0)](y_1 - y_2)\|_{L_p([0,a],X)} \\
&\leq C \sup_{t\in[0,a]} \|A(t) - A(0)\|_{B(D,X)}\|y_1 - y_2\|_{L_p([0,a],D)}.
\end{aligned}
$$

Hence, if a is small enough then $C \cdot \sup\{\|A(t) - A(0)\| : t \leq a\} < 1$ and we can apply Banach's fixed point theorem to $L : L_p([0, a], D) \to L_p([0, a], D)$ to obtain a solution for (1.10) on $[0, a]$. Since our constants do not depend on t and $A(t)$ is uniformly continuous on $[0, T]$ we repeat the argument with initial value $x = y(a)$, and after finitely many repetitions we obtain a solution on $[0, T]$. Here, of course, we have to guarantee that $x = y(a)$ belongs again to the space Y we introduced above. Fortunately, this is the case and follows from the general embedding result

$$W_p^1([0, T], X) \cap L_p([0, T], D) \hookrightarrow BUC([0, T], Y), \tag{1.12}$$

again we refer to the notes for further explanations. $\square$

Not every generator of an analytic semigroup has maximal regularity. We will present some counterexamples in 1.15. The main result of this section will be a characterization of maximal L_p-regularity of A in terms of the resolvent of A (see Theorem 1.11). The following discussion will motivate the conditions that will appear in Theorem 1.11 and also explain the connection between the maximal regularity problem and Fourier multiplier theorems.

1.5 Discussion. Let A be the generator of a bounded analytic semigroup and consider the Cauchy problem (1.1) with $x_0 = 0$. To obtain a formula for y' we differentiate the variation of constant formula (1.6). Using $\frac{d}{dt}T_t = AT_t$ we obtain

$$y'(t) = \int_0^t AT_{t-s}f(s)ds + f(t) \tag{1.13}$$

which certainly makes sense for $f \in C_c(\mathbb{R}_+, D(A))$. Since this space is dense in $L_p(\mathbb{R}_+, X)$, A has maximal L_p-regularity if and only if the operator

$$Kf(t) = \int_0^t AT_{t-s}f(s)ds, \qquad f \in C_c(\mathbb{R}_+, D(A)), \tag{1.14}$$

extends to a bounded operator $K : L_p(\mathbb{R}_+, X) \longrightarrow L_p(\mathbb{R}_+, X)$. The operator K is a convolution operator with a singular kernel since $\|AT_t\| \sim \frac{1}{t}$. Therefore it is tempting to apply the Fourier transform

$$\mathcal{F}(f)(t) = \hat{f}(t) = \int e^{-it \cdot s} f(s)\, ds \tag{1.15}$$

for integrable f. As in the case of scalar convolution we obtain from (1.14)

$$\widehat{Kf}(u) = (AT_t)\widehat{}(u)[\hat{f}(u)], \qquad u \in \mathbb{R}.$$

By (1.5) we can identify $(AT_t)\widehat{}$ as

$$m(u) := (AT_t)\widehat{}(u) = AR(iu, A) = iuR(iu, A) - I. \tag{1.16}$$

Since A is the generator of an analytic semigroup, the function m must be bounded on $\mathbb{R} \setminus \{0\}$ and the same is true for

$$um'(u) = -iuAR(iu, A)^2 = [uR(iu, A)]^2 + iuR(iu, A). \tag{1.17}$$

The last two conditions (1.16) and (1.17) remind us of Mihlin's multiplier theorem, but we need a version for operator valued multiplier functions m and Bochner spaces $L_p(\mathbb{R}, X)$. Indeed, J. Schwartz proved such a theorem if X is a Hilbert space.

1.6 Theorem. *Let X be a Hilbert space. Assume that for the function $m \in C^1(\mathbb{R} \setminus \{0\}, B(X))$ the sets*

$$\{m(u) : u \in \mathbb{R} \setminus \{0\}\} \quad and \quad \{u\, m'(u) : u \in \mathbb{R} \setminus \{0\}\}$$

are bounded in $B(X)$. Then the Fourier multiplier operator

$$T_m f = \mathcal{F}^{-1}(m(\cdot)\hat{f}(\cdot)), \qquad f \in \mathcal{S}(\mathbb{R}, X) \tag{1.18}$$

extends to a bounded operator T_m on $L_p(\mathbb{R}, X)$ for $1 < p < \infty$.

By $\mathcal{S}(\mathbb{R}, X)$ we denote here the rapidly decreasing smooth functions with values in a Banach space X; if $f \in \mathcal{S}(\mathbb{R}, X)$, then $m(\cdot)\hat{f}(\cdot)$ is in $L_1(\mathbb{R}, X)$, $T_m f$ belongs at least to $C(\mathbb{R}, X)$ and one may ask whether $T_m f \in L_p(\mathbb{R}, X)$.

Applying this theorem for a Hilbert space X to the situation discussed in 1.5 above, we can conclude that the operator K in (1.14) is bounded on $L_p(\mathbb{R}, X)$ and we find the following classical result of De Simon.

1.7 Corollary. *Every generator of a bounded analytic semigroup on a Hilbert space X has maximal L_p-regularity.*

Unfortunately, Theorem 1.6 is only true for Hilbert spaces. G. Pisier observed that if Theorem 1.6 holds in the stated form for some Banach space X, then X is already isomorphic to a Hilbert space. So if we want a multiplier theorem, that can be applied to maximal regularity for a larger class of Banach spaces, we need additional assumptions on the Banach space X and the multiplier function m. What assumptions will work?

1.8 UMD-spaces. A useful multiplier theorem should cover the basic multiplier function $m(u) = i\pi\mathrm{sign}(u)$, which corresponds to the Hilbert transform

$$Hf(t) = PV - \int \frac{1}{t-s} f(s)ds, \qquad f \in \mathcal{S}(\mathbb{R}, X). \qquad (1.19)$$

A Banach space X is called a *UMD-space* if H extends to bounded operator on $L_p(\mathbb{R}, X)$ for some (or equivalently, for each) $p \in (1, \infty)$.

All subspaces and quotient spaces of $L_q(\Omega, \mu)$ for $1 < q < \infty$ have the UMD property but $L_1(\Omega, \mu)$ or spaces of continuous functions $C(K)$ do not. As a rule of thumb we can say that Sobolev spaces, Hardy spaces and other well known spaces of analysis are UMD if they are reflexive.

1.9 R-boundedness. The boundedness assumption for the multiplier function in Mihlin's theorem will be replaced by the following stronger boundedness concept: A set of operators $\tau \subset B(X, Y)$ is called R-*bounded*, if there is a constant $C < \infty$ such that, for all $T_1, ..., T_m \in \tau$ and $x_1, ..., x_m \in X$ with $m \in \mathbb{N}$, we have

$$\left\| \sum_{n=1}^{m} r_n T_n x_n \right\|_{L_2([0,1],Y)} \leq C \left\| \sum_{n=1}^{m} r_n x_n \right\|_{L_2([0,1],X)}$$

where $r_n(t) = \mathrm{sign}\, \sin(2^n \pi t)$ are the Rademacher functions on $[0, 1]$. This notion will be discussed in detail in Section 2.

For the moment we note that in a Hilbert space H every bounded set is R-bounded. Indeed, $L_2([0, 1], H)$ is a Hilbert space and $(r_n x_n)$ and $(r_n T_n x_n)$ are orthogonal sequences in $L_2([0, 1], H)$ so that

$$\begin{aligned}
\left\| \sum_n r_n T_n x_n \right\|_{L_2([0,1],H)}^2 &= \sum_n \|T_n x_n\|_H^2 \leq (\sup_n \|T_n\|^2) \sum_n \|x_n\|_H^2 \\
&= (\sup_n \|T_n\|^2) \left\| \sum_n r_n x_n \right\|_{L_2([0,1],H)}^2 .
\end{aligned} \qquad (1.20)$$

Using these notions we will prove in Section 3 the following Fourier multiplier theorem due to L. Weis.

1.10 Theorem. *Let X and Y be UMD spaces. Assume that for $m \in C^1(\mathbb{R} \setminus \{0\}, B(X,Y))$ the sets*

$$\{m(t) : t \in \mathbb{R} \setminus \{0\}\} \quad and \quad \{tm'(t) : t \in \mathbb{R} \setminus \{0\}\}$$

are R-bounded. Then the operator

$$Tf = \mathcal{F}^{-1}[m(\cdot)\widehat{f}(\cdot)], \qquad f \in \mathcal{S}(\mathbb{R}, X)$$

extends to a bounded operator $T : L_p(\mathbb{R}, X) \longrightarrow L_p(\mathbb{R}, Y)$ for all $p \in (1, \infty)$.

Note that, if X is a Hilbert space, this theorem reduces to Theorem 1.6. Now we apply this more general Fourier multiplier theorem to the maximal regularity problem. If we replace the boundedness in (1.16) and (1.17) by R-boundedness, the discussion in 1.5 and Theorem 1.10 show that A has maximal L_p-regularity if the set $\{uR(iu, A) : u \in \mathbb{R} \setminus \{0\}\}$ is R-bounded. This is the main implication of the following characterization of maximal-L_p-regularity, due to L. Weis.

1.11 Theorem. *Let A be the generator of a bounded analytic semigroup in a UMD-space X. Then A has maximal L_p-regularity for one (all) $p \in (1, \infty)$ on $\mathbb{R}_+$ if and only if one of the following equivalent conditions is fulfilled.*

(i) $\{\lambda R(\lambda, A) : \lambda \in \Sigma_\sigma\}$ *is R-bounded for some $\sigma \geq \frac{\pi}{2}$;*

(ii) $\{T_z : z \in \Sigma_\delta\}$ *is R-bounded for some $\delta > 0$;*

(iii) $\{T_t, \, tAT_t : t > 0\}$ *is R-bounded.*

It is instructive to compare Theorem 1.11 with Theorem 1.1: If we replace in the equivalent conditions for analyticity boundedness by R-boundedness we obtain characterizations of maximal L_p-regularity. Therefore we call an analytic semigroup R-*analytic* if it satisfies the conditions of Theorem 1.11.

Condition (i) is usually expressed in terms of R-*sectoriality*: Let B be a closed operator on X with dense domain. We call B *sectorial of angle $\omega \in (0, \pi)$*, if $\sigma(B) \subset \overline{\Sigma_\omega}$ and for all $\nu > \omega$ we have that $\{\lambda R(\lambda, B) : \nu \leq |\arg(\lambda)| \leq \pi\}$ is bounded. If this set is even R-bounded then B is called R-*sectorial of angle ω.* Hence (i) states that $B = -A$ is R-sectorial for some angle $\omega < \frac{\pi}{2}$.

The remaining implications of Theorem 1.11 will be discussed in Section 2 after we will have developed the basic properties of R-boundedness and have seen how the Laplace transform (1.5) and its inverse (1.2) map R-bounded sets into R-bounded sets. There we will also justify the following remarks:

1.12 Remarks.

a) It is sufficient for (i) to show that $\{tR(it, A) : t \in \mathbb{R} \setminus \{0\}\}$ is R-bounded.

b) If $0 \in \varrho(A)$ then it suffices for maximal L_p-regularity on $\mathbb{R}_+$ that the sets

$$\{tR(it, A) : |t| \geq \delta\} \text{ or } \{T_t, tAT_t : 0 < t < \varepsilon\}$$

are R-bounded for some $0 < \varepsilon$ or $\delta < \infty$.

c) A has maximal L_p-regularity on a finite interval $[0, T]$ if and only if one of the conditions in b) holds. (Indeed, without loss of generality we can consider $A - \varepsilon I$ in place of A and apply b).)

Theorem 1.11 can be used to derive in a unified way the known criteria for maximal L_p-regularity mostly by checking the R-boundedness of the resolvent (i) in Theorem 1.11. Here are some examples that will be considered in later sections.

1.13 Examples. The following generators have maximal L_p-regularity (for the numerous contributions to this list see the notes).

a) All analytic generators on a Hilbert space H (see Corollary 1.7).

b) All analytic generators on $L_p(\Omega, \mu)$ which have Gaussian bounds (see Section 5) or satisfy certain weighted norm estimates (see Section 8).

c) Elliptic differential operators on $L_p(\Omega)$ with general boundary conditions where Ω is a region in $\mathbb{R}^n$ (see Sections 6 and 7).

d) All A that generate a positive and contractive semigroup on $L_p(\Omega, \mu)$ spaces (see Sections 5 and 10). This criteria applies in particular to many generators of stochastic processes.

e) All analytic generators with a bounded H^∞-functional calculus (or more generally, with bounded imaginary powers) on a UMD-space X (see Section 12).

Viewed together the criteria in 1.13 show that classical differential operators on $L_p(\Omega)$-spaces for $1 < p < \infty$, are R-sectorial if they are sectorial. This lead Brézis to the question whether all analytic semigroups on $L_p(\Omega, \mu)$ are R-analytic. Recently, G. Lancien and N. J. Kalton provided a complete answer to his question:

1.14 Theorem. *If X has an unconditional basis and all analytic semigroups on X are R-analytic then X is isomorphic to a Hilbert space.*

Within the class of Banach spaces with an unconditional basis which includes $L_p(\Omega)$-spaces, this is a complete converse to Corollary 1.7. However, the existence of counterexamples on non-Hilbert spaces follows only indirectly from the theory of bases in Banach spaces. It is still an open question whether there are "natural" differential operators on $L_p(\mathbb{R}^n)$ which generate analytic but not R-analytic semigroups.

We would also like to point out that the UMD-assumption in the Characterization Theorem 1.11 cannot be dropped.

1.15 Counterexamples. E.g., the Laplace operator on $L_1(\mathbb{R}^n)$ does not have maximal L_p-regularity. However, the need for the UMD-condition in Theorem 1.11 becomes even more apparent from the following result of T. Couhlon and D. Lamberton: If we extend the Poisson semigroup (1.8) to $L_2(\mathbb{R}, X)$ by

$$T_t f(u) = \frac{1}{\pi} \int \frac{t}{t^2 + (u - v)^2} f(v)dv, \qquad f \in L_2(\mathbb{R}, X),$$

then its generator $A = (H\frac{d}{dx}) \otimes Id_X$ has maximal L_p-regularity if and only if X is a UMD-space. This follows from the fact that the Hilbert transform appears in the formulation of A as well as in the definition of UMD-spaces.

There is a second approach to the maximal regularity problem initiated by DaPrato and Grisvard which uses operator theory instead of multiplier theorems. To prove the Characterization Theorem 1.11 along these lines, we first introduce a functional calculus for sectorial operators.

1.16 H^∞-calculus. By $H^\infty(\Sigma_\sigma)$, $\sigma \in (0, \pi)$, we denote the Banach space of all bounded analytic functions defined on Σ_σ with norm $\|f\|_{H^\infty(\Sigma_\sigma)} = \sup_{z \in \Sigma_\sigma} |f(z)|$. If A is a bounded operator on X and $\sigma(A) \subset \Sigma_\sigma$, there is a classical way to define a functional calculus $f \in H^\infty(\Sigma_\sigma) \mapsto f(A) \in B(X)$ for A which

(i) is linear and multiplicative,

(ii) is norm bounded,

(iii) satisfies $f_\mu(A) = R(\mu, A)$ for $f_\mu(\lambda) = (\mu - \lambda)^{-1}$ with $\mu \notin \overline{\Sigma_\sigma}$.

Namely, we can employ the Dunford integral

$$f(A) := \frac{1}{2\pi i} \int_\Gamma f(\lambda) R(\lambda, A) d\lambda, \tag{1.21}$$

with $\Gamma = \partial(\Sigma_\sigma \cap \{\lambda : |\lambda| \leq r\})$ and $r = 2\|A\|$.

If A is an unbounded sectorial operator of angle ω and $f \in H^\infty(\Sigma_\sigma)$ with $\sigma > \nu > \omega$, then it is tempting to consider the limit $r \to \infty$ and to replace the Dunford integral (1.21) by

$$f(A) = \frac{1}{2\pi i} \int_{\partial \Sigma_\nu} f(\lambda) R(\lambda, A) d\lambda. \tag{1.22}$$

But note that $\|R(\lambda, A)\| \sim \frac{1}{|\lambda|}$ on $\partial \Sigma_\nu$, and therefore (1.22) is a singular integral, which may not always exist. Of course the integral (1.22) is finite if

$$\int_{\delta \Sigma_\nu} |f(\lambda)| \; |\lambda|^{-1} d|\lambda| < \infty. \tag{1.23}$$

Following A. McIntosh we say that A has a bounded $H^\infty(\Sigma_\sigma)$-*calculus* if we can extend the map $f \mapsto f(A) \in B(X)$ first defined for $f \in H^\infty(\Sigma_\sigma)$ with the additional property (1.23) to a bounded map on all of $H^\infty(\Sigma_\sigma)$, satisfying the properties (i), (ii), (iii) above. The details of this construction will be given in Section 9, and in Sections 10, 11, 13 and 14 we will show that many differential operators have such a bounded H^∞-calculus.

To discuss maximal regularity we will need in particular the following example.

1.17 Example. Let X be a UMD-space. Then the operator $\tilde{B} = \frac{d}{dt}$ on $\tilde{X} = L_p(\mathbb{R}, X)$ with $1 < p < \infty$ has a bounded $H^\infty(\Sigma_\sigma)$-calculus for all $\sigma > \frac{\pi}{2}$.

We sketch how to derive this fact from the Multiplier Theorem 1.10. For $g \in \mathcal{S}(\mathbb{R}, X)$ we have $\tilde{B}g = \mathcal{F}^{-1}[it\hat{g}(t)]$. Therefore, for $f \in H^\infty(\Sigma_\sigma)$ the operator $f(\tilde{B})g$ should be the Fourier multiplier operator $\mathcal{F}^{-1}[f(it)\hat{g}(t)]$. To show that this operator is bounded on $L_p(\mathbb{R}, X)$, consider Cauchy's formula

$$tf'(it) = \frac{1}{2\pi i} \int_{\partial\Sigma_\nu} \frac{t}{(\lambda - it)^2} f(\lambda)d\lambda.$$

Since the integrand is uniformly bounded in $L_1(\partial\Sigma_\nu)$, $\sigma < \nu < \frac{\pi}{2}$, for $t \in \mathbb{R} \setminus \{0\}$ we obtain

$$\sup_{t\in\mathbb{R}} |tf'(it)| \leq C \sup_{z\in\Sigma_\sigma} |f(z)| \leq C \|f\|_{H^\infty(\Sigma_\sigma)}.$$

Now one can check that $f \in H^\infty(\Sigma_\sigma) \mapsto f(\tilde{B}) \in B(L_p(X))$ satisfies (i), (ii) and (iii) of 1.16 and that

$$\|f(\tilde{B})\| \leq C \sup_{t\in\mathbb{R}}\{|f(it)|, |tf'(it)|\} \leq C' \|f\|_{H^\infty(\Sigma_\sigma)}.$$

$\square$

1.18 Discussion. Now we can describe the approach of Da Prato and Grisvard to the maximal regularity problem for the Cauchy problem

$$y'(t) = Ay(t) + f(t), \qquad y(0) = 0.$$

First we rewrite the maximal regularity estimate (1.9)

$$\|y'\|_{L_p(\mathbb{R}_+,X)} + \|Ay\|_{L_p(\mathbb{R}_+,X)} \leq C\|f\|_{L_p(\mathbb{R}_+,X)}$$

as an inequality for operators on $L_p(\mathbb{R}_+, X)$ for $1 < p < \infty$. If we denote by $\tilde{B}$ the derivative $\frac{d}{dt}$ on $\tilde{X} = L_p(\mathbb{R}_+, X)$ and by $\tilde{A}$ the extended operator $(\tilde{A}f)(t) = -A(f(t))$, then (1.9) is equivalent to

$$\|\tilde{A}y\|_{\tilde{X}} + \|\tilde{B}y\|_{\tilde{X}} \le C\|(\tilde{A} + \tilde{B})y\|_{\tilde{X}}, \qquad y \in D(\tilde{A}) \cap D(\tilde{B}). \qquad (1.24)$$

Note that $\tilde{A}$ and $\tilde{B}$ have commuting resolvents. If X is a UMD-space then $\tilde{B}$ has a bounded H^∞-calculus on $\tilde{X} = L_p(\mathbb{R}_+, X)$ for all $\sigma > \frac{\pi}{2}$ by 1.17. If we assume now that A is R-sectorial with angle $\omega < \frac{\pi}{2}$, then the following 'sum theorem' due to N. J. Kalton and L. Weis, implies (1.24) and therefore maximal regularity for A (which is again the main result of our Characterization Theorem 1.11).

1.19 Theorem. *Let A and B be closed operators on X whose resolvents commute. In particular, $D(A) \cap D(B)$ is dense in X. Assume that B has a bounded $H^\infty(\Sigma_\sigma)$-calculus and A is R-sectorial with angle ω such that $\sigma + \omega < \pi$. Then $A + B$ is closed on $D(A) \cap D(B)$ and*

$$\|Ax\| + \|Bx\| \le C\|Ax + Bx\|, \qquad x \in D(A) \cap D(B). \qquad (1.25)$$

Let us just indicate how the bounded H^∞-calculus enters into the proof of 1.19 (for details see Section 12): The estimate (1.25) holds if and only if $A(A + B)^{-1}$ and $B(A + B)^{-1}$ can be defined as bounded operators on X. Since formally $B(A + B)^{-1} = I - A(A + B)^{-1}$ we can concentrate on the first operator. We may think of $A(B + A)^{-1}$ as $f(B)$, where $f(\lambda) = A(\lambda + A)^{-1} = -AR(-\lambda, A)$. f is analytic on Σ_σ since $\sigma < \pi - \omega$. So the boundedness of $A(B + A)^{-1} = f(B)$ follows if B has a functional calculus for operator valued analytic functions $\lambda \mapsto f(\lambda)$ with R-bounded range. In Section 12 we will show that a bounded H^∞-calculus can always be extended to such an operator valued calculus. For an operator valued f satisfying the additional assumption (1.23) this extended calculus will again be defined by (1.22). $\square$

Notes on Section 1:

N 1.1 - N 1.2: For the standard material on analytic semigroups see books on semigroup theory, e.g. [88, II.4.a], [186, 2.5] or [12], [172], for Gaussian bounds see [63]. A short introduction to semigroups can also be found in [197] and [222]. See also 9.8.

An alternative definition of maximal L_p-regularity requires in addition that $y \in W_p^1([0, T), X)$. For $T < \infty$ this definition is equivalent to ours, since the convolution (1.6) gives a function in $L_p([0, T])$ thanks to the boundedness of the semigroup. But for $T = \infty$ the alternative definition already implies that $0 \in \rho(A)$ (see e.g. [70]). It seems desirable to us, not to exclude the possibility that $0 \in \sigma(A)$.

For the theory of maximal regularity in the scale of Hölder continuous functions (instead of the L_p-scale) see the monograph [172]. This theory does not require assumptions on the Banach space X.

N 1.3: It is shown in [70] that maximal L_p-regularity of (1.1) already implies that A generates an analytic semigroup. For the applications of maximal regularity mentioned in 1.3, we refer to Amann [5], [8], [9], Clément and Li [47], Clément and Prüss [49], [192], [89], Lunardi [172] and Monniaux [179].

N 1.4: The result on the non-autonomous equation in 1.4 is a special case of known results (see e.g. [193], [8]). The space Y we introduced here is a special case of the spaces $D_A(\alpha, p)$ where A is the generator of a bounded analytic semigroup in X and $\alpha \in (0, 1)$. By definition, $D_A(\alpha, p)$ is the space of all $x \in X$ such that

$$[x]_{\alpha,p} := \|s \mapsto s^{1-\alpha} A e^{sA} x\|_{L_p(\mathbb{R}_+, \frac{ds}{s}, X)} < \infty.$$

It is a Banach space for $\|x\| + [x]_{\alpha,p}$. If $0 \in \rho(A)$, then $x \in D_A(1 - 1/p, p)$ is equivalent to $s \mapsto e^{sA} x \in W_p^1(\mathbb{R}_+, X)$ and equivalent to $s \mapsto e^{sA} x \in W_p^1([0, T], X)$ (by the semigroup property). This means that the space Y we introduced in 1.4 is the space $D_{A(t)}(1 - 1/p, p)$. It may be shown, cf. [172, Prop. 2.2.2], that $D_A(\alpha, p) = (X, D)_{\alpha,p}$ with equivalent norms where $(X, D)_{\alpha,p}$ is the real interpolation space between X and the domain D of A equipped with the graph norm. For the embedding (1.12) we refer to [5, Thm. 4.10.2].

N 1.5: In his proof of Corollary 1.7, De Simon [64] may have been the first who applied methods from the theory of singular intergal operators to maximal regularity estimates. This approach is also used in [203] and [44] to show that maximal L_p-regularity does not depend on $p \in (1, \infty)$.

N 1.6 - N 1.7: A proof of J. Schwartz's result 1.6 using the Calderon-Zygmund method can be found in [29], Section 6.1. Corollary 1.7 is from [64]. Pisier's observation on the operator-valued Mihlin multiplier theorem was a folklore result for many years and is now contained in [14].

N 1.8: There are many important statements in vector-valued harmonic analysis and probability theory that are equivalent to the UMD-property. For this we refer to [42]. The significance of the UMD-property for vector-valued multiplier theorems was recognized in Bourgain's paper [39], in which Theorem 1.10 was proven for scalar valued multiplier functions in the periodic case.

N 1.9: The notion of R-boundedness is also implicit in this paper; it was named "Riesz-property" in [30], "Randomized boundedness" in [46] and is also known as "Rademacher boundedness". Luckily enough everybody can agree on the 'R'.

N 1.10 - N 1.12: The operator-valued Multiplier Theorem 1.10 is from [221]. For more information on R-boundedness and multiplier theorems see the notes of Sections 2 to 5. The Characterization Theorem 1.11 and Remark 1.12 are also from [221]. This theorem has already inspired characterizations of maximal L_p-regularity for evolution equations different from (1.1), e.g. for Volterra equations in [49], [220] and for periodic equations in [14]. The term 'R-sectorial' was introduced in [48] and [220].

N 1.13: Maximal L_p-regularity for generators of semigroups with Gaussian bounds (or even Poisson bounds) were shown first by Hieber and Prüss in [119], see also [52]. In [48] and [220] maximal functions from harmonic analysis were used to show that these estimates imply R-sectoriality. In Section 5 we will present a more elementary approach from [105]. For differential operators, see the notes to Sections 6 and 7. For analytic semigroups which are contractive on the whole L_p-scale, $1 \leq p \leq \infty$, maximal regularity was established by L. Lamberton in [151]. Using Theorem 1.11 it is shown in [220] that it is sufficient if the semigroup is positive and contractive on L_p for some $p \in (1, \infty)$.

N 1.14 - N 1.15: G. Lancien and N. J. Kalton's solution to the maximal regularity problem is in [132] and the result in 1.15 is due to T. Coulhon and D. Lamberton [54]. For further counterexamples see [157] and [153].

N 1.16 - N 1.17: For Hilbert spaces the H^∞-calculus was developed by A. McIntosh [176]; for the Banach space setting see [56]. For further information we refer to the notes of Sections 9, 10, 11 and 12.

N 1.18 - N 1.19: The 'method' of operator sums discussed in 1.18 goes back to G. Da Prato and P. Grisvard [59], who considered mainly real interpolation spaces. Dore and Venni ([73], [70]) took a major step forward by proving the sum theorem for resolvent commuting operators with bounded imaginary powers (BIP) on UMD-spaces. While the assumption in the Dore-Venni theorem are symmetric in A and B (both have BIP), Theorem 1.19 from [134] makes a stronger assumption on B (bounded H^∞-calculus) but a weaker assumption on A (R-sectoriality). Since B is $\frac{d}{dt}$ or Δ in many applications, this asymmetry allows to reach the Characterization Theorem 1.11.

For the joint functional calculus of resolvent commuting operators and its applications to sum theorems see [152], [154], [158] and [134].

I. Fourier Multiplier Theorems and Maximal Regularity

2 *R*-boundedness

In 1.9 we defined R-boundedness in terms of Rademacher series $\sum_n r_n x_n$. First we take a closer look at such series and discuss Kahane's inequality and the contraction principle. Then we explore R-boundedness and show how it can be expressed in L_p-spaces. A convexity result leads to various criteria for R-boundedness, which should give a feeling how R-bounded sets appear in analysis. Their usefulness will be demonstrated by the proof of Theorem 1.11 at the end of the section.

2.1 Rademacher Functions. The Rademacher functions, which are defined by $r_n(t) = \operatorname{sign} \sin(2^n \pi t)$, $n \in \mathbb{N}$, are an orthonormal sequence in $L_2[0,1]$ as you can see from their graphs:

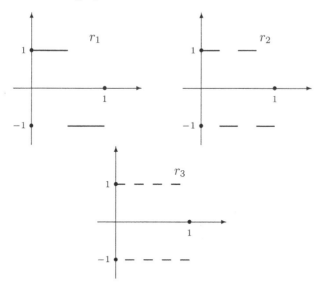

We can also look at them as a sequence of identically distributed, stochastically independent functions since for the Lebesgue-measure μ on $[0,1]$ and $\delta_1, \ldots, \delta_m \in \{-1, 1\}$ we have

$$\mu(\{r_{n_1} = \delta_1, \ldots, r_{n_m} = \delta_m\}) = \frac{1}{2^m} = \prod_{j=1}^{m} \mu(\{r_{n_j} = \delta_j\}).$$

Therefore, for $a_1, \ldots, a_m \in \mathbb{C}$

$$\int_0^1 \left| \sum_{n=1}^{m} r_n(u) a_n \right|^p du = \mathbb{E} \left| \sum_{n=1}^{m} r_n(u) a_n \right|^p = \frac{1}{2^m} \sum_{\delta_n \in \{-1,1\}} \left| \sum_{n=1}^{m} \delta_n a_n \right|^p.$$

This formula shows that the integral above remains unchanged if we replace the Rademacher functions by any sequence (ε_n) of independent functions with the distribution $\mu(\{\varepsilon_n = 1\}) = \mu(\{\varepsilon_n = -1\}) = \frac{1}{2}$. We will use this observation occasionally.

2.2 Khinchine's Inequality. This inequality allows to evaluate L_p-norms of Rademacher sums: For $1 \leq p < \infty$ there is a constant C_p such that for all $a_n \in \mathbb{C}$

$$\frac{1}{C_p} \left(\sum_n |a_n|^2 \right)^{\frac{1}{2}} \leq \left\| \sum_n r_n a_n \right\|_{L_p[0,1]} \leq C_p \left(\sum_n |a_n|^2 \right)^{\frac{1}{2}}. \qquad (2.1)$$

Proof. For $p = 2$ the claim follows of course directly from the fact that (r_n) is an orthonormal sequence in $L_2[0,1]$ with $C_2 = 1$.

If all the L_p-norms are supposed to be equivalent for $f(t) = \sum_{n=1}^{m} a_n r_n(t)$, then, intuitively, f should not 'grow' too fast. Indeed, we show first for $\sum_{n=1}^{m} |a_n|^2 = 1$, that

$$\int_0^1 e^{|f(t)|} dt \leq 2e^{\frac{1}{2}}. \qquad (2.2)$$

Since the r_n are independent

$$\int_0^1 e^{f(t)} dt = \prod_{n=1}^{m} \int_0^1 e^{a_n r_n(t)} dt = \prod_{n=1}^{m} \frac{1}{2} (e^{a_n} + e^{-a_n}).$$

Since $e^{\beta} + e^{-\beta} \leq 2e^{\beta^2/2}$ (e.g. compare the coefficients of the power series on both sides) we get

$$\int_0^1 e^{f(t)} dt \leq \prod_{n=1}^{m} e^{\frac{a_n^2}{2}} = \exp\left(\sum_{n=1}^{m} \frac{a_n^2}{2} \right) = e^{\frac{1}{2}}.$$

The same estimate holds for $\int_0^1 e^{-f(t)} dt$ and hence

$$\int_0^1 e^{|f(t)|} dt \leq \int_0^1 e^{f(t)} dt + \int_0^1 e^{-f(t)} dt \leq 2e^{\frac{1}{2}}.$$

Now if $p \in \mathbb{N}$ we have for $y \in \mathbb{R}$ that

$$|y|^p < p!(1 + |y|^p/p!) \leq p! e^{|y|}$$

and therefore, still assuming $\|f\|_{L_2} = \left(\sum_{n=1}^m |a_n|^2\right)^{\frac{1}{2}} = 1$, by (2.2)

$$\int_0^1 |f(t)|^p dt \leq p! \int_0^1 e^{|f(t)|} dt \leq 2p! e^{\frac{1}{2}}.$$

Hence $\|f\|_{L_p} \leq C_p$ for $\|f\|_{L_2} = 1$ and $C_p = (2e^{\frac{1}{2}} p!)^{1/p}$. By homogeneity and monotonicity of the $L_p[0,1]$-norm we get for $p \in \mathbb{N}$, $p \geq 2$,

$$\|f\|_{L_2} \leq \|f\|_{L_p} \leq C_p \|f\|_{L_2}.$$

For arbitrary $p \in (2, \infty)$ we put $q = [p] + 1$ and conclude that $\|f\|_{L_2} \leq \|f\|_{L_p} \leq \|f\|_{L_q} \leq C_q \|f\|_{L_2}$.
If $1 \leq p < 2$, we use Hölder's inequality. Choose $\theta \in (0,1)$ so that $2 = \theta p + (1 - \theta)4$, i.e. $\theta = (2 - \frac{p}{2})^{-1}$. Then

$$\int_0^1 |f(t)|^2 dt = \int_0^1 |f(t)|^{p\theta} \cdot |f(t)|^{4(1-\theta)} dt \leq \left(\int_0^1 |f(t)|^p dt\right)^\theta \left(\int_0^1 |f(t)|^4 dt\right)^{1-\theta}.$$

Since $\|f\|_4 \leq C_4 \|f\|_2$ we can divide by $\|f\|^{4(1-\theta)}$ and obtain

$$C_4^{-4(1-\theta)} \|f\|_{L_2}^{p\theta} \leq \|f\|_{L_p}^{p\theta}.$$

Since $\|f\|_{L_p} \leq \|f\|_{L_2}$ we have shown again that $\|f\|_{L_p}$ is equivalent to $\|f\|_{L_2} = \left(\sum |a_n|^2\right)^{\frac{1}{2}}$. $\qquad\square$

Remark. By inequality (2.1), Rad_p, the space of all series $\sum_{n=1}^\infty r_n a_n$ convergent in $L_p[0,1]$, is isomorphic to l_2 in all L_p-spaces, and this has two interesting consequences:

(i) Rad_p is isomorphic to Rad_2 for $1 \leq p < \infty$;

(ii) For all $(\alpha_n) \subset \mathbb{C}$ with $|\alpha_n| \leq 1$

$$\left\|\sum_{n=1}^\infty r_n \alpha_n a_n\right\|_{L_p} \leq C \left\|\sum_{n=1}^\infty r_n a_n\right\|_{L_p}.$$

If we consider now vector valued sums $\sum_n r_n x_n$ with x_n in a Banach space X, then (2.1) fails. As a matter of fact, the inequality

$$\frac{1}{C} \left(\sum_n \|x_n\|^2\right)^{\frac{1}{2}} \leq \left\|\sum_n r_n x_n\right\|_{L_2([0,1],X)} \leq C \left(\sum_n \|x_n\|^2\right)^{\frac{1}{2}}$$

holds for a fixed constant C and arbitrary $x_n \in X$ if and only if X is a Hilbert space. But it is important for us, that the conditions (i) and (ii) which were derived from (2.1) have vector valued analogues.

2.3 Notation. By $Rad_p(X)$ we denote the Banach space of the sums

$$\sum_{n=1}^{\infty} r_n x_n \qquad \text{with } x_n \in X$$

that converge in the Bochner space $L_p([0,1], X)$ for $1 \leq p < \infty$.
If $\sum_n x_n r_n = 0$, then $\sum_n x'(x_n) r_n = 0$ for all $x' \in X'$ and the orthogonality
of (r_n) implies that $x_n = 0$ for all n. Hence for every $f \in Rad_p(X)$ there is a
unique sequence (x_n) in X such that $\sum_n x_n r_n$ converges to f in $L_p([0,1], X)$
and we can identify $Rad_p(X)$ with a space of sequences (x_n) in X with norm
$\|(x_n)\|_{Rad_p(X)} = \|\sum_n r_n x_n\|_{L_p(X)}$:

$$Rad_p(X) = \{(x_n) : x_n \in X, \|(x_n)\|_{Rad_p(X)} < \infty\}.$$

If X is reflexive, then $(x_n)_{n \in \mathbb{N}} \in Rad_p(X)$ is equivalent to

$$M = \sup_N \left\| \sum_{n=1}^{N} r_n x_n \right\|_{L_p([0,1], X)} < \infty. \tag{2.3}$$

$Rad_p^m(X)$ is the subspace of finite sums $\sum_{n=1}^{m} r_n x_n$. Our first theorem extends
(i) in the remark after 2.2:

2.4 Theorem. (Kahane's inequality) *The spaces $Rad_p(X)$ are isomorphic
for $1 \leq p < \infty$, i.e. there is a constant $C_p < \infty$, such that for $x_n \in X$*

$$\frac{1}{C_p} \left\| \sum_n r_n x_n \right\|_{L_2(X)} \leq \left\| \sum_n r_n x_n \right\|_{L_p(X)} \leq C_p \left\| \sum_n r_n x_n \right\|_{L_2(X)}.$$

One can show that there is an analogue of (2.2) also in the vector valued
case, i.e. there are constants $\alpha > 0$ and $C < \infty$ such that for all $f(t) = \sum_{i=1}^{n} r_i(t) x_i$ with $\|f\|_{L_2(X)} = 1$ we have

$$\int_0^1 e^{\|f(t)\|} dt \leq C. \tag{2.4}$$

With this inequality, we can repeat the proof of Khinchine's inequality. But
(2.4) is a lot harder to prove than (2.2) in the scalar case (see the notes).
There is also an extension of (ii) in the remark after 2.2:

2.5 Proposition. (Kahane's contraction principle) *A convergent series
in $Rad_p(X)$ converges unconditionally, i.e. for all $(a_n) \subset \mathbb{C}$ with $|a_n| \leq 1$,*

$$\left\| \sum_n r_n a_n x_n \right\|_{L_p(X)} \leq 2 \left\| \sum_n r_n x_n \right\|_{L_p(X)}.$$

Proof. We go step by step. First, if $a_n \in \{-1, 1\}$, then $\|\sum_{n=1}^{m} r_n a_n x_n\|_{L_p(X)} = \|\sum_{n=1}^{m} r_n x_n\|_{L_p(X)}$ since both equal

$$\left(\frac{1}{2^m} \sum_{\sigma_n \in \{-1,1\}} \left\| \sum_{n=1}^{m} \sigma_n x_n \right\|^p \right)^{\frac{1}{p}}.$$

If $a_n \in \{0, 1\}$ we write $\sum_n r_n a_n x_n = \frac{1}{2} \sum_n r_n x_n + \frac{1}{2} \sum_n r_n (2a_n - 1) x_n$. Since $2a_n - 1 \in \{-1, 1\}$ we can conclude that $\|\sum_n r_n a_n x_n\| \leq \|\sum_n r_n x_n\|$.

Next we consider coefficients a_n in $[0, 1]$. Let $a_n = \sum_{k=1}^{\infty} 2^{-k} a_{nk}$ be the dyadic expansion of a_n with $a_{nk} \in \{0, 1\}$. Then by the last step

$$\left\| \sum_n r_n a_n x_n \right\| \leq \sum_k 2^{-k} \left\| \sum_n r_n a_{nk} x_n \right\| \leq \left(\sum_{k=1}^{\infty} 2^{-k} \right) \left\| \sum_n r_n x_n \right\| = \left\| \sum_n r_n x_n \right\|.$$

The case of real coefficients a_n in $[-1, 1]$ can be reduced to the case $a_n \in [0, 1]$ by our first step. If the a_n are complex with $|a_n| \leq 1$ we consider the real part and the imaginary part of a_n separately. Note that for real a_n we can replace the constant 2 in our claim by 1. $\qquad\square$

2.6 Remark. Our results on Rademacher sums have now some very useful consequences for R-boundedness.

a) We can replace the $L_2(X)$-norm in our earlier definition 1.9 by the $L_p(X)$-norm, i. e. a set $\tau \subseteq B(X, Y)$ is *R-bounded* if and only if for one (or all) $p \in [1, \infty)$ there is a constant C such that for all $T_1, \ldots, T_m \in \tau$ and $x_1, \ldots, x_m \in X$

$$\left\| \sum_n r_n T_n x_n \right\|_{L_p([0,1],Y)} \leq C \left\| \sum_n r_n x_n \right\|_{L_p([0,1],X)}. \tag{2.5}$$

But note that the constant in (2.5) depends on p. We denote by $R_p(\tau)$ the best constant such that (2.5) holds. With C_p from Theorem 2.4 we have $(C_p)^{-2} R_2(\tau) \leq R_p(\tau) \leq C_p^2 R_2(\tau)$. Usually we will not distinguish between the $R_p(\tau)$'s and incorporate C_p into the constant at hand. For $R_2(\tau)$ we write $R(\tau)$.

b) We may replace the Rademacher functions in (2.5) by any sequence of independent, $\{-1, 1\}$-valued sequence (ε_n) with mean value 0, since (cf. 2.1)

$$\left(\mathbb{E} \left\| \sum_{n=1}^{m} \varepsilon_n x_n \right\|_X^p \right)^{1/p} = \left(\frac{1}{2^m} \sum_{\sigma_j \in \{-1,1\}} \left\| \sum_{j=1}^{m} \sigma_j x_j \right\|^p \right)^{\frac{1}{p}}.$$

Such random variables (ε_n) will be referred to as Bernoulli variables.

c) If we introduce in the definition of R-boundedness the additional restriction that the chosen $T_1, \ldots, T_n \in \tau$ are always different operators we obtain an equivalent definition (see the notes).

d) Kahane's contraction principle says that $\{aI : |a| \leq 1\}$ is R-bounded. Hence we may view R-boundedness as a contraction principle for operators instead of scalars which may or may not hold for a given subset τ of $B(X, Y)$.

Sometimes, it is useful, at least on an intuitive level, to think of an R-bounded sequence of operators as *one* bounded operator on the larger space $Rad(X)$ and of the R-boundedness estimate (2.5) as a norm estimate on $Rad(X)$. Indeed, looking back at our definitions we obtain

2.7 Proposition. *a) A sequence of operators* $(T_n) \subseteq B(X, Y)$ *is R-bounded if and only if the map* $\mathbf{T}(x_n) = (T_n x_n)$ *extends to a bounded linear operator from* $Rad(X)$ *to* $Rad(Y)$ *and* $\|\mathbf{T}\| = R(\{T_n : n \in \mathbb{N}\})$.

b)For a set $\tau \subseteq B(X, Y)$ *we have* $R(\tau) \leq C$ *if and only if for all* $T_1, \ldots, T_m \in \tau$ *the map* $\mathbf{T}(x_n) = (T_n x_n)$ *extends to a bounded operator* $\mathbf{T} : Rad_2^m(X) \to Rad_2^m(X)$ *and* $\|\mathbf{T}\| \leq C$.

The next statements replace the norm inequalities $\|S + T\| \leq \|T\| + \|S\|$ and $\|S \circ T\| \leq \|S\| \cdot \|T\|$ in estimates of R-bounded sets.

2.8 Fact. If τ and σ are R-bounded subsets of $B(X, Y)$ and ρ is an R-bounded subset of $B(Y, Z)$, then

$$\sigma + \tau = \{S + T : S \in \sigma, T \in \tau\} \quad \text{and} \quad \rho \circ \tau = \{R \circ T : R \in \rho, T \in \tau\}$$

are R-bounded. Moreover, we have

$$R(\sigma + \tau) \leq R(\sigma) + R(\tau) \quad \text{and} \quad R(\rho \circ \tau) \leq R(\rho)R(\tau).$$

Proof. If $S_n \in \sigma$, $T_n \in \tau$, and $R_n \in \rho$, then the claim follows from

$$\left\| \sum_n r_n (T_n + S_n) x_n \right\|_{L_1(Y)} \leq \left\| \sum_n r_n T_n x_n \right\|_{L_1(Y)} + \left\| \sum_n r_n S_n x_n \right\|_{L_1(Y)}$$

$$\left\| \sum_n r_n R_n T_n x_n \right\|_{L_1(Z)} \leq R(\rho) \left\| \sum_n r_n T_n x_n \right\|_{L_1(Y)}. \quad \square$$

We noticed already in 1.9 that in a Hilbert space H we have

$$\left\| \sum_n r_n x_n \right\|_{L_2(H)} = \left(\sum_n \|x_n\|_H^2 \right)^{1/2}.$$

Therefore every bounded subset of $B(H)$ is R-bounded. Kahane's and Khinchine's inequalities allow us now to characterize R-bounded sets of operators on L_p-spaces in terms of square functions estimates known from harmonic analysis.

2.9 Remark. Let $X = L_q(\Omega, \mu)$ on a σ-additive measure space (Ω, μ) with $1 \leq q < \infty$. We can identify each $x_1, \ldots, x_m \in X$ with a function $x_j(\cdot)$ on Ω and by Kahane's inequality 2.4, Fubini's theorem and Khinchine's inequality

2.2 applied to the scalar sequence $x_n(\omega)$, $n \in \mathbb{N}$, we can reformulate the R-boundedness condition

$$\Big\| \sum_{n=1}^{m} r_n x_n \Big\|_{L_2([0,1],X)} \sim \Big\| \sum_{n=1}^{m} r_n x_n \Big\|_{L_q([0,1],L_q(\Omega))}$$

$$= \left(\int_\Omega \int_0^1 \Big| \sum_{n=1}^{m} r_n(u) x_n(\omega) \Big|^q du\, d\mu(\omega) \right)^{\frac{1}{q}} \qquad (2.6)$$

$$\sim \left(\int_\Omega \Big(\sum_{n=1}^{m} |x_n(\omega)|^2 \Big)^{\frac{q}{2}} d\mu(\omega) \right)^{\frac{1}{q}} = \Big\| \Big(\sum_{n=1}^{m} |x_n|^2 \Big)^{\frac{1}{2}} \Big\|_{L_q(\Omega)}.$$

Now the R-boundedness condition (2.5) turns into the square function estimate

$$\Big\| \Big(\sum_{n=1}^{m} |T_n x_n|^2 \Big)^{\frac{1}{2}} \Big\|_{L_q} \le C \Big\| \Big(\sum_{n=1}^{m} |x_n|^2 \Big)^{\frac{1}{2}} \Big\|_{L_q}. \qquad (2.7)$$

This reformulation will allow us to connect with the wealth of square function estimates known in harmonic analysis and will guide us – at least intuitively – when we treat Fourier multiplier theorems in the next section. Again by (2.6) we see that $Rad_2^m(X)$ for $m \in \mathbb{N} \cup \{\infty\}$ is isomorphic to the Banach lattice $L_q(\Omega, l_2^m)$ with norm

$$\|(x_n)\|_{L_q(l_2^m)} = \Big\| \Big(\sum_{n=1}^{m} |x_n|^2 \Big)^{\frac{1}{2}} \Big\|_{L_q(\Omega)}.$$

Elements of $L_q(\Omega, l_2^m)$ can be identified with sequences $(x_n)_{n=1}^m$ of functions in $L_q(\Omega)$ and as in 2.7 we note that a set $\tau \subseteq B(L_q(\Omega))$ is R-bounded with $R(\tau) \le C$ if and only if for all $T_1, \ldots, T_m \in \tau$ the operator $\mathbf{T}(x_n) = (T_n x_n)$ is bounded on $L_q(\Omega, l_2)$ with $\|\mathbf{T}\| \le C$.

Hence in an L_q-space R-boundedness reduces to an estimate in the Banach lattice $L_q(\Omega, l_2)$ and this will allow for special results not available in the general Banach space setting. Here is a simple example.

2.10 Example. a) Let $\tau \subseteq B(L_q(\Omega))$. Assume that the operators in τ are dominated by one positive operator $T \in B(L_q(\Omega))$ in the sense that

$$|Sx| \le T|x| \qquad \text{for all } S \in \tau, \ x \in L_q(\Omega).$$

Then τ is R-bounded.

Proof. Choose $S_1, \ldots, S_m \in \tau$, and $x_1, \ldots, x_n \in L_q(\Omega)$. Then with (2.6)

$$\Big\| \Big(\sum_n |S_n x_n|^2 \Big)^{\frac{1}{2}} \Big\|_{L_q} \le \Big\| \Big(\sum_n (T|x_n|)^2 \Big)^{\frac{1}{2}} \Big\|_{L_q} \le C \Big\| \sum_n r_n T |x_n| \Big\|_{L_q([0,1],L_q(\Omega))}$$

$$\le C\|T\| \Big\| \sum_n r_n |x_n| \Big\|_{L_q([0,1],L_q(\Omega))}$$

$$\le C^2 \|T\| \Big\| \Big(\sum_n |x_n|^2 \Big)^{\frac{1}{2}} \Big\|_{L_q}. \qquad \square$$

b) The remark in a) may be applied e.g. to pointwise estimates of integral operators

$$T_k f(t) = \int_\Omega k(t, s) f(s) ds.$$

If T_{k_0} is bounded on $L_q(\Omega)$ with a positive kernel k_0, then the set of integral operators $\{T_k : |k| \leq k_0\}$, whose kernels are dominated by k_0, is R-bounded in $B(L_q(\Omega))$.

c) Also, for a positive semigroup T_t on $L_q(\Omega)$ it is well known ([180]), that for $\lambda_0 \in \rho(A) \cap \mathbb{R}$ the resolvent operator $R(\lambda_0, A)$ is positive and

$$|R(\lambda, A)f| \leq R(\lambda_0, A)|f|$$

for $\mathrm{Re}\,\lambda \geq R_0$. Hence the set $\{R(\lambda, A) : \mathrm{Re}\,\lambda \geq \lambda_0\}$ is R-bounded.   □

On $L_q(\Omega)$ with $q \in (1, \infty)$ we also have complete duality for R-boundedness.

2.11 Corollary. Let $X = L_q(\Omega, \mu)$ with $1 < q < \infty$. Then $\tau \subseteq B(X)$ is R-bounded if and only if $\tau' = \{T' : T \in \tau\} \subseteq B(X')$ is R-bounded and $R(\tau') = R(\tau)$.

Proof. We have $L_q(\Omega, l_2^m)' = L_{q'}(\Omega, l_2^m)$ for $\frac{1}{q} + \frac{1}{q'} = 1$ with respect to the duality

$$\langle (x_n), (y_n) \rangle = \sum_{n=1}^m \int_\Omega x_n(\omega) y_n(\omega) d\mu(\omega)$$

for $(x_n) \subset L_q(\Omega, l_2^m)$ and $(y_n) \subset L_{q'}(\Omega, l_2^m)$. For all $T_1, \ldots, T_m \in \tau$ the operators $\mathbf{T}(x_n) = (T_n x_n)$ on $L_q(\Omega, l_2^m)$, $\mathbf{S}(y_n) = (T_n' y_n)$ on $L_{q'}(\Omega, l_2^m)$ are dual to each other, i.e. $\mathbf{S} = \mathbf{T}'$, and the claim follows from $\|\mathbf{S}\| = \|\mathbf{T}'\| = \|\mathbf{T}\|$.   □

Before we give some criteria for R-boundedness, we present a simple example of a set of operators, which is not R-bounded.

2.12 Example. Let $X = L_p(\mathbb{R})$ with $1 \leq p < \infty$ and $p \neq 2$. The set of translation operators $T_n f(\cdot) = f(\cdot - n)$, for $n \in \mathbb{N} \cup \{0\}$, is not R-bounded on X. Indeed, for $f_n = \chi_{[0,1]}$ we have

$$\left\| \left(\sum_{n=0}^{m-1} |T_n f_n|^2 \right)^{\frac{1}{2}} \right\|_X = \|\chi_{[0,m]}\|_X = m^{\frac{1}{p}},$$

$$\left\| \left(\sum_{n=0}^{m-1} |f_n|^2 \right)^{\frac{1}{2}} \right\|_X = \|m^{\frac{1}{2}} \chi_{[0,1]}\|_X = m^{\frac{1}{2}}.$$

So for $1 \leq p < 2$ it is impossible to have (2.7) with a constant C for all m. For $2 < p < \infty$ we argue similarly, or use 2.11.

Many practical criteria for R-boundedness are based on the following important convexity result.

2.13 Theorem. *Let $\tau \subseteq B(X,Y)$ be R-bounded. Then the convex hull $co(\tau)$, the absolute convex hull $absco(\tau) = \{\sum_{k=1}^{n} \lambda_k T_k : n \in \mathbb{N}, T_k \in \tau, \lambda_k \in \mathbb{C}$ with $\sum_{k=1}^{n} |\lambda_k| = 1\}$ of τ and their closure in the strong operator topology are also R-bounded and*

$$R(\overline{co(\tau)}^s) \leq R(\tau), \qquad R(\overline{absco(\tau)}^s) \leq 2R(\tau).$$

Observation: Note that, for subsets A, B of a complex vector space V,

$$co(A \times B) = co(A) \times co(B).$$

The inclusion "$\subset$" is easily checked.

Conversely, take an element $(\sum_{i=1}^{l} \alpha_i x_i, \sum_{j=1}^{m} \beta_j y_j)$ of $co(A) \times co(B)$. Here $x_i \in A$, $y_j \in B$, and $\alpha_i, \beta_j \geq 0$ are scalars such that $\sum_{i=1}^{l} \alpha_i = \sum_{j=1}^{m} \beta_j = 1$. Then

$$\left(\sum_{i=1}^{l} \alpha_i x_i, \sum_{j=1}^{m} \beta_j y_j \right) = \sum_{i=1}^{l} \sum_{j=1}^{m} \alpha_i \beta_j (x_i, y_i) \in co(A \times B)$$

since $\sum_{i,j} \alpha_i \beta_j = 1$. Applying this observation repeatedly, we obtain for $A_1, \ldots, A_n \subset V$ that

$$co(A_1) \times \ldots \times co(A_n) = co(A_1 \times \ldots \times A_n).$$

Proof (of Theorem 2.13). Let $T_1, \ldots, T_n \in co(\tau)$ be a sequence. Using our observation above with $V = B(X,Y)$ and $A_1 = \ldots = A_n = \tau$, we can find $\lambda_1, \ldots, \lambda_N$ with $\lambda_j \geq 0$, $\sum_{j=1}^{N} \lambda_j = 1$ and operators $T_{kj} \in \tau$, $k = 1, \ldots, n$, $j = 1, \ldots, N$ such that

$$T_k = \sum_{j=1}^{N} \lambda_j T_{kj}, \qquad k = 1, \ldots, n.$$

For $x_1, \ldots, x_n \in X$ we then have

$$\left\| \sum_{k=1}^{n} r_k T_k x_k \right\|_{L_2(X)} \leq \sum_{j=1}^{N} \lambda_j \left\| \sum_{k=1}^{n} r_k T_{kj} x_k \right\|_{L_2(X)}$$

$$\leq \left(\sum_{j=1}^{N} \lambda_j \right) R(\tau) \left\| \sum_{k=1}^{n} r_k x_k \right\|_{L_2(X)}.$$

Since a complex sequence (a_k) with $\sum_{k=1}^{n} |a_k| \leq 1$ can be written as $a_k = \alpha_k \lambda_k$ with $|\alpha_k| \leq 1$, $\lambda_k \geq 0$ and $\sum_{k=1}^{n} \lambda_k = 1$, we obtain from the contraction principle 2.5 that $R(absco(\tau)) \leq 2R(co(\tau)) \leq 2R(\tau)$. Furthermore, $R(\overline{co(\tau)}^s) \leq R(\tau)$ follows directly from the definition. $\qquad \square$

2.13 is often used in the following form.

2.14 Corollary. *Let τ be an R-bounded set in $B(X,Y)$. For every strongly measurable $N : \Omega \to B(X,Y)$ on a σ-finite measure space Ω with values in τ and every $h \in L_1(\Omega, \mu)$ we define an operator $T_{N,h} \in B(X,Y)$ by*

$$T_{N,h}x = \int_\Omega h(\omega)N(\omega)x \, d\mu(\omega), \qquad x \in X. \tag{2.8}$$

Then $\sigma = \{T_{N,h} : \|h\|_{L_1} \le 1, N \text{ as above}\}$ is R-bounded and $R(\sigma) \le 2R(\tau)$.

Proof. Given are $h \in L_1(\Omega, \mu)$, $x_1, \ldots, x_n \in X$ and $\varepsilon > 0$. Approximating the measurable function $M : \Omega \to X^n, M(\omega) = (N(\omega)x_k)_{k=1}^n$ by step functions in $L_\infty(\Omega, X^n)$, we find a measurable partition $(V_j)_{j=1,\ldots,m}$ of Ω, where $m \in \mathbb{N} \cup \{\infty\}$, and $\omega_j \in V_j$ such that for all j:

$$\|N(\omega)x_k - N(\omega_j)x_k\| \le \varepsilon \qquad \text{for almost all } \omega \in V_j, \quad k = 1, \ldots, n.$$

Put

$$S = \sum_{j=1}^m \left(\int_{V_j} h(\omega)d\mu(\omega) \right) N(\omega_j).$$

Then by (2.8), $\|T_{N,h}x_k - Sx_k\| < \varepsilon$ for $k = 1, \ldots, n$ and S belongs to the neighborhood of $T_{N,h}$ in the strong operator topology determined by $x_1, \ldots, x_n$ and $\varepsilon > 0$. Since $S \in \overline{absco}^s(\tau)$, we have $\sigma \subseteq \overline{absco}^s(\tau)$, and we can appeal to 2.13. $\qquad \square$

We note two immediate consequences of 2.14. They will be crucial when we work with semigroups and their resolvents.

2.15 Example. Consider the Laplace transform of a strongly continuous and bounded function $t \in \mathbb{R}_+ \mapsto N(t) \in B(X,Y)$, i.e.

$$\widehat{N}(\lambda) := \int_0^\infty e^{-\lambda t} N(t)dt, \qquad Re\lambda > 0$$

If $\sigma = \{N(t) : t \ge 0\}$ is R-bounded, then $\tau = \{\lambda\widehat{N}(\lambda) : \lambda \in \Sigma_\theta\}$ is R-bounded for all $\theta < \frac{\pi}{2}$ and $R(\tau) \le 2R(\sigma)$.

Proof. Clearly $\lambda\widehat{N}(\lambda) = \int h_\lambda(t)N(t)dt$ with $h_\lambda(t) = \lambda e^{-\lambda t}$. Since $\theta < \frac{\pi}{2}$ we have that $\sup\{\|h_\lambda\|_{L_1(\mathbb{R}_+)} : \lambda \in \Sigma_\theta\} < \infty$. $\qquad \square$

2.16 Example. Let $\lambda \mapsto N(\lambda) \in B(X,Y)$ be analytic on $\Sigma_{\theta'}$ and assume that $\sigma = \{N(\lambda) : \lambda \in \partial\Sigma_\theta, \lambda \ne 0\}$ is R-bounded for some $\theta < \theta'$. Then

a) $\tau = \{N(\lambda) : \lambda \in \Sigma_\theta\}$ is R-bounded.

b) $\tau = \{\lambda\frac{d}{d\lambda}N(\lambda) : \lambda \in \Sigma_{\theta_1}\}$ is R-bounded for each $\theta_1 < \theta$.

If we have the slightly weaker assumption that $R(\{N(e^{\pm i\theta}a^k t) : k \in \mathbb{Z}\}) \le C$ for all $t \ge 0$ and some $a > 0$, we still can conclude that $\{N(\lambda) : \lambda \in \Sigma_{\theta_1}\}$ is R-bounded for all $\theta_1 < \theta$.

Proof. If we replace N by the function $\lambda \in \mathbb{C}_+ \mapsto M(\lambda) = N(\lambda^\alpha), \alpha = \frac{2\theta}{\pi}$, we can assume that $\theta = \frac{\pi}{2}$.

a) follows now from 2.14 and Poisson's formula

$$N(\alpha + i\beta) = \int_{-\infty}^{\infty} h_{\alpha,\beta}(s)N(is)ds, \qquad \alpha > 0,$$

with $h_{\alpha,\beta}(s) = \frac{1}{\pi} \frac{\alpha}{\alpha^2 + (s-\beta)^2}$ since $\|h_{\alpha,\beta}\|_{L_1} = 1$.

b) follows from 2.14 and Cauchy's formula

$$\lambda \frac{d}{d\lambda} N(\lambda) = \int_{\partial \Sigma(\theta)} h_\lambda(\mu)N(\mu)d\mu, \qquad \lambda \in \Sigma_{\theta_1},$$

with $h_\lambda(\mu) = \frac{1}{2\pi i}\lambda(\mu - \lambda)^{-2}$ since $\sup\{\|h_\lambda\|_{L_1} : \lambda \in \Sigma_{\theta_1}\} < \infty$.

We now prove the last claim. Again we may assume that $\theta = \frac{\pi}{2}$ and we consider points $z_k = \alpha_k + i\beta\alpha_k$ with $\alpha_k > 0$, $\beta = \tan\theta_1$ for $|\theta_1| < \theta$. Choose $m_k \in \mathbb{Z}$ so that $a^{m_k} < \alpha_k < a^{m_k+1}$. Then

$$N(z_k) = \int_{-\infty}^{\infty} N(it)h_{z_k}(t)dt = \int_{-\infty}^{\infty} N(ia^{m_k}t)h_{z_k}(a^{m_k}t)a^{m_k}dt$$

with

$$|h_{z_k}(a^{m_k}t)a^{m_k}| = \frac{1}{\pi} \frac{\alpha_k a^{-m_k}}{(\alpha_k a^{-m_k})^2 + (t - \alpha_k a^{-m_k}\beta)^2} \leq \frac{C_1}{1+t^2}.$$

Hence for $x_1, \ldots, x_n \in X$,

$$\Big\| \sum_{k=1}^{n} r_k N(z_k)x_k \Big\|_{L_1(X)} \leq C_1 \int_{-\infty}^{\infty} \Big\| \sum_{k=1}^{n} r_k N(ita^{m_k})x_k \Big\|_{L_1(X)} \frac{1}{1+t^2} dt$$

$$\leq C_1 C \Big\| \sum_{k=1}^{n} r_k x_k \Big\|_{L_1(X)}.$$

Hence $\{N(\lambda) : \lambda \in \partial\Sigma_{|\theta_1|}\}$ is R-bounded and we apply a) to complete the proof. $\qquad\square$

As a consequence of Kahane's contraction principle we obtain a companion to 2.14, which allows to create R-bounded sets without any R-boundedness assumptions

2.17 Corollary. *Let $N : \Omega \mapsto B(X,Y)$ be strongly integrable on the measure space (Ω, Σ, μ) such that for some constant A*

$$\int \|N(\omega)x\|_Y d\mu(\omega) \leq A\|x\|_X \qquad \text{for all } x \in X.$$

For $h \in L_\infty(\Omega, \mu)$ define an operator $T_h \in B(X,Y)$ as in 2.14. Then the set $\tau = \{T_h : \|h\|_{L_\infty(\Omega)} \leq 1\}$ is R-bounded and $R(\tau) \leq 2A$.

Proof. For $h_1, \ldots, h_m \in L_\infty(\Omega, \mathbb{C})$ with $\|h_j\|_\infty \leq 1$ and $x_1, \ldots, x_m \in X$ we have

$$\left\|\sum_n r_n T_{h_n}(x_n)\right\|_{L_1(Y)} = \left\|\int_\Omega \left[\sum_n r_n h_n(\omega) N(\omega) x_n\right] d\mu\right\|_{L_1(Y)}$$

$$\leq \int_\Omega \left\|\sum_n r_n h_n(\omega) N(\omega) x_n\right\|_{L_1(Y)} d\mu$$

$$\leq 2\int_\Omega \left\|N(\omega) \sum_n r_n x_n\right\|_{L_1(Y)} d\mu$$

$$\leq 2A\left\|\sum_n r_n x_n\right\|_{L_1(X)}$$

by Kahane's contraction principle and our assumption. □

This criterion implies that operator functions of bounded variation are R-bounded.

2.18 Example. Assume that $t \in (a, b) \mapsto N(t) \in B(X)$ has an integrable derivative. Then $\tau = \{N(t) : t \in (a, b)\}$ is R-bounded and

$$R(\tau) \leq \int_a^b \|\tfrac{d}{dt} N(t)\| dt + \|N(a)\|.$$

Proof. By the fundamental theorem of calculus we have, with $h_t(s) = \chi_{[0,t]}(s)$,

$$N(t) = N(a) + \int_a^t \tfrac{d}{ds} N(s) ds = N(a) + \int_a^b h_t(s) \tfrac{d}{ds} N(s) ds. \qquad \square$$

Finally we state a result, which allows us to neglect compact sets when we deal with R-sectoriality.

2.19 Example. Let $G \subseteq \mathbb{C}$ be open and let $\lambda \in G \mapsto N(\lambda) \in B(X)$ be an analytic function. Then $\tau = \{N(\lambda) : \lambda \in K\}$ is R-bounded for every compact set $K \subseteq G$.

Proof. Choose a contour Γ in $G \setminus K$ so that by Cauchy's formula

$$N(\lambda) = \int_\Gamma h_\lambda(\mu) N(\mu) d\mu$$

with $h_\lambda(\mu) = \frac{1}{2\pi i}(\mu - \lambda)^{-1}$. The function $\lambda \in \Gamma \longrightarrow N(\lambda)$ is integrable and

$$\|h_\lambda\|_{L_\infty(\Gamma)} \leq \left(\inf\{|\lambda - \mu| : \lambda \in \Gamma, \mu \in K\}\right)^{-1}$$

Now 2.17 applies. □

Now we are in a position to show the equivalence of the various R-boundedness conditions stated in Theorem 1.11 and Remark 1.12. The proof of Theorem 1.11 will be completed in the next section by proving the Mihlin Multiplier Theorem.

2.20 Theorem. *Let A be the generator of a bounded analytic semigroup in a Banach space X. Then the following conditions are equivalent:*

(i) *$(-A)$ is R-sectorial for some angle larger than $\frac{\pi}{2}$, i.e., $\{\lambda R(\lambda, A) : \lambda \in \Sigma_{\frac{\pi}{2}+\delta}\}$ is R-bounded for some $\delta > 0$;*

(ii) *For some $n \in \mathbb{N}$, the set $\{t^n R(it, A)^n : t \in \mathbb{R} \setminus \{0\}\}$ is R-bounded.*

(iii) *A is R-analytic, i.e., $\{T_z : z \in \Sigma_\delta\}$ is R-bounded for some $\delta > 0$;*

(iv) *$\{T_t, tAT_t : t > 0\}$ is R-bounded.*

If $\{T_t\}_{t \geq 0}$ is exponentially stable, i.e., $0 \in \rho(A)$, then we can add two more equivalent conditions:

(v) *$\{T_t, tAT_t : t \in [0, T]\}$ is R-bounded for some $0 < T < \infty$.*

(vi) *$\{tR(it, A) : |z| > T\}$ is R-bounded for some $0 < T < \infty$.*

The following lemma will be useful.

2.21 Lemma. *Let A be the generator of a bounded analytic semigroup $\{T_z\}$ in a Banach space X.*

a) *If $\{\lambda R(\lambda, A) : \lambda \in \overline{\Sigma_\omega} \setminus \{0\}\}$ is R-bounded with $\omega \in [0, \pi)$, then $\{\lambda R(\lambda, A) : \lambda \in \Sigma_{\omega_1}\}$ is R-bounded for some $\omega_1 > \omega$.*

b) *If $\{T_z, zAT_z : z \in \overline{\Sigma_\delta}\}$ is R-bounded for some $\delta \in [0, \frac{\pi}{2})$, then $\{T_z : z \in \Sigma_{\delta_1}\}$ is R-bounded for some $\delta_1 > \delta$.*

Proof. a) We use the power expansion

$$\lambda R(\lambda, A) = \sum_{m=0}^{\infty} \lambda (te^{i\omega} - \lambda)^m R(te^{i\omega}, A)^{m+1}.$$

If $C = R(\{tR(te^{i\omega}, A) : t \neq 0\})$, we choose $\omega_1 > \omega$ so small that $\Sigma_{\omega_1} \subset \rho(A)$ and $|e^{i\omega} - e^{i\omega_1}| < \frac{1}{2C}$. For $\lambda = e^{i\varphi}t$ with $\omega < \varphi < \omega_1$ we have

$$\lambda R(\lambda, A) = \sum_{m=0}^{\infty} e^{i\varphi}(e^{i\omega} - e^{i\varphi})^m [tR(te^{i\omega}, A)]^{m+1}$$

and therefore

$$R(\{\lambda R(\lambda, A) : \omega \leq \arg \lambda \leq \omega_1\}) \leq \sum_{m=0}^{\infty} \left(\frac{1}{2C}\right)^m R(\{tR(e^{i\omega}t, A)\})^{m+1}$$

$$\leq 2C \sum_{m=0}^{\infty} \frac{1}{2^{m+1}}.$$

The same argument applies on the line $\{te^{-i\omega} : t > 0\}$.

b) We use the Taylor expansion of T_z in a point $te^{i\delta}$, $t > 0$ (cf. the proof of [186, Thm. 5.2]).

$$T_z = T_{e^{i\delta}t} + \sum_{n=1}^{\infty} \frac{1}{n!} A^n T_{e^{i\delta}t}(z - e^{i\delta}t)^n = T_{e^{i\delta}t} + \sum_{n=1}^{\infty} \frac{1}{n!}(tAT_{e^{i\delta}t}/n)^n(\tfrac{z}{t} - e^{i\delta})^n.$$

Choose $\delta_1 > \delta$ so small that $|e^{i\delta} - e^{i\delta_1}| < \frac{1}{2eC}$, where C denotes the R-bound of $\{T_{e^{i\delta}t}, tAT_{e^{i\delta}t} : t > 0\}$. For $z = e^{i\varphi}t$ with $\delta < |\varphi| < \delta_1$, we obtain

$$R(\{T_z : \delta < \arg\lambda < \delta_1\}) \le C + \sum_{n=1}^{\infty} \frac{n^n C^n}{n!}|e^{i\varphi} - e^{i\delta}|^n \le C + \sum_{n=1}^{\infty} 2^{-n} < \infty$$

since $n^n \le n!e^n$. $\square$

Proof (of Theorem 2.20).

(i)$\Rightarrow$(ii) is clear by 2.8.

(ii)$\Rightarrow$(i) First we show that if (ii) holds for some $n \in \mathbb{N}$, it also holds for $n = 1$. We have

$$R(it, A)^{n-1} = (n-1)i \int_t^{\infty} R(is, A)^n ds$$

or

$$(it)^{n-1}R(it, A)^{n-1} = \int_0^{\infty} h_t(s)[(is)^n R(is, A)^n]ds$$

where $h_t(s) = (n-1)t^{n-1}s^{-n}\chi_{[t,\infty)}(s)$ with $\int_0^{\infty} h_t(s)ds = 1$ for $t > 0$ and $n \ge 2$. Hence the R-boundedness of $\{\lambda^{n-1}R(\lambda, A)^{n-1} : \lambda \in i\mathbb{R}\}$ follows from the R-boundedness of $\{\lambda^n R(\lambda, A)^n : \lambda \in i\mathbb{R}\}$ by 2.14, and we can iterate this step. By 2.16 we conclude that $\{\lambda R(\lambda, A) : Re\,\lambda \ge 0, \lambda \ne 0\}$ is R-bounded. To reach a larger sector we use Lemma 2.21 a).

(iii)$\Rightarrow$(i) follows from 2.15 since the resolvent is the Laplace transform of the semigroup as in (1.5).

(i)$\Rightarrow$(iii) follows also from 2.15 by the usual representation of the semigroup operators as an inverse Laplace transform (see e.g. (5.11) in the proof of [186, Thm. 5.2])

$$T_z = \frac{1}{2\pi i} \int_{\Gamma_t} e^{\lambda z} R(\lambda, A)\, d\lambda$$

for $z = te^{i\varphi}$, where $|\varphi| \le \delta_1 < \delta$ and Γ_t is given by $\{se^{\pm(\frac{\pi}{2}+\delta)} : |s| > \frac{1}{t}\} \cup \{\frac{1}{t}e^{is} : |s| \le \frac{\pi}{2} + \delta\}$ where $\sigma(A)$ lies to the left. Then

$$T_z = \frac{1}{2\pi i} \int_{\Gamma_1} h_z(\lambda)[\tfrac{1}{t}\lambda R(\tfrac{1}{t}\lambda, A)]d\lambda,$$

where $h_z(\lambda) = \frac{1}{\lambda}e^{\lambda z}$. Since $\arg(\lambda z) > \frac{\pi}{2} + \delta - \delta'$, the integrals $\int_{\Gamma_1} |h_z(\lambda)||d|\lambda|$ are uniformly bounded for $z \in \Sigma_{\delta'}$ and by 2.14 the set $\{T_z : z \in \Sigma_\delta\}$ is R-bounded.

(iv)$\Rightarrow$(iii) follows from Lemma 2.21 b) with $\delta = 0$.

(iii)$\Rightarrow$(iv) follows by 2.16.

We assume now in addition that $0 \in \rho(A)$.

(v)$\Leftrightarrow$(iii) We argue as in the proof of Lemma 2.21 b) and obtain R-boundedness of $\{T_z : |z| \leq T/2, |\arg z| \leq \delta\}$ for some $\delta > 0$. In this case $\{T_{te^{\pm i\delta}} : t \geq T/2\}$ is automatically R-bounded since $\frac{d}{dt} T_{te^{\pm i\delta}} = A T_{te^{\pm i\delta}}$ is integrable on $[T/2, \infty)$ in the operator norm. Now we use 2.16.

(vi)$\Leftrightarrow$(ii) $\{R(it, A) : t \in [-T, T]\}$ is R-bounded by 2.19. $\square$

2.22 Remark. a) For a sectorial (R-sectorial) operator B let $\omega(B)$ ($\omega_R(B)$) be the infimum over all ω so that B is (R-)sectorial of angle ω.

For an (R-)analytic operator A let $\delta(A)$ ($\delta_R(A)$) be the supremum over all δ so that A is (R-)analytic.

It is well-known that $\omega(-A) = \frac{\pi}{2} - \delta(A)$ and our proof shows that $\omega_R(-A) = \frac{\pi}{2} - \delta_R(A)$. (The suprema over δ appearing in (i) or (iii), respectively, are identical.

b) We will also need that the R-boundedness of the set

$$\{\tfrac{1}{z} \int_0^z T_\lambda d\lambda : z \in \Sigma_\delta\}$$

is equivalent to the conditions (i) to (iv) of 2.20.

Proof. b) The R-boundedness of this set follows from (iii) and the fact that $\overline{absco}\{T_z : z \in \Sigma_\delta\}$ is R-bounded by 2.13.

For the converse, note that $M(t) = \frac{1}{t} \int_0^t T_{e^{i\varphi}s} ds$ with $|\varphi| < \delta$ satisfies

$$T_{e^{i\varphi}s} = M(t) + t\frac{d}{dt}M(t).$$

Furthermore $\{t\frac{d}{dt}M(t) : t > 0\}$ is R-bounded by 2.16. $\square$

For C_0-semigroups which are not analytic we have the following version of the Hille-Yosida Theorem.

2.23 Theorem. *A C_0-semigroup T_t with generator A on a Banach space X is R-bounded with $R(\{T_t : t > 0\}) \leq C$ if and only if the set $\tau = \{\lambda^n R(\lambda, A)^n : \lambda > 0\}$ satisfies $R(\tau) \leq C$.*

Proof. On the one hand, differentiating the Laplace representation (1.5) gives (cf. [186])

$$\lambda^{n+1} R(\lambda, A)^{n+1} x = \int_0^\infty h_{n,\lambda}(t) T_t x \, dt$$

where $h_{n,\lambda}(t) = \frac{1}{n!}(\lambda t)^{n+1}\frac{1}{t}e^{-\lambda t}$ is uniformly bounded in $L_1(\mathbb{R}_+)$. So if T_t is R-bounded we can apply 2.14 to see that τ is R-bounded.

For the converse we use the exponential formula

$$(\tfrac{n}{t})^n R(\tfrac{n}{t}, A)^n x = (I - \tfrac{t}{n}A)^{-n} x \to T_t x \qquad \text{as } n \to \infty$$

for $t > 0$ and $x \in X$. Hence $T_t \in \overline{\tau}^s$. $\square$

Notes on Section 2:

N 2.1: A nice discussion of the many lives of Rademacher functions can be found in the classical book [128].

N 2.2: Khinchine's inequality as stated here appears for the first time in [227], Chap IV, §5.2, although all ingredients for the classical proof given there (see also [165]) go back to work of Rademacher in 1922 and Khinchine in 1923. Our proof is from [206], Appendix D, as presented in [69] Sect. 1. The Hilbert space characterization quoted in the remark is due to Kwapien [148].

N 2.3: We give a proof of the last statement for $p = 2$; the general case follows from 2.4. Let $\{x_n\}_{n \in \mathbb{N}}$ satisfy (2.3). Since X is reflexive, $L_2([0,1], X)$ is reflexive with dual $L_2([0,1], X^*)$. For every $x^* \in X^*$ we have

$$\left(\sum_{n=1}^{N} |x^*(x_n)|^2 \right)^{1/2} \leq C \left(\int_0^1 \left| x^* \left(\sum_{n=1}^{N} r_n(u) x_n \right) \right|^2 du \right)^{1/2}$$

$$\leq C \|x^*\| \cdot \left\| \sum_{n=1}^{N} r_n(u) x_n \right\|_{L_2(X)} \leq CM \|x^*\| .$$

For each $F = x^* \otimes f$ with $x^* \in X^*$ and $f \in L_2([0,1])$, Bessel's inequality gives, for each $N \in \mathbb{N}$,

$$\sum_{n=1}^{N} |\langle F, r_n \otimes x_n \rangle_{L_2(X)}| = \sum_{n=1}^{N} |x^*(x_n) \int_0^1 r_n(u) f(u) \, du|$$

$$\leq \left(\sum_{n=1}^{N} |x^*(x_n)|^2 \right)^{1/2} \left(\sum_{n=1}^{N} \int_0^1 |r_n(u) f(u)|^2 \, du \right)^{1/2}$$

$$\leq CM \|x^*\| \cdot \|f\|_{L_2[0,1]} .$$

Put $g_n = \sum_{m=1}^{n} r_m x_m$. We have shown that for all subsequences (n_k) and F of the form $F = \sum_{j=1}^{m} x_j^* \otimes f_j$ with $x_j^* \in X^*$ and $f_j \in L_2[0,1]$ the sequence $\langle F, g_{n_k} \rangle$ is a Cauchy sequence. Since such F's are dense in $L_2([0,1], X^*)$ and (g_n) is bounded, the same is true for all $F \in L_2([0,1], X^*)$.

However, the reflexive space $L_2([0,1], X)$ is weakly sequentially complete, so that $\sum_k r_{n_k} x_{n_k}$ converges weakly in $L_2([0,1], X)$ and therefore $\sum_n r_n x_n$ converges (even unconditionally) in $L_2([0,1], X)$ by Orticz's convergence theorem [69, §1]. □

The statement we just proved also holds in spaces which do not contain c_0, cf. [149].

The space $RadX$ is at the center of many important notions in the geometry of Banach spaces. A Banach space has type $p \in [1,2]$ [cotype $q \in [2,\infty]$] if there is a constant C such that for all $f = \sum_{n=1}^{m} r_n x_n$ in $Rad(X)$

$$\|f\|_{Rad(X)} \leq C \left(\sum_{n=1}^{m} \|x_n\|^p \right)^{\frac{1}{p}} \left[\left(\sum_{n=1}^{m} \|x_n\|^q \right)^{\frac{1}{q}} \leq C\|f\|_{Rad(X)} \right].$$

Kwapien's results in 2.2 say now, that X is isomorphic to a Hilbert space if X has type 2 and cotype 2. G.Pisier has shown that a Banach space X has type p with $p > 1$ if and only if $Rad(X)$ is complemented in $L_p([0,1], X)$ for $1 < p < \infty$.
For these results and many related ones, see e.g. [189],[190] and [69].

    N 2.4: A proof of (2.4) can be found in [129] Section 2.5, 2.7. A clear presentation of Kahane's original proof is given in [69], Sect. 11, which contains further references. We also recommend the approach in [191]. For $1 \leq p \leq 2$ there is an elementary proof in [155] which we reproduce here for the curious reader:

Proof. Because of the monotonicity of the $L_p([0,1], X)$-norm we only have to show that for all $x_1, \ldots, x_n \in X$

$$\left\| \sum_{i=1}^{n} r_i x_i \right\|_{L_2([0,1],X)} \leq \sqrt{2} \left\| \sum_{i=1}^{n} r_i x_i \right\|_{L_1([0,1],X)}. \tag{2.9}$$

Now we fix $x_1, \ldots, x_n \in X$ and put for $\varepsilon = (\varepsilon_i) \in \{-1,1\}^n$

$$F_\varepsilon = \left\| \sum_{i=1}^{n} \varepsilon_i x_i \right\|.$$

As in 2.6.b) (see also 2.1) we can restate our claim (2.9) as

$$\frac{1}{2^n} \sum_{\varepsilon \in \{-1,1\}^n} F_\varepsilon^2 \leq 2 \left(\frac{1}{2^n} \sum_{\varepsilon \in \{-1,1\}^n} F_\varepsilon \right)^2. \tag{2.10}$$

For real $\beta_1, \ldots, \beta_n$ consider the identity

$$\prod_{i=1}^{n} (1 + \beta_i) = \sum_{\sigma \in \{0,1\}^n} \prod_{i=1}^{n} \beta_i^{\sigma_i}.$$

Replacing β_i by $t^{-1}\alpha_i$ and multiplying by t^2 for $t > 0$ we find

$$t^2 \prod_{i=1}^{n} (1 + t^{-1}\alpha_i) = \sum_{\sigma \in \{0,1\}^n} t^{2-|\sigma|} \prod_{i=1}^{n} \alpha_i^{\sigma_i}$$

with $|\sigma| = \sum_{i=1}^{n} \sigma_i$. Differentiating this equality with respect to t we get for $t = 1$:

$$2\prod_{i=1}^{n}(1+\alpha_i) - \sum_{j=1}^{n}\alpha_j\prod_{\substack{i=1\\i\neq j}}^{n}(1+\alpha_i) = \sum_{\sigma\in\{0,1\}^n}(2-|\sigma|)\prod_{i=1}^{n}\alpha_i^{\sigma_i}.$$

Next we set $\alpha_i = \varepsilon_i\eta_i$ with $\varepsilon = (\varepsilon_i)$ and $\eta = (\eta_i)$ in $\{-1,1\}^n$ and sum over all possible ε and η:

$$\sum_{\varepsilon,\eta\in\{-1,1\}^n}\left[2\prod_{i=1}^{n}(1+\varepsilon_i\eta_i) - \sum_{j=1}^{n}\varepsilon_j\eta_j\prod_{\substack{i=1\\i\neq j}}^{n}(1+\varepsilon_i\eta_i)\right]F_\varepsilon F_\eta$$

$$= \sum_{\varepsilon,\eta\in\{-1,1\}^n}\left[\sum_{\sigma\in\{0,1\}^n}(2-|\sigma|)\left(\prod_{i=1}^{n}\varepsilon_i^{\sigma_i}\right)\left(\prod_{i=1}^{n}\eta_i^{\sigma_i}\right)\right]F_\varepsilon F_\eta \quad (2.11)$$

$$= \sum_{\sigma\in\{0,1\}^n}(2-|\sigma|)\left(\sum_{\varepsilon\in\{-1,1\}^n}\left(\prod_{i=1}^{n}\varepsilon_i^{\sigma_i}\right)F_\varepsilon\right)^2.$$

If $|\sigma| = 1$ with, say $\sigma_j = 1$, then

$$\sum_{\varepsilon\in\{-1,1\}^n}\left(\prod_{i=1}^{n}\varepsilon_i^{\sigma_i}\right)F_\varepsilon = \sum_{\substack{\varepsilon\in\{-1,1\}^n\\\varepsilon_j=1}}(F_\varepsilon - F_{-\varepsilon}) = 0$$

since by definition we have that $F_\varepsilon = F_{-\varepsilon}$. Hence if we omit all summands with $|\sigma| \geq 2$ we are left with $|\sigma| = 0$ and

$$(2.11) \leq 2\left(\sum_{\varepsilon\in\{-1,1\}^n}F_\varepsilon\right)^2.$$

To estimate (2.11) from below, observe that $\prod_{i=1}^{n}(1+\varepsilon_i\eta_i) \neq 0$ if and only if $\varepsilon = \eta$, and $\prod_{\substack{i=1\\i\neq j}}^{n}(1+\varepsilon_i\eta_i) \neq 0$ if and only if $\varepsilon_i = \eta_i$ for $i \neq j$. Hence, with the notation $d(\varepsilon,\eta) = \text{card}\{i : \varepsilon_i \neq \eta_i\}$, we only have to consider pairs (ε,η) with $\varepsilon = \eta$ or $d(\varepsilon,\eta) = 1$, and the left-hand side of (2.11) equals

$$\sum_{\varepsilon\in\{-1,1\}^n}2\cdot 2^n F_\varepsilon^2 - \sum_{\varepsilon\in\{-1,1\}^n}n2^{n-1}F_\varepsilon^2 + \sum_{\substack{\varepsilon,\eta\in\{-1,1\}^n\\d(\varepsilon,\eta)=1}}2^{n-1}F_\varepsilon F_\eta$$

$$= 2^n\sum_{\varepsilon\in\{-1,1\}^n}F_\varepsilon^2 + 2^{n-1}\sum_{\varepsilon\in\{-1,1\}^n}F_\varepsilon\left[\sum_{\substack{\eta\in\{-1,1\}^n\\d(\varepsilon,\eta)=1}}F_\eta - (n-2)F_\varepsilon\right]. \quad (2.12)$$

We want to show that $[\cdots]$ is positive. Indeed, for fixed ε,

$$\sum_{\substack{\eta\in\{-1,1\}^n\\d(\varepsilon,\eta)=1}}F_\eta \geq \left\|\sum_{j=1}^{n}\left(\sum_{\substack{\eta\in\{-1,1\}^n\\d(\varepsilon,\eta)=1}}\eta_j\right)x_j\right\| = (n-2)F_\varepsilon, \quad (2.13)$$

since

$$\sum_{\substack{\eta\in\{-1,1\}^n \\ d(\varepsilon,\eta)=1}} \eta_j = (n-1)\varepsilon_j - \varepsilon_j = (n-2)\varepsilon_j.$$

By (2.13), we can estimate (2.11) from below, by omitting the last terms of (2.12). This gives

$$2^n \sum_{\eta\in\{-1,1\}^n} F_\varepsilon^2 \leq (2.11) \leq 2 \left(\sum_{\eta\in\{-1,1\}^n} F_\varepsilon \right)^2.$$

Dividing by 2^{2n} we get (2.10) and we are through. □

N 2.5 and its proof are from [129], Sect. 2.5. For a more general statement, see e. g. [69], 12.2.

N 2.6 - N 2.8: Basic properties of R-boundedness were studied in [46], where in particular Remark 2.6 c) was justified by the following argument: Suppose that (2.5) is satisfied for a set τ with a constant C and operators $S_1, \ldots, S_m \in \tau$ with $S_1 \neq S_j$ for $i \neq j$. Given an arbitrary selection $T_1, \ldots, T_n$ from τ we denote by $S_1, \ldots, S_m$ the distinct operators among $T_1, \ldots, T_n$. Let $F_k = \{j : T_j = S_k\}$. Furthermore, let (ε_j) and (ε_k') be two independent sequences of Bernoulli variables on Ω and Ω', respectively. Then for $x_1, \ldots, x_n \in X$ and every fixed $\omega' \in \Omega'$

$$\left\| \sum_{j=0}^n \varepsilon_j T_j x_j \right\|_{L_2(\Omega,X)} = \left\| \sum_{k=0}^m \sum_{j\in F_k} \varepsilon_j S_j x_j \right\|_{L_2(\Omega,X)}$$

$$= \left\| \sum_{k=0}^m \sum_{j\in F_k} \varepsilon_k'(\omega') \varepsilon_j S_j x_j \right\|_{L_2(\Omega,X)}.$$

Since $\{\varepsilon_k'(\omega')\varepsilon_j : j \in F_k, k = 1, \ldots, m\}$ and $\{\varepsilon_j : j = 1, \ldots, n\}$ are both Bernoulli random variables. By Fubini's theorem and our assumption we obtain

$$\left\| \sum_{j=0}^n \varepsilon_j T_j x_j \right\|_{L_2(\Omega,X)} \leq \left(\int_{\Omega'} \left\| \sum_{k=0}^m \varepsilon_k'(\omega') \varepsilon_j S_j x_j \right\|_{L_2(\Omega,X)}^2 d\omega' \right)^{\frac{1}{2}}$$

$$= \left(\int_\Omega \left\| \sum_{k=0}^m \varepsilon_k' S_k \left(\sum_{j\in F_k} \varepsilon_j(\omega) x_j \right) \right\|_{L_2(\Omega',X)}^2 d\omega \right)^{\frac{1}{2}}$$

$$\leq C \left(\int_\Omega \left\| \sum_{k=0}^m \varepsilon_k' \left(\sum_{j\in F_k} \varepsilon_j(\omega) x_j \right) \right\|_{L_2(\Omega',X)}^2 d\omega \right)^{\frac{1}{2}}.$$

Reversing our steps we find the last integral to be equal to $\left\| \sum_{j=1}^n \varepsilon_j x_j \right\|_{L_2(\Omega,X)}$.

N 2.9: The reformulation of $\|\sum_n r_n x_n\|_{L_2(X)}$ as a square sum in 2.9 is a folklore result, that goes back at least to the proof of a classical theorem of Marcinkiewicz and Zygmund, see theorem V.2.7 in [101]. Square function estimates are equivalent to R-boundedness not only in L_q spaces but in every Banach function space which is q-concave for some $q < \infty$. See [165] Theorem 1.16 for the equivalent of (2.6). This fact was already used in [152].

N 2.10 - N 2.11: These observations and further results on R-boundedness in L_p-spaces are contained in the last section of [220]. If τ is an R-bounded subset of $B(X, Y)$ for arbitrary Banach spaces X and Y, then $\tau'' = \{T'' : T \in \tau\}$ is R-bounded in $B(X'', Y'')$ by the principle of local reflexivity, but in order to conclude that $\tau' = \{T' : T \in \tau\}$ is R-bounded in $B(Y', X')$ we have to assume that X and Y have non-trivial type (see [135]).

N 2.12: It follows from Kwapien's theorem mentioned in N 2.3 that in every Banach space not isomorphic to a Hilbert space, there is a sequence of one-dimensional operators which is not R-bounded (see [14] for details). Further non-trivial examples are provided by [132], [133].

N 2.13: The convexity theorem is implicit in [39], [226] and spelled out in [46].

N 2.14 - N 2.19 are special cases of results in [221]. An exception is 2.17 which appears in [135].

N 2.20 - N 2.22: Theorem 2.20 and Remark 2.22 are also adapted from [221].

3 Fourier Multiplier Theorem on $\mathbb{R}$

In this section we prove the vector-valued multiplier theorem announced in 1.10 by first proving a vector-valued version of Steklin's multiplier theorem in 3.6 and then extending it to a vector-valued Mihlin theorem (see 3.12) via R-boundedness (see 3.8) and the Paley-Littlewood decomposition (see 3.11). We will also show that the R-boundedness of the multiplier function is necessary for boundedness of the Fourier multiplier operator (see 3.13). Finally, in 3.14 we explain why a closed subspace of $L_q(\Omega, \mathbb{C})$, with $1 < q < \infty$, has the UMD property. But first we fix our notations.

3.1 Notation.
Fourier Transform: Recall that the Schwartz class $\mathcal{S}(\mathbb{R}^N, X)$ of rapidly decreasing smooth functions from $\mathbb{R}^N$ into X is norm dense in $L_p(\mathbb{R}^N, X)$ for $1 \leq p < \infty$. The Fourier transform $\mathcal{F} \colon \mathcal{S}(\mathbb{R}^N, X) \to \mathcal{S}(\mathbb{R}^N, X)$, defined by

$$(\mathcal{F}f)(t) \equiv \widehat{f}(t) := \int_{\mathbb{R}^N} e^{-it \cdot s} f(s) \, ds \,, \tag{3.1}$$

is a bijection whose inverse is given by

$$(\mathcal{F}^{-1}f)(t) \equiv \check{f}(t) := (2\pi)^{-N} \int_{\mathbb{R}^N} e^{it \cdot s} f(s) \, ds \,, \tag{3.2}$$

where $f \in \mathcal{S}(\mathbb{R}^N, X)$ and $t \in \mathbb{R}^N$. Note that the formula in (3.1) (resp. (3.2)) defines a mapping $\mathcal{F}$ (resp. $\mathcal{F}^{-1}$) in $B\left(L_1(\mathbb{R}^N, X), L_\infty(\mathbb{R}^N, X)\right)$. Recall that

$$\mathcal{S}_0(\mathbb{R}^N, X) := \left\{ f \in \mathcal{S}(\mathbb{R}^N, X) \colon \text{supp } \widehat{f} \text{ is compact}, 0 \notin \text{supp } \widehat{f} \right\}$$

is norm dense in $L_p(\mathbb{R}^N, X)$ for $1 < p < \infty$.

Fourier Multipliers: Consider a bounded measurable function $m : \mathbb{R}^N \to B(X, Y)$. It induces a map T_m where

$$\mathcal{S}(\mathbb{R}^N, X) \ni f \quad \xrightarrow{T_m} \quad \mathcal{F}^{-1}(m(\cdot)[\widehat{f}(\cdot)]) \in L_\infty(\mathbb{R}^N, Y) . \tag{3.3}$$

We call m an L_p-*Fourier multiplier*, $1 < p < \infty$, if there is a constant c_p so that

$$\|T_m f\|_{L_p(\mathbb{R}^N, Y)} \leq c_p \|f\|_{L_p(\mathbb{R}^N, X)} , \qquad \text{for each } f \in \mathcal{S}(\mathbb{R}^N, X); \tag{3.4}$$

in which case the map T_m in (3.3) extends uniquely to an operator $T_m \in B\left(L_p\left(\mathbb{R}^N, X\right), L_p\left(\mathbb{R}^N, Y\right)\right)$, which is called the L_p-*Fourier multiplier operator corresponding to* m and whose operator norm is the smallest constant c_p for which (3.4) holds. Let

$$\mathcal{M}_p(\mathbb{R}^N, X, Y) = \left\{ m \colon \mathbb{R}^N \to B(X, Y) \mid m \text{ is an } L_p\text{-Fourier multiplier} \right\} .$$

We sometimes denote $\mathcal{M}_p(\mathbb{R}^N, X, X)$ by just $\mathcal{M}_p(\mathbb{R}^N, X)$, or just $\mathcal{M}_p(X)$ if the dimension is understood. If $m \in \mathcal{M}_p(\mathbb{R}^N, X, Y)$, then we define

$$\|m\|_{\mathcal{M}_p(\mathbb{R}^N, X, Y)} \equiv \|m\|_p := \|T_m\|_{B(L_p(\mathbb{R}^N, X), L_p(\mathbb{R}^N, Y))} . \tag{3.5}$$

As an elementary example, note that if $m(\cdot) := m_0 \in B(X, Y)$, then $m \in \mathcal{M}_p(\mathbb{R}^N, X, Y)$ with $(T_m f)(\cdot) = m_0(f(\cdot))$ for $f \in \mathcal{S}(\mathbb{R}^N, X)$ and $\|m\|_{\mathcal{M}_p(\mathbb{R}^N, X, Y)} = \|m_0\|_{B(X,Y)}$. A map $m \colon \mathbb{R}^N \to \mathbb{C}$ induces a map $\widetilde{m} \colon \mathbb{R}^N \to B(X)$ via $\widetilde{m}(\cdot) := m(\cdot)\text{Id}_X$; we often identify m with $\widetilde{m}$.

3.2 Discussion. In the scalar case $X = Y = \mathbb{C}$ one usually proceeds as follows to prove multiplier theorems such as Mihlin's theorem. First observe that each $m \in L_\infty(\mathbb{R}^N, \mathbb{C})$ is already an L_2-Fourier multiplier since $\mathcal{F}$ is an isomorphism on $L_2(\mathbb{R}^N, \mathbb{C})$. Next prove a weak L_1-estimate for T_m. Then interpolate and use duality to obtain the L_p-result. This approach breaks down if X is infinite dimensional since the Fourier transform is not bounded on $L_2(\mathbb{R}^N, X)$ unless X is isomorphic to a Hilbert space. So there is no *easy* space that can serve as the *endpoint* for the interpolation procedure. Therefore Bourgain derived a new strategy to prove multiplier theorems. Assume that the Hilbert transform is bounded on $L_p(\mathbb{R}, X)$, i. e. $m(\cdot) = -i\pi\text{sign}(\cdot) \in \mathcal{M}_p(\mathbb{R}, X, X)$. Then try to build more complicated multipliers step by step from variants of this function until one reaches multipliers with the Mihlin condition. This makes UMD-spaces (see 1.8) a natural starting point. Here are some elementary operations on multiplier functions that we will use in this process.

3.3 Facts.

a) (algebra of multipliers)

If $m_i \in \mathcal{M}_p(\mathbb{R}^N, X, Y)$ and $c \in \mathbb{C}$, then $cm_1, m_1 + m_2 \in \mathcal{M}_p(\mathbb{R}^N, X, Y)$ with $T_{cm_1} = cT_{m_1}$ and $T_{m_1+m_2} = T_{m_1} + T_{m_2}$.

If $m_1 \in \mathcal{M}_p(\mathbb{R}^N, X, Y)$ and $m_2 \in \mathcal{M}_p(\mathbb{R}^N, Y, Z)$ then $m_2 \cdot m_1 \in \mathcal{M}_p(\mathbb{R}^N, X, Z)$ with $T_{m_2 \cdot m_1} = T_{m_2} \circ T_{m_1}$.

Thus $\mathcal{M}_p(\mathbb{R}^N, X, Y)$, endowed with the norm $\| \cdot \|_p$ given in (3.5), is a normed algebra.

b) (shifting multiplier)

If $m \in \mathcal{M}(\mathbb{R}^N, X, Y)$, then $m_a(t) = m(t - a)$ is an L_p-Fourier multiplier function for each $a \in \mathbb{R}^N$ and $\|m_a\|_p = \|m\|_p$, i.e. $\mathcal{M}_p(\mathbb{R}^N, X, Y)$ is translation invariant.

c) (convergence of multipliers)

Let $m_n \in \mathcal{M}_p(\mathbb{R}^N, X, Y)$ for each $n \in \mathbb{N}$ and a measurable function $m \colon \mathbb{R}^N \to B(X, Y)$ be such that, for each $t \in \mathbb{R}^N$, $n \in \mathbb{N}$, and $x \in X$,

$$m(t)\, x = \lim_{k \to \infty} m_k(t)\, x \quad , \quad \|m_n(t)\|_{B(X,Y)} \leq D \quad , \quad \|m_n\|_p \leq C \,.$$

Then $m \in \mathcal{M}_p(\mathbb{R}^N, X, Y)$ with $\|m\|_p \leq C$.

Proof. a) Follows from the definitions; note that

$$T_{m_1 \cdot m_2} f = \mathcal{F}^{-1}(m_1[\mathcal{F}^{-1}(m_2 \cdot \hat{f})]^{\wedge}) \quad \text{for } f \in \mathcal{S}(\mathbb{R}^N, X).$$

b) Since $[e^{-ia \cdot (\cdot)} f(\cdot)]^{\wedge}(s) = \hat{f}(s + a)$ and $e^{ia \cdot t}(\mathcal{F}^{-1}g)(t) = \mathcal{F}^{-1}(g(\cdot - a))(t)$ we have for $f \in \mathcal{S}(\mathbb{R}^N, X)$

$$(T_{m_a} f)(t) = e^{ia \cdot t} T_m(e^{-ia \cdot (\cdot)} f(\cdot))(t)$$

and therefore $\|T_{m_a} f\|_{L_p(\mathbb{R}^N, Y)} \leq \|T_m\|_{B(L_p(\mathbb{R}^N, X), L_p(\mathbb{R}^N, Y))} \|f\|_{L_p(\mathbb{R}^N, X)}$.

c) Let $f \in \mathcal{S}(\mathbb{R}^N, X)$. Then $m_n \hat{f} \to m\hat{f}$ in $L_1(\mathbb{R}^N, Y)$ by Lebesgue's theorem and so $T_{m_n} f \to T_m f$ in $L_\infty(\mathbb{R}^N, Y)$. Thus

$$\|T_m f\|_{L_p(\mathbb{R}^N, Y)}^p = \int_{\mathbb{R}^N} \|T_m f(t)\|_Y^p \, dt$$

$$\leq \lim_{n \to \infty} \int_{\mathbb{R}^N} \|T_{m_n} f(t)\|_Y^p \, dt \leq C^p \|f\|_{L^p(\mathbb{R}^N, X)}^p$$

by Fatou's lemma. □

By 3.3 a) and 3.3 c), if $\mathcal{M} \subset \mathcal{M}_p(\mathbb{R}^N, X, Y)$ satisfies

$$\sup_{m \in \mathcal{M}} \|m\|_p < \infty \quad \text{and} \quad \sup_{m \in \mathcal{M}} \sup_{t \in \mathbb{R}^N} \|m(t)\|_{B(X,Y)} < \infty \,,$$

then the measurable functions in the sequential closure of absco($\mathcal{M}$) with respect to pointwise convergence in the strong operator topology are again multipliers in $\mathcal{M}_p(\mathbb{R}^N, X, Y)$. We will use this observation in the following integral form.

3.4 Facts. (on averaging of multiplier functions)
Let S be a measurable subset of $\mathbb{R}^M$.

a) Consider a collection $\{m(s,\cdot) \in \mathcal{M}_p(\mathbb{R}^N, X, Y): s \in S\}$ of multiplier functions that satisfies

$$\|m(s,\cdot)\|_p \leq C \qquad \text{for each } s \in S \qquad (3.6)$$

and $m \in L_\infty(S \times \mathbb{R}^N, B(X,Y))$. Let $h \in L_1(S,\mathbb{C})$ and define $m_h \colon \mathbb{R}^N \to B(X,Y)$ by

$$m_h(t) := \int_S m(s,t)h(s)\,ds \qquad \text{for a.e. } t \in \mathbb{R}^N .$$

(Note that one can think of $m_h(\cdot)$ as a weighted average of the $m(s,\cdot)$'s.)
Then $m_h \in \mathcal{M}_p(\mathbb{R}^N, X, Y)$ with $\|m_h\|_p \leq C\|h\|_{L_1(S,\mathbb{C})}$ and

$$T_{m_h} f = \int_S (T_{m(s,\cdot)} f)h(s)\,ds , \qquad \text{for } f \in L_p(\mathbb{R}^N, X) ,$$

where the integral is a Bochner integral in $L_p(\mathbb{R}^N, Y)$.

b) To obtain an R-bounded result, we now assume that $\{T_{m(s,\cdot)}\}_{s \in S}$ is not only uniformly norm bounded but also R-bounded. More precisely, let $\sigma \subset B(L_p(\mathbb{R}^N, X), L_p(\mathbb{R}^N, Y))$ be R-bounded and

$$S_\sigma := \{m \in L_\infty(S \times \mathbb{R}^N, B(X,Y)): \text{ for each } s \in S$$
$$m(s,\cdot) \in \mathcal{M}_p(\mathbb{R}^N, X, Y) \text{ and } T_{m(s,\cdot)} \in \sigma\} .$$

Then the set

$$\tilde{\sigma} := \{T_{m_h}: m \in S_\sigma \text{ and } \|h\|_{L_1(S,\mathbb{C})} \leq 1\}$$

is R-bounded in $B(L_p(\mathbb{R}^N, X), L_p(\mathbb{R}^N, Y))$ with $R_p(\tilde{\sigma}) \leq 2R_p(\sigma)$.

Proof a) Note that, by the assumptions, $m_h \in L_\infty(\mathbb{R}^N, B(X,Y))$.
   Fix $f \in S(\mathbb{R}^N, X)$. By the assumptions,

$$S \ni s \to m(s,\cdot)\widehat{f}(\cdot)h(s) \in L_1(\mathbb{R}^N, Y)$$

is an integrable function. Thus, since $\mathcal{F}^{-1} \in B(L_1(\mathbb{R}^N, Y), L_\infty(\mathbb{R}^N, Y))$,

$$T_{m_h} f = \mathcal{F}^{-1}\left[\int_S m(s,\cdot)\widehat{f}(\cdot)h(s)\,ds\right]$$
$$= \int_S \mathcal{F}^{-1}\left(m(s,\cdot)\widehat{f}(\cdot)\right) h(s)\,ds = \int_S \left(T_{m(s,\cdot)} f\right) h(s)\,ds$$

where the first Bochner integral is in $L_1(\mathbb{R}^N, Y)$ and the latter two are in $L_\infty(\mathbb{R}^N, Y)$. Thus

$$S \ni s \to \left(T_{m(s,\cdot)}f\right) h(s)$$

is measurable from S into $L_\infty(\mathbb{R}^N, Y)$ and thus also into $L_p(\mathbb{R}^N, Y)$. Also

$$\int_S \|(T_{m(s,\cdot)}f) h(s)\|_{L_p(\mathbb{R}^N,Y)} \, ds \le C\|f\|_{L_p(\mathbb{R}^N,Y)} \|h\|_{L_1(S,\mathbb{C})} \,.$$

Thus

$$T_{m_h} f = \int_S \left(T_{m(s,\cdot)}f\right) h(s) \, ds \tag{3.7}$$

where the Bochner integral is in $L_p(\mathbb{R}^N, Y)$ and so

$$\|T_{m_h}f\|_{L_p(\mathbb{R}^N,Y)} \le \int_S \| \left(T_{m(s,\cdot)}f\right) h(s)\|_{L_p(\mathbb{R}^N,Y)} \, ds$$
$$\le C\|f\|_{L_p(\mathbb{R}^N,X)} \|h\|_{L_1(S,\mathbb{C})} \,.$$

Thus $m_h \in \mathcal{M}_p(\mathbb{R}^N, X, Y)$ and $\|m_h\|_p \le C\|h\|_{L_1(S,\mathbb{C})}$. Also, since $\mathcal{S}(\mathbb{R}^N, X)$ is norm dense in $L_p(\mathbb{R}^N, X)$, thanks to (3.6), the equality in (3.7) holds for each $f \in L_p(\mathbb{R}^N, X)$.

Part b) follows from a) and 2.14. □

For the remainder of Section 3 we restrict our attention to the case $\mathbb{R}^N = \mathbb{R}$.

3.5 Discussion. What kind of multipliers can we construct from the basic multiplier function

$$m_H = -\chi_{(-\infty,0]} + \chi_{(0,\infty)}$$

by applying the operations of 3.3 or 3.4? We assume that X is a UMD space so that $m_H \in \mathcal{M}_p(\mathbb{R}, X, X)$. But Y is an arbitrary Banach space.

i) Note that $\|m_H\|_{\mathcal{M}_p(X)} \ge 1$ since $1_{\mathbb{R}} = m_H^2$. By 3.3 a) and b), each of the following functions

$$\chi_{(0,\infty)} = \frac{1}{2}(1 + m_H) \qquad\qquad \chi_{(-\infty,0]} = \frac{1}{2}(1 - m_H)$$
$$\chi_{(a,\infty)}(t) = \chi_{(0,\infty)}(t-a) \qquad \chi_{(-\infty,b]}(t) = \chi_{(-\infty,0]}(t-b)$$

is in $\mathcal{M}_p(X)$ with $\| \cdot \|_{\mathcal{M}_p(X)}$-norm at most $\|m_H\|_{\mathcal{M}_p(X)}$. Thus, $\chi_{(a,b]} \equiv \chi_{(-\infty,b]} \cdot \chi_{(a,\infty]}$ is in $\mathcal{M}_p(X)$ with norm at most $\|m_H\|_{\mathcal{M}_p(X)}^2$.

ii) Let $m: \mathbb{R} \to B(X,Y)$ have the form $m(\cdot) = \chi_I(\cdot)B$ where I is an interval of $\mathbb{R}$ and $B \in B(X,Y)$. Then $(T_m f)(t) = B[(T_{\chi_I}f)(t)]$ for $f \in \mathcal{S}(\mathbb{R}, X)$. Therefore, $m \in \mathcal{M}_p(\mathbb{R}, X, Y)$ and $\|m\|_p \le \|B\|_{B(X,Y)} \|\chi_I\|_{\mathcal{M}_p(\mathbb{R},X,X)}$.

iii) Now we explore what kind of multipliers we can obtain by averaging the multipliers we built up so far. To this end consider a continuously differentiable function $m : \mathbb{R} \to B(X,Y)$ with an integrable derivative. Then $m_{-\infty} := \lim_{u \to -\infty} m(u)$ exists since $m(t) - m(u) = \int_u^t m'(s) \, ds$ for $u, t \in \mathbb{R}$. Thus for each $t \in \mathbb{R}$ we can write

$$m(t) = m_{-\infty} + \int_{-\infty}^{t} m'(s)\, ds = m_{-\infty} + \int_{-\infty}^{\infty} \chi_{(s,\infty)}(t) m'(s)\, ds$$

$$= m_{-\infty} + \int_{\mathbb{R}} m_s(t) h(s)\, ds$$

where $h(s) := \|m'(s)\|_{B(X,Y)}$ and

$$m_s(\cdot) := \chi_{(s,\infty)}(\cdot) \frac{m'(s)}{\|m'(s)\|_{B(X,Y)}} \chi_{\{u:\, m'(u)\neq 0\}}(s) \,,$$

following the convention that $\frac{0}{0}$ is 0. By i) and ii), $m_s \in \mathcal{M}_p(\mathbb{R}, X, Y)$ with $\|m_s\|_{\mathcal{M}_p(\mathbb{R},X,Y)} \leq \|m_H\|_{\mathcal{M}_p(X)}$. Now 3.4 a) implies a first serious multiplier theorem, Proposition 3.6.

3.6 Proposition. *Let X be a UMD-space. If $m \in C^1(\mathbb{R}, B(X,Y))$ has an integrable derivative and $\lim_{t\to-\infty} m(t) = 0$, then $m \in \mathcal{M}_p(\mathbb{R}, X, Y)$ and*

$$\|m\|_{\mathcal{M}_p(\mathbb{R},X,Y)} \leq C \int_{\mathbb{R}} \|m'(t)\|_{B(X,Y)}\, dt$$

where $C = \|m_H\|_{\mathcal{M}_p(X)}$ with m_H as in 3.5.

Since a C^1 function with integrable derivative is of bounded variation, we can view this result as an operator-valued version of Steklin's multiplier theorem.

Reexamining the proof of Proposition 3.6 we also find a criterion for the R-boundedness of sets of such multipliers of Steklin type. First we look at multipliers of the form considered in 3.5 i) and 3.5 ii).

Let (Ω, μ) be a σ-finite measure space. For $n \in L_\infty(\Omega, \mathbb{C})$ denote by M_n the multiplication operator $(M_n f)(\omega) = n(\omega) f(\omega)$ on $L_p(\Omega, X)$. Also, for $T \in B(X,Y)$ define the extension operator $\widetilde{T} \in B(L_p(\Omega, X), L_p(\Omega, Y))$ by $(\widetilde{T} f)(\omega) = T(f(\omega))$.

3.7 Lemma. *Let τ be an R-bounded subset of $B(X,Y)$ and $\mathcal{I}$ be the set of all (possibly unbounded) interval of $\mathbb{R}$.*

a) *The set of multiplication operators $\sigma = \{M_n : \|n\|_{L_\infty(\Omega,\mathbb{C})} \leq 1\}$ is R-bounded in $B(L_p(\Omega, X))$ with $R_p(\sigma) \leq 2$.*

b) *$\widetilde{\tau} := \{\widetilde{T} : T \in \tau\}$ is an R-bounded subset of $B(L_p(\Omega, X), L_p(\Omega, Y))$ with $R_p(\widetilde{\tau}) \leq R_p(\tau)$.*

c) *If X is an UMD space, then the set of Fourier multiplier operators $\sigma_{\mathcal{I}} := \{T_{\chi_I} : I \in \mathcal{I}\}$ is an R-bounded subset of $B(L_p(\mathbb{R}, X))$.*

d) *If X is an UMD space, then the set of Fourier multiplier operators $\{T_{B\chi_I} : I \in \mathcal{I} \text{ and } B \in \tau\}$ is R-bounded in $B(L_p(\mathbb{R}, X), L_p(\mathbb{R}, Y))$ with R_p-bound at most $C\, R_p(\tau)$.*

Here, $C = R_p(\sigma_{\mathcal{I}}) \leq 32 \, \|m_H\|^2_{\mathcal{M}_p(X)}$ where m_H is as in 3.5.

Proof. Towards a), choose $n_j \in L_\infty(\Omega, \mathbb{C})$ with $\|n_j\|_{L_\infty} \leq 1$ and $f_j \in L_p(\Omega, X)$. Then by Fubini's theorem and Kahane's contraction principle

$$\|\sum_{j=1}^m r_j M_{n_j} f_j\|^p_{L_p([0,1], L_p(\Omega, X))} = \int_\Omega \int_0^1 \|\sum_{j=1}^m r_j(u) n_j(\omega) f_j(\omega)\|^p_X \, du \, d\omega$$

$$\leq 2^p \int_\Omega \int_0^1 \|\sum_{j=1}^m r_j(u) f_j(\omega)\|^p_X \, du \, d\omega$$

$$= 2^p \|\sum_{j=1}^m r_j f_j\|^p_{L_p([0,1], L_p(\Omega, X))} \, .$$

Towards b), choose $T_j \in \tau$ and $f_j \in L_p(\Omega, X)$. Then by Fubini's theorem

$$\|\sum_{j=1}^m r_j \widetilde{T}_j f_j\|^p_{L_p([0,1], L_p(\Omega, Y))} = \int_\Omega \int_0^1 \|\sum_{j=1}^m r_j(u) T_j(f_j(\omega))\|^p_Y \, du \, d\omega$$

$$\leq R_p^p(\tau) \int_\Omega \int_0^1 \|\sum_{j=1}^m r_j(u) f_j(\omega)\|^p_X \, du \, d\omega$$

$$= R_p^p(\tau) \|\sum_{j=1}^m r_j f_j\|^p_{L_p([0,1], L_p(\Omega, X))} \, .$$

Towards c), consider the multiplication operators $M_s(\cdot) := M_{e^{is(\cdot)}}$ in $B(L_p(\mathbb{R}, X))$ and the (since X is UMD) Fourier multiplier operators $T_{\chi_{(0,\infty)}}$ and $T_{\chi_{(-\infty,0)}}$ in $B(L_p(\mathbb{R}, X))$. If $f \in \mathcal{S}(\mathbb{R}, X)$, then

$$(T_{\chi_{(s,\infty)}} f)(t) = e^{ist} (T_{\chi_{(0,\infty)}} (e^{-is(\cdot)} f(\cdot)))(t) \, .$$

Thus by 3.3 a), $T_{\chi_{(s,\infty)}} = M_s T_{\chi_{(0,\infty)}} M_{-s}$ and so

$$T_{\chi_{(-\infty,r)}} = I - T_{\chi_{(r,\infty)}} = M_r(I - T_{\chi_{(0,\infty)}}) M_{-r} = M_r T_{\chi_{(-\infty,0)}} M_{-r}$$

$$T_{\chi_{(s,r)}} = T_{\chi_{(-\infty,r)}} T_{\chi_{(s,\infty)}} = M_r T_{\chi_{(-\infty,0)}} M_{s-r} T_{\chi_{(0,\infty)}} M_{-s} \, .$$

Thus $\sigma_{\mathcal{I}} \subset \sigma \circ \{T_{\chi_{(-\infty,0)}}, I\} \circ \sigma \circ \{T_{\chi_{(0,\infty)}}, I\} \circ \sigma$ where σ is as in part a) with $\Omega = \mathbb{R}$. Thus by part a), 3.5 i), and 2.8, $\sigma_{\mathcal{I}}$ is R-bounded with $R_p(\sigma_{\mathcal{I}}) \leq 2^3 \left[2 \|m_H\|_{\mathcal{M}_p(X)} \right]^2$.

Towards d), note that, as shown in 3.5 ii), $T_{B\chi_I} \equiv \widetilde{B} \circ T_{\chi_I}$ is indeed an L_p-Fourier multiplier operator. Apply part b), part c) and 2.8 to see that $R_p(\{T_{B\chi_I} : I \in \mathcal{I} \text{ and } B \in \tau\}) \leq R_p(\tau) R_p(\sigma_{\mathcal{I}})$. $\square$

Now we consider sets of Steklin type multipliers of the following form. For a bounded subset τ of $B(X, Y)$, define

$$S(\tau) := \{ m \in C^1(\mathbb{R}, B(X, Y)): \lim_{t \to -\infty} m(t) = 0 \text{ and}$$

$$m'(\cdot) = h(\cdot)n(\cdot) \text{ for some} \tag{3.8}$$

$$h \in L_1(\mathbb{R}, \mathbb{C}) \text{ with } \|h\|_{L_1} \leq 1 \text{ and } \tau\text{-valued } n \in L_\infty(\mathbb{R}, B(X, Y)) \} .$$

3.8 Corollary. *Let X have the UMD property and $\tau \subset B(X, Y)$ be R-bounded. Then $\sigma = \{T_m : m \in S(\tau)\}$ is an R-bounded set of Fourier multiplier operators in $B(L_p(\mathbb{R}, X), L_p(\mathbb{R}, Y))$ and $R_p(\sigma) \leq C R_p(\tau)$, where $C = 64 \|m_H\|^2_{\mathcal{M}_p(X)}$ with m_H as in 3.5.*

Proof. Fix $m \in S(\tau)$ and find a decomposition $m'(\cdot) = h(\cdot)n(\cdot)$ as in (3.8). As in 3.5 iii),

$$m(t) = \int_{\mathbb{R}} \chi_{(s,\infty)}(t) m'(s) \, ds = \int_{\mathbb{R}} m_s(t) h(s) \, ds ,$$

where $m_s(\cdot) = \chi_{(s,\infty)}(\cdot)n(s)$. For each $s \in \mathbb{R}$, $m_s \in \mathcal{M}_p(\mathbb{R}, X, Y)$ by 3.5 i) and ii). Also, for each $s \in \mathbb{R}$, T_{m_s} is in $\{T_{\chi_I B} : I \text{ is an interval of } \mathbb{R} , B \in \tau\}$, which, by 3.7 d), is an R-bounded subset of $B(L_p(\mathbb{R}, X), L_p(\mathbb{R}, Y))$ with R_p bound at most $32R_p(\tau)\|m_H\|^2_{\mathcal{M}_p(X)}$.

Thus, by 3.4 b), σ is R-bounded with $R_p(\sigma) \leq 64R_p(\tau)\|m_H\|^2_{\mathcal{M}_p(X)}$.   □

We return now to Mihlin multipliers.

3.9 Discussion. Let $m \in C^1(\mathbb{R} \setminus \{0\}, B(X, Y))$ satisfy the Mihlin's conditions of 1.10, i.e.,

$$\{m(t) : t \in \mathbb{R} \setminus \{0\}\} \qquad \text{and} \qquad \{tm'(t) : t \in \mathbb{R} \setminus \{0\}\}$$

are R-bounded (and so also norm bounded). Thus, since $\|m'(t)\|_{B(X,Y)} \leq \frac{C}{t}$, m' is not necessarily integrable, but we do have

$$\int_{2^n}^{2^{n+1}} \|m'(t)\|_{B(X,Y)} \, dt \leq C \log 2, \quad n \in \mathbb{Z}.$$

This motivates us to decompose functions f in $L_p(\mathbb{R}, X)$ into formal series $f = \sum_{n \in \mathbb{Z}} P_n f$ with

$$\text{supp}(\widehat{P_n f}) \subset I_n := \{t \in \mathbb{R}: 2^{n-1} \leq |t| \leq 2^n\}, \quad n \in \mathbb{Z} .$$

To be precise, we assume that X has UMD so that, by 3.5 i),

$$P_n := T_{\chi_{I_n}} \in B(L_p(\mathbb{R}, X), L_p(\mathbb{R}, X)) \qquad \text{for } n \in \mathbb{Z} . \tag{3.9}$$

Note that $\sum_{n \in \mathbb{Z}} P_n f$ converges to f in $L_p(\mathbb{R}, X)$; indeed, it obviously does so for $f \in S_0(\mathbb{R}, X)$ and $\sum_{|n| \leq N} P_n$ is uniformly (in N) bounded in operator norm. Next we can apply T_m to each $P_n f$, which amounts to applying to f the operator $T_m P_n \equiv T_{m\chi_{I_n}}$, which is a bounded operator on $L_p(\mathbb{R}, Y)$ by 3.3 c) and 3.6. Next one wants that $\sum_{n\in\mathbb{Z}} T_m P_n f$ converges in $L_p(\mathbb{R}, Y)$. This is not easy. It is not enough that $\sum_{n\in\mathbb{Z}} P_n f$ converges in $L_p(\mathbb{R}, X)$. Rather we need a stronger convergence: in functional analytic terms we want that $\sum_{n\in\mathbb{Z}} P_n f$ converges unconditionally (this section's endnotes discuss unconditional convergence) while in the terminology of harmonic analysis we are looking for the Paley-Littlewood decomposition (by which we mean the Paley-Littlewood estimate (3.13) holds). However, both viewpoints are equivalent, as shown by Proposition 3.10, which we state in a general form for later use.

3.10 Proposition. *Let X be reflexive and $1 < p < \infty$. Let $\mathcal{Q} := \{H_n\}_{n\in\mathbb{Z}}$ be a decomposition of $\mathbb{R}^N$ with $P_n := T_{\chi_{H_n}} \in B(L_p(\mathbb{R}^N, X))$ that satisfies the geometric conditions:*

- *if $H_n \in \mathcal{Q}$ then $-H_n \in \mathcal{Q}$*

- *if $0 < r < R < \infty$ then there are finitely many $H_{n_1}, \ldots, H_{n_k}$ so that $B^\infty(R) \setminus B^\infty(r) \subset \cup_{j=1}^k H_{n_j}$ where $B^\infty(a)$ is as in (4.1).*

Then the following are equivalent.

i) (unconditional convergence of $\sum_{n\in\mathbb{Z}} P_n f$)
There is a constant C_{Xp} such that, for each $f \in L_p(\mathbb{R}^N, X)$,

$$\sup_{\substack{\delta_n \in \mathbb{C} \\ |\delta_n| \leq 1}} \| \sum_{n\in\mathbb{Z}} \delta_n P_n f \|_{L_p(\mathbb{R}^N, X)} \leq C_{Xp} \|f\|_{L_p(\mathbb{R}^N, X)} . \qquad (3.10)$$

ii) For each (or equivalently, for some) choice of $q_1, q_2 \in [1, \infty)$, there is a constant E (depending on X, p, q_1, q_2) such that

$$\| \sum_{n\in\mathbb{Z}} r_n P_n f \|_{L_{q_1}([0,1], L_p(\mathbb{R}^N, X))} \leq E \|f\|_{L_p(\mathbb{R}^N, X)} \qquad (3.11)$$

$$\| \sum_{n\in\mathbb{Z}} r_n P_n^* g \|_{L_{q_2}([0,1], L_{p'}(\mathbb{R}^N, X^*))} \leq E \|g\|_{L_{p'}(\mathbb{R}^N, X^*)} \qquad (3.12)$$

for each $f \in L_p(\mathbb{R}^N, X)$ and $g \in L_{p'}(\mathbb{R}^N, X^*)$.

iii) (Paley-Littlewood decomposition for $L_p(\mathbb{R}^N, X)$)
For each (or equivalently, for some) $q \in [1, \infty)$ there is a constant D_{Xpq} such that, for each $f \in L_p(\mathbb{R}^N, X)$,

$$\frac{1}{D_{Xpq}} \|f\|_{L_p(\mathbb{R}^N, X)} \leq \| \sum_{n\in\mathbb{Z}} r_n P_n f \|_{L_q([0,1], L_p(\mathbb{R}^N, X))} \leq D_{Xpq} \|f\|_{L_p(\mathbb{R}^N, X)} .$$
$$(3.13)$$

Here, $\{r_n\}_{n\in\mathbb{Z}}$ is an enumeration of the Rademacher functions.

In application, we usually use i) or iii). But to show that these hold, we usually show that (3.11) holds and then use a duality argument to show that (3.12) holds. To avoid distraction from the main ideas, the proof of Proposition 3.10 is in this section's endnotes.

3.11 Theorem. *Let X be a UMD space and $I_n := \{t \in \mathbb{R}: 2^{n-1} < |t| \leq 2^n\}$ for $n \in \mathbb{Z}$. Then the Paley-Littlewood estimate (3.13) holds with $P_n = T_{\chi_{I_n}}$ for each $1 < p < \infty$.*

Remark. For $X = \mathbb{C}$, by 2.9, (3.13) takes the form

$$\|f\|_{L_p(\mathbb{R},\mathbb{C})} \sim \|(\sum_{n \in \mathbb{Z}} |P_n f|^2)^{\frac{1}{2}}\|_{L_p(\mathbb{R},\mathbb{C})} ,$$

which is the classical Paley-Littlewood theorem.

Proof (of Theorem 3.11). That UMD implies (3.13) is a difficult result of Bourgain and will not be presented here. Note that $\mathcal{Q} = \{I_n\}_{n \in \mathbb{Z}}$ satisfies the hypotheses of Proposition 3.10 and so i) and ii) of 3.10 also hold.   □

We are ready now for the multiplier theorem announced in 1.10.

3.12 Theorem. *Let X and Y be UMD-spaces and $1 < p < \infty$.*
Let $m \in C^1(\mathbb{R} \setminus \{0\}, B(X,Y))$ satisfy the following conditions:

i) $\tau_1 := \{m(t) : t \in \mathbb{R} \setminus \{0\}\}$ is R-bounded with R_p-bound C_1,

ii) $\tau_2 := \{tm'(t) : t \in \mathbb{R} \setminus \{0\}\}$ is R-bounded with R_p-bound C_2.

Then $m \in \mathcal{M}_p(\mathbb{R}, X, Y)$ and $\|T_m\|_{B(L_p(\mathbb{R},X),L_p(\mathbb{R},Y))} \leq C\max(C_1, C_2)$ for some constant C depending only on X, Y, and p.

Proof. Let $\tau = \text{absco}\ \{m(t), tm'(t): t \in \mathbb{R} \setminus \{0\}\}$; so, $R_p(\tau) \leq 4\max(C_1, C_2)$.
Choose a $\varphi \in C^\infty(\mathbb{R}, \mathbb{R})$ with supp $\varphi \subset \{t : 2^{-2} \leq |t| \leq 2^1\}$ and $\varphi \equiv 1$ on $\{t : 2^{-1} \leq |t| \leq 2^0\}$. Let $C_\varphi := \sup\{|\varphi(t)|, |t\varphi'(t)| : t \in \mathbb{R}\}$.
For each $k \in \mathbb{Z}$ let

$$\varphi_k(t) = \varphi(2^{-k}t) \qquad\qquad I_k = \{t \in \mathbb{R}: 2^{k-1} \leq |t| \leq 2^k\}$$
$$m_k = \varphi_k m \qquad\qquad\qquad J_k = \{t \in \mathbb{R}: 2^{k-2} \leq |t| \leq 2^{k+1}\}$$
$$h_k(t) = \frac{1}{t}\chi_{J_k}(t) \qquad\qquad n_k(t) = 2^{-k}t\varphi'(2^{-k}t)m(t) + \varphi(2^k t)tm'(t) .$$

Thus, for each $k \in \mathbb{Z}$

$$\varphi_k \equiv 1\ \ \text{on}\ I_k \qquad\qquad \text{supp}\ \varphi_k \subset J_k$$
$$\|h_k\|_{L_1(\mathbb{R},\mathbb{R})} = 6\log 2 \qquad\quad n_k(t) \in 2C_\varphi\, \tau$$
$$m_k'(t) = h_k(t)n_k(t) \qquad\quad m_k \in S(12(\log 2)C_\varphi\, \tau)$$

where $S(\cdot)$ is as in (3.8) .
By 3.8, $\sigma := \{T_{m_k} : k \in \mathbb{Z}\} \subset B(L_p(\mathbb{R}, X), L_p(\mathbb{R}, Y))$ is an R-bounded set of Fourier multipliers with $R_p(\sigma) \leq D\max(C_1, C_2)$ for some constant D depending only on X and p. Since $T_{\chi_{I_n}}T_m = T_{m\chi_{I_n}\varphi_n} = T_{m_n}T_{\chi_{I_n}}$, we can now deliver our punchline (using Theorem 3.11 and (3.13))

$$\|T_m f\|_{L_p(\mathbb{R},Y)} \le D_{Ypp} \|\sum_{n\in\mathbb{Z}} r_n P_n T_m f\|_{L_p([0,1],L_p(\mathbb{R},Y))}$$

$$= D_{Ypp} \|\sum_{n\in\mathbb{Z}} r_n T_{m_n} P_n f\|_{L_p([0,1],L_p(\mathbb{R},Y))}$$

$$\le D_{Ypp} R_p(\sigma) \|\sum_{n\in\mathbb{Z}} r_n P_n f\|_{L_p([0,1],L_p(\mathbb{R},X))}$$

$$\le D_{Xpp} D_{Ypp} R_p(\sigma) \|f\|_{L_p(\mathbb{R},X)}$$

for each $f \in \mathcal{S}(\mathbb{R},X)$. $\qquad\qquad\qquad\qquad\qquad\qquad\qquad\qquad\square$

3.13 Necessity of R-boundedness. We show now that condition i) in Theorem 3.12 is necessary. Condition ii) on the other hand is a convenient sufficient condition (together with i)), which is not necessary even in the scalar-valued case. More precisely, the following is true.

Let $m \in L_\infty(\mathbb{R}, B(X,Y))$ for some Banach spaces X and Y and let $L(m)$ be the set of continuity points (or even Lebesgue points) of m. If T_m defines a bounded operator from $L_p(\mathbb{R},X)$ to $L_p(\mathbb{R},Y)$ for some $p \in (1,\infty)$, then the set $\{m(t) : t \in L(m)\}$ must be R-bounded in $B(X,Y)$.

Proof. We have to find a constant C such that for all choices $t_1,\ldots,t_k \in L(m)$ and $x_1,\ldots,x_k \in X$ we have

$$\|\sum_{j=1}^k r_j m(t_j) x_j\|_{L_p([0,1],Y)} \le C \|\sum_{j=1}^k r_j x_j\|_{L_p([0,1],X)}.$$

To relate the left hand side to T_m we choose a symmetric function $\psi \in \mathcal{D}(\mathbb{R},\mathbb{R})$ with $0 \le \psi \le 1$ and $\int \psi^2(t)\, dt = 1$. Put $\psi_n(t) = \psi(nt)$ and for $\phi = \check\psi$ let $\phi_n(t) = \phi(\frac{1}{n}t)$. Observe that

$$\widehat{\phi_n}(t) = n\psi_n(t), \quad \int_{\mathbb{R}} \widehat{\phi_n}(t)\psi_n(t)\, dt = 1.$$

Hence for every $t_j \in L(m)$

$$\lim_{n\to\infty} \int_{\mathbb{R}} m(t)\widehat{\phi_n}(t-t_j)\psi_n(t-t_j)\, dt = \lim_{n\to\infty} \left[m * \left(\widehat{\phi_n}\psi_n\right)\right](t_j) = m(t_j)$$

in $B(X,Y)$ and therefore, by Fatou's lemma,

$$\int_0^1 \|\sum_{j=1}^k r_j(u) m(t_j) x_j\|_Y^p\, du$$

$$\le \varliminf_{n\to\infty} \int_0^1 \|\sum_{j=1}^k r_j(u) \int_{\mathbb{R}} m(t) x_j \widehat{\phi_n}(t-t_j)\psi_n(t-t_j)\, dt\|_Y^p\, du \ . \tag{3.14}$$

Note that $m(t) x_j \widehat{\phi_n}(t-t_j) = \mathcal{F} T_m[e^{it_j(\cdot)}\phi_n(\cdot)x_j](t)$ and

$$\int_{\mathbb{R}} \mathcal{F}[T_m(e^{it_j(\cdot)}\phi_n(\cdot)x_j)](t)\psi_n(t-t_j)\,dt$$

$$= \int_{\mathbb{R}} T_m(e^{it_j(\cdot)}\phi_n(\cdot)x_j)(s)\mathcal{F}[\psi_n(\cdot-t_j)](s)\,ds \quad = \int_{\mathbb{R}} T_j[\phi_n(\cdot)x_j](s)\widehat{\psi}_n(s)\,ds$$

where $(T_jf)(s) = e^{-it_j s}T_m(e^{it_j(\cdot)}f(\cdot))(s)$ for $f \in L_p(\mathbb{R}, X)$. By Lemma 3.7 a), the set $\{T_j: j = 1,\ldots,k\}$ is R-bounded in $B(L_p(\mathbb{R}, X), L_p(\mathbb{R}, Y))$ with R_p-bound at most $D := 4\|T_m\|_{B(L_p(\mathbb{R}, X), L_p(\mathbb{R}, Y))}$. Now coming back to (3.14), we have, with the help of Hölder's inequality,

$$\|\sum_{j=1}^{k} r_j m(t_j)x_j\|_{L_p([0,1],Y)}^p$$

$$\leq \lim_{n\to\infty} \int_0^1 \left\|\sum_{j=1}^{k} r_j(u) \int_{\mathbb{R}} T_j[\phi_n(\cdot)x_j](s)\widehat{\psi}_n(s)\,ds\right\|_Y^p du$$

$$\leq \lim_{n\to\infty} \int_0^1 \left[\int_{\mathbb{R}} \left\|\sum_{j=1}^{k} r_j(u)T_j(\phi_n(\cdot)x_j)(s)\right\|_Y \left|\widehat{\psi}_n(s)\right|_{\mathbb{C}} ds\right]^p du$$

$$\leq \lim_{n\to\infty} \int_0^1 \left[\left\|\sum_{j=1}^{k} r_j(u)T_j(\phi_n x_j)\right\|_{L_p(\mathbb{R},Y)} \left\|\widehat{\psi}_n\right\|_{L_{p'}(\mathbb{R},\mathbb{C})}\right]^p du$$

$$\leq \lim_{n\to\infty} \|\sum_{j=1}^{k} r_j T_j(\phi_n x_j)\|_{L_p([0,1],L_p(\mathbb{R},Y))}^p \|\widehat{\psi}_n\|_{L_{p'}(\mathbb{R},\mathbb{C})}^p$$

$$\leq D^p \lim_{n\to\infty} \|\sum_{j=1}^{k} r_j \phi_n x_j\|_{L_p([0,1],L_p(\mathbb{R},X))}^p \|\widehat{\psi}_n\|_{L_{p'}(\mathbb{R},\mathbb{C})}^p$$

$$= D^p \|\sum_{j=1}^{k} r_j x_j\|_{L_p([0,1],X)}^p \left(\lim_{n\to\infty} \|\phi_n\|_{L_p}\|\widehat{\psi}_n\|_{L_{p'}}\right)^p.$$

Since $\|\phi_n\|_{L_p} = n^{\frac{1}{p}}\|\phi\|_{L_p}$ and $\|\widehat{\psi}_n\|_{L_{p'}} = n^{\frac{-1}{p}}\|\widehat{\psi}\|_{L_{p'}}$, we are done. □

Now we can also complete the proof of our characterization Theorem 1.11 of maximal regularity for bounded analytic semigroups. In 1.5 we have seen that the generator A has maximal regularity if and only if $m(t) = tR(it, A)$ is a bounded Fourier-multiplier on $L_p(\mathbb{R}, X)$ for $1 < p < \infty$. We have just shown that for this it is necessary that $\{tR(it, A) : t \neq 0\}$ is R-bounded. Hence maximal regularity implies the equivalent conditions i), ii), iii) in 1.11.

So far we have not discussed how to check the UMD property for a concrete Banach space X nor how to prove Bourgain's result on the existence of a vector-valued Paley-Littlewood estimate (3.13). It is easier to treat these questions for the special case of $X = L_q(\Omega, \mathbb{C})$-spaces, because here we can use many tools from classical harmonic analysis instead of martingale transforms. Also many important spaces in analysis (e. g. Sobolev- and Hardy spaces) are closed subspaces of some L_q-space. Therefore it seems reasonable to concentrate on L_q-spaces for the last part of this section. (See the notes for further information).

3.14 UMD and Paley Littlewood for $L_q(\Omega)$. Let X be a closed subspace of an $L_q(\Omega, \mathbb{C})$-space, for a σ-finite measure space (Ω, μ) and $1 < q < \infty$. Then

a) X has the UMD property

b) $L_p(\mathbb{R}, X)$, for $1 < p < \infty$, has a Paley-Littlewood estimate (3.13) for $P_n = T_{\chi_{I_n}}$ where $I_n := \{t \in \mathbb{R} : 2^{n-1} < |t| \leq 2^n\}$.

Proof. a) It shown in many books (see e. g. [86], [101], [205]) that the Hilbert transform

$$H\phi(t) = PV - \int_{\mathbb{R}} \frac{1}{t-s} \phi(s) ds, \qquad \phi \in \mathcal{D}(\mathbb{R}, \mathbb{C})$$

is bounded on $L_q(\mathbb{R}, \mathbb{C})$. For a function $f : \Omega \to L_q(\mathbb{R}, \mathbb{C})$ of the form $f(\omega) = \sum_{j=1}^{n} \psi_j(\omega)\phi_j$ with $\psi_j \in L_q(\Omega, \mathbb{C})$ and $\phi_j \in \mathcal{D}(\mathbb{R}, \mathbb{C})$ we define $(H \otimes I)f(\omega) = H(f(\omega)) = \sum_{j=1}^{n} \psi_j(\omega)H(\phi_j)$, so that

$$\|(H \otimes I)f\|_{L_q(\Omega, L_q(\mathbb{R}, \mathbb{C}))}^q = \int_{\Omega} \|H(f(\omega))\|_{L_q(\mathbb{R}, \mathbb{C})}^q d\mu(\omega)$$
$$\leq \|H\|_{B(L_q(\mathbb{R}, \mathbb{C}))}^q \|f\|_{L_q(\Omega, L_q(\mathbb{R}, \mathbb{C}))}^q .$$

Hence $H \otimes I$ extends to a bounded operator on $L_q(\Omega, L_q(\mathbb{R}, \mathbb{C}))$ and by Fubini's theorem also on $L_q(\mathbb{R}, L_q(\Omega, \mathbb{C})) = L_q(\Omega, L_q(\mathbb{R}, \mathbb{C}))$. Clearly, $L_q(\mathbb{R}, X)$ is an invariant subspace for $H \otimes I$. To see that $H \otimes I$ is also bounded on $L_p(\mathbb{R}, X)$ for $1 < p < \infty$, we employ Theorem 3.15 below with $k(t) = \frac{1}{t}\mathrm{Id}_X$.

b) Without loss of generality, $X = L_q(\Omega, \mathbb{C})$. As in the proof of Theorem 3.12, choose a $\varphi \in C^{\infty}(\mathbb{R}, \mathbb{R})$ with $\mathrm{supp}\,\varphi \subset \{t : 2^{-2} \leq |t| \leq 2^1\}$ and $\varphi \equiv 1$ on $\{t : 2^{-1} \leq |t| \leq 2^0\}$ and put $\varphi_j(t) = \varphi(2^{-j}t)$ for $j \in \mathbb{Z}$.

Let Γ be the collection of all finitely-supported $\{0, 1, -1\}$-valued sequences $\{\varepsilon_j\}_{j \in \mathbb{Z}}$. For an $\varepsilon = (\varepsilon_j)_{j \in \mathbb{Z}} \in \Gamma$, define a (potential) multiplier function

$$m_\varepsilon(t) = \sum_{j \in \mathbb{Z}} \varepsilon_j \varphi_j(t) .$$

Note that

$$k_\varepsilon(t) := \check{m}_\varepsilon(t) = \sum_{j \in \mathbb{Z}} \varepsilon_j \check{\varphi}_j(t) .$$

So for $f \in \mathcal{S}_0(\mathbb{R}, \mathbb{C})$

$$T_{m_\varepsilon} f(t) = [m_\varepsilon \hat{f}]^{\vee}(t) = (k_\varepsilon * f)(t) = \int_{\mathbb{R}} k_\varepsilon(t-s) f(s)\, ds .$$

There is a constant C, depending only on the choice of φ, such that

$$|m_\varepsilon(t)| \le C \qquad\qquad\qquad |t m'_\varepsilon(t)| \le C \qquad\qquad (3.15)$$

$$k_\varepsilon \in L_{1,\text{loc}}(\mathbb{R} \setminus \{0\}, \mathbb{C}) \qquad\qquad |k'_\varepsilon(t)| \le \frac{C}{|t|^2}\ . \qquad\qquad (3.16)$$

The estimates (3.15) follow from the support property of φ. Towards (3.16), note that $\frac{d}{dt}\check{\varphi}_j(t) = i(2^j)^2 [(\cdot)\varphi(\cdot)]^\vee (2^j t)$ and, since $[(\cdot)\varphi(\cdot)]^\vee \in \mathcal{S}(\mathbb{R}, \mathbb{C})$, there is a constant C_1 so that $|[(\cdot)\varphi(\cdot)]^\vee (t)| \le \frac{C_1|t|}{1+|t|^4}$ for each $t \in \mathbb{R}$. So for each $t \in \mathbb{R} \setminus \{0\}$, say $2^{n-1} \le |t| < 2^n$ where $n \in \mathbb{Z}$,

$$|t^2 k'_\varepsilon(t)| \le \sum_{j\in\mathbb{Z}} |t|^2 \left|\frac{d}{dt}\check{\varphi}_j(t)\right| \ \le \sum_{j\in\mathbb{Z}} |2^j t|^2\, |[(\cdot)\varphi(\cdot)]^\vee (2^j t)|$$

$$\le C_1 \sum_{j\in\mathbb{Z}} \frac{|2^j t|^3}{1+|2^j t|^4} \ \le C_1 \sum_{j\in\mathbb{Z}} \frac{|2^{n+j}|^3}{1+|2^{n-1+j}|^4}$$

$$= 2C_1 \sum_{j\in\mathbb{Z}} \frac{|2^j|^3}{1+|2^j|^4} \ := C_2 < \infty\ .$$

Now (3.15) implies that T_{m_ε} is a bounded Fourier-multiplier on $L_q(\mathbb{R}, \mathbb{C})$ by the classical Mihlin multiplier theorem. As in a) we can show with Fubini's theorem that $T_{m_\varepsilon} \otimes I$ is bounded on $L_q(\mathbb{R}, X)$. By (3.16) we can apply Theorem 3.15 to $k(t) = k_\varepsilon(t)\mathrm{Id}_X$ and obtain that $T_{m_\varepsilon} \otimes I$ is bounded on $L_p(\mathbb{R}, X), 1 < p < \infty$, and $\|T_{m_\varepsilon} \otimes I\| \le C_p$ for all such ε's.

Now we are ready to prove (3.13). Fix $1 < p < \infty$. Let $f \in \mathcal{S}_0(\mathbb{R}, X)$. Note that $P_j f = T_{I_j} f = P_j(\check{\varphi}_j * f)$ since $\varphi_j \equiv 1$ on I_j. Also the set $\{P_j : j \in \mathbb{Z}\}$ is R-bounded on $L_p(\mathbb{R}, X)$ by part a) and Lemma 3.7c). Hence, for some $J \in \mathbb{N}$ sufficiently large,

$$\int_0^1 \Big\| \sum_{j\in\mathbb{Z}} r_j(u) P_j f \Big\|_{L_p(\mathbb{R},X)}^p du = \int_0^1 \Big\| \sum_{\substack{j\in\mathbb{Z}\\|j|\le J}} r_j(u) P_j (\check{\varphi}_j * f) \Big\|_{L_p(\mathbb{R},X)}^p du$$

$$\le R_p^p(\{P_j\}_{j\in\mathbb{Z}}) \int_0^1 \Big\| \sum_{\substack{j\in\mathbb{Z}\\|j|\le J}} r_j(u) \check{\varphi}_j * f \Big\|_{L_p(\mathbb{R},X)}^p du$$

$$\le R_p^p(\{P_j\}_{j\in\mathbb{Z}}) \sup_{\varepsilon\in\Gamma} \|T_{m_\varepsilon} f\|_{L_p(\mathbb{R},X)}^p$$

$$\le C_p^p\, R_p^p(\{P_j\}_{j\in\mathbb{Z}}) \|f\|_{L_p(\mathbb{R},X)}^p.$$

Thus, for each $p \in (1,\infty)$ and each $q \in (1,\infty)$, for $X = L_q(\Omega, \mathbb{C})$, there is a constant D, depending on X and p, so that

$$\Big\| \sum_{j\in\mathbb{Z}} r_j P_j f \Big\|_{L_p([0,1], L_p(\mathbb{R},X))} \le D \|f\|_{L_p(\mathbb{R},X)} \qquad \text{for each } f \in L_p(\mathbb{R}, X)\ .$$

But $X^* = L_{q'}(\Omega, \mathbb{C})$ and $P_n^* = T_{\chi_{I_n}}$. So ii) of 3.10 holds for $X = L_q(\Omega, \mathbb{C})$.

$\square$

Finally, we quote an important extrapolation result, which we used already in 3.14.

3.15 Theorem. *Let X and Y be Banach spaces and let $T : L_q(\mathbb{R}^n, X) \to L_q(\mathbb{R}^n, Y)$ be a bounded linear operator for <u>one</u> $q \in (1, \infty)$. Assume that there is a locally integrable kernel $k : \mathbb{R}^n \setminus \{0\} \to B(X, Y)$ such that*

$$Tf(t) = \int k(t - s)f(s)\, ds, \qquad t \notin \operatorname{supp} f$$

for $f \in L_\infty(\mathbb{R}^n, X)$ with compact support.
If the kernel satisfies

$$\|\nabla k(t)\| \leq C|t|^{-n-1} \text{ for } t \neq 0$$

then T extends to a bounded operator from $L_p(\mathbb{R}, X)$ to $L_p(\mathbb{R}, Y)$ for <u>all</u> $p \in (1, \infty)$.

This theorem highlights the point we made in the beginning of this section. If X is not a Hilbert space there is often no easy way to show the boundedness of a Fourier multiplier or singular integral operator T on $L_2(\mathbb{R}, X)$. Indeed, the main effort will be to find *one* q such that T is bounded on $L_q(\mathbb{R}, X)$.

Notes on Section 3:

N 3.1 - N 3.4: The classical theory of Fourier multipliers on $L_p(\mathbb{R}^N, \mathbb{C})$ and the basic facts about vector-valued multipliers considered here are contained in many books, e. g. [86], [101] and [205].

N 3.5 - N 3.6: Proposition 3.6 is a special case of a more general Steklin multiplier theorem for operator-valued multipliers in [210], Theorem 3.4.

N 3.7 - N 3.8: Part a) of Lemma 3.7 is the birth place of R-boundedness. It is implicit in [39] and stated explicitly in [30]. Part b) was observed in [221]. Corollary 3.8 is new.

N 3.9: A fundamental notion is that of unconditional convergence of a formal series $\sum_{n=1}^{\infty} x_n$ where $x_n \in X$. As a definition of a *unconditional convergence series* we may take any of the following equivalent statements, which we quote here for the convenience of the reader.

- $\sum_{n=1}^{\infty} x_{\pi(n)}$ converges for each permutation π of $\mathbb{N}$.
- $\sum_{n=1}^{\infty} x_{n_k}$ converges weakly for each subsequence $(n_k)_{k \in \mathbb{N}}$.
- $\sum_{n=1}^{\infty} \delta_n x_n$ converges for each choice of signs $\delta_n \in \{\pm 1\}$.
- $\sum_{n=1}^{\infty} \alpha_n x_n$ converges for each sequence $(\alpha_n)_{n \in \mathbb{N}}$ in $\mathbb{C}$ with $|\alpha_n| \leq 1$.

If X is reflexive (and therefore weakly sequentially complete) one may add

- $\sum_{n=1}^{\infty} |x^*(x_n)| < \infty$ for each $x^* \in X^*$.

Proofs of the equivalence of these (and more) statements can be found in many books, e.g. [69].

If X is a Hilbert space with an orthonormal sequence $(h_n)_{n \in \mathbb{N}}$, e.g. the Rademacher functions $(r_n)_{n \in \mathbb{N}}$ in $L_2[0,1]$, then the series

$$\sum_{n=1}^{\infty} \langle x, h_n \rangle \, h_n$$

converges unconditionally. Although we loose the concept of orthogonality in a general Banach space, some of the important "convergence properties" of orthogonal series are captured in the notion of unconditional convergence, which therefore will play an important role in these notes.

Note that Kahane's contraction principle can be rephrased in these terms: for each $(x_n)_{n \in \mathbb{N}} \in Rad(X)$, the series $\sum_{n=1}^{\infty} r_n x_n$ converges unconditionally in $L_p([0,1], X)$ for $1 \leq p < \infty$.

N 3.10: As promised, here is the proof of Proposition 3.10.

Proof. Note that $[L_p(\mathbb{R}^N, X)]^* = L_{p'}(\mathbb{R}^N, X^*)$ and $P_n^* = T_{\chi_{-H_n}}$.

In Proposition 3.10, our initial understanding of convergence for expressions of the form $\sum_{n \in \mathbb{Z}} g_n$ is: the limit, as $N \to \infty$, of $\sum_{n=-N}^{N} g_n$ in the underlying function space. To see that we can interpret the summand differently, consider

$$S_Q(\mathbb{R}^N, Z) := \left\{ f \in S(\mathbb{R}^N, Z) : \text{supp } \widehat{f} \subset \cup_{|n| \leq N} H_n \text{ for some } N \in \mathbb{N} \right\} .$$

$S_Q(\mathbb{R}^N, Z)$ is dense in $L_q(\mathbb{R}^N, Z)$, for $1 < q < \infty$, since it contains $S_0(\mathbb{R}^N, Z)$.

Let $f \in S_Q(\mathbb{R}^N, X)$ and $g \in S_Q(\mathbb{R}^N, X^*)$. Then $P_n f \equiv 0$ and $P_n^* g \equiv 0$ for $|n|$ sufficiently large (thus the summands appearing in Proposition 3.10 are actually finite summands); also, $f = \sum_{n \in \mathbb{Z}} P_n f$ and $g = \sum_{n \in \mathbb{Z}} P_n^* g$ (again, these are actually finite sums). Furthermore, if an inequality in Proposition 3.10 holds for each $f \in S_Q(\mathbb{R}^N, X)$ (resp. $g \in S_Q(\mathbb{R}^N, X^*)$) then it holds for each $f \in L_p(\mathbb{R}^N, X)$ ((resp. $g \in L_{p'}(\mathbb{R}^N, X^*)$); this is clear in i) while in ii) and iii), since X is reflexive, this follows from 2.3. Also note that, in ii) and iii), for any choice of signs $\delta_n \in \{\pm 1\}$, we can replace $\{r_n\}_{n \in \mathbb{Z}}$ by $\{\delta_n r_n\}_{n \in \mathbb{Z}}$ since these two sequences are identically distributed. Thus we can replace our initial understanding of convergence by unconditional convergence (see the endnotes for 3.9) in the underlying function space. But for this proof, we keep with the initial understanding.

i) $\Rightarrow$ ii). Assume i) holds. For $f \in S_Q(\mathbb{R}^N, X)$,

$$\left\| \sum_{n \in \mathbb{Z}} r_n P_n f \right\|_{L_{q_1}([0,1], L_p(\mathbb{R}^N, X))} \leq \left\| \sum_{n \in \mathbb{Z}} r_n P_n f \right\|_{L_\infty([0,1], L_p(\mathbb{R}^N, X))}$$

$$\leq C_{Xp} \|f\|_{L_p(\mathbb{R}^N, X)} .$$

Fix $g \in \mathcal{S}_Q(\mathbb{R}^N, X^*)$. By duality, i) implies

$$\sup_{\substack{\delta_n \in \mathbb{C} \\ |\delta_n| \leq 1}} \left\| \sum_{n \in \mathbb{Z}} \delta_n P_n^* g \right\|_{L_{p'}(\mathbb{R}^N, X^*)} \leq C_{Xp} \|g\|_{L_{p'}(\mathbb{R}^N, X^*)}$$

and so, similar to the calculation above for f, (3.12) holds. So ii) holds.

ii) $\Rightarrow$ iii). Assume ii) holds. Fix $f \in \mathcal{S}_Q(\mathbb{R}^N, X)$. For $g \in \mathcal{S}_Q(\mathbb{R}^N, X^*)$

$$\langle f, g \rangle = \left\langle \sum_{n \in \mathbb{Z}} P_n f, \sum_{m \in \mathbb{Z}} P_m^* g \right\rangle = \sum_{n \in \mathbb{Z}} \langle P_n f, P_n^* g \rangle$$

$$= \int_0^1 \left\langle \sum_{n \in \mathbb{Z}} r_n(u) P_n f, \sum_{m \in \mathbb{Z}} r_m(u) P_m^* g \right\rangle du$$

$$\leq E \, \|g\|_{L_{p'}(\mathbb{R}^N, X^*)} \left\| \sum_{n \in \mathbb{Z}} r_n P_n f \right\|_{L_2([0,1], L_p(\mathbb{R}^N, X))}.$$

Taking the supremum over all such g's with $\|g\|_{L_{p'}(\mathbb{R}^N, X^*)} \leq 1$ gives the lower inequality in iii).

iii) $\Rightarrow$ i). Assume that iii) holds. Fix an admissible sequence $\{\delta_n\}_{n \in \mathbb{Z}}$. If $f \in \mathcal{S}_Q(\mathbb{R}^N, X)$ then

$$\left\| \sum_{n \in \mathbb{Z}} \delta_n P_n f \right\|_{L_p(\mathbb{R}^N, X)} \leq D_{Xp1} \left\| \sum_{n \in \mathbb{Z}} r_n P_n \left(\sum_{k \in \mathbb{Z}} \delta_k P_k f \right) \right\|_{L_1([0,1], L_p(\mathbb{R}^N, X))}$$

$$= D_{Xp1} \left\| \sum_{n \in \mathbb{Z}} \delta_n r_n P_n f \right\|_{L_1([0,1], L_p(\mathbb{R}^N, X))}$$

$$\leq 2 D_{Xp1} \left\| \sum_{n \in \mathbb{Z}} r_n P_n f \right\|_{L_1([0,1], L_p(\mathbb{R}^N, X))}$$

$$\leq 2 D_{Xp1}^2 \|f\|_{L_p(\mathbb{R}^N, X)}$$

by Kahane's contraction principle and iii).   □

N 3.11: That $L_p(\mathbb{R}, X)$ has a Paley-Littlewood estimate (3.13) for UMD spaces X is the break-through contribution of J. Bourgain [39]. He stated the result originally for $L_p([0, 2\pi], X)$ and Zimmermann([226]) extended his proof to $L_p(\mathbb{R}, X)$. The connection of the Littlewood Paley decomposition with unconditional Schauder decompositions (as in part i) of 3.10) was studied in [46].

N 3.12: Theorem 3.12 is from [221]. We give here a new proof by reducing the claim to the R-boundedness version 3.8 of Steklin's theorem. The case of scalar multiplier functions is already contained in [226]. For variants and extensions of this theorem see the notes of Section 4.

N 3.13: That R-boundedness of the multiplier function m is necessary for boundedness of T_m is shown in [48] with the argument reproduced here. A slightly weaker statement was already contained in [221]. This result was strong enough to imply that the R-boundedness conditions i), ii), and iii) in 1.11 are necessary.

N 3.14 is inspired by standard techniques in harmonic analysis, which are detailed e. g. in [86], [101], [205]. That $L_q(\Omega, \mathbb{C}), 1 < q < \infty$, has the UMD-property is already implicitly in [28], maybe also in earlier papers. The same proof as for 3.14 actually shows that for a closed subspace X of $L_q(\Omega, \mu)$ and every scalar function $m \in C^1(\mathbb{R} \setminus \{0\})$ with $|m(t)| \leq C$ and $|tm'(t)| \leq C$ the Fourier multiplier T_m is bounded on $L_p(\mathbb{R}, X)$ for $1 < p < \infty$. (To check the Hörmander condition given in N 3.15 below, use the argument in [101], proof of II.6.3.)

N 3.15: For this result of [27] see also [101], Theorem V.3.4. The result holds under a weaker assumption on K, namely the so called Hörmander condition:

$$\int_{|t|>2|s|} \|k(t - s) - k(t)\|_{B(X,Y)} dt \leq C \text{ for all } s \in \mathbb{R}^N .$$

Furthermore, [101] contains information on the 'endpoint' estimates $L_1 \rightarrow$ weak L_1, $H_1 \rightarrow L_1$ and $L_\infty \rightarrow BMO$.

We take this opportunity to summarize some further results, which make it clear that UMD-spaces are the right class of Banach spaces for the study of vector-valued harmonic analysis and stochastics. (For a survey and detailed references see [42] and [196]).

If X has the UMD property, then X^*, closed subspaces of X, and quotient spaces of X have UMD. Also $L_p(\Omega, X)$ inherits the UMD property for $1 < p < \infty$. A UMD space is always uniformly convex, but not all uniformly convex spaces have UMD. In particular, a UMD space is reflexive and has a type $p > 1$ and cotype $q < \infty$. Besides the L_p-spaces we mention the Schatten classes $\mathcal{S}_p$ with $1 < p < \infty$ as examples of UMD-spaces.

Each of the following conditions is equivalent to the UMD property (see [42], [196], [38]).

- For one (all) $p \in (1, \infty)$ the Hilbert transform is bounded on $L_p(\mathbb{R}, X)$.

- X-valued martingale difference sequences (d_j) on (Ω, μ) converges unconditionally, i. e. for one (all) $p \in (1, \infty)$ there is a constant C_p only dependent on X and p such that

$$\sup_{\varepsilon_j = \pm 1} \| \sum_j \varepsilon_j d_j \|_{L_p(\Omega, X)} \leq C_p \| \sum_j d_j \|_{L_p(\Omega, X)} .$$

- X is ζ-convex, i. e. there is a function $\zeta : X \times X \rightarrow \mathbb{R}$ such that $\zeta(0,0) > 0$ and $\zeta(x, y) = \zeta(y, x)$ for all $(x, y) \in X \times X$;
  $\zeta(x, \cdot)$ is a convey function for any $x \in X$;
  $\zeta(\cdot, y)$ is a convex function for any $y \in X$;
  $\zeta(x, y) \leq \|x + y\|$ when $\|x\| = \|y\| = 1$.

The second property gave rise to the name UMD.

4 Fourier Multipliers on $\mathbb{R}^N$

In this section we present a vector-valued version of the Mihlin multiplier theorem (in 4.6) and of the Marcinkiewicz multiplier theorem (in 4.13). The latter theorem requires an additional assumption on the Banach space X, Pisier's property (α), which we discuss in 4.9.

4.1 Discussion. We would like to extend the argument of the last section to the multidimensional case. This will require the following steps.

i) Use the UMD property of the space X to show that the characteristic functions of cubes define bounded Fourier multipliers on $L_p(\mathbb{R}^N, X)$ for $1 < p < \infty$.

ii) Show that functions of bounded variation on $\mathbb{R}^N$ define bounded Fourier multipliers by representing them as "convex combinations" of characteristic functions of cubes.

iii) Establish a Littlewood-Paley decomposition for $L_p(\mathbb{R}^N, X)$, $1 < p < \infty$.

iv) Use iii) to reduce general multiplier functions satisfying Mihlin's condition to step ii).

While i) and ii) will be done quickly in Lemma 4.3 and Proposition 4.4, the Littlewood-Paley decomposition will require some thought. Among the many possibilities to decompose $\mathbb{R}^N$ in *cubes* (generalizing the dyadic intervals $(2^{k-1}, 2^k]$ we used in Section 3) we consider the following two (see Figure 1). The first way decomposes $\mathbb{R}^N$ into *dyadic cubical annulars*:

$$
\begin{aligned}
&\mathbb{R}^N \setminus \{0\} = \bigcup_{j \in \mathbb{Z}} I_j \\
&I_j = B^\infty(2^j) \setminus B^\infty(2^{j-1}) \\
&B^\infty(a) = \{(t_1, \ldots, t_N) \in \mathbb{R}^N : |t_i| \le a \text{ for each } i \in \{1, \ldots, N\}\}.
\end{aligned}
\tag{4.1}
$$

The second decomposition consists of all possible products of dyadic intervals; it refines (4.1) by further decomposing each *dyadic cubical annular* into *dyadic cubes*:

$$
\begin{aligned}
&\mathbb{R}^N \setminus \{\text{coordinate axes}\} \quad = \bigcup_{(j_1, \ldots, j_N) \in \mathbb{Z}^N} (J_{j_1} \times J_{j_2} \times \cdots \times J_{j_N}) \\
&J_{2j} = \{t \in \mathbb{R} : 2^{j-1} < t \le 2^j\} \\
&J_{2j+1} = \{t \in \mathbb{R} : 2^{j-1} < -t \le 2^j\}.
\end{aligned}
\tag{4.2}
$$

In Proposition 4.5 we will show that, if X is a UMD space and $1 < p < \infty$, then the decomposition (4.1) defines a Littlewood-Paley decomposition of

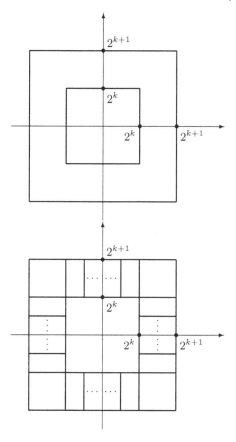

Fig. 1. The two decompositions

$L_p(\mathbb{R}^N, X)$ as in 3.11. The finer decomposition (4.2) will allow us to formulate a multiplier theorem with a weaker Mihlin assumption on the multiplier function than for decomposition (4.1). But we will see that the price for this improvement is an additional assumption on X, Pisier's property (α) (which is satisfied for all closed subspaces of $L_p(\Omega, \mathbb{C})$ spaces with $1 < p < \infty$, see 4.9 and 4.10). So the geometry of the decompositions (4.1) and (4.2) is reflected in the assumptions we have to put on the space X! Having settled iii), step iv) is routine again.

Let us start this program with a simple observation.

4.2 Fact. Fix $j \in \{1, \ldots, N\}$.

a) If $m \colon \mathbb{R} \to \mathbb{C}$ defines a bounded Fourier multiplier on $L_p(\mathbb{R}, X)$, then so does $\tilde{m}(t_1, \ldots, t_N) := m(t_j)$ on $L_p(\mathbb{R}^N, X)$; also, for each $f \in L_p(\mathbb{R}^N, X)$,

$$(T_{\tilde{m}} f)(t_1, \ldots, t_N) = T_m(f(t_1, \ldots, t_{j-1}, \cdot, t_{j+1}, \ldots, t_N))(t_j) \qquad (4.3)$$

for a.e. $t \in \mathbb{R}^N$ and $\|T_{\tilde{m}}\|_{B(L_p(\mathbb{R}^N, X))} \leq \|T_m\|_{B(L_p(\mathbb{R}, X))}$.

b) If $\mathcal{M}$ is a set of multiplier functions $m \colon \mathbb{R} \to \mathbb{C}$ such that $\{T_m \colon m \in \mathcal{M}\}$ is R-bounded in $B(L_p(\mathbb{R}, X))$, then $\{T_{\tilde{m}} \colon m \in \mathcal{M}\}$ is R-bounded in $B(L_p(\mathbb{R}^N, X))$ with $R_p(\{T_{\tilde{m}} \colon m \in \mathcal{M}\}) \leq R_p(\{T_m \colon m \in \mathcal{M}\})$.

Proof It is easy to check that (4.3) holds for $f \in \mathcal{S}(\mathbb{R}^N, X)$ of the form $f(t_1, \ldots, t_N) = x \prod_{k=1}^{N} \varphi_k(t_k)$ where $x \in X$ and $\varphi_k \in \mathcal{S}(\mathbb{R}, \mathbb{C})$. Since finite sums of such functions are norm dense in $L_p(\mathbb{R}^N, X)$, the first part holds.

For finite sequences $\{m_k\}_k$ in $\mathcal{M}$ and $\{f_k\}_k$ in $L_p(\mathbb{R}^N, X)$, by Fubini's theorem

$$
\int_0^1 \|\sum_k r_k(u) T_{\tilde{m}_k} f_k\|_{L_p(\mathbb{R}^N, X)}^p \, du
$$

$$
= \int_{\mathbb{R}^{N-1}} \int_0^1 \|\sum_k r_k(u) T_{m_k}(f_k(t_1, \ldots, t_{j-1}, \cdot, t_{j+1}, \ldots, t_N))\|_{L_p(\mathbb{R}, X)}^p \, du \, ds
$$

$$
\leq R_p^p(\{T_{m_k}\}_k) \int_0^1 \|\sum_k r_k(u) f_k\|_{L_p(\mathbb{R}^N, X)}^p \, du
$$

where $ds = d(t_1, \ldots, t_{j-1}, t_{j+1}, \ldots, t_N)$. □

4.3 Lemma. *Let X be a UMD-space and $1 < p < \infty$. Let $\mathcal{Q}$ be the set of all (possibly unbounded) cubes $\prod_{j=1}^{N}(a_j, b_j)$ in $\mathbb{R}^N$. Then $\{T_{\chi_Q} \colon Q \in \mathcal{Q}\}$ is an R-bounded set of multipliers on $L_p(\mathbb{R}^N, X)$.*

Proof. Let $\mathcal{Q}_k$ be the set of all cubes $\prod_{j=1}^{N}(a_j, b_j)$ with $(a_j, b_j) = \mathbb{R}$ for $j \neq k$. Then $\{T_{\chi_Q} \colon Q \in \mathcal{Q}_k\}$ is R-bounded by 3.7 and 4.2. Since, by 3.3a), $\{T_{\chi_Q} \colon Q \in \mathcal{Q}\} = \prod_{j=1}^{N}\{T_{\chi_{Q_j}} \colon Q_j \in \mathcal{Q}_j\}$, the claim follows from 2.8. □

To formulate a Steklin-type multiplier theorem we extend a notion from Section 3 (see (3.8)). For a bounded subset τ of $B(X, Y)$, let

$$
S_N(\tau) := \{m \in C^N(\mathbb{R}^N, B(X, Y)) \colon \lim_{|t| \to \infty} m(t) = 0 \text{ and}
$$

$$
\frac{\partial^N}{\partial t_1 \cdots \partial t_N} m(\cdot) = h(\cdot) n(\cdot) \text{ for some} \tag{4.4}
$$

$h \in L_1(\mathbb{R}^N, \mathbb{C})$ with $\|h\|_{L_1} \leq 1$ and τ-valued $n \in L_\infty(\mathbb{R}^N, B(X, Y))\}$.

Now we can extend 3.6 and 3.8.

4.4 Proposition. *Let X have the UMD-property and $1 < p < \infty$.*

a) If $m \in C^N(\mathbb{R}^N, B(X, Y))$ with $\lim_{|t| \to \infty} m(t) = 0$ and $\frac{\partial^N}{\partial t_1 \cdots \partial t_N} m$ is integrable on $\mathbb{R}^N$, then $m \in \mathcal{M}_p(\mathbb{R}^N, X, Y)$.

b) *If* $\tau \subset B(X,Y)$ *is R-bounded then* $\sigma = \{T_m : m \in \mathcal{S}_N(\tau)\}$ *is an R-bounded set of Fourier multiplier operators in* $B(L_p(\mathbb{R}^N, X), L_p(\mathbb{R}^N, Y))$ *with* $R_p(\sigma) \le C\, R_p(\tau)$ *for some constant* C *depending on* $X, N,$ *and* p.

Proof. Below, C_1 and C_2 are constants depending on $X, N,$ and p.

a) By the fundamental theorem of calculus, for $t = (t_1, \ldots, t_N) \in \mathbb{R}^N$,

$$m(t) = \int_{-\infty}^{t_1} \cdots \int_{-\infty}^{t_N} \frac{\partial^N}{\partial s_1 \cdots \partial s_N} m(s_1, \ldots, s_N) ds_1 \cdots ds_N$$

$$= \int_{\mathbb{R}^N} \prod_{j=1}^{N} \chi_{(s_j, \infty)}(t_j) \frac{\partial^N}{\partial s_1 \ldots \partial s_N} m(s)\, ds = \int_{\mathbb{R}^N} m_s(t) k(s)\, ds$$

with $k(s) := \|\frac{\partial^N}{\partial s_1 \cdots \partial s_N} m(s)\|_{B(X,Y)}$ and $m_s(t) = \chi_{Q(s)}(t) \cdot k(s)^{-1} \frac{\partial^N}{\partial s_1 \cdots \partial s_N} m(s)$ with $Q(s) = \prod_{j=1}^{N}(s_j, \infty)$ if $k(s) \ne 0$ and $m_s(t) = 0$ if $k(s) = 0$.

For each fixed $s \in \mathbb{R}^N$, $m_s(\cdot)$ defines a bounded Fourier multiplier on $L_p(\mathbb{R}^N, X)$ with $\sup_{s \in \mathbb{R}^N} \|m_s\|_{\mathcal{M}_p(\mathbb{R}^N, X, Y)} \le C_1$ by 3.5ii) and 4.3. Since $k \in L_1(\mathbb{R}^N, \mathbb{C})$ by assumption, Fact 3.4a) implies that $m \in \mathcal{M}_p(\mathbb{R}^N, X, Y)$.

b) Fix $m \in \mathcal{S}_N(\tau)$ and find the decomposition $\frac{\partial^N}{\partial t_1 \cdots \partial t_N} m(\cdot) = h(\cdot) n(\cdot)$ as given in (4.4). Similar to the proof of part a), $m(t) = \int_{\mathbb{R}^N} n_s(t) h(s)\, ds$ where $n_s(\cdot) = \chi_{Q_s}(\cdot) n(s)$ where $Q_{(s_1, \ldots, s_N)} = \prod_{j=1}^{N}(s_j, \infty)$. For each $s \in \mathbb{R}^N$, $n_s \in \mathcal{M}_p(\mathbb{R}^N, X, Y)$ by 3.5ii) and 4.3. For each $s \in \mathbb{R}^N$, T_{n_s} is in

$$\{T_{B\chi_Q} : B \in \tau\, ,\ Q \text{ is a cube in } \mathbb{R}^N\} = \tilde\tau \circ \{T_{\chi_Q} : Q \text{ is a cube in } \mathbb{R}^N\}\, ,$$

which is an R-bounded subset of $B(L_p(\mathbb{R}^N, X), L_p(\mathbb{R}^N, Y))$ with R_p-bound at most $C_2 R_p(\tau)$ by 2.8, 3.7b, and 4.3.

Thus, by 3.4b), σ is an R-bounded set of Fourier multiplier operators with $R_p(\sigma) \le 2C_2\, R_p(\tau)$. $\qquad\square$

Now for the Paley-Littlewood decomposition corresponding to (4.1).

4.5 Proposition. *Let X have the UMD property and $\{I_n\}_{n \in \mathbb{Z}}$ be the decomposition of $\mathbb{R}^N$ in (4.1). Then the Paley-Littlewood estimate (3.13) holds with $P_n = T_{\chi_{I_n}}$ for each $1 < p < \infty$.*

Proof. Note that $P_j = T_{\chi_{B^\infty(2^j)}} - T_{\chi_{B^\infty(2^{j-1})}}$ and so, since X is a UMD space, $P_j \in \mathcal{B}(L_p(\mathbb{R}^N, X))$. Thus the decomposition $\mathcal{Q} := \{I_n\}_{n \in \mathbb{Z}}$ of $\mathbb{R}^N$ satisfies the hypotheses of Proposition 3.10. We shall show that, from Proposition 3.10, ii) holds; for then iii) and i) also hold.

Put $I_{j,k} = \{t \in \mathbb{R}^N : 2^{j-1} < |t_k| \le 2^j\}$ and $B_{j,k} = \{t \in \mathbb{R}^N : |t_k| \le 2^j\}$. We can decompose I_j into a disjoint union according to the first component of t which is larger that 2^{j-1}:

$$I_j = \bigcup_{k=1}^{N} [B_{j-1,1} \cap \ldots \cap B_{j-1,k-1} \cap I_{j,k} \cap B_{j,k+1} \ldots \cap B_{j,N}].$$

Let $Q_{j,k}$ and $P_{j,k}$ be the Fourier multiplier operators on $L_p(\mathbb{R}^N, X)$ corresponding to $\chi_{B_{j,k}}$ and $\chi_{I_{j,k}}$, respectively. Then

$$P_j = \sum_{k=1}^{N} R_{j,k} P_{j,k}, \quad R_{j,k} := Q_{j-1,1} \cdots Q_{j-1,k-1} Q_{j,k+1} \cdots Q_{j,N} .$$

For each $k \in \{1, \ldots, N\}$, $P_{j,k}$ can be obtained from its one-dimensional analogue by the extension procedure of Fact 4.2 (namely, $P_{j,k} = T_{\tilde{m}_j}$ where $m_j = \chi_{I_j}$ and $\tilde{m}_j(t_1, \ldots, t_N) = m_j(t_k)$) and so

$$\sup_{\substack{\delta_j \in \mathbb{C} \\ |\delta_j| \leq 1}} \| \sum_{j \in \mathbb{Z}} \delta_j P_{j,k} \|_{B(L_p(\mathbb{R}^N, X))} \leq C_{Xp} ,$$

where C_{Xp} is from (3.10) for Theorem 3.11; indeed, this is a direct computation. Furthermore $\{R_{j,k} : j \in \mathbb{Z}, k \in \{1, \ldots, N\}\}$ is an R-bounded subset of $B(L_p(\mathbb{R}^N, X)))$. So, by Theorem 3.11, for each $f \in \mathcal{S}_0(\mathbb{R}^N, X)$

$$\int_0^1 \| \sum_{j \in \mathbb{Z}} r_j(u) P_j f \|_{L_p(\mathbb{R}^N, X)} \, du \leq \sum_{k=1}^{N} \int_0^1 \| \sum_{j \in \mathbb{Z}} r_j(u) R_{j,k} P_{j,k} f \|_{L_p(\mathbb{R}^N, X)} \, du$$

$$\leq R_1(\{R_{j,k}\}_{j,k}) \sum_{k=1}^{N} \int_0^1 \| \sum_{j \in \mathbb{Z}} r_j(u) P_{j,k} f \|_{L_p(\mathbb{R}^N, X)} \, du$$

$$\leq R_1(\{R_{j,k}\}_{j,k}) \sum_{k=1}^{N} \sup_{\delta_j = \pm 1} \| \sum_{j \in \mathbb{Z}} \delta_j P_{j,k} f \|_{L_p(\mathbb{R}^N, X)}$$

$$\leq R_1(\{R_{j,k}\}_{j,k}) \, N \, C_{Xp} \|f\|_{L_p(\mathbb{R}^N, X)} .$$

So, if X has UMD, then for each $1 < p < \infty$

$$\| \sum_{j \in \mathbb{Z}} r_j(u) P_j f \|_{L_1([0,1], L_p(\mathbb{R}^N, X))} \leq E \|f\|_{L_p(\mathbb{R}^N, X)} , \quad \text{for } f \in L_p(\mathbb{R}^N, X) ,$$

for some constant E depending on X, N and p. But if X has UMD, then so does X^*. Also, $P_j^* = T_{\chi_{I_j}}$. So ii) of 3.10 holds. $\square$

Next comes a multiplier theorem based on the decomposition (4.1).

4.6 Theorem. *Let X and Y have the UMD property.*
Let $m \in C^N(\mathbb{R}^N \setminus \{0\}, B(X,Y))$ and assume that

$$\tau := \{|t|^{|\alpha|} D^\alpha m(t) : t \in \mathbb{R}^N \setminus \{0\}, \alpha \leq (1,1, \ldots, 1)\} \tag{4.5}$$

is R-bounded in $B(X,Y)$. Then $m \in \mathcal{M}_p(\mathbb{R}^N, X, Y)$ for $1 < p < \infty$ and
$\|T_m\|_{B(L_p(\mathbb{R}^N, X), L_p(\mathbb{R}^N, Y))} \leq C R_p(\tau)$ for some constant C depending only on X, Y, N and p.

Proof. Let $\{I_n\}_{n\in\mathbb{Z}}$ be the decomposition of $\mathbb{R}^N$ in (4.1) and $P_n = T_{\chi_{I_n}}$.

Choose $\varphi \in C^\infty(\mathbb{R}^N, \mathbb{R})$ with $\operatorname{supp} \varphi \subset B^\infty(2) \setminus B^\infty(2^{-2}) \equiv I_{-1} \cup I_0 \cup I_1$ and $\varphi \equiv 1$ on $B^\infty(1) \setminus B^\infty(2^{-1}) \equiv I_0$. For $j \in \mathbb{Z}$, put $\varphi_j(t) := \varphi(2^{-j}t)$ and so $m_j := \varphi_j m$ has support in $J_j := B^\infty(2^{j+1}) \setminus B^\infty(2^{j-2})$ and $m_j = m$ on I_j.

The constants C_i's below depend on, at most, X, N, p, and φ.

With $\beta = (1, 1, \ldots, 1)$, we have

$$\frac{\partial^N}{\partial t_1 \cdots \partial t_N} m_j(t) = \frac{1}{|t|^N} \sum_{\alpha \leq \beta} \binom{\beta}{\alpha} \left[|t|^{N-|\alpha|} D^{\beta-\alpha}\varphi_j(t) \right] \left[|t|^{|\alpha|} D^\alpha m(t) \right]$$

$$= h_j(t) \, n_j(t)$$

where $h_j(\cdot) := |\cdot|^{-N} \chi_{J_j}(\cdot)$ has $L_1(\mathbb{R}^N, \mathbb{C})$-norm at most some constant C_1 and

$$n_j(t) := \sum_{\alpha \leq \beta} \binom{\beta}{\alpha} \left[|2^{-j}t|^{|\beta-\alpha|} D^{\beta-\alpha}\varphi(2^{-j}t) \right] \left[|t|^{|\alpha|} D^\alpha m(t) \right] \in C_2(\text{absco } \tau)$$

for some constant C_2. Hence $m_j \in \mathcal{S}_N (C_1 C_2 (\text{absco } \tau))$ and Proposition 4.4 tells us that $\{T_{m_j} : j \in \mathbb{Z}\}$ is an R-bounded set of Fourier multiplier operators in $B(L_p(\mathbb{R}^N, X), L_p(\mathbb{R}^N, Y))$ with R_p-bound at most $C_3 C_2 C_1 R_p(\tau) := C_4 R_p(\tau)$ for some constant C_3.

Using $P_j T_m = T_{m_j} P_j$ and the Paley-Littlewood estimate (3.13) from Proposition 4.5, similar to the one-dimensional case, we can complete the proof:

$$\|T_m f\|_{L_p(\mathbb{R}^N, Y)} \leq D_{Ypp} \| \sum_{j\in\mathbb{Z}} r_j P_j T_m f \|_{L_p([0,1], L_p(\mathbb{R}^N, Y))}$$

$$= D_{Ypp} \| \sum_{j\in\mathbb{Z}} r_j T_{m_j} P_j f \|_{L_p([0,1], L_p(\mathbb{R}^N, Y))}$$

$$\leq D_{Ypp} \, C_4 \, R_p(\tau) \| \sum_{j\in\mathbb{Z}} r_j P_j f \|_{L_p([0,1], L_p(\mathbb{R}^N, X))}$$

$$\leq D_{Ypp} \, C_4 \, R_p(\tau) \, D_{Xpp} \|f\|_{L_p(\mathbb{R}^N, X)}$$

for each $f \in \mathcal{S}(\mathbb{R}^N, X)$. $\square$

4.7 Example. In connection with parabolic equations, multipliers of the following form appear

$$m(t_1, t_2) = \frac{t_1}{t_1 + it_2^2} \, .$$

This multiplier does <u>not</u> satisfy the condition

$$|t|^{|\alpha|} |D^\alpha m(t)| \leq C \qquad \text{for } \alpha \leq (1,1) \, , \, t \in \mathbb{R}^2 \setminus \{0\}$$

(consider $\alpha = (1,0)$). But it does satisfy the weaker condition

$$|t^\alpha D^\alpha m(t)| \leq C \qquad \text{for } \alpha \leq (1,1) , \ t \in \mathbb{R}^2 \setminus \{0\} \qquad (4.6)$$

where $t^\alpha = t_1^{\alpha_1} t_2^{\alpha_2}$ for $\alpha = (\alpha_1, \alpha_2)$.

Next we will show that the finer decomposition (4.2) allows us to consider the weaker condition (4.6). We start by investigating the Paley-Littlewood decomposition (i.e. 3.10iii)) for the decomposition (4.2).

A simple observation is needed. We obtained the Paley-Littlewood estimate for the decomposition (4.1) by reducing the N-dimensional case to the 1-dimensional case and then using Theorem 3.11 for the symmetric decomposition $\{I_n\}_{n \in \mathbb{Z}}$ of $\mathbb{R}$ that corresponds to (4.1) with $N = 1$. But in Theorem 3.11 we could have used the non-symmetric decomposition $\{J_n\}_{n \in \mathbb{Z}}$ that corresponds to (4.2) with $N = 1$. Indeed,

$$\Big\| \sum_{n \in \mathbb{Z}} r_n T_{\chi_{I_n}} f \Big\|_{L_q([0,1], L_p(\mathbb{R}, X))} \sim \Big\| \sum_{n \in \mathbb{Z}} r_n T_{\chi_{J_n}} f \Big\|_{L_q([0,1], L_p(\mathbb{R}, X))}$$

for $f \in L_p(\mathbb{R}, X)$; to see this, use the decomposition of the identity operator on $L_p(\mathbb{R}, X)$ into the sum of the two projections $T_{\chi_{[0,\infty)}}$ and $T_{\chi_{(-\infty,0)}}$.

4.8 Discussion. Let X have the UMD property.

For $k \in \{1, \ldots, N\}$, $j \in \mathbb{Z}$, and $\{J_j\}_{j \in \mathbb{Z}}$ as in (4.2), let

$$J_{j,k} = \{t = (t_1, \ldots, t_N) \in \mathbb{R}^N : t_k \in J_j\} \qquad \text{and} \qquad P_{j,k} = T_{\chi_{J_{j,k}}} \ ;$$

so $P_{j,k} \in B(L_p(\mathbb{R}^N, X))$ by 4.2. For $j = (j_1, \ldots, j_N) \in \mathbb{Z}^N$, let

$$P_j := \prod_{k=1}^N P_{j_k,k} \qquad \text{and so} \qquad P_j = T_{\chi_{J_{j_1} \times \cdots \times J_{j_N}}} \qquad (4.7)$$

and $\{P_j\}_{j \in \mathbb{Z}^N}$ corresponds to the decomposition in (4.2).

Let $f \in L_p(\mathbb{R}^N, X)$. By (3.13) for Theorem 3.11, for each $k \in \{1, \ldots, N\}$,

$$\Big\| \sum_{j \in \mathbb{Z}} r_j P_{j,k} f \Big\|_{L_1([0,1], L_p(\mathbb{R}^N, X))} \leq C_{Xp} \|f\|_{L_p(\mathbb{R}^N, X)} \qquad (4.8)$$

for some constant C_{Xp} depending on X and p (this is a direct computation if you replace L_1 with L_p). For notational simplicity, let $N = 2$ now. By applying (4.8) twice, we get

$$\int_0^1 \int_0^1 \Big\| \sum_{(j_1, j_2) \in \mathbb{Z}^2} r_{j_1}(u) r_{j_2}(v) P_{(j_1, j_2)} f \Big\|_{L_p(\mathbb{R}^2, X)} \, du \, dv$$

$$= \int_0^1 \int_0^1 \Big\| \Big(\sum_{j_1 \in \mathbb{Z}} r_{j_1}(u) P_{(j_1, 1)} \Big) \Big(\sum_{j_2 \in \mathbb{Z}} r_{j_2}(v) P_{(j_2, 2)} \Big)(f) \Big\|_{L_p(\mathbb{R}^2, X)} \, du \, dv \qquad (4.9)$$

$$\leq C_{Xp}^2 \|f\|_{L_p(\mathbb{R}^2, X)} \ .$$

If the left hand side of (4.9) were equivalent to

$$\int_0^1 \|\sum_{(j_1,j_2)\in\mathbb{Z}^2} r_{(j_1,j_2)}(u)P_{(j_1,j_2)}f\|_{L_p(\mathbb{R}^2,X)}\,du$$

where $\{r_{(i,j)}\}_{(i,j)\in\mathbb{Z}^2}$ is a Bernoulli sequence, then the Paley-Littlewood estimate for the decomposition (4.2) that we aim for would follow from 3.10. The two integrals would be equivalent if $\{(u,v)\to r_i(u)\cdot r_j(v)\}_{(i,j)\in\mathbb{Z}^2}$ were an independent sequence of random variables. But this is not the case (just consider $\{r_1r_1, r_1r_2, r_2r_1, r_2r_2\}$) and so we need an additional property of the Banach space X.

4.9 Property (α). A Banach space X has property (α) if there is a constant C_X such that for each $n\in\mathbb{N}$, subset $\{x_{ij}\}_{i,j=1}^n$ of X, and subset $\{\alpha_{ij}\}_{i,j=1}^n$ of $\mathbb{C}$ with $|\alpha_{ij}|\leq 1$ we have

$$\int_0^1\int_0^1\|\sum_{i,j=1}^n r_i(u)r_j(v)\alpha_{ij}x_{ij}\|_X\,du\,dv$$
$$\leq C_X\int_0^1\int_0^1\|\sum_{i,j=1}^n r_i(u)r_j(v)x_{ij}\|_X\,du\,dv\;; \qquad (4.10)$$

i.e. we require that the contraction principle holds for the random sequence $\{r_ir_j\}_{i,j\in\mathbb{N}}$. Note that in (4.10) we could replace $\{r_i\}_{i\in\mathbb{N}}$ and $\{r_j\}_{j\in\mathbb{N}}$ by two mutually independent Bernoulli sequences $\{\varepsilon_i\}_{i\in\mathbb{N}}$ and $\{\varepsilon'_j\}_{j\in\mathbb{N}}$ on a probability space (Ω,μ); for example, $\varepsilon_i(u,v)=r_i(u)$ and $\varepsilon'_j(u,v)=r_j(v)$ on $\Omega=[0,1]^2$. Also, in the definition of property (α), one can replace $\alpha_{ij}\in\mathbb{C}$ with $|\alpha_{ij}|\leq 1$ by $\alpha_{ij}\in\{\pm 1\}$ (with a change in the constant C_X); indeed, just use ideas from the proof of Kahane's contraction principle to see this.

4.10 Remark. a) By applying Kahane's inequality repeatedly and by Fubini's theorem we can show that (4.10) can be replaced, for each $p\in[1,\infty)$, by the condition

$$\left(\int_0^1\int_0^1\|\sum_{i,j=1}^n r_i(u)r_j(v)\alpha_{ij}x_{ij}\|_X^p\,du\,dv\right)^{\frac{1}{p}}$$
$$\leq C_{Xp}\left(\int_0^1\int_0^1\|\sum_{i,j=1}^n r_i(u)r_j(v)x_{ij}\|_X^p\,du\,dv\right)^{\frac{1}{p}}. \qquad (4.11)$$

b) Clearly, property (α) is inherited by closed subspaces. $\mathbb{C}$ has property (α); indeed, Khintchine's inequality gives that, for $\beta_{ij}\in\mathbb{C}$,

$$\int_0^1\int_0^1|\sum_{i,j=1}^n r_i(u)r_j(v)\beta_{ij}|^2\,du\,dv\sim\sum_{i,j=1}^n|\beta_{ij}|^2.$$

Thus, by c) to come, closed subspaces of $L_p(\Omega, \mathbb{C})$ have property (α), where $1 \le p < \infty$ and (Ω, μ) is a σ-finite measure space.

c) If X has property (α), then $L_p(\Omega, X)$ has property (α) for $1 \le p < \infty$ and a σ-finite measure space (Ω, μ). This follows easily from a) and Fubini's theorem; indeed, if (4.11) holds for X with constant C_{Xp}, then (4.11) holds for $L_p(\Omega, X)$ with the same constant C_{Xp}.

d) The next result shows that property (α) allows a natural identification $J : Rad(X) \to Rad(Rad(X))$ by $J((x_{ij})_{(i,j) \in \mathbb{Z}^2}) = ((x_{ij})_{j \in \mathbb{Z}})_{i \in \mathbb{Z}}$ for finitely supported sequences $(x_{ij})_{(i,j) \in \mathbb{Z}^2}$ in X; note that

$$\|(x_{ij})_{(i,j) \in \mathbb{Z}^2}\|_{Rad_1(X)} = \int_0^1 \| \sum_{(i,j) \in \mathbb{Z}^2} r_{ij}(u) x_{ij} \|_X \, du$$

$$\|((x_{ij})_{j \in \mathbb{Z}})_{i \in \mathbb{Z}}\|_{Rad_1(Rad_1 X)} = \int_0^1 \int_0^1 \| \sum_{(i,j) \in \mathbb{Z}^2} r_i(u_1) r_j(u_2) x_{ij} \|_X \, du_1 \, du_2$$

where $\{r_{ij}\}_{(i,j) \in \mathbb{Z}^2}$ (resp. $\{r_i\}_{i \in \mathbb{Z}}$) is a Bernoulli sequence.

4.11 Lemma. *Let X and Y have property (α) and C_X be the constant in (4.10). Let $(\varepsilon_j)_{j \in \mathbb{Z}^N}$ be a Bernoulli sequence on (Ω, μ).*

a) For all finitely supported sequences $(x_j)_{j \in \mathbb{Z}^N}$ in X

$$\frac{1}{C_X^N} \int_\Omega \| \sum_{j \in \mathbb{Z}^N} \varepsilon_j(\omega) x_j \|_X \, d\mu(\omega)$$

$$\le \int_0^1 \cdots \int_0^1 \| \sum_{j \in \mathbb{Z}^N} [r_{j_1}(u_1) \cdots r_{j_N}(u_N)] x_j \|_X \, du_1 \cdots du_N$$

$$\le C_X^N \int_\Omega \| \sum_{j \in \mathbb{Z}^N} \varepsilon_j(\omega) x_j \|_X \, d\mu(\omega) \, .$$

b) If $\tau \subset B(X, Y)$ is R-bounded, then for all finitely supported sequences $(x_j)_{j \in \mathbb{Z}^N}$ in X and $(T_j)_{j \in \mathbb{Z}^N}$ in τ

$$\int_0^1 \cdots \int_0^1 \| \sum_{j \in \mathbb{Z}^N} r_{j_1}(u_1) \cdots r_{j_N}(u_N) T_j x_j \|_Y \, du_1 \cdots du_N$$

$$\le C_X^N C_Y^N R_1(\tau) \int_0^1 \cdots \int_0^1 \| \sum_{j \in \mathbb{Z}^N} r_{j_1}(u_1) \cdots r_{j_N}(u_N) x_j \|_X \, du_1 \cdots du_N \, .$$

As customary, $j = (j_1, \ldots, j_n)$ for $j \in \mathbb{Z}^N$.

Proof. a) We prove the lower inequality by induction on N. For $N = 1$ there is nothing to prove.

Now assume that the lower inequality holds for $N \in \mathbb{N}$ and all spaces with property (α). Let (ε'_{jk}) (resp. (ε_j), (r_k)) be a Bernoulli sequence on Ω' (resp. Ω, $[0,1]$) indexed by $(j,k) \in \mathbb{Z}^N \times \mathbb{Z}$ (resp. $j \in \mathbb{Z}^N$, $k \in \mathbb{Z}$). Choose a finitely supported sequence $\{x_{jk}\}_{(j,k) \in \mathbb{Z}^N \times \mathbb{Z}}$ in X. For fixed $u \in [0,1]$ and $\omega \in \Omega$ we apply the contraction principle and then integrate over $[0,1] \times \Omega$:

$$\int_{\Omega'} \| \sum_{(j,k) \in \mathbb{Z}^N \times \mathbb{Z}} \varepsilon'_{jk}(\omega') x_{jk} \|_X \, d\mu(\omega')$$

$$\leq \int_{\Omega'} \left(\int_0^1 \int_{\Omega} \| \sum_{(j,k) \in \mathbb{Z}^N \times \mathbb{Z}} \varepsilon'_{jk}(\omega') r_k(u) \varepsilon_j(\omega) x_{jk} \|_X \, du \, d\omega \right) d\omega' \, .$$

Applying assumption (4.10) for each fixed $\omega' \in \Omega'$ and then integrating over Ω', we can continue with:

$$\leq C_X \int_0^1 \int_{\Omega} \| \sum_{j \in \mathbb{Z}^N} \sum_{k \in \mathbb{Z}} r_k(u) \varepsilon_j(\omega) x_{jk} \|_X \, du \, d\omega$$

$$= C_X \int_{\Omega} \| \sum_{j \in \mathbb{Z}^N} \varepsilon_j(\omega) (\sum_{k \in \mathbb{Z}} r_k x_{jk}) \|_{L_1([0,1],X)} \, d\omega \, .$$

By 4.10c) $L_1([0,1],X)$ has property (α) so we can apply the inductive hypothesis to $x_j = \sum_{k \in \mathbb{Z}} r_k x_{jk} \in L_1([0,1],X)$ and continue with:

$$\leq C_X C_X^N \int_0^1 \cdots \int_0^1 \| \sum_{j \in \mathbb{Z}^N} r_{j_1}(u_1) \cdots r_{j_N}(u_N) x_j \|_{L_1([0,1],X)} \, du_1 \cdots du_N$$

$$= C_X^{N+1} \int_0^1 \cdots \int_0^1 \| \sum_{(j,k) \in \mathbb{Z}^{N+1}} r_{j_1}(u_1) \cdots r_{j_N}(u_N) r_k(u) x_{j,k} \|_X \, du \, du_1 \cdots du_N \, .$$

So the lower inequality in a) is proven for all $N \in \mathbb{N}$. Reversing these steps we can prove the upper inequality in a) in a similar way.
b) If we replace x_j in a) by $T_j x_j$ we can cancel T_j in the left hand side at the expense of the constant $R_1(\tau)$. □

Now we return to the Paley-Littlewood estimate for decomposition (4.2). Note that Proposition 3.10 still holds if we replace $n \in \mathbb{Z}$ by $j \in \mathbb{Z}^N$ and replace the Rademacher functions $\{r_n\}_{n \in \mathbb{Z}}$ on $[0,1]$ by a Bernoulli sequence $\{\varepsilon_j\}_{j \in \mathbb{Z}^N}$ on (Ω, μ).

4.12 Proposition. *Let X have UMD and property (α).*
Let $\{J_{j_1} \times \ldots \times J_{j_N}\}_{(j_1,\ldots,j_N) \in \mathbb{Z}^N}$ be the decomposition of $\mathbb{R}^N$ in (4.2). Then the Paley-Littlewood estimate (3.13), modified as mentioned above, holds with $P_{(j_1,\ldots,j_N)} = T_{\chi_{J_{j_1} \times \cdots \times J_{j_N}}}$ for each $1 < p < \infty$.

Proof. Clearly, the decomposition $\mathcal{Q} = \{J_{j_1} \times \ldots \times J_{j_N}\}_{(j_1,\ldots,j_N) \in \mathbb{Z}^N}$ of $\mathbb{R}^N$ satisfies the hypotheses of Proposition 3.10. We shall show that, from Proposition 3.10, ii) holds; for then iii) and i) also hold.

In Discussion 4.8, we explored the Fourier multiplier operators $\{P_j\}_{j \in \mathbb{Z}^N}$; so we will keep with the notation (and ideas) from this discussion.

Since X have UMD and property (α), say with constant C_X in (4.10), so does $L_p(\mathbb{R}^N, X)$. So by Discussion 4.8 and Lemma 4.11, with C_{Xp} as in (4.8),

$$\int_\Omega \|\sum_{j \in \mathbb{Z}^N} \varepsilon_j(\omega) P_j f\|_{L_p(\mathbb{R}^N,X)} \, d\mu(\omega)$$

$$\leq C_X^N \int_0^1 \cdots \int_0^1 \|\sum_{j \in \mathbb{Z}^N} r_{j_1}(u_1) \cdots r_{j_N}(u_N)(P_{j_1,1} \cdots P_{j_N,N})(f)\| \, du_1 \cdots du_N$$

$$= C_X^N \int_0^1 \cdots \int_0^1 \|\prod_{k=1}^N \left(\sum_{j_k \in \mathbb{Z}} r_{j_k}(u_k) P_{j_k,k}\right) f\|_{L_p(\mathbb{R}^N,X)} \, du_1 \cdots du_N$$

$$\leq C_X^N C_{Xp}^N \|f\|_{L_p(\mathbb{R}^N,X)}$$

for each $f \in \mathcal{S}_0(\mathbb{R}^N, X)$.

Thus, if a Banach space Z has UMD and property (α), then for each $1 < p < \infty$ there is a constant C (depending on Z, N, and p) so that

$$\|\sum_{j \in \mathbb{Z}^N} \varepsilon_j P_j f\|_{L_1(\Omega, L_p(\mathbb{R}^N, Z))} \leq C\|f\|_{L_p(\mathbb{R}^N, Z)} \quad \text{for} \ f \in L_p(\mathbb{R}^N, Z) .$$

Since X has UMD and property (α), so does X^*. Also, $P^*_{(j_1,\ldots,j_N)} = T_{\chi_{-(J_{j_1} \times \cdots \times J_{j_N})}}$. Thus, there is a constant $\widetilde{C}$ (depending on X^*, N, and p') so that

$$\|\sum_{j \in \mathbb{Z}^N} \varepsilon_j P_j^* g\|_{L_1(\Omega, L_{p'}(\mathbb{R}^N, X^*))} \leq \widetilde{C}\|g\|_{L_{p'}(\mathbb{R}^N, X^*)} \quad \text{for} \ g \in L_{p'}(\mathbb{R}^N, X^*) .$$

Thus 3.10ii) holds. $\square$

This leads to the following multiplier theorem.

4.13 Theorem. *Let X and Y be UMD-spaces with property (α). Let $m \in C^N(\mathbb{R}^N \setminus \{0\}, B(X,Y))$ and assume that*

$$\tau := \{t^\alpha D^\alpha m(t) : t \in \mathbb{R}^N \setminus \{0\} , \ \alpha \leq (1,1,\ldots,1)\}$$

is R-bounded in $B(X,Y)$. Then $m \in \mathcal{M}_p(\mathbb{R}^N, X, Y)$ for $1 < p < \infty$ and $\|T_m\|_{B(L_p(\mathbb{R}^N,X),L_p(\mathbb{R}^N,Y))} \leq C \, R_p(\tau)$ for some constant C depending only on X, Y, N, and p.

Proof. The proof closely parallels the proof of 3.12 and of 4.6.

We choose $\varphi \in C^\infty(\mathbb{R}, \mathbb{R})$ with supp $\varphi \subset \{t \in \mathbb{R}: 2^{-2} < t \le 2\}$ and $\varphi \equiv 1$ on $\{t \in \mathbb{R}: 2^{-1} < t \le 1\}$. For $i \in \mathbb{Z}$, put

$$\varphi_{2i}(t) := \varphi(2^{-i}t) \quad \text{and} \quad \varphi_{2i+1}(t) := \varphi(-2^{-i}t) .$$

For $j = (j_1, \ldots, j_N) \in \mathbb{Z}^N$ and $t = (t_1, \ldots, t_N) \in \mathbb{R}^N$, let

$$m_j(t) := \prod_{k=1}^N \varphi_{j_k}(t_k) m(t) .$$

Note that, following the notation from (4.2), $m_j = m$ on $\prod_{k=1}^N J_{j_k}$ and

$$\text{supp } m_j \subset \prod_{k=1}^N \left(\cup_{l=-1}^1 J_{2l+j_k} \right) := M_j .$$

Fix $j = (j_1, \ldots, j_N) = (2i_1 + \delta_1, \ldots, 2i_N + \delta_N) \in \mathbb{Z}^N$ where $i_k \in \mathbb{Z}$ and $\delta_k \in \{0,1\}$ for each $k \in \{1, \ldots, N\}$. Let $\varepsilon_k = 1$ if $\delta_k = 0$ and $\varepsilon_k = -1$ if $\delta_k = 1$. With $\beta = (1, \ldots, 1)$ we have

$$\frac{\partial^N}{\partial t_1 \cdots \partial t_N} m_j(t)$$

$$= \frac{1}{t^\beta} \sum_{\alpha \le \beta} \binom{\beta}{\alpha} \left[t^{\beta-\alpha} \left(\prod_{k=1}^N D^{\beta_k - \alpha_k} \varphi_{j_k}(t_k) \right) \right] [t^\alpha D^\alpha m(t)]$$

$$= \frac{1}{t^\beta} \sum_{\alpha \le \beta} \binom{\beta}{\alpha} \left[\prod_{k=1}^N \left(\varepsilon_k 2^{-i_k} t_k \right)^{\beta_k - \alpha_k} \varphi^{\beta_k - \alpha_k} \left(\varepsilon_k 2^{-i_k} t_k \right) \right] [t^\alpha D^\alpha m(t)]$$

$$= h_j(t) n_j(t)$$

where $h_j(t) := \frac{1}{t^\beta} \chi_{M_j}(t)$ and

$$n_j(t) := \sum_{\alpha \le \beta} \binom{\beta}{\alpha} \left[\prod_{k=1}^N \left(\varepsilon_k 2^{-i_k} t_k \right)^{\beta_k - \alpha_k} \varphi^{\beta_k - \alpha_k} \left(\varepsilon_k 2^{-i_k} t_k \right) \right] [t^\alpha D^\alpha m(t)] .$$

Note that h_j has $L_1(\mathbb{R}^N, \mathbb{C})$-norm $(3 \log 2)^N := C_1$ and $n_j(t) \in C_2$ (absco τ) for some constants C_2 (depending on N and φ). Thus, following the notation in (4.4), $m_j \in S_N(C_1 C_2 (\text{absco } \tau))$.

So, by Proposition 4.4, $\{T_{m_j}: j \in \mathbb{Z}^N\}$ is an R-bounded set of Fourier multiplier operators in $B(L_p(\mathbb{R}^N, X), L_p(\mathbb{R}^N, Y))$ with R_p-bound at most $C_3 R_p(\tau)$ for some constant C_3 (depending on X, N, p, and φ). For P_j as in 4.12, by construction, $P_j T_m = T_{m_j} P_j$. We can conclude this proof as we concluded the proof of 4.6, but using the Paley-Littlewood estimate from 4.12 instead of 4.5. □

Notes on Section 4:

N 4.2 - N 4.3 are basic observations implicitly contained in [226].

N 4.4: Proposition 4.4 is new. A related result for Fourier multipliers of bounded variation is Theorem 3.4 in [210].

N 4.5: Proposition 4.5 is a special case of the Littlewood-Paley decomposition in [226], which generalizes the one-dimensional result of [39].

N 4.6: Theorem 4.6 was first shown in [210], Theorem 4.4, by transference to the discrete case $L_p([0, 2\pi]^N, X)$. More direct and streamlined proofs appeared in [114] and [66], Section 3.4. The proof given here, based on 4.4, is new and probably the most transparent argument so far. [210] and [114] also contain a more general Marcinkiewicz multiplier theorem. For scalar multiplier functions these results were proved much earlier in [175] and [226]. There are a number of variants and improvements of this theorem. In [104] the optional smoothness of the multiplier function was determined and found to depend again on the geometry of the Banach spaces X and Y.

A Banach space X has **Fourier type** $p \in [1, 2]$ if the Fourier transform defines a bounded operator $T : L_p(\mathbb{R}^N, X) \to L_{p'}(\mathbb{R}^N, X)$ with $\frac{1}{p} + \frac{1}{p'} = 1$ (vector-valued Hausdorff-Young inequality). Every Banach space has Fourier type 1, but the space X has Fourier type 2 (Plancherel's inequality) if and only if X is isomorphic to a Hilbert space ([148]). A space $L_q(\Omega, \mathbb{C})$ has Fourier type $p = \min(q, q')$, but not better ([187]). Every uniformly convex space (in particular every UMD-space) has a non-trivial Fourier type $p > 1$. For a survey see [100]. Now if X and Y have Fourier type $p \in [1, 2]$, then it suffices to assume (cf. [104]) that the sets

$$\{|t|^{|\alpha|} D^\alpha m(t) : t \in \mathbb{R}^N \setminus \{0\} , \ |\alpha| \le [N/p] + 1\}$$

are R-bounded, i. e. $\frac{N}{p}$ is the required order of smoothness (which can be made more precise using Besov-norms), see [104]. Such results can also be obtained as a corollary of boundedness theorems for more general singular integral operators with operator-valued kernels (see [126], [127]). In [125] the number of partial derivatives one has to take into account is reduced further and Hytönen can even improve the classical Fourier multiplier theorems of Mihlin and Marcinkiewicz in $L_p(\mathbb{R}^N)$ (as presented e. g. in [205]). $H^1(X)$- and $\mathrm{BMO}(X)$-estimates for operator-valued Fourier multipliers can be found in [104] and [123].

Versions of Theorem 4.6 on $L_p([0, 2\pi]^N, X)$, i. e. in the discrete case, are considered in [210], Section 3, and for $N = 1$ also in [14]. A version on $l_p(\mathbb{Z}, X)$ is given in [31]. For operator-valued Fourier multiplier theorems on Besov spaces $B^s_{pq}(\mathbb{R}^N, X)$, see [6, 11, 103].

N 4.7: This example is from [205], 6.2.4, where further examples can be found.

N 4.8 - 4.10: Property (α) and its elementary properties were introduced in [188]. It is shown there, that every subspace of a space X with local unconditional structure (e. g. X a Banach function space) with finite cotype has property (α). Conversely if X has property (α), than X has finite cotype. The spaces S_p of Schatten class operators on a Hilbert space have UMD for $1 < p < \infty$, but they fail property (α) unless $p = 2$ ([188]).

N 4.11: Lemma 4.11 a) is in [188] and part b) in [46] for $N = 2$. The latter paper contains interesting applications of property (α) to unconditional Schauder decompositions.

N 4.12: Proposition 4.12 is from [226]. The connection of property (α) with multiplier theorems was noticed first by Zimmermann and he also gave counterexamples, showing that property (α) cannot be dropped. In the proof we used the fact that, if X is a <u>UMD</u> space with property (α), then X^* has property (α).

N 4.13: For scalar m, Theorem 4.13 was shown in [226], and extended to operator-valued multipliers in [210], Section 4, using transference from $L_p([0,1]^N, X)$. A more direct proof is given in [114]. The present proof is new again.

5 R-bounded Sets of Classical Operators

In this section we will prove the R-boundedness of large classes of classical operators on $L_p(\Omega, X)$-spaces, such as: convolution operators, integral operators, and positive contractions. In particular, they include the Gaussian and the Poisson semigroups $(t > 0)$

$$T_t f(v) = (4\pi t)^{\frac{-N}{2}} \int_{\mathbb{R}^N} \exp\left(-\frac{|u|^2}{4t}\right) f(v - u)\, du \qquad (5.1)$$

$$P_t f(v) = C_N \int_{\mathbb{R}^N} \frac{t}{(t^2 + |u|^2)^{(N+1)/2}} f(v - u)\, du$$

$$\text{with } C_N = \Gamma\left(\frac{N+1}{2}\right) \pi^{-\frac{N+1}{2}}. \qquad (5.2)$$

A first result for convolution operators (in the form of Fourier multipliers) was already given in 3.8 and 4.4.

5.1 Example. Consider the set of multiplier functions

$$\mathcal{M} = \{m \in C^N(\mathbb{R}^N, \mathbb{C}): \lim_{|t| \to \infty} |m(t)| = 0, \left\|\frac{\partial^N}{\partial t_1 \cdots \partial t_n} m\right\|_{L_1(\mathbb{R}^N, \mathbb{C})} \leq 1\}.$$

If X has the UMD property, then the set of Fourier multiplier operators $\{T_m: m \in \mathcal{M}\}$ is R-bounded on $L_p(\mathbb{R}^N, X)$ for $1 < p < \infty$.

Since $\mathcal{F}[\exp(-\frac{|\cdot|^2}{4t})](u) = (16\pi)^{\frac{N}{2}} t^N \exp(-4t|u|^2)$ it is easy to check that the Gaussian semigroup (5.1) (considered as Fourier multipliers) is <u>not</u> covered by 5.1.

Therefore we need a stronger result which we obtain by *bootstrapping* our multiplier Theorems 4.6 and 4.13: given a sequence m_j of multiplier functions, we form one vector-valued multiplier function $m = (m_j)_j$, identify the Fourier multiplier operator T_m on $L_p(RadX)$ with the operator (T_{m_j}) on $Rad(L_p(X))$, and conclude that the sequence (T_{m_j}) is R-bounded on $L_p(X)$ by 2.7. In this way, one can use boundedness theorems for a single operator with a vector-valued kernel to prove R-boundedness results for sets of operators with scalar-valued kernels. Here is a first example.

5.2 Fourier Multipliers. Let X and Y be UMD spaces with property (α) and $1 < p < \infty$.

a) Consider the set of multiplier functions

$$\mathcal{M} := \{m \in C^N(\mathbb{R}^N \setminus \{0\}, \mathbb{C}):$$
$$|t^\alpha||D^\alpha m(t)| \le A \text{ for } \alpha \le (1, \ldots, 1), t \in \mathbb{R}^N \setminus \{0\}\} .$$

Then $\tau := \{T_m : m \in \mathcal{M}\}$ is an R-bounded set of Fourier multiplier operators on $L_p(\mathbb{R}^N, X)$ and $R_p(\tau) \le CA$ for some constant C depending on X, N and p.

b) We will need the following vector-valued version. Let $\sigma \subset B(X, Y)$ be an R-bounded set and

$$\mathcal{M}_\sigma := \{m \in C^N(\mathbb{R}^N \setminus \{0\}, B(X, Y)):$$
$$t^\alpha D^\alpha m(t) \in \sigma \text{ for } \alpha \le (1, \ldots, 1), t \in \mathbb{R}^N \setminus \{0\}\} .$$

Then $\tau := \{T_m : m \in \mathcal{M}_\sigma\}$ is an R-bounded set of Fourier multiplier operators in $B(L_p(\mathbb{R}^N, X), L_p(\mathbb{R}^N, Y))$ with $R_p(\tau) \le CR_p(\sigma)$ for some constant C depending on X, Y, N and p.

Proof. Clearly b) implies a). So let's just show b).

Let $(m_j)_{j \in \mathbb{N}}$ be a finitely supported sequence from $\mathcal{M}_\sigma$. By Proposition 2.7, we can define a function $m \in C^N(\mathbb{R}^N \setminus \{0\}, B(Rad_p X, Rad_p Y))$ via

$$[m(t)]\left((x_j)_{j \in \mathbb{N}}\right) := (m_j(t) x_j)_{j \in \mathbb{N}}$$

for $t \in \mathbb{R}^N \setminus \{0\}$ and $(x_j)_{j \in \mathbb{N}} \in Rad_p X$. Note that

$$R_p(\{t^\alpha D^\alpha m(t): t \in \mathbb{R}^N \setminus \{0\}, |\alpha| \le (1, \ldots, 1)\}) \le C_1 R_p(\sigma)$$

for some constant C_1 (depending on X, Y and p); indeed, for $k \in \{1, \ldots, n\}$, if $x_k = (x_{kj})_{j \in \mathbb{N}} \in Rad_1 X$ and $t_k \in \mathbb{R}^N \setminus \{0\}$ and $|\alpha_k| \le (1, \ldots, 1)$ then

$$\int_0^1 \| \sum_{k=1}^n r_k(u)[(t_k)^{\alpha_k} D^{\alpha_k} m(t_k)](x_k) \|_{Rad_1 Y} \, du$$

$$= \int_0^1 \int_0^1 \| \sum_{k=1}^n \sum_{j \in \mathbb{N}} r_k(u) r_j(v)[(t_k)^{\alpha_k} D^{\alpha_k} m_j(t_k)] x_{kj} \|_Y \, dv \, du$$

$$\leq C_X^2 C_Y^2 R_1(\sigma) \int_0^1 \int_0^1 \| \sum_{k=1}^n \sum_{j \in \mathbb{N}} r_k(u) r_j(v) x_{kj} \|_X \, dv \, du$$

$$= C_X^2 C_Y^2 R_1(\sigma) \int_0^1 \| \sum_{k=1}^n r_k(u) x_k \|_{Rad_1 X} \, du.$$

by 4.11, where C_X (resp. C_Y) is the property (α) constant from (4.10). Both UMD and property (α) lift from Z to $Rad_p Z$. Thus, by Theorem 4.13, T_m is a Fourier multiplier operator in $B(L_p(\mathbb{R}^N, Rad_p X), L_p(\mathbb{R}^N, Rad_p Y))$ with operator norm at most $C_2 R_p(\sigma)$ for some constant C_2 (depending on X, Y, N and p).

By Fubini's theorem, we have a natural isometry J_Z where

$$L_p(\mathbb{R}^N, Rad_p Z) \ni f(\cdot) = (f_j(\cdot))_{j \in \mathbb{N}} \xrightarrow{J_Z} (f_j)_{j \in \mathbb{N}} \in Rad_p(L_p(\mathbb{R}^N, Z)) .$$

By Theorem 4.13, $T_{m_j} \in B((L_p(\mathbb{R}^N, X), L_p(\mathbb{R}^N, Y))$ for each j. It is easy to check that

$$Rad_p(L_p(\mathbb{R}^N, X)) \ni (f_j)_{j \in \mathbb{N}} \xrightarrow{J_Y T_m J_X^{-1}} (T_{m_j} f_j) \in Rad_p(L_p(\mathbb{R}^N, Y)) .$$

So, by Proposition 2.7, the claim follows. $\qquad\qquad\square$

As a corollary we obtain an improvement of 5.1.

5.3 Convolution Integrals. Consider the subset of integrable functions

$$\mathcal{K} = \left\{ k \in L_1(\mathbb{R}^N, \mathbb{C}): \int_{\mathbb{R}^N} |t^\alpha| |D^\alpha k(t)| \, dt \leq A \text{ for } \alpha \leq (1, 1, \ldots, 1) \right\} .$$

For $k \in L_1(\mathbb{R}^N, \mathbb{C})$ define the convolution operator

$$K_k f(t) = \int_{\mathbb{R}^N} k(t - s) f(s) \, ds , \quad \text{for } f \in L_p(\mathbb{R}^N, \mathbb{C}) , \text{ a.e. } t \in \mathbb{R}^N .$$

If X has UMD and property (α) then $\tau = \{K_k : k \in \mathcal{K}\}$ is an R-bounded set of operators on $L_p(\mathbb{R}^N, X)$ for $1 < p < \infty$ and $R_p(\tau) \leq CA$ for some constant C depending only on X, p and N.

Proof. Let $k \in \mathcal{K}$. Put $m = \hat{k}$. Then for $\alpha \leq (1, \ldots, 1)$ we have

$$t^\alpha D^\alpha m(t) = \int_{\mathbb{R}^N} e^{-is\cdot t}(-1)^{|\alpha|} D^\alpha[s^\alpha k(s)]\,ds$$

$$= \int_{\mathbb{R}^N} e^{-is\cdot t}(-1)^{|\alpha|} \sum_{\beta\leq\alpha} \binom{\alpha}{\beta} D^{\alpha-\beta}(s^\alpha)(D^\beta k)(s)\,ds$$

$$= \sum_{\beta\leq\alpha} c_{\alpha,\beta} \int_{\mathbb{R}^N} e^{-is\cdot t}[s^\beta D^\beta k(s)]\,ds \ .$$

Hence $\|(\cdot)^\alpha D^\alpha m(\cdot)\|_{L_\infty(\mathbb{R}^N,\mathbb{C})} \leq C_N \sup_{\beta\leq\alpha} \|(\cdot)^\beta D^\beta k(\cdot)\|_{L_1(\mathbb{R}^N,\mathbb{C})}$. So, since $k \in \mathcal{K}$, m belongs to the set $\mathcal{M}$ in 5.2. Since $K_k = T_m$ the claim now follows from 5.2. $\qquad\square$

5.4 Gauß- and Poisson Semigroup. The semigroups $\{T_t : t \geq 0\}$ of (5.1) and $\{P_t : t \geq 0\}$ of (5.2) are R-bounded (R-analytic) on $L_p(\mathbb{R}^N, X)$, for $1 < p < \infty$, if X is UMD and has property (α).

Proof. For any $k \in \mathcal{K}$ (as in 5.3) the kernels

$$k_\gamma f(u) = \gamma^N k(\gamma u) , \qquad \gamma \in \mathbb{R}_+$$

also belong to $\mathcal{K}$ by a simple substitution and therefore $\{K_{k_\gamma} : \gamma > 0\}$ is R-bounded. For

$$k(s) = (4\pi)^{-\frac{N}{2}} \exp\left(-\frac{|s|^2}{4}\right) \in \mathcal{K}$$

we get the Gauß-semigroup $T_t = K_{k_\gamma}$ with $\gamma = t^{-\frac{1}{2}}$. For

$$k(s) = C_N(1 + |s|^2)^{-\frac{N+1}{2}} \in \mathcal{K}$$

we obtain the Poisson semigroup $P_t = K_{k_\gamma}$ with $\gamma = t^{-1}$. (Of course we could have applied 5.2 directly since $\pi^{-\frac{N}{2}}\mathcal{F}[\exp(-|\cdot|^2)](s) = \exp(-\frac{|s|^2}{4})$ and $\mathcal{F}[C_N(1 + |\cdot|^2)^{-\frac{N+1}{2}}](s) = e^{-|s|}$.) $\qquad\square$

We give now a far-reaching extension of 2.10b).

5.5 Integral Operators. Let X and Y be Banach spaces and (Ω, μ) a σ-finite measure space. Let $\mathcal{K}$ be a class of measurable, operator-valued kernels $k : \Omega \times \Omega \to B(X, Y)$ such that, for each $k \in \mathcal{K}$,

$$K_k f(t) = \int_\Omega k(t, s) f(s) d\mu(s), \ t \in \Omega$$

is defined in a dense subset $\mathcal{D}$ of $L_p(\Omega, X)$, $1 \leq p < \infty$, and extends to a bounded operator $K_k : L_p(\Omega, X) \to L_p(\Omega, Y)$. Assume that

$$R_p(\{k(t, s) \in B(X, Y) : k \in \mathcal{K}\}) \leq r(t, s) \text{ for } s, t \in \Omega,$$

and that $R_0 f(t) = \int_\Omega r(t, s) f(s) ds$ also defines a bounded operator R_0 on $L_p(\Omega, \mathbb{C})$. Then $\tau = \{K_k : k \in \mathcal{K}\} \subset B(L_p(\Omega, X), L_p(\Omega, Y))$ is R-bounded and $R_p(\tau) \leq \|R_0\|_{B(L_p(\Omega,\mathbb{C}),L_p(\Omega,\mathbb{C}))}$.

Proof. For $k_1, \ldots, k_n \in \mathcal{K}$ and $f_1, \ldots, f_n \in \mathcal{D}$ we have, by Fubini's theorem,

$$\| \sum_{j=1}^{n} r_j K_{k_j} f_j \|_{L_p([0,1],L_p(\Omega,Y))}^{p}$$

$$= \int_{\Omega} \| \sum_{j=1}^{n} r_j \int_{\Omega} k_j(t,s) f_j(s) ds \|_{L_p([0,1],Y)}^{p} dt$$

$$\leq \int_{\Omega} \left(\int_{\Omega} \| \sum_{j=1}^{n} r_j k_j(t,s) f_j(s) \|_{L_p([0,1],Y)} ds \right)^{p} dt$$

$$\leq \int_{\Omega} \left(\int_{\Omega} r(t,s) \| \sum_{j=1}^{n} r_j f_j(s) \|_{L_p([0,1],X)} ds \right)^{p} dt$$

$$= \| R_0 g \|_{L_p(\Omega,\mathbb{C})}^{p} \leq \| R_0 \|_{B(L_p(\Omega,\mathbb{C}),L_p(\Omega,\mathbb{C}))}^{p} \| g \|_{L_p(\Omega,\mathbb{C})}^{p}$$

with $g(\cdot) = \| \sum_{j=1}^{n} r_j f_j(\cdot) \|_{L_p([0,1],X)}$, a function belonging to $L_p(\Omega,\mathbb{C})$. $\quad\square$

We turn now to contractive semigroups (T_t) on $L_q(\mathbb{R}^N)$ (i. e. $\|T_t\| \leq 1$ for all t). For $k_t(s) = (4\pi t)^{-\frac{N}{2}} \exp(-\frac{|s|^2}{4t})$ we have $\|k_t\|_{L_1} = 1$ for all $t > 0$. Hence by Young's inequality $\|k_t * f\|_{L_q(\mathbb{R}^N)} \leq \|k_t\|_{L_1(\mathbb{R}^N)} \|f\|_{L_q(\mathbb{R}^N)}$ and the Gauß semigroup $T_t f = k_t * f$ from (5.1) is contractive on $L_q(\mathbb{R}^N)$ for all $1 \leq q \leq \infty$. The same reasoning applies to the Poisson semigroup in (5.2). Hence the next theorem gives another proof for the R-boundedness of these semigroups (see 5.4).

Further examples of this kind are semigroups generated by Dirichlet forms (see [173]), and, more generally, many semigroups generated by stochastic processes.

5.6 Theorem. *Assume that $(T_{t,q})_{t \geq 0}$ are contractive semigroups on $L_p(\Omega,\mu)$ for all $1 \leq q \leq \infty$ such that, for all $q_1, q_2 \in [1,\infty]$ and $t \geq 0$, the operators T_{t,q_1} and T_{t,q_2} agree on $L_{q_1}(\Omega) \cap L_{q_2}(\Omega)$. If $T_{t,q}$ is analytic on $L_q(\Omega)$ for some $1 < q < \infty$ then it is even R-analytic.*

Our proof is based on the following maximal estimate from ergodic theory. Consider the semigroup averages on $L_q(\Omega)$

$$N(t)f = \frac{1}{t} \int_0^t T_{s,q} f ds, \ f \in L_q(\Omega) \ .$$

If $(T_{t,q})_{t \geq 0}$ is contractive on all $L_q(\Omega)$ for $1 \leq q \leq \infty$, then for each $1 < q < \infty$ and $f_1, \ldots, f_m \in L_q(\Omega)$ we have

$$\| \sup_j \sup_{t > 0} |N(t)f_j| \|_{L_q} \leq 2 \left(\frac{q}{q-1} \right)^{\frac{1}{q}} \| \sup_j |f_j| \|_{L_q} \tag{5.3}$$

(see, [77], Theorem VIII 7.7; in ergodic theory this estimate is used to show that $T_{t,q}f \rightarrow f$ a. e. for $t \rightarrow 0$).

The next lemma shows how such maximal estimates can be used to prove R-boundedness.

5.7 Lemma. *Let $\lambda \in \Sigma_\delta \longrightarrow N(\lambda) \in B(L_q(\Omega, \mu))$ be an analytic, operator-valued function which is strongly continuous and bounded on $\overline{\Sigma_\delta}$ with $\|N(\lambda)\| \leq C_q$. Assume that $q \in [1, 2)$ and that $N(t), t > 0$, satisfies a maximal estimate: There is a constant C_∞ such that, for any $t_1, \ldots, t_m \in \mathbb{R}_+, f_1, \ldots, f_m \in L_q$,*

$$\| \sup_{j=1}^{m} |N(t_j)f_j| \|_{L_q} \leq C_\infty \| \sup_{j=1}^{m} |f_j| \|_{L_q}. \tag{5.4}$$

Then $\{N(\lambda) : \lambda \in \Sigma_{\frac{q}{2}\delta}\}$ is R-bounded with R-bound smaller than $C_q^{\frac{q}{2}} C_\infty^{1-\frac{q}{2}}$.

The proof of Theorem 5.6 is a simple consequence of this lemma and (5.3).

Proof (of 5.6). Fix $q \in (1, 2)$ and assume that the semigroup $(T_{t,q})$ extends analytically to a bounded semigroup $T_\lambda, \lambda \in \Sigma_{\delta_1}$, for some $\delta_1 > 0$. We also extend $N(t)$ by

$$N(\lambda)f = \frac{1}{\lambda} \int_0^\lambda T_\mu d\mu, \quad \lambda \in \Sigma_{\delta_1}.$$

Fix some δ with $0 < \delta < \delta_1$ and observe that the ergodic estimate (5.3) implies (5.4) for the analytic function $N(\lambda)$. By Lemma 5.7 we conclude that for some $\delta' > 0$ the set $\{N(\lambda) : \lambda \in \Sigma_{\delta'}\}$ is R-bounded. Since $T_\lambda = N(\lambda) + \lambda \frac{d}{d\lambda} N(\lambda)$ and 2.15 b) implies the R-boundedness of $\{\lambda \frac{d}{d\lambda} N(\lambda) : \lambda \in \Sigma_{\delta''}\}$ for $0 < \delta'' < \delta'$, the claim follows for $q \in (1, 2)$. If $q \in (2, \infty)$, we know that the dual semigroup $T_\lambda, \lambda \in \Sigma_{\delta''}$, is R-bounded on $L_{q'}(\Omega), \frac{1}{q} + \frac{1}{q'} = 1$, and we can apply Corollary 2.11. $\square$

So the work is in proving the lemma.

Proof (of 5.7). The boundedness of $N(\lambda)$ on $\overline{\Sigma_\delta}$ implies by Fubini's theorem, that for $\lambda_1, \ldots, \lambda_m$ with $|\arg \lambda_j| = \delta$ and $g_1, \ldots, g_m \in L_q(\Omega)$ we have

$$\| \left(\sum_{j=1}^{m} |N(\lambda_j)g_j|^q \right)^{\frac{1}{q}} \|_{L_q}^q = \sum_{j=1}^{m} \|N(\lambda_j)g_j\|_{L_q}^q$$

$$\leq C_q^q \sum_{j=1}^{m} \|g_j\|^q = C_q^q \| \left(\sum_{j=1}^{m} |g_j|^q \right)^{\frac{1}{q}} \|_{L_q}^q \tag{5.5}$$

On the other hand, by 2.15 a) and Remark 2.8 the claim is equivalent to the following square function estimate: for all $\lambda_1, \ldots, \lambda_m$ with $|\arg \lambda_j| = \frac{q}{2}\delta$ and $g_1, \ldots, g_m \in L_q(\Omega)$

$$\left\|\left(\sum_{j=1}^{m}|N(\lambda_j)g_j|^2\right)^{\frac{1}{2}}\right\|_{L_q} \leq C_q^{\frac{q}{2}}C_\infty^{1-\frac{q}{2}}\left\|\left(\sum_{j=1}^{m}|g_j|^2\right)^{\frac{1}{2}}\right\|_{L_q}. \qquad (5.6)$$

We want to show that we can obtain (5.6) by "interpolating" the inequalities (5.4) and (5.5). To make this precise we fix $t_1,\ldots,t_m \in \mathbb{R}_+, m \in \mathbb{N}$, and define a strongly continuous mapping

$$\lambda \in \overline{\Sigma_\delta} \longrightarrow M(\lambda) \in B(L_q(\Omega, l_q^m))$$
$$M(\lambda)(g_1,\ldots,g_m) = (N(t_1\lambda)g_1,\ldots,N(t_m\lambda)g_m)$$

which is analytic on Σ_δ. (5.5) says now that

$$\|M(\lambda)\|_{B(L_q(\Omega,l_q^m))} \leq C_q \text{ for } |\arg\lambda| = \delta \qquad (5.7)$$

and (5.4) implies

$$\|M(\lambda)\|_{B(L_q(\Omega,l_\infty^m))} \leq C_\infty \text{ for } \arg\lambda = 0. \qquad (5.8)$$

Since $\frac{1}{2} = \theta\frac{1}{q} + (1-\theta)\frac{1}{\infty}$ for $\theta = \frac{q}{2}$, the interpolation lemma 5.8 below gives (with $\sigma_0 = 0, \sigma_1 = \delta, q_0 = q_1 = q, \ p_0 = \infty, p_1 = 2, \theta = \frac{q}{2}$)

$$\|M(\lambda)\|_{B(L_q(\Omega,l_2^m))} \leq C_q^{\frac{q}{2}}C_\infty^{1-\frac{q}{2}} \quad \text{for } |\arg\lambda| = \frac{q}{2}\delta \qquad (5.9)$$

which in turn implies (5.6) for $\lambda_j = t_j e^{\pm i\frac{q}{2}\delta}$. Since the bound in (5.9) does not depend on t_j or m, we conclude that (5.6) holds for all λ_j with $|\arg\lambda_j| = \frac{q}{2}\delta$, and the claim follows. $\qquad\qquad\qquad\square$

We will use the notation $\Sigma(\sigma_0,\sigma_1) = \{\lambda \in \mathbb{C} : \sigma_0 < \arg\lambda < \sigma_1\}$.

5.8 Lemma. *Let $q_0, q_1, p_0, p_1 \in [1,\infty]$ and $m \in \mathbb{N}$. Consider a family of operators*

$$N(\lambda) : L_{q_0}(\Omega, l_{p_0}^m) \cap L_{q_1}(\Omega, l_{p_1}^m) \to L_{q_0}(\Omega, l_{p_0}^m) + L_{q_1}(\Omega, l_{p_1}^m)$$

such that

$$\lambda \in \overline{\Sigma(\sigma_0,\sigma_1)} \to N(\lambda)F \in L_{q_0}(l_{p_0}^m) + L_{q_1}(l_{p_1}^m)$$

is continuous and analytic in $\Sigma(\sigma_0,\sigma_1)$ for all $F \in L_{q_0}(l_{p_0}^m) \cap L_{q_1}(l_{p_1}^m)$. Assume that, for $j = 0, 1$,

$$\|N(\lambda)\|_{B(L_{q_j}(l_{p_j}^m))} \leq C_j \quad \text{for } \arg\lambda = \sigma_j. \qquad (5.10)$$

If for some $\theta \in (0,1)$ we put

$$\frac{1}{q} = (1-\theta)\frac{1}{q_0} + \theta\frac{1}{q_1}, \ \frac{1}{p} = (1-\theta)\frac{1}{p_0} + \theta\frac{1}{p_1}, \ \sigma = (1-\theta)\sigma_0 + \theta\sigma_1$$

then

$$\|N(\lambda)\|_{B(L_q(l_p^m))} \leq C_0^{1-\theta}C_1^\theta \text{ for } \arg(\lambda) = \sigma.$$

Proof. Let $S = \{\lambda : 0 < Re\,\lambda < 1\}$ and put $M(\lambda) = N(\exp[i(\sigma_0 + (\sigma_1 - \sigma_0)\lambda))])$. Then for $j = 0, 1$ we have by (5.10)

$$\|M(\lambda)\|_{B(L_{q_j}(l_{p_j}^m))} \le C_j \quad \text{for } Re\,\lambda = j$$

To give a short argument we use now the well known fact, that for the complex interpolation method

$$L_q(\Omega, l_p^m) = [L_{q_0}(\omega, l_{p_0}^m), L_{q_1}(\Omega, l_{p_1}^m)]_\theta . \tag{5.11}$$

For $F \in L_q(\Omega, l_p^m)$ with $\|F\| < 1$ there is by definition of the complex method a continuous function $\mathcal{F} : \overline{S} \to L_{q_0}(l_{p_0}^m) + L_{q_1}(l_{p_1}^m)$, analytic on S, with $\mathcal{F}(\theta) = F$ and

$$\|\mathcal{F}(\lambda)\|_{L_{q_j}(l_{p_j}^m)} \le 1 \text{ for } Re\,\lambda = j \text{ where } j = 0, 1.$$

For a fixed $\mu = re^{i\sigma}$ define now $\mathcal{G}(\lambda) = M(\lambda - i\frac{\log \sigma}{\sigma_1 - \sigma_0})F(\lambda)$. Then $\|\mathcal{G}(\lambda)\|_{L_{q_j}(l_{p_j}^m)} \le C_j$ for $j = 0, 1$ and $\mathcal{G}(\theta) = N(re^{i\sigma})\mathcal{F}(\theta) = N(\mu)F$. Again by the definition of the complex method we have

$$\|N(\mu)F\|_{L_q(l_p^m)} \le C_0^{1-\theta}C_1^\theta$$

and this holds now for all μ with $\arg \mu = \sigma$. Since $F \in L_q(l_p^m)$ with $\|F\| < 1$ was arbitrary, the claim follows. $\qquad\square$

Notes on Section 5:

N 5.1 recalls Proposition 4.4.

N 5.2: Statement 5.2 appears in [215] (with a different proof) and also in [105], Theorem 3.2, where other families of operator-valued multiplier functions are considered. For example, for an R-bounded set $\sigma \subset B(X, Y)$ for X and Y having UMD, property (α), and Fourier type p (see N 4.6) one can replace $\mathcal{M}_\sigma$ in 5.2 by

$$\mathcal{M}(\sigma) = \{m \in C^N(\mathbb{R}^N \setminus \{0\}, B(X, Y)):$$
$$|t|^{|\alpha|}D^\alpha m(t) \in \sigma \text{ for } t \in \mathbb{R}^N \setminus \{0\} \text{ and } |\alpha| \le [N/p] + 1\} .$$

N 5.3: Statement 5.3 appears in [105], again for operator valued kernels. Let

$$\mathcal{K}_\sigma = \{k : \mathbb{R}^N \to B(X, Y) : \int_{\mathbb{R}^N} \|t^\alpha D^\alpha k(t)\|_\sigma \, dt \le 1 \text{ for } \alpha \le (1, \ldots, 1)\}$$

where $\|T\|_\sigma = \inf\{r \ge 0 : T \in r\sigma\}$ for $T \in B(X, Y)$ and an absolutely convex, strongly closed, R-bounded subset $\sigma \subset B(X, Y)$. Let X and Y be

UMD spaces with property (α). Then $\tau = \{K_k : k \in \mathcal{K}_\sigma\}$ is an R-bounded set in $B(L_p(\mathbb{R}^N, X), L_p(\mathbb{R}^N, Y))$ for $1 < p < \infty$.

It is clear that the set of convolution operators $\{T_k : \|k\|_{L_1(\mathbb{R}^N)} \leq 1\}$ cannot be R-bounded since the strong closure of this set contains the translations operators, which do not form an R-bounded set (by Example 2.12). Hence some additional conditions an k such as in 5.3 are needed. R-boundedness results for large classes of singular integral operators with operator-valued kernels are contained in [126], Section 6. For scalar kernels see also [101], V Theorem 3.11.

N 5.4: The R-analyticity of the Gauß semigroup was proved independently in [48] and [220] using (different but non-trivial) maximal estimates from harmonic analysis. The present more elementary proof is from [105]. In [48] it is also shown that $\{K_{k_\gamma} : \gamma > 0\}$ is R-bounded if k is of the form $k(t) = p(|t|)$ with continuous, nonincreasing function $p : (0,\infty) \to (0,\infty)$ with $\int_0^\infty p(r) r^{N-1} dr < \infty$.

N 5.5 is from [66], Proposition 4.12. Criteria for boundedness of such integral operators with operator-valued kernels are given in [107].

N 5.6: Theorem 5.6 implies of course that $T_{t,q}$ has maximal L_p regularity. This result was already proved in [151] by constructing a special functional calculus for $T_{t,q}$. Theorem 5.6 and its proof are from [221], section 4.d) where actually a stronger theorem is proved. It is enough that T_t is positive, contractive and analytic on $L_q(\mathbb{R}^N)$ for one $q \in (1,\infty)$. We shall reprove this result in Section 10 (5.6 follows since $T_{t,q}$ can be dominated by a contractive, positive semigroup under the assumptions of 5.6, see [139]). For a discussion of the optimal angle of R-boundedness, see [144].

N 5.7 is a special case of the proposition in [221], section 4.b. We shall take up similar arguments in Section 8.

N 5.8 is a variant of the Stein interpolation theorem. For a more general approach see [217]. Formula (5.11) can be found, e.g. in [29, Theorem 5.1.2].

6 Elliptic Systems on $\mathbb{R}^n$

This section is the first of three devoted to the study of maximal regularity for classes of elliptic differential operators. Here we shall study L_p-realizations of elliptic systems A on $\mathbb{R}^n$ for $p \in (1,\infty)$. In the following Section 7, we study elliptic boundary value problems on subsets of $\mathbb{R}^n$ where we confine ourselves to the case of scalar equations of second order to limit technicalities. Finally, in Section 8, we sketch another approach, particularly useful for operators in divergence form.

Before starting the investigation of differential operators we introduce some terminology in the abstract situation. For an (elliptic) operator A in a Banach space X, we are interested in maximal L_q-regularity for the corresponding (parabolic) evolution equation

$$u'(t) + Au(t) = f(t), \quad t > 0,$$
$$u(0) = 0.$$

If X is a UMD-space (which is the case for L_p if $p \in (1, \infty)$), then we know by the characterization of maximal regularity in terms of R-boundedness in Theorem 1.11 and by Theorem 2.20 that A is R-sectorial with $\omega_R(A) < \pi/2$.

Recall (cf. Section 1 after Theorem 1.11) that a densely defined linear operator A in a Banach space X is called $(R$-$)$sectorial of angle $\omega \in [0, \pi)$ if $\rho(A) \subset \overline{\Sigma_\omega}$ and, for all $\nu > \omega$ the set

$$\{\lambda R(\lambda, A) : \nu \leq |\arg(\lambda)| \leq \pi\}$$

is $(R$-$)$bounded in $B(X)$. The infimum of all such angles ω is denoted $\omega(A)$ $(\omega_R(A))$, cf. Remark 2.22.

It shall turn out useful to introduce some more notation for $(R$-$)$sectorial operators. If A is a sectorial operator and $\pi > \theta > \omega(A)$ then we let

$$M_\theta(A) := \sup\{\|\lambda R(\lambda, A)\| : \lambda \neq 0, \pi \geq |\arg \lambda| > \theta\},$$
$$\tilde{M}_\theta(A) := \sup\{\|A R(\lambda, A)\| : \lambda \neq 0, \pi \geq |\arg \lambda| > \theta\}.$$

If A is an R-sectorial operator and $\pi > \theta > \omega_R(A)$ then we replace boundedness in operator-norm by R-boundedness and define

$$R_\theta(A) := R(\{\lambda R(\lambda, A) : \lambda \neq 0, \pi \geq |\arg \lambda| > \theta\}),$$
$$\tilde{R}_\theta(A) := R(\{A R(\lambda, A) : \lambda \neq 0, \pi \geq |\arg \lambda| > \theta\}).$$

Note that we always have $M_\theta(A) \leq R_\theta(A)$ and $\tilde{M}_\theta(A) \leq \tilde{R}_\theta(A)$.

6.1 Elliptic Operators.

We now introduce the class of operators to be studied in this section. For $n, N \in \mathbb{N}$ and $1 < p < \infty$, we consider realizations in $L_p(\mathbb{R}^n, \mathbb{C}^N)$ of systems of the following form

$$(\mathcal{A}(D)u)(x) = \sum_{|\alpha| \leq 2m} a_\alpha(x)(D^\alpha u)(x), \quad x \in \mathbb{R}^n \tag{6.1}$$

where $m \in \mathbb{N}$ and $a_\alpha : \mathbb{R}^n \to \mathbb{C}^{N \times N}$, $|\alpha| \leq 2m$. We have used the usual multiindex notation, i. e. $D^\alpha := D_1^{\alpha_1} \cdot \ldots \cdot D_n^{\alpha_n}$ where $D_j := -i\frac{\partial}{\partial x_j}$ denotes $-i$ times the partial derivative in the j-th coordinate direction, and $|\alpha| = \alpha_1 + \ldots + \alpha_n$ for $\alpha = (\alpha_1, \ldots, \alpha_n) \in \mathbb{N}_0^n$.

For $\alpha, \beta \in \mathbb{N}_0^n$ we also write $\beta \leq \alpha$ if $\beta_1 \leq \alpha_1, \ldots, \beta_n \leq \alpha_n$ and define $\binom{\alpha}{\beta} := \binom{\alpha_1}{\beta_1} \cdot \ldots \cdot \binom{\alpha_n}{\beta_n}$ in case $\beta \leq \alpha$.

A decisive role is played by the symbol $\mathcal{A}_\pi(x, \xi)$ of the principle part $\mathcal{A}_\pi(D) := \sum_{|\alpha|=2m} a_\alpha D^\alpha$ of $\mathcal{A}(D)$ which is obtained from $\mathcal{A}_\pi(D)$ by replacing D_j with ξ_j, i.e.,

$$\mathcal{A}_\pi(x,\xi) = \sum_{|\alpha|=2m} a_\alpha(x)\xi^\alpha, \qquad (x,\xi) \in \mathbb{R}^n \times \mathbb{R}^n, \qquad (6.2)$$

where $\xi^\alpha := \xi_1^{\alpha_1} \cdot \ldots \cdot \xi_n^{\alpha_n}$ for $\xi = (\xi_1, \ldots, \xi_n) \in \mathbb{R}^n$ and $\alpha = (\alpha_1, \ldots, \alpha_n) \in \mathbb{N}_0^n$. We have, formally,

$$\mathcal{A}_\pi(D)u(x) = \mathcal{F}_\xi^{-1}(\xi \mapsto \mathcal{A}_\pi(x,\xi)\mathcal{F}u(\xi))(x),$$

and that is where the relevance of the symbol $\mathcal{A}_\pi(x,\xi)$ stems from.

We assume that the system $\mathcal{A}(D)$ is (M,ω_0)-elliptic in the sense of [10], [79] where $M > 0$ and $\omega_0 \in [0,\pi)$, i.e., we assume

$$\sum_{|\alpha|=2m} \|a_\alpha\|_\infty \leq M, \qquad (6.3)$$

and that, for all $x \in \mathbb{R}^n$, $|\xi| = 1$, the principle symbol satisfies

$$\sigma(\mathcal{A}_\pi(x,\xi)) \subset \overline{\Sigma_{\omega_0}} \setminus \{0\}, \qquad (6.4)$$

$$|(\mathcal{A}_\pi(x,\xi))^{-1}| \leq M. \qquad (6.5)$$

The condition (6.4) is a natural generalization to systems (with $N > 1$) of the sectoriality condition

$$\mathcal{A}_\pi(x,\xi) \subset \overline{\Sigma_{\omega_0}}, \qquad (x,\xi) \in \mathbb{R}^n \times \mathbb{R}^n,$$

which is usually imposed in the case $N = 1$. For $N = 1$, (6.4) and (6.5) together imply the usual ellipticity assumption

$$|\mathcal{A}_\pi(x,\xi)| \geq \frac{1}{M}|\xi|^{2m}, \qquad (x,\xi) \in \mathbb{R}^n \times \mathbb{R}^n.$$

In the present section we show R-sectoriality of (M,ω_0)-elliptic systems under the following regularity assumption on the coefficients of the principle part:

$$a_\alpha \in BUC(\mathbb{R}^n, \mathbb{C}^{N\times N}), \qquad |\alpha| = 2m, \qquad (6.6)$$

where BUC denotes the space of bounded uniformly continuous functions. For the lower order coefficients we assume

$$a_\alpha \in L_\infty(\mathbb{R}^n, \mathbb{C}^{N\times N}), \qquad |\alpha| \leq 2m - 1, \qquad (6.7)$$

for the moment. This condition may be relaxed (cf. N 6.8 in the notes). The main difficulty, however, are homogeneous operators without lower order terms.

We shall proceed by the classical three-steps-approach to problems of this kind. In Step I we study the case of homogeneous systems with constant coefficients, i.e., the a_α do not depend on the variable x. This is done by an application of the R-boundedness version of Mihlin's multiplier theorem 5.2.

In Step II we study homogeneous systems with variable coefficients, but we assume that they are small perturbations of systems with constant coefficients. This is done via a perturbation result for R-sectorial operators (Theorem 6.5).

In Step III we study systems satisfying (6.6) by a localization procedure, but one which uses an infinite partition of unity. It is in this step where lower order terms are included. R-sectoriality for a suitable translate of A will follow from a natural corollary to Theorem 6.5.

Step I: Constant Coefficients

Here we study homogeneous (M, ω_0)-elliptic systems with constant coefficients. We show that they are R-sectorial where R-bounds may be estimated by constants only depending on M and ω_0 but not on the special operator under consideration. Precisely, we shall prove the following:

6.2 Theorem. *Let $M > 0$, $\omega_0 \in [0, \pi)$ and $1 < p < \infty$. Then, for each $\omega_0 < \theta < \pi$, the set of all operators $\lambda(\lambda - A)^{-1}$ where $\lambda \neq 0$, $|\arg \lambda| \geq \theta$, and A is the L_p-realization of a homogeneous (M, ω_0)-elliptic operator with constant coefficients, is R-bounded in $L_p(\mathbb{R}^n, \mathbb{C}^N)$.*

Proof. We fix $M > 0$, $\omega_0 \in [0, \pi)$, $\theta \in (\omega_0, \pi)$ and $1 < p < \infty$. The proof of Theorem 6.2 is done by an application of the R-boundedness version of Mihlin's theorem in 5.2 a). All we have to do is to check that Mihlin's condition holds uniformly for $\xi \mapsto \lambda(\lambda - a(\xi))^{-1}$ where $\lambda \notin \Sigma_\theta \cup \{0\}$, $a(\xi) = \sum_{|\alpha|=2m} a_\alpha \xi^\alpha$

and

$$\sum_{|\alpha|=2m} |a_\alpha| \leq M, \quad \sigma(a(\xi)) \subset \overline{\Sigma_{\omega_0}} \setminus \{0\}, \quad |a(\xi)^{-1}| \leq M \text{ for } |\xi| = 1. \qquad (6.8)$$

Writing $\lambda = e^{i\omega}|\sigma|^{2m}$ where $|\omega| \in [\theta, \pi]$ and $\sigma \in \mathbb{R}$, we have that

$$(\sigma, \xi) \mapsto m(\sigma, \xi) := e^{i\omega}|\sigma|^{2m}(e^{i\omega}|\sigma|^{2m} - a(\xi))^{-1}$$

is a C^∞-function on $(\mathbb{R} \times \mathbb{R}^n) \setminus \{0\}$ which is positively homogeneous of degree 0, i.e., $m(\rho\sigma, \rho\xi) = m(\sigma, \xi)$ for all $(\sigma, \xi) \in \mathbb{R} \times \mathbb{R}^n$ and $\rho > 0$. We use the following lemma whose proof is elementary.

6.3 Lemma. *If m is a differentiable function on $\mathbb{R}^k \setminus \{0\}$ which is positively homogeneous of degree $r \in \mathbb{R}$, i.e., $m(\rho\zeta) = \rho^r m(\zeta)$ for $\rho > 0$, $\zeta \neq 0$, then each partial derivative of m is positively homogeneous of degree $r - 1$.*

By Lemma 6.3 we have that, for any $\alpha \in \mathbb{N}_0^n$, the function

$$(\sigma, \xi) \mapsto |\xi|^{|\alpha|} D_\xi^\alpha m(\sigma, \xi)$$

is also C^∞ on $(\mathbb{R} \times \mathbb{R}^n) \setminus \{0\}$ and positively homogeneous of degree 0. This implies that $\xi \mapsto m(\sigma, \xi)$ satisfies Mihlin's condition uniformly in $\sigma \neq 0$.

Since we also need uniformity in $|\omega| \in [\theta, \pi]$ and tuples $(a_\alpha)_{|\alpha|=2m}$ satisfying (6.8), we have to extend the argument. Let $I := \{\omega : |\omega| \in [\theta, \pi]\}$, and let K denote the set of all tuples $(a_\alpha)_{|\alpha|=2m}$ of matrices $a_\alpha \in \mathbb{C}^{N \times N}$ satisfying (6.8) where $a(\xi) := \sum\limits_{|\alpha|=2m} a_\alpha \xi^\alpha, \xi \in \mathbb{R}^n$. Moreover, let $\Gamma := \{(\sigma, \xi) \in \mathbb{R} \times \mathbb{R}^n : \sigma^2 + |\xi|^2 = 1\}$. Then I and Γ are compact, and we show now that also K is compact.

By $\sum\limits_{|\alpha|=2m} |a_\alpha| \leq M$ the set K is bounded, and we show that it is closed. If (a_α) belongs to the closure of K then $\sum\limits_{|\alpha|=2m} |a_\alpha| \leq M$ and $|a(\xi)^{-1}| \leq M$ for $|\xi| = 1$ by continuity of $(a_\alpha) \mapsto a(\xi)$ (where ξ is fixed) and continuity of inversion in $\mathbb{C}^{N \times N}$.

Moreover, $\sigma(a(\xi)) \subset \overline{\Sigma_{\omega_0}} \setminus \{0\}$ by a spectral projection argument or by Hurwitz' theorem on zeros of analytic functions applied to $\lambda \mapsto \det(\lambda - a(\xi))$. (Observe that (M, ω_0)-ellipticity implies actually $\sigma(a(\xi)) \subset \{z \in \mathbb{C} : \frac{1}{M} \leq |z| \leq M, |\arg z| \leq \omega_0\} =: B$ for $|\xi| = 1$ and that B is a closed set.)

Denoting now, for $(a_\alpha) \in K$, $\omega \in I$, $(\sigma, \xi) \in (\mathbb{R} \times \mathbb{R}^n) \setminus \{0\}$,

$$m((a_\alpha), \omega, \sigma, \xi) := e^{i\omega}|\sigma|^{2m}(e^{i\omega}|\sigma|^{2m} - a(\xi))^{-1},$$

we have as before that, for any $\beta \in \mathbb{N}_0^n$, $(a_\alpha) \in K$, $\omega \in I$, the function

$$(\sigma, \xi) \mapsto |\xi|^{|\beta|} D_\xi^\beta m((a_\alpha), \omega, \sigma, \xi)$$

is C^∞ on $(\mathbb{R} \times \mathbb{R}^n) \setminus \{0\}$ and positively homogeneous of degree 0. For fixed $\beta \in \mathbb{N}_0^n$, the map

$$((a_\alpha), \omega, \sigma, \xi) \mapsto |\xi|^\beta D_\xi^\beta m((a_\alpha), \omega, \sigma, \xi)$$

is continuous on $K \times I \times \Gamma$. We conclude that

$$\sup\{|\xi|^{|\beta|} D_\xi^\beta m((a_\alpha), \omega, \sigma, \xi) : \sigma \neq 0, \xi \neq 0, \omega \in I, (a_\alpha) \in K\}$$
$$= \sup\{|\xi|^{|\beta|} D_\xi^\beta m((a_\alpha), \omega, \sigma, \xi) : (\sigma, \xi) \in \Gamma, \omega \in I, (a_\alpha) \in K\} < \infty,$$

since $K \times I \times \Gamma$ is compact.

This means that Mihlin's condition holds uniformly in $\sigma \neq 0$, $|\omega| \in [\theta, \pi]$ and tuples $(a_\alpha) \in K$, and the assertion is proved. $\square$

Step II: Small Perturbations

Now we study homogeneous (M, ω_0)-elliptic systems on $\mathbb{R}^n$ with bounded measurable coefficients which are sufficiently close to systems with constant coefficients. This is done via a perturbation result for R-sectorial operators (Theorem 6.5 below). Precisely, we shall show the following.

6.4 Theorem. *Let $M > 0$, $\omega_0 \in [0, \pi)$, $1 < p < \infty$, and $\theta \in (\omega_0, \pi)$. Then there are constants $K, \epsilon > 0$ such that, for any (M, ω_0)-elliptic system $\mathcal{A}(D) = \sum_{|\alpha|=2m} a_\alpha D^\alpha$ with constant coefficients and any differential operator*

$$\mathcal{B}(D) := \sum_{|\alpha|=2m} b_\alpha D^\alpha$$

with $b_\alpha \in L_\infty(\mathbb{R}^n, \mathbb{C}^N)$ and

$$\sum_{|\alpha|=2m} \|b_\alpha\|_\infty < \epsilon, \tag{6.9}$$

we have that

$$R(\{\lambda(\lambda - (A + B))^{-1} : \lambda \neq 0, |\arg \lambda| \geq \theta\}) \leq K \tag{6.10}$$

where $A+B$ denotes the L_p-realization of the differential operator $\mathcal{A}(D)+\mathcal{B}(D)$ with domain $W_p^{2m}(\mathbb{R}^n, \mathbb{C}^N)$.

In the proof we shall use the following perturbation result which is an R-boundedness version of the usual perturbation theorem for sectorial operators.

6.5 Theorem. *Let A be an R-sectorial operator in the Banach space X and $\pi \geq \theta > \omega_R(A)$. Let B be a linear operator satisfying $D(B) \supset D(A)$ and*

$$\|Bx\| \leq a\|Ax\| \qquad , x \in D(A).$$

If $a < 1/\tilde{R}_\theta(A)$ then $A + B$ is again R-sectorial and

$$R_\theta(A + B) \leq \frac{R_\theta(A)}{1 - a\tilde{R}_\theta(A)}.$$

In particular, we have $\omega_R(A + B) \leq \theta$.

Proof. For $|\arg \lambda| \geq \theta$ we obtain that

$$\|BR(\lambda, A)x\| \leq a\|AR(\lambda, A)x\| \leq a\tilde{M}_\theta(A)\|x\| \quad , x \in X.$$

Hence $I + BR(\lambda, A)$ is invertible by the assumption and

$$R(\lambda, A+B) = R(\lambda, A)(I+BR(\lambda, A))^{-1} = R(\lambda, A) \sum_{k=0}^{\infty} (-BR(\lambda, A))^k. \tag{6.11}$$

This representation implies

$$M_\theta(A + B) \leq \frac{M_\theta(A)}{1 - a\tilde{M}_\theta(A)},$$

hence $A + B$ is a sectorial operator and $\sigma(A+B) \subset \Sigma_\theta$. From the definition of R-boundedness and the assumption it is clear that, for any subset Λ of $\rho(A)$, we have

$$R(\{BR(\lambda, A) : \lambda \in \Lambda\}) \leq a \cdot R(\{AR(\lambda, A) : \lambda \in \Lambda\}).$$

Applying this to the series representation of $R(\lambda, A + B)$ in (6.11) we obtain

$$R_\theta(A + B) \leq \frac{R_\theta(A)}{1 - a\tilde{R}_\theta(A)}$$

as asserted.   □

Later on, in Step III, we shall derive the usual corollary for perturbations B satisfying

$$\|Bx\| \leq a\|Ax\| + b\|x\|, \quad x \in D(A),$$

for some $a, b \geq 0$ where a has to be small and b is arbitrary. But first we make up for the proof of Theorem 6.4.

Proof (of Theorem 6.4). For any $\epsilon > 0$ and all $f \in W_p^{2m}(\mathbb{R}^n, \mathbb{C}^N)$, we have

$$
\begin{aligned}
\|Bf\|_p &\leq \sum_{|\alpha|=2m} \|b_\alpha\|_\infty \|D^\alpha f\|_p \\
&\leq \epsilon \max_{|\alpha|=2m} \|D^\alpha f\|_p \\
&\leq \epsilon \max_{|\alpha|=2m} \|D^\alpha A^{-1}\|_{p \to p} \|Af\|_p
\end{aligned}
$$

if (b_α) satisfies (6.9). For $|\alpha| = 2m$, the symbol of $D^\alpha A^{-1}$ is given by

$$m(\xi) := \xi \mapsto \xi^\alpha \Big(\sum_{|\beta|=2m} a_\beta \xi^\beta \Big)^{-1}$$

which is C^∞ on $\mathbb{R}^n \setminus \{0\}$ and positively homogeneous of degree 0. By an argument similar to the one used in the proof of Theorem 6.2 we see that $\xi \mapsto m(\xi)$ satisfies Mihlin's conditions uniformly in $(a_\beta) \in K$ where K is as in the proof of Theorem 6.2. Hence

$$\|D^\alpha A^{-1}\|_{p \to p} \leq C$$

for all $|\alpha| = 2m$ and all (M, ω_0)-elliptic operators A of the given form where C only depends on M, ω_0 and p. Letting

$$R_\theta := R\Big(\{\lambda(\lambda - A)^{-1} : A = \sum_{|\beta|=2m} a_\beta D^\beta, (a_\beta) \in K, \lambda \neq 0, |\arg \lambda| \geq \theta\} \Big),$$

which is finite by Theorem 6.2, and choosing $0 < \epsilon < \frac{1}{C(R_\theta + 1)}$ we have by Theorem 6.5 that (6.10) holds for $K := \frac{R_\theta}{1 - \epsilon C(R_\theta + 1)}$. Observe that the constants ϵ and K we obtained depend only on M, ω_0, θ and p.   □

Step III: Localization and the General Case

Now we study (M, ω_0)-elliptic systems of the form (6.1) where the coefficients of the principle part satisfy (6.6). Let us fix M, ω_0 and the system. For fixed $p \in (1, \infty)$ and $\theta \in (\omega_0, \pi)$, the BUC-assumption on the top order coefficients allows for a "uniform" localization: Recall that Theorem 6.4 gives constants $\epsilon, K > 0$ which allow to handle ϵ-small perturbations of (M, ω_0)-elliptic systems with constant coefficents where the precise meaning of ϵ-small is given by (6.9). We now choose $\delta > 0$ according to uniform continuity such that

$$\sum_{|\alpha|=2m} |a_\alpha(x) - a_\alpha(y)| < \epsilon \quad \text{for } |x - y| \le \delta.$$

Then, locally, the situation is clear by Step II. All we have to do is to set up a procedure that allows to pass from the general situation to a family of localized problems and back again to the original problem. This procedure should respect uniform estimates of the constants we are interested in.

We fix $r \in (0, \delta)$ and a C^∞-function φ with $0 \le \varphi \le 1$ and $\operatorname{supp} \varphi \subset Q := (-r, r)^n$ such that

$$\sum_{l \in r\mathbb{Z}^n} \varphi_l^2(x) = 1, \quad x \in \mathbb{R}^n, \tag{6.12}$$

where $\varphi_l(x) := \varphi(x - l)$. We let $Q_l := l + Q$ for $l \in r\mathbb{Z}^n =: \Gamma$. We also fix a C^∞-function ψ with $\operatorname{supp} \psi \subset Q$, $0 \le \psi \le 1$, and $\psi = 1$ on $\operatorname{supp} \varphi$ and define $\psi_l(x) := \psi(x - l)$ for $l \in \Gamma$. For any $l \in \Gamma$ and $|\alpha| = 2m$ we define coefficients $a_\alpha^l : \mathbb{R}^n \to \mathbb{C}^{N \times N}$ by

$$a_\alpha^l(x) := \begin{cases} a_\alpha(x), x \in Q_l \\ a_\alpha(l) \text{ otherwise.} \end{cases}$$

We define operators A_l with domain $W_p^{2m}(\mathbb{R}^n, \mathbb{C}^N)$ by

$$A_l u(x) := \sum_{|\alpha|=2m} a_\alpha^l(x) D^\alpha u(x).$$

Writing $A = A_\pi + A_{\text{low}}$ where $A_\pi = \sum_{|\alpha|=2m} a_\alpha$ denotes the principle part of A, and A_{low} denotes the part containing the lower order terms of A, we have that $A_\pi u = A_l u$ if $u \in W_p^{2m}(\mathbb{R}^n, \mathbb{C}^N)$ is supported in Q_l.

6.6 The Localization Scheme. We now set up the localization scheme. First, we define the space in which the family of localized operators will act. For a Banach space $(X, \|\cdot\|_X)$ we denote by $l_p(\Gamma, X)$ the space of all families $(x_l)_{l \in \Gamma}$ in X such that $\|(x_l)_l\|_p := (\sum_{l \in \Gamma} \|x_l\|_X^p)^{1/p} < \infty$, which is a Banach space for $\|\cdot\|_p$. We let

$$\mathbb{X}_p := l_p(\Gamma, L_p(\mathbb{R}^n, \mathbb{C}^N)).$$

Denoting the elements of $\mathbb{X}_p$ by $(u_l)_{l\in\Gamma}$ we define the operator $\mathbb{A}(u_l) := (A_l u_l)$ with domain

$$\mathbb{X}_p^{2m} := l_p(\Gamma, W_p^{2m}(\mathbb{R}^n, \mathbb{C}^N)).$$

The operators A_l are small perturbations of (M, ω_0)-elliptic operators with constant coefficients. By Theorem 6.4 we know that the operators A_l are R-sectorial and that we have $R_\theta(A_l) \le K$ for any $l \in \Gamma$. Since $\mathbb{A}$ is a diagonal operator in $\mathbb{X}_p$ this clearly implies that $\mathbb{A}$ is a sectorial operator of type $\omega(\mathbb{A}) \le \theta$.

But $\mathbb{A}$ is also R-sectorial of R-type $\omega_R(\mathbb{A}) \le \theta$. To see this recall that, in the definition of R-boundedness of a family of operators in a Banach space X with the Rademachers on $[0,1]$, one may use any $L_q([0,1], X)$-norm, $q \in [1,\infty)$, and that the resulting R-bounds are equivalent (cf. Theorem 2.4). For the present purpose we choose $q = p$. Suppose that (T_j) is a family of diagonal operators $\mathbb{X}_p \to \mathbb{X}_p$, each T_j being given by $T_j(f_l) := (T_l^{(j)} f_l)$ where $T_l^{(j)} : L_p \to L_p$ for $l \in \Gamma$. Then we have for elements $x_j := (f_l^{(j)})_l \in \mathbb{X}_p$:

$$\left(\int_0^1 \| \sum_j r_j(t) T_j x_j \|_{\mathbb{X}_p}^p \, dt \right)^{1/p}$$

$$= \left(\int_0^1 \sum_l \| \sum_j r_j(t) T_l^{(j)} f_l^{(j)} \|_{L_p}^p \, dt \right)^{1/p}$$

$$= \left(\sum_l \int_0^1 \| \sum_j r_j(t) T_l^{(j)} f_l^{(j)} \|_{L_p}^p \, dt \right)^{1/p}$$

$$\le \sup_l R(\{T_l^{(j)} : j\}) \left(\sum_l \int_0^1 \| \sum_j r_j(t) f_l^{(j)} \|_{L_p}^p \, dt \right)^{1/p}$$

$$= \sup_l R(\{T_l^{(j)} : j\}) \left(\int_0^1 \| \sum_j r_j(t) x_j \|_{\mathbb{X}_p}^p \, dt \right)^{1/p},$$

which means that the family (T_j) is R-bounded in $\mathbb{X}_p$ if, for each $l \in \Gamma$, the set $\{T_l^{(j)} : j\}$ is R-bounded and if the R-bounds are bounded uniformly in $l \in \Gamma$. This argument shows that the operator $\mathbb{A}$ is an R-sectorial operator in $\mathbb{X}_p$ and that $\omega_R(\theta) \le K$.

The operation of "localization" itself is represented by the operator $J : L_p(\mathbb{R}^n, \mathbb{C}^N) \to \mathbb{X}_p$ that takes a function f to the sequence $(\varphi_l f)$. By

$$\sum_{l\in\Gamma} \|\varphi_l f\|_p^p \le C_\varphi \sum_{l\in\Gamma} \|1_{Q_l} f\|_p^p = 2C_\varphi \|f\|_p^p$$

the operator J is continuous and it is clearly injective. A similar argument shows that J maps $W_p^{2m}(\mathbb{R}^n, \mathbb{C}^N)$ continuously into $\mathbb{X}_p^{2m}$.

"Patching together again" is done via the operator $P : \mathbb{X}_p \to L_p(\mathbb{R}^n, \mathbb{C}^N)$ that takes a sequence (f_l) to $\sum_l \varphi_l f_l$. The sum is locally finite here. Note that, writing more precisely J_p for the operator J defined above, we have $P = (J_{p'})'$. The operator P is continuous $\mathbb{X}_p \to L_p$ and $\mathbb{X}_p^{2m} \to W_p^{2m}$. By the definition of the φ_l and the choice of φ we have $PJf = f$ for all $f \in L_p(\mathbb{R}^n, \mathbb{C}^N)$. Actually, these properties imply that $\|Jf\|_{\mathbb{X}_p} \sim \|f\|_p$ and that $\|Ju\|_{\mathbb{X}_p^{2m}} \sim \|u\|_{W_p^{2m}}$, but we shall not use this fact explicitly.

Now we want to relate the operator A in L_p and the operator $\mathbb{A}$ in $\mathbb{X}_p$. To this end we calculate $JA - \mathbb{A}J$ on W_p^{2m} and $AP - P\mathbb{A}$ on $\mathbb{X}_p^{2m}$. We define $k \bowtie l$ for $k, l \in \Gamma$ if $Q_k \cap Q_l \neq \emptyset$. For $u \in W_p^{2m}$ and $l \in \Gamma$ we have

$$\varphi_l Au = A\varphi_l u + [\varphi_l A - A\varphi_l]u = A_l \varphi_l u + A_{\text{low}} \varphi_l u + \sum_{k \bowtie l}[\varphi_l A - A\varphi_l]\varphi_k \varphi_k u,$$

which leads to

$$JA = (\mathbb{A} + \mathbb{B})J \tag{6.13}$$

where $\mathbb{B}(u_l) := (A_{\text{low}} \psi_l u_l + \sum_{k \bowtie l}[\varphi_l A - A\varphi_l]\varphi_k u_k)_l$ for $(u_l) \in \mathbb{X}_p^{2m}$. Observe that $\mathbb{B}$ is a matrix of differential operators of order $\leq 2m - 1$ and that the number of non-zero entries in the l-th row is bounded by $\#\{k \in \Gamma : k \bowtie l\} \in \mathbb{N}$ which is independent of $l \in \Gamma$. For a sequence $(u_l) \in \mathbb{X}_p^{2m}$ we obtain

$$(AP - P\mathbb{A})(u_l) = A(\sum_l \varphi_l u_l) - \sum_l \varphi_l A_l u_l$$

$$= \sum_l \varphi_l A_{\text{low}} u_l + \sum_k [A\varphi_k - \varphi_k A]u_k$$

$$= \sum_l \varphi_l A_{\text{low}} u_l + \sum_k \sum_{l \bowtie k} \varphi_l \varphi_l [A\varphi_k - \varphi_k A]u_k$$

$$= \sum_l \varphi_l \left(A_{\text{low}} u_l + \sum_{k \bowtie l} \varphi_l [A\varphi_k - \varphi_k A]u_k\right)$$

which means

$$AP = P(\mathbb{A} + \mathbb{D}) \tag{6.14}$$

where $\mathbb{D}(u_l) := (A_{\text{low}} u_l + \sum_{k \bowtie l} \varphi_l [A\varphi_k - \varphi_k A]u_k)$.

Note that the structure of the operator $\mathbb{D}$ is similar to the structure of $\mathbb{B}$ and that the operators $A\varphi_k - \varphi_k A$ are of order $\leq 2m - 1$. Since the coefficients in A_{low}, i.e., the a_α for $|\alpha| \leq 2m - 1$, are L_∞-functions, the operators $\mathbb{B}$ and $\mathbb{D}$ act boundedly $\mathbb{X}_p^{2m-1} \to \mathbb{X}_p$. This implies that, for any $\epsilon > 0$, we have an estimate

$$\|\mathbb{B}(u_l)\|_{\mathbb{X}_p} + \|\mathbb{D}(u_l)\|_{\mathbb{X}_p} \leq \epsilon \|\mathbb{A}(u_l)\|_{\mathbb{X}_p} + C_\epsilon \|(u_l)\|_{\mathbb{X}_p}, \quad (u_l) \in \mathbb{X}_p^{2m}.$$

Hence the following corollary to Theorem 6.5 applies.

6.7 Corollary. *Let A be an R-sectorial operator in X and $\theta > \omega_R(A)$. Let B be a linear operator satisfying $D(B) \supset D(A)$ and*

$$\|Bx\| \leq a\|Ax\| + b\|x\| \qquad , x \in D(A) \tag{6.15}$$

for some $a, b \geq 0$. If $a < (\tilde{M}_\theta(A)\tilde{R}_\theta(A))^{-1}$ then $A + B + \nu$ is R-sectorial for any

$$\nu > \frac{bM_\theta(A)\tilde{R}_\theta(A)}{(1 - a\tilde{M}_\theta(A)\tilde{R}_\theta(A))}. \tag{6.16}$$

Moreover, we have $\omega_R(A + B + \nu) \leq \theta$ for all such ν.

Proof. For any $\nu > 0$ and $x \in X$ we have

$$\begin{aligned}\|BR(-\nu, A)x\| &\leq a\|AR(-\nu, A)x\| + b\|R(-\nu, A)x\| \\ &\leq (a\tilde{M}_\theta(A) + bM_\theta(A)/\nu)\|x\| =: c(\nu)\|x\|.\end{aligned}$$

This means that B satisfies the assumptions of Theorem 6.5 for $A + \nu$ in place of A if $c(\nu) < \tilde{R}_\theta(A)^{-1}$ (note that $\tilde{R}_\theta(A + \nu) \leq \tilde{R}_\theta(A)$ for $\nu > 0$). Since $a\tilde{M}_\theta(A)\tilde{R}_\theta(A) < 1$ by assumption, the condition $c(\nu)\tilde{R}_\theta(A) < 1$ is equivalent to (6.16). □

By Corollary 6.7 we find $\nu \geq 0$ such that $\nu + \mathbb{A} + \mathbb{B}$ and $\nu + \mathbb{A} + \mathbb{D}$ are both R-sectorial operators of R-type $\leq \theta$. Let now $|\arg(\lambda + \nu)| > \theta$. If $u \in W_p^{2m}$ and f satisfy $(\lambda - A)u = f$ then (6.13) yields $Jf = (\lambda - (\mathbb{A} + \mathbb{B}))Ju$ and

$$u = PJu = P(\lambda - (\mathbb{A} + \mathbb{B}))^{-1}Jf.$$

In particular, $\lambda - A$ is injective. On the other hand, (6.14) yields, for any $f \in L_p$,

$$\begin{aligned}f = PJf &= P(\lambda - (\mathbb{A} + \mathbb{D}))(\lambda - (\mathbb{A} + \mathbb{D}))^{-1}Jf \\ &= (\lambda - A)P(\lambda - (\mathbb{A} + \mathbb{D}))^{-1}Jf\end{aligned}$$

which means that $\lambda - A : W_p^{2m} \to L_p$ is surjective. Together we have that any λ satisfying $|\arg(\lambda + \nu)| > \theta$ belongs to $\rho(A)$ and

$$(\lambda - A)^{-1} = P(\lambda - (\mathbb{A} + \mathbb{B}))^{-1}J = P(\lambda - (\mathbb{A} + \mathbb{D}))^{-1}J.$$

Since P and J are bounded operators this means that $A + \nu$ is an R-sectorial operator of R-type $\leq \theta$. Moreover, the proof shows that $R_\theta(A + \nu)$ only depends on M, ω_0, θ, p, the δ from the BUC-assumption for the highest order coefficients, and the L_∞-norm of the lower order coefficients.

Thus we have proved the following.

6.8 Theorem. *Let $M > 0$, $\omega_0 \in [0, \pi)$, and $p \in (1, \infty)$. Let A be the L_p-realization of an (M, ω_0)-elliptic system $\mathcal{A}(D)$ satisfying (6.6) and (6.7). Then, for any $\theta \in (\omega_0, \pi)$, there exists a $\nu \geq 0$ such that $A + \nu$ is R-sectorial of R-type $\leq \theta$.*

Notes on Section 6:

N 6.1: The notion of (M, ω_0)-elliptic systems was introduced in [10]. It was also used in [79], [146], [66], [85], [113], [116].

N 6.2: The fact that the constant coefficient case in Step I relies on Mihlin's theorem is standard. In [195] it was shown that second order operators have bounded imaginary powers (cf. also [201]). In [10] it was shown that (M, ω_0)-elliptic systems with constant coefficients have a bounded H^∞-calculus. In a much more general context this is also contained in [66]. New is the use of the R-boundedness version 5.2 of Mihlin's theorem in this context.

N 6.4 is from [146]. The proof given in the present section for maximal regularity of (M, ω_0)-elliptic systems whose highest order coefficients are BUC was sketched there.

N 6.5 is from [146]. An earlier version of this perturbation result has been used in [141]. From the application in Step II it is clear how the regularity of the coefficients depends on the assumptions in perturbation results. Before the characterization of maximal regularity via R-sectoriality was established, maximal regularity for elliptic operators was proved using the Dore-Venni-approach which meant that one had to prove that the operators had bounded imaginary powers (BIP) or – even stronger – a bounded H^∞-calculus (cf. Chapter II). The perturbation theorem for BIP in [195] (which was also used in [204]) makes a stronger assumption on the perturbing operator. This is also the case for the perturbation result for the H^∞-calculus which is proved in [10]. For a perturbation result for the H^∞-calculus and further details we refer to Section 13.

In [70, Thm. 6.1], a perturbation theorem for maximal regularity is given via the operator sum method: suppose that A has maximal regularity in X. Define $\mathcal{A}$ on $L_p(I, X)$ by $(\mathcal{A}f)(t) := A(f(t))$ and denote the inverse of $d/dt + \mathcal{A}$ by $\mathcal{M}$. Then $A + B$ has maximal regularity if, in addition to (6.15), we have $b\|\mathcal{M}\| + a\|\mathcal{A}\mathcal{M}\| < 1$ which holds in particular if $b = 0$ and a is sufficiently small. The smallness condition $a\|\mathcal{A}\mathcal{M}\| < 1$ should be compared with the assumption $a\tilde{R}_\theta(A) < 1$ in Theorem 6.5. Of course, in UMD-spaces, Theorem 6.5 yields a perturbation theorem for maximal L_p-regularity, but if the space X is not UMD, the results are not comparable.

Also [70, Thm. 6.1] can be applied to prove maximal regularity for (M, ω_0)-systems satisfying (6.6) and (6.7) (or (6.17) below) if $\omega_0 < \pi/2$. The assertion on the R-type in Theorem 6.8 might also be proved but that would involve considerations of maximal regularity for rotated operators. Apparently such an approach has never been carried out.

N 6.6: The localization scheme is taken from [10]. It was also used in [79], [141], [113], [116]. A different localization with a finite partition of unity was used in [195]. [66] also uses a finite partition of unity. Localizing on $\mathbb{R}^n$ with a *finite* partition of unity means that one has to impose conditions on the highest order coefficients at infinity.

N 6.7 is again from [146].

N 6.8 is the main result of this section. The proof presented here has been sketched in [146]. The result was known before, in [79] it was shown that those systems even have a bounded H^∞-calculus. The proof is very involved and relies on techniques related to the $T(1)$-Theorem from harmonic analysis. It also relies on the special structure of the operators whereas the proof given here relies on the general perturbation results 6.5 and 6.7.

The prior proofs of maximal regularity for such systems used the operator sum method, i.e., the Dore–Venni approach, which means that those operators were shown to have BIP or a bounded H^∞-calculus first. For systems with C^∞-coefficients boundedness of imaginary powers was shown by Seeley ([201]). For second order operators with Hölder continuous coefficients in the principle part boundedness of imaginary powers was studied in [195] and also in [204].

In [10, Thm. 9.4] it was shown that L_p-realizations A of (M, ω_0)-elliptic systems satisfying (6.6) have the property that, for a suitable $\nu \geq 0$, $A + \nu$ is a sectorial operator. The assumptions on the lower order coefficients in [10] are more general than (6.7) (see (6.17) below). Under an additional assumption on the modulus of continuity of the highest order coefficients (which holds in particular if they are Hölder-continuous) it was shown in [10, Thm. 9.6] that $A + \nu$ has even a bounded H^∞-calculus for a suitable $\nu \geq 0$ (we refer to Section 13 in Chapter II where we shall give a different proof of this result). This property implies that $A + \nu$ is R-sectorial.

Duong and Simonett ([79]) were able to remove the additional assumption on the modulus of continuity. They proved that, under the assumption (6.6) and the assumption (6.17) below, the (M, ω_0)-elliptic operator $A + \nu$ has a bounded H^∞-calculus for a suitable $\nu \geq 0$. Their proof, however, is considerably more involved (cf. the remarks above). As far as R-sectoriality in concerned, the proof given in the present section is much simpler and seems more natural.

The assertion of Theorem 6.8 was also shown in [113] by a different proof that relies on the connection of R-boundedness to certain weighted norm inequalities for Muckenhoupt weights (cf. also [101]). By the same approach (but technically much more involved) the assertion of Theorem 6.8 was shown in [116] for operators whose highest order coeffcients belong to $L_\infty \cap VMO$. On the other hand, it was proved in [85] that second order elliptic operators with second order coefficients in $L_\infty \cap VMO$ even have a bounded H^∞-calculus. The proof of this result makes extensive use of wavelet techniques.

The assumptions (6.7) on the lower order coefficients enters in Step III. It may be weakened as long as the arguments used to show that $\mathbb{B}$ and $\mathbb{D}$ are $\mathbb{A}$-small perturbations still work. One might, e.g., use Hölder's inequality and Sobolev embedding for $k := |\alpha| \leq 2m - 1$ to obtain

$$\|a_\alpha D^\alpha u\|_p \leq \|a_\alpha\|_{q_k} \|D^\alpha u\|_{r_k} \leq \|a_\alpha\|_{q_k} \|u\|_{W_p^s}$$

where $1/q_k + 1/r_k = 1/p$ and s is such that $W_p^s \hookrightarrow W_{r_k}^k$. Observe that, by the localization scheme, we only need to use this argument locally. The following condition is used in [10], [79]:

$$a_\alpha \in L_{q_{|\alpha|},\text{loc,unif}}(\mathbb{R}^n, \mathbb{C}^{N \times N}), \quad |\alpha| \le 2m - 1, \tag{6.17}$$

where $q_k := p$ if $k < 2m - n/p$ and $q_k > p_k := n/(2m - k)$ if $2m - n/p \le k < 2m$, and

$$L_{q,\text{loc,unif}} := \{g \in L_{q,\text{loc}} : \sup_{l \in \Gamma} \|g\|_{L_q(Q_l)} < \infty\}.$$

Observe that this definition does not depend on the δ in the definition of the cubes Q_l and the grid Γ. We refer to [10, Lem. 9.3]. In [66] a version of the assertion in Theorem 6.8 has been shown for the case of an infinite dimensional state space which is assumed to have the UMD-property.

7 Elliptic Boundary Value Problems

In this section we study elliptic boundary value problems in open subsets Ω of $\mathbb{R}^n$. Already the case $n = 1$ makes it clear that, if we want to obtain a sectorial operator, then the elliptic system – as it was studied in Section 6 but defined in Ω in place of $\mathbb{R}^n$ – has to be complemented by conditions on the boundary $\partial\Omega$. These boundary conditions will also be given by differential operators. We will impose the assumptions that are known form the classical theory of elliptic boundary value problems, i.e., ellipticity for the system in Ω (as we have done in Section 6) and the so-called *Lopatinskij-Shapiro condition* (in the form it takes for parabolic or parameter-elliptic problems). For an introduction to the classical theory of elliptic boundary value problems we refer to [224] and [87].

In order to limit technicalities we restrict ourselves here to scalar equations (the case $N = 1$) of second order (the case $m = 1$). Although this is the simplest case it already shows much of the structure that one encounters in more general situations. For the most general situation studied so far (systems of arbitrary order with general boundary conditions and infinite dimensional state space) we refer to [66].

The operators we shall study have the following form

$$\mathcal{A}(D) = \sum_{|\alpha| \le 2} a_\alpha D^\alpha$$

where

$$a_\alpha \in BUC(\Omega, \mathbb{C}), \quad |\alpha| = 2,$$

$$a_\alpha \in L_\infty(\Omega, \mathbb{C}), \quad |\alpha| < 2,$$

and $\Omega \subset \mathbb{R}^n$ is an open subset.

We keep the multiindex notation from Section 6, i.e., we denote $D_j := -i\frac{\partial}{\partial x_j}, j = 1, \ldots, n$, and $D^\alpha := D_1^{\alpha_1} \cdot \ldots \cdot D_n^{\alpha_n}$ for $\alpha = (\alpha_1, \ldots, \alpha_n) \in \mathbb{N}_0^n$. This may seem artificial for a second order operator but, on the one hand, we want to be consistent with Section 6 and, on the other hand, we prefer a notation that makes generalizations to higher orders easily accessible. The symbol $\mathcal{A}(x, \xi)$ of $\mathcal{A}(D)$ is then given by

$$\mathcal{A}(x, \xi) = \sum_{|\alpha| \leq 2} a_\alpha(x) \xi^\alpha, \quad x \in \Omega, \xi \in \mathbb{R}^n,$$

and the principle symbol $\mathcal{A}_\pi(x, \xi)$ is given by

$$\mathcal{A}_\pi(x, \xi) = \sum_{|\alpha| = 2} a_\alpha(x) \xi^\alpha, \quad x \in \Omega, \xi \in \mathbb{R}^n.$$

The ellipticity assumption now reads

$$\mathcal{A}_\pi(x, \xi) = \sum_{|\alpha| = 2} a_\alpha(x) \xi^\alpha \subset \overline{\Sigma_{\omega_0}}, \quad x \in \Omega, |\xi| = 1,$$

$$|\mathcal{A}_\pi(x, \xi)| \geq \delta |\xi|^2, \quad x \in \Omega, \xi \in \mathbb{R}^n.$$

for some $\omega_0 \in [0, \pi), \delta > 0$. Note that this corresponds exactly to the notion of (M, ω_0)-ellipticity for $N = 1$ (cf. Section 6) where, e.g., $M := \max\{\sum_{|\alpha|=2} \|a_\alpha\|_\infty, \delta^{-1}\}$.

For an elliptic operator of order $2m$ we need m boundary conditions, each one given by a differential operator $\mathcal{B}_j(D)$ of order $m_j < 2m$ (see [224] or [87]). This means that, for an elliptic second order operator, we need only one boundary condition, i.e., one differential operator $\mathcal{B}(D)$ of order < 2 on the boundary. Hence $\mathcal{B}(D)$ has order zero or order one. For $p \in (1, \infty)$, we are interested in showing R-sectoriality for the operator A_B in $L_p(\Omega)$ or a suitable translate $\nu + A_B$ of A_B, where A_B is given by

$$A_B u(x) := \mathcal{A}(D) u(x)$$
$$\text{for} \quad u \in D(A_B) := \{v \in W_p^2(\Omega) : \mathcal{B}(D) v = 0 \text{ on } \partial\Omega\}.$$

If $\mathcal{B}(D)$ has order zero and satisfies the Lopatinskij-Shapiro condition then, as we shall see, the homogeneous boundary condition $\mathcal{B}(D)v = 0$ on $\partial\Omega$ is equivalent to the Dirichlet boundary condition $v = 0$ on $\partial\Omega$. This is a case that allows for several simplifications compared to the general situation. If $\mathcal{B}(D)$ has order one, we shall see that we have an oblique derivative problem. We concentrate here on the oblique derivative problem, which gives a good impression of the arguments that are employed in more general situations. We comment on the changes and simplifications for Dirichlet boundary conditions later on.

Again, we shall use the classical three-step-method: constant coefficients in the half space, small perturbations, localization and coordinate transformation.

Apart from additional coordinate transformations which locally flatten the boundary, the localization procedure in the last step is as in the boundaryfree case in Section 6. Hence we concentrate on the first two steps for the moment.

Oblique Derivative Condition

Step I: Constant Coefficients in the Half Space

In this subsection we study boundary value problems for homogeneous constant coefficient operators in the half space. We shall use the following notation: $x = (x', y) \in \mathbb{R}^{n+1}_+$ where $x' = (x_1, \ldots, x_n) \in \mathbb{R}^n$ and $y > 0$. We recall $D_j := -i\frac{\partial}{\partial x_j}$ and write $D = (D', D_{n+1})$ where $D' = (D_1, \ldots, D_n)$ and $D_{n+1} = D_y = -i\frac{\partial}{\partial y}$.

In $\mathbb{R}^{n+1}_+$, we study the boundary value problem for the homogeneous second order operator

$$\mathcal{A}(D) = \sum_{|\alpha|=2} a_\alpha D^\alpha$$

and the homogeneous first order boundary operator

$$\mathcal{B}(D) = \sum_{|\beta|=1} b_\beta D^\beta.$$

We assume (M, ω_0)-ellipticity of $\mathcal{A}(D)$, i.e.,

$$\mathcal{A}(\xi) \subset \overline{\Sigma_{\omega_0}} \text{ and } |\mathcal{A}(\xi)| \geq \frac{1}{M}|\xi|^2 \text{ for } \xi \in \mathbb{R}^{n+1}, \quad \sum_{|\alpha|=2} |a_\alpha| \leq M,$$

where $\mathcal{A}(\xi) = \sum_{|\alpha|=2} a_\alpha \xi^\alpha, \xi \in \mathbb{R}^{n+1}$, denotes the symbol of the operator $\mathcal{A}(D)$, and $M > 0$, $\omega_0 \in [0, \pi)$ are fixed constants.

7.1 The Boundary Value Problem. It has the following form: Given $\theta > \omega_0, \lambda \in \Sigma_{\pi-\theta}, f \in L_p(\mathbb{R}^{n+1}_+), g \in W^1_p(\mathbb{R}^{n+1}_+)$, find $u \in W^2_p(\mathbb{R}^{n+1}_+)$ such that

$$\begin{aligned} \lambda u + \mathcal{A}(D)u &= f \quad \text{in } \mathbb{R}^{n+1}_+ \\ \mathcal{B}(D)u &= g \quad \text{on } \partial\mathbb{R}^{n+1}_+. \end{aligned} \tag{7.1}$$

Here we have fixed $p \in (1, \infty)$. We postpone our assumption on $\mathcal{B}(D)$, i.e., the formulation of the Lopatinskij-Shapiro condition, for the moment.

In this section, we are – of course – primarily interested in the case $g = 0$: Defining the operator A_B in $L_p(\mathbb{R}^{n+1}_+)$ by

$$\begin{aligned} D(A_B) &:= \{u \in W^2_p(\mathbb{R}^{n+1}_+) : \mathcal{B}(D)u = 0 \text{ on } \partial\mathbb{R}^{n+1}_+\} \\ A_B u &:= \mathcal{A}(D)u \end{aligned}$$

we have $u = (\lambda + A_B)^{-1}f$ if u is a solution of (7.1) where $g = 0$.

One reason to study the full inhomogeneous problem (7.1) is that we need a representation of its solution in the next step where we study small perturbations of constant coefficient operators.

Another reason is that, due to our knowledge on elliptic operators on the whole space $\mathbb{R}^{n+1}$, we can reduce the problem (7.1) to the case $f = 0$.

7.2 Reduction to $f = 0$. Denote by A_{n+1} the $L_p(\mathbb{R}^{n+1})$-realization of the operator $\mathcal{A}(D)$ with domain $W_p^2(\mathbb{R}^{n+1})$ and by $E_0 f$ the zero-extension of $f \in L_p(\mathbb{R}_+^{n+1})$ to a function on $\mathbb{R}^{n+1}$. By the results of Section 6 we know that $-\lambda \in \rho(A_{n+1})$. Hence $v := (\lambda + A_{n+1})^{-1} E_0 f \in W_p^2(\mathbb{R}^{n+1})$ is a solution to

$$\lambda v + \mathcal{A}(D)v = E_0 f \quad \text{in } \mathbb{R}^{n+1}.$$

If we denote by Ph the restriction to $\mathbb{R}_+^{n+1}$ of a function h defined on $\mathbb{R}^{n+1}$, then $w := P(\lambda + A_{n+1})^{-1} E_0 f \in W_p^2(\mathbb{R}_+^{n+1})$ is a solution to

$$\lambda w + \mathcal{A}(D)w = f \quad \text{in } \mathbb{R}_+^{n+1}.$$

This means that we obtain the solution to (7.1) as $u = w + z$ where $z \in W_p^2(\mathbb{R}_+^{n+1})$ is a solution to

$$\lambda z + \mathcal{A}(D)z = 0 \qquad\qquad \text{in } \mathbb{R}_+^{n+1}$$
$$\mathcal{B}(D)z = g - \mathcal{B}(D)w \qquad\qquad \text{on } \partial\mathbb{R}_+^{n+1}.$$

Hence we concentrate on the case $f = 0$ in (7.1) for the moment.

7.3 Partial Fourier Transform. We use the standard approach (see, e.g., [224], [87]) via partial Fourier transform with respect to the variable $x' \in \mathbb{R}^n$, which we denote by

$$\widehat{u}(\xi', y) = \int_{\mathbb{R}^n} e^{-ix'\xi'} u(x', y)\, dx'.$$

Then

$$(\widehat{\mathcal{A}(D)u})(\xi', y) = \sum_{k=0}^{2} \sum_{\beta \in \mathbb{N}_0^n, |\beta|=k} a_{(\beta, 2-k)}(\xi')^\beta (D_y^{2-k}\widehat{u})(\xi', y)$$

$$= \sum_{k=0}^{2} a_k(\xi')(D_y^{2-k}\widehat{u})(\xi', y)$$

where

$$a_k(\xi') := \sum_{|\beta|=k} a_{(\beta, 2-k)}(\xi')^\beta \qquad\qquad (7.2)$$

is a homogeneous polynominal of degree k for $k = 0, 1, 2$. In particular, $a_0 = a_{(0,\dots,0,2)}$ is independent of ξ'. Similarly, we have for the boundary operator

$$(\widehat{\mathcal{B}(D)u})(\xi', y) = \sum_{k=0}^{1} \sum_{\gamma \in \mathbb{N}_0^n, |\gamma|=k} b_{(\gamma, 1-k)}(\xi')^\gamma (D_y^{1-k}\widehat{u})(\xi', y)$$

$$= \sum_{k=0}^{1} b_k(\xi')(D_y^{1-k}\widehat{u})(\xi', y)$$

where

$$b_k(\xi') := \sum_{|\gamma|=k} b_{(\gamma,1-k)}(\xi')^\gamma$$

is a homogeneous polynominal of degree k for $k = 0, 1$. Precisely, $b_1(\xi') = \sum_{j=1}^n b_{(e_j,0)}\xi_j$ is homogeneous of degree 1 and $b_0 = b_{(0,\dots,0,1)}$ is independent of ξ'.

Thus we have transformed (7.1) with $f = 0$ into the equation

$$\lambda \widehat{u}(\xi', y) + \sum_{k=0}^{2} a_k(\xi')(D_y^{2-k}\widehat{u}(\xi', y)) = 0, \qquad (\xi', y) \in \mathbb{R}^{n+1}_+$$

$$b_0 D_y \widehat{u}(\xi', 0) + b_1(\xi')\widehat{u}(\xi', 0) = \widehat{g}(\xi', 0), \qquad \xi' \in \mathbb{R}^n.$$

For fixed $\xi' \in \mathbb{R}^n$ this is an ordinary differential equation on the half line $[0, \infty)$ for the function $y \mapsto \phi(y) := \widehat{u}(\xi', y)$:

$$(\lambda + a_2(\xi'))\phi(y) + a_1(\xi')D_y\phi(y) + a_0 D_y^2\phi(y) = 0, \quad y > 0$$
$$b_0 D_y\phi(0) + b_1(\xi')\phi(0) = \zeta \tag{7.3}$$

where $\zeta := \widehat{g}(\xi', 0)$.

7.4 The Lopatinskij-Shapiro Condition. Our second assumption besides (M, ω_0)-ellipticity of the operator $\mathcal{A}(D)$, i.e., our assumption on the boundary operator $\mathcal{B}(D)$, which we have postponed so far, is the so-called *Lopatinskij-Shapiro condition* in the following form:

> For any $\xi' \in \mathbb{R}^n$ and $\lambda \in \Sigma_\theta$ satisfying $|\xi'| + |\lambda| \neq 0$
> and for any $\zeta \in \mathbb{C}$ the problem (7.3) has a unique
> solution $\phi \in L_p(0, \infty)$. $\tag{7.4}$

Condition (7.4) implies

$$|b_0| + |b_1(\xi')| \neq 0 \text{ for all } \xi' \in \mathbb{R}^n,$$

since otherwise we would not get solutions for $\zeta \neq 0$. In particular, taking $\xi' = 0$, we obtain $b_0 \neq 0$ which means (recall $b_0 = b_{(0,\dots,0,1)}$) that $\mathcal{B}(D)$ is *non-tangential*, i.e., the component orthogonal to $\partial\mathbb{R}^{n+1}_+$ does not vanish. This is where the name oblique derivative stems from. Under the assumption $b_0 \neq 0$ the condition (7.4) is equivalent to the fact that the characteristic polynomial

$$a_0\mu^2 + a_1(\xi')\mu + a_2(\xi') + \lambda = 0 \tag{7.5}$$

of the equation (7.3) has two distinct roots $\mu_\pm$ with $Im\mu_+ > 0 > Im\mu_-$. Recall that $D_y = -i\frac{\partial}{\partial y}$ and that (7.5) is obtained by writing μ for D_y. Hence any solution of (7.3) in $L_p(0, \infty)$ is given by

$$y \mapsto c\, e^{iy\mu_+}$$

where $c \in \mathbb{C}$ is a constant. Here, of course, $\mu_+ = \mu_+(\xi', \lambda)$ depends on ξ' and λ, and c has to be determined by the boundary condition in (7.3) where we recall $\zeta = \widehat{g}(\xi', 0)$. This yields

$$c = \frac{\widehat{g}(\xi', 0)}{b_0 \mu_+(\xi', \lambda) + b_1(\xi')}.$$

Actually, writing $\lambda = e^{i\omega}\sigma^2$ where $\sigma \neq 0$, we have that $(\xi', \sigma) \mapsto \mu_+(\xi', e^{i\omega}\sigma^2)$ is positively homogeneous of degree 1: if $\rho > 0$ then $\mu_+(\rho\xi', e^{i\omega}\rho^2\sigma^2)$ is the unique root with positive imaginary part of the polynomial

$$a_0\mu^2 + a_1(\rho\xi')\mu + a_2(\rho\xi') + e^{i\omega}\rho^2\sigma^2 = \rho^2(a_0(\mu/\rho)^2 + a_1(\xi')(\mu/\rho) + e^{i\omega}\sigma^2)$$

where we used the homogeneity properties of a_1 and a_2. This equality yields $\mu_+(\xi', e^{i\omega}\sigma^2) = \rho^{-1}\mu_+(\rho\xi', e^{i\omega}\rho^2\sigma^2)$. For later purposes we note here that $(\xi', \sigma) \mapsto \mu_+(\xi', e^{i\omega}\sigma^2)$ is C^∞ on $(\mathbb{R}^n \times \mathbb{R}) \setminus \{(0, 0)\}$.

7.5 Representation of the Solution. Coming back to the ordinary differential equation, we have shown that the unique solution $\phi \in L_p(0, \infty)$ of (7.3) with $\zeta = \widehat{g}(\xi', 0)$ is given by

$$\phi(y) = \frac{e^{iy\mu_+(\xi', \lambda)}}{b_1(\xi') + b_0\mu_+(\xi', \lambda)}\widehat{g}(\xi', 0) =: m(\xi', \lambda, y)\widehat{g}(\xi', 0), \quad y > 0. \tag{7.6}$$

Recalling $\phi(y) = \widehat{u}(\xi', y)$ we thus have

$$\widehat{u}(\xi', y) = m(\xi', \lambda, y)\widehat{g}(\xi', 0), \quad (\xi', y) \in \mathbb{R}_+^{n+1}.$$

Taking the inverse Fourier transform with respect to ξ' yields the following representation for the solution u of (7.1) in case $f = 0$:

$$u(x', y) = \frac{1}{(2\pi)^n} \int_{\mathbb{R}^n} e^{ix'\xi'} m(\xi', \lambda, y)\widehat{g}(\xi', 0)d\xi' \tag{7.7}$$

for $(x', y) \in \mathbb{R}_+^{n+1}$. We write this as $u = M(\lambda)g$.

But the representation (7.7) is not suited to establish bounds on the resolvents of A_B. We use the following trick (borrowed from [66])

$$m(\xi', \lambda, y)\widehat{g}(\xi', 0) = -\int_0^\infty \partial_{\tilde{y}}(m(\xi', \lambda, y + \tilde{y})\widehat{g}(\xi', \tilde{y}))d\tilde{y}$$

$$= -\int_0^\infty (\partial_{\tilde{y}}m(\xi', \lambda, y + \tilde{y}))\widehat{g}(\xi', \tilde{y})d\tilde{y} - \int_0^\infty m(\xi', \lambda, y + \tilde{y})\widehat{(\partial_{\tilde{y}}g)}(\xi', \tilde{y})d\tilde{y}.$$

We denote the Laplacian in $\mathbb{R}^n$ by Δ_n and rewrite the first integral as

$$\int\limits_0^\infty \partial_{\tilde{y}} m(\xi',\lambda,y+\tilde{y})(|\lambda|+|\xi'|^2)^{-1/2}\big((|\lambda|-\Delta_n)^{\frac{1}{2}}g\big)\widehat{\ }(\xi',\tilde{y})d\tilde{y}.$$

By Fubini, we may write $L_p(\mathbb{R}^{n+1}_+) = L_p(\mathbb{R}^n, L_p(0,\infty))$ and obtain finally

$$M(\lambda)g = -K(\lambda)(|\lambda|-\Delta_n)^{\frac{1}{2}}g - L(\lambda)(\partial_y g),$$

where, denoting Fourier transform in $\mathbb{R}^n$ with respect to the variables $x' \to \xi'$ by $\mathcal{F}_n$,

$$K(\lambda) = \mathcal{F}_n^{-1}k(\xi',\lambda)\mathcal{F}_n, \quad L(\lambda) = \mathcal{F}_n^{-1}l(\xi',\lambda)\mathcal{F}_n,$$

and $k(\xi',\lambda)$, $l(\xi',\lambda)$ are operators $L_p(0,\infty) \to L_p(0,\infty)$ given by

$$(k(\xi',\lambda)h)(y) = \int\limits_0^\infty \partial_{\tilde{y}} m(\xi',\lambda,y+\tilde{y})(|\lambda|+|\xi'|^2)^{-1/2}h(\tilde{y})\,d\tilde{y}$$

$$(l(\xi',\lambda)h)(y) = \int\limits_0^\infty m(\xi',\lambda,y+\tilde{y})h(\tilde{y})\,d\tilde{y}.$$

The following proposition is essentially what we need to establish sectoriality and R-sectoriality in L_p for the operator A_B and small perturbations of it.

7.6 Lemma. *For $T \in \{K,L\}$ and $j,j' \in \{1,\ldots,n+1\}$ the sets*

$$\{\lambda T(\lambda) : \lambda \in \Sigma_{\pi-\theta}\}, \{|\lambda|^{1/2}\partial_{x_j}T(\lambda) : \lambda \in \Sigma_{\pi-\theta}\}, \{\partial_{x_j}\partial_{x_{j'}}T(\lambda) : \lambda \in \Sigma_{\pi-\theta}\}$$

are R-bounded in $B(L_p(\mathbb{R}^n, L_p(0,\infty)))$.

Proof. We write $T(\lambda) = \mathcal{F}_n^{-1}t(\xi',\lambda)\mathcal{F}_n$. Once again we apply the R-boundedness version of Mihlin's theorem, this time for operator-valued multipliers on $L_p(0,\infty)$, i.e., in the form 5.2 b). We write $\lambda = e^{i\omega}\sigma^2$. The operators $e^{i\omega}\sigma^2 t(\xi',e^{i\omega}\sigma^2)$ are integral operators on $L_p(0,\infty)$ having kernels of the form

$$\tau(\xi',\sigma)e^{-\varphi(\xi',\sigma)(y+\tilde{y})} \tag{7.8}$$

where $\varphi,\tau : \mathbb{R}^{n+1} \setminus \{0\} \to \mathbb{C}$ are C^∞-functions which are positively homogeneous of degree one. To see this, observe that

$$(\xi',\sigma) \mapsto \mu_+(\xi',e^{i\theta}\sigma^2), \quad (\xi',\sigma) \mapsto b_1(\xi') + b_0\mu_+(\xi',e^{i\theta}\sigma^2),$$

$$\text{and} \quad (\xi',\sigma) \mapsto (\sigma^2 + |\xi'|^2)^{1/2}$$

are C^∞ on $\mathbb{R}^{n+1} \setminus \{0\}$ and positively homogeneous of degree one, and recall the definition of m, k, and l.

Representing the operators $|\lambda|^{1/2}\partial_{x_j}T(\lambda)$ and $\partial_{x_j}\partial_{x_{j'}}T(\lambda)$ in the same way as Fourier multiplier operators $\sigma\partial_{x_j}T(e^{i\omega}\sigma^2)$, $\partial_{x_j}\partial_{x_{j'}}T(e^{i\omega}\sigma^2)$, we show that also their kernels have the structure (7.8) with the stated properties of τ and φ. Indeed, observe that an occurrence of $\partial_{x_{n+1}} = \partial_y$ means that we have to derive the kernel of the integral operator with respect to y which yields multiplication of the kernel by $i\mu_+(\xi', e^{i\omega})$ and this factor is positively homogeneous of degree one. Partial derivatives ∂_{x_j} for $j = 1, \ldots, n$ yield a multiplication of the kernel by $i\xi_j$ and this factor is also homogeneous of degree one.

The maps $\xi' \mapsto \lambda k(\xi', \lambda), \xi' \mapsto \lambda l(\xi', \lambda)$ are C^∞ on $\mathbb{R}^n \setminus \{0\}$ and we obtain derivatives $\partial_{\xi'}^\alpha$ by taking derivatives of the kernels. By induction on $|\alpha|$, it is not hard to see that

$$|\xi'|^{|\alpha|} \partial_{\xi'}^\alpha (\tau(\xi', \sigma) e^{-\varphi(\xi', \sigma)(y + \tilde{y})}) = \sum_{k=0}^{|\alpha|} \tau_k(\xi', \sigma)(y + \tilde{y})^k e^{-\varphi(\xi', \sigma)(y + \tilde{y})}$$

where $(\xi', \sigma) \mapsto \tau_k(\xi', \sigma)$ is C^∞ on $\mathbb{R}^{n+1} \setminus \{0\}$ and positively homogeneous of degree $k + 1$ for $k = 0, \ldots, |\alpha|$. Then the proof will be finished if we have shown that, for each $k \in \mathbb{N}_0$, the set of integral operators with kernels

$$\tau_k(\xi', \sigma)(y + \tilde{y})^k e^{-\varphi(\xi', \sigma)(y + \tilde{y})}, \quad (\xi', \sigma) \in \mathbb{R}^{n+1} \setminus \{0\}$$

is R-bounded on $L_p(0, \infty)$ where $\tau_k : \mathbb{R}^{n+1} \setminus \{0\} \mapsto \mathbb{C}$ is a C^∞-function which is positively homogeneous of degree $k + 1$. Fixing $k \in \mathbb{N}_0$ and letting

$$Z := \{\varphi(\xi', \sigma) : |\xi'|^2 + \sigma^2 = 1\}$$
$$W := \{\tau_k(\xi', \sigma) : |\xi'|^2 + \sigma^2 = 1\}$$

we see that Z, W are compact, and that Z is contained in the open right half plane.

Hence we consider the set of integral operators on $L_p(0, \infty)$ with kernels

$$r^{k+1} w(y + \tilde{y})^k e^{-rz(y + \tilde{y})}, \quad r > 0, w \in W, z \in Z.$$

For any $r > 0, w \in W, z \in Z$, and $h \in L_p(0, \infty)$ we have

$$\left| \int_0^\infty r^{k+1} w(y + \tilde{y})^k e^{-rz(y + \tilde{y})} h(\tilde{y}) \, d\tilde{y} \right|$$

$$\leq \int_{-\infty}^\infty r^{k+1} |w| |y - \tilde{y}|^k e^{-r(Rez)|y - \tilde{y}|} |\tilde{h}(-\tilde{y})| \, d\tilde{y}$$

where $\tilde{h}(y) := h(-y)$ for $y < 0$, $\tilde{h}(y) := 0$ for $y > 0$. Letting $b := \inf\{Rez : z \in Z\}$, $a := \sup\{|w| : w \in W\}$ we have $b > 0$, $a < \infty$, and it suffices to show that the set of integral operators on $L_p(\mathbb{R})$ with kernels

$$ar^{k+1} |y - \tilde{y}|^k e^{-br|y - \tilde{y}|}, \quad r > 0,$$

are R-bounded. Since

$$c_k := \sup_{y,\tilde{y}\in\mathbb{R}, r>0} (r|y-\tilde{y}|)^k e^{-br|y-\tilde{y}|/2} < \infty,$$

it is by Example 2.10 sufficient to show that the set of integral operators with kernels

$$re^{-\frac{b}{2}r|y-\tilde{y}|}, \qquad r > 0,$$

is R-bounded on $L_p(\mathbb{R})$. Here we may take $\frac{b}{2} = 1$ without loss of generality. We observe that

$$re^{-r|\cdot|} * h = \mathcal{F}^{-1}(\eta \mapsto \psi(r,\eta)\mathcal{F}h(\eta))$$

where $\psi(r,\eta) = \frac{cr^2}{r^2+\eta^2}$. The functions $\psi(r,\cdot)$ are C^∞ and satisfy

$$|\psi(r,\eta)| \le c,$$

$$|\eta\partial_\eta\psi(r,\eta)| = \left|\frac{2cr^2\eta^2}{(r^2+\eta^2)^2}\right| = 2c\frac{r^2}{r^2+\eta^2}\cdot\frac{\eta^2}{r^2+\eta^2} \le 2c.$$

This means that $\psi(r,\cdot)$ satisfies Mihlin's condition uniformly in $r > 0$. We use the R-boundedness version of Mihlin's theorem for scalar-valued multipliers 5.2 a) to conclude that the set $\{re^{r|\cdot|}* : r > 0\}$ is R-bounded on $L_p(\mathbb{R})$. The proof of the lemma is complete.   □

We recall that the solution of the full inhomogeneous problem (7.1) was obtained as

$$u = P(\lambda + A_{n+1})^{-1}E_0 f + M(\lambda)g - M(\lambda)\mathcal{B}(D)P(\lambda + A_{n+1})^{-1}E_0 f$$

where

$$M(\lambda) = -K(\lambda)(|\lambda| - \Delta_n)^{\frac{1}{2}} - L(\lambda)\partial_y.$$

Now we are ready to prove the following.

7.7 Theorem. *Let $\mathcal{A}(D)$ be a homogeneous second order operator with constant coefficients which is (M,ω_0)-elliptic. Let $\mathcal{B}(D)$ be a homogeneous first order operator with constant coefficients such that (A,B) satisfies the Lopatinskij-Shapiro condition (7.4). Let $\theta \in (\omega_0,\pi)$. For any $f \in L_p(\mathbb{R}^{n+1}_+)$, $g \in W_p^1(\mathbb{R}^{n+1}_+)$, and $\lambda \in \Sigma_{\pi-\theta}$ there exists a unique solution $u \in W_p^2(\mathbb{R}^{n+1}_+)$ of the boundary value problem*

$$(\lambda + A)u = f \qquad \text{in } \mathbb{R}^{n+1}_+$$
$$Bu = g \qquad \text{on } \partial\mathbb{R}^{n+1}_+.$$

Moreover, there is a constant $C > 0$ only depending on n, p, M, ω_0, θ such that, for all such finite families (f_j) in $L_p(\mathbb{R}^{n+1}_+)$, (g_j) in $W_p^1(\mathbb{R}^{n+1}_+)$, (λ_j) in $\Sigma_{\pi-\theta}$ and corresponding solutions (u_j) in $W_p^2(\mathbb{R}^{n+1}_+)$, we have

$$\sum_{k=0}^{2}\left\|\sum_{j}r_j\lambda_j^{1-k/2}\nabla^k u_j\right\|_{X_p^{(n+1)k}}$$

$$\leq C\left(\left\|\sum_{j}r_j f_j\right\|_{X_p} + \left\|\sum_{j}r_j\lambda_j^{1/2}g_j\right\|_{X_p} + \left\|\sum_{j}r_j\nabla g_j\right\|_{X_p^{n+1}}\right) \tag{7.9}$$

where $X_p := L_2([0,1], L_p(\mathbb{R}_+^{n+1}))$, the r_j are the Rademacher functions on $[0,1]$, and $\nabla_2 := (\partial_j\partial_k)_{j,k=1}^{n+1}$.

Proof. Let $k \in \{0,1,2\}$ and $\alpha \in \mathbb{N}_0^{n+1}$ such that $|\alpha| = k$. We set $l := 1 - k/2$. Take finite sequences (λ_j) in $\Sigma_{\pi-\theta}$, (f_j) in $L_p(\mathbb{R}_+^{n+1})$ and (g_j) in $W_p^1(\mathbb{R}_+^{n+1})$. For any j, we let

$$h_j := \mathcal{B}(D)P(\lambda_j + A_{n+1})^{-1}E_0 f_j.$$

Denoting the solution of (7.1) with data (λ_j, f_j, g_j) in place of (λ, f, g) by u_j we have

$$\begin{aligned}
\lambda_j^l D^\alpha u_j &= P(\lambda_j^l D^\alpha(\lambda_j + A_{n+1})^{-1})E_0 f_j \\
&\quad + \lambda_j^l D^\alpha K(\lambda_j)(|\lambda_j| - \Delta_n)^{\frac{1}{2}}(g_j - h_j) \\
&\quad - \lambda_j^l D^\alpha L(\lambda_j)\partial_y(g_j - h_j).
\end{aligned}$$

By using arguments similar to those in Section 6 the set of operators

$$\{\lambda^l D^\alpha(\lambda + A_{n+1})^{-1} : \lambda \in \Sigma_{\pi-\theta}\} \tag{7.10}$$

is R-bounded on $L_p(\mathbb{R}_+^{n+1})$. Indeed, the symbol

$$(\xi, \sigma) \mapsto i\eta\,\mathcal{B}(\xi)(e^{i\theta}\sigma^2 + \mathcal{A}(\xi))^{-1}$$

is C^∞ on $\mathbb{R}^{n+1} \setminus \{0\}$ and homogeneous of degree 0, and we apply the R-boundedness version of Mihlin's theorem 5.2 once again, this time for scalar-valued multipliers on $\mathbb{R}^{n+1}$.

By R-boundedness of (7.10) and Lemma 7.6 we obtain

$$\begin{aligned}
\left\|\sum_{j}r_j\lambda_j^l D^\alpha u_j\right\|_{X_p} &\leq C_1\left(\left\|\sum_{j}r_j f_j\right\|_{X_p} + \left\|\sum_{j}r_j(|\lambda_j| - \Delta_n)^{1/2}(g_j - h_j)\right\|_{X_p}\right. \\
&\quad \left. + \left\|\sum_{j}r_j\partial_y(g_j - h_j)\right\|_{X_p}\right),
\end{aligned}$$

recall that r_j are the Rademacher functions and $X_p := L_2([0,1], L_p(\mathbb{R}_+^{n+1}))$. Now we use the definition of the h_j and obtain

$$(|\lambda_j| - \Delta_n)^{1/2}h_j = P(|\lambda_j| - \Delta_n)^{1/2}\mathcal{B}(D)(\lambda_j + A_{n+1})^{-1}E_0 f_j.$$

The set of operators

$$\{(|\lambda| - \Delta_n)^{1/2}\mathcal{B}(D)(\lambda + A_{n+1})^{-1} : \lambda \in \Sigma_{\pi-\theta}\}$$

is R-bounded on $L_p(\mathbb{R}^{n+1})$ by arguments as in Section 6. Indeed, the symbol

$$(\xi, \sigma) \mapsto (\sigma^2 + |\xi'|^2)^{\frac{1}{2}}\mathcal{B}(\xi)(e^{i\theta}\sigma^2 + \mathcal{A}(\xi))^{-1}$$

is C^∞ on $\mathbb{R}^{n+1} \setminus \{0\}$ and homogeneous of degree 0, and we apply 5.2 again. Moreover,

$$\partial_y h_j = P(\partial_y \mathcal{B}(D)(\lambda_j + A_{n+1})^{-1}E_0 f_j),$$

and the set of operators

$$\{\partial_y \mathcal{B}(D)(\lambda + A_{n+1})^{-1} : \lambda \in \Sigma_{\pi-\theta}\}$$

is R-bounded on $L_p(\mathbb{R}^{n+1})$, again by arguments similar to those used above. Thus we obtain the estimate

$$\left\|\sum_j r_j \lambda_j^l D^\alpha u_j\right\|_{X_p} \le C_2\left(\left\|\sum_j r_j f_j\right\|_{X_p} + \left\|\sum_j r_j(|\lambda_j| - \Delta_n)^{1/2}g_j\right\|_{X_p}\right.$$
$$\left. + \left\|\sum_j r_j \partial_y g_j\right\|_{X_p}\right).$$

$$(7.11)$$

For a simplification of the terms involving the functions g_j we apply the following lemma.

7.8 Lemma. *The set of operators*

$$\{(|\lambda| - \Delta_n)^{1/2}(|\lambda|^{1/2} + (-\Delta_n)^{1/2})^{-1} : \lambda \in \Sigma_{\pi-\theta}\}$$

is R-bounded on $L_p(\mathbb{R}_+^{n+1})$.

Proof. We write again $L_p(\mathbb{R}_+^{n+1}) = L_p(\mathbb{R}^n, L_p(0, \infty))$ and use the R-boundedness version of Mihlin's multiplier theorem in the form 5.2 b). We thus have to show that the family of multipliers

$$m_\lambda : \xi' \mapsto \frac{(|\lambda| + |\xi'|^2)^{1/2}}{|\lambda|^{1/2} + |\xi'|}$$

satisfies Mihlin's conditions uniformly in $\lambda \in \Sigma_{\pi-\theta}$. Writing again $\lambda = e^{i\omega}\sigma^2$ we note that $m(\xi', \sigma) := m_{e^{i\omega}\sigma^2}(\xi')$ is positively homogeneous of degree zero. In contrast to the situations we encountered so far, this time m is not C^∞ on $\mathbb{R}^{n+1}\setminus\{0\}$. Thus we have to do a little calculation. Clearly, $m(\xi', \sigma) = k(\sigma, |\xi'|)$ where

$$k : (\sigma, \tau) \mapsto \frac{(\sigma^2 + \tau^2)^{1/2}}{\sigma + \tau}$$

is C^∞ on $[0, \infty)^2 \setminus \{(0, 0)\}$ and positively homogeneous of degree zero. By induction on $k = |\alpha|$ we obtain that, for any multiindex $\alpha \in \mathbb{N}_0^n \setminus \{0\}$, we have

$$\partial_\xi^\alpha m(\xi', \sigma) = \sum_{\nu=1}^{k} \partial_\tau^\nu k(\sigma, |\xi'|) \phi_{\alpha,\nu}(\xi')$$

where $\xi' \mapsto \phi_{\alpha,\nu}(\xi')$ is a sum of products of partial derivatives of $\xi' \mapsto |\xi'|$ which is positively homogeneous of degree $k - \nu$. Moreover $\phi_{\alpha,\nu}$ is C^∞ on $\mathbb{R}^n \setminus \{0\}$. Hence

$$|\xi'|^{|\alpha|} \partial_\xi^\alpha m(\xi', \sigma) = \sum_{\nu=1}^{k} |\xi'|^\nu \partial_\tau^\nu k(\sigma, |\xi'|) \cdot |\xi'|^{k-\nu} \phi_{\alpha,\nu}(\xi') \tag{7.12}$$

where each factor $|\xi'|^\nu \partial_\tau^\nu k(\sigma, |\xi'|)$ is positively homogeneous of degree zero and continuous on $\mathbb{R}^{n+1} \setminus \{0\}$, and each factor $|\xi'|^{k-\nu} \phi_{\alpha,\nu}(\xi')$ is positively homogeneous of degree zero and continuous on $\mathbb{R}^n \setminus \{0\}$. When taking the sup in (7.12) over $\xi' \in \mathbb{R}^n \setminus \{0\}$ and $\sigma > 0$ we thus may take the sup over $|\xi'|^2 + \sigma^2 = 1$ for the first factors and the sup over $|\xi'| = 1$ for the second factors. By continuity all those sup's are finite and we conclude that $m(\cdot, \sigma)$ satisfies Mihlin's conditions uniformly in $\sigma > 0$. □

We continue the proof of Theorem 7.7 by applying the lemma to the estimate (7.11). This yields

$$\left\| \sum_j r_j \lambda_j^l D^\alpha u_j \right\|_{X_p} \le C_3 \left(\left\| \sum_j r_j f_j \right\|_{X_p} + \left\| \sum_j r_j |\lambda_j|^{1/2} g_j \right\|_{X_p} \right.$$
$$\left. + \left\| \sum_j r_j (-\Delta_n)^{1/2} g_j \right\|_{X_p} + \left\| \sum_j r_j \partial_y g_j \right\|_{X_p} \right).$$

Using now that $\|(-\Delta_n)^{1/2} g\|_{L_p} \sim \|\nabla_n g\|_{L_p^n}$ where ∇_n denotes the gradient with respect to x' we get

$$\left\| \sum_j r_j \lambda_j^l D^\alpha u_j \right\|_{X_p}$$
$$le\ C_4 \left(\left\| \sum_j r_j f_j \right\|_{X_p} + \left\| \sum_j r_j |\lambda_j|^{1/2} g_j \right\|_{X_p} + \left\| \sum_j r_j \nabla g_j \right\|_{X_p^{n+1}} \right).$$

as asserted. □

Letting $g_j = 0$ in Theorem 7.7 and recalling the definition of the operator A_B we immediately get the following.

7.9 Corollary. *The operator A_B is R-sectorial in $L_p(\mathbb{R}_+^{n+1})$ and, for all $k = 0, 1, 2$ and $\alpha \in \mathbb{N}_0^{n+1}$ with $|\alpha| = k$, the set*

$$\{|\lambda|^{1-k/2} D^\alpha (\lambda + A_B)^{-1} : \lambda \in \Sigma_{\pi-\theta}\}$$

is R-bounded in $L_p(\mathbb{R}_+^{n+1})$. In particular, $\omega_R(A_B) \le \theta$ for any $\theta \in (\omega_0, \pi)$, i.e., $\omega_R(A_B) \le \omega_0$.

Step II: Small Perturbations in the Half Space

The following lemma will provide the technical key for a substitute of the perturbation theorem 6.5 for sectorial operators we used in the previous section. As we shall see below, the lemma fits exactly the situation we have for boundary value problems.

7.10 Lemma. *Let X, Y, Z be Banach spaces, and let $S : X \to Y$, $Q : Y \to Z$ be bounded linear operators. Assume that, for every $g \in Y$, there exists a unique $u \in X$ such that*

$$QSu = Qg. \tag{7.13}$$

Assume that we have, for some constant $C > 0$, the estimate

$$\|u\|_X \le C\|g\|_Y, \quad g \in Y. \tag{7.14}$$

If $P : X \to Y$ is a bounded linear operator satisfying $\|P\| < 1/C$, then, for every $h \in Y$ there is a unique $v \in X$ such that

$$Q(S + P)v = Qh. \tag{7.15}$$

Moreover, we have the estimate

$$\|v\|_X \le \frac{C}{1 - \|P\|C} \|h\|_Y, \quad h \in Y. \tag{7.16}$$

Proof. Define the operator $T : Y \to X$, $g \mapsto Tg := u$, where u is the unique solution of (7.13). Then T is a bounded linear operator, $QST = Q$, $\|T\| \le C$, and $\|PT\| \le \|P\|C < 1$. Hence $V := T \sum_{k=0}^{\infty}(-PT)^k$ defines a bounded linear operator in X, and we have $\|V\| \le C(1 - \|P\|C)^{-1}$. Moreover,

$$Q(S + P)V = QST \sum_{k=0}^{\infty}(-PT)^k + QPT \sum_{k=0}^{\infty}(-PT)^k$$

$$= Q\left(\sum_{k=0}^{\infty}(-PT)^k - \sum_{k=1}^{\infty}(-PT)^k \right)$$

$$= Q$$

which means that, for any $h \in Y$, the vector $v := Vh$ is a solution of (7.15). This solution satisfies (7.16).

To prove uniqueness, we take $v \in X$ such that $Q(S + P)v = 0$. Then $QSv = -QPv$, and by $-Pv \in Y$ and the assumed uniqueness for (7.13) we obtain $v = -TPv$. Hence $\|TP\| \le C\|P\| < 1$ yields $v = 0$.

Thus, for any $h \in Y$, the vector $v = Vh$ is the unique solution of (7.15). $\qquad\square$

To make our approach more transparent, we first study boundedness and sectoriality in L_p, and then we study R-boundedness and R-sectoriality. By Theorem 7.7 we know the following for homogeneous constant coefficient operators A_0, B_0 where A_0 is (M, ω_0)–elliptic and (A_0, B_0) satisfy the Lopatinskij-Shapiro condition:

For any $f \in L_p(\mathbb{R}_+^{n+1})$, $g \in W_p^1(\mathbb{R}_+^{n+1})$, and $\lambda \in \Sigma_{\pi-\theta}$ there exists a unique solution $u \in W_p^2(\mathbb{R}_+^{n+1})$ of the boundary value problem

$$(\lambda + A_0)u = f \quad \text{in } \mathbb{R}_+^{n+1}$$
$$B_0 u = g \quad \text{on } \partial\mathbb{R}_+^{n+1}$$

and we have the estimate

$$\sum_{k=0}^{2} \|\lambda^{1-k/2}\nabla^k u\|_{L_p^{(n+1)k}} \leq C(\|f\|_p + \|\lambda^{1/2}g\|_p + \|\nabla g\|_p) \tag{7.17}$$

for a constant $C > 0$ only depending on n, p, M, ω_0, θ. We simply get (7.17) by considering one-element families in (7.9). To avoid clumsy notation we shall write $\|\cdot\|_p$ in place of $\|\cdot\|_{L_p^\nu}$, $\nu \in \mathbb{N}$.

We now want to perturb this situation by considering operators $A = A_0 + A_\epsilon$, $B = B_0 + B_\epsilon$ in place of A_0, B_0, respectively, where A_ϵ, B_ϵ have the same form as A_0, B_0 but with coefficients depending on the space variable x.

To this end we apply the lemma above and let $X := W_p^2(\mathbb{R}_+^{n+1})$ equipped with the norm (depending on λ!) on the left hand side of (7.17), i.e., with the norm

$$|u|_{p,\lambda} := \sum_{k=0}^{2} \|\lambda^{1-k/2}\nabla^k u\|_p.$$

We let $Y := L_p(\mathbb{R}_+^{n+1}) \times W_p^1(\mathbb{R}_+^{n+1})$ equipped with the norm

$$\|(f, g)\|_{p,\lambda} := \|f\|_p + \|\lambda^{1/2}g\|_p + \|\nabla g\|_p.$$

Moreover, Q is the restriction $(f, g) \mapsto g|_{\partial\mathbb{R}_+^{n+1}}$ and Z is some space (no matter which space precisely) such that $Q : Y \to Z$ is bounded. The operator $S : X \to Y$ is given by $Su := ((\lambda + A_0)u, B_0 u)$, and P is given by $Pu := (A_\epsilon u, B_\epsilon u)$. We write $A_\epsilon = \sum_{|\alpha|=2} a_{\epsilon,\alpha} D^\alpha$, $B_\epsilon = \sum_{|\beta|=1} b_{\epsilon,\beta} D^\beta = b_\epsilon^t \nabla$ and let $a_\epsilon := (a_{\epsilon,\alpha})_\alpha$. Furthermore we let $\|a_\epsilon\|_\infty = \sum_\alpha \|a_{\epsilon,\alpha}\|_\infty$, $\|b_\epsilon\|_\infty := \sum_j \|b_{\epsilon,j}\|_\infty$ and $\|\nabla b_\epsilon\|_\infty := \sum_{j,k} \|\partial_k b_{\epsilon,j}\|_\infty$.

It rests to show that $\|P\|_{X \to Y}$ is small, uniformly in λ. We have

$$\|Pu\|_{p,\lambda} = \|A_\epsilon u\|_p + |\lambda|^{1/2}\|B_\epsilon u\|_p + \|\nabla B_\epsilon u\|_p$$
$$\leq \|a_\epsilon\|_\infty \|\nabla^2 u\|_p + \|b_\epsilon\|_\infty |\lambda|^{1/2}\|\nabla u\|_p + \|\nabla B_\epsilon u\|_p.$$

For the third term in the last line we recall $B_\epsilon u = b_\epsilon^t \nabla u$ which yields

$$\|\nabla B_\epsilon u\|_p \leq \|\nabla b_\epsilon\|_\infty \|\nabla u\|_p + \|b_\epsilon\|_\infty \|\nabla^2 u\|_p.$$

Hence we have

$$\|Pu\|_{p,\lambda} \leq \left(\|a_\epsilon\|_\infty + \|b_\epsilon\|_\infty + \frac{\|\nabla b_\epsilon\|_\infty}{|\lambda|^{1/2}} \right) |u|_{p,\lambda}.$$

Application of Lemma 7.10 now yields

7.11 Theorem. *There is a constant $\epsilon > 0$ (simply take $\epsilon := (6C)^{-1}$ with C from (7.17)) such that, whenever $\|a_\epsilon\|_\infty \leq \epsilon$, $\|b_\epsilon\|_\infty \leq \epsilon$, $\lambda \in \Sigma_{\pi-\theta}$, and $|\lambda| \geq (\|\nabla b_\epsilon\|_\infty/\epsilon)^2$, then the boundary value problem*

$$\begin{aligned} (\lambda + A_0 + A_\epsilon)u &= f &&\text{in } \mathbb{R}^{n+1}_+ \\ (B_0 + B_\epsilon)u &= g &&\text{on } \partial\mathbb{R}^{n+1}_+ \end{aligned}$$

has for any $f \in L_p(\mathbb{R}^{n+1}_+)$, $g \in W^1_p(\mathbb{R}^{n+1}_+)$ a unique solution $u \in W^2_p(\mathbb{R}^{n+1}_+)$, and that we have the estimate

$$\sum_{k=0}^{2} \|\lambda^{1-k/2}\nabla^k u\|_p \leq 2C(\|f\|_p + \|\lambda^{1/2}g\|_p + \|\nabla g\|_p) \tag{7.18}$$

where C is the same constant as in (7.17).

7.12 Corollary. *Taking $g = 0$ we obtain that the operator $\nu + A_B$, where $\nu := (\|\nabla b_\epsilon\|/\epsilon)^2$ and A_B is given by the restriction of $A = A_0 + A_\epsilon$ to the domain*

$$D(A_B) := \{u \in W^2_p(\mathbb{R}^{n+1}_+) : Bu = (B_0 + B_\epsilon)u = 0 \text{ on } \partial\mathbb{R}^{n+1}_+\},$$

is a sectorial operator in $L_p(\mathbb{R}^{n+1}_+)$ satisfying $\omega(\nu + A_B) \leq \theta$.

Now we turn to R-sectoriality. This time we let $X := Rad_2^m(W^2_p(\mathbb{R}^{n+1}_+))$ equipped with the norm – which now depends on $\Lambda := (\lambda_1, \ldots, \lambda_m)!$ – on the left hand side of (7.9), i.e., with the norm

$$|(u_j)|_{p,\Lambda} := \sum_{k=0}^{2} \| \sum_j r_j \lambda_j^{1-k/2} \nabla^k u_j \|$$

where we write $\| \cdot \|$ for the norm in $L_2([0,1], L_p(\mathbb{R}^{n+1}_+))$ or in

$$L_2([0,1], L_p(\mathbb{R}^{n+1}_+, \mathbb{C}^{(n+1)^k})) = L_2([0,1], L_p(\mathbb{R}^{n+1}_+))^{(n+1)^k}.$$

We let $Y := Rad_2^m(L_p(\mathbb{R}^{n+1}_+) \times W^1_p(\mathbb{R}^{n+1}_+))$ equipped with the norm

$$\|(f_j, g_j)\|_{p,\Lambda} := \| \sum_j r_j f_j \| + \| \sum_j r_j \lambda_j^{1/2} g_j \| + \| \sum_j r_j \nabla g_j \|.$$

Moreover, Q is now the restriction $((f_j, g_j))_j \mapsto (g_j|_{\partial\mathbb{R}^{n+1}_+})_j$ and Z is some space such that $Q : Y \to Z$ is bounded (we may take $Z = Rad_2^m(\tilde{Z})$ where

$\widetilde{Z}$ may be any space such that the restriction $(f,g) \mapsto g|_{\partial \mathbb{R}^{n+1}_+}$ is continuous $L_p(\mathbb{R}^{n+1}_+) \times W^1_p(\mathbb{R}^{n+1}_+) \to \widetilde{Z})$.

The operator $S : X \to Y$ is now given by $S((u_j)_j) := \big(((\lambda_j + A_0)u_j, B_0 u_j)\big)_j$, and P is given by $P((u_j)_j) := ((A_\epsilon u_j, B_\epsilon u_j))_j$.

It rests to show that $\|P\|_{X \to Y}$ is small, uniformly in $\Lambda \in \Sigma^m_{\pi-\theta}$ and $m \in \mathbb{N}$. We have

$$\|P((u_j)_j)\|_{p,\Lambda} = \|\sum_j r_j A_\epsilon u_j\| + \|\sum_j r_j \lambda_j^{1/2} B_\epsilon u_j\| + \|\sum_j r_j \nabla B_\epsilon u_j\|$$

$$= \|A_\epsilon \sum_j r_j u_j\| + \|B_\epsilon \sum_j r_j \lambda_j^{1/2} u_j\| + \|\nabla B_\epsilon \sum_j r_j u_j\|$$

$$\leq \|a_\epsilon\|_\infty \|\sum_j r_j \nabla^2 u_j\| + \|b_\epsilon\|_\infty \|\sum_j r_j \lambda_j^{1/2} \nabla u_j\|$$

$$+ \|\nabla B_\epsilon \sum_j r_j u_j\|.$$

For the third term in the last line we again observe $B_\epsilon u = b_\epsilon^t \nabla u$ which yields

$$\|\nabla B_\epsilon \sum_j r_j u_j\| \leq \|\nabla b_\epsilon\|_\infty \|\sum_j r_j \nabla u_j\| + \|b_\epsilon\|_\infty \|\sum_j r_j \nabla^2 u_j\|.$$

By the contraction principle Proposition 2.5 we have

$$\|\sum_j r_j \nabla u_j\| = \|\sum_j r_j \lambda_j^{-1/2} \lambda_j^{1/2} \nabla u_j\| \leq \frac{2}{(\inf_j |\lambda_j|)^{1/2}} \|\sum_j r_j \lambda_j^{1/2} \nabla u_j\|.$$

Hence we have

$$\|P(u_j)\|_{p,\Lambda} \leq \left(\|a_\epsilon\|_\infty + \|b_\epsilon\|_\infty + \frac{2\|\nabla b_\epsilon\|_\infty}{\inf_j(|\lambda_j|^{1/2})} \right) |(u_j)|_{p,\Lambda}.$$

Application of Lemma 7.10 now yields

7.13 Theorem. *There is a constant $\epsilon > 0$ (simply take $\epsilon := (6C)^{-1}$) such that, whenever $\|a_\epsilon\|_\infty \leq \epsilon$, $\|b_\epsilon\|_\infty \leq \epsilon$, (λ_j) is a finite sequence in $\Sigma_{\pi-\theta} \cap \{|\lambda| \geq (2\|\nabla b_\epsilon\|_\infty/\epsilon)^2\}$, (f_j) and (g_j) are finite sequences in $L_p(\mathbb{R}^{n+1}_+)$ and $W^1_p(\mathbb{R}^{n+1}_+)$, respectively, then the solutions (u_j) to the boundary value problems*

$$(\lambda + A_0 + A_\epsilon)u_j = f_j \qquad \text{in } \mathbb{R}^{n+1}_+$$
$$(B_0 + B_\epsilon)u_j = g_j \qquad \text{on } \partial\mathbb{R}^{n+1}_+$$

satisfy

$$\sum_{k=0}^2 \|\sum_j r_j \lambda_j^{1-k/2} \nabla^k u_j\| \leq 2C(\|\sum_j r_j f_j\| + \|\sum_j r_j \lambda_j^{1/2} g_j\| + \|\sum_j r_j \nabla g_j\|)$$

$$(7.19)$$

where C is the same constant as in (7.9) and $\|\cdot\|$ denotes the norm in $L_2([0,1], L_p(\mathbb{R}^{n+1}_+))^\nu$, $\nu \in \mathbb{N}$.

7.14 Corollary. *Taking $g_j = 0$ we obtain that the operator $\nu + A_B$ is R-sectorial in $L_p(\mathbb{R}^{n+1}_+)$ and $\omega_R(\nu + A_B) \leq \theta$ where $\nu := (2\|\nabla b_\epsilon\|/\epsilon)^2$ and A_B is defined in Corollary 7.12.*

Step III: Localization and Coordinate Transformation

Let Ω be an open subset of $\mathbb{R}^{n+1}$. The main idea is to use the same localization scheme we used in Section 6. Let $r \in (0, \delta)$ and φ be a C^∞-function satisfying supp $\varphi \subset Q := (-r, r)^{n+1}$, $0 \leq \varphi \leq 1$ and

$$\sum_{l \in r\mathbb{Z}^{n+1}} \varphi_l^2(x) = 1, \quad x \in \mathbb{R}^{n+1}$$

where $\varphi_l(x) := \varphi(x - l)$. We let $Q_l := l + Q$ for $l \in r\mathbb{Z}^{n+1}$, and $\Gamma := \{l \in r\mathbb{Z}^{n+1} : Q_l \cap \Omega \neq \emptyset\}$. Then Γ splits into the set Γ_0 of l such that $Q_l \subset \Omega$ and the set Γ_1 of l such that $Q_l \cap \partial\Omega \neq \emptyset$.

For $l \in \Gamma_0$ the situation is as in Section 6. The new feature is the situation for $l \in \Gamma_1$. We restrict ourselves here to the case that $\partial\Omega$ is compact. Then Γ_1 is finite and Ω is either bounded or has bounded complement. For each $l \in \Gamma_1$, we have to apply a coordinate transformation $\Phi_l : Q_l \to \mathbb{R}^{n+1}$ that maps $\Omega \cap Q_l$ into $\mathbb{R}^{n+1}_+$, $\partial\Omega \cap Q_l$ into $\partial\mathbb{R}^{n+1}_+$ and $Q_l \setminus \overline{\Omega}$ into $\mathbb{R}^{n+1}_-$. The coordinate transformation Φ_l also transforms the operator in $\Omega \cap Q_l$ and the boundary condition on $\partial\Omega \cap Q_l$.

In the following we shall study this transformation for fixed l which we shall drop in notation. Moreover, it is sufficient to study the situation up to a rotation of $\mathbb{R}^{n+1}$. In order to apply the localization scheme from Section 6 we transform the boundary condition *including lower order terms*. After we have done that, it is sufficient to apply perturbation theorems for R-sectorial operators and we do not have to resort to Lemma 7.10 again.

7.15 The Bended Halfspace. Let $\Omega \subset \mathbb{R}^{n+1}$ be such that there is a C^2-diffeomorphism Φ mapping Ω onto $\mathbb{R}^{n+1}_+$ and $\partial\Omega$ onto $\partial\mathbb{R}^{n+1}_+$. Suppose that $\Phi(x) = x$ outside a compact set and that Φ is an admissible coordinate transformation in the sense of [224, Def.2.9] for a point $x_0 \in \partial\Omega$, i.e., the outer normal of Ω in x_0 is mapped to an outer normal of $\mathbb{R}^{n+1}_+$.

Suppose that $A := \sum_{j,k} a_{jk}(x)\partial_j\partial_k$ is a homogeneous second order operator which is (M, ω_0)-elliptic and satisfies

$$\sum_{j,k} |a_{jk}(x) - a_{jk}(y)| \leq \eta, \quad x, y \in \Omega,$$

where η is to be specified later. The boundary condition is given by a first order operator $B := \sum_j b_j\partial_j + c$ where we assume that $b := (b_1, \ldots, b_{n+1})$ and c are C^1_b-functions on $\overline{\Omega}$ and that

$$\sum_j |b_j(x) - b_j(y)| \leq \eta, \quad x, y \in \overline{\Omega}.$$

Moreover, we assume that the Lopatinskij-Shapiro condition is satisfied in each boundary point. For details we refer to [224, pp.149-157].

The operator $J : f \mapsto f \circ \Phi$ maps functions on $\mathbb{R}_+^{n+1}$ to functions on Ω and its inverse is the operator $J^{-1} : g \mapsto g \circ \Phi^{-1}$. We denote $\Phi(x) = (\Phi_1(x), \ldots, \Phi_{n+1}(x))$ and calculate

$$\partial_j(u \circ \Phi)(x) = \sum_\nu (\partial_\nu u)(\Phi(x))\partial_j\Phi_\nu(x),$$

$$\partial_j\partial_k(u \circ \Phi)(x) = \sum_{\nu,\mu} (\partial_\nu\partial_\mu u)(\Phi(x))\partial_j\Phi_\mu(x)\partial_k\Phi_\nu(x)$$
$$+ \sum_\nu (\partial_\nu u)(\Phi(x))\partial_j\partial_k\Phi_\nu(x).$$

This means, on the one hand, that J is maps isomorphically $L_p(\mathbb{R}_+^{n+1}) \to L_p(\Omega)$, $W_p^1(\mathbb{R}_+^{n+1}) \to W_p^1(\Omega)$, and $W_p^2(\mathbb{R}_+^{n+1}) \to W_p^2(\Omega)$. On the other hand, we see that the operator $\widetilde{A} := J^{-1}AJ$ in $L_p(\mathbb{R}_+^{n+1})$ is given by

$$\widetilde{A}u(x) = \sum_{\nu,\mu} \widetilde{a}_{\nu\mu}(x)\partial_\nu\partial_\mu u(x) + \sum_\nu \widetilde{a}_\nu(x)\partial_\nu u(x)$$

where

$$\widetilde{a}_{\nu\mu}(x) = \left(\sum_{j,k} a_{jk}\partial_j\Phi_\mu\partial_k\Phi_\nu\right)(\Phi^{-1}(x)) = \left((\nabla\Phi_\mu)^t(a_{jk})(\nabla\Phi_\nu)\right)(\Phi^{-1}(x))$$

$$\widetilde{a}_\nu = \left(\sum_{j,k} a_{jk}\partial_j\partial_k\Phi_\nu\right)(\Phi^{-1}(x)) = (A\Phi_\nu)(\Phi^{-1}(x)).$$

The boundary operator $\widetilde{B} := J^{-1}BJ$ is given by

$$\widetilde{B}u(x) = \sum_\nu \widetilde{b}_\nu(x)\partial_\nu u(x) + c(\Phi^{-1}(x))u(x)$$

where

$$\widetilde{b}_\nu = \left(\sum_j b_j\partial_j\Phi_\nu\right)(\Phi^{-1}(x)).$$

This means that $J^{-1}A_BJ = \widetilde{A}_{\widetilde{B}}$ and this operator may be studied as in Step II. The only difference is the first order term in $\widetilde{A}$ and the zero order term in $\widetilde{B}$. The first order term in $\widetilde{A}$ may be treated by the perturbation theorem for R-sectorial operators 6.7, but the zero order term in $\widetilde{B}$ changes the domain of the operator. We let $\widetilde{x}_0 := \Phi(x_0)$ where $x_0 \in \partial\Omega$ is the point fixed at the beginning, and define the constant coefficient operators

$$\widetilde{A}_0 u(x) := \sum_{\nu,\mu} \widetilde{a}_{\nu\mu}(\widetilde{x}_0)\partial_\nu\partial_\mu u(x)$$

and

$$\widetilde{B}_0 u(x) := \sum_\nu \widetilde{b}_\nu(\widetilde{x}_0) \partial_\nu u(x).$$

Moreover, we let $\widetilde{A}_\epsilon := \widetilde{A} - \widetilde{A}_0$, $\widetilde{B}_\epsilon := \widetilde{B} - \widetilde{B}_0$. Recall that, by assumption and [224, Thm.11.3], $\widetilde{B}_0$ satisfies the Lopatinskij-Shapiro condition with respect to $\widetilde{A}_0$. Observe that $\widetilde{A}_0$ is $(\widetilde{M}, \omega_0)$-elliptic where $\widetilde{M}$ is a suitable constant. We denote the first order coefficient of $\widetilde{B}_\epsilon$ by b_ϵ and the zero order coefficient by c_ϵ $(= c \circ \Phi^{-1})$. We apply Lemma 7.10 in the same setting as in Step II: We let $S((u_j)_j) := ((\widetilde{A}_0 u_j, \widetilde{B}_0 u_j))_j$ and have the required estimate by (7.9) for $\widetilde{A}_0$, $\widetilde{B}_0$. We let $P((u_j)_j) := ((\widetilde{A}_\epsilon u_j, \widetilde{B}_\epsilon u_j))_j$ and have to show that $\|P\|_{X \to Y}$ is small, uniformly in Λ. Similar as before, we have

$$\|P((u_j)_j)\|_{p,\Lambda} \leq \left(\|(\widetilde{a}_{\nu\mu} - \widetilde{a}_{\nu\mu}(\widetilde{x}_0))\|_\infty + \|(\widetilde{b}_\nu - \widetilde{b}_\nu(\widetilde{x}_0))\|_\infty \right) |(u_j)_j|_{p,\Lambda}$$

$$+ \left(\|(\widetilde{a}_\nu)\|_\infty + \|\nabla b_\epsilon\|_\infty + \|c_\epsilon\|_\infty \right) \| \sum_j r_j \nabla u_j \|$$

$$+ \|\nabla c_\epsilon\|_\infty \| \sum_j r_j u_j \|.$$

An application of the contraction principle as before yields by our assumptions

$$\|P((u_j)_j)\|_{p,\Lambda} \leq 2 \left(\eta + \frac{\|(\widetilde{a}_\nu)\|_\infty + \|\nabla b_\epsilon\|_\infty + \|c_\epsilon\|_\infty}{\inf_j |\lambda_j|^{1/2}} + \frac{\|\nabla c_\epsilon\|_\infty}{\inf_j |\lambda_j|} \right) |(u_j)_j|_{p,\Lambda}.$$

We thus have proved

7.16 Theorem. *There is a constant η (we may take $\eta := (12\widetilde{C})^{-1}$ where $\widetilde{C}$ is the constant from (7.9) for the operators $\widetilde{A}_0$ and $\widetilde{B}_0$) such that, for*

$$\widetilde{\nu} := \max \left\{ \left(\frac{2\|(\widetilde{a}_\nu)\|_\infty}{\eta} \right)^2, \left(\frac{2\|\nabla b_\epsilon\|_\infty}{\eta} \right)^2, \left(\frac{2\|c_\epsilon\|_\infty}{\eta} \right)^2, \frac{2\|\nabla c_\epsilon\|}{\eta} \right\}$$

the operator $\widetilde{\nu} + \widetilde{A}_{\widetilde{B}}$ is R-sectorial of type $\leq \omega_0$. Consequently, also the operator $\widetilde{\nu} + A_B$ is R-sectorial of type $\leq \omega_0$.

Moreover, it is clear that we also have an estimate for A_B that corresponds to the estimate (7.9).

7.17 The Case of Compact Boundary. Suppose that $\Omega \subset \mathbb{R}^{n+1}$ is an open subset with a compact boundary of class C^2. Assume that A is a second order (M, ω_0)-elliptic operator on Ω whose second order coefficients belong to BUC whereas the lower order coefficients are in L_∞. Suppose that B is a first order boundary operator with C^1-coefficients and that (A, B) satisfies the Lopatinskij-Shapiro condition at each boundary point of Ω. First we find η (by Theorem 7.16 and Section 6) and then we set up the localization scheme with suitable fineness r as described above. This is possible since we may obtain uniform constants by compactness of $\partial\Omega$.

We may proceed as in Section 6 where we use Theorem 7.16 for $l \in \Gamma_1$ and obtain that, for a suitable $\nu \geq 0$, $\nu + A_B$ is R-sectorial with $\leq \omega_0$. Moreover, we have estimates like (7.9) for the operator A_B.

7.18 Dirichlet Boundary Conditions. We only sketch the changes that have to be made for a boundary operator $\mathcal{B}(D)$ of order zero, i.e., of the form

$$(\mathcal{B}(D)u)(x) = b_0(x)u(x). \tag{7.20}$$

As before, the operator $\mathcal{A}(D)$ is assumed to be of second order and (M, ω_0)-elliptic. We comment on the three steps separately.

Step I: Here $\mathcal{A}(D)$ and $\mathcal{B}(D)$ are homogeneous differential operators with constant coefficients. We consider the full inhomogeneous problem (7.1), but here we take $g \in W_p^2(\mathbb{R}_+^{n+1})$. As before, the full inhomogeneous problem (7.1) may be reduced to the case $f = 0$. For the case $f = 0$, the same procedure as before, i.e., taking partial Fourier transform with respect to the variable $x' \in \mathbb{R}^n$ and freezing $\xi' \in \mathbb{R}^n$, leads to

$$(\lambda + a_2(\xi'))\phi(y) + a_1(\xi')D_y\phi(y) + a_0D_y^2\phi(y) = 0, \quad y > 0 \\ b_0\phi(0) = \zeta \tag{7.21}$$

where $\zeta := \widehat{g}(\xi', 0)$, ϕ is the function $y \mapsto \widehat{u}(\xi', y)$, the $a_k(\xi')$ are given by (7.2), and b_0 is the constant value of the function in (7.20) (observe that b_0 has not the same meaning as in the oblique derivative problem). The Lopatinskij-Shapiro condition now reads

> For any $\xi' \in \mathbb{R}^n$ and $\lambda \in \Sigma_\theta$ satisfying $|\xi'| + |\lambda| \neq 0$
> and for any $\zeta \in \mathbb{C}$ the problem (7.21) has a unique
> solution $\phi \in L_p(0, \infty)$. $\tag{7.22}$

This condition clearly implies $b_0 \neq 0$, and under the assumption $b_0 \neq 0$ it is equivalent to the fact that the characteristic polynomial (7.5) has two distinct roots $\mu_\pm$ with $Im\mu_+ > 0 > Im\mu_-$ (observe that this is the same condition as for the oblique derivative problem, but recall that b_0 has a different meaning). Then the solution $\phi(y) = \widehat{u}(\xi', y)$ of (7.21) is given by

$$\widehat{u}(\xi', y) = \frac{e^{iy\mu_+(\xi', \lambda)}}{b_0}\widehat{g}(\xi', 0) =: m(\xi', \lambda, y)\widehat{g}(\xi', 0), \quad y > 0, \xi' \in \mathbb{R}^n.$$

As before we write

$$m(\xi', \lambda, y)\widehat{g}(\xi', 0) = -\int_0^\infty (\partial_{\tilde{y}}m(\xi, \lambda, y + \tilde{y}))\widehat{g}(\xi', \tilde{y})d\tilde{y}$$

$$-\int_0^\infty m(\xi', \lambda, y + \tilde{y})\widehat{(\partial_{\tilde{y}}g)}(\xi', \tilde{y})d\tilde{y}.$$

In order to get the right degree of homogeneity for the arguments used in the proof of Lemma 7.6 we rewrite this as

$$= - \int_0^\infty (\partial_{\tilde{y}} m(\xi, \lambda, y + \tilde{y}))(|\lambda| + |\xi'|^2)^{-1}((|\lambda| - \Delta_n)g)\widehat{\ }(\xi', \tilde{y})d\tilde{y}$$

$$- \int_0^\infty m(\xi', \lambda, y + \tilde{y})(|\lambda| + |\xi'|^2)^{-1/2}((|\lambda| - \Delta_n)^{1/2}\partial_y g)\widehat{\ }(\xi', \tilde{y})d\tilde{y}$$

and obtain, for the solution u of (7.1) in the case $f = 0$,

$$u = M(\lambda)g = -K(\lambda)(|\lambda| - \Delta_n)g - L(\lambda)(|\lambda| - \Delta_n)^{1/2}\partial_y g.$$

Here, as far as homogeneity properties are concerned, the kernels of the Fourier multipliers

$$K(\lambda) = \mathcal{F}_n^{-1} k(\xi', \lambda)\mathcal{F}_n \quad L(\lambda) = \mathcal{F}_n^{-1} l(\xi', \lambda)\mathcal{F}_n$$

have the same structure as before (cf. (7.8) but recall that $\lambda = e^{i\omega}\sigma^2$ and that the kernels have been multiplied by a factor which is positively homogeneous of degree two before having the form (7.8)). Thus we obtain the same assertion as in Lemma 7.6, and arguments similar to those used for the oblique derivative problem lead to R-sectoriality of the operator A_B in $L_p(\mathbb{R}_+^{n+1})$ and to $\omega_R(A_B) \leq \omega_0$.

We have seen here that we may take $b_0 = 1$ without loss of generality for the case of constant coefficients. Moreover, an L_∞-small perturbation of $b_0 \equiv 1$ in (7.20) does not affect the domain of the operator A_B. This means that, as far as R-sectoriality is concerned, we do not need an estimate for solutions to the *full* inhomogeneous problem 7.1.

Step II and Step III: Since small perturbations and coordinate transformations do not affect the domain of the operator one may simply apply the perturbation results 6.5 and 6.7 and there is no need to resort to Lemma 7.10. In fact, we may take $b_0 \equiv 1$ without loss of generality (in the general case b_0 should be a C_b^2-function on $\overline{\Omega}$ which is bounded away from zero on $\partial\Omega$, recall that, if A is an elliptic operator of order $2m$, then the usual regularity required for the coefficients of a boundary operator B_j of order $m_j \leq 2m - 1$ is $C_b^{2m-m_j}$, cf. [66]).

Notes on Section 7:

N 7.1 - 7.3: The presentation owes a lot to the corresponding section in [66]. As already mentioned, the procedure, in particular the use of partial Fourier transform, is standard and we refer to [224] and [87] for the classical background on boundary value problems.

N 7.4: Actually, this is the form the Lopatinskij-Shapiro condition takes for parameter-elliptic problems. We refer, e. g., to [87] p.112 where (7.4) is a part of the *Agranovich-Vishik condition* and (7.4) is compared to its *elliptic* counterpart (which is the same as (7.4) but with $\lambda = 0$). In [66] the term *parameter-elliptic* is used.

N 7.5: We took the representation of the solution for constant coefficient operators from [66] where the case studied here is just a very special one (cf. below).

N 7.6: The proof given here is quite different from the one given in [66] where it is based on kernel estimates for the operators $K(\lambda)$ and $L(\lambda)$. The new feature in our proof is the consequent use of the R-boundedness version of Mihlin's multlpier theorem 5.2, in particular of 5.2 b) for operators $L_p(0, \infty) \to L_p(0, \infty)$.

N 7.7 - N 7.9: Theorem 7.7 contains the crucial estimate for boundary value problems with constant coefficient operators. It should be compared to the corresponding result in [66] where the presentation is different (and, of course, much more general, cf. N 7.17). The difference is due, on the one hand, to the application of Lemma 7.8 and, on the other hand, to the requirements of Lemma 7.10 which we apply in Step II. The assertion of 7.9 is in any case immediate.

N 7.10: This lemma is the technical key which makes perturbation arguments very transparent (for both boundedness and R-boundedness in L_p). It also gives the right estimates for solutions to the full inhomogeneous perturbed problem. We do not know a reference for this lemma but (due to its simplicity) we hesitate to claim it as new.

N 7.11 - N 7.14: We chose the presentation given here to show the close analogy in the proofs of uniform boundedness and R-boundedness in L_p. The use of Lemma 7.10 in the proofs is new.

N 7.15 - N 7.16: The intermediate step of the bended half space is also used, e.g., in [195], [204], [183].

N 7.17: The assertion is a special case of the main result in part II of DHP1. The situation studied there is much more general: $\mathcal{A}(D)$ is an elliptic system of order $2m$ on an infinite dimensional state space X (which is assumed to have the UMD-property) and $\mathcal{B}_1, \ldots, \mathcal{B}_m$ are boundary operators satisfying a suitable generalization of the Lopatinskij-Shapiro condition.

In [195], second order operators with Dirichlet boundary conditions were studied under the assumption that the second order coefficients are Hölder continuous. Those operators were shown to have bounded imaginary powers (BIP). In [204], second order operators were studied under Neumann boundary conditions, i.e., under conormal boundary conditions of the form

$$\frac{\partial}{\partial \nu}(Au)(x) = 0 \quad \text{on } \partial\Omega$$

where ν denotes the outer normal on $\partial\Omega$. Under suitable smoothness assumptions those operators were shown to have BIP. In the case of conormal boundary conditions the regularity of the coefficients of A enters also in the boundary operator which means that one has to require more smoothness than BUC (cf. [204]). [141] studied second order operators with Dirichlet boundary conditions whose second order coefficients are BUC and allowed for a non-compact boundary. To this end one has to keep track of the precise constants and to impose a certain uniformity on the constants that enter into the localization scheme via coordinate transformations.

N 7.18: Dirichlet boundary conditions for real and symmetric second order coefficients (a_{jk}) may also be studied by "odd" reflection on the boundary. This method has been employed in [195] and [141]. Reflection allows to reduce the half space case $\mathbb{R}^{n+1}_+$ to $\mathbb{R}^{n+1}$. "Even" reflection on the boundary has been used in [204] for the case of conormal boundary conditions. We refer to [102] for more details on the reflection principle.

8 Operators in Divergence Form

In this section we present some of the methods that may be used for operators in divergence form. We start with a brief review on the functional analytic approach to such operators.

8.1 Review. Let H be Hilbert space with inner product $(\cdot|\cdot)_H$ and let V be a dense subspace of H. Assume that $\mathfrak{a} : V \times V \to \mathbb{C}$ is a sesquilinear form. Define $\mathfrak{a}^*(u,v) := \overline{\mathfrak{a}(v,u)}$ and $\operatorname{Re}\mathfrak{a} := \frac{1}{2}(\mathfrak{a} + \mathfrak{a}^*)$, $\operatorname{Im}\mathfrak{a} := \frac{1}{2i}(\mathfrak{a} - \mathfrak{a}^*)$. Write $\mathfrak{b}(u) := \mathfrak{b}(u,u)$ if $\mathfrak{b}$ is a sesquilinear form. The form $\mathfrak{a}$ is called *sectorial of angle* $\omega \in [0, \pi/2)$ if

$$|\operatorname{Im}\mathfrak{a}(u)| \leq (\tan\omega)\operatorname{Re}\mathfrak{a}(u), \quad u \in V.$$

The form $\mathfrak{a}$ is called *closed* if V is a Hilbert space with respect to the inner product

$$(u|v)_V := \operatorname{Re}\mathfrak{a}(u,v) + (u|v)_H, \quad u,v \in V.$$

Now suppose that $\mathfrak{a}$ is a closed sectorial form of angle ω. We associate an operator A in H to the form $\mathfrak{a}$ by letting

$$u \in D(A) \text{ and } Au = f \iff u \in V \text{ and } \forall v \in V : (f|v)_H = \mathfrak{a}(u,v).$$

Observe that A is well defined by denseness of V in H. The operator A is called *associated* to the form $\mathfrak{a}$. Now we show that the operator A is sectorial and $\omega(A) \leq \omega$.

Proof. We consider V equipped with the inner product $(\cdot|\cdot)_V$. Then $\mathfrak{a}$ and $\mathfrak{a}+1$ are continuous $V \times V \to \mathbb{C}$ where we write 1 for the sesquilinear form given by the inner product $(\cdot|\cdot)_H$. We identify H with its antidual space H^* and denote the antidual of $(V, (\cdot|\cdot))$ by V^*. Then we have continuous and dense injections $V \subset H = H^* \subset V^*$. The continuous sesquilinear form $\mathfrak{a} + 1$ gives rise to a bounded operator $\mathcal{A}+1 : V \to V^*$ via $((\mathcal{A}+1)u, v)_{V^*,V} = (\mathfrak{a}+1)(u, v)$. By Lax-Milgram $\mathcal{A} + 1$ is an isomorphism $V \to V^*$ with inverse $(\mathcal{A} + 1)^{-1} : V^* \to V$. We let R denote the restriction of $(\mathcal{A} + 1)^{-1}$ to H. Observe that $R \in B(H)$ is injective. We define the operator B in H by $B := R^{-1} - 1$. Clearly, B is a closed operator in H. We have, for $u, f \in H$,

$$u \in D(B), Bu = f \Leftrightarrow R(u + f) = u \Leftrightarrow u \in V, (\mathcal{A} + 1)u = u + f$$
$$\Leftrightarrow u \in V, \ \forall v \in V : (f|v)_H = \mathfrak{a}(u, v).$$

The last line is the definition of the operator A from above. Note that we have $R = (A+1)^{-1}$ hence $-1 \in \rho(A)$. Sectoriality of $\mathfrak{a}$ now implies $(Au|u)_H \subset \overline{\Sigma_\omega}$, and together with $-1 \in \rho(A)$ this yields $\mathbb{C} \setminus \overline{\Sigma_\omega} \subset \rho(A)$ and

$$\|R(\lambda, A)\|_{H \to H} \le \frac{1}{d(\lambda, \overline{\Sigma_\omega})}, \quad \text{for } \lambda \notin \overline{\Sigma_\omega}. \tag{8.1}$$

In particular, A is a sectorial operator and $\omega(A) \le \omega$. $\qquad\qquad\square$

Hence operators associated to densely defined, closed and sectorial forms in $L_2(\Omega)$ are closed and sectorial in $L_2(\Omega)$, and since $L_2(\Omega)$ is a Hilbert space, they are R-sectorial of the same type. They even have a bounded H^∞-calculus of optimal angle in $L_2(\Omega)$ (we refer to Section 11 in Chapter II).

One can already see from these comments that, at least in $L_2(\Omega)$, the study of divergence form operators is easier than the study of operators in non-divergence form. Recall, e.g., that for the operators studied in Sections 6 and 7 sectoriality had to be proven and was *not* clear from the beginning. Moreover, it is not known whether elliptic operators in non-divergence form always have a bounded H^∞-calculus in $L_2(\Omega)$ (we refer to Sectin 13 in Chapter II for further details).

So, roughly speaking and for the properties that we investigate in these notes, one can say that, for operators in divergence form, the L_2-theory is clear. But what about the situation in L_p for $p \ne 2$?

First of all one has to "extend" or "extrapolate" the operator A, which originally is given by a form in L_2, to other L_p-spaces. Due to the unboundedness of A this leads to delicate domain problems which are further complicated by the fact that, in general, even the domain of A in L_2 is unknown and one only has information about the domain of the form. The usual way to get around these problems is to extrapolate the semigroup operators $T(t) := e^{-tA}$, $t > 0$, to other L_p-spaces. If this is possible and $(T(t))$ induces a C_0-semigroup $(T_p(t))$ in L_p then one defines the operator $-A_p$ to be the generator of $(T_p(t))$ in L_p. Extrapolation means here that the semigroups are *consistent*, i.e.,

$$T(t)f = T_p(t)f, \quad t > 0, f \in L_2 \cap L_p.$$

By this method one may investigate properties of the given divergence form operator in spaces L_p, in which the semigroup acts as a C_0-semigroup.

There are basically two methods which allow to extrapolate semigroups from L_2 to other L_p-spaces: contractivity (which may be combined with positivity) and kernel bounds (or, more general, weighted norm bounds). To give an idea of what is meant by all that, we illustrate both methods by the simplest example available, i.e., by the Laplacian on $\mathbb{R}^n$. Of course, the domain of Δ in $L_2(\mathbb{R}^n)$ is the Sobolev space $H^2(\mathbb{R}^n) = W_2^2(\mathbb{R}^n)$, but Δ is also associated to the following form

$$\mathfrak{a}(u,v) := \int_{\mathbb{R}^n} \overline{\nabla v}^t \nabla u \, dx, \quad u, v \in V := H^1(\mathbb{R}^n) = W_2^1(\mathbb{R}^n).$$

The form $\mathfrak{a}$ is sesquilinear, symmetric and closed on the form domain V. The associated operator A in $L_2(\mathbb{R}^n)$ is given by

$$u \in D(A) \text{ and } Au = f$$
$$:\Longleftrightarrow u \in H^1(\mathbb{R}^n) \text{ and } \forall v \in H^1(\mathbb{R}^n) : (f, v) = \mathfrak{a}(u, v).$$

By partial integration we obtain $A = -\Delta$. The semigroup generated by $-A = \Delta$ is of course the Gaussian semigroup, but let us forget for the moment that we know a lot about it.

8.2 Extrapolation by Contractivity. The well-known Beurling-Deny criteria (see, e.g., [63], [184]) allow to check L_∞-contractivity (and positivity) of (e^{-tA}) *in terms of the form* $\mathfrak{a}$. Here, L_∞-contractivity means

$$\|e^{-tA}f\|_\infty \leq \|f\|_\infty, \quad t > 0, f \in L_2 \cap L_\infty.$$

By Riesz-Thorin interpolation we obtain that any operator e^{-tA} extends to a contraction $T_p(t)$ on L_p for $p \in [2, \infty)$. The semigroup $(T_p(t))$ is weakly continuous in L_p for $p \in [2, \infty)$, hence strongly continuous. By symmetry of A we also obtain a consistent C_0-semigroup $(T_p(t))$ for $p \in (1, 2)$. The semigroup also acts as semigroup of contractions $(T_1(t))$ in L_1. By positivity of e^{-tA}, $(T_1(t))$ is a C_0-semigroup in L_1 ([218]). Thus we may define $-A_p$ to be the generator of $(T_p(t))$ for all $p \in [1, \infty)$. For operators of this kind, Theorem 5.6 gives maximal regularity for A_p.

There are much more refined techniques for such operators where L_∞-contractivity is replaced by a kind of Moser-Nash-iteration (see, e.g., [58], [167], [202]). Also symmetry is not essential, but the method is restricted to (though quite general!) *second* order operators with *real* coefficients. For those operators one can use the result mentioned in N 5.6.

8.3 Extrapolation by Kernel Bounds. We forget about the previous arguments, but recall that the Gaussian semigroup is given by kernels $k(t, x, y)$ satisfying a bound

$$|k(t, x, y)| \leq C t^{-n/2} e^{-b\frac{|x-y|^2}{t}} =: g_t(x - y), \quad t > 0, \ x, y \in \mathbb{R}^n,$$

where $C, b > 0$. For obvious reasons this is called a *Gaussian* kernel bound, and, of course, we even have equality for a suitable choice of b and C, but we shall not need this fact. All that we need is that the kernel bound implies

$$|e^{-tA}f| \leq g_t * f, \quad t > 0, \ f \in L_1 \cap L_\infty$$

which by $\sup_t \|g_t\|_1 < \infty$ yields that (e^{-tA}) acts as a bounded semigroup $(T_p(t))$ in all L_p, $p \in [1, \infty)$. Again, for $p \in (1, \infty)$, the semigroup $(T_p(t))$ is weakly continuous, hence strongly continuous. The kernel bound allows to prove strong continuity also in L_1 (see, e.g., [218]). Gaussian kernel bounds allow to extrapolate maximal regularity from L_2 to all spaces L_p, $1 < p < \infty$ (cf. [119], [52]).

Of course, now the problem is to establish Gaussian kernel bounds for the semigroup (e^{-tA}), which is not an easy matter and, moreover, is not always possible. But, fortunately enough, in this method Gaussian kernel bounds may be replaced by so-called weighted norm bounds and we shall concentrate on those for the rest of Section 8. For uniformly elliptic operators the validity of suitable weighted norm bounds basically relies on the "right" Sobolev embedding $V \hookrightarrow L_q$ of the form domain V. Hence weighted norm bounds do *naturally* hold for those operators. The method applies to quite general operators of higher order and with complex coefficients, but it might fail for "bad" domains where the Sobolev embedding does not hold (cf. the Notes).

After this very sketchy introduction, the section is organized as follows. We shall present an abstract result (Theorem 8.6) that generalizes the mentioned result in [119], [52] and allows to extrapolate maximal regularity and R-sectoriality from L_2 to other L_p-spaces provided certain so-called weighted norm bounds hold for the semigroup operators.

The main idea behind this result is domination and we shall give full details on that. In fact, the arguments are somehow similar to those that were used to prove Theorem 5.6. First we show that suitable weighted norm bounds (and hence in particular suitable kernel bounds) imply a certain maximal estimate (Proposition 8.9) which leads to Proposition 8.14. The assertion of Proposition 8.14 now plays the role of (5.3). Afterwards we use Stein interpolation (Lemma 5.8) in a way similar to the argument in the proof of Lemma 5.7. But the proof also relies on boundedness properties of maximal functions known from harmonic analysis. For those properties of the maximal function we refer to the literature.

Having proved the abstract result, we study a class of elliptic operators in divergence form and show how weighted norm bounds may be derived.

We start with some preparations for the abstract result.

8.4 Spaces of Homogeneous Type. Suppose that (Ω, d, μ) is a measured metric space, i.e., (Ω, d) is a metric space and μ is a Borel measure. The space

(Ω, d, μ) is said to satisfy the *doubling property* (cf. [205]) if there is a constant $c > 0$, called the *doubling constant*, such that

$$|B(x, 2\rho)| \le c|B(x, \rho)|, \quad x \in \Omega, \ \rho > 0,$$

where $|B(x, \rho)| := \mu(B(x, \rho))$ denotes the volume of the open ball $B(x, \rho)$ with radius $\rho > 0$ and center $x \in \Omega$. This means that the volume of large balls can be estimated by the volume of small balls with the same center, uniformly over the whole space.

It is not hard to see that the doubling property implies

$$|B(x, \lambda\rho)| \le C\lambda^n|B(x, \rho)|, \quad x \in \Omega, \ \rho > 0, \ \lambda \ge 1, \qquad (8.2)$$

for $C := c$ and $n := \log c / \log 2$. This means that (Ω, d, μ) is a *space of homogeneous type* and *dimension* n. An example is of course $\Omega = \mathbb{R}^n$ with Euclidean distance and Lebesgue measure. But observe that, in general, n is not uniquely defined by (8.2).

If (Ω, d, μ) is a space of homogeneous type and dimension n, and $x, y \in \Omega$, $\rho > 0$ then $B(x, \rho) \subset B(y, d(x, y) + \rho)$ which yields

$$\frac{|B(x, \rho)|}{|B(y, \rho)|} \le C\Big(1 + \frac{d(x, y)}{\rho}\Big)^n. \qquad (8.3)$$

This estimate allows comparison of volume of balls with same radius but different centers. Of course, the bound in (8.3) is worst possible for a space of dimension n. In $\Omega = \mathbb{R}^n$, e.g., we have $|B(x, \rho)| = |B(y, \rho)|$ for all x, y and ρ.

To motivate the kind of weighted norm bounds assumed in Theorem 8.6 we prove the following lemma. Throughout this section we shall use the notation $\|T\|_{p \to q}$ to denote the operator norm of linear operators $T : L_p(\Omega) \to L_q(\Omega)$.

8.5 Lemma. *Let (Ω, d, μ) be a space of homogeneous type and doubling constant c. Let $m > 1$ and $(K_t)_{t>0}$ be a family of integral operators with kernels $k(t, x, y)$ satisfying*

$$|k(t, x, y)| \le |B(x, t^{1/m})|^{-1} g\Big(\frac{d(x, y)}{t^{1/m}}\Big), \quad t > 0, x, y \in \Omega,$$

for a non-increasing function $g : [0, \infty) \to [0, \infty)$. Then we have, for all $1 \le p \le q \le \infty$,

$$\|1_{B(x, t^{1/m})} K_t 1_{B(y, t^{1/m})}\|_{p \to q}$$
$$\le |B(x, t^{1/m})|^{-1+1/q}|B(y, t^{1/m})|^{1-1/p}\tilde{g}\Big(\frac{d(x,y)}{t^{1/m}}\Big), \quad t > 0, x, y \in \Omega, \qquad (8.4)$$

where $\tilde{g}(s) := cg((s - 2)_+), s \ge 0$.

Conversely, if this bound holds for $(p, q) = (1, \infty)$ then the kernel bound holds for $\tilde{g}$ in place of g.

Here, the action of the operator $1_{B(x, t^{1/m})} K_t 1_{B(y, t^{1/m})}$ is given by

$$(1_{B(x, t^{1/m})} K_t 1_{B(y, t^{1/m})} f)(u) = \int_\Omega 1_{B(x, t^{1/m})}(u) k(t, u, v) 1_{B(y, t^{1/m})}(v) f(v) \, d\mu(v).$$

Proof. We start with the case $p = 1$, $q = \infty$. Then

$$\|1_{B(x,t^{1/m})} K_t 1_{B(y,t^{1/m})}\|_{1 \to \infty}$$

$$\leq \sup_{d(x,u) < t^{1/m}, d(y,v) < t^{1/m}} |B(u,t^{1/m})|^{-1} g\left(\frac{d(u,v)}{t^{1/m}}\right).$$

For $d(x,u), d(y,v) < t^{1/m}$ we have $d(u,v) \geq d(x,y) - 2t^{1/m}$. Since g is non-increasing we obtain the claim by the doubling property.

In the general case we use Hölder's inequality

$$\|1_{B(x,t^{1/m})} K_t 1_{B(y,t^{1/m})} f\|_q$$
$$\leq \|1_{B(x,t^{1/m})}\|_q \|1_{B(x,t^{1/m})} K_t 1_{B(y,t^{1/m})}\|_{1 \to \infty} \|1_{B(y,t^{1/m})}\|_{p'} \|f\|_p$$

and the case just proved to obtain the result. The converse assertion follows from the natural isometry $B(L_1, L_\infty) \cong L_\infty(\Omega \times \Omega)$. □

Recalling the bound (8.3) it is clear that we can estimate the left hand side of (8.4) by

$$C^{1-1/p} |B(x,t^{1/m})|^{-(1/p-1/q)} \left(1 + \frac{d(x,y)}{t^{1/m}}\right)^{n(1-1/p)} \tilde{g}\left(\frac{d(x,y)}{t^{1/m}}\right).$$

Letting $h(s) := (1+s)^{n(1-1/p)} \tilde{g}(s)$, $s \geq 0$, or (independent of p) $h(s) := (1+s)^n \tilde{g}(s)$ we then have an estimate of the left hand side of (8.4) against

$$C^{1-1/p} |B(x,t^{1/m})|^{-(1/p-1/q)} h\left(\frac{d(x,y)}{t^{1/m}}\right).$$

We take this as a motivation for the assumption in the following theorem.

8.6 Theorem. *Let (Ω, d, μ) be a space of homogeneous type, and let $-A$ be the generator of a bounded analytic semigroup in $L_2(\Omega)$. Let $1 \leq p < 2 < q \leq \infty$. Assume that the following weighted norm bound holds*

$$\|1_{B(x,t^{1/m})} e^{-tA} 1_{B(y,t^{1/m})}\|_{p \to q} \leq |B(x,t^{1/m})|^{-(1/p-1/q)} g\left(\frac{d(x,y)}{t^{1/m}}\right),$$

for all $t > 0$, $x, y \in \Omega$, where $g : [0, \infty) \to [0, \infty)$ is a non-increasing function that decays faster than any polynomial. Then, for all $r \in (p,q)$, the semigroup (e^{-tA}) induces a consistent and bounded analytic semigroup $(T_r(t))$ in $L_r(\Omega)$. If $-A_r$ denotes the generator of $(T_r(t))$ in $L_r(\Omega)$ then A_r is R-sectorial in $L_r(\Omega)$ for $r \in (p,q)$, in particular A_r has maximal regularity in $L_r(\Omega)$ for $r \in (p,q)$.

We start the proof with a domination property that is implied by the weighted norm bound. Due to the fact that we do not necessarily have kernels for our operators we cannot hope for pointwise domination. But we can use normalized L_r-norms on balls instead. To this end we introduce the following notation. For $r \in [1, \infty]$, $x \in \Omega$, $\rho > 0$, and a measurable function $f : \Omega \to \mathbb{C}$ we let

$$N_{r,\rho} f(x) := |B(x, \rho)|^{-1/r} \|1_{B(x,\rho)} f\|_r.$$

Observe that the Hardy-Littlewood maximal function M_1 (see, e.g., [205, Ch.I]) is given by

$$M_1 f(x) = \sup_{\rho > 0} N_{1,\rho} f(x), \quad x \in \Omega.$$

Moreover, if $r \in (1, \infty]$, and M_r is defined via

$$M_r f(x) := \sup_{\rho > 0} N_{r,\rho} f(x), \quad x \in \Omega,$$

then $M_r f = (M_1(|f|^r))^{1/r}$. We call M_r the r-*maximal function*. Just as M_1 is, also the operators $N_{r,\rho}$ and M_r are sublinear. The following lemma collects some basic properties of the operators $N_{r,\rho}$.

8.7 Lemma. *For all* $1 \le r \le s \le \infty$, *all* $\rho > 0$ *and all measurable functions* f *we have*

$$N_{r,\rho} f \le N_{s,\rho} f$$

and

$$c^{-1/r} \|N_{r,\rho} f\|_r \le \|f\|_r \le c^{1/r} \|N_{r,\rho} f\|_r$$

where c *denotes the doubling constant.*

Proof. The first assertion follows by Hölder's inequality. For the second assertion we observe that, by Fubini,

$$\|f\|_r^r = \int_\Omega |f(x)|^r \, dx = \int_\Omega \int_{B(x,\rho)} \frac{dy}{|B(x,\rho)|} |f(x)|^r \, dx$$

$$= \int_\Omega \int_{B(y,\rho)} \frac{|f(x)|^r \, dx}{|B(x,\rho)|} \, dy$$

By the doubling property we have $|B(x,\rho)|/|B(y,\rho)| \le c$ for $d(x,y) < \rho$ which ends the proof. □

For the following important property (a) of the maximal function we refer to [205, Ch.I].

8.8 Theorem. (a) *The sublinear operator* M_1 *is bounded on* $L_p(\Omega)$ *for* $1 < p \le \infty$, *but not on* $L_1(\Omega)$.

(b) *For* $p \in [1, \infty]$, *the sublinear operator* M_p *is bounded on* $L_r(\Omega)$ *for* $p < r \le \infty$, *but not on* $L_p(\Omega)$.

Proof. We refer to [205, Ch.I] for (a). By the representation of M_r in terms of M_1, (b) is an immediate consequence of (a). □

We now turn to the announced domination property that is fundamental for the proof of Theorem 8.6.

8.9 Proposition. *Under the assumptions of Theorem 8.6 we have*

$$N_{q,t^{1/m}}(e^{-tA}f)(x) \le C_1 M_p f(x), \quad t > 0, x \in \Omega,$$

for all $f \in L_p(\Omega) \cap L_q(\Omega)$.

Proof. We first use the assumption in Theorem 8.6 to derive a weighted norm bound of annular type. Let $k \in \mathbb{N}_0$ and

$$A(x, t^{1/m}, k) := B(x, (k+1)t^{1/m}) \setminus B(x, kt^{1/m}).$$

We aim to bound

$$\|1_{B(x,t^{1/m})} e^{-tA} 1_{A(x,t^{1/m},k)}\|_{p \to q}.$$

To this end we choose a maximal subset F of centers $y \in A(x, t^{1/m}, k)$ such that the balls $B(y, t^{1/m}/2)$ are disjoint. Then the balls $B(y, t^{1/m})$, $y \in F$, cover $A(x, t^{1/m}, k)$, which yields

$$\|1_{B(x,t^{1/m})} e^{-tA} 1_{A(x,t^{1/m},k)}\|_{p \to q} \le \sum_{y \in F} \|1_{B(x,t^{1/m})} e^{-tA} 1_{B(y,t^{1/m})}\|_{p \to q}.$$

By $F \subset A(x, t^{1/m}, k)$ we have $d(x, y) \ge kt^{1/m}$ for $y \in F$ which means that any summand on the right hand side is bounded by

$$|B(x, t^{1/m})|^{-(1/p-1/q)} g(k).$$

By the volume doubling property we obtain an estimate on the cardinality of F which we denote by α: For any $y \in F$ we have, by $d(x, y) < (k+1)t^{1/m}$,

$$|B(x, t^{1/m}/2)| \le C(1 + \frac{2d(x,y)}{t^{1/m}})^n |B(y, t^{1/m}/2)|$$
$$\le C(2k+3)^n |B(y, t^{1/m}/2)|.$$

On the other hand we have by construction

$$\sum_{y \in F} |B(y, t^{1/m}/2)| = |\bigcup_{y \in F} B(y, t^{1/m}/2)|$$
$$\le |B(x, (k+1)t^{1/m} + t^{1/m}/2)| \le C(2k+3)^n |B(x, t^{1/m}/2)|$$

which together yields

$$\alpha C^{-1}(2k+3)^{-n}|B(x, t^{1/m}/2)| \le C(2k+3)^n |B(x, t^{1/m}/2)|,$$

i.e., $\alpha \le C^2(2k+3)^{2n}$. We finally obtain the annular type weighted norm bound

$$\|1_{B(x,t^{1/m})} e^{-tA} 1_{A(x,t^{1/m},k)}\|_{p\to q}$$
$$\leq |B(x,t^{1/m})|^{-(1/p-1/q)} h(k), \quad t > 0, \ x \in \Omega, \ k \in \mathbb{N}_0$$

where $h(k) := C^2(2k+3)^{2n} g(k)$, $k \in \mathbb{N}_0$, decays faster in k than any polynomial.

We now use this bound to prove the assertion.

$$N_{q,t^{1/m}}(e^{-tA}f)(x)$$
$$= |B(x,t^{1/m})|^{-1/q} \|1_{B(x,t^{1/m})} e^{-tA} f\|_q$$
$$\leq |B(x,t^{1/m})|^{-1/q} \sum_{k=0}^{\infty} \|1_{B(x,t^{1/m})} e^{-tA} 1_{A(x,t^{1/m},k)} f\|_q$$
$$\leq |B(x,t^{1/m})|^{-1/p} \sum_{k=0}^{\infty} h(k) \|1_{A(x,t^{1/m},k)} f\|_p$$
$$\leq C^{1/p} \sum_{k=0}^{\infty} (k+1)^{n/p} |B(x,(k+1)t^{1/m})|^{-1/p} h(k) \|1_{B(x,(k+1)t^{1/m})} f\|_p$$
$$\leq M_p f(x) \cdot C^{1/p} \sum_{k=0}^{\infty} (k+1)^{n/p} h(k).$$

We have thus obtained the assertion with $C_1 := C^{1/p} \sum_k (k+1)^{n/p} h(k)$ which is finite by the decay of h.   □

8.10 Corollary. *Under the assumptions of Theorem 8.6 we have, for all $r \in (p,q]$, the following bound*

$$\|e^{-tA}f\|_r \leq C(r)\|f\|_r, \quad t > 0, f \in L_p(\Omega) \cap L_q(\Omega).$$

Proof. We have by Lemma 8.7

$$\|e^{-tA}f\|_r \leq c^{1/r} \|N_{r,t^{1/m}}(e^{-tA}f)\|_r \leq c^{1/r} \|N_{q,t^{1/m}}(e^{-tA}f)\|_r$$
$$\leq c^{1/r} C_1 \|M_p f\|_r$$

and we finish the proof by the boundedness of M_p on $L_r(\Omega)$.   □

8.11 Corollary. *Under the assumptions of Theorem 8.6, for any $r \in (p,q)$, the semigroup (e^{-tA}) induces a bounded analytic C_0-semigroup $(T_r(t))$ in $L_r(\Omega)$.*

Proof. Since (e^{-tA}) is a C_0-semigroup in $L_2(\Omega)$ we obtain weak continuity in $L_r(\Omega)$ by the previous corollary. Hence (e^{-tA}) induces a C_0-semigroup $(T_r(t))$ in $L_r(\Omega)$. Since (e^{-tA}) is a bounded analytic semigroup in $L_2(\Omega)$, we obtain by Stein interpolation that $(T_s(t))$ extends to bounded analytic semigroup for any s strictly between r and 2. This proves the assertion.   □

By now, we have proved the first assertion of Theorem 8.6, and we now use Proposition 8.9 and boost our arguments to prove the second assertion.

We shall prove that, for any $r \in (p,q)$, there is a $\delta > 0$ such that the set

$$\{T_r(z) : z \neq 0, |\arg z| < \delta\}$$

is an R-bounded set of operators $L_r(\Omega) \to L_r(\Omega)$. Then maximal L_p-regularity follows by Theorem 1.11. Recall (cf. Remark 2.9) that R-boundedness of this set is equivalent to the existence of a constant $C > 0$ such that, for all finite families (z_j) in Σ_δ and (f_j) in $L_r(\Omega)$, the following square function estimate holds:

$$\left\|\left(\sum_j |T_r(z_j)f_j|^2\right)^{1/2}\right\|_r \leq C \left\|\left(\sum_j |f_j|^2\right)^{1/2}\right\|_r$$

This opens the door for interpolation arguments, and we shall use the following notion.

8.12 Definition. A subset τ of bounded (sub)linear operators $L_r(\Omega) \to L_r(\Omega)$ is called R_s-*bounded in* $L_r(\Omega)$ if there exists a constant $C > 0$ such that, for all finite families (T_j) in τ and (f_j) in $L_r(\Omega)$, we have

$$\left\|\left(\sum_j |T_j f_j|^s\right)^{1/s}\right\|_r \leq C \left\|\left(\sum_j |f_j|^s\right)^{1/s}\right\|_r, \quad \text{if } 1 \leq s < \infty$$

$$\left\|\sup_j |T_j f_j|\right\|_r \leq C \left\|\sup_j |f_j|\right\|_r, \quad \text{if } s = \infty.$$

The infimum of all such C is denoted $R_s(\tau; L_r)$.

We call the subset τ R_s-*bounded from below* if there is a constant $C > 0$ such that the reverse inequalities hold for all finite families (T_j) in τ and (f_j) in $L_r(\Omega)$. The supremum over $1/C$ where C is a such constant is denoted $R'_s(\tau; L_r)$.

By Fubini, a subset τ of bounded linear operators $T : L_r \to L_r$ is R_r-bounded in L_r if and only if τ is uniformly bounded.

Known properties of the maximal function M_1 give us

8.13 Proposition. *For* $1 \leq p < r, s \leq \infty$ *the set* $\{M_p\}$ *is* R_s-*bounded in* $L_r(\Omega)$. *For* $1 \leq r, s < q \leq \infty$ *the set* $\{N_{q,\rho} : \rho > 0\}$ *is* R_s-*bounded from below in* $L_r(\Omega)$.

Proof. For the first assertion we refer to [93] in case $p = 1$. In fact, it suffices to prove the case $p = 1$ since

$$\left\|\left(\sum_j |M_p f_j|^s\right)^{1/s}\right\|_r = \left\|\left(\sum_j (M_1|f_j|^p)^{s/p}\right)^{s/p}\right\|_{r/p}^{1/p}.$$

We turn to the second assertion. Taking $1 \leq r, s < q$ the first assertion yields that $\{N_{q',\rho} : \rho > 0\}$ is $R_{s'}$-bounded in $L_{r'}(\Omega)$. We use a duality argument to show that this implies R_s-boundedness of $\{N_{q,\rho} : \rho > 0\}$ in $L_r(\Omega)$ from below.

For finite families (f_j) in $L_r(\Omega)$, (g_j) in $L_{r'}(\Omega)$, and (ρ_j) in $(0, \infty)$ we have, by Fubini and Hölder,

$$
\begin{aligned}
\int_\Omega \sum_j g_j f_j \, dx &= \int_\Omega \sum_j g_j(x) f_j(x) |B(x, \rho_j)|^{-1} \int_{B(x,\rho_j)} dy \, dx \\
&= \int_\Omega \sum_j \int_{B(y,\rho_j)} g_j(x) f_j(x) \frac{dx}{|B(x, \rho_j)|} \, dy \\
&\leq c \int_\Omega \sum_j \int_{B(y,\rho_j)} g_j(x) f_j(x) \frac{dx}{|B(y, \rho_j)|} \, dy \\
&\leq c \int_\Omega \sum_j N_{q',\rho_j} g_j(y) \, N_{q,\rho_j} f_j(y) \, dy \\
&\leq c \| (\sum_j |N_{q',\rho_j} g_j|^{s'})^{1/s'} \|_{r'} \| (\sum_j |N_{q,\rho_j} f_j|^s)^{1/s} \|_r \\
&\leq cM \| (\sum_j |g_j|^{s'})^{1/s'} \|_{r'} \| (\sum_j |N_{q,\rho_j} f_j|^s)^{1/s} \|_r.
\end{aligned}
$$

The claim follows by taking the sup over all families (g_j) with

$$
\| (\sum_j |g_j|^{s'})^{1/s'} \|_{r'} \leq 1
$$

which is a norming set for $L_r(l_s)$. The proof works with obvious modifications in notation also in the limit cases $s = \infty$, $s' = \infty$. $\qquad\square$

Now we use the assertion of Proposition 8.9 to prove

8.14 Proposition. *Under the assumptions of Theorem 8.6, for any $r, s \in (p, q)$, the set $\{T_r(t) : t \geq 0\}$ is R_s-bounded in $L_r(\Omega)$.*

Proof. Let (t_j) and (f_j) be finite families in $(0, \infty)$ and $L_r(\Omega)$, respectively. Then we have

$$
\| (\sum_j |T_r(t_j) f_j|^s)^{1/s} \|_r
$$

$$
\leq R'_s(\{N_{q,\rho} : \rho > 0\}; L_r) \| (\sum_j N_{q, t_j^{1/m}} (T_r(t_j) f_j)^s)^{1/s} \|_r
$$

$$
\leq C_{q,r,s} \| (\sum_j M_p(f_j)^s)^{1/s} \|_r
$$

$$
\leq C_{q,r,s} R_s(\{M_p\}; L_r) \| (\sum_j |f_j|^s)^{1/s} \|_r
$$

which proves the claim by Proposition 8.13. $\qquad\square$

Observe that the assertion corresponds to the assertion of Corollary 8.11. Now we use again Stein interpolation as in the form of Leema 5.8 to show

8.15 Proposition. *For any $r \in (p,q)$ there is a $\delta > 0$ such that $\{T_r(z) : z \neq 0, |\arg z| < \delta\}$ is R_2-bounded in $L_r(\Omega)$.*

which ends the proof of Theorem 8.6.

Proof. Fix $r \in (p,q) \setminus \{2\}$. Find $\tilde{r} \in (p,q)$ such that r is strictly between 2 and $\tilde{r}$. In $L_{\tilde{r}}(\Omega)$, the semigroup is R_2-bounded on $(0,\infty)$, and in $L_2(\Omega)$ it is bounded analytic of some angle ω. Defining $\mathbb{T}_t(f_j) := (T_{tt_j} f_j)$ where t_j runs through the positive dyadic rational numbers we obtain a C_0-semigroup which is bounded in $L_{\tilde{r}}(\Omega; l_2)$ and bounded analytic in $L_2(\Omega; l_2)$ of angle ω. Stein interpolation as in Lemma 5.8 now yields that $(\mathbb{T}_t)$ is bounded analytic in $L_r(\Omega; l_2)$ of angle $\delta := \theta\omega$ where $\theta \in (0,1)$ is given by $r^{-1} = \theta/2 + (1-\theta)/\tilde{r}$. Hence $\{T_r(z) : z \neq 0, |\arg z| < \delta\}$ is R_2-bounded in $L_r(\Omega)$. □

Having proved the abstract result on extrapolation of maximal regularity, we now prove weighted norm bounds as they are assumed in Theorem 8.6 for a particular class of elliptic operators. Consequently, those operators have maximal regularity n a certain scale of L_r-spaces (cf. Theorem 8.24).

8.16 Divergence Form Operators. We shall study elliptic operators A in divergence form with bounded measurable complex-valued coefficients on $\mathbb{R}^n$ and derive weighted norm bounds for the semigroup (e^{-tA}) which are of the type used in Theorem 8.6. We study operators of arbitrary order. The case of second order operators is also included but the emphasis is on the non-symmetric complex situation here. This means that one cannot hope for positivity or L_∞-contractivity of the semigroup which one usually has for operators with real coefficients. A priori, second order operators with complex coefficients do not behave better than operators of higher order.

The operators we study are given by forms $\mathfrak{a}$ of the type

$$\mathfrak{a}(u,v) = \int_{\mathbb{R}^n} \sum_{|\alpha|=|\beta|=m} a_{\alpha\beta}\, \partial^\alpha u\, \overline{\partial^\beta v}\, dx, \quad u,v \in H_2^m(\mathbb{R}^n), \qquad (8.5)$$

where the $a_{\alpha\beta} : \mathbb{R}^n \to \mathbb{C}$ are bounded and measurable, and H_2^m is the usual Sobolev space of order m in $L_2(\mathbb{R}^n)$, which is also a Bessel potential space.

We assume sectoriality of the form $\mathfrak{a}$, i.e.,

$$|Im\,\mathfrak{a}(u)| \leq \tan\psi \cdot Re\,\mathfrak{a}(u), \quad u \in H_2^m,$$

where $\psi \in [0, \pi/2)$. Moreover, we assume ellipticity in the form of the following Gårding inequality

$$Re\,\mathfrak{a}(u) \geq \delta\|(-\Delta)^{m/2}u\|_2^2, \quad u \in H_2^m, \qquad (8.6)$$

for some $\delta > 0$. Then the form $\mathfrak{a}$ is closed and sectorial on H_2^m. The associated operator A is defined via $Au = g$ if and only if $u \in H_2^m$ and $(g, v) = \mathfrak{a}(u, v)$ for all $v \in H_2^m$. The operator A is closed and sectorial in $L_2(\mathbb{R}^n)$.

Following Davies ([61]) we define the twisted forms

$$\mathfrak{a}_{\lambda\phi}(u, v) := \mathfrak{a}(e^{\lambda\phi} u, e^{-\lambda\phi} v)$$

where $\lambda \in \mathbb{R}$ and ϕ is a real-valued C^∞-function with compact support satisfying $\|\partial^\alpha \phi\|_\infty \leq 1$ for all $1 \leq |\alpha| \leq m$. The space of all such functions is denoted by $\mathcal{E}_m$. Observe that the functions $e^{\lambda\phi}$ are pointwise multipliers in H_2^m.

The following lemma relates the class $\mathcal{E}_m$ to the Euclidean distance in $\mathbb{R}^n$.

8.17 Lemma. *There is a constant $c > 0$ such that, for all $x, y \in \mathbb{R}^n$,*

$$c^{-1}|x - y| \leq \sup\{|\phi(x) - \phi(y)| : \phi \in \mathcal{E}_m\} \leq c|x - y|.$$

Proof. The upper bound with constant $\sqrt{n}$ is immediate from the mean value theorem and the boundedness of the first derivatives, uniformly in $\phi \in \mathcal{E}_m$. For the lower bound we fix $x \neq y$ and let $\phi_0(z) := (x-y, z)/|x-y|$. The function ϕ_0 satisfies the desired bounds but is not of compact support. Moreover, $|\phi_0(z)| \leq |z|$, $z \in \mathbb{R}^n$. We choose a function $\psi \in C_c^\infty(B(0, 2))$ such that $0 \leq \psi \leq 1$ and $\psi = 1$ on $B(0, 1)$, and let $\psi_k := \psi(\cdot/k)$, $k \in \mathbb{N}$. For any multiindex $\alpha \in \mathbb{N}_0^n$ with $|\alpha| \geq 2$ we have

$$\partial^\alpha(\psi_k \phi_0) = (\partial^\alpha \psi_k)\phi_0 + \sum_{e_j \leq \alpha} \alpha_j (\partial^{\alpha - e_j} \psi_k)\partial_j \phi_0.$$

This leads to the bound

$$|\partial(\psi_k \phi_0)| \leq 2k^{1-|\alpha|}\|\partial^\alpha \psi\|_\infty + k^{1-|\alpha|} \sum_{e_j \leq \alpha} \alpha_j \|\partial^{\alpha - e_j}\psi\|_\infty,$$

and the right hand side can be made arbitrarily small by choosing k sufficiently large. We turn to the case $|\alpha| = 1$. If $j \in \{1, \ldots, n\}$ then, for any k,

$$\begin{aligned}
|\partial_j(\psi_k \phi_0)| &\leq |\partial_j \psi_k| 1_{B(0,2k)}|\phi_0| + |\psi_k||\partial_j \phi_0| \\
&\leq k^{-1}\|\partial_j \psi\|_\infty \cdot 2k + 1 \\
&\leq 2 \max_\nu \|\partial_\nu \phi_0\|_\infty + 1 =: c.
\end{aligned}$$

Hence we have for k large that $\phi := c^{-1}\psi_k \phi_0 \in \mathcal{E}_m$ and $|\phi(x) - \phi(y)| = c^{-1}|x - y|$. Observe that c is independent of x and y. $\qquad\square$

We shall derive estimates for the twisted forms. To this end we start with the following lemma, whose proof can be done by induction on $|\beta|$.

8.18 Lemma. *There are constants* $c_{\sigma_1,\ldots,\sigma_k}$ *such that, for* $|\beta| \geq 1$, $\lambda \in \mathbb{R}$, *and any smooth function* ϕ, *we have*

$$\partial^\beta(e^{\lambda\phi}) = \left[\sum_{k=1}^{|\beta|} \lambda^k \sum_{\substack{\sigma_1+\ldots+\sigma_k=\beta \\ \sigma_\nu \neq 0}} c_{\sigma_1,\ldots,\sigma_k} \prod_{\nu=1}^{k} \partial^{\sigma_\nu}\phi\right] e^{\lambda\phi}.$$

This formula yields the following estimate for $\phi \in \mathcal{E}_m$ and $\beta > 0$:

$$|\partial^\beta(e^{\lambda\phi})| \leq \sum_{k=1}^{|\beta|} |\lambda|^k c_{\beta,k}\, e^{\lambda\phi} \quad, \lambda \in \mathbb{R}.$$

By Leibniz' formula we obtain

$$|\partial^\alpha(e^{-\lambda\phi}u)\,\partial^\beta(e^{\lambda\phi}v) - \partial^\alpha u\,\partial^\beta v|$$

$$\leq C_{\alpha,\beta} \sum_{\substack{0\leq\gamma\leq\alpha;\,0\leq\delta\leq\beta \\ |\gamma+\delta|<2m}} (1+|\lambda|)^{2m-|\gamma|-|\delta|}|\partial^\gamma u||\partial^\delta v|$$

which yields

$$|\mathfrak{a}_{\lambda\phi}(u,v) - \mathfrak{a}(u,v)|$$

$$\leq C \sum_{|\alpha|=|\beta|=m} \|a_{\alpha\beta}\|_\infty \sum_{\substack{0\leq l,k\leq m \\ l+k<2m}} (1+|\lambda|)^{2m-(l+k)}\|\nabla^k u\|_2\|\nabla^l v\|_2.$$

Using the estimate $\|\nabla^\nu u\|_2 \leq C\|\nabla^m u\|_2^{\nu/m}\|u\|_2^{1-\nu/m}$ for $\nu = 0,\ldots,m$ (which may be reduced to a special case of b) in Theorem 15.14) and the elementary inequality $x^a y^b \leq \epsilon x^2 + C(\epsilon)y^2$ for $a,b,x,y \geq 0$, $a+b < 2$, we obtain

$$|\mathfrak{a}_{\lambda\phi}(u,u) - \mathfrak{a}(u,u)| \leq \epsilon\|\nabla^m u\|_2^2 + C(\epsilon,\|(a_{\alpha\beta})\|_\infty)(1+\lambda^{2m})\|u\|_2^2, \qquad (8.7)$$

for $u \in H_2^m$ and each $\epsilon > 0$.

We fix $\epsilon \in (0,1)$ and write $C_\epsilon = C(\epsilon,\|(a_{\alpha\beta})\|_\infty)$. If $\omega \geq C_\epsilon$ we conclude that

$$Re\,\mathfrak{a}_{\lambda\phi} \geq Re\,\mathfrak{a} - |\mathfrak{a}_{\lambda\phi} - \mathfrak{a}| \geq (1-\epsilon)Re\,\mathfrak{a} - \omega(1+\lambda^{2m}).$$

Putting $\mathfrak{a}_{\lambda\phi\omega} := \mathfrak{a}_{\lambda\phi} + \omega(1+\lambda^{2m})$ we hence have, by $\epsilon \in (0,1)$,

$$Re\,\mathfrak{a}_{\lambda\phi\omega} \geq (1-\epsilon)Re\,\mathfrak{a} \geq 0.$$

This implies that $\mathfrak{a}_{\lambda\phi\omega}$ is a closed form with form domain H_2^m. Denoting the operator associated to the form $\mathfrak{a}_{\lambda\phi\omega}$ by $A_{\lambda\phi\omega}$ it also implies that

$$\|e^{-tA_{\lambda\phi\omega}}\|_{2\to2} \leq 1 \qquad (t \geq 0, \lambda \in \mathbb{R}, \phi \in \mathcal{E}_m, \omega \geq C_\epsilon).$$

We now aim for estimates like this on a sector. To this end we consider the "rotated" operator $e^{i\theta}A$ where $0 \leq |\theta| < \pi/2 - \psi$. This operator is associated to the form $\mathfrak{b}_\theta := e^{i\theta}\mathfrak{a}$. Then

$$Re\, \mathfrak{b}_\theta = Re\,(e^{i\theta}\mathfrak{a})$$
$$= \cos\theta\, Re\,\mathfrak{a} - \sin\theta\, Im\,\mathfrak{a}$$
$$\geq \cos\theta\cdot Re\,\mathfrak{a}\cdot(1 - \tan|\theta|\cdot\tan\psi)$$
$$= \cos\theta\left(1 - \frac{\tan|\theta|}{\tan(\pi/2 - \psi)}\right)Re\,\mathfrak{a}$$
$$=: c_{\theta,\psi}\, Re\,\mathfrak{a}$$

and

$$|Im\,\mathfrak{b}_\theta| = |\cos\theta\cdot Im\,\mathfrak{a} + \sin\theta\cdot Re\,\mathfrak{a}|$$
$$\leq \cos\theta|Im\,\mathfrak{a}| + \sin|\theta|\, Re\,\mathfrak{a}$$
$$\leq \cos\theta(\tan\psi + \tan|\theta|)\, Re\,\mathfrak{a}$$
$$\leq \frac{\cos\theta(\tan\psi + \tan|\theta|)}{c_{\theta,\psi}}\, Re\,\mathfrak{b}_\theta.$$

Hence $\mathfrak{b}_\theta$ is again closed and sectorial with form domain H_2^m.

We also consider the rotated operators $e^{i\theta}A_{\lambda\phi}$ which are associated to the forms $\mathfrak{b}_{\theta\lambda\phi} := e^{i\theta}\mathfrak{a}_{\lambda\phi}$. The assumption (8.7) and the estimate obtained for $Re\,\mathfrak{b}_\theta$ then imply

$$|\mathfrak{b}_{\theta\lambda\phi} - \mathfrak{b}_\theta| \leq \epsilon\, Re\,\mathfrak{a} + C_\epsilon(1 + \lambda^{2m}) \leq \frac{\epsilon}{c_{\theta,\psi}}\, Re\,\mathfrak{b}_\theta + C_\epsilon(1 + \lambda^{2m})$$

for all $\lambda \in \mathbb{R}$, $\phi \in \mathcal{E}_m$, $|\theta| < \pi/2 - \psi$. We now use the arguments from above to conclude that

$$\|e^{-t(e^{i\theta}A_{\lambda\phi} + \omega(1+\lambda^{2m}))}\|_{2\to2} \leq 1, \quad t \geq 0, \lambda \in \mathbb{R}, \phi \in \mathcal{E}_m, \omega \geq C_\epsilon, \qquad (8.8)$$

for any $|\theta| \leq \theta_0 < \pi/2 - \psi$ provided $\epsilon/c_{\theta_0,\psi} < 1$ which holds for sufficiently small $\epsilon > 0$ (for a fixed θ_0 one has to choose ϵ accordingly to obtain (8.8) for $|\theta| \leq \theta_0$).

We now write $z = te^{i\theta}$ where $|\theta| \leq \theta_0$ and obtain

$$\|e^{-zA_{\lambda\phi}}\|_{2\to2} \leq e^{|z|C_\epsilon(1+\lambda^{2m})} \leq e^{Re\, z(\cos\theta_0)^{-1}C_\epsilon(1+\lambda^{2m})}$$

for all $\lambda \in \mathbb{R}$, $\phi \in \mathcal{E}_m$, and $|\arg z| \leq \theta_0$. Letting now $\omega := C_\epsilon/\cos\theta_0$ we conclude

$$\|e^{-zA_{\lambda\phi\omega}}\|_{2\to2} \leq 1 \qquad (|\arg z| \leq \theta_0, \lambda \in \mathbb{R}, \phi \in \mathcal{E}_m).$$

Fixing $t > 0$ and letting $r := t\sin\theta_0$ the Cauchy formula yields

$$\|A_{\lambda\phi\omega}e^{-tA_{\lambda\phi\omega}}\| = \|\frac{1}{2\pi i}\int_{|z-t|=r}\frac{e^{-zA_{\lambda\phi\omega}}}{(z-t)^2}\,dz\|$$
$$\leq \frac{1}{2\pi}\cdot 2\pi r\cdot\frac{1}{r^2}$$
$$= \frac{1}{r} = \frac{1}{t\sin\theta_0}$$

for all $\lambda \in \mathbb{R}$ and $\phi \in \mathcal{E}_m$.

We shall now use the Gårding inequality $Re\,\mathfrak{a}(u) \geq \delta \|(-\Delta)^{m/2}u\|_2^2$, $u \in H_2^m$, where $\delta > 0$ (see (8.6) above). For $f \in L_2$, $\lambda \in \mathbb{R}$, $\phi \in \mathcal{E}_m$, and $t > 0$ we let $v(t) := e^{-tA_{\lambda\phi\omega}}f$. Then $v(t) \in H_2^m$ and

$$
\begin{aligned}
\|(-\Delta)^{m/2}v(t)\|_2^2 &\leq \delta^{-1}\,Re\,\mathfrak{a}(v(t)) \\
&\leq (\delta(1-\epsilon))^{-1}\,Re\,\mathfrak{a}_{\lambda\phi\omega}(v(t)) \\
&\leq (\delta(1-\epsilon))^{-1}\,|\langle A_{\lambda\phi\omega}v(t), v(t)\rangle| \\
&\leq (\delta(1-\epsilon))^{-1}\,\|A_{\lambda\phi\omega}v(t)\|_2\|v(t)\|_2 \\
&\leq (\delta(1-\epsilon)\sin\theta_0)^{-1}\,t^{-1}\,\|f\|_2^2.
\end{aligned}
$$

We conclude

$$
\|e^{-tA_{\lambda\phi}}\|_{L_2 \to \dot{H}_2^m} \leq (\delta(1-\epsilon)\sin\theta_0)^{-1/2}\,t^{-1/2}\,e^{\omega(1+\lambda^{2m})t}
$$

for all $t > 0$, $\lambda \in \mathbb{R}$, and $\phi \in \mathcal{E}_m$, where $\dot{H}_2^m$ denotes the homogeneous Sobolev space (a Riesz potential space) whose norm is given by $\|u\|_{\dot{H}_2^m} := \|(-\Delta)^{m/2}u\|_2$.

We claim that

$$
e^{-tA_{\lambda\phi}} = e^{-\lambda\phi}\,e^{-tA}\,e^{\lambda\phi} \quad (t \geq 0, \lambda \in \mathbb{R}, \phi \in \mathcal{E}_m), \tag{8.9}
$$

and hence have to show $A_{\lambda\phi} = e^{-\lambda\phi}Ae^{\lambda\phi}$. But

$$
A_{\lambda\phi}u = f \Leftrightarrow u \in H_2^m \text{ and } \forall v \in H_2^m : (f, v) = \mathfrak{a}_{\lambda\phi}(u, v) = \mathfrak{a}(e^{\lambda\phi}u, e^{-\lambda\phi}v),
$$

$$
e^{-\lambda\phi}Ae^{\lambda\phi}u = f \Leftrightarrow A(e^{\lambda\phi}u) = e^{\lambda\phi}f
$$

$$
\Leftrightarrow e^{\lambda\phi}u \in H_2^m \text{ and } \forall v \in H_2^m : (e^{\lambda\phi}f, v) = \mathfrak{a}(e^{\lambda\phi}u, v),
$$

and the claim boils down to the fact that $e^{\pm\lambda\phi}H_2^m = H_2^m$.

Now (8.9) is a weighted estimate for the semigroup acting $L_2(\mathbb{R}^n) \to \dot{H}_2^m(\mathbb{R}^n)$, i.e., we have proved

8.19 Proposition. *For all $t > 0$, $\lambda \in \mathbb{R}$, and $\phi \in \mathcal{E}_m$ we have*

$$
\|e^{-\lambda\phi}\,e^{-tA}\,e^{\lambda\phi}\|_{L_2 \to \dot{H}_2^m} \leq (\delta(1-\epsilon)\sin\theta_0))^{-1/2}\,t^{-1/2}\,e^{\omega(1+\lambda^{2m})t}.
$$

We next derive weighted $(2, q)$-estimates where we distinguish the two cases $n > 2m$ and $n < 2m$ (in [25, p. 59] the case $n = 2m$ is treated: using Sneiberg's Lemma it is essentially reduced to a situation where the arguments used for the case $n > 2m$ apply. We shall not go into details here).

We treat the case $n > 2m$ first. By Sobolev embedding we have $\|u\|_q \leq C\|u\|_{\dot{H}_2^m}$ where $q := 2n/(n-2m)$, and Proposition 8.19 thus yields

$$
\|e^{-\lambda\phi}\,e^{-tA}\,e^{\lambda\phi}\|_{2 \to q} \leq C\,(\delta(1-\epsilon)\sin\theta_0)^{-1/2}\,t^{-1/2}\,e^{\omega(1+\lambda^{2m})t}
$$

for all $t > 0$, $\lambda \in \mathbb{R}$, $\phi \in \mathcal{E}_m$.

If $n < 2m$ we use the following.

8.20 Lemma. *If $n < 2m$ then*

$$\|u\|_\infty \le C \|u\|_{\dot{H}_2^m}^{n/(2m)} \|u\|_2^{1-n/(2m)}.$$

for $u \in H_2^m(\mathbb{R}^n)$.

Proof. We have $\|u\|_\infty \le c\|\hat{u}\|_1$ and, for $\gamma > 0$,

$$\left(\int |\hat{u}(\xi)| \, d\xi \right)^2 \le \int (\gamma^m + |\xi|^{2m}) |\hat{u}(\xi)|^2 \, d\xi \cdot \int (\gamma^m + |\xi|^{2m})^{-1} \, d\xi$$

$$= (\gamma^m \|\hat{u}\|_2^2 + \| \xi \mapsto |\xi|^m \hat{u}\|_2^2) \gamma^{(n-2m)/2} \int (1 + |\eta|^{2m})^{-1} \, d\eta$$

$$\le c'(\gamma^{n/2} \|u\|_2^2 + \gamma^{(n-2m)/2} \|u\|_{\dot{H}_2^m}^2).$$

Now we optimize with respect to γ, taking $\gamma := ((2m-n)\|u\|_{\dot{H}_2^m}^2)/(n\|u\|_2^2))^{1/m}$. $\qquad\square$

Lemma 8.20 and the estimate above yields

$$\|e^{-\lambda\phi} e^{-tA} e^{\lambda\phi}\|_{2\to\infty} \le C \left(\delta(1-\epsilon)\sin\theta_0 \right)^{-n/(4m)} t^{-n/(4m)} e^{\omega(1+\lambda^{2m})t}$$

for all $t > 0$, $\lambda \in \mathbb{R}$, and $\phi \in \mathcal{E}_m$.

Letting $q := \infty$ in the case $n < 2m$, we hence have for $n \ne 2m$

$$\|e^{-\lambda\phi} e^{-tA} e^{\lambda\phi}\|_{2\to q} \le K \, t^{-n/(2m)\cdot(1/2-1/q)} e^{\omega(1+\lambda^{2m})t}$$

for all $t > 0$, $\lambda \in \mathbb{R}$, $\phi \in \mathcal{E}_m$, where K is a fixed constant, depending only on δ, θ_0, ϵ and the Sobolev embedding constants.

An inspection of the proof shows that, for $|\theta| < \theta_0$, we may obtain in the same way

$$\|e^{-\lambda\phi} e^{-te^{i\theta}A} e^{\lambda\phi}\|_{2\to q} \le K_\theta \, t^{-n/(2m)\cdot(1/2-1/q)} e^{\omega(1+\lambda^{2m})t}$$

for all $t > 0$, $\lambda \in \mathbb{R}$, and $\phi \in \mathcal{E}_m$.

Using the arguments for the dual operator A' of A (which satisfies the same assumptions!) we obtain a weighted $(2,q)$-estimate for $(e^{-te^{i\theta}A'})$. Dualization of this estimate leads to a weighted $(p,2)$-estimate for $(e^{-te^{i\theta}A})$ where $p := q'$ is the exponent dual to q. We write

$$e^{-\lambda\phi} e^{-te^{i\theta}A} e^{\lambda\phi} = \left(e^{-\lambda\phi} e^{-\frac{t}{2}e^{i\theta}A} e^{\lambda\phi} \right) \left(e^{-\lambda\phi} e^{-\frac{t}{2}e^{i\theta}A} e^{\lambda\phi} \right)$$

and put together the weighted $(p,2)$-estimate for $t/2$ and the weighted $(2,q)$-estimate for $t/2$. We thus have proved

8.21 Proposition. *Let $n \ne 2m$ and define $q := 2n/(n-2m)$ if $n > 2m$ and $q := \infty$ if $n < 2m$. Let $p := q'$. Then, for any $|\theta| < \pi/2 - \psi$, there are constants $M_\theta, \omega_\theta > 0$ such that*

$$\|e^{-\lambda\phi} e^{-te^{i\theta}A} e^{\lambda\phi}\|_{p\to q} \le M_\theta \, t^{-n/(2m)\cdot(1/p-1/q)} e^{\omega_\theta(1+\lambda^{2m})t} \qquad (8.10)$$

for all $t > 0$, $\lambda \in \mathbb{R}$, and $\phi \in \mathcal{E}_m$.

We take the Euclidean distance $|\cdot - \cdot|$ in $\mathbb{R}^n$ as metric d and denote the open ball with radius $\rho > 0$ and center $x \in \mathbb{R}^n$ by $B(x, \rho)$. We derive the estimates we need in order to apply Theorem 8.6.

8.22 Proposition. *Let (8.10) hold for some θ, $1 \leq p \leq q \leq \infty$, and all $t > 0$, $\lambda \in \mathbb{R}$, $\phi \in \mathcal{E}_m$. Then there is a constant $b_\theta > 0$ such that*

$$\begin{aligned}
&\left\| 1_{B(x,t^{1/(2m)})}\, e^{-te^{i\theta}A}\, 1_{B(y,t^{1/(2m)})} \right\|_{p \to q} \\
&\leq M_\theta t^{-n/(2m)\cdot(1/p-1/q)}\, e^{(\omega_\theta+2)t}\, e^{-b_\theta\left(\frac{|x-y|^{2m}}{t}\right)^{1/(2m-1)}}
\end{aligned}
\tag{8.11}$$

for all $t > 0$ and $x, y \in \mathbb{R}^n$.

Proof. Fix $t > 0$ and $x, y \in \mathbb{R}^n$. For any $\lambda \in \mathbb{R}$ and $\phi \in \mathcal{E}_m$ we have

$$\begin{aligned}
&\left\| 1_{B(x,t^{1/(2m)})}\, e^{-tA}\, 1_{B(y,t^{1/(2m)})} \right\|_{p \to q} \\
&\leq \left\| 1_{B(x,t^{1/(2m)})}\, e^{\lambda\phi} \right\|_\infty \left\| e^{-\lambda\phi}\, e^{-tA}\, e^{\lambda\phi} \right\|_{p \to q} \left\| 1_{B(y,t^{1/(2m)})} e^{-\lambda\phi} \right\|_\infty \\
&\leq e^{\lambda\phi(x)} e^{|\lambda|t^{1/(2m)}} M_\theta t^{-n/(2m)\cdot(1/p-1/q)} e^{\omega_\theta(1+\lambda^{2m})t} e^{-\lambda\phi(y)} e^{|\lambda|t^{1/(2m)}}.
\end{aligned}$$

We observe that $|\lambda|t^{1/(2m)} \leq (1+\lambda^{2m})t$. Taking, for fixed λ, the infimum over all $\phi \in \mathcal{E}_m$ we obtain by [61, Lem.4]

$$\begin{aligned}
&\left\| 1_{B(x,t^{1/(2m)})}\, e^{-tA}\, 1_{B(y,t^{1/(2m)})} \right\|_{p \to q} \\
&\leq M_\theta t^{-n/(2m)\cdot(1/p-1/q)} e^{-c|\lambda||x-y|+(\omega_\theta+2)(1+\lambda^{2m})t}.
\end{aligned}$$

Now optimizing with respect to $|\lambda| \geq 0$, i.e., taking

$$|\lambda| = \left(\frac{c|x-y|}{2mt(\omega_\theta+2)} \right)^{-1/(2m-1)},$$

yields

$$\leq M_\theta t^{-n/(2m)\cdot(1/p-1/q)} e^{(\omega_\theta+2)t} e^{-b_\theta\left(\frac{|x-y|^{2m}}{t}\right)^{1/(2m-1)}}$$

where $b_\theta = \frac{2m-1}{2m}(c^{2m}/(2t(\omega_\theta+2)))^{1/(2m-1)}$. $\square$

8.23 Remark. For an operator of the form (8.5) on $\mathbb{R}^n$ we can get rid of the factor $e^{(\omega_\theta+2)t}$ in (8.11) by scaling and thus obtain

$$\begin{aligned}
&\left\| 1_{B(x,t^{1/(2m)})}\, e^{-te^{i\theta}A}\, 1_{B(y,t^{1/(2m)})} \right\|_{p \to q} \\
&\leq M_\theta\, t^{-n/(2m)\cdot(1/p-1/q)}\, e^{-b_\theta\left(\frac{|x-y|^{2m}}{t}\right)^{1/(2m-1)}}
\end{aligned}
\tag{8.12}$$

for all $t > 0$, $x, y \in \Omega$, $|\theta| < \pi/2 - \psi$, and some constants $M_\theta, b_\theta > 0$ where p and $q = p'$ are as in Proposition 8.21.

Hence the assumptions of Theorem 8.6 are satisfied for all operators $e^{i\theta}A$, $|\theta| < \pi/2 - \psi$. We obtain that A_r is R-sectorial in $L_r(\mathbb{R}^n)$ with $\omega_R(A_r) \leq \psi$ for all $r \in (p, p')$. This yields

8.24 Theorem. *For an operator A of the form (8.5) and p as in 8.21 we have for all $r \in (p, p')$ that A_r is R-sectorial and $\omega_R(A_r) = \omega(A_2)$.*

Here we just have to observe that the optimal ψ equals the type $\omega(A_2)$ of the operator A in L_2. In other words, for $r \in (p, p')$, the operator A_r is R-sectorial in $L_r(\mathbb{R}^n)$ with optimal angle.

Notes on Section 8:

N 8.1: For full details and further information we refer to [137, Ch. 6 §1-2]. The Hilbert space dual A^* of A is associated to the form $\mathfrak{a}^*$; the operator A is self-adjoint if and only if the form $\mathfrak{a}$ is symmetric (i.e., $\mathfrak{a} = \operatorname{Re} \mathfrak{a}$). Notice, however, that Kato's notion of a sectorial operator, which is based on (8.1), is different from ours. We shall come back to (8.1) in Section 11.

N 8.2 recalls 5.6. The same arguments apply to the Laplacian with Dirichlet or Neumann boundary conditions on arbitrary open subsets Ω of $\mathbb{R}^n$. Those operators are also given by the sesquilinear form $\mathfrak{a}$ but with form domain $H_0^1(\Omega) := \overline{C_c^\infty(\Omega)}^{\|\cdot\|_{H^1}}$ or $H^1(\Omega)$, respectively. The semigroup generated by the Dirichlet Laplacian always satisfies Gaussian kernel bounds. For the semigroup generated by the Neumann Laplacian this property depends on the domain Ω and is closely related to Sobolev embeddings of $H^1(\Omega)$. We refer to, e.g., [63], [198] and the literature cited in these references. Concrete examples of Neumann Laplacians without Gaussian bounds are in [142]. For applications to Schrödinger operators with general potentials we refer to [167].

N 8.4: Spaces of homogeneous type are designed to provide a frame for a variety of arguments in harmonic analysis that have first been used on $\mathbb{R}^n$ or the torus $\mathbb{T}^n$. In particular, the properties of the maximal function we use in this section rely on the doubling property. Bounded open subsets of $\mathbb{R}^n$ with a Lipschitz boundary satisfy the doubling property with respect to Lebesgue measure, but, e.g., exponential horns $\Omega := \{(x, y) \in \mathbb{R}^2 : x > 0, |y| < e^{-x}\}$ as in [142] do not.

N 8.5 serves as a motivation for the weighted norm bounds we consider here. In the form given in (8.4), i.e., with projection on the two balls $B(x, t^{1/m})$, $B(y, t^{1/m})$ they were introduced in [143]. Weighted norm bounds of Gaussian type were first introduced in a form corresponding to (8.10) in [200]. They were used to show p-independence of the L_p-spectrum of Schrödinger operators with potentials from the pseudo-Kato class. The technique of those weighted norm estimates was further developed in [199], [61], [171], [145], [170]. The weighted norm bounds of annular type used in the proof of 8.9 were introduced in [33].

N 8.6 is a simplified version of the main result in [33] (cf. also [143]). In the case $(p, q) = (1, \infty)$ it was first proved in [119], and improved in [52]. The proofs given there relied on the technique from [78] and are quite involved.

Simplified proofs involving domination by maximal functions have been given in [48], [32], [220] (cf. also N 8.9).

N 8.8 is well-known from harmonic analysis.

N 8.9 is from [33]. The proof given here is slightly different. In the special case $(p, q) = (1, \infty)$ the estimate implies

$$e^{-tA} f(x) \leq C_1 M_1 f(x), \quad t > 0, \text{a.e. } x \in \Omega.$$

This estimate was used in [48] to give a simplified proof of the main results in [119], [52], which correspond to the assertion of Theorem 8.6 for the case $(p, q) = (1, \infty)$.

N 8.10: By estimating more carefully in Proposition 8.9 one can obtain in Corollary 8.10 a constant C' which is independent of $r \in (p, q]$ in place of $C(r)$. Then the estimate holds also in the limit case $r = p$.

N 8.11: One can show that (e^{-tA}) induces bounded analytic semigroups also in the limit spaces $L_p(\Omega)$ and $L_q(\Omega)$ provided $q < \infty$. Moreover, the type of the negative generator $-A_r$ does not depend on $r \in [p, q] \setminus \{\infty\}$. (see, e.g., [185], [170]).

N 8.12 is taken from [220].

N 8.13: The second assertion was first proved in [33] (cf. also [143]).

N 8.14 - N 8.15 are from [33]. The proof of 8.15 is taken from [143].

N 8.16 - N 8.24: The class of elliptic operators studied here is a complex analog of the class of self-adjoint operators Davies introduced in [61]. We refer also to [25] and [170, Sect.6]. The proof of weighted norm bounds given here parallels [61]. It is taken from [143]. Lemma 8.17 is [61, Lem.4] where the lower bound is proved for $c = 1$. Lemma 8.20 and its proof are [61, Lem.16] Theorem 8.24 was given in [33].

As stated in 8.21 one has $p = 1$ which corresponds to Gaussian kernel bounds if $n < 2m$ (this was shown in [61]). It has been shown in [25] that this still holds for $n = 2m$. If $n > 2m$ then Gaussian bounds do not hold in general, even in the self-adjoint case: By [62] the range $[p, p']$ in 8.21 is optimal in the following sense: for every $r \notin [p, q]$ there is an operator in the given class such that the corresponding semigroup does not extend to L_r.

For the case $n > 2m$ there is a characterization of the operators for which the semigroup satisfies Gaussian bounds in terms of elliptic regularity, see [25] and [24].

Since the operators are given by a sectorial form in L_2 they have a bounded H^∞-calculus in L_2 (cf. Section 11). The weighted norm bounds may be used to extrapolate a bounded H^∞-calculus to L_p-spaces with $p \neq 2$ (we refer to Section 14, see also [78], [82] for the case $(p, q) = (1, \infty)$ and [34] for the general case).

II. The H^∞-calculus

9 Construction of the H^∞-calculus

In this section, we would like to construct a functional calculus for sectorial operators looking towards the Dunford calculus for inspiration. If $A \in B(X)$ and $\sigma(A) \subset \Sigma_\omega$, then we can define a map

$$f \in H^\infty(\Sigma_\sigma) \mapsto f(A) := \frac{1}{2\pi i} \int_{\partial \Gamma_n} f(\lambda) R(\lambda, A) d\lambda, \tag{9.1}$$

where $\Gamma_n := \Sigma_{\sigma'} \cap B(0, n)$ with $\omega < \sigma' < \sigma < \pi$ and $n > \|A\|$. $\partial \Gamma_n$ is oriented counterclockwise. It is well known that this map is a bounded algebra homomorphism from $H^\infty(\Sigma_\sigma)$ with sup norm $\|f\|_{H^\infty(\Sigma_\sigma)} = \sup_{\lambda \in \Sigma_\sigma} |f(\lambda)|$ to $B(X)$. It also has a convergence property corresponding to Lebesgue's Convergence Theorem:
If $f_n, f \in H^\infty(\Sigma_\sigma)$, $\|f_n\| \leq C$ and $\lim_{n \to \infty} f_n(\lambda) = f(\lambda)$ for $\lambda \in \Sigma_\sigma$, then $\lim_{n \to \infty} f_n(A)x = f(A)x$ for all $x \in X$.
These properties allow us to form many operators important in semigroup theory and evolution equations such as

$$e^{-tA}, \quad t^n A^n e^{-tA}, \quad A^n R(\lambda, A)^n, \quad \lambda^n A^{\alpha n} R(\lambda, A^\alpha)^n, \quad A^{it}$$

($\lambda \notin \Gamma_n, t \in \mathbb{R}_+, n \in \mathbb{N}$) by simply plugging A into the corresponding analytic functions and also to estimate their norm by the sup of these functions on Γ_n (cf. Illustration 9.9).

Furthermore, algebraic rules for functions translate directly into operator equations, e.g.

$$(\lambda^\alpha e^{-t\lambda})(\lambda^\beta e^{-s\lambda}) = \lambda^{\alpha+\beta} e^{-(t+s)\lambda} \mapsto (A^\alpha e^{-tA})(A^\beta e^{-sA}) = A^{\alpha+\beta} e^{-(t+s)A}$$

and formulas like

$$\mu^{-\alpha} = \frac{1}{\Gamma(\alpha)} \int_0^\infty t^{\alpha-1} e^{-\mu t} dt$$

can be used to show that (cf. Lemma 9.12)

$$A^{-\alpha} = \frac{1}{\Gamma(\alpha)} \int_0^\infty t^{\alpha-1} e^{-tA} dt.$$

In order to move towards unbounded sectorial operators, assume now that f in (9.1) has some decay at 0 and ∞, say

$$|f(\lambda)| \leq C|\varrho(\lambda)|^{\varepsilon}, \qquad \text{for all } \lambda \in \Sigma_{\sigma} \tag{9.2}$$

for some $C, \varepsilon > 0$, where $\varrho(\lambda) := \frac{\lambda}{(1+\lambda)^2}$. Then we can take the limit $n \to \infty$ and obtain

$$f(A) = \frac{1}{2\pi i} \int_{\partial \Sigma_{\sigma'}} f(\lambda) R(\lambda, A) d\lambda.$$

This formula also makes sense for a sectorial operator of angle ω, since we have the growth estimate $\|R(\lambda, A)\| \sim \frac{1}{\lambda}$ on $\partial \Sigma_{\sigma'}$. Recall that an operator A on X with dense domain is *sectorial* of angle $\omega \in (0, \pi)$, if $\sigma(A) \subset \overline{\Sigma_{\omega}}$ and if for $\omega' > \omega$ we have that the set $\{\lambda R(\lambda, A) : \omega' \leq |\arg \lambda| \leq \pi\}$ is bounded. $\omega(A)$ is the infimum over all such ω. In this section we will find it convenient to work with operators that are also injective and have dense range. (This is not really a loss of generality, since a sectorial operator has restrictions with these additional properties; see §15.) We introduce the notation $\mathcal{S}(X)$ for the class of sectorial operators on X that are injective and have dense range.

Of course we cannot hope to define $f(A)$ in terms of the Dunford integral above for arbitrary $f \in H^{\infty}(\Sigma_{\sigma})$ (since $\frac{1}{\lambda}$ is not integrable on $\mathbb{R}_+$), but we can use it to define a first "auxiliary" calculus that we will extend to a larger class of functions later on in 9.6 and 9.10.

9.1 Notation. For $0 < \sigma < \pi$ put

$$H_0^{\infty}(\Sigma_{\sigma}) = \{f \in H^{\infty}(\Sigma_{\sigma}) : \text{ there are } C, \varepsilon > 0 \text{ such that } (9.2) \text{ holds}\}.$$

We define now a functional calculus on $H_0^{\infty}(\Sigma_{\sigma})$.

9.2 Theorem. *Let X be a Banach space and $A \in \mathcal{S}(X)$ a sectorial operator of angle ω. For $\omega < \sigma' < \sigma$ and $f \in H_0^{\infty}(\Sigma_{\sigma})$ put*

$$\Phi_A(f) = f(A) = \frac{1}{2\pi i} \int_{\partial \Sigma_{\sigma'}} f(\lambda) R(\lambda, A) d\lambda. \tag{9.3}$$

Then $\Phi_A : H_0^{\infty}(\Sigma_{\sigma}) \to B(X)$ defines a linear and multiplicative map with the following properties:

(i) *Let $f_n, f \in H^{\infty}(\Sigma_{\sigma})$ be uniformly bounded and $f_n(\lambda) \to f(\lambda)$ for $\lambda \in \Sigma_{\sigma}$. Then for all $g \in H_0^{\infty}(\Sigma_{\sigma})$*

$$\lim_{n \to \infty} \Phi_A(f_n \cdot g) = \Phi_A(f \cdot g)$$

in $B(X)$.

(ii) *If $f(\lambda) = \frac{\lambda}{(\mu_1 - \lambda)(\mu_2 - \lambda)}$ with $\mu_1, \mu_2 \notin \overline{\Sigma_{\sigma}}$, then*

$$f(A) = A R(\mu_1, A) R(\mu_2, A).$$

(iii) *$\|f(A)\| \leq \frac{c}{2\pi} \int_{\partial \Sigma_{\sigma'}} |f(\lambda)| \frac{d|\lambda|}{|\lambda|}$ for some positive constant c.*

9.3 Remark on Contour Integrals. a) Let $\Sigma(\sigma_1, \sigma_2) = \{\lambda \in \mathbb{C} \setminus \{0\} : \sigma_1 < \arg \lambda < \sigma_2\}$. If $\partial\Sigma(\sigma_1, \sigma_2)$ appears in a contour integral such as in (9.3) we always choose the *orientation* on $\partial\Sigma(\sigma_1, \sigma_2)$ in such a way that the interior of $\Sigma(\sigma_1, \sigma_2)$ lies to the left as we move along the boundary.

b) The integral (9.3) is independent of the particular choice of $\sigma' \in (\sigma, \omega)$. Indeed, let G be analytic on $\Sigma(\sigma_1, \sigma_2)$ with values in a Banach space Y. If in addition $\|G(\lambda)\| \leq C|\lambda|^{-1}|\varrho(\lambda)|^\varepsilon$ on $\Sigma(\nu_1, \nu_2)$ for some $\varepsilon > 0$ and $\sigma_1 < \nu_1 < \nu_2 < \sigma_2$, then as an extension of *Cauchy's theorem* we have

$$\int_{\partial\Sigma(\nu_1,\nu_2)} G(\lambda)d\lambda = 0.$$

Indeed, with $\Sigma_n = \partial\Sigma(\nu_1, \nu_2) \cap \{\lambda : \frac{1}{n} < |\lambda| < n\}$ and $\Gamma_n = \Sigma(\nu_1, \nu_2) \cap \{\lambda : |\lambda| = n, |\lambda| = \frac{1}{n}\}$ we have by Cauchy's theorem

$$\int_{\partial\Sigma(\nu_1,\nu_2)} G(\lambda)d\lambda = \lim_{n\to\infty} \int_{\Sigma_n} G(\lambda)d\lambda$$

$$= \lim_{n\to\infty} \int_{\Sigma_n \cup \Gamma_n} G(\lambda)d\lambda - \lim_{n\to\infty} \int_{\Gamma_n} G(\lambda)d\lambda = 0,$$

since $\|\int_{\Gamma_n} G(\lambda)d\lambda\| \leq 2\pi C \sup_{\lambda\in\Gamma_n} |\varrho(\lambda)|^\varepsilon \to 0$ for $n \to \infty$.

c) Similarly we can extend *Cauchy's formula*: If $\|G(\lambda)\| \leq C|\varrho(\lambda)|^\varepsilon$ on $\Sigma(\nu_1, \nu_2)$ for some $\varepsilon > 0$, then for $\mu \in \Sigma(\nu_1, \nu_2)$

$$G(\mu) = \frac{1}{2\pi i} \int_{\partial\Sigma(\nu_1,\nu_2)} \frac{1}{\lambda - \mu} G(\lambda)d\lambda.$$

Proof of 9.2. $f(A)$ belongs to $B(X)$, since $\|f(\lambda)R(\lambda, A)\| \leq C|\lambda|^{-1}|\varrho(\lambda)|^\varepsilon$ for $\lambda \in \partial\Sigma_{\sigma'}$ so that the integral (9.3) exists in $B(X)$. Linearity is clear. To prove multiplicativity we choose $\omega < \sigma' < \sigma'' < \sigma$. Then, for $f, g \in H_0^\infty(\Sigma_\sigma)$,

$$\Phi_A(f)\Phi_A(g) = \left(\frac{1}{2\pi i}\right)^2 \left(\int_{\partial\Sigma_{\sigma''}} f(\mu)R(\mu, A)d\mu\right)\left(\int_{\partial\Sigma_{\sigma'}} g(\lambda)R(\lambda, A)d\lambda\right)$$

$$= \left(\frac{1}{2\pi i}\right)^2 \int_{\partial\Sigma_{\sigma''}}\int_{\partial\Sigma_{\sigma'}} f(\mu)g(\lambda)R(\mu, A)R(\lambda, A)d\lambda\, d\mu$$

$$= \left(\frac{1}{2\pi i}\right)^2 \int_{\partial\Sigma_{\sigma''}}\int_{\partial\Sigma_{\sigma'}} \frac{f(\mu)g(\lambda)}{\lambda - \mu}[R(\mu, A) - R(\lambda, A)]d\lambda\, d\mu$$

by the resolvent equation and

$$\Phi_A(f)\Phi_A(g) = \frac{1}{2\pi i}\int_{\partial\Sigma_{\sigma'}} g(\lambda)\left[\frac{1}{2\pi i}\int_{\partial\Sigma_{\sigma''}} \frac{f(\mu)}{\mu - \lambda}d\mu\right] R(\lambda, A)d\lambda$$

$$+ \frac{1}{2\pi i}\int_{\partial\Sigma_{\sigma''}} f(\mu)\left[\frac{1}{2\pi i}\int_{\partial\Sigma_{\sigma'}} \frac{g(\lambda)}{\lambda - \mu}d\lambda\right] R(\mu, A)d\lambda$$

$$= \frac{1}{2\pi i}\int_{\partial\Sigma_{\sigma'}} g(\lambda)f(\lambda)R(\lambda, A)d\lambda = \Phi_A(f \cdot g),$$

using Remark 9.3. Indeed, $\frac{1}{2\pi i}\int_{\partial\Sigma_{\sigma''}}f(\mu)(\mu-\lambda)^{-1}d\mu = f(\lambda)$ by Cauchy's formula and $\frac{1}{2\pi i}\int_{\partial\Sigma_{\sigma'}}g(\lambda)(\lambda-\mu)^{-1}d\lambda = 0$ since $\lambda\mapsto g(\lambda)(\lambda-\mu)^{-1}$ is analytic on $\Sigma_{\sigma'}$.

(i) Since $|f_n(\lambda)g(\lambda)|\leq C|g(\lambda)|$ for all $\lambda\in\Sigma_\sigma$ we obtain with Lebesgue's Convergence Theorem:

$$\Phi_A(f\cdot g) = \frac{1}{2\pi i}\int_{\partial\Sigma_{\sigma'}}f(\lambda)g(\lambda)R(\lambda,A)d\lambda$$

$$= \lim_{n\to\infty}\frac{1}{2\pi i}\int_{\partial\Sigma_{\sigma'}}f_n(\lambda)g(\lambda)R(\lambda,A)d\lambda$$

$$= \lim_{n\to\infty}\Phi_A(f_n\cdot g).$$

(ii) First we consider $g(\lambda) = (\mu-\lambda)^{-1}$ with $\mu\notin\overline{\Sigma_{\sigma'}}$. If $x\in R(A)$, say $x = Ay$ with $y\in D(A)$, then

$$R(\lambda,A)x = R(\lambda,A)Ay = \lambda R(\lambda,A)y - y$$

is bounded in $\{\lambda\in\mathbb{C}\setminus\Sigma_{\sigma'} : 0<|\lambda|<1\}$. Put $\Sigma(\varepsilon) = \Sigma_{\sigma'}\cup B(0,\varepsilon)$. Then

$$\int_{\partial\Sigma_{\sigma'}}\frac{1}{\mu-\lambda}R(\lambda,A)x\,d\lambda = \lim_{\varepsilon\to 0}\int_{\partial\Sigma(\varepsilon)}\frac{1}{\mu-\lambda}R(\lambda,A)x\,d\lambda$$

$$= \int_{\partial\Sigma(\delta)}\frac{1}{\mu-\lambda}R(\lambda,A)x\,d\lambda$$

for a fixed δ with $0<\delta<|\mu|$ by Cauchy's theorem. Applying Cauchy's theorem to $\tilde{\Sigma} = \mathbb{C}\setminus\overline{\Sigma(\delta)}$ and $\mu\in\tilde{\Sigma}$ (we have no singularity at 0 here!) we obtain for $x\in R(A)$

$$\frac{1}{2\pi i}\int_{\partial\Sigma_{\sigma'}}\frac{1}{\mu-\lambda}R(\lambda,A)x\,d\lambda = \frac{1}{2\pi i}\int_{\partial\tilde{\Sigma}}\frac{1}{\lambda-\mu}R(\lambda,A)x\,d\lambda \tag{9.4}$$

$$= R(\mu,A)x.$$

If $\mu_1\neq\mu_2$ we write f as

$$f(\lambda) = a(\mu_1-\lambda)^{-1} + b(\mu_2-\lambda)^{-1},\quad a = \frac{\mu_1}{\mu_2-\mu_1},\ b = \frac{\mu_2}{\mu_1-\mu_2}.$$

For $x\in R(A)$ we get from (9.4)

$$\Phi_A(f)x = \frac{1}{2\pi i}\int_{\partial\Sigma_{\sigma'}}[a(\mu_1-\lambda)^{-1} + b(\mu_2-\lambda)^{-1}]R(\lambda,A)x\,d\lambda$$

$$= aR(\mu_1,A)x + bR(\mu_2,A)x$$

$$= (\mu_2-\mu_1)^{-1}[\mu_1 R(\mu_1,A) - \mu_2 R(\mu_2,A)]x$$

$$= (\mu_2-\mu_1)^{-1}A[R(\mu_1,A) - R(\mu_2,A)]x$$

$$= AR(\mu_1,A)R(\mu_2,A)x$$

by the resolvent equation. Since $R(A)$ is dense in X the claim follows.

If $\mu_1 = \mu_2 = \mu$, we choose a sequence ν_n in $\mathbb{C} \setminus \Sigma_\sigma$ such that $\nu_n \neq \mu$ and $\nu_n \to \mu$. Let $g(\lambda) := \frac{\lambda^{1/2}}{\mu - \lambda}$ and $g_n(\lambda) := \frac{\lambda^{1/2}}{\nu_n - \lambda}$. Then $g_n, g \in H_0^\infty(\Sigma_\sigma)$ are uniformly bounded and $g_n(\lambda) \to g(\lambda)$ for $\lambda \in \Sigma_\sigma$. Now by (i)

$$\Phi_A(f) = \Phi_A(g^2) = \lim_{n \to \infty} \Phi_A(g_n \cdot g).$$

We have proved above that $\Phi_A(g_n \cdot g) = AR(\nu_n, A)R(\mu, A)$. This concludes the proof of (ii).

(iii) follows by a direct estimate of (9.3).   □

Of course, Theorem 9.2 does not give yet a satisfying functional calculus since it does not apply to standard functions such as $(\mu - \lambda)^{-1}$, $e^{t\lambda}$ or λ^{it}. But Theorem 9.2 can be improved by an approximation procedure based on an "approximate unity" ϱ_n that we introduce now.

9.4 Proposition. *Let $A \in \mathcal{S}(X)$ be a sectorial operator.*

(a) For all $x \in X$ we have

$$\lim_{t \to \infty} t(t + A)^{-1}x = x, \quad \lim_{t \to 0} t(t + A)^{-1}x = 0$$

$$\lim_{t \to \infty} A(t + A)^{-1}x = 0, \quad \lim_{t \to 0} A(t + A)^{-1}x = x$$

(b) For $n \geq 2$ define $\varrho_n(\lambda) := \frac{n}{n+\lambda} - \frac{1}{1+n\lambda}$. Then

$$\varrho_n(A) = n(n + A)^{-1} - \frac{1}{n}(\frac{1}{n} + A)^{-1}.$$

For $k \in \mathbb{N}$ the range of $\varrho_n(A)^k$ equals $D(A^k) \cap R(A^k)$. Furthermore,

$$\lim_{n \to \infty} \varrho_n(A)^k x = x \quad \text{for all } x \in X.$$

(c) $D(A^k) \cap R(A^k)$ is dense in X for all $k \in \mathbb{N}$.

(d) $\varrho(A) = A(1 + A)^{-2}$ has range $R(A) \cap D(A)$. Recall that $\varrho(\lambda) = \frac{\lambda}{(1 + \lambda)^2}$.

The operator sequence $\varrho_n(A)$ will be used as an approximate identity.

Proof. (a) Everything follows from

$$I - t(t + A)^{-1} = A(t + A)^{-1} \tag{9.5}$$

and the fact that $\tau = \{t(t + A)^{-1}, A(t + A)^{-1} : t > 0\}$ is bounded in $B(X)$. If $x \in D(A)$ then by (9.5)

$$t(t + A)^{-1}x - x = -\frac{1}{t}[t(t + A)^{-1}](Ax) \to 0 \quad \text{for } t \to \infty.$$

Similarly if $x \in R(A)$, i.e. $x = Ay$ for some $y \in D(A)$, then

$$A(t + A)^{-1}x - x = -t[(t + A)^{-1}A](y) \to 0 \quad \text{for } t \to 0.$$

Hence the first and fourth claim hold for x in a dense subset of X. Since τ is bounded they hold for all $x \in X$ by a 3ε-argument. The remaining two claims follow from these and (9.5).

(b) $\varrho_n(\lambda) = (\frac{n^2-1}{n})\frac{\lambda}{(-\frac{1}{n}-\lambda)(-n-\lambda)}$ is in $H_0^\infty(\Sigma_\sigma)$ and we can apply Theorem 9.2(ii) so that

$$\varrho_n(A) = \frac{n^2-1}{n}A(n + A)^{-1}(\tfrac{1}{n} + A)^{-1} = n(n + A)^{-1} - \tfrac{1}{n}(\tfrac{1}{n} + A)^{-1}.$$

The boundedness of the sequence $\varrho_n(A)$ is clear since $A \in \mathcal{S}(X)$. For $k \in \mathbb{N}$,

$$\varrho_n(A)^k = (\tfrac{n^2-1}{n})^k \underbrace{(n + A)^{-k}}_{=:U} \underbrace{A^k(\tfrac{1}{n} + A)^{-k}}_{=:V}$$

$$= (\tfrac{n^2-1}{n})^k A^k \underbrace{(n + A)^{-k}(\tfrac{1}{n} + A)^{-k}}_{=:W}$$

where V is bounded and U and W map X into $D(A^k)$. Hence the range of $\varrho_n(A)^k$ is contained in $D(A^k) \cap R(A^k)$. For the reverse inclusion, take $x \in D(A^k) \cap R(A^k)$ and put $y = (\frac{n}{n^2-1})^k(\frac{1}{n} + A)^k(n + A)^k A^{-k}x$. Then $\varrho_n(A)^k y = x$, i.e., x is in the range of $\varrho_n(A)^k$.

By (a) we have $\lim_{n\to\infty} \varrho_n(A)x = x$ and then also

$$\lim_{n\to\infty} \varrho_n(A)^k x = x.$$

(c) follows from the last part of (b), and (d) can be proved as (b).   □

One can think of Φ_A as a linear, possibly unbounded operator from $(H^\infty(\Sigma_\sigma), \|\cdot\|_{H^\infty(\Sigma_\sigma)})$ to $B(X)$ with domain $H_0^\infty(\Sigma_\sigma)$. For this operator we now will construct a closed extension whose domain $H_A^\infty(\Sigma_\sigma)$ will contain all $f \in H^\infty(\Sigma_\sigma)$ for which $f(A)$ can be defined as a bounded operator on X. Of course the graph norm of Φ_A will be involved in this procedure.

9.5 Definition. *Let $A \in \mathcal{S}(X)$ and $\sigma > \omega(A)$. For $f \in H_0^\infty(\Sigma_\sigma)$ define*

$$\|f\|_A = \|f\|_{H^\infty(\Sigma_\sigma)} + \|f(A)\|_{B(X)}.$$

Then $H_A^\infty(\Sigma_\sigma)$ is the class of functions $f \in H^\infty(\Sigma_\sigma)$, for which there exists a sequence $f_n \in H_0^\infty(\Sigma_\sigma)$ with $f_n(\lambda) \to f(\lambda)$ for all $\lambda \in \Sigma_\sigma$ and $\sup_n \|f_n\|_A < \infty$.

It is easy to check that $H_A^\infty(\Sigma_\sigma)$ is a subalgebra of $H^\infty(\Sigma_\sigma)$.

Let $f \in H_A^\infty(\Sigma_\sigma)$ and $f_n \in H_0^\infty(\Sigma_\sigma)$ with $f_n(\lambda) \to f(\lambda)$ on Σ_σ and $\|f_n\|_A \leq C$. In the next theorem we will show that the limit

$$\overline{\Phi}_A(f)x := \lim_{n \to \infty} \Phi_A(f_n)x, \quad x \in X \tag{9.6}$$

always exists, and that this definition produces a functional calculus on $H_A^\infty(\Sigma_\sigma)$.

9.6 Theorem. *Let $A \in \mathcal{S}(X)$ and $\sigma > \omega(A)$. Then there exists an extension $\overline{\Phi}_A : H_A^\infty(\Sigma_\sigma) \to B(X)$ of Φ_A (in Theorem 9.2) with the following properties:*

(i) $\overline{\Phi}_A$ *is linear and multiplicative.*

(ii) $\tau_\mu(\lambda) = (\mu - \lambda)^{-1}$ *belongs to $H_A^\infty(\Sigma_\sigma)$ for $\mu \notin \overline{\Sigma_\sigma}$ and $\overline{\Phi}_A(\tau_\mu) = R(\mu, A)$.*

(iii) *If $f_n \in H_A^\infty(\Sigma_\sigma)$ and $f \in H^\infty(\Sigma_\sigma)$ with $f_n(\lambda) \to f(\lambda)$ for $\lambda \in \Sigma_\sigma$ and $\|f_n\|_A \leq C$, then $f \in H_A^\infty(\Sigma_\sigma)$, $\lim \overline{\Phi}_A(f_n)x = \overline{\Phi}_A(f)x$ for $x \in X$ and $\|\overline{\Phi}_A(f)\| \leq C$.*

In particular, $\overline{\Phi}_A$ is closed as an operator from $H^\infty(\Sigma_\sigma)$ to $B(X)$ with domain $H_A^\infty(\Sigma_\sigma)$.

Remark. Note the following consequence of (iii): If M is a closed subset of $H^\infty(\Sigma_\sigma)$ (with respect to uniform convergence) and contained in $H_A^\infty(\Sigma_\sigma)$, then $\|\overline{\Phi}_A(f)\| \leq C\|f\|_{H^\infty(\Sigma_\sigma)}$ for some $C < \infty$ and all $f \in M$.

Proof of Theorem 9.6. Let $f \in H_A^\infty(\Sigma_\sigma)$ and $f_n \in H_0^\infty(\Sigma_\sigma)$ with $f_n(\lambda) \to f(\lambda)$ and $\|f_n\|_A \leq C$. By the convergence assertion in Theorem 9.2(i) we get with ϱ from (9.2)

$$\lim_{n \to \infty} f_n(A)(\varrho(A)x) = \lim_{n \to \infty} (f_n \cdot \varrho)(A)x = (f \cdot \varrho)(A)x \tag{9.7}$$

for all $x \in X$. Hence for all y in the dense subset $R(\varrho(A))$ we can define

$$\overline{\Phi}_A(f)y = \lim_{n \to \infty} f_n(A)\varrho(A)[\varrho(A)^{-1}y] = (f \cdot \varrho)(A)[\varrho(A)^{-1}y]. \tag{9.8}$$

Since $\|f_n(A)\| \leq C$ this limit exists for all $x \in X$ by a 3ε–argument and defines a bounded operator $\overline{\Phi}_A(f)$ with $\|\overline{\Phi}_A(f)\| \leq C$. (9.7) and (9.8) also show that $\overline{\Phi}_A(f)$ does not depend on a particular choice of the approximating sequence (f_n). Furthermore, by choosing the stationary sequence $f, f, f, \ldots$, we see

$$\overline{\Phi}_A(f) = \Phi_A(f) \quad \text{for } f \in H_0^\infty(\Sigma_\sigma).$$

(i) follows from $\overline{\Phi}_A(f)x = \lim_{n \to \infty} \Phi_A(f_n)x$.

(ii) Given $\mu \in \mathbb{C} \setminus \overline{\Sigma_\sigma}$, choose $f_n(\lambda) = \frac{-\lambda}{(\mu-\lambda)(-\frac{1}{n}-\lambda)}$. Then $\lim_{n \to \infty} f_n(\lambda) = (\mu - \lambda)^{-1}$, $\|f_n\|_{H^\infty(\Sigma_\sigma)} \leq C < \infty$ and $f_n(A) = -AR(-\frac{1}{n}, A)R(\mu, A)$ (by Theorem 9.2 (ii)) is also a bounded sequence in $B(X)$. Hence $f \in H_A^\infty(\Sigma_\sigma)$ and $\lim_n f_n(A)x = R(\mu, A)x$ by Proposition 9.4 a).

(iii) Let f_n and f be as in (iii). We want to show that $f \cdot \varrho_m \in H_0^\infty(\Sigma_\sigma)$, $m \in \mathbb{N}$, $m \geq 2$ is an approximating sequence for f. Of course $(f\varrho_m)(\lambda) \to f(\lambda)$ for

$\lambda \in \Sigma_\sigma$ (by the definition of ϱ_m in 9.4). To see that $\sup \|(f\varrho_m)(A)\| < \infty$, we observe that

$$(f\varrho_m)(A) = \lim_{n\to\infty} (f_n\varrho_m)(A) \tag{9.9}$$

by Theorem 9.2(i). Hence

$$\|(f\varrho_m)(A)\| \leq \sup_n \|(f_n\varrho_m)(A)\| = \sup_n \|f_n(A)\varrho_m(A)\|$$
$$\leq C \sup_m \|\varrho_m(A)\| < \infty$$

by Proposition 9.4, and we have shown that $f \in H_A^\infty(\Sigma_\sigma)$. (9.9) implies

$$\overline{\Phi}_A(f)(\varrho_m(A)x) = \lim_{n\to\infty} \overline{\Phi}_A(f_n)(\varrho_m(A)x) \quad \text{for } x \in X.$$

Since $\|\overline{\Phi}_A(f_n)\| \leq C$ and $R(\varrho_m(A))$ is dense in X, it follows that $\overline{\Phi}_A(f)x = \lim \overline{\Phi}_A(f_n)x$ for all $x \in X$ and $\|\overline{\Phi}_A(f)\| \leq C$.

The closedness of $\overline{\Phi}_A$ is a special case of (iii). If M is as in the remark, then M is complete with respect to the norm of H^∞ and H_A^∞. By the open mapping theorem these norms must be equivalent on M. □

As usual we use for $\overline{\Phi}_A(f)$ also the classical notation $f(A)$.

9.7 Remark (Uniqueness of the H^∞-calculus). Let $A \in \mathcal{S}(X)$ be a sectorial operator on X and $\sigma > \omega(A)$. If $\Psi : H_A^\infty(\Sigma_\sigma) \to B(X)$ satisfies the properties of the H^∞ functional calculus, namely (i), (ii), and (iii) in Theorem 9.6, then $\Psi = \overline{\Phi}_A$ on $H_A^\infty(\Sigma_\sigma)$.

Proof. If $\phi \in H_0^\infty(\Sigma_{\sigma_1})$ for $\sigma_1 > \sigma$, we can interpret Cauchy's formula as a Bochner integral in $H_0^\infty(\Sigma_\sigma)$:

$$\phi(\cdot) = \frac{1}{2\pi i} \int_{\partial\Sigma_\omega} \frac{\phi(\mu)}{\mu - \cdot} d\mu = \int_{\partial\Sigma_\omega} g_\mu(\cdot)h(\mu)d\mu$$

where $\omega \in (\sigma, \sigma_1)$, $g_\mu(\cdot) = \mu(\mu - \cdot)^{-1}$ is bounded in $H_0^\infty(\Sigma_\sigma)$ and $h(\mu) = \frac{1}{2\pi i} \frac{\phi(\mu)}{\mu}$ is integrable. Ψ is a closed operator by (iii) and therefore

$$\Psi(\phi) = \int_{\partial\Sigma_\omega} \Psi(g_\mu)h(\mu)d\mu = \int_{\partial\Sigma_\omega} \mu R(\mu, A)h(\mu)d\mu$$
$$= \frac{1}{2\pi i} \int_{\partial\Sigma_\omega} \phi(\mu)R(\mu, A)d\mu = \overline{\Phi}_A(\phi).$$

If $\phi \in H_0^\infty(\Sigma_\sigma)$, consider the function $\phi_n(\lambda) = \phi(\lambda^{1-\frac{1}{n}})$, which belongs to $H_0^\infty(\Sigma_{\sigma\frac{n}{n-1}})$ for $n \geq 2$. We just saw that $\Psi(\phi_n) = \overline{\Phi}_A(\phi_n)$. Furthermore $\phi_n(\lambda) \to \phi(\lambda)$ for $n \to \infty$ and $\lambda \in \Sigma_\sigma$ and

$$\limsup_{n\to\infty} \|\overline{\Phi}_A(\phi_n)\| \leq \int_{\partial\Sigma_\sigma} |\phi(\lambda)|\, \|R(\lambda, A)\| d\lambda < \infty.$$

By the convergence property (iii) we have $\Psi(\phi) = \overline{\Phi}_A(\phi)$.

For a general $f \in H_A^\infty(\Sigma_\sigma)$ there are $f_n \in H_0^\infty(\Sigma_\sigma)$ with $f_n(\lambda) \to f(\lambda)$ for $\lambda \in \Sigma_\sigma$, $\|f_n\|_{H^\infty(\Sigma_\sigma)} \leq C$ and $\|\Psi(f_n)\| = \|\overline{\Phi}_A(f_n)\| \leq C$ since $\Psi(f_n) = \overline{\Phi}_A(f_n)$ by the last step. However, Ψ and $\overline{\Phi}_A$ satisfy both the convergence property (iii) and we conclude that $\Psi(f) = \overline{\Phi}_A(f)$.   □

For every $A \in \mathcal{S}(X)$ the class $H_A^\infty(\Sigma_\sigma)$ is large enough to contain many basic functions, in any case enough functions to derive the usual theory of analytic semigroups from our functional calculus.

9.8 Examples. Let $A \in \mathcal{S}(X)$ on a Banach space X and $\sigma > \omega(A)$.

a) Assume that $f \in H^\infty(\Sigma_\sigma)$ is analytic in a neighborhood of 0 and satisfies $|f(\lambda)| \leq C(1 + |\lambda|)^{-\varepsilon}$ in Σ_σ for some $\varepsilon > 0$. Then $f \in H_A^\infty(\Sigma_\sigma)$ and

$$\overline{\Phi}_A(f) = \frac{1}{2\pi i} \int_{\partial \Sigma(\delta)} f(\lambda) R(\lambda, A) d\lambda$$

where $\Sigma(\delta) = \Sigma_{\sigma'} \cup B(0, \delta)$, $\sigma' \in (\omega(A), \sigma)$ and $\delta > 0$ so small that f is analytic on $\Sigma(\delta)$. Also note the norm estimate

$$\|\overline{\Phi}_A(f)\| \leq \frac{1}{2\pi} \sup\{\|\lambda R(\lambda, A)\| : \lambda \in \Sigma(\delta)\} \|f\|_{\Sigma(\delta)}$$

with $\|f\|_{\Sigma(\delta)} = \int_{\partial \Sigma(\delta)} |f(\lambda)| \frac{d|\lambda|}{|\lambda|}$.

Proof. Put $f_n(\lambda) = f(\lambda)\varrho(\lambda)^{1/n}$. As in the proof of Theorem 9.2(ii) we can deform the contour $\partial \Sigma_{\sigma'}$ into $\partial \Sigma(\varepsilon)$ and obtain

$$\frac{1}{2\pi i} \int_{\partial \Sigma(\varepsilon)} f_n(\lambda) R(\lambda, A) d\lambda = \frac{1}{2\pi i} \int_{\partial \Sigma_{\sigma'}} f_n(\lambda) R(\lambda, A) d\lambda = f_n(A)$$

for all n. A direct estimate of the first integral shows that $\|f_n\|_A$ is bounded. Since $f_n \to f$ pointwise we have $f \in H_A^\infty(\Sigma_\sigma)$.

b) $1 \in H_A^\infty(\Sigma_\sigma)$ and $\Phi_A(1) = I_X$.

Proof. $\varrho_m(\lambda) \to 1$ for $\lambda \in \Sigma_\sigma$ and $\varrho_m(A)x \to x$ for $x \in X$ by Proposition 9.4.

c) For a rational function $f(\lambda) = \prod_{i=1}^n \frac{a_i - \lambda}{\mu_i - \lambda}$ with $a_i \in \mathbb{C}, \mu_i \notin \overline{\Sigma_\sigma}$ we have $f(A) = \prod_{i=1}^n (a_i - A) R(\mu_i, A)$.

Proof. For a single factor we have the decomposition $f(\lambda) = \frac{a - \lambda}{\mu - \lambda} = \frac{a - \mu}{\mu - \lambda} + 1$. By a) and b) we obtain that $f \in H_A^\infty(\Sigma_\sigma)$ and (9.5) yields

$$\overline{\Phi}_A(f) = \overline{\Phi}_A\left(\frac{a - \mu}{\mu - \lambda} + 1\right) = (a - \mu) R(\mu, A) + I$$

$$= (a - A) R(\lambda, A).$$

The general case follows from the multiplicativity of $\overline{\Phi}_A$.

9.9 Illustration. In books on semigroup theory, the usual theory of analytic semigroups such as formulas (1.2) to (1.5) are proved with methods that we now can recognize as special cases of the functional calculus approach. We will see that if A is a sectorial operator with $\omega(A) < \frac{\pi}{2}$, then $-A$ generates an analytic semigroup. Indeed, if $z \in \mathbb{C}$ with $|\arg z| < \frac{\pi}{2} - \omega(A)$, then $e_z(\lambda) = e^{-\lambda z}$ belongs to $H_A^\infty(\Sigma_\sigma)$ for $\omega(A) < \sigma < \frac{\pi}{2} - |\arg z|$ by Example 9.8 a). The representation

$$T_z := e_z(A) = \frac{1}{2\pi i} \int_{\partial \Sigma(\epsilon)} e^{-\lambda z} R(\lambda, A) d\lambda = \frac{1}{2\pi i} \int_{-\partial \Sigma(\epsilon)} e^{\lambda z} R(\lambda, -A) d\lambda$$

from the proof of 9.8 corresponds to (1.2). Property (1.4), i.e. the semigroup property $T_{z_1} T_{z_2} = T_{z_1+z_2}$, follows directly from $e_{z_1} e_{z_2} = e_{z_1+z_2}$. If we think of $\frac{d}{dz} e_z(\lambda) = -\lambda e_z(\lambda)$ as a Fréchet derivative with respect to the norm $\| \cdot \|_{\Sigma(\epsilon)}$ of 9.8 a), then

$$\lim_{h \to 0} \tfrac{1}{h}(T_{z+h} - T_z) = \lim_{h \to 0} \overline{\Phi}_A[\tfrac{1}{h}(e_{z+h} - e_z)] = \overline{\Phi}_A[\tfrac{d}{dz} e_z] = -AT_z.$$

(For a different argument see 9.12 below.) Similarly we get (1.5) if we read the Laplace integral in

$$r_\mu(\lambda) = \int_0^\infty e^{-\mu t} e_t(\lambda) dt$$

as a Bochner integral in $H^\infty(\Sigma_\omega)$ and use the closedness of Φ_A:

$$R(\mu, A) = \overline{\Phi}_A\left[\int_0^\infty e^{-\mu t} e_t dt\right] = \int_0^\infty e^{-\mu t} \overline{\Phi}_A[e_t] dt = \int_0^\infty e^{-\mu t} T_t dt.$$

(See also 9.12 for a more detailed explanation.) Finally we obtain the boundedness of $\{t^n A^n e^{tA} : t > 0\} = \{\varphi(tA) : t > 0\}$ with $\varphi(\lambda) = \lambda^n e^{-\lambda}$ from 9.2 (iii) since $\int_{\partial \sigma_\omega} |\varphi(t\lambda)| \frac{d|\lambda|}{|\lambda|}$ is finite and independent of t. Similarly we can show the boundedness of $\{t^{1/2} A^{\alpha/2} R(t, A^\alpha) : t > 0\} = \{\phi(t^{-1/\alpha} A) : t > 0\}$ where $\phi(\lambda) = \lambda^{\alpha/2}(1 - \lambda^\alpha)^{-1}$. On the other hand, there are generators of analytic semigroups even in Hilbert space for which $\mu_t(\lambda) = \lambda^{it}$ does not belong to $H_A^\infty(\Sigma_\sigma)$.

Often it is not easy to decide whether or not a concrete function belongs to $H_A^\infty(\Sigma_\sigma)$. A case in point is the function $g_t(\lambda) = \lambda^{it}$ for $t \in \mathbb{R}$. If $g_t \in H_A^\infty(\Sigma_\sigma)$ one says that A has property BIP, i.e., A has *bounded imaginary powers*.

Therefore, operators with $H_A^\infty(\Sigma_\sigma) = H^\infty(\Sigma_\sigma)$ are of particular interest.

9.10 Definition. *A sectorial operator* $A \in \mathcal{S}(X)$ *has a bounded* H^∞*-calculus of angle* σ, $\sigma > \omega(A)$, *if* $H^\infty(\Sigma_\sigma) = H_A^\infty(\Sigma_\sigma)$, *i.e., there exists a bounded algebra homomorphism* $f \in H^\infty(\Sigma_\sigma) \mapsto f(A) \in B(X)$ *With the convergence property (iii) of 9.6 such that*

$$f(A) = \frac{1}{2\pi i} \int_{\partial \Sigma_\nu} f(\lambda) R(\lambda, A) d\lambda$$

for $f \in H_0^\infty(\Sigma_\sigma)$ and $\sigma > \nu > \omega(A)$.

The infimum over all σ for which A has a bounded H^∞–calculus is called the H^∞–angle $\omega_{H^\infty}(A)$ of A.

The following remark is very useful in checking that a concrete operator A has a bounded H^∞-calculus.

9.11 Remark (Uniform boundedness of the H^∞-calculus). If $\sigma > \omega(A)$, then $A \in \mathcal{S}(X)$ has a bounded $H^\infty(\Sigma_\sigma)$–calculus if and only if there is a $C < \infty$ with

$$\|\Phi_A(f)\|_{B(X)} \leq C\|f\|_{H^\infty(\Sigma_\sigma)} \quad \text{for all } f \in H_0^\infty(\Sigma_\sigma).$$

In this case, we also have $\|f\|_A \sim \|f\|_{H^\infty(\Sigma_\sigma)}$.

Proof. Indeed, let $f \in H^\infty(\Sigma_\sigma)$ and $f_n := f\varrho_n$. Then $f_n \to f$ pointwise and $\sup_n \|f_n\|_A < \infty$ since the latter condition implies that $\|f\|_A \sim \|f\|_{H^\infty(\Sigma_\sigma)}$. The converse follows from the closed graph theorem. □

Many examples of operators with a bounded H^∞–calculus will be presented in the following sections 10, 11, 13 and 14. We close this section with two results that will simplify calculations with the H^∞-calculus in coming sections. First we formalize the idea of "plugging" A into integral or differential formula. We have applied these ideas already in 9.7 and 9.9.

9.12 Lemma. *Let $A \in \mathcal{S}(X)$ and $\sigma > \omega(A)$. Assume that $f : (a,b) \times \Sigma_\sigma \to \mathbb{C}$ satisfies*

(i) $f(\cdot, \lambda)$ is continuous on (a,b) for all $\lambda \in \Sigma_\sigma$,

(ii) $f(t, \cdot) \in H_A^\infty(\Sigma_\sigma)$ for all $t \in (a,b)$,

(iii) $t \in (a,b) \mapsto \|f(t, \cdot)\|_A$ is locally bounded and integrable.

Then we have:

a) The function $g(\lambda) = \int_a^b f(t, \lambda) dt$ belongs to $H_A^\infty(\Sigma_\sigma)$ and

$$g(A)x = \int_a^b f(t, A)x \, dt, \qquad x \in X.$$

b) For each $s \in (a,b)$ the function $g(s, \lambda) = \int_a^s f(t, \lambda) dt$ belongs to $H_A^\infty(\Sigma_\sigma)$ and $s \mapsto g(s, A)$ is strongly differentiable with

$$\frac{\partial}{\partial s} g(s, A)x = f(s, A)x$$

for $s \in (a,b)$ and $x \in X$.

If A has a bounded H^∞-calculus, then $\|\cdot\|_A$ can be replaced by $\|\cdot\|_{H^\infty(\Sigma_\sigma)}$.

Proof. By the convergence property 9.6 (iii) and condition (iii) we have for all $x \in X$ that the function $t \in (a, b) \mapsto f(t, A)x$ is continuous and integrable.
a) Since $|f(t, \cdot)| \le \|f(t, \cdot)\|_A$, g is bounded on Σ_σ and it follows from Lebesgue's convergence theorem that g is continuous on Σ_σ. Applying first Fubini's theorem and then Cauchy's theorem to a simple closed curve Γ in Σ_σ we obtain that

$$\int_\Gamma g(\lambda)d\lambda = \int_a^b \left(\int_\Gamma f(t, \lambda)d\lambda \right) dt = 0.$$

Hence Morera's theorem implies that g is analytic on Σ_σ.

Put $g_n(\lambda) = g(\lambda)\varrho_n(\lambda)$, where ϱ_n is as in Proposition 9.4. Then $g(\lambda) = \lim_{n\to\infty} g_n(\lambda)$, $\|g_n\|_{H^\infty(\Sigma_\sigma)} \le \|g\|_{H^\infty(\Sigma_\sigma)}$ and for $\sigma' \in (\omega(A), \sigma)$ and $x \in X$

$$g_n(A)x = \frac{1}{2\pi i} \int_{\partial\Sigma_{\sigma'}} \varrho_n(\lambda)g(\lambda)R(\lambda, A)x\, d\lambda$$

$$= \int_a^b \left(\frac{1}{2\pi i} \int_{\partial\Sigma_{\sigma'}} \varrho_n(\lambda)f(t, \lambda)R(\lambda, A)x\, d\lambda \right) dt$$

$$= \int_a^b \varrho_n(A)f(t, A)x\, dt \longrightarrow \int_a^b f(t, A)x\, dt.$$

Also $\sup_n \|g_n(A)\| \le \sup_n \|\varrho_n(A)\| \int_a^b \|f(t, A)\|_A\, dt < \infty$. Hence $g \in H^\infty_A(\Sigma_\sigma)$ and the claim follows.
b) By a), $g(s, \cdot)$ is in $H^\infty_A(\Sigma_\sigma)$ for $s \in (a, b)$ and

$$\frac{1}{h}[g(s + h, \lambda) - g(s, \lambda)] = \frac{1}{h} \int_s^{s+h} f(t, \lambda)dt \to f(s, \lambda)$$

for $h \to 0$ and all $\lambda \in \Sigma_\sigma$. But then

$$\frac{1}{h}[g(s + h, A) - g(s, A)]x \to f(s, A)x$$

by the convergence property 9.6(iii).   □

Finally, we use the H^∞-calculus to construct rather general "partitions of unity".

9.13 Lemma. *Let $A \in S(X)$ be sectorial and $\psi \in H^\infty_0(\Sigma_\sigma)$ with $\pi > \sigma > \omega(A)$.*
a) Let $\int_0^\infty \psi(t)\frac{dt}{t} = 1$. Then for $x \in D(A) \cap R(A)$ we have

$$x = \int_0^\infty \psi(tA)x\, \frac{dt}{t} \qquad \text{(as an improper integral)}. \tag{9.10}$$

b) Let $\omega \in (\omega(A), \sigma)$ and $C := \frac{1}{2\pi} \int_{\partial\Sigma_\omega} |\psi(\lambda)| \frac{d|\lambda|}{|\lambda|} < \infty$. Then

$$\sup_{t>0} \|\psi(tA)\|_{\mathcal{B}(X)} \le C \sup \left\{ \|\lambda R(\lambda, A)\| : \pi \ge |\arg \lambda| \ge \omega \right\}.$$

Proof. a) By the uniqueness of analytic continuation we have $\int_0^\infty \psi(t\lambda)x \frac{dt}{t} = 1$ for all $\lambda \in \Sigma_\sigma$. Put $\varphi_n(\lambda) = \int_{1/n}^n \psi(\lambda t)\frac{dt}{t}$. Then $\varphi_n \in H_0^\infty(\Sigma_\sigma)$, $\|\varphi_n\|_\infty \leq \sup_{|\theta|\leq\sigma} \int_0^\infty |\psi(e^{i\theta}t)| \frac{dt}{t} < \infty$ and $\varphi_n(\lambda) \to 1$ for $n \to \infty$. For $x = \varrho(A)y$ with $y \in X$ we obtain by Corollary 9.12 and Theorem 9.2

$$\int_{1/n}^n \psi(tA)x \frac{dt}{t} = \varphi_n(A)\varrho(A)y = (\varphi_n\varrho)(A)y \to \varrho(A)y = x.$$

b) Again by Theorem 9.2 we have for all $t > 0$:

$$\|\psi(tA)\| \leq \frac{1}{2\pi} \int_{\partial\Sigma_\omega} |\psi(t\lambda)| \frac{d|\lambda|}{|\lambda|} \sup_{\lambda\in\partial\Sigma_\omega} \|\lambda R(\lambda, A)\|$$

$$= \frac{1}{2\pi} \int_{\partial\Sigma_\omega} |\lambda^{-1}\psi(\lambda)| \frac{d|\lambda|}{|\lambda|} \sup\{\|\lambda R(\lambda, A)\| : \pi \geq |\arg\lambda| \geq \omega\} \quad \square$$

$\square$

9.14 Remark. One can even show that the operators $J_{a,b} = \int_a^b \psi(tA) \frac{dt}{t}$, $0 < a < b < \infty$, are uniformly bounded in $B(X)$. Hence (9.10) holds for all $x \in X$. This is easy to see in the special case $\psi(\lambda) = \frac{\lambda}{(1-\lambda)^2}$. Indeed, $\psi(\frac{1}{t}A) = tAR(t, A)^2$ and for $x \in D(A)$:

$$\int_a^b \psi(tA)x \frac{dt}{t} = \int_{b^{-1}}^{a^{-1}} \psi(\frac{1}{t}A)x \frac{dt}{t} = \int_{b^{-1}}^{a^{-1}} AR(t, A)^2 x\, dt$$

$$= AR(a^{-1}, A)x - AR(b^{-1}, A)x \quad \square$$

Notes on Section 9

The Dunford functional calculus for bounded operators can be found in many functional analysis text books, e.g. in [57] or [77].

N 9.1 - N 9.13: We give a version of McIntosh's approach to the H^∞-calculus (see e. g. [176], [56] and its references). We use H_A^∞ to frame the convergence lemma instead of an "extended" functional calculus. For another approach to H_A^∞, see [KW]. We also use H_A^∞ to state the uniqueness of the H^∞-calculus. Lemma 9.12 is part of the folklore of the H^∞-calculus, here formalized in the framework of H_A^∞. For a proof of Remark 9.14, see [176], or [112], Lemma 1.30.

10 First Examples for the H^∞-functional Calculus

In this section we will use the harmonic analysis tools from Chapter I to show that many classical operators do have a bounded H^∞-calculus. As a warm up we consider first some basic examples such as multiplication operators,

$\frac{d}{dx}$ and the Laplace operator Δ. Then we discuss the connection between the H^∞–calculus and the existence of dilations for a semigroup in UMD spaces. An important tool here is a vector–valued version of the Coifman–Weiss transference method. We will apply this to generators of contraction semigroups in Hilbert space and to positive contraction semigroups in $L_q(\Omega)$–spaces. To show the limitations of the theory we finally consider some sectorial operators without a bounded H^∞–calculus. Along the way we give a preview of some of the results presented in the following sections.

10.1 Example. Let $X = L_2(\Sigma_\sigma, \mu)$ with Lebesgue measure μ. The multiplication operator $(Ax)(\lambda) = \lambda x(\lambda)$ has spectrum $\sigma(A) = \overline{\Sigma_\sigma}$ and is sectorial. Indeed, $(R(\mu, A)x)\lambda = (\mu - \lambda)^{-1}x(\lambda)$ for $\mu \notin \overline{\Sigma_\sigma}$. Moreover, for $|\arg \mu| = \omega \in (\sigma, \pi]$,

$$\|R(\mu, A)\| = \sup\{|\mu - \lambda|^{-1} : \lambda \in \Sigma_\sigma\} = \frac{C_{\sigma,\omega}}{|\mu|},$$

where $C_{\sigma,\omega} = [\sin(\omega - \sigma)]^{-1}$ if $\omega \leq \sigma + \frac{\pi}{2}$ and $C_{\sigma,\omega} = 1$ else. Clearly, $(f(A)x)(\lambda) = f(\lambda)x(\lambda)$ defines an algebra homomorphism $C(\Sigma_\sigma) \to B(X)$ with (i), (ii), (iii) of 9.6. Hence the restriction of this map $f \mapsto f(A)$ to the subspace $H^\infty(\Sigma_\sigma)$ of $C(\Sigma_\sigma)$ gives the H^∞-calculus by 9.7.

While in this case the H^∞-calculus is only the restriction of the much richer function calculus for normal operators, the H^∞-calculus is the *best* we can get if we consider the operator A not on $L_2(\Sigma_\sigma)$ but on $X = H^\infty(\Sigma_\sigma)$. Here $f(A)$ maps X into X only if f is analytic.

Since differential operators with constant coefficients become multiplication operators under the Fourier transform we combine now the idea of Example 10.1 with the Fourier multiplier theorem from Section 3.

10.2 Example. a) Let X be a UMD space and $A = \frac{d}{dt}$ on $L_p(\mathbb{R}, X)$ with $p \in (1, \infty)$. For $f \in H^\infty(\Sigma_\sigma)$ with $\sigma > \frac{\pi}{2}$ we put

$$\Phi(f)g = \mathcal{F}^{-1}[f(i\cdot)\widehat{g}(\cdot)]$$

for $g \in \mathcal{S}(\mathbb{R}, X)$. We have seen in Example 1.17 (using the Mihlin Multiplier Theorem 3.15) that $\Phi(f)$ defines a bounded operator on $L_p(\mathbb{R}, X)$ and $\|\Phi(f)\| \leq C\|f\|_{H^\infty(\Sigma_\sigma)}$. One can check directly that Φ satisfies (i), (ii), and (iii) of 9.6. (E.g., the multiplicativity follows from the properties of the convolution product. For $g \in \mathcal{S}(\mathbb{R}, X)$, the convergence property (iii) follows from Lebesgue's Convergence Theorem since $\widehat{g} \in L_1(\mathbb{R}, X)$ and $\mathcal{S}(\mathbb{R}, X)$ is dense in $L_p(\mathbb{R}, X)$.)

b) The same technique works if $A = -\Delta$ on $L_p(\mathbb{R}^n, X)$ with $p \in (1, \infty)$ and a UMD space X. For $f \in H^\infty(\Sigma_\sigma)$ with $\sigma > 0$ we put now

$$\Phi(f)g = \mathcal{F}^{-1}[f(|\cdot|^2)\widehat{g}(\cdot)]$$

for $g \in \mathcal{S}(\mathbb{R}^n, X)$. The boundedness of $\Phi(f)$ will follow from the Mihlin Multiplier Theorem 4.6. To apply 4.6, we have to bound for $\alpha \leq (1, \ldots, 1)$

$$|t|^{|\alpha|} D^\alpha f(|t|^2) = |t|^{|\alpha|} 2^{|\alpha|} t^\alpha f^{(|\alpha|)}(|t|^2) = \left(\frac{2t}{|t|}\right)^\alpha (|t|^2)^{|\alpha|} f^{(|\alpha|)}(|t|^2)$$

But the boundedness of $t^n f^{(n)}(t)$ for $t > 0$ follows from the Cauchy formula since $f \in H^\infty(\Sigma_\sigma)$ for $\sigma > 0$. Hence $\Phi(f)$ defines a bounded operator on $L_p(\mathbb{R}^n, X)$. The conditions (i), (ii), and (iii) of 9.6 can be checked as in part a) and Φ must be the bounded H^∞-calculus of A by 9.7. $\square$

10.3 Remark. The UMD property is not just sufficient for the boundedness of the H^∞ functional calculus of $\frac{d}{dt}$ on $L_p(\mathbb{R}, X)$ but also necessary. Hence the UMD property can also be characterized in terms of spectral theory.

More sophisticated differential operators will be treated in sections 13 and 14. As a first step in this direction we formulate a vector-valued version of the well known transference principle from harmonic analysis.

10.4 Notation. Let X be a Banach space. For an operator-valued kernel function $k \in L_1(\mathbb{R}, B(X))$ we define a convolution operator $K_k : L_p(\mathbb{R}, X) \to L_p(\mathbb{R}, X)$ for $1 \leq p < \infty$ by the Bochner integrals

$$K_k f(t) = \int_{-\infty}^\infty k(s)\big[f(t-s)\big]\, ds\,, \qquad f \in L_p(\mathbb{R}, X)\,.$$

As in the scalar case one shows with Young's inequality that

$$\|K_k\|_p := \|K_k\|_{B(L_p(\mathbb{R},X))} \leq \|k\|_{L_1(B(X))}\,. \tag{10.1}$$

10.5 Theorem. Let $U(t)$ be a c_0-group of operators on X such that

$$C_U := \sup_{t \in \mathbb{R}} \|U(t)\|_{B(X)} < \infty \tag{10.2}$$

Suppose that $k \in L_1(\mathbb{R}, B(X))$ satisfies $k(t)U(s) = U(s)k(t)$ for all $t, s \in \mathbb{R}$. Then for all $x \in X$ the Bochner integral

$$T_k x = \int_{-\infty}^\infty k(t)\big[U(t)x\big]\, dt \tag{10.3}$$

defines a bounded operator T_k on X such that

$$\|T_k\|_{B(X)} \leq C_U^2 \|K_k\|_{B(L_p(X))}\,.$$

10.6 Discussion. The point of the theorem is to "transfer" a good estimate on $\|K_k\|$ to a good estimate on $\|T_k\|$. Of course we always have the elementary estimate

$$\|T_k x\|_X \le \int_{-\infty}^{\infty} \|U(t)\| \cdot \|k(t)\| \cdot \|x\| \, dt \,,$$

or

$$\|T_k\| \le C_U \|k\|_{L_1(\mathbb{R}, B(X))} \,. \tag{10.4}$$

But there may be better estimates available for $\|K_k\|_p$. E. g., if X is a Hilbert space, $p = 2$ and k is scalar. Then using Parseval's identity, we get $\|K_k\|_2 \le \sup_{s \in \mathbb{R}} \|\widehat{k}(s)\|$ and the theorem gives $\|T_k\| \le C_U^2 \sup \|\widehat{k}(\cdot)\|$. If X is a UMD space, the vector-valued Mihlin multiplier theorem 3.15 and (10.3) imply for $1 < p < \infty$

$$\|T_k\| \le C_U^2 C_p R\left(\left\{ \widehat{k}(t), \, t\tfrac{d}{dt}\widehat{k}(t) : t \in \mathbb{R} \right\}\right). \tag{10.5}$$

These improvements over (10.4) are essential if we want to estimate the norm of the functional calculus

Proof of Theorem 10.5. The elementary estimate (10.4) shows that for $\varepsilon > 0$ and $N > 0$ large enough the operator

$$T_N x = \int_{|t|>N} U(t)\big[k(t)x\big] \, dt$$

satisfies

$$\|T_N\|_{B(L_p(X))} \le \int_{|s|>N} \|k(s)\| \, ds < \varepsilon \,.$$

Therefore we can assume without loss of generality that k has support in a fixed interval $[-N, N]$. Given N and $\varepsilon > 0$, we choose M so large that $\frac{M+N}{M} \le 1 + \varepsilon$. Since $U(t)$ is a group and $U(t) = U(-t)$ we have for $x \in X$

$$\|x\| \le C_U \|U(t)x\| \,, \qquad \|U(t)x\| \le C_U \|x\| \,, \qquad t \in \mathbb{R} \,. \tag{10.6}$$

Hence for $x \in X$

$$\|T_k x\|^p \le \frac{C_U^p}{2M} \int_{-M}^{M} \|U(-s)T_k x\|^p \, ds$$

$$= \frac{C_U^p}{2M} \int_{-M}^{M} \left\| \int_{-\infty}^{\infty} U(t-s)\big[k(t)x\big] \, dt \right\|^p ds$$

by the definition of T_k. Also $\chi_{[-M-N,M+N]}(t-s) = 1$ if $t \in \operatorname{supp} k \subset [-N, N]$ and $s \in [-M, M]$, so that with $\chi := \chi_{[-M-N,M+N]}$ the last expression can be rewritten as

$$= \frac{C_U^p}{2M} \int_{-M}^{M} \left\| \int_{-\infty}^{\infty} k(t)\big[U(t-s)x\,\chi(t-s)\big] \, dt \right\|^p ds$$

and by the definition of K_k

$$= \frac{C_U^p}{2M} \int_{-M}^{M} \left\| K_k \left[U(-\cdot) x \, \chi(-\cdot) \right] (s) \right\|^p ds$$

$$\leq \frac{C_U^p}{2M} \left\| K_k \left[U(\cdot) x \, \chi(\cdot) \right] \right\|_{L_p(\mathbb{R},X)}^p$$

$$\leq \frac{C_U^p}{2M} \|K_k\|_p^p \int_{-N-M}^{N+M} \|U(s) x \, \chi(s)\|^p ds$$

$$\leq \frac{C_U^p}{2M} \|K_k\|_p^p \, 2(N+M) C_U^p \|x\|^p$$

$$\leq (1+\varepsilon) C_U^{2p} \|K_k\|_p^p \|x\|^p$$

since $\frac{N+M}{N} \leq 1+\varepsilon$. Now the claim follows.   □

Next we apply the transference principle to the H^∞-functional calculus.

10.7 Theorem. *Let $U(t)$ be a bounded c_0-group on a UMD-space X with generator $-A$. Then $A \in \mathcal{S}(X)$ has a bounded $H^\infty(\Sigma_\sigma)$-calculus for all $\sigma \in (\frac{\pi}{2}, \pi]$.*

Proof. **1. step**: We construct first a functional calculus for the functions in the Mihlin-multiplier space

$$\mathfrak{M}(\mathbb{R}) := \left\{ m \in C^1(\mathbb{R}\backslash\{0\}) : \|m\|_{\mathfrak{M}} = \sup_{t\neq 0} |m(t)| + |t m'(t)| < \infty \right\}$$

It is easy to see that $\mathfrak{M}$ is a Banach algebra under pointwise multiplication and $\|m(a\cdot)\|_{\mathfrak{M}} = \|m(\cdot)\|_{\mathfrak{M}}$ for all $a \in \mathbb{R}\backslash\{0\}$. Put

$$\mathfrak{M}_1(\mathbb{R}) := \{m \in \mathfrak{M}(\mathbb{R}) : m = \widehat{k} \text{ for some } k \in L_1(\mathbb{R})\}.$$

For $m = \widehat{k} \in \mathfrak{M}_1$ we can define for $x \in X$

$$\Phi(A)x := \int_{-\infty}^{\infty} k(t) U(t) x \, dt.$$

The transference principle 10.5 gives us immetiately $\|\Phi(A)\| \leq C_p \|m\|_{\mathfrak{M}}$, where the constant C_p is determined by the Mihlin multiplier theorem 3.15 on $L_p(\mathbb{R},X)$ for some $1 < p < \infty$. For $k_\lambda(t) = exp(\lambda t)\chi_{\mathbb{R}_+}(t)$ with $Re\lambda < 0$ we have that $r_\lambda(t) := \widehat{k}_\lambda(t) = (it - \lambda)^{-1}$ so that

$$r_\lambda \in \mathfrak{M}, \qquad \Phi(r_\lambda) = R(\lambda, A) \tag{i}$$

If $m_1 = \widehat{k}_1, m_2 = \widehat{k}_2 \in \mathfrak{M}_1(\mathbb{R})$, then $m_1 \cdot m_2 = (k_1 * k_2)\widehat{} \in \mathfrak{M}_1(\mathbb{R})$ and for $x \in X$

$$\Phi(m_1 \cdot m_2)x = \int U(t)x \left(\int k_1(t-s) k_2(s) \, ds \right) dt$$

$$= \int U(s) k(s) \left(\int k_1(t-s) U(t-s) x \, dt \right) ds \tag{ii}$$

$$= \Phi(m_1)\Phi(m_2)x$$

To extend Φ from $\mathfrak{M}_1(\mathbb{R})$ to $\mathfrak{M}(\mathbb{R})$ fix the function $\varrho(t) = \frac{(it)^2}{(1+it)^4}$ with $\varrho(A) = -A^2(1+iA)^{-4}$. Note that for every $m \in \mathfrak{M}(\mathbb{R})$ the product $m\varrho$ is in $W_2^1(\mathbb{R})$, so that by the classical Bernstein inequality $(m\varrho)^{\cdot} \in L_1(\mathbb{R})$ and $m\varrho \in L_1(\mathbb{R})$. Furthermore we chose a function $\psi \in \mathcal{S}(\mathbb{R})$ such that $0 \leq \psi(t) \leq 1$, $\psi(t) = 1$ for $|t| \leq 1$ and $\psi(t) = 0$ for $|t| \geq 2$. For $m \in \mathfrak{M}$ put

$$m_n(t) := m(t)\psi(t/n)\big(1 - \psi(nt)\big).$$

Since m_n has compact support in $\mathbb{R}\backslash\{0\}$ we have $m_n \in W_2^1(\mathbb{R})$ and $m_n \in \mathfrak{M}_1(\mathbb{R})$. Also $\|m_n(t)\|_{\mathfrak{M}} \leq \|\psi\|_{\mathfrak{M}}(1 + \|\psi\|_{\mathfrak{M}})\|m\|_{\mathfrak{M}}$ for all n and $m_n(t) \to m(t), tm_n(t) \to tm(t)$ for $n \to \infty$. In particular $\varrho m_n \to \varrho m$ in $W_2^1(\mathbb{R})$ and by Bernstein's inequality $(\varrho m)^{\cdot} \to (\varrho m)^{\cdot}$ in $L^1(\mathbb{R})$. Hence

$$\Phi(m_n)\Phi(\varrho)x = \Phi(m_n\varrho)x \to \Phi(m\varrho)x \text{ for } n \to \infty.$$

Furthermore, $\|\Phi(m_n)\| \leq C\|m\|_{\mathfrak{M}}$ and $R(\Phi(\varrho)) = R(A^2) \cap D(A^2)$ is dense in X. We can conclude that $Sx := \lim_{n\to\infty} \Phi(m_n)x$ exists for all $x \in X$ and $\|S\|_{B(X)} \leq C\|m\|_{\mathfrak{M}}$. Now define $\Phi(m) := S$.

In a similar way one can show that $\Phi(m)$ is well defined and that one has the convergence property:

> If $m, m_n \in \mathfrak{M}$ with $\|m_n\|_{\mathfrak{M}} \leq C$ and $m_n(t) \to m(t)$, $m_n'(t) \to m'(t)$ then $\Phi(m_n)x \to \Phi(m)x$ for all $x \in X$. $\qquad$ (iii)

2.step: From the algebra homomorphism $\Phi : \mathfrak{M}(\mathbb{R}) \to B(X)$ we obtain a bounded $H^\infty(\Sigma_\sigma)$–calculus for all $\sigma > \frac{\pi}{2}$. Indeed, if $f \in H^\infty(\Sigma_\sigma)$ then $m(t) = f(it)$ belongs to $\mathfrak{M}(\mathbb{R})$ and $\|m\|_{\mathfrak{M}} \leq \tilde{C}\|f\|_{H^\infty(\Sigma_\sigma)}$ (compare the estimate in 2.16 b). By (i), (ii),(iii) above and the uniqueness result 9.7 we have that $f(A) = \Phi(m)$ agrees with the H^∞–calculus of A and $\|f(A)\| \leq C\|f\|_{H^\infty(\Sigma_\sigma)}$. $\qquad\square$

10.8 Remark. If we use the vector valued Mihlin multiplier theorem 3.15 in connection with the transference principle we can repeat the construction of $\Phi(m)$ above (with minor changes) also for functions $m : \mathbb{R} \setminus \{0\} \to B(X)$ with

$$R(\{m(t), tm'(t) : t \neq 0\}) < \infty$$

and for which each $m(t)$ commutes with T_t. We will study such an operator-valued functional calculus in greater generality in section 12.

Theorem 10.7 can be applied to semigroups via dilations. A bounded c_0-group $U(t)$ on a Banach space $\tilde{X}$ is a **dilation** of the semigroup T_t on a Banach space X, if there exists an isomorphic embedding $J : X \to \tilde{X}$ and a projection $P : \tilde{X} \to J(X)$ such that

$$JT_t = PU_tJ \qquad \text{for } t \geq 0. \tag{10.7}$$

If $-A$ is the generator of T_t and $-\tilde{A}$ the generator of $U(t)$ then for $\operatorname{Re}\lambda > 0$

$$JR(\lambda, -A)x = J\left[\int_0^\infty e^{-\lambda t}T_t x\, dt\right]$$

$$= \int_0^\infty e^{-\lambda t}\left(JT_t\right)x\, dt = \int_0^\infty e^{-\lambda t}\left(PU_t J\right)x\, dt$$

$$= PR(\lambda, -\tilde{A})Jx\,,$$

so that for $f \in H_0^\infty(\Sigma_\sigma)$ with $\sigma > \nu > \omega(A)$

$$Jf(A) = \frac{1}{2\pi i}\int_{\partial\Sigma_\nu} f(\lambda)JR(\lambda, -A)d\lambda = Pf(\tilde{A})J\,.$$

Together with Theorem 10.7 we have shown the following corollary.

10.9 Corollary. *Let X be a UMD-space. If the c_0-semigroup T_t with generator $-A$ has a dilation to a bounded c_0-group on a UMD space $\tilde{X}$, then A has a bounded H^∞-calculus (possibly with $\omega_{H^\infty}(A) > \frac{\pi}{2}$).*

For generators of R-analytic semigroups, the reverse implication is also true.

10.10 Theorem. *Let X be a UMD space and $A \in \mathcal{S}(X)$ be R-sectorial with $\omega_R(A) < \frac{\pi}{2}$. Then A has a bounded H^∞-calculus with $\omega_{H^\infty}(A) = \omega_R(A)$ if and only if the semigroup T_t generated by $-A$ has a dilation to a bounded c_0-group on $\tilde{X} = Rad(X)$.*

In these notes we only prove this statement fully in the Hilbert space case in Section 11. Corollary 10.9 has some important consequences. E.g., in Hilbert spaces one has the classical dilation theorem of Sz.-Nagy.

10.11 Theorem. *Let T_t be a contractive c_0-semigroup on a Hilbert space H. Then there exists a Hilbert space $\tilde{H}$ and an unitary group $U(t)$ on $\tilde{H}$ such that*

$$JT_t = PU(t)J$$

where J is an isometric embedding and $P : \tilde{H} \to J(H)$ the orthogonal projection.

Now Corollary 10.9, or more directly, the spectral theory of normal operators on Hilbert spaces, gives

10.12 Corollary. *Suppose that $A \in \mathcal{S}(H)$ and $-A$ generate a contractive semigroup on a Hilbert space H. Then A has a bounded H^∞-calculus and $\omega(A) = \omega_{H^\infty}(A)$.*

The last two results will be treated in detail for analytic semigroups in the next section. In a similiar way one can treat positive contractive semigroups on $L_q(\Omega, \mu)$-spaces. Again, there is a powerful dilation theorem, which we only quote here.

10.13 Theorem. *Let T_t be a c_0-semigroup of positive contractions on $L_q(\Omega, \mu)$ with $1 < q < \infty$. Then there exists a measure space $(\widetilde{\Omega}, \widetilde{\mu})$ and a c_0-group of positive isometries on $L_q(\widetilde{\Omega}, \widetilde{\mu})$ such that*

$$JT_t = PU(t)J, \quad t \geq 0,$$

where $P : L_q(\Omega, \mu) \to L_q(\widetilde{\Omega}, \widetilde{\mu})$ is an isometric, positive embedding and $P : L_q(\widetilde{\Omega}, \widetilde{\mu}) \to L_q(\widetilde{\Omega}, \widetilde{\mu})$ is a positive and contractive projection.

We are also interested in operators on $L_p(\Omega, X)$ and for this purpose we need an extention theorem for positive operators on $L_p(\Omega)$.

10.14 Lemma. *Let $S : L_p(\Omega_1, \mu_1) \to L_p(\Omega_2, \mu_2)$ be a positive operator and X a Banach space. We define the tensor extention $\overline{S} = S \otimes Id.$ of S for functions $F(\omega) = \sum_{j=1}^{m} f_j(\omega) x_j$ with stepfunctions $f_j \in L_1(\Omega_1)$ and $x_j \in X$ by*

$$\left(\overline{S}F \right)(\cdot) = \sum_{j=1}^{m} (Sf_j)(\cdot) x_j. \tag{10.8}$$

Then $\overline{S}$ extends to a bounded operator $\overline{S} : L_p(\Omega_1, X) \to L_p(\Omega_2, X)$ and $\|\overline{S}\| = \|S\|$.

Proof. Given $F = \sum_{j=1}^{m} f_j \otimes x_j$ we choose a sequence x_n^* in the unit sphere of X^* that norms span $(x_1 \cdots x_m)$. For each $x^* \in X^*$ we have

$$\left\langle \left(\overline{S}F \right)(\cdot), x^* \right\rangle = S \left(\sum_{j=1}^{m} \langle x_j, x^* \rangle f_j \right)(\cdot)$$

For allmost all $\omega \in \Omega_2$ we have

$$\left\| \left(\overline{S}F \right)(\omega) \right\| = \sup_n \left| \left\langle \left(\overline{S}F \right)(\omega), x_n^* \right\rangle \right|$$

$$= \sup_n \left| S \left(\sum_{j=1}^{m} \langle x_j, x_n^* \rangle f_j(\cdot) \right)(\omega) \right|$$

and by the positivity of S

$$\leq S\left(\sup_n \left|\sum_{j=1}^m \langle x_j, x_n^* \rangle f_j(\cdot)\right|\right)(\omega)$$

$$= S\left(\sup_n |\langle F(\cdot), x_n^* \rangle|\right)(\omega)$$

$$= S(\,\|F(\cdot)\|_X\,)(\omega)$$

Taking the L_p-norm we get

$$\|\overline{S}F\|_{L_p(\Omega_2,X)} \leq \|S\|\,\|F\|_{L_p(\Omega_1,X)}.$$

Since functions of the form F are dense in $L_p(\Omega_1, X)$, the claim follows.   □

Bringing Corollary 10.9 and Theorem 10.13 together we obtain

10.15 Corollary. *Let X be a UMD Banach space. Suppose that $-A$ generates a positive contraction semigroup T_t on $L_p(\Omega, \mu)$ with $1 < p < \infty$ and let $\overline{T}_t = T_t \otimes I$ be its tensor extension to $L_p(\Omega, X)$ (as in 10.14) with generator $-\overline{A}$. Then A and $\overline{A}$ have a bounded H^∞-calculus.*

Proof. If the positive operators $U(t) : L_p(\widetilde{\Omega}, \widetilde{\mu})$ and $P : L_1(\widetilde{\Omega}, \widetilde{\mu}) \to Y(L_1(\Omega, \mu))$ define a dilation for the semigroup T_t as in Theorem 10.13 then the extensions $\overline{U}(t) : L_p(\widetilde{\Omega}, X) \to L_p(\widetilde{\Omega}, X), \overline{J} : L_p(\Omega, \mu) \to L_1(\widetilde{\Omega}, X)$ and $\overline{P} : L_p(\Omega, X) \to \overline{J}(L_p(\Omega, x))$ define a dilation for $\overline{T}_t$ on $L_p(\Omega, X)$. Indeed, $\overline{J}\,\overline{T}_t = \overline{P}\,\overline{U}(t)\,\overline{J}$ can be checked directly with 10.13. Now apply Corollary 10.9 to A and $\overline{A}$.   □

10.16 Corollary. *Assume $(T_{t,q})_{t\geq 0}$ are contractive and positive semigroups on $L_q(\Omega, \mu)$ for all $1 \leq q \leq \infty$ such that for all $q_1, q_2 \in [1, \infty)$ and $t \geq 0$ the operators T_{t,q_1} and T_{t,q_2} agree on $L_{q_1}(\Omega) \cap L_{q_2}(\Omega)$. If $T_{t,q}$ is also analytic on L_q for some $1 < q < \infty$ with generator $-A_q$ then A_q has a bounded H^∞-calculus on $L_q(\Omega, \mu)$ with $\omega_{H^\infty}(A_q) < \frac{\pi}{2}$.*

Proof. A_q has a bounded H^∞-calculus by Corollary 10.15 , but we could have $\omega_{H^\infty}(A_q) > \frac{\pi}{2}$. On the other hand, we have shown in Theorem 5.6 that $\omega_R(A) < \frac{\pi}{2}$. The Corollary follows now from Theorem 12.8 where we will show that $\omega_R(A) = \omega_{H^\infty}(A)$ in our situation.   □

We close this section with an example showing that the notions of R–sectoriality, BIP and bounded H^∞–calculus are all different, and that even in Hilbert space there are sectorial operators without a bounded H^∞–calculus.

10.17 Example. For $p \in (1, \infty)$ let $w(t) = |t|^\alpha$ with $-1 < \alpha < p - 1$. Put $X = L_p(\mathbb{R}, w(t)\, dt)$ and consider the unbounded Fourier multiplier operator

$$Ax = \mathcal{F}^{-1}(m \cdot \hat{x})$$

with $m(t) = e^t$ and maximal domain in X, i.e., $D(A) = \{x \in X : Ax \in X\}$. Note that for $\lambda = e^{i\theta} e^r$ with $0 < |\theta| \leq \pi$, $r, s \in \mathbb{R}$

$$\lambda R(\lambda, A)x = \mathcal{F}^{-1}(r_\lambda \hat{x}) \qquad \text{with } r_\lambda(t) = \frac{\lambda}{\lambda - e^t} = \frac{e^{i\theta}}{e^{i\theta} - e^{t-r}},$$

$$A^{is}x = \mathcal{F}^{-1}(a_s\hat{x}) \qquad \text{with } a_s(t) = e^{ist}.$$

So by the basic properties of the Fourier transform

$$(A^{is}x)(t) = x(s+t) \qquad \text{are translation operators} \tag{10.9}$$

and

$$\lambda R(\lambda, A)x(t) = e^{irt}[e^{i\theta}R(e^{i\theta}, A)(e^{-ir(\cdot)}x(\cdot))](t). \tag{10.10}$$

From this we can draw the following conclusions

a) A is always R-sectorial with $\omega_R(A) = 0$.
b) If $\alpha = 0$, then A has BIP, but A has a bounded H^∞-calculus only for $p = 2$.
c) If $\alpha \neq 0$, then A has not BIP, in particular A has not a bounded H^∞-calculus even in the Hilbert space case $p = 2$.

Proof. a) For $\lambda_j = e^{i\theta}e^{r_j}$ with $r_1, \ldots, r_m \in \mathbb{R}_+$ and $x_1, \ldots, x_m \in X$ we have by Kahane's contraction principle

$$\|\sum_{j=1}^m r_j\lambda_j R(\lambda_j, A)x_j\|_{L_1([0,1],X)}$$

$$\leq 2\|\sum_{j=1}^m r_j R(e^{i\theta}, A)[e^{-ir_j\cdot}x_j(\cdot)]\|_{L_1([0,1],X)}$$

$$\leq 4\|R(e^{i\theta}, A)\|_{B(X)}\|\sum_{j=1}^m r_j x_j\|_{L_1([0,1],X)}.$$

Since $r_{e^{i\theta}}(t) = \frac{e^{i\theta}}{e^{i\theta}-e^t}$ satisfies the assumption of Mihlin's theorem and w is a so-called A_p-weight, the boundedness of $R(e^{i\theta}, A)$ follows from standard results in harmonic analysis. b) If $\alpha = 0$ the translation operators A^{ist} are clearly bounded on $L_p(\mathbb{R})$ with norm 1. We will see in 11.9 that for the Hilbert space $L_2(\mathbb{R})$ this already implies the boundedness of the H^∞-calculus. In 12.11 we will show that $\{A^{is} : s \in [-1, 1]\}$ is even R-bounded if A has a bounded H^∞-calculus. But we saw already in 2.12 that the translation group cannot be R-bounded for $p \neq 2$. c) If $\alpha \neq 0$ translation operators cannot be bounded on $L_p(\mathbb{R}, w(t)\,dt)$ since $w(t) = |t|^\alpha$. For $p = 2$ it follows then from 11.9 that A has no bounded H^∞-calculus. $\qquad\square$

Inspite of this example we will see that classical operators on $L_p(\Omega)$ with $1 < p < \infty$ usually have a bounded H^∞-calculus. But in $L_1(\Omega)$ or $C(K)$ there are very few examples. This corresponds to the breakdown of the theory of singular integral operators in $L_1(\mathbb{R}^n)$ and $L_\infty(\mathbb{R}^n)$.

10.18 Remark. Let T_t be a semigroup on $L_1(\Omega, \mu)$ for a non-atomic measure space (Ω, μ) so that each T_t has a representation as an integral operator. If $-A$ is the generator of T_t, one can show that A does not have a bounded H^∞–calculus. This applies in particular to $A = -\Delta$ on $L_1(\mathbb{R}^n)$.

Notes on Section 10:

N 10.2 - 10.3: The connection between UMD and the bounded H^∞-calculus for $\frac{d}{dt}$ and Δ on $L_p(X)$ was pointed out in [120].

N 10.4 - 10.6: The proof of the transference principle follows closely the classical proof in [50]. Vector-valued transference was treated in [45, 120, 48].

N 10.7: is also from [120]. The proof actually gives a bounded H^∞–calculus on double cones $\Sigma_\sigma \cap (-\Sigma_\sigma)$ for $\sigma > \frac{\pi}{2}$, but we do not consider this case here. See e. g. [75].

N 10.8: Such a vector-valued calculus is worked out in [48].

N 10.9: See [120] and [48].

N 10.10: is from [99], which also contains more general characterizations of the bounded H^∞-calculus in terms of dilations, where it is only assumed that X has finite cotype, not UMD.

N 10.11 -10.12: These results go back to [181] see also [60, Sect. 6.3]

N 10.13: is due to G. Fendler [90]. Instead of this more recent dilation theorem, an older and somewhat simpler result about dilations of the powers of one positive contraction [1] is often used in the literature on the transference principle. But then tedious 'discretization steps' are necessary (see e. g. [50], Sect. 4)

N 10.14: A folklore result.

N 10.15: The scalar case goes back to [84], the vector-valued case is in [120].

N 10.16: is shown in [134] in a more general setting : It is enough that $-A$ generates a bounded analytic semigroup $T(z)$ such that $T(t)$ is contractive and positive for one $q \in (1, \infty)$.

N 10.17: For $\alpha = 0$ this example is already in [56]. The use of A_p-weights may be new. The harmonic analysis result quoted in the text may be found in [101, VI.3].

There is a systematic way to construct oparators without a bounded H^∞-calculus starting from a Schauder basis $(e_n)_{n \in \mathbb{Z}}$ of a Banach space X. (e_n) is called a Schauder basis if for every $x \in X$ there is a unique sequence $(\alpha_n)_{n \in \mathbb{Z}}$ of scalars with

$$x = \sum_{n \in \mathbb{Z}} \alpha_n e_n := \lim_{N \to \infty} \sum_{n=-N}^{N} \alpha_n e_n.$$

For the operator $A(\sum \alpha_n e_n) = \sum_{n \in \mathbb{Z}} 2^n \alpha_n e_n$ with its maximal domain in X one can show (see, e.g., [214] and [76]):

i) A is a sectorial operator with $\omega(A) = 0$.

ii) For $\varphi \in H_0^\infty(\Sigma_\sigma)$ with $\sigma \in (0, \pi)$

$$\varphi(A)x = \sum_{n \in \mathbb{Z}} \varphi(2^n)a_n e_n, \qquad x = \sum_{n \in \mathbb{Z}} a_n e_n \in X.$$

iii) A has a bounded H^∞-calculus if and only if the expansion $x = \sum a_n e_n$ converges unconditionally for all $x \in X$. (the latter statement follows from the fact that 2^n, $n \in \mathbb{Z}$, is an interpolating sequence for analytic functions on Σ_σ)

If we choose $X = L_p([-\pi, \pi], w(t) dt)$ with $w(t) = |t|^\alpha$ and $e_n(t) = e^{int}$, then this point of view gives us a discrete version of our example (see [161] for $\alpha = 0$), where we can use the fact that e_n is only an unconditional basis if $p = 2$ and $\alpha = 0$.

N 10.18 This is a sample of the results proved for $L_1(\Omega)$ and $C(K)$ spaces in [134, 121, 140].

11 H^∞-calculus for Hilbert Space Operators

In this section, H is a complex Hilbert space with scalar product $(\cdot \mid \cdot)$. As in Sections 9 and 10 we restrict our attention to sectorial operators A which are injective and have dense range, i.e. $A \in \mathcal{S}(H)$. As a justification we refer again to the decomposition 15.2. Central for our discussion is the following class of operators.

11.1 Accretive Operators. For $\omega \in [0, \frac{\pi}{2}]$ we call a linear operator A with dense domain $D(A)$ on H ω-accretive if

$$\sigma(A) \subset \overline{\Sigma}_\omega, \qquad (Ax \mid x) \in \overline{\Sigma}_\omega \quad \text{for all } x \in D(A).$$

We also assume that $A : D(A) \to H$ is injective and then, by 15.2, $R(A)$ is dense in H. If $\omega = \frac{\pi}{2}$, an ω-accretive operator is called maximal accretive. (Note that in many books injectivity of A is not included in these definitions.)

Remark. An ω-accretive operator is ω-sectorial with

$$\|R(\mu, A)\| \leq d(\mu, \Sigma_\omega)^{-1} \quad \text{for } \mu \notin \overline{\Sigma}_\omega. \tag{11.1}$$

Indeed, for $x \in D(A)$ with $\|x\| = 1$

$$\|(\mu - A)x\| \geq |((\mu - A)x \mid x)| = |\mu - (Ax \mid x)| \geq d(\mu, \Sigma_\omega).$$

This class includes many interesting operators of applied analysis, and, somewhat surprisingly, it describes the class of operators on a Hilbert space that have a bounded H^∞-calculus in a very satisfactory way. In this section we will show that for a sectorial operator A on H with $\omega(A) < \frac{\pi}{2}$ the following conditions are equivalent (some of them were already announced in the last section):

- A has a bounded H^∞-*calculus.*
- A is *accretive* with respect to an inner product equivalent to the given one on H (see 11.5, 11.13).
- There is an equivalent inner product on H such that A is *associated to a closed sectorial form* (see 11.3).
- A generates an analytic semigroup which is *contractive* with respect to an equivalent Hilbert space norm (see 11.13).
- A is a *dilation* of a normal operator. (This is a variant of the famous Nagy dilation theorem, see 10.11 and 11.15).
- A has *BIP*, i.e. bounded imaginary powers (see 11.9).
- The norm on H is equivalent (see 11.9) to *square function norms* of the form

$$|||x||| = \left(\int_0^\infty \|A^{1/2}(t+A)^{-1}x\|^2 \, dt \right)^{1/2}.$$

The last statement actually provides the technical tool that is essential in the proofs of most of the other statements.

We start with some basic examples of accretive operators:

11.2 Example. Consider the Hilbert space $L_2(\mathbb{R}, H)$ with the inner product $(f|g) = \int_{-\infty}^\infty (f(t)|g(t)) \, dt$. For $\alpha \in (0, 2)$ we denote by M_α the multiplication operator

$$(M_\alpha g)(t) = (it)^\alpha g(t), \quad g \in L_2(\mathbb{R}, H).$$

Clearly, M_α is a normal operator with $\sigma(M_\alpha) = \{\lambda : |\arg \lambda| = \alpha \frac{\pi}{2}\}$. It is $(\alpha \frac{\pi}{2})$-accretive since

$$(M_\alpha g|g) = e^{i\alpha \frac{\pi}{2}} \int_0^\infty t^\alpha |g(t)|^2 \, dt + e^{-i\alpha \frac{\pi}{2}} \int_0^\infty t^\alpha |g(-t)|^2 \, dt \in \Sigma_{\alpha \frac{\pi}{2}}.$$

In the last theorems of this section we will see that every operator with a bounded H^∞-calculus is the dilation of M_α for $\alpha > \omega(A)$.

Originally, accretive operators were introduced to study differential operators given by forms. We have seen in 8.1 that an operator A which is associated to a closed sectorial form $\mathfrak{a}$ of angle $\omega \in [0, \pi/2)$ is sectorial. Moreover, the arguments that lead to (8.1) show that A is ω-accretive. Now we show the converse.

11.3 Theorem. *We now show that an ω-accretive operator A is associated to a closed sectorial form $\mathfrak{a}$ if $\omega \in [0, \pi/2)$.*

Proof. We define the form $\mathfrak{a}$ on $D(A)$ by $\mathfrak{a}(u, v) := (Au|v)$. Then $\mathfrak{a}$ is sectorial of angle ω (cf. 11.1 and 8.1). We have an inner product $(u|v)_V := (\operatorname{Re} \mathfrak{a})(u, v) + (u|v)$ on $D(A)$ and want to consider V as completion of $D(A)$ with respect to $\|\cdot\|_V$. Then $\mathfrak{a}$ has a continuous extension to V which is also sectorial of angle ω. We only have to make sure that V may be realized as a subspace of H. To this end let (u_n) be a $\|\cdot\|_V$-Cauchy sequence in $D(A)$ such that

$\|u_n\|_H \to 0$. We have to show $\|u_n\|_V \to 0$, i.e., $\operatorname{Re} \mathfrak{a}(u_n) \to 0$. Now, for fixed n and arbitrary m,

$$
\begin{aligned}
\operatorname{Re} \mathfrak{a}(u_n) &= \operatorname{Re}\left(Au_n | u_n - u_m\right) + \operatorname{Re}\left(Au_n | u_m\right) \\
&\le |\mathfrak{a}(u_n, u_n - u_m)| + |(Au_n | u_m)| \\
&\le C\|u_n\|_V \|u_n - u_m\|_V + \|Au_n\|_H \|u_m\|_H.
\end{aligned}
$$

Letting $m \to \infty$ we obtain

$$
\operatorname{Re} \mathfrak{a}(u_n) \le C \sup_k \|u_k\|_V \limsup_{m\to\infty} \|u_n - u_m\|_V
$$

which tends to 0 as $n \to \infty$. We denote the extension of $\mathfrak{a}$ to V again by $\mathfrak{a}$ and the operator associated to $\mathfrak{a}$ (cf. 8.1) by B. Then, by definition of $\mathfrak{a}$ on $D(A)$ and continuity, B is an extension of A. Since both operators are sectorial this yields $A = B$. $\qquad\square$

Now we recall the connection between accretive operators and contraction semigroups.

11.4 Theorem (Lumer Phillips). *The injective operator $-A$ generates a contractive semigroup on H if and only if A is maximal accretive. Moreover, $-A$ generates a contractive analytic semigroup $T(z)$, $z \in \overline{\Sigma}_\delta$ with $\delta \in (0, \frac{\pi}{2})$ (i. e. $\|T(z)\| \le 1$ for all $z \in \overline{\Sigma}_\delta$), if and only if A is $(\frac{\pi}{2} - \delta)$-accretive.*

Proof. We concentrate on the case of analytic semigroups where $\omega(A) < \frac{\pi}{2}$, so that we can use [186, Theorem II.5.2]. If we assume that $\|T(z)\| \le 1$ for $z \in \overline{\Sigma}_\delta$ then $\sigma(-A) \subset \overline{\Sigma}_{\frac{\pi}{2}-\delta}$ by [186, Theorem II.5.2]. Furthermore

$$
\operatorname{Re}(T(z)x|x) \le \|T(z)x\| \cdot \|x\| \le \|x\|^2,
$$

so that

$$
\begin{aligned}
\operatorname{Re}\left[e^{i\delta}(-Ax|x)\right] &= \operatorname{Re}\left(\lim_{t\to 0} \frac{T(e^{i\delta}t)x - x}{t} \,\Big|\, x\right) \\
&= \lim_{t\to 0} \frac{1}{t}\left[\operatorname{Re}(T(e^{i\delta}t)x|x) - \|x\|^2\right] \le 0.
\end{aligned}
$$

We can replace δ by $-\delta$ in this line and obtain

$$
\operatorname{Re}\left[e^{\pm i\delta}(Ax|x)\right] \ge 0,
$$

or $(Ax, x) \in \overline{\Sigma}_{\frac{\pi}{2}-\delta}$. Conversely, if $\mu = te^{i\omega}$ with $|\omega| \le \delta$, we have

$$
\begin{aligned}
\|\mu x + Ax\| \cdot \|x\| &\ge |(tx + e^{-i\omega}Ax|x)| \\
&\ge \operatorname{Re}(tx + e^{-i\omega}Ax|x) = \operatorname{Re}\left[e^{-i\omega}(Ax|x)\right] + |\mu| \cdot \|x\|^2.
\end{aligned}
$$

Since $(Ax|x) \in \overline{\Sigma}_{\frac{\pi}{2}-\delta}$ and $|\omega| \leq \delta$ we obtain $\mathrm{Re}[e^{-i\omega}(Ax|x)] \geq 0$ and therefore $\|\mu(\mu + A)^{-1}\| \leq 1$ for $\mu \in \overline{\Sigma}_\delta$.

Hence for $f_{z,n}(\lambda) = (1 + n^{-1}\lambda z)^{-n}$, where $z \in \overline{\Sigma}_\delta$ and $n \in \mathbb{N}$,

$$\|f_{z,n}(A)\| = \left\|\left(\frac{n}{z}\right)^n \left(\frac{n}{z} + A\right)^{-n}\right\| \leq 1 \quad \text{for all } n, z.$$

Since $f_{z,n}(\lambda) \xrightarrow{n \to \infty} e^{-z\lambda}$, the convergence property of the H^∞-calculus gives $f_{z,n}(A)x \to T(z)x$ for all x and the contractivity of $T(z)$ follows. □

The last argument already indicates that an accretive operator has a good functional calculus. In the following we prove this fact not relying on Nagy's dilation theorem as in Corollary 10.12 but with a more direct argument.

11.5 Theorem. *Suppose that A is maximal accretive on H. Then A has a bounded $H^\infty(\Sigma_\sigma)$-functional calculus for all $\sigma > \frac{\pi}{2}$ and*

$$\|f(A)\| \leq \|f\|_{H^\infty(\Sigma_{\pi/2})}.$$

Proof. a) First we assume in addition that A is a bounded operator and $\mathrm{Re}(Ax|x) \geq c\|x\|^2$ for all $x \in H$ and some fixed $c > 0$. Note that

$$((A + A^*)x|x) = (Ax|x) + \overline{(Ax|x)} = 2\,\mathrm{Re}(Ax, x) \geq 2c\|x\|^2,$$

so that $A + A^*$ is selfadjoint and positive, and $(A + A^*)^{1/2}$ exists.

Choose $\sigma > \frac{\pi}{2}$ and $f \in H_0^\infty(\Sigma_\sigma)$. Then $f(\lambda)(A^* + \lambda)^{-1}$ is analytic in a neighborhood of $\mathbb{C}_+$ and

$$\int_{i\mathbb{R}} f(\lambda)(A^* + \lambda)^{-1}\,d\lambda = 0.$$

Hence

$$\begin{aligned}
f(A) &= \frac{1}{2\pi i} \int_{i\mathbb{R}} f(\lambda)R(\lambda, A)\,d\lambda \\
&= -\frac{1}{2\pi i} \int_{i\mathbb{R}} f(\lambda)\big[(A - \lambda)^{-1} + (A^* + \lambda)^{-1}\big]\,d\lambda \\
&= \frac{1}{2\pi} \int_{\mathbb{R}} f(it)(A^* + it)^{-1}(A^* + A)(A - it)^{-1}\,dt,
\end{aligned}$$

and for all $x, y \in H$

$$(f(A)x|y) \tag{11.2}$$
$$= \frac{1}{2\pi} \int_{\mathbb{R}} f(it)\big((A^* + A)^{1/2}(A - it)^{-1}x \mid (A^* + A)^{1/2}(A - it)^{-1}y\big)\,dt.$$

Since $1 \in H_A(\Sigma_\sigma)$, we can choose $f \equiv 1$ and $x = y$ in (11.2). Then

$$\|x\|^2 = \frac{1}{2\pi} \int_{\mathbb{R}} \|(A^* + A)^{1/2}(A - it)^{-1}x\|^2 \, dt \,. \tag{11.3}$$

With Hölder's inequality it follows now from (11.2) and (11.3) that

$$|(f(A)x, y)| \le \|f\|_{H^\infty(\Sigma_{\pi/2})} \left(\frac{1}{2\pi} \int_{\mathbb{R}} \|(A^* + A)^{1/2}(A - it)^{-1}x\|^2 dt \right)^{1/2}$$

$$\cdot \left(\frac{1}{2\pi} \int_{\mathbb{R}} \|(A^* + A)^{1/2}(A - it)^{-1}y\|^2 dt \right)^{1/2}$$

$$= \|f\|_{H^\infty(\Sigma_{\pi/2})} \|x\| \cdot \|y\| \,.$$

b) Now we reduce the general case to a). For any $\varepsilon > 0$ put

$$A_\varepsilon = \left[(A + \varepsilon)^{-1} + \varepsilon \right]^{-1} \,.$$

Then A_ε is bounded and $\sigma(A_\varepsilon) \subset \overline{\Sigma}_{\pi/2}$ by the spectral mapping theorem. For $x \in H$ let $y = A_\varepsilon x$ and $z = (A + \varepsilon)^{-1}y$. Then

$$\mathrm{Re}(A_\varepsilon x | x) = \mathrm{Re}(y | (A + \varepsilon)^{-1}y + \varepsilon y) = \mathrm{Re}(Az | z) + \varepsilon(z | z) + \varepsilon(y | y)$$

$$\ge \varepsilon \|y\|^2 \ge \varepsilon \|A_\varepsilon^{-1}\|^{-2} \|x\|^2 \,.$$

Hence A_ε satisfies the assumption of part a). Moreover, for a fixed ω with $\frac{\pi}{2} < \omega < \sigma$ we have by (11.1) a uniform estimate

$$\|R(\lambda, A_\varepsilon)\| \le C_\omega |\lambda|^{-1} \quad \text{for all } \lambda \in \partial\Sigma_\omega, \ \varepsilon > 0 \,.$$

Therefore, for each $f \in H_0^\infty(\Sigma_\sigma)$ we obtain by the uniform convergence theorem

$$f(A) = \int_{\partial\Sigma_\omega} f(\lambda)R(\lambda, A) \, d\lambda = \lim_{\varepsilon \to 0} \int_{\partial\Sigma_\omega} f(\lambda)R(\lambda, A_\varepsilon) \, d\lambda$$

$$= \lim_{\varepsilon \to 0} \int_{i\mathbb{R}} f(\lambda)R(\lambda, A_\varepsilon) \, d\lambda$$

in the operator norm. By a)

$$\|f(A)\| \le \|f\|_{H^\infty(\Sigma_{\pi/2})}, \quad f \in H_0^\infty(\Sigma_\sigma) \,,$$

and the claim follows from the convergence property of the functional calculus.

$\square$

We have just seen in (11.3) that square function estimates are useful in establishing a bounded H^∞-calculus. Since they are even more important in establishing a converse to Theorem 11.5, we study them now in some detail.

11.6 Square Functions. Let $A \in \mathcal{S}(X)$ be a sectorial operator on H. For simplicity we only consider the square functions

$$\left(\int_0^\infty \|A^{1/2} R(te^{i\omega}, A)x\|^2 \, dt \right)^{1/2}, \quad x \in D(A) \cap R(A), \tag{11.4}$$

with $|\omega| > \omega(A)$, and, if $\omega(A) < \frac{\pi}{2}$ and $|\delta| < \frac{\pi}{2} - \omega(A)$,

$$\left(\int_0^\infty \|A^{1/2} T(te^{i\delta})x\|^2 \, dt \right)^{1/2}, \quad x \in D(A) \cap R(A). \tag{11.5}$$

By $T(z)$ we denote the analytic semigroup $T(z) = e^{-zA}$ generated by $-A$ (see [186, Theorem II.5.2]). These are finite for $x \in D(A) \cap R(A)$, since for $y = A^{1/2}x$ and $z = A^{-1/2}x$ we have

$$\|R(te^{i\omega}, A)y\| \leq \frac{C}{t} \quad \text{for } t \geq 1, \qquad \|AR(te^{i\omega}, A)z\| \leq C \quad \text{for } t \leq 1,$$

$$\|AT(te^{i\delta})z\| \leq \frac{C}{t} \quad \text{for } t \geq 1, \qquad \|T(te^{i\delta})y\| \leq C \quad \text{for } t \leq 1.$$

One can consider more general square functions of the form

$$\|\psi(\cdot A)x\|_{L_2(\mathbb{R}_+, dt/t, H)} \quad \text{for } \psi \in H_0^\infty;$$

but we will discuss this only in the notes.

11.7 Example. The square functions (11.4) and (11.5) are easy to compute if A is a selfadjoint operator. Indeed, for $x \in D(A) \cap R(A)$

$$\int_0^\infty \|A^{1/2} R(te^{i\omega}, A)x\|^2 \, dt = \int_0^\infty (AR(te^{i\omega}, A)^2 x | x) \, dt$$

$$= - \lim_{\substack{a \to 0 \\ b \to \infty}} \left[(AR(te^{i\omega}, A)x | x) \right]_a^b = (x|x) = \|x\|^2$$

and

$$\int_0^\infty \|A^{1/2} T(te^{i\delta})x\|^2 \, dt = - \int_0^\infty ((-A)T(te^{i\delta})^2 x | x) \, dt$$

$$= - \lim_{\substack{a \to 0 \\ b \to \infty}} - \left[(-T(2te^{i\delta})x | x) \right]_a^b$$

$$= \lim_{a \to 0} (T(2ae^{i\delta})x | x) - \lim_{b \to \infty} (AT(2te^{i\delta})A^{-1}x | x) = (x|x) = \|x\|^2.$$

11.8 A Classical Square Function. Square functions originated in harmonic analysis and the theory of harmonic functions. E. g., it is well known that a function $x \in L_2(i\mathbb{R})$ has an harmonic extension to the right half plane $\Sigma_{\frac{\pi}{2}}$ given by

$$\tilde{x}(t + is) = (P_t x)(is), \qquad t > 0, \ s \in \mathbb{R},$$

where P_t is the Poisson semigroup from 5.4 with generator $H\frac{d}{dt}$. Since the Hilbert transform is an isometry on $L_2(\mathbb{R})$ we can apply our basic estimates for analytic semigroups to obtain

$$\left\| t\frac{d}{dt}\tilde{x}(t + i\cdot) \right\|_{L_2} = \left\| t(H\frac{d}{dt})P_t x \right\|_{L_2} \le C\|x\|_{L_2}.$$

Paley and Littlewood have shown now that we can recover $\|x\|$ from this estimate by integrating over t:

$$\|x\|^2 \sim \int_0^\infty \left(t\|\frac{d}{dt}\tilde{x}(t + i\cdot)\| \right)^2 \frac{dt}{t} = \int_0^\infty t\|(H\frac{d}{dt})P_t x\|^2 \, dt, \tag{11.6}$$

which corresponds to a variant of (11.5), namely $\int_0^\infty t\|AT_t x\|^2 dt$. Since $\frac{d}{dt}$ has a bounded H^∞-calculus (see 10.2) the classical result (11.6) is a special case of Theorem 11.13 below.

The main point of the next theorem will be that the square functions in (11.4) are equivalent to the given norm if and only if the operator A has a bounded H^∞-calculus. In our later considerations it will be very helpful that we can replace the given norm by the seemingly more complicated norm (11.4).

11.9 Theorem. *For a sectorial operator $A \in \mathcal{S}(H)$ on a Hilbert space H the following conditions are equivalent.*

(H1) A has a bounded $H^\infty(\Sigma_\sigma)$-calculus for one (all) $\sigma \in (\omega(A), \pi]$.

(H2) A has bounded imaginary powers (BIP) and for one (all) $\nu \in (\omega(A), \pi]$ there is a constant C_1 with $\|A^{is}\| \le C_1 e^{\nu|s|}$.

(H3) For one (all) ω with $|\omega| \in (\omega(A), \pi]$ we have for $x \in H$

$$\|A^{1/2}R(\cdot e^{i\omega}, A)x\|_{L_2(\mathbb{R}_+, H)} \le C\|x\|,$$
$$\|(A^*)^{1/2}R(\cdot e^{i\omega}, A^*)x\|_{L_2(\mathbb{R}_+, H)} \le C\|x\|.$$

(H4) For one (all) ω with $|\omega| \in (\omega(A), \pi]$ there is a C with

$$\frac{1}{C}\|x\| \le \|A^{1/2}R(\cdot e^{i\omega}, A)x\|_{L_2(\mathbb{R}_+, H)} \le C\|x\|.$$

11.10 Remarks. a) Condition (H2) states that A has a bounded $H^\infty(\Sigma_\sigma)$-calculus if only the functions $f_t(\lambda) = \lambda^{it}$ for $t \in [-1, 1]$ belong to $H_A(\Sigma_\sigma)$.

b) An important part of the statements is that these conditions do not depend on the particular choice of the angle σ, ν, or ω in the given interval. In particular, if one can show that A has a "bad" H^∞-functional calculus or a "bad" estimate on A^{is} (for some large σ or ν), then A has automatically a "good" H^∞-calculus or estimate for A^{is} (for a small σ or ν close to the sectoriality angle $\omega(A)$.)

For the proof of 11.9 we need an extension procedure for operators on $L_2(\mathbb{R}^N)$ to $L_2(\mathbb{R}^N, H)$.

11.11 Lemma. *Let H be a Hilbert space and let T be a bounded operator on $L_2(\mathbb{R}^N)$. Then it can be extended to a bounded operator $T \otimes I$ on $L_2(\mathbb{R}^N, H)$ by*

$$T\left(\sum_{j=1}^{m} f_j x_j\right) = \sum_{j=1}^{m} T(f_j) \cdot x_j$$

and completion. Furthermore, $\|T \otimes I\| \leq \|T\|$.

Proof. Since measurable functions have a separable range, we can assume that H is separable and has an orthonormal basis (e_n). Then functions of the form

$$f(t) = \sum_{j=1}^{m} f_j(t) e_j, \qquad f_j \in \mathcal{S}(\mathbb{R}^N), \ m \in \mathbb{N},$$

are dense in $L_2(\mathbb{R}^N, H)$. For such f:

$$\|T(f)(\cdot)\|_{L_2(H)}^2 = \left\|\sum_{j=1}^{m} T(f_j)(\cdot) e_j\right\|_{L_2(H)}^2$$

$$= \int_{\mathbb{R}} \left\|\sum_{j=1}^{m} T(f_j)(t) e_j\right\|_H^2 dt = \sum_{j=1}^{m} \int_{\mathbb{R}} |T(f_j)(t)|^2 \, dt$$

$$\leq \|T\| \sum_{j=1}^{m} \int_{\mathbb{R}} |f_j(t)|^2 \, dt = \|T\| \cdot \|f(\cdot)\|_{L_2(H)},$$

where the last equality follows by reversing our steps. □

Now, we apply this procedure to obtain the Fourier transform and the Hilbert transform for H-valued functions.

11.12 Hilbert Space Valued Integral Transforms. a) If we apply the lemma to

$$\mathcal{F}_0 f(t) = (2\pi)^{-N/2} \int_{\mathbb{R}^N} e^{-is \cdot t} f(s) \, ds$$

we obtain a vector-valued Plancherel identity:

$$\|\mathcal{F}_0 f(\cdot)\|_{L_2(\mathbb{R}^N, H)} = \|f(\cdot)\|_{L_2(\mathbb{R}^N, H)}.$$

b) Also the following variant of the Hilbert transform is bounded on $L_2(\partial\Sigma_\theta, H)$:

$$\mathcal{H}_\theta g(\mu) = \text{PV-}\int_{\partial\Sigma_\theta} \frac{1}{\mu - \lambda} g(\lambda) \, d\lambda.$$

Proof of b). By Lemma 11.11 it is enough to show that $\mathcal{H}_\theta$ is bounded on the scalar functions $L_2(\partial\Sigma_\theta)$. Using the parameterization $t \in \mathbb{R} \mapsto \gamma(t) = |t|e^{i(\operatorname{sign} t)\theta}$ of $\partial\Sigma_\theta$ we write

$$\mathcal{H}_\theta g(\gamma(t)) = \int_{-\infty}^\infty \frac{(\operatorname{sign} s)e^{i(\operatorname{sign} s)\theta}}{e^{i\operatorname{sign} t\theta}|t| - e^{i\operatorname{sign} s\theta}|s|} g(\gamma(s))\, ds$$

$$= \chi_{\mathbb{R}_+}(t)\cdot H[g(\gamma(\cdot))\chi_{\mathbb{R}_+}(\cdot)](t) + \chi_{\mathbb{R}_-}(t)\cdot H[g(\gamma(\cdot))\chi_{\mathbb{R}_-}(\cdot)](t)$$

$$+ \chi_{\mathbb{R}_+}(t)\cdot K_\theta[g(\gamma(-\cdot))](t) - \chi_{\mathbb{R}_-}(t)K_{-\theta}[g(\gamma(\cdot))](-t)\,,$$

where H is the Hilbert transform, and K_θ is a variant of the Carleman transform:

$$K_\theta g(t) = \int_0^\infty \frac{1}{s - e^{i2\theta}t} g(s)\, ds, \qquad t \in \mathbb{R}_+\,.$$

We know that the Hilbert transform is bounded on $L_2(\mathbb{R})$ and K_θ is bounded on $L_2(\mathbb{R}_+)$ by [205, A.3], since the kernel $k(t,s) = (s-e^{i2\theta}t)^{-1}$ is homogeneous of degree -1 and $\int_0^\infty |s - e^{i2\theta}t|^{-1}s^{-1/2}\, ds < \infty$ for $|\theta| \in (0,\pi)$. □

Now, we are ready for the proof of 11.9

Proof of 11.9. We first give a simple argument showing that the conditions (H3) and (H4) hold for all ω with $|\omega| \in (\omega(A),\pi]$ if they hold for one such ω. Indeed the resolvent equation implies that

$$A^{1/2}R(te^{i\omega}, A) = [I + (e^{i\nu} - e^{i\omega})tR(te^{i\omega}, A)]A^{1/2}R(te^{i\nu}, A)\,.$$

Since the expression in brackets $[\dots]$ is bounded on $\mathbb{R}_+$, we obtain

$$\int_0^\infty \|A^{1/2}R(te^{i\omega}, A)x\|^2\, dt \le C\int_0^\infty \|A^{1/2}R(te^{i\nu}, A)x\|^2\, dt\,.$$

The same argument applies to A^*.

(H1) $\Rightarrow$ (H2) follows from the elementary estimate

$$|\lambda^{it}| = e^{-(\arg\lambda)t} \le e^{\sigma|t|} \qquad \text{for } \lambda \in \Sigma_\sigma\,.$$

(H2) $\Rightarrow$ (H3) We show the estimate for A; the same arguments apply to A^*. Fix ν and ω with $\omega(A) < \nu < \omega$. For $y \in R(A)$ we use a representation formula for the fractional powers of A (cf. Theorem 15.18)

$$A^{is-\frac{1}{2}}y = -\frac{\sin(\pi(is - \frac{1}{2}))}{\pi}\int_0^\infty t^{is-\frac{1}{2}}(t+A)^{-1}y\, dt\,.$$

For $x \in A(D(A^2))$ and $y = A^{1/2}x$ it follows that

$$A^{is}x = \frac{\cosh(\pi s)}{\pi}\int_0^\infty t^{is}[t^{1/2}(t+A)^{-1}A^{1/2}x]\frac{dt}{t}\,,$$

and if we replace A by the sectorial operator $e^{-i\theta}A$ with $|\theta| < \pi - \nu$

$$\frac{\pi}{\cosh(\pi s)}e^{\theta s}A^{is}x = \int_0^\infty t^{is}\left[e^{i\frac{\theta}{2}}t^{1/2}A^{1/2}(e^{i\theta}t + A)^{-1}x\right]\frac{dt}{t}\,.$$

With the substitution $t = e^u$ we get

$$\frac{\pi e^{\theta s}}{\cosh(\pi s)}A^{is}x = \int_{-\infty}^\infty e^{ius}\left[e^{i\frac{\theta}{2}}e^{u/2}(e^{i\theta}e^u + A)^{-1}A^{1/2}x\right]du\,.$$

Since $\cosh(\pi s) \sim e^{\pi|s|}$ the left hand side is in $L_2(\mathbb{R}, H)$; the Plancherel identity for H-valued functions, cf. 11.12 a), gives

$$\left(\int_0^\infty \|A^{1/2}(e^{i\theta}t + A)^{-1}x\|^2\,dt\right)^{1/2}$$

$$= \left(\int_{-\infty}^\infty \|e^{u/2}(e^{i\theta}e^u + A)^{-1}A^{1/2}x\|^2\,du\right)^{1/2}$$

$$= \frac{\pi}{\sqrt{2\pi}}\left(\int_{-\infty}^\infty \left\|\frac{e^{\theta s}}{\cosh(\pi s)}A^{is}x\right\|^2\,ds\right)^{1/2} \leq C\|x\|$$

by the assumption on A^{is}. For $\omega = \pi - \theta$ we have $(e^{i\theta}t + A)^{-1} = -R(e^{-i\omega}t, A)$ and (H3) follows since $|e^{\theta s}\cosh(\pi s)^{-1}| \leq e^{-\omega|s|}$.

(H3) $\Rightarrow$ (H4) For $x \in R(A) \cap D(A)$ we have

$$e^{i\omega}\int_0^\infty AR(e^{i\omega}t, A)^2 x\,dt = \lim_{R \to \infty}\left[-AR(e^{i\omega}t, A)x\right]_{-1/R}^R = x\,.$$

Hence for any $y \in H$

$$(x|y) = e^{i\omega}\int_0^\infty (AR(e^{i\omega}t, A)^2 x|y)\,dt$$

$$= e^{i\omega}\int_0^\infty (A^{1/2}R(e^{i\omega}t, A)x \mid (A^*)^{1/2}R(e^{-i\omega}t, A^*)y)\,dt\,,$$

and using Hölder's inequality

$$|(x|y)| \leq \left(\int_0^\infty \|A^{1/2}R(e^{i\omega}t, A)x\|^2\,dt\right)^{1/2}$$

$$\cdot\left(\int_0^\infty \|(A^*)^{1/2}R(e^{-i\omega}t, A^*)y\|^2\,dt\right)^{1/2}$$

$$\leq \left(\int_0^\infty \|A^{1/2}R(e^{i\omega t}, A)x\|^2\,dt\right)^{1/2}C\|y\|\,.$$

Together with (H3) we obtain

$$\frac{1}{C}\|x\| \leq \left(\int_0^\infty \|A^{1/2}R(e^{i\omega}t, A)x\|^2 dt\right)^{1/2} \leq C\|x\|.$$

(H4) $\Rightarrow$ (H1) Choose σ and ω with $\omega(A) < \omega < \sigma$. For $f \in H_0^\infty(\Sigma_\sigma)$ we show now how to estimate $f(A)$ by "pushing" it through the square function. For $|\arg \mu| = \omega$ we calculate using the resolvent equation

$$\begin{aligned}
A^{1/2}R(\mu, A)f(A) &= \frac{1}{2\pi i}\int_{\partial\Sigma_\omega} f(\lambda)A^{1/2}R(\mu, A)R(\lambda, A)\, d\lambda \\
&= \left(\frac{1}{2\pi i}\,\text{PV-}\!\!\int_{\partial\Sigma_\omega} \frac{f(\lambda)}{\lambda - \mu}\, d\lambda\right) A^{1/2}R(\mu, A) \\
&\quad - \frac{1}{2\pi i}\,\text{PV-}\!\!\int_{\partial\Sigma_\omega} \frac{f(\lambda)A^{1/2}R(\lambda, A)}{\lambda - \mu}\, d\lambda \\
&= f(\mu)A^{1/2}R(\mu, A) - K[f(\cdot)A^{1/2}R(\cdot, A)](\mu)
\end{aligned} \tag{11.7}$$

by Cauchy's theorem and using the notation

$$KG(\mu) = \frac{1}{2\pi i}\,\text{PV-}\!\!\int_{\partial\Sigma_\omega} \frac{G(\lambda)}{\lambda - \mu}\, d\lambda, \qquad \mu \in \partial\Sigma_\omega.$$

K is a variant of the Hilbert transform and it is a bounded operator on $L_2(\partial\Sigma_\omega)$. By 11.12 b) we have for $g \in L_2(\partial\Sigma_\omega, H)$

$$\int_{\partial\Sigma_\omega} \|Kg(\mu)\|^2 d|\mu| \leq M^2 \int_{\partial\Sigma_\omega} \|g(\lambda)\|^2 d|\mu|.$$

Then for $x \in D(A)$ is follows from (11.7) and (H4) that

$$\begin{aligned}
\|f(A)x\|_H &\leq C\|A^{1/2}R(\cdot, A)[f(A)x]\|_{L_2(\partial\Sigma_\omega, H)} \\
&\leq C\|f(\cdot)A^{1/2}R(\cdot, A)x\|_{L_2(\partial\Sigma_\omega, H)} \\
&\quad + CM\|f(\cdot)A^{1/2}R(\cdot, A)x\|_{L_2(\partial\Sigma_\omega, H)} \\
&\leq C(1+M)\|f\|_{H^\infty(\partial\Sigma_\omega)}\|A^{1/2}R(\cdot, A)x\|_{L_2(\partial\Sigma_\omega, H)} \\
&\leq C_1\|x\|\,\|f\|_\infty
\end{aligned}$$

again by (H4) and the first part of the proof. $\qquad\square$

Now we have the tools to prove our main theorem:

11.13 Theorem. *For a sectorial operator $A \in \mathcal{S}(H)$ on a Hilbert space H with $\omega(A) < \frac{\pi}{2}$ the following conditions are equivalent.*

(H1) A has a bounded $H^\infty(\Sigma_\sigma)$ functional calculus for one (all) $\sigma \in (\omega(A), \pi]$.

(H5) For one (all) $\frac{\pi}{2} \geq \omega > \omega(A)$ there is an equivalent scalar product $(\cdot|\cdot)_{A,\omega}$ on H (depending on A and ω) so that A is $(\frac{\pi}{2} - \omega)$-accretive with respect to $(\cdot|\cdot)_{A,\omega}$.

(H6) For one (all) $\omega \in (\omega(A), \frac{\pi}{2}]$ there is an equivalent Hilbert space norm $\|\cdot\|_{A,\omega}$ on H (depending on A and ω) such that $-A$ generates an analytic semigroup $T(z) = e^{-zA}, z \in \Sigma_{\frac{\pi}{2}-\omega}$, which is contractive with respect to the new norm: $\|T(z)\|_{A,\omega} \leq 1$ for $z \in \Sigma_{\frac{\pi}{2}-\omega}$.

(H7) For one (all) $\omega \in (\omega(A), \frac{\pi}{2}]$ there is a constant C such that

$$\frac{1}{C}\|x\| \leq \left(\int_{\Sigma_{\pi/2-\omega}} \|AT(z)x\|^2\, dz\right)^{1/2} \leq C\|x\|.$$

Remarks. a) Again it is part of the statements that if they hold for one ω then they hold for all ω in the possible range.

b) It is important that we consider equivalent *Hilbert space* norms in (H5) and (H6). It is easy to construct for every bounded analytic semigroup $T(z)$, $z \in \Sigma_\sigma$, on H an equivalent norm $\||\cdot\||$, so that $T(z)$ is contractive with respect to $\||\cdot\||$, just take

$$\||x\|| = \sup_{z \in \Sigma_\sigma} \|T(z)x\|.$$

Since there are analytic generators on H without a bounded H^∞-calculus, the norm $\||\cdot\||$ cannot always be given by an inner product. This shows again how subtle the notion of an H^∞-calculus is.

Proof of Theorem 11.13. (H1) $\Rightarrow$ (H7) By Theorem 11.9 (H3), there is a constant $C < \infty$ such that for $x \in D(A) \cap R(A)$

$$\|A^{1/2}R(e^{i\theta}\cdot, -A)x\|_{L_2(\mathbb{R}_+, H)} \leq C\|x\|$$

for all θ with $\omega \leq |\theta| \leq \pi$. Since $-e^{i\theta}A$ is the generator of $T(e^{i\theta}t)$ we have for $|\delta| < \frac{\pi}{2} - \omega$

$$\int_0^\infty e^{-its}[A^{1/2}T(se^{i\delta})]\, ds = A^{1/2}R(it, -e^{i\delta}A)$$

$$= -e^{-i\delta}A^{1/2}R(te^{i(-\frac{\pi}{2}-\delta)}, A)$$

and by the vector-valued Plancherel formula 11.12 a)

$$\|A^{1/2}T(e^{i\delta}\cdot)x\|_{L_2(\mathbb{R}_+, H)} \leq C\|x\| \qquad \text{for } |\delta| < \frac{\pi}{2} - \omega.$$

Now with $\Sigma = \Sigma_{\frac{\pi}{2}-\omega}$

$$\|AT(\cdot)x\|_{L_2(\Sigma, H)}^2 = \int_{-(\frac{\pi}{2}-\omega)}^{\frac{\pi}{2}-\omega} \int_0^\infty \|AT(e^{i\theta}t)x\|^2 t\, dt\, d\theta$$

$$\leq (\pi - 2\omega) \max_{|\theta| \leq \frac{\pi}{2}-\omega} \int_0^\infty \left\|[t^{1/2}A^{1/2}T(e^{i\theta}\tfrac{t}{2})]A^{1/2}T(e^{i\theta}\tfrac{t}{2})x\right\|^2 dt$$

$$\leq 2(\pi - 2\omega)C \max_{|\theta| \leq \frac{\pi}{2}-\omega} \int_0^\infty \|A^{1/2}T(e^{i\theta}t)x\|^2\, dt \leq C_1\|x\|^2$$

since the operator family in $[\dots]$ is bounded by (H1). The same argument applies to the dual operators and (with $\Sigma = \Sigma_{\frac{\pi}{2}-\omega}$)

$$\|A^*T(\lambda)^*x\|_{L_2(\Sigma,H)} \leq C_1\|x\|.$$

Then, for $x, y \in D(A) \cap R(A)$,

$$\left|\int_\Sigma (A^2T(2\lambda)x|y)\,d\lambda\right| \leq \|AT(\lambda)x\|_{L_2(\Sigma,H)}\,\|A^*T(\lambda)^*y\|_{L_2(\Sigma,H)}$$

$$\leq C_1\|AT(\lambda)x\|_{L_2(\Sigma,H)} \cdot \|y\|.$$

To obtain the lower estimate in (H7), it remains to calculate the integral on the left hand side (with $\delta = \tan(\frac{\pi}{2} - \omega)$):

$$\int_\Sigma A^2T(2\lambda)x\,d\lambda = \frac{1}{4i}\int_0^\infty \int_{-\delta t}^{\delta t} \frac{\partial}{\partial t}\frac{\partial}{\partial s}T(t+is)x\,ds\,dt$$

$$= \frac{1}{4i}\int_0^\infty \left[\frac{\partial}{\partial t}T(t+is)x\Big|_{s=\delta t} - \frac{\partial}{\partial t}T(t+is)x\Big|_{s=-\delta t}\right]dt$$

$$= \frac{1}{4i}\lim_{a\to 0}\left[(1+i\delta)^{-1}T((1+i\delta)a)x - (1-i\delta)^{-1}T((1-i\delta)a)x\right]$$

$$= \frac{1}{2}\frac{\delta}{1+\delta^2}\,x$$

since the upper limit disappears for $x \in R(A)$. Hence

$$|(x|y)| \leq C\|AT(\lambda)x\|_{L_2(\Sigma,H)} \cdot \|y\|$$

and the lower estimate in (H7) follows.

(H7) $\Rightarrow$ (H6) For a fixed $\frac{\pi}{2} \geq \omega > \omega(A)$ and $\delta = \frac{\pi}{2} - \omega$ the norm

$$\|x\|_{A,\omega} = \left(\int_{\Sigma_\delta} \|AT(z)x\|^2\,dz\right)^{1/2}$$

is equivalent to $\|\cdot\|$ on H by (H7), and is given by the inner product

$$(x|y)_{A,\omega} = \int_{\Sigma_\delta} (A^{1/2}T(\lambda)x|A^{1/2}T(\lambda)y)\,d\lambda.$$

$T(\mu)$ is contractive in the new norm for $\mu \in \Sigma_\delta$ since by the semigroup property

$$\|T(\mu)x\|^2_{A,\omega} = \int_{\Sigma_\delta} \|AT(\lambda+\mu)x\|^2_H\,d\lambda$$

$$= \int_{\mu+\Sigma_\delta} \|AT(\lambda)x\|^2_H\,d\lambda \leq \|x\|^2_{A,\omega}.$$

(H6) $\Rightarrow$ (H5) follows from Example 11.4 applied to A on $(H, \|\cdot\|_{A,\omega})$.

(H5) $\Rightarrow$ (H1) By Theorem 11.5 we get a bounded $H^\infty(\Sigma_\sigma)$ calculus for $\sigma > \frac{\pi}{2}$. By Theorem 11.9 we can improve the angle. $\square$

One of the most attractive results of spectral theory is certainly that an operator A on H is normal if and only if A is similar to a multiplication operator on some space $L_2(\Omega, \mu)$. The class of operators with a bounded H^∞-calculus is of course much larger than the class of normal operators, but we can show that it is precisely the class of operators which have a dilation to certain multiplication operators.

11.14 Theorem. *A sectorial operator $A \in \mathcal{S}(H)$ on the Hilbert space H has a bounded H^∞-calculus if and only if A has a dilation to a multiplication operator*

$$M f(t) = (it)^\alpha f(t) \qquad on \ L_2(\mathbb{R}, H)$$

for one (all) $\alpha > \frac{2\omega(A)}{\pi}$, i.e. for an equivalent scalar product $(\cdot | \cdot)_A$ there is an isometric embedding $J : (H, (\cdot | \cdot)_A) \to L_2(\mathbb{R}, H)$, such that $P = JJ^*$ is an orthogonal projection from $L_2(\mathbb{R}, H)$ onto $J(H)$ and for μ with $\omega(A) < |\arg \mu| < \alpha \frac{\pi}{2}$

$$J^* R(\mu, M) J = R(\mu, A).$$

In this case, we have for all $f \in H^\infty(\Sigma_\sigma)$ with $\omega(A) < \sigma < \alpha \frac{\pi}{2}$

$$J^* f(M) J = f(A), \quad or \quad P f(M) J = J f(A). \tag{H8}$$

If $\omega(A) < \frac{\pi}{2}$, $B = -A$ generates a semigroup T_t and we can choose $\alpha = 1$. If N_t denotes the unitary group $N_t f(t) = e^{-it} f(t)$, then

$$P N_t J = J T_t. \tag{H9}$$

As an immediate corollary we obtain a version of the dilation theorem of Sz.-Nagy, which also proves 10.10 in the Hilbert space case.

11.15 Corollary. *The generator $-A$ of an analytic semigroup T_t on a Hilbert space H has a bounded H^∞-calculus for one (all) $\sigma \in (\omega(A), \pi)$ if and only if it is a dilation of a unitary group N_t (i.e. (H9) holds with the notation of 11.14).*

Proof of Theorem 11.14. Assume that A has a bounded H^∞-functional calculus.

1st Step: We assume that $\omega(A) < \frac{\pi}{2}$ and consider $M f(t) = it f(t)$ on $L_2(\mathbb{R}, H)$. We define a new inner product for $x, y \in D(A) \cap R(A)$ by

$$(x|y)_A = \int_{-\infty}^{\infty} (A^{1/2} R(it, A) x | A^{1/2} R(it, A) y) \, dt.$$

Since A has a bounded H^∞-calculus and $\omega(A) < \frac{\pi}{2}$, the norm

$$\|x\|_A := \|A^{1/2} R(i \cdot, A) x\|_{L_2(\mathbb{R}, H)}$$

is then equivalent to the given norm on H. If we equip H with the new norm $\| \cdot \|_A$, then the map

$$J : (H, \|\cdot\|_A) \to L_2(\mathbb{R}, H), \quad x \mapsto Jx = A^{1/2}R(i\cdot, A)x$$

becomes an isometric embedding. For the Hilbert space dual J^* we note that for $x, y \in H$

$$(J^*Jx|y)_A = (Jx|Jy)_{L_2(H)} = \int_{-\infty}^{\infty} (A^{1/2}R(it, A)x|A^{1/2}R(it, A)y)\,dt$$

$$= (x|y)_A,$$

so that $J^*J = I_H$ and $P = JJ^*$ is a contractive projection of $L_2(\mathbb{R}, H)$ onto $J(H)$. To check (H8) we choose μ with $\mathrm{Re}\,\mu < 0$. Then by the resolvent equation

$$JR(\mu, A)x = A^{1/2}R(i\cdot, A)R(\mu, A)x$$

$$= (\mu - i\cdot)^{-1}A^{1/2}R(i\cdot, A)x - (\mu - i\cdot)^{-1}A^{1/2}R(\mu, A)x,$$

and for $y \in H$

$$(R(\mu, A)x|y)_A = (JR(\mu, A)x|Jy)_{L_2(H)}$$

$$= ((\mu - i\cdot)^{-1}A^{1/2}R(i\cdot, A)x|Jy)$$

$$\quad + ((\mu - i\cdot)^{-1}A^{1/2}R(\mu, A)x|A^{1/2}R(i\cdot, A)y)$$

$$= (R(\mu, M)Jx|Jy)$$

$$\quad + \int_{-\infty}^{\infty} (\mu - it)^{-1}(A^{1/2}R(\mu, A)x, A^{1/2}R(it, A)y)\,dt$$

$$= (J^*R(\mu, M)Jx|y)_A$$

since the last integral is zero by Cauchy's theorem. Indeed,

$$\lambda \mapsto (\mu - \lambda)^{-1} \quad \text{and} \quad \lambda \mapsto R(-\bar{\lambda}, A)^* = R(-\lambda, A^*)$$

are both analytic on Σ_δ for $\frac{\pi}{2} < \delta < |\arg\mu|$. Furthermore, if $\sigma > \frac{\pi}{2}$ and $f \in H_0^\infty(\Sigma_\sigma)$, then

$$(f(A)x|y)_A = \int_{\partial\Sigma_\mu} f(\lambda)(R(\lambda, A)x|y)_A\,d\lambda$$

$$= \int_{\partial\Sigma_\mu} f(\lambda)(J^*R(\mu, A)Jx|y)_A\,d\lambda$$

$$= (J^*[f(A)Jx]|y)_A$$

and the dilation property is justified.

   2nd Step: To reduce the general case to the first step we choose an $\alpha > \frac{2\omega(A)}{\pi}$. Then $\tilde{A} = A^{1/\alpha}$ has $\omega(\tilde{A}) = \omega(A)/\alpha < \frac{\pi}{2}$ by Theorem 15.16. Choose a σ with $\omega(A) < \sigma < \alpha\frac{\pi}{2}$. For every $f \in H_0^\infty(\Sigma_\sigma)$ the function $\tilde{f}(\lambda) = f(\lambda^\alpha)$

belongs to $H_0^\infty(\Sigma_{\sigma/\alpha})$ with $\omega(\widetilde{A}) < \sigma/\alpha < \frac{\pi}{2}$. Proposition 15.11 gives then that $\widetilde{f}(\widetilde{A}) = f(A)$. In particular it follows that $\widetilde{A}$ has a bounded $H^\infty(\Sigma_{\sigma/\alpha})$ functional calculus. Step 1 of the proof applies to $\widetilde{A}$ now and gives operators $\widetilde{M}, \widetilde{J}$ such, that for all $f \in H_0^\infty(\Sigma_\sigma)$

$$\widetilde{J}^* \widetilde{f}(\widetilde{M}) \widetilde{J} = \widetilde{f}(\widetilde{A}) = f(A).$$

So for $Mg(t) = (it)^\alpha g(t)$ and $J = \widetilde{J}$ we obtain our claim since

$$[\widetilde{f}(\widetilde{M})g](t) = \widetilde{f}(it)g(t) = f((it)^\alpha)g(t) = [f(M)g](t). \qquad \square$$

Notes on Section 11:

N 11.1 – N 11.3: Accretive operators are a classical topic discussed e. g. in [137] Chap. V, §3.10. Their introduction was motivated by the study of differential operators such as in Section 8. The point in the proof in 11.3 is that the form $\mathfrak{a}$ defined on $D(A)$ is *closable* (we refer to [137, Ch. VI §1.4] for this notion). Not every sectorial form is closable, but every sectorial form induced by an operator is closable (cf. [137, Ch. VI §1.5]). For the fact that A is associated to a form (for an equivalent inner product) if and only if A satisfies the equivalent statements after 11.1 we also refer to [16].

N 11.4: The classical Lumer-Phillips theorem applies to general c_0 semigroups, not just analytic semigroups (see e. g. [88] or [186]).

N 11.5: Our proof of theorem 10.5 is due to E. Franks and appears in [4], Theorem G.

N 11.6 – N 11.10: Square functions originate in Harmonic Analysis, as indicated in 10.8 (see [207] for an in-depth discussion). McIntosh and Yagi introduced them as a tool in operator theory [176], [225] and considered square functions of the form

$$\|x\|_\psi = \int_0^\infty \|\psi(tA)x\|^2 \frac{dt}{t}$$

for general $\psi \in H_0^\infty(\Sigma_r)$. Our theorem 10.9 is a special case of McIntosh's result in [176], which was stated for general square functions. We made the choice to give an introduction to the square function method by limiting us to very concrete square functions and to give more direct proofs of $(H2) \Rightarrow (H3)$ and $(H4) \Rightarrow (H1)$, which are taken from [135]. The original approach of McIntosh in [176] is based on the equivalence of all $\|\cdot\|_\psi$ for $\psi \in H_0^\infty(\Sigma_\nu)$. For a related result see the end of section 12.

N 11.11 – N 11.12 are well known classical facts.

N 11.13: Implications $(H1) \Rightarrow (H5), (H6)$, which is not covered by the previous results was first shown in [162] in the form: BIP implies that A is maximal accretive for an equivalent Hilbert space norm. In [160] LeMerdy observed the simple proof using the square functions $\|A^{1/2}T_tx\|$. That A is actually ω-accretive for $0 < \omega$ for an equivalent Hilbert space norm was shown in [163], using completely bounded maps. Condition $(H7)$ is new and simplifies the proof of [163].

N 11.14 is a special case of a similar dilation theorem for operators with a bounded H^∞-calculus on a Banach space with finite cotype in [99].

N 11.15: This version of the dilation theorem for contractive semigroups in an immediate consequence of the results in [162].

12 The Operator-valued H^∞-calculus and Sums of Closed Operators

In this section we give a characterization of the bounded H^∞-calculus on Banach spaces, which will be our basic tool in developing the theory of the H^∞-calculus (Theorem 12.2). In particular, it will allow us to construct a functional calculus for certain operator-valued analytic functions (Theorem 12.7). From this in turn we obtain a joint functional calculus for commuting sectorial operators (Theorem 12.12) and a theorem on the closedness of the sum of two sectorial operators (Theorem 12.13). As we explained already in Section 1.18, this theorem allows to give an operator-theoretic proof of the characterization theorem 1.11 of maximal regularity.

In the last section we used square function estimates to prove characterizations of the bounded H^∞-calculus in Hilbert spaces. Our first task is therefore to find a substitute for these square functions in the Banach space setting. We will do this using Rademacher averages, and hence it will be no surprise that R-boundedness will play again an important role in this section. In particular, in Theorem 12.8 we show how a bounded H^∞-calculus creates large sets of R-bounded operators. As a consequence we can clarify the connection between a bounded H^∞-calculus and BIP: A sectorial operator has a bounded H^∞-calculus if and only if the imaginary powers are bounded and form an R-bounded set (see Corollary 12.11).

Finally we extend a further Hilbert space result ($H4$ in Theorem 11.9) to the Banach space setting: An R–sectorial operator A has a bounded H^∞-calculus if and only if its square function norms are equivalent to the given norm on X.

12.1 Discussion. Looking for square functions, it is, of course, tempting to use again the norms from Section 10

$$\|x\|_A = \left(\int\limits_0^\infty \|A^{\frac{1}{2}} R(-t, A)x\|^2 dt \right)^{\frac{1}{2}}, \ x \in D(A)$$

also in a general Banach space X. But they do not give the desired results. If X is not a Hilbert space, $\| \cdot \|_A$ is usually not equivalent to the given norm on X and there is no characterization of the bounded H^∞-calculus in terms of $\|\cdot\|_A$. To find a workable replacement for $\|\cdot\|_A$ we first make a calculation in the Hilbert space case. Using the notation

$$\psi_\nu(\lambda) = \lambda^{\frac{1}{2}}(e^{i\nu} - \lambda)^{-1} \tag{12.1}$$

we have

$$\int_0^\infty \|A^{\frac{1}{2}}R(-t,A)x\|^2 dt = \int_0^\infty \|\psi_\pi(tA)x\|^2 \frac{dt}{t}$$

$$= \sum_{k\in\mathbb{Z}} \int_{2^k}^{2^{k+1}} \|\psi_\pi(tA)x\|^2 \frac{dt}{t} = \int_1^2 \sum_k \|\psi_\pi(2^k tA)x\|^2 \frac{dt}{t}$$

$$= \int_1^2 \left(\int_0^1 \|\sum_k r_k(u)\psi_\pi(2^k tA)x\|^2 du \right) \frac{dt}{t} ,$$

where we use the fact that $r_k(\cdot)\psi_\pi(2^k tA)x$, $k \in \mathbb{Z}$, is an orthogonal sequence in the Hilbert space $L_2(X) := L_2([0,1], X)$ for the Bernoulli sequence $(r_k)_{k\in\mathbb{Z}}$.

Hence one might hope that

$$\left(\int_0^\infty \|A^{\frac{1}{2}}R(t,A)x\|^2 dt \right)^{\frac{1}{2}} \sim \sup_{t\in[1,2]} \|\sum_{k\in\mathbb{Z}} r_k \psi_\nu(2^k tA)x\|_{L_2(X)}$$

and that the last expression is the right one for the general Banach space setting. For $\varphi \in H_0^\infty(\Sigma_\sigma)$ with $\sigma > \omega(A)$ we will use the notation

$$\|x\|_\varphi := \sup_{t>0} \left\|\sum_k r_k\varphi(t2^k A)x\right\|_{L_p(X)}, \quad x \in X,$$

$$\|x^*\|_\varphi := \sup_{t>0} \left\|\sum_k r_k\varphi(t2^k A)^* x^*\right\|_{L_p(X^*)}, \quad x^* \in X^*. \tag{12.2}$$

In particular for $\varphi = \psi_\theta$ we have

$$\|x\|_{\psi_\theta} = \sup_{t>0} \left\|\sum_k r_k A^{1/2} R(t2^k e^{i\theta}, A)x\right\|_{L_p(X)}.$$

Now we can formulate our basic characterization of the bounded H^∞-calculus for sectorial operators. As in Sections 9 and 10 we will work with sectorial operators $A \in \mathcal{S}(X)$, i.e. A is also assumed to be injective with dense range.

12.2 Theorem. *Let $A \in \mathcal{S}(X)$ be a sectorial operator on a Banach space X. Consider the following conditions (with ψ_θ and $\|\cdot\|_{\psi_\theta}$ as above).*

i) A has a bounded $H^\infty(\Sigma_\sigma)$-calculus.

ii) We have

$$\sup_{N} \sup_{t>0} \sup_{\varepsilon_k=\pm 1} \| \sum_{k=-N}^{N} \varepsilon_k \psi_{\pm\theta}(2^k tA)\|_{B(X)} < \infty.$$

iii) There is a constant $C < \infty$ such that for all $x \in X$ and $x^* \in X^*$ we have

$$\|x\|_{\psi_\theta} + \|x\|_{\psi_{-\theta}} \leq C\|x\|,$$

$$\|x^*\|_{\psi_\theta} + \|x^*\|_{\psi_{-\theta}} \leq C\|x^*\|.$$

iv) The functions $t \in \mathbb{R}_+ \to AR(te^{\pm i\theta}, A)x$ are weakly of bounded variation, i.e. they have weakly integrable derivatives and there exists a constant $C < \infty$ such that, for all $x \in X$ and $x^* \in X^*$, we have

$$\int_0^\infty |\langle AR(te^{\pm i\theta}, A)^2 x, x^*\rangle| dt \leq C\|x\| \cdot \|x^*\|.$$

Then i) $\Rightarrow$ ii) $\Rightarrow$ iii) $\Rightarrow$ iv) holds for $\theta > \sigma > \omega(A)$, and iv) $\Rightarrow$ i) holds if $\sigma > \theta > \omega(A)$.

12.3 Remark. More general square functions can be used in this theorem: our proof will show that i) $\Rightarrow$ ii) $\Rightarrow$ iii) holds for an arbitrary $\varphi \in H_0^\infty(\Sigma_\sigma) \setminus \{0\}$, $\sigma > \omega(A)$, in place of ψ_θ. In Proposition 12.15 and Theorem 12.17 we will show that for an R-sectorial operator $\| \cdot \|_\varphi$ and $\| \cdot \|_\psi$ are equivalent for all $\varphi, \psi \in H_0^\infty(\Sigma_\sigma) \setminus \{0\}$ with $\sigma > \omega_R(A)$. Hence we can replace ψ_θ by any $\varphi \in H_0^\infty(\Sigma_\sigma) \setminus \{0\}$ with $\sigma > \omega_R(A)$ also in iii).

For the proof of Theorem 12.2 we need some preparations.

12.4 Lemma. *Let $A \in \mathcal{S}(X)$ be a sectorial operator on X. Suppose that $F : \Sigma_\sigma \to B(X)$ is a bounded analytic function, such that $F(\lambda)$ commutes with $R(\mu, A)$ for $\lambda \in \Sigma_\sigma$ and $\|F(\lambda)\| \leq |\lambda|^\varepsilon(1 + |\lambda|)^{-2\varepsilon}$ for some $\varepsilon > 0$ on Σ_σ. Then for any $\omega(A) < \nu < \sigma$ and $s \in (0, 1)$*

$$F(A) := \int_{\partial\Sigma_\nu} F(\lambda)R(\lambda, A)d\lambda = \int_{\partial\Sigma_\nu} F(\lambda)\lambda^{-s} A^s R(\lambda, A)d\lambda. \tag{12.3}$$

Proof. First note that $A^s R(\lambda, A)$ is a bounded operator for $\lambda \in \partial\Sigma_\nu$, which is given by the integral

$$A^s R(\lambda, A) = \frac{1}{2\pi i} \int_{\partial\Sigma_{\nu'}} \mu^s (\lambda - \mu)^{-1} R(\mu, A) d\mu$$

if $\omega(A) < \nu' < \nu$. This gives the estimate

$$\|A^s R(\lambda, A)\| \leq C \int_{\partial \Sigma_\nu} |\mu|^{s-1} |\lambda - \mu|^{-1} d|\mu| \leq C_1 |\lambda|^{s-1}$$

and shows that the second integral in (12.3) exists as a Bochner integral. We consider now elements in X of the form $x = \varrho_m(A)y$ where $y \in X$ and ϱ_n is the approximating sequence $\varrho_n(z) = \frac{n}{n+z} - \frac{1}{1+nz}$ from 9.4. Then

$$F(A)\varrho_n(A)x = F(A)\varrho_m(A)\varrho_n(A)y$$
$$= [A^s \varrho_m(A)][F(A)A^{-s}\varrho_n(A)]y$$

where $A^s \varrho_n(A)$ and $A^{-s}\varrho_n(A)$ are bounded operators. Hence, by the multiplicativity of the Dunford integral, which also holds for operator-valued integrals,

$$F(A)\varrho_n(A)x = \frac{1}{2\pi i}[A^s \varrho_m(A)] \int_{\partial \Sigma_\nu} \lambda^{-s} \varrho_n(\lambda)F(\lambda)R(\lambda, A)y \, d\lambda$$

$$= \frac{1}{2\pi i} \int_{\partial \Sigma_\nu} \lambda^{-s} \varrho_n(\lambda)F(\lambda)[A^s \varrho_m(A)]R(\lambda, A)y \, d\lambda$$

$$= \frac{1}{2\pi i} \int_{\partial \Sigma_\nu} \lambda^{-s} \varrho_n(\lambda)F(\lambda)A^s R(\lambda, A)x \, d\lambda.$$

For $n \to \infty$, Lebesgue's convergence theorem and $\lim_n \varrho_n(\lambda) = 1$ yield (12.3) for all x in the dense subspace $R(\varrho_m(A)) = D(A) \cap R(A)$.   □

12.5 Lemma. *For every $\psi \in H_0^\infty(\Sigma_\omega)$ there is a $\varphi \in H_0^\infty(\Sigma_\omega)$ and a constant γ such that for $\omega > \nu > \omega(A)$ and $t > 0$*

$$\psi(tA) = \frac{1}{2\pi i} \int_{\partial \Sigma_\nu} \varphi(\lambda)[tAR(\lambda, tA)^2]d\lambda + \gamma tA(1 + tA)^{-2}. \qquad (12.4)$$

Proof. Let $\bar\psi$ be an antiderivative of $\lambda^{-1}\psi(\lambda)$ which vanishes at 0. Define $\varphi(\lambda) = \bar\psi(\lambda) + \gamma\lambda(1 + \lambda)^{-1}$, where $\gamma = -\int_0^\infty t^{-1}\psi(t)dt$. Then

$$\varphi'(\lambda) = \lambda^{-1}\psi(\lambda) + \gamma(1 + \lambda)^{-2}, \qquad (12.5)$$

and we can show that $\varphi \in H_0^\infty(\Sigma_\omega)$. Hence for $\omega > \nu > \omega(A)$

$$\varphi(tA) = \frac{1}{2\pi i} \int_{\partial \Sigma_\nu} \varphi(t\lambda)R(\lambda, A)d\lambda = \frac{1}{2\pi i} \int_{\partial \Sigma_\nu} \varphi(\lambda)R(\lambda, tA)d\lambda.$$

Therefore

$$\frac{d}{dt}\varphi(tA) = \frac{1}{2\pi i} \int_{\partial \Sigma_\nu} \lambda\varphi'(t\lambda)R(\lambda, A)d\lambda = A\varphi'(tA)$$

and

$$\frac{d}{dt}\varphi(tA) = \frac{1}{2\pi i}\int_{\partial\Sigma_\nu}\varphi(\lambda)[AR(\lambda, tA)^2]d\lambda$$

Applying the functional calculus to (12.5) we get all together

$$\psi(tA) + \gamma tA(1 + tA)^{-2} = (tA)\varphi'(tA) = \frac{1}{2\pi i}\int_{\partial\Sigma_\nu}\varphi(\lambda)[tAR(\lambda, tA)^2]d\lambda. \quad \square$$

The estimates of the next and last lemma are often referred to as "randomization" of series and integrals.

12.6 Lemma. *a) Let $\psi_k, \phi_k, M_k \in B(X)$, $k = 1, \ldots, N$. For $x_k \in X$ and $x_k^* \in X^*$ we have*

$$\sum_{k=1}^{N}|\langle\psi_k M_k\phi_k x_k, x_k^*\rangle| \le 2R(\{M_k : k\})\cdot\Big\|\sum_k\epsilon_k\phi_k x_k\Big\|_{L_2(X)}\cdot\Big\|\sum_k\epsilon_k\psi_k^* x_k^*\Big\|_{L_2(X^*)}.$$

b) For strongly measurable, locally bounded operator valued functions ψ, ϕ and M on $\mathbb{R}_+$ we have for $x \in X$ and $x^* \in X^*$

$$\lim_{R\to\infty}\int_{1/R}^{R}|\langle\psi(t)M(t)\phi(t)x, x^*\rangle|\frac{dt}{t}$$

$$\le 2R(\{M(t) : t > 0\})\Big(\sup_{t>0}\sup_{N}\Big\|\sum_{|k|\le N}r_k\phi(a^k t)x\Big\|_{L_2(X)}$$

$$\le \qquad \cdot\Big\|\sum_{|k|\le N}r_k\psi^*(a^k t)x^*\Big\|_{L_2(X^*)}.$$

Proof. a) We have

$$\sum_{k=1}^{N}|\langle\psi_k M_k\phi_k x_k, x_k^*\rangle| \le \sup_{|a_k|=1}\Big|\sum_{k=1}^{N}a_k\langle M_k\phi_k x_k, \psi_k^* x_k^*\rangle\Big|$$

$$= \sup_{|a_k|=1}\Big\|\sum_{k=1}^{N}r_k^2\langle a_k M_k\phi_k x_k, \psi_k^* x_k^*\rangle\Big\|_{L_1(0,1)}$$

$$= \sup_{|a_k|=1}\Big\|\langle\sum_k r_k a_k M_k\phi_k x_k, \sum_k r_k\psi_k^* x_k^*\rangle\Big\|_{L_1(0,1)}$$

$$\le \sup_{|a_k|=1}\Big\|r_k\epsilon_k a_k M_k\phi_k x_k\Big\|_{L_2(X)}\cdot\Big\|\sum_k r_k\psi_k^* x_k^*\Big\|_{L_2(X^*)}.$$

Now we use the R–boundedness of $\{a_k M_k : k = 1, \ldots, N\}$.

b) We have

$$\int_{a^{-N}}^{a^{N+1}} |\langle \psi(t)M(t)\phi(t)x, x^*\rangle| \frac{dt}{t} = \sum_{k=-N}^{N} \int_{a^k}^{a^{k+1}} |\ldots| \frac{dt}{t}$$

$$\leq \sup_{t>0} \sum_{k=-N}^{N} |\langle \psi(a^k t)M(2^k t)\phi(a^k t)x, x^*\rangle|$$

and apply part a). □

Now we are ready for the proof of Theorem 12.2.

Proof of Theorem 12.2. i) ⇒ ii) Let φ be any function in $H_0^\infty(\Sigma_\sigma)$ (e. g. ψ_θ with $\theta > \sigma$). Choose constants $C < \infty$ and $\varepsilon > 0$ such hat $|\varphi(\lambda)| \leq C|\lambda|^\varepsilon(1 + |\lambda|)^{-2\varepsilon}$ on Σ_σ. Then the functions $g(\lambda) = \sum_{k=-N}^{N} \varepsilon_k \psi_\theta(2^k t\lambda)$ with $\varepsilon_k = \pm 1, t > 0$, satisfy the estimate

$$|g(\lambda)| \leq \sum_{k=-\infty}^{\infty} \frac{|2^k t\lambda|^\varepsilon}{(1 + |2^k t\lambda|)^{2\varepsilon}}$$

$$\leq C \sum_{k=-\infty}^{\infty} \frac{(2^{k+l+1})^\varepsilon}{(1 + 2^{k+l})^{2\varepsilon}} = C_1 < \infty \qquad \text{if } |\lambda t| \in [2^l, 2^{l+1}],$$

where C_1 is independent of $\lambda \in \Sigma_\sigma, t > 0, N$ and ε_k. By the boundedness of the H^∞-calculus we get for all $N, \varepsilon_k = \pm 1, t > 0$,

$$\|\sum_{k=-N}^{N} \varepsilon_k \varphi(2^k tA)\| = \|g(A)\| \leq C \cdot C_1.$$

ii) ⇒ iii) If C denotes the sup in ii) we have e. g.

$$\int_0^1 \|\sum_k r_k(u)\psi_\theta(2^k tA)^* x^*\| du$$

$$\leq \sup_{\varepsilon_k = \pm 1} \|[\sum_k \varepsilon_k \psi(2^k tA)]^*\| \, \|x^*\| \leq C\|x^*\|.$$

iii) ⇒ vi) For $x \in X, x^* \in X^*$ we estimate with Lemma 12.6

$$\int_0^\infty |\langle AR(te^{i\theta}, A)^2 x, x^*\rangle| dt = \int_0^\infty |\langle \psi_\theta(tA)x, \psi_\theta(tA)^* x^*\rangle| \frac{dt}{t}$$

$$\leq \sup_{t>0} \|\sum_k r_k a_k(t)\psi_\theta(2^k tA)x\|_{L_2(X)} \cdot \|\sum_l r_l \psi_\theta(2^l tA)^* x^*\|_{L_2(X^*)}$$

$$\leq C^2 \|x\| \cdot \|x^*\|$$

by assumption iii).

iv) $\Rightarrow$ i) First we want to show that

$$\int_0^\infty |\langle A^{\frac{1}{2}} R(te^{\pm i\theta}, A)x, x^* \rangle| t^{-\frac{1}{2}} dt \le C\|x\| \cdot \|x^*\|. \tag{12.6}$$

Indeed, by Lemma 12.5 there is a $\varphi \in H_0^\infty(\Sigma_\sigma)$ and a constant γ such that

$$\psi_\theta(tA) = \frac{1}{2\pi i} \int_{\partial \Sigma_\theta} \varphi(\lambda)[tAR(\lambda, tA)^2]d\lambda + \gamma tA(1 + tA)^{-2}$$

where $\sigma > \theta$. Now the integral in (12.6) equals

$$\int_0^\infty |\langle \psi_\theta(tA)x, x^* \rangle| \frac{dt}{t}$$

$$\le \int_{\partial \Sigma_\theta} |\varphi(\lambda)| \int_0^\infty |\langle tAR(\lambda, tA)^2 x, x^* \rangle| \frac{dt}{t} d|\lambda|$$

$$+ |\gamma| \int_0^\infty |\langle tA(1 + tA)^{-2} x, x^* \rangle| \frac{dt}{t}$$

$$\le \left(\int_{\partial \Sigma_\theta} |\varphi(\lambda)| \frac{d|\lambda|}{|\lambda|} \right) \sup_{\lambda \in \partial \Sigma_\theta} \int_0^\infty |\langle \lambda t^{-1} AR(\lambda t^{-1}, A)^2 x, x^* \rangle| \frac{dt}{t}$$

$$+ |\gamma| \int_0^\infty |\langle AR(-t, A)^2 x, x^* \rangle| dt$$

$$\le C \sup_{j=\pm 1} \int_0^\infty |\langle AR(e^{ij\theta} t, A)^2 x, x^* \rangle| dt.$$

To justify the estimate of the second integral we may assume that $\theta = \frac{\pi}{2}$ (indeed, we may replace λ by λ^α). Then the Poisson representation formula applied to the function $g(\lambda) = \langle AR(\lambda, A)^2 x, x^* \rangle$ gives

$$g(t) = \frac{1}{\pi} \int_{-\infty}^\infty \frac{t}{t^2 + u^2} g(iu) \, du$$

and therefore

$$\int_0^\infty |g(t)| \, dt \le \int_{-\infty}^\infty \left[\frac{1}{\pi} \int_0^\infty \frac{t}{t^2 + u^2} \, dt \right] |g(iu)| \, du \le \int_{-\infty}^\infty |g(iu)| \, du.$$

Together with iv) we have shown

$$\int\limits_0^\infty | < \psi_0(tA)x, x^* > | \, dt \leq C_1 \|x\| \cdot \|x^*\|.$$

Now we finish using Lemma 12.4 again. For any $f \in H_0^\infty(\Sigma_\sigma)$ we get with (12.6)

$$|\langle f(A)x, x^* \rangle| = \left| \frac{1}{2\pi i} \int\limits_{\partial\Sigma_\theta} f(\lambda)\lambda^{-\frac{1}{2}} \langle A^{\frac{1}{2}} R(\lambda, A)x, x^* \rangle d\lambda \right|$$

$$\leq C_\theta \sum_{j=\pm 1} \frac{1}{2\pi} \int\limits_0^\infty |f(e^{ij\theta}t)| \, |\langle A^{\frac{1}{2}} R(e^{ij\theta}t, A)x, x^* \rangle| t^{-\frac{1}{2}} dt$$

$$\leq C\|f\|_{H^\infty(\Sigma_\sigma)} \|x\| \cdot \|x^*\|.$$

Hence $\|f(A)\| \leq C\|f\|_{H^\infty(\Sigma_\sigma)}$ with a constant C independent of $f \in H_0^\infty(\Sigma_\sigma)$ and we can appeal to 9.12. $\qquad\square$

As a consequence of the characterization theorem we can extend a bounded H^∞-calculus to a **calculus for operator-valued functions** of the following kind:

Given a sectorial operator $A \in \mathcal{S}(X)$ on X, we denote by $\mathcal{A}$ the subalgebra of $B(X)$ of all bounded operators, that commute with resolvents of A. Then we denote by $RH^\infty(\Sigma_\sigma, \mathcal{A})$ the space of all bounded analytic functions $F : \Sigma_\sigma \to \mathcal{A}$ with the additional property that the range $\{F(\lambda) : \lambda \in \Sigma_\sigma\}$ is R-bounded in $B(X)$. The norm for an element $F \in RH^\infty(\Sigma_\sigma, \mathcal{A})$ is given by

$$\|F\| = R(\{F(\lambda) : \lambda \in \Sigma_\sigma\})$$

By $RH_0^\infty(\Sigma_\sigma, \mathcal{A})$ we denote elements of $RH^\infty(\Sigma_\sigma, \mathcal{A})$ which are dominated in operator norm by a multiple of $|\lambda|^\varepsilon(1 + |\lambda|)^{-2\varepsilon}$ for some $\varepsilon > 0$. For $F \in RH_0^\infty(\Sigma_\sigma, \mathcal{A})$ the integral

$$F(A) = \frac{1}{2\pi i} \int\limits_{\partial\Sigma_\nu} F(\lambda) R(\lambda, A) d\lambda \qquad (12.7)$$

exists as a Bochner integral for $\omega(A) < \nu < \sigma$. We show now that (12.7) can be extended to a bounded functional calculus on $RH^\infty(\Sigma_\sigma, \mathcal{A})$, if A has a bounded H^∞-calculus:

12.7 Theorem. *Assume that $A \in \mathcal{S}(X)$ has a bounded $H^\infty(\Sigma_{\sigma'})$-calculus. Then for all $\sigma > \sigma'$ there is a bounded algebra homomorphism $\Phi_A : RH^\infty(\Sigma_\sigma, \mathcal{A}) \to B(X)$ with $\Phi_A(F) = F(A)$ as in (12.7) for $F \in RH_0^\infty(\Sigma_\sigma, \mathcal{A})$ and the following convergence property:*

If (F_n) is a bounded sequence in $RH^\infty(\Sigma_\sigma, \mathcal{A})$
and $F_n(\lambda)x \to F(\lambda)x$ for some $F \in RH^\infty(\Sigma_\sigma, \mathcal{A})$
and all $\lambda \in \Sigma_\sigma$ and $x \in X$, then
$$\Phi_A(F_n)x \xrightarrow{n\to\infty} \Phi_A(F)x \quad \text{for all } x \in X.$$

(12.8)

Notation. It is often convenient to write $F(A)$ for $\Phi_A(F)$.

Proof. First we estimate (12.7) for $F \in RH_0^\infty(\Sigma_\sigma, \mathcal{A})$ using Lemma 12.4 and Lemma 12.6:

$$|\langle F(A)x, x^*\rangle| = \frac{1}{2\pi}\Big| \int\limits_{\partial\Sigma_\nu} \langle \lambda^{-\frac{1}{2}}F(\lambda)A^{\frac{1}{2}}R(\lambda, A)x, x^*\rangle d\lambda\Big|$$

$$\leq \frac{1}{2\pi} \sum_{j=\pm 1} \int\limits_0^\infty |\langle F(te^{ij\nu})\psi_{j\nu}^{\frac{1}{2}}(tA)x, \psi_{j\nu}^{\frac{1}{2}}(tA)^*x^*\rangle| \frac{dt}{t}$$

$$\leq \frac{1}{2\pi} \sup_{j=\pm 1} \sup_{t>0} \Big\|\sum_k r_k a_k(t)F(2^k te^{ij\nu})\psi_{j\nu}^{\frac{1}{2}}(2^k tA)x\Big\|_{L_2(X)}$$

$$\cdot \Big\|\sum_k r_k \psi_{j\nu}^{\frac{1}{2}}(2^k t)^*x^*\Big\|_{L_2(X^*)}$$

$$\leq C\|F\|_{RH^\infty}\Big(\|x\|_{\psi_\nu^{1/2}} + \|x\|_{\psi_{-\nu}^{1/2}}\Big)\Big(\|x^*\|_{\psi_\nu^{1/2}} + \|x^*\|_{\psi_{-\nu}^{1/2}}\Big).$$

By the R-boundedness of $\{a_k(t)F(2^k te^{ij\nu}) : t > 0, j = \pm 1, k \in \mathbb{Z}\}$ and 12.2. iii) for $\psi_\nu^{\frac{1}{2}}$ in place of ψ_ν (see the remark after 12.2) we obtain

$$|\langle F(A)x, x^*\rangle| \leq C\|F\|_{RH^\infty} \cdot \|x\| \cdot \|x^*\|$$

with a constant C independent of F. Now we proceed as in the proof of 9.2 to show the multiplicativity of Φ_A on $RH_0^\infty(\Sigma_\sigma, \mathcal{A})$ and the convergence property in 9.2.i). This in turn allows to adapt the proof of 9.6 to check the required properties of Φ_A on $RH^\infty(\Sigma_\sigma, \mathcal{A})$. □

As a first consequence of the extended H^∞-calculus we show that the set of operators produced by a bounded H^∞-calculus (from a uniformly bounded set of functions) is R-bounded. However, we need property (α) from 4.8 again. (Recall that all subspaces of $L_q(\Omega), 1 < q < \infty$, have this property).

12.8 Theorem. *Let X be a Banach space with property (α) and $A \in \mathcal{S}(X)$ be a sectorial operator with a bounded $H^\infty(\Sigma_{\sigma'})$-calculus. Then for $\sigma > \sigma', C > 0$, the set*

$$\{f(A) : \|f\|_{H^\infty(\Sigma_\sigma)} \leq C\}$$

(12.9)

is R-bounded in $B(X)$. In particular we have

a) A is R-sectorial and $\omega_{H^\infty}(A) = \max\{\omega_R(A), \omega_R(A^*)\}$

b) A is R-analytic if $\omega_{H^\infty}(A) < \frac{\pi}{2}$.

12.9 Remarks. We say that A has an *R-bounded H^∞-functional calculus* if the set (12.9) is R-bounded.

a) The equality of $\omega_R(A)$ and $\omega_{H^\infty}(A)$ is very useful in establishing an optimal H^∞-calculus. We already used it in Corollary 10.16; first we showed that if $-A$ generates a contraction semigroup on all $L_q(\mathbb{R}^n)$ for $1 \le q \le \infty$ then A has a bounded $H^\infty(\Sigma_\sigma)$–calculus for $\sigma > \frac{\pi}{2}$. Since $-A$ also generates an analytic semigroup we concluded from theorem 5.6 that $\omega_R(A) < \frac{\pi}{2}$. Now 12.8a) gives $\omega_{H^\infty}(A) < \frac{\pi}{2}$. This argument applies to many generators A of stochastic processes.

b) If $X = L_q(\Omega)$ with $1 < q < \infty$, then $\omega_{H^\infty}(A) = \omega_R(A)$ since $\{ \lambda R(\lambda, A) : \lambda \in \Sigma_\omega \}$ is R-bounded in $B(L_q)$ if and only if $\{ \lambda R(\lambda, A)^* : \lambda \in \Sigma_\omega \}$ is R-bounded in $B(L_{q'})$ (cf. 2.11) and therefore $\omega_R(A) = \omega_{H^\infty}(A^*)$. The same is true for a large class of Banach spaces including all uniformly convex Banach spaces.

c) By 12.8 b) and 1.11 we have in particular, that boundedness of the H^∞-calculus implies maximal regularity.

12.10 Remark. We will also need the following R-boundedness result for the vector-valued H^∞-calculus: Let X have property (α) and $A \in \mathcal{S}(X)$. For an R-bounded subset $\tau \subset B(X)$ put

$$RH^\infty(\Sigma_\sigma, \tau) = \{ F \in RH^\infty(\Sigma_\sigma, A) : F(\lambda) \in \tau \text{ for } \lambda \in \Sigma_\sigma \}.$$

If A has a bounded $H^\infty(\Sigma_{\sigma'})$-calculus, then for $\sigma > \sigma'$ the set $\{ F(A) : F \in RH^\infty(\Sigma_\sigma, \tau) \}$ is R-bounded in $B(X)$.

Proof of 12.8 and 12.10. We prove 12.8 in the more general form 12.10. (In the scalar case we choose τ as the unit ball in $\mathbb{C}$.)

It is enough to show that $\{F_n(A) : n \in \mathbb{N}\}$ is R-bounded for every sequence $F_n \in RH_0^\infty(\Sigma_\sigma, \tau)$ or that $T(x_n) = (F_n(A)x_n)_n$ defines a bounded operator on $\operatorname{Rad} X$ (by 2.7). To show this we want to find a sectorial operator $\tilde{A}$ on $\operatorname{Rad} X$ with a bounded $H^\infty(\Sigma_\sigma)$-calculus and an $F \in RH^\infty(\Sigma_\sigma, \tilde{\mathcal{A}})$ with $T = F(\tilde{A})$ and apply Theorem 12.7. Here $\tilde{\mathcal{A}}$ denotes the subalgebra of $B(\operatorname{Rad} X)$ of operators commuting with the resolvent of $\tilde{A}$.

We define $\tilde{A}$ by $D(\tilde{A}) = \{(x_n)_n \in \operatorname{Rad} X : x_n \in D(A)\}$ and $\tilde{A}[(x_n)] = (Ax_n)_n$ for $(x_n) \in D(\tilde{A})$. We can extend the bounded H^∞-calculus of A to $\tilde{A}$ by $f(\tilde{A})[(x_n)_n] = (f(A)x_n)_n$ for all $f \in H^\infty(\Sigma_\sigma)$. (Use the uniqueness of the H^∞-calculus, cf. 9.8). Our candidate for F is $F(\lambda) = (F_n(\lambda))_n$. Since τ is R-bounded we have $F(\lambda) \in B(\operatorname{Rad} X)$ and clearly, $F(\lambda)$ belongs to $\tilde{\mathcal{A}}$. Property (α) is now used to show that the range of F in $B(\operatorname{Rad} X)$ is R-bounded. Indeed, let $(\varepsilon_j), (\varepsilon'_j)$ two independent copies of Bernoulli random variables on $[0, 1]$ and use on $\operatorname{Rad} X$ the norm $\|(y_n)\| = \int \| \sum_n \varepsilon'_n(u) y_n \| du$. Then, by Lemma 4.11 and the R-boundedness of τ we have for all $\lambda_j \in \Sigma_\sigma$ and $\tilde{x}_j = (x_{j,n})_n \in \operatorname{Rad} X$

$$\|\sum_j r_j F(\lambda_j)\tilde{x}_j\|_{L_1(RadX)} = \int_0^1\int_0^1 \|\sum_n\sum_j \varepsilon_n'(u)\varepsilon_j(v)F_n(\lambda_j)x_{jn}\|_X\,dudv$$

$$\leq C\int_0^1\int_0^1 \|\sum_n\sum_j \varepsilon_n'(u)\varepsilon_j(v)x_{jn}\|_X\,dudv$$

$$= C\|\sum_j r_j\tilde{x}_j\|_{L_1(RadX)}.$$

Now we can apply 12.7 to F and $\tilde{A}$ and obtain the boundedness of T on Rad X since for $\tilde{x} = (x_n) \in \text{Rad}\,X$.

$$T\tilde{x} = \left(\int_{\partial\Sigma_\omega} F_n(\lambda)R(\lambda, A)x_n\,d\lambda\right)_n = \int_{\partial\Sigma_\omega} F(\lambda)R(\lambda,\tilde{A})\tilde{x}d\lambda,$$

where $\sigma > \omega > \omega(A)$. This proves the first claim of 12.8 and 12.10.

a) Since $f_\mu(\lambda) = \lambda(\mu-\lambda)^{-1}$ is uniformly bounded in $H^\infty(\Sigma_\sigma)$ for $|\arg(\mu)| > \omega$ where $\omega > \sigma$, (12.9) implies that A is R-bounded and $\omega_R(A) \leq \omega_{H\infty}(A)$. For the converse inequality we use the characterization theorem 12.2 again. Choose $\omega > \omega_R(A)$ and $\nu > \omega_{H\infty}(A)$. By the resolvent equation we have

$$R(te^{i\omega}, A) = R(te^{i\nu}, A) + (e^{i\nu} - e^{i\omega})tR(te^{i\omega}, A)R(te^{i\nu}, A),$$

and therefore

$$A^{\frac{1}{2}}R(te^{i\omega}, A) = [Id + (e^{i\nu} - e^{i\omega})tR(te^{i\omega}, A)]A^{\frac{1}{2}}R(te^{i\nu}, A).$$

By the choice of ω the operators $T_k = Id + (e^{i\nu} - e^{i\omega})2^k tR(2^k te^{i\omega}, A)$ with $k \in \mathbb{Z}$ form an R-bounded set. Hence for $x \in X$

$$\|x\|_{\psi_\omega} = \|\sum_k r_k A^{\frac{1}{2}}R(2^k te^{i\omega}, A)x\|_{L_p(X)}$$

$$= \|\sum_k r_k T_k A^{\frac{1}{2}}R(2^k te^{i\nu}, A)x\|_{L_p(X)}$$

$$\leq R(T_k)\|\sum_k r_k A^{\frac{1}{2}}R(2^k te^{i\nu}, A)x\|_{L_p(X)} \leq C\|x\|_{\psi_\nu} \leq C_1\|x\|,$$

since $\nu > \omega_{H\infty}(A)$ by 12.2.

If $\omega > \omega_R(A^*)$ we can repeat this argument for A^* and obtain for $x^* \in X^*$

$$\|x^*\|_{\psi_\omega} = \|\sum_k r_k(A^*)^{\frac{1}{2}}R(2^k te^{i\omega}, A^*)x^*\|_{L_p(X^*)} \leq C\|x^*\|_{\psi_\nu} \leq C_1\|x^*\|.$$

Now 12.2 gives $\omega_{H\infty}(A) \leq \nu$ and $\nu > \max\{\omega_R(A), \omega_R(A^*)\}$ was arbitrary.

b) The functions $e_\nu(\lambda) = e^{-\nu\lambda}$ are uniformly bounded in $H^\infty(\Sigma_\sigma)$ for $|\arg\nu| < \omega$ where $\omega < \frac{\pi}{2} - \sigma$, and $\omega_{H\infty}(A) < \sigma < \frac{\pi}{2}$. □

Every sectorial operator with a bounded H^∞ functional calculus has bounded imaginary powers (BIP), since the functions $f_s(\lambda) = e^{-d|s|}\lambda^{is}$ with $d > \sigma$ are uniformly bounded in $H^\infty(\Sigma_\sigma)$ for all $s \in \mathbb{R}$. In Theorem 11.9 we showed the converse if X is a Hilbert space. But even in $L_q(\Omega)$-spaces with $q \neq 2$ there are operators with BIP but without a bounded H^∞-functional calculus. The following result shows what additional properties the imaginary powers should have so that the H^∞-calculus is bounded. It reduces to 11.9 in the Hilbert space case.

12.11 Corollary. *Let X have property (α). Then $A \in \mathcal{S}(X)$ has a bounded H^∞-calculus if and only if A has R-BIP of type $< \pi$, i.e. A has bounded imaginary powers and the set $\tau_d = \{ e^{-d|s|}A^{is} : s \in \mathbb{R} \}$ is R-bounded for some $0 \le d < \pi$.*
Furthermore, $\omega_{H^\infty}(A) = \inf\{ d : \tau_d$ is R-bounded$\}$.

Proof. The boundedness of the H^∞-calculus implies R-BIP by Theorem 12.8. The proof of the reverse implication we do not present here.   □

We used this corollary already in Example 10.17 to discuss operators without a bounded H^∞-calculus.

As a second application of the operator-valued functional calculus we construct the **joint functional calculus** of resolvent commuting operators each having a bounded H^∞-calculus. Denote by $H^\infty(\Sigma_{\sigma_1} \times \cdots \times \Sigma_{\sigma_n})$ the space of all bounded analytic functions of n variables on $\Sigma_{\sigma_1} \times \cdots \times \Sigma_{\sigma_n}$ and by $H_0^\infty(\Sigma_{\sigma_1} \times \cdots \times \Sigma_{\sigma_n})$ the subspace of functions $f \in H^\infty(\Sigma_{\sigma_1} \times \cdots \times \Sigma_{\sigma_n})$ which obey an estimate of the form

$$\|f(\lambda_1, \ldots, \lambda_n)\| \le C \prod_{k=1}^{n} |\lambda_k|^\varepsilon (1 + |\lambda_k|)^{-2\varepsilon}$$

for some $\varepsilon > 0$. For $f \in H_0^\infty(\Sigma_{\sigma_1} \times \cdots \times \Sigma_{\sigma_n})$ we define the joint functional calculus by

$$f(A_1, \ldots, A_n) = \left(\frac{1}{2\pi i}\right)^n \int_{\partial \Sigma_{\nu_n}} \cdots \int_{\partial \Sigma_{\nu_1}} f(\lambda_1, \ldots, \lambda_n) \prod_{j=1}^{n} R(\lambda_j, A_j) \, d\lambda_1 \ldots d\lambda_n$$

(12.10)

where $\omega(A_i) < \nu_i < \sigma_i$ for $i = 1, \ldots, n$. If $n = 2$ and Φ_2 denotes the bounded $H^\infty(\Sigma_{\sigma_2})$-calculus of A_2 and Φ_1 denotes the operator valued functional calculus of A_1 then we can rewrite (12.10) by Fubini's theorem as

$$f(A_1, A_2) = \frac{1}{2\pi i} \int\limits_{\partial \Sigma_{\nu_1}} \left(\frac{1}{2\pi i} \int\limits_{\partial \Sigma_{\nu_2}} f(\lambda_1, \lambda_2) R(\lambda_2, A_2) d\lambda_2 \right) R(\lambda_1, A_1) d\lambda_1$$

$$= \frac{1}{2\pi i} \int\limits_{\partial \Sigma_{\nu_1}} f(\lambda_1, A_2) R(\lambda_1, A_1) d\lambda_1 \qquad (12.11)$$

$$= \Phi_1(f(\cdot, A_2)).$$

This indicates already how to extend (12.10) to $H^\infty(\Sigma_{\sigma_1} \times \cdots \times \Sigma_{\sigma_n})$, using induction.

12.12 Theorem. *Let X have property (α) and assume that $A_1, \ldots, A_n$ are resolvent commuting operators with a bounded $H^\infty(\Sigma_{\sigma_j'})$ functional calculus for $j = 1, \ldots, n$, respectively. Then for $\sigma_j > \sigma_j'$, (12.10) extends to an algebra homomorphism $\Phi : H^\infty(\Sigma_{\sigma_1} \times \cdots \times \Sigma_{\sigma_n}) \to B(X)$ with the following convergence property:*

If $\{f_m, f\}$ is uniformly bounded in $H^\infty(\Sigma_{\sigma_1} \times \cdots \times \Sigma_{\sigma_n})$
and $f_m(\lambda_1, \ldots, \lambda_n) \xrightarrow{m \to \infty} f(\lambda_1, \ldots, \lambda_n)$ for all $\lambda_i \in \Sigma_{\sigma_i}$, then $\qquad (12.12)$
$f_m(A_1, \ldots, A_n)x \xrightarrow{m \to \infty} f(A_1, \ldots, A_n)x$ for all $x \in X$.

Furthermore, the set of operators

$$\left\{ f(A_1, \ldots, A_n) : f \in H^\infty(\Sigma_{\sigma_1} \times \cdots \times \Sigma_{\sigma_n}), \; \|f\|_{H^\infty} \le 1 \right\}$$

is R-bounded in $B(X)$.

Notation. In place of $\Phi(f)$ we may write $f(A_1, \ldots, A_n)$.

Proof. By $\mathcal{A}_j$ we denote the subalgebra of all operators in $B(X)$ that commute with A_j and put $\mathcal{A} = \mathcal{A}_1 \cap \cdots \cap \mathcal{A}_n$. Note that $R(\lambda, A_j) \in \mathcal{A}$ for all $j = 1, \ldots, n$. The case $n = 1$ is covered by Theorem 12.7. Assume now that $A_2, \ldots, A_n$ have a joint functional calculus $\Psi : H^\infty(\Sigma_{\sigma_2} \times \cdots \times \Sigma_{\sigma_n}) \to B(X)$ with the required properties. In particular, we have that

$$\tau = \left\{ g(A_2, \ldots, A_n) : g \in H^\infty(\Sigma_{\sigma_2} \times \cdots \times \Sigma_{\sigma_n}), \; \|g\|_{H^\infty} \le 1 \right\}$$

is an R-bounded subset of $\mathcal{A} \subset \mathcal{A}_1$. By Φ_1 we denote the operator-valued functional calculus of A_1 defined on $RH^\infty(\Sigma_{\sigma_1}, \mathcal{A}_1)$ as constructed in Theorem 12.7. Given $f \in H^\infty(\Sigma_{\sigma_1} \times \cdots \times \Sigma_{\sigma_n})$ with $\|f\|_{H^\infty} \le 1$ the set

$$\left\{ f(\lambda_1, \cdot, \ldots, \cdot) : \lambda_1 \in \Sigma_{\sigma_1} \right\}$$

is uniformly bounded in $H^\infty(\Sigma_{\sigma_2} \times \cdots \times \Sigma_{\sigma_n})$. Hence

$$\Psi\big[f(\lambda_1, \cdot, \ldots, \cdot)\big] = f(\lambda_1, A_2, \ldots, A_n) \in \tau \qquad \text{for all } \lambda_1 \in \Sigma_{\sigma_1}.$$

Furthermore, for all $k \in \mathbb{N}$, the functions

$$\lambda_1 \in \Sigma_{\sigma_1} \mapsto \prod_{j=2}^{n} \rho_k(A_j) f(\lambda_1, A_2, \dots, A_n)$$

$$= \left(\frac{1}{2\pi i}\right)^{n-1} \int_{\partial \Sigma_{\nu_2}} \cdots \int_{\partial \Sigma_{\nu_n}} \prod_{j=2}^{n} \rho_k(\lambda_j) f(\lambda_1, \dots, \lambda_n) R(\lambda_j, A_j) \, d\lambda_2 \dots d\lambda_n$$

(here $\sigma_j' < \nu_j < \sigma_j$) are analytic, and so is their pointwise limit for $k \to \infty$

$$\lambda_1 \in \Sigma_{\sigma_1} \mapsto f(\lambda_1, A_2, \dots, A_n) \in \tau.$$

So we have $f(\cdot, A_2, \dots, A_n) \in RH^\infty(\Sigma_{\sigma_1}, \mathcal{A}_1)$ and we can define

$$\Phi(f) = \Phi_1\big(f(\cdot, A_2, \dots, A_n)\big).$$

The required properties of Φ follow now from the corresponding properties of Ψ and Φ_1. E. g., by Fubini's theorem we get that Φ extends (12.10). To check the multiplicativity, choose $f, g \in H^\infty(\Sigma_{\sigma_1} \times \cdots \times \Sigma_{\sigma_n})$. Then

$$\Phi(f \cdot g) = \Phi_1\big((f \cdot g)(\cdot, A_2, \dots, A_n)\big)$$
$$= \Phi_1\big(f(\cdot, A_2, \dots, A_n) \cdot g(\cdot, A_2, \dots, A_n)\big)$$
$$= \Phi_1\big(f(\cdot, A_2, \dots, A_n)\big)\Phi_1\big(g(\cdot, A_2, \dots, A_n)\big) = \Phi(f)\Phi(g).$$

Also, if f_m, f are bounded in $H^\infty(\Sigma_{\sigma_1} \times \cdots \times \Sigma_{\sigma_n})$ and $f_m(\lambda_1, \dots, \lambda_n) \to f(\lambda_1, \dots, \lambda_n)$ for $m \to \infty$, then by the convergence property of Ψ we have

$$f_m(\lambda_1, A_2, \dots, A_n)x \xrightarrow{m \to \infty} f(\lambda_1, A_2, \dots, A_n)x.$$

for every fixed $\lambda_1 \in \Sigma_{\sigma_1}$ and $x \in X$. Now apply the convergence property of Φ_1 to $F_m(\lambda) = f_m(\lambda, A_2, \dots, A_n)$ and $F(\lambda) = f(\lambda, A_2, \dots, A_n)$. The R-boundedness of $\Phi(U_{H^\infty})$ follows directly from 12.10.   □

Finally we derive a theorem on the closedness of the sum of two resolvent commuting operators A and B. It stands in the tradition of a well known theorem of Dore and Venni, who assume that X is a UMD-space and A and B have BIP. In the following version, we assume more on B (boundedness of the H^∞-calculus) but less on A (R-sectoriality). We have already seen in 1.18 how such a theorem can be used to prove the characterization theorem 1.11. for maximal regularity, where $B = \frac{d}{dx}$ on $L_p(\mathbb{R}_+, X)$ has a bounded H^∞-calculus by 10.2.

12.13 Theorem. *Let $A, B \in \mathcal{S}(x)$ be two resolvent commuting sectorial operators on a Banach space X. Assume that B has a bounded H^∞-calculus and that A is R-sectorial with $\omega_{H^\infty}(B) + \omega_R(A) < \pi$. Then $A + B$ is closed on $D(A) \cap D(B)$ and*

$$\|Ax\| + \|Bx\| \le C\|Ax + Bx\| \quad \text{for } x \in D(A) \cap D(B). \tag{12.13}$$

If X has property (α), then $A + B$ is again R-sectorial with $\omega_R(A + B) \le \max\{\omega_R(A), \omega_{H^\infty}(B)\}$.

Proof. The idea is to define $A(A + B)^{-1}$ and $B(A + B)^{-1}$ as bounded operators with the help of the operator-valued functional calculus for B. To this end denote by $\mathcal{B}$ the subalgebra of $B(X)$ of all operators commuting with resolvents of B. Choose σ, ω with $\sigma > \omega_{H^\infty}(B), \omega > \omega_R(A)$ and $\sigma + \omega < \pi$. By assumption we have that the function

$$F(\lambda) = A(\lambda + A)^{-1} = -AR(-\lambda, A)$$

belongs to $RH^\infty(\Sigma_\sigma, \mathcal{B})$ since $-\Sigma_\sigma = \{\lambda : |\arg \lambda| > \pi - \sigma\}$ and $\omega < \pi - \sigma$. Furthermore, the function

$$G(\lambda) = (\lambda + A)\varrho_n(\lambda)^2 \varrho_n(A)^2$$

with (ϱ_n) from 9.4 belongs to $H_0^\infty(\Sigma_\sigma, \mathcal{B})$ and even to $RH_0^\infty(\Sigma_\sigma, \mathcal{B})$ since $\lambda^{-1}\varrho_n(\lambda)$ is bounded in λ for a fixed n. We have

$$G(\lambda)F(\lambda) = \varrho_n(A)^2 A\varrho_n(\lambda)^2$$

and therefore by the multiplicativity of the functional calculus

$$(B + A)\varrho_n(B)^2 \varrho_n(A)^2 F(B) = G(B)F(B) = A\varrho_n(B)^2 \varrho_n(A)^2$$

and by its boundedness for $x \in D(A) \cap D(B)$

$$\|\varrho_n(B)^2 \varrho_n(A)^2 Ax\| \leq C\|\varrho_n(B)^2 \varrho_n(A)^2 (B + A)x\|$$

with $C = \|F(B)\|$. For $n \to \infty$ this gives

$$\|Ax\| \leq C\|(A + B)x\|.$$

Furthermore,

$$\|Bx\| \leq \|Ax\| + \|Ax + Bx\| \leq (C + 1)\|Ax + Bx\|$$

and (12.13) follows. To check the closedness of $A + B$ on $D(A) \cap D(B)$ choose $x_n \in D(A) \cap D(B)$ with $x_n \to x$ and $Ax_n + Bx_n \to y$ in X. By (12.13), (Ax_n) and (Bx_n) are Cauchy sequences, so that $x \in D(A) \cap D(B)$ and $y = \lim(Ax_n + Bx_n) = Ax + Bx$ by closedness of A and B.

Now we assume property (α) so that we can apply 12.8. For $|\arg \mu| \geq \theta$ for a fixed θ with $\max(\sigma, \omega) < \theta < \pi$ consider the analytic functions

$$\lambda \in \Sigma_\sigma \to F_\mu(\lambda) = \mu R(\mu - \lambda, A) \in \mathcal{B}$$

which are well defined since $\arg(\mu - \lambda) \geq \omega$. Furthermore they have values in the R-bounded set

$$\tau = C\{\nu R(\nu, A) : |\arg \nu| \geq \omega\}, \quad C = \sup\left\{\left|\frac{\mu}{\mu - \lambda}\right| : \lambda \in \Sigma_\sigma, |\arg \mu| \geq \theta\right\}.$$

Hence $F_\mu \in RH^\infty(\Sigma_\sigma, \tau)$ for $|\arg \mu| \geq \theta$. An approximation argument with $\varrho_n(A)$ as above shows that

$$\mu R(\mu, A + B) = F_\mu(A).$$

Now 12.10 implies that $\omega_R(A + B) \leq \theta$. $\qquad \square$

12.14 Remark. Let X have property (α). If B has a bounded $H^\infty(\Sigma_\sigma)$-calculus and A has a bounded $H^\infty(\Sigma_\omega)$-calculus with $\omega + \sigma < \pi$, then we can derive (12.13) also from the joint calculus of A and B. Indeed, the function $f(w, z) = w(w + z)^{-1}$ belongs to $H^\infty(\Sigma_\sigma \times \Sigma_\omega)$ if $\sigma + \omega < \pi$. Then $f(B, A)$, which formally equals $B(A + B)^{-1}$, is bounded and for $x \in D(A) \cap D(B)$

$$f(A, B)(A + B)\varrho_n(A)^2 \varrho_n(B)^2 x = \varrho_n(A)^2 \varrho_n(B)^2 Ax$$

so that with $C = \|f(A, B)\|$

$$\|\varrho_n(A)^2 \varrho_n(B)^2 Ax\| \leq C\|\varrho_n(A)^2 \varrho_n(B)^2 (A + B)x\|$$

and the claim follows for $n \to \infty$.   □

Motivated by (H4) in Section 11 we want to characterize the H^∞–calculus in terms of its square functions

$$\|x\|_\varphi = \sup_{t \in [1,2]} \left\| \sum_k r_k \varphi(2^k tA)x \right\|_{L_p(X)},$$

where $\varphi \in H_0^\infty(\Sigma_\sigma)$ with $\sigma > \omega_R(A)$. As a first step we show the equivalence of such square functions, a result that will also be useful in the next section

12.15 Proposition. *Let $A \in \mathcal{S}(X)$ on a Banach space X. For every φ and ψ in $H_0^\infty(\Sigma_\sigma) \setminus \{0\}, \sigma > \omega_R(A)$, there is a constant C such that for all $x \in X$*

$$\frac{1}{C}\|x\|_\varphi \leq \|x\|_\psi \leq C\|x\|_\varphi.$$

The proof will be based on the following lemma.

12.16 Lemma. *Let $M, N : \mathbb{R}_+ \to B(X)$ be strongly measurable, bounded functions and $h \in L_1(\mathbb{R}_+, \frac{dt}{t})$. If*

$$M(t) = \int_0^\infty h(ts)N(s)\frac{ds}{s}, \quad t > 0,$$

then for all $x \in X$

$$\sup_{t>0} \left\| \sum_{k \in \mathbb{Z}} r_k M(2^k t)x \right\|_{L_p(X)} \leq C \sup_{t>0} \left\| \sum_k r_k N(2^k t)x \right\|_{L_p(X)},$$

where $C = 2\int_0^\infty |h(s)| \frac{ds}{s}$.

Proof. For every $k \in \mathbb{Z}$ and $t > 0$

$$M(t2^k)x = \sum_{j \in \mathbb{Z}} \int_1^2 h(t2^k s2^j)N(s2^j)x\frac{ds}{s}.$$

Hence

$$\left\| \sum_k r_k(\cdot) M(t2^k) x \right\| \le \int_1^2 \left\| \sum_j \sum_k r_k(\cdot) h(ts2^{k+j}) N(s2^j) x \right\| \frac{ds}{s}$$

$$\le \sum_l \int_1^2 \left\| \sum_j r_{l-j}(\cdot) h(ts2^l) N(s2^j) x \right\| \frac{ds}{s},$$

and therefore by Kahane's contraction principle and Fubini

$$\left\| \sum_k r_k M(t2^k) x \right\|_{L_1(X)}$$

$$\le 2 \sum_l \int_1^2 |h(ts2^l)| \frac{ds}{s} \left(\sup_{s \in [1,2]} \left\| \sum_j r_{l-j} N(s2^j) x \right\|_{L_1(X)} \right)$$

$$\le \int_0^\infty |h(ts)| \frac{ds}{s} \sup_{s \in [1,2]} \left\| \sum_j r_j N(s2^j) x \right\|_{L_1(X)} \quad \square$$

Proof of Proposition 12.15. We choose an auxiliary function $g \in H_0^\infty(\Sigma_\sigma)$ such that

$$\int_0^\infty g(t)\varphi(t) \frac{dt}{t} = 1.$$

By Lemma 9.13 we then have for all $x \in R(A) \cap D(A)$

$$\int_0^\infty g(tA)\varphi(tA) x \frac{dt}{t} = x.$$

Hence for $s > 0$ and $\omega \in (\omega(A), \sigma)$ we get by Fubini's theorem

$$\psi(sA)x = \int_0^\infty [\psi(sA)g(tA)]\varphi(tA) \frac{dt}{t}$$

$$= \int_0^\infty \left[\frac{1}{2\pi i} \int_{\partial \Sigma_\omega} \psi(s\lambda) g(t\lambda) R(\lambda, A) d\lambda \right] \varphi(tA) \frac{dt}{t}$$

$$= \frac{1}{2\pi i} \int_{\partial \Sigma_\omega} \psi(s\lambda)[\lambda R(\lambda, A)] \left(\int_0^\infty g(t\lambda)\varphi(tA) \frac{dt}{t} \right) \frac{d\lambda}{\lambda}$$

$$= \frac{1}{2\pi i} \int_{\partial \Sigma_\omega} \psi(s\lambda) M(\lambda) \frac{dt}{t}$$

where $M(\lambda) = \lambda R(\lambda, A)N(\lambda)$ and $N(\lambda) = \int_0^\infty g(t\lambda)\varphi(tA)\frac{dt}{t}$. Since

$$\psi(sA)x = \sum_{\delta=\pm 1} \frac{-\delta e^{i\delta\omega}}{2\pi i} \int_0^\infty \psi(e^{i\delta\omega}st)M(e^{i\delta\omega}t)x\frac{dt}{t}$$

we conclude with Lemma 12.16 that

$$\|x\|_\psi \le C \sup_{\sigma=\pm 1} \sup_{s>0} \|\sum_j r_j M(e^{i\delta\omega}s2^j)x\|_{L_p(X)}$$

$$\le C \sup_{\sigma=\pm 1} \sup_{s>0} \|\sum_j r_j N(e^{i\delta\omega}s2^j)x\|_{L_p(X)}.$$

Since $\{\lambda R(\lambda, A) : \lambda \in \Sigma_\sigma\}$ is R-bounded, the last expression can be estimated by $C \cdot \|x\|_\varphi$ using again Lemma 12.16.

Of course, we can exchange φ and ψ in these arguments to obtain the equivalence of the square function norms. $\square$

12.17 Theorem. *Let $A \in \mathcal{S}(X)$ be R–sectorial on the Banach space X. If A has a bounded H^∞–functional calculus and $\phi \in H_0^\infty(\Sigma_\omega)$ with $\omega > \omega_{H^\infty}(A)$ then there is a constant C such that*

$$\frac{1}{C}\|x\| \le \|x\|_\phi \le C\|x\| \qquad \text{for } x \in X. \tag{12.14}$$

Conversely, if (12.14) holds for some $\phi \in H_0^\infty(\Sigma_\omega)$ with $\omega > \omega_R(A)$ then A has a bounded H^∞–calculus and $\omega_{H^\infty}(A) \le \omega$.

Proof. Let A have a bounded $H^\infty(\Sigma_\sigma)$–calculus. Given $\phi \in H_0^\infty(\Sigma_\omega)$ with $\omega > \sigma$ we choose $g \in H_0^\infty(\Sigma_\omega)$ so that by Lemma 9.13

$$\int_0^\infty \phi(tA)g(tA)x\,\frac{dt}{t} = x \qquad \text{for } x \in D(A) \cap R(A).$$

For $x^* \in X^*$ we have by Lemma 12.6b)

$$\langle x, x^* \rangle \le \int_0^\infty |\langle \phi(tA)x, g(tA)^*x^* \rangle| \frac{dt}{t} \le \|x\|_\phi \|x^*\|_g \le C\|x\|_\phi \|x^*\|$$

by Theorem 12.2 and Remark 12.3 applied to g. Hence $\|x\| \le C\|x\|_\phi$ and the reverse inequality follows again from Theorem 12.2.

For the converse statement we note first that for $\sigma > \omega$ there is a constant $C > 0$ such that

$$R(\{f(A)\phi(tA) : t > 0\}) \le C\|f\|_{H^\infty(\Sigma_\sigma)} \tag{12.15}$$

for all $f \in H_0^\infty(\Sigma_\sigma)$. Indeed, for $\nu \in (\omega_R(A), \omega)$ we have

$$f(A)\phi(tA) = \frac{1}{2\pi i}\int_{\partial\Sigma_\nu} \frac{1}{\lambda}\phi(t\lambda)[f(\lambda)\lambda R(\lambda, A)]\,d\lambda$$

so that (12.15) follows from the R-boundedness of $\{f(\lambda)\lambda R(\lambda, A) : \lambda \in \partial\Sigma_\nu\}$ and Corollary 2.14. Using Proposition 12.15 and (12.15) we get for all $f \in H^\infty(\Sigma_\sigma)$

$$\|f(A)x\|_X \le C\|f(A)x\|_\phi \le C_1\|f(A)x\|_{\phi^2}$$
$$= C_1 \sup_{t>0} \|\sum_k r_k[f(A)\phi(t2^k)]\phi(t2^k)x\|_{L_p(X)}$$
$$\le C_2\|f\|_{H^\infty}\|x\|_\phi \le C_3\|f\|_{H^\infty}\|x\|_X$$

and we can appeal to Remark 9.7   □

Notes on Section 12:

N 12.1 – 12.3: The oldest part of the characterization theorem is i) $\Longleftrightarrow$ iv) which was first shown in [41]. It is shown in [56] that $\psi^2_{1/2}$ can be replaced in iv) by a large class of $\varphi \in H_0^\infty$. The proof given here is more elementary than the original ones. For $X = L_q(\Omega)$, $1 < q < \infty$, i) $\Longleftrightarrow$ iii) was demonstrated also in [56], where $\|\cdot\|_\varphi$ was expressed in terms of classical square functions. i) $\Longleftrightarrow$ iii) for the general case and i) $\Longleftrightarrow$ ii) are from [134].

N 12.4 appears in [134] and [135].

N 12.5, 12.6 are from [131].

N 12.7: The idea of an operator-valued H^∞-calculus goes back to D. Albrecht, A. McIntosh, F. Lancien, G. Lancien and Ch. LeMerdy. For Hilbert spaces X and in the special case $X \subset L_q(\Omega, H)$, $\mathcal{A} \subset \mathrm{Id}_{L_q} \otimes B(H)$, where $1 < q < \infty$ and H is a Hilbert space, Theorem 12.7 is shown in [152] and [154]. The general case is from [134]. Note that one cannot avoid the R-boundedness assumption. It is shown in [152] that if 12.7 holds for all $A \in \mathcal{S}(X)$ and $F \in H^\infty(\Sigma_\sigma, \mathcal{A})$, then X is isometric to a Hilbert space.

Applications to elliptic operators on a half space with general boundary conditions are given in [74].

N 12.8, 12.9 are from [134], but the proof in [134] is somewhat more direct. Part a) and b) can be shown under weaker assumptions. That the R-boundedness of the H^∞-calculus implies R-sectoriality was shown for UMD-spaces in [48] and for the larger class of spaces with *property* (Δ) in [134]. A Banach space has property (Δ) if for all $x_{ij} \in X$

$$\int_0^1\int_0^1 \left\|\sum_{i=1}^\infty\sum_{j=i}^\infty r_i(u)r_j(v)x_{ij}\right\| du\,dv \le C \int_0^1\int_0^1 \left\|\sum_{i=1}^\infty\sum_{j=1}^\infty r_i(u)r_j(v)x_{ij}\right\| du\,dv.$$

All spaces with property (α), UMD or analytic UMD satisfy this property (cf. [134]).

N 12.8: a) For Hilbert spaces X the equality $\omega_{H\infty}(A) = \omega(A)$ was observed by McIntosh [176], but for more general Banach spaces we may have $\omega_{H\infty}(A) > \omega(A)$ (see [130]). Hence $\omega_{H\infty}(A) = \omega_R(A)$ is a reasonable extension of the Hilbert space result to Banach spaces with property (Δ) (cf. [134]). To avoid assumptions on the Banach space X we introduced in [131] *almost R-bounded* operators, i.e. sectorial operators A such that $\{\lambda R(\lambda, A)^2 : |\arg \lambda| > \omega\}$ is R-bounded for some $\omega \in (\omega(A), \pi)$. The infimum over all such ω is denoted by $\tilde{\omega}_R(A)$. It is shown in [131] that $\omega_{H\infty}(A) = \tilde{\omega}_R(A)$ for all Banach spaces X. Every sectorial operator with BIP is almost R-bounded, but not necessarily R-bounded. Also R-boundedness implies almost R-boundedness, but not conversely (cf. [131]). A useful property of almost R-bounded operators is the following one (see [131]): The set $\{\varphi(tA) : t > 0\}$ is R-bounded for all $\varphi \in H_0^\infty(\Sigma_\sigma)$ with $\sigma > \tilde{\omega}_R(A)$.

b) $\tau \subset B(X,Y)$ is R-bounded if and only if $\tau' = \{T' : T \in \tau\}$ is R-bounded whenever X has finite type (cf. [131]).

N 12.11: R-BIP implies the boundedness of the H^∞-calculus whenever X has finite cotype. This is shown in [135] using a different and more powerful extension of the Hilbert space square functions (from section 10) in place of the expressions $\|\cdot\|_\varphi$ from (12.2). It is sufficient for a bounded H^∞-calculus if $\{A^{it} : t \in [-\varepsilon, \varepsilon]\}$ is R-bounded for some $\varepsilon > 0$.

Examples of operators in $L_q(\Omega)$, $q \neq 2$, with BIP but without a bounded H^∞-calculus can be found in [56].

N 12.12: For $n = 2$, Theorem 12.12 (aside from the R-boundedness statement) was shown for $L_q(\Omega)$ spaces with $1 < q < \infty$ in [2], and for spaces with property (α) in [152]. The proof given here is essentially the one in [134]. The extension to n operators is observed in [135]. As a consequence of the joint functional calculus one can derive that the closure of $A_1 + \cdots + A_n$ has a bounded H^∞-calculus. The latter result can be shown under the weaker assumption that X has property (Δ), cf. [159]. In [75] an operator-valued extension of the joint functional calculus of commuting sectorial operators is constructed (adapting the methods of 12.7) with properties analogous to 12.7.

N 12.13: Theorem 12.13 was shown in [134]. For Hilbert spaces it already appears in [73], some further special situations are contained in [152], [154] and [48].

The Dore-Venni theorem is shown in [73] for sectorial operators A with $0 \in \rho(A)$. The latter assumption was removed in [194].

N 12.16 - 12.17: These results on square functions are proven in a more general form in [131]. In the Hilbert space case, such equivalence theorems go back to McIntosh, see [176], and they were extended to L_p-spaces (in a somewhat different form) in [161].

13 Perturbation Theorems and Elliptic Operators

In this section we address the problem how to "transfer" the property of having a bounded H^∞–calculus from one operator to another. We approach this problem in two ways. First we compare (fractional) domain and range spaces of sectorial operators and show that the infomation on the H^∞-calculus is "encoded" in the natural norms on these spaces. An interesting faeture of this approach is that no "smallness" assumptions are involved. Secondly, we present theorems on "small" perturbations. Here the classical relative boundedness assumptions, which were still good enough for perturbation theorems for R-sectorial operators (cf. Section 6), are not sufficient anymore. Stronger assumtptions must be imposed unless the perturbation is of lower order, and this is the case we shall start out with (Proposition 13.1).

In this section we assume throughout – with exception of Proposition 13.1 – that X is a uniformly convex Banach space. This assumption is chosen only to simplify duality arguments: a uniformly convex space is reflexive and has non-trivial type. In particular, an operator A in a unifomrly convex space is R-sectorial if and only if A^* is R-sectorial (cf. Notes on Section 2). On the other hand, the class of uniformly convex spaces is large enough to contain the spaces relevant for the differential operators we will consider: closed subspaces and quotient spaces of $L_p(\Omega)$-spaces with $1 < p < \infty$ are uniformly convex.

As announced we start with a result on lower order perturbation where X may be an arbitrary Banach space.

13.1 Proposition. *Let $A \in \mathcal{S}(X)$ have a bounded $H^\infty(\Sigma_\sigma)$-calculus in X and assume $0 \in \rho(A)$. Let $\delta \in (0,1)$ and suppose that B is a linear operator in X satisfying $D(B) \supset D(A)$, and*

$$\|Bx\| \le C\|A^{1-\delta}x\|, \quad x \in D(A),$$

where $C > 0$. Then $\nu + A + B$ has a bounded $H^\infty(\Sigma_\sigma)$-calculus in X for $\nu \ge 0$ sufficiently large.

Proof. We assume withuout loss of generality that $A + B$ is again sectorial. By the resolvent equation, we can write

$$R(\lambda, A + B) = R(\lambda, A) + R(\lambda, A + B)BR(\lambda, A)$$
$$= R(\lambda, A) + R(\lambda, A + B)[BA^{\delta-1}]A^{1-\delta}R(\lambda, A)$$
$$= R(\lambda, A) + M(\lambda).$$

For $\psi \in H_0^\infty(\Sigma_\sigma)$ this yields

$$\psi(A + B) = \psi(A) + \frac{1}{2\pi i} \int_{\Gamma_\nu} \psi(\lambda)M(\lambda)\,d\lambda,$$

where the integrand may be estimated

$$\|\psi(\lambda)M(\lambda)\| \leq M_{A+B}\|\psi\|_{H^\infty}|\lambda|-1\|BA^{\delta-1}\|M_A|\lambda|^{-\delta},$$

since $M_{A+B} := \sup\{\|\lambda R(\lambda, A + B)\| : \pi \geq |\arg\lambda| \geq \nu\}$ and $M_A := \sup\{\|\lambda^\delta A^{1-\delta}R(\lambda, A)\| : \pi \geq |\arg\lambda| \geq \nu\}$ are finite. Hence the integral converges absolutely and its norm is $\leq C_1\|\psi\|_{H^\infty}$.   □

We now present the comparison result.

13.2 Theorem. *Let $A \in \mathcal{S}(X)$ be R-sectorial in a uniformly convex space X. Suppose that A has a bounded $H^\infty(\Sigma_\sigma)$-calculus in X. Let $B \in \mathcal{S}(X)$ be R-sectorial in X with $\omega_R(B) \leq \sigma$.*
Assume that for some $u, v \in (0, 3/2)$ and some $C > 0$ we have

$$D(A^u) = D(B^u) \ and \ \frac{1}{C}\|A^u x\| \leq \|B^u x\| \leq C\|A^u x\| \ for \ x \in D(A^u).$$

$$R(A^v) = R(B^v) \ and \ \frac{1}{C}\|A^{-v}x\| \leq \|B^{-v}x\| \leq C\|A^{-v}x\| \ for \ x \in R(A^v).$$

Then B has a bounded $H^\infty(\Sigma_\nu)$-calculus for $\nu > \sigma$.

Before we give the proof of the theorem we state a lemma that reformulates a part of the hypotheses.

13.3 Lemma. *Let A and B be injective operators on a reflexive space X with dense domain and range. Then the following two conditions are equivalent:*

(i) $D(A^{-1}) \subset D(B^{-1})$ and $\|B^{-1}x\| \leq C\|A^{-1}x\|$ *for $x \in R(A)$*
(ii) $D(B^*) \subset D(A^*)$ and $\|A^*x^*\| \leq C\|B^*x^*\|$ *for $x^* \in D(B^*)$*

Proof. $(i) \Longrightarrow (ii)$: For $z \in D(A)$ we have $Az \in D(B^{-1})$. For $y^* \in D((B^{-1})^*)$ we obtain

$$\langle Az, (B^{-1})^* y^*\rangle = \langle B^{-1}Az, y^*\rangle \leq C\|z\|\|y^*\|.$$

Hence $(B^{-1})^*y^* \in D(A^*)$, $\|A^*(B^*)^{-1}y^*\| \leq C\|y^*\|$ and the claim follows.
$(ii) \Longrightarrow (i)$: Let $x = Ay \in D(A^{-1})$. For $x^* = B^*y^* \in R(B^*)$ we have $y^* \in D(B^*) \subset D(A^*)$ and

$$|\langle(B^*)^{-1}x^*, x\rangle| = |\langle y^*, Ay\rangle| = |\langle A^*y^*, y\rangle| \leq \|A^*y^*\|\|A^{-1}x\|$$
$$\leq C\|B^*y^*\|\|A^{-1}x\| = C\|x^*\|\|A^{-1}x\|.$$

By reflexivity of X we conclude $x \in D(((B^*)^{-1})^*) = D(B^{-1})$, (i) follows since $R(B^*)$ is dense in X^*.   □

Actually, we shall use this lemma for fractional powers of A and B, namely, for A^u, B^u and for A^{-v}, B^{-v}.

Proof (of Theorem 13.2). Denote by M and N the bounded extensions of $B^u A^{-u}$ and $B^{-v} A^v$ to X, respectively. Here we use the inequalities on the right of our assumptions. We want to apply Theorem 12.2. To this end we compare expressions of the form $\lambda^{n-u} T^u R(\lambda, T)^n$ for $T = A$ and $T = B$ via the generalized resolvent equation

$$R(\lambda, B) = R(\lambda, A) + (BR(\lambda, B) - R(\lambda, B)A)R(\lambda, A).$$

Multiplying with $R(\lambda, B)^2$ from the left we get

$$
\begin{aligned}
&R(\lambda, B)^3 \\
&= R(\lambda, B)^2 (\lambda - A) R(\lambda, A)^2 \\
&\quad + R(\lambda, B)^2 (BR(\lambda, B) - R(\lambda, B)A)(\lambda - A)R(\lambda, A)^2 \\
&= R(\lambda, B)^2 (\lambda - B)R(\lambda, B)\lambda R(\lambda, A)^2 - R(\lambda, B)^2 AR(\lambda, A)^2 \\
&\quad + R(\lambda, B)^2 BR(\lambda, B)\lambda R(\lambda, A)^2 - BR(\lambda, B)^3 AR(\lambda, A)^2 \\
&\quad - \lambda R(\lambda, B)^3 AR(\lambda, A)^2 + R(\lambda, B)^3 A^2 R(\lambda, A)^2 \\
&= [\lambda^2 R(\lambda, B)^3] R(\lambda, A)^2 - [R(\lambda, B)^2 + \lambda R(\lambda, B)^3 + BR(\lambda, B)^3] AR(\lambda, A)^2 \\
&\quad + [R(\lambda, B)^3] A^2 R(\lambda, A)^2.
\end{aligned}
$$

For $\frac{3}{2} > s > u$ this yields

$$
\begin{aligned}
&\lambda^{3-s} B^s R(\lambda, B)^3 \\
&= [\lambda^{3-s+u} B^{s-u} R(\lambda, B)^3 M]\{\lambda^{2-u} A^u R(\lambda, A)^2\} \\
&\quad - [\lambda^{2-s} B^s R(\lambda, B)^2 + \lambda^{3-s} B^s R(\lambda, B)^3 + \lambda^{2-s} B^{1+s} R(\lambda, B)^3] \times \\
&\quad \times \{\lambda AR(\lambda, A)^2\} + [\lambda^{3-s-v} B^{s+v} R(\lambda, B)^3 N]\{\lambda^v A^{2-v} R(\lambda, A)^2\}.
\end{aligned}
$$

By R–sectoriality of B and boundedness of M and N the three sets in $[\,\ldots\,]$ are R–bounded for $|\arg(\lambda)| \geq \nu, \nu > \sigma$. Observe that, indeed, all exponents are strictly positive. Hence, for $\psi_{s,k}(\lambda) = \lambda^s (e^{i\nu} - \lambda)^{-k}$ we obtain for $x \in X$, $t > 0$

$$
\left\| \sum_k r_k \psi_{s,3}(t^{-1} 2^{-k} B) x \right\|_{L_2(X)} \leq D \sum_{s=u, 1, 2-v} \left\| \sum_k r_k \psi_{s,2}(t^{-1} 2^{-k} A) x \right\|_{L_2(X)}
$$
$$
\leq D_1 \|x\|,
$$

since A has a $H^\infty(\Sigma_\sigma)$–calculus (cf. also 12.17).

We now use the estimates on the left of our assumptions translating them by 13.3 into estimates for the dual operators A^* and B^*.

Since B^* is again R-sectorial and A^* has a bounded H^∞-calculus we can repeat the above argument on X^* and obtain

$$
\left\| \sum_k r_k \psi_{s,3}(t^{-1} 2^{-k} B^*) x^* \right\|_{L_2(X^*)} \leq D_2 \|x^*\|, \quad x^* \in X^*.
$$

Now the claim follows from Theorem 12.2 in combination with 12.15.   □

When we use Proposition 15.27 we obtain the following.

13.4 Corollary. *Let $A \in \mathcal{S}(X)$ be R-sectorial. Suppose that A has a bounded $H^\infty(\Sigma_\sigma)$-calculus in X. Let $B \in \mathcal{S}(X)$ be R-sectorial in X with $\omega_R(B) \leq \sigma$.*
Assume that for some $0 < u_1 < u_2 < 3/2$, some $C > 0$, and $j = 0,1$ we have

$$D(A^{u_j}) = D(B^{u_j}) \text{ and } \frac{1}{C}\|A^{u_j}x\| \leq \|B^{u_j}x\| \leq C\|A^{u_j}x\| \text{ for } x \in D(A^{u_j}).$$

Then B has a bounded $H^\infty(\Sigma_\nu)$-calculus for $\nu > \sigma$.

Proof. We let $Y := \dot{X}_{u_1,A} = \dot{X}_{u_1,B}$ (cf. Section 15). By Proposition 15.27 we have, for $T \in \{A, B\}$, $X = \dot{Y}_{-u_1, \dot{T}_{u_1}}$ and $\dot{X}_{u_2} = \dot{Y}_{u_2-u_1, \dot{T}_{u_1}}$. The operator $\dot{A}_{u_1}$ is similar to A and $\dot{B}_{u_1}$ is similar to B. Hence the assumptions of Theorem 13.2 are satisfied for $\dot{A}_{u_1}, \dot{B}_{u_1}$ in Y and $u = u_2 - u_1$, $v = u_1$. We conclude that $\dot{B}_{u_1}$ has a bounded $H^\infty(\Sigma_\nu)$-calculus in Y for $\nu > \sigma$. By similarity, the same holds for B in X.   □

We now give a version of the perturbation theorem for R–sectoriality 6.5 that asserts persistence of a bounded H^∞-calculus under an additional assumption.

13.5 Theorem. *Let $A \in \mathcal{S}(X)$ be an R-sectorial operator having a bounded $H^\infty(\Sigma_\sigma)$-calculus in X. Let $\delta \in (0,1)$ and suppose that B is a linear operator in X satisfying $D(B) \supset D(A)$, $R(B) \subset R(A^\delta)$ and*

$$\|Bx\| \leq C_0\|Ax\|, \quad x \in D(A),$$
$$\|A^{-\rho}Bx\| \leq C_1\|A^{1-\rho}x\|, \quad x \in D(A), \rho \in (0,\delta)$$

where $C_0, C_1 < 1/R_0$ and $R_0 := R(\{AR(\lambda, A) : |\arg \lambda| \geq \sigma\})$. Then $A + B$ has a bounded $H^\infty(\Sigma_\nu)$-calculus in X for any $\nu > \sigma$.

Proof. Using the smallness of C_0 we obtain by Theorem 6.5 that $A + B$ is R–sectorial in X where

$$R(\lambda, A + B) = R(\lambda, A) \sum_{k=0}^{\infty}(BR(\lambda, A))^k, \quad |\arg \lambda| > \sigma.$$

We shall apply Theorem 13.2 with $A + B$ in place of B and $u = 1$, $v \in (0, \delta)$. Since we have in particular $C_0 < 1$, coincidence of the domains $D(A + B) = D(A)$ and equivalence of norms $\|(A+B)x\| \sim \|Ax\|$ are immediate, and $A+B$ is injective. Since $A + B = (I + BA^{-1})A$, A has dense range, and $I + BA^{-1}$ extends to an isomorphism of X, we obtain that $A + B$ has dense range.

We fix $\rho \in (0, \delta)$ and denote by $L \in B(X)$ the bounded extension of $A^{-\rho}BA^{\rho-1}$. Then

$$R(\lambda, A)BR(\lambda, A) = R(\lambda, A)A^\rho L A^{1-\rho}R(\lambda, A) \supset A^\rho R(\lambda, A)LAR(\lambda, A)A^{-\rho}$$

which yields

$$R(\lambda, A + B) = R(\lambda, A) + A^\rho R(\lambda, A) \sum_{k=0}^{\infty} (LAR(\lambda, A))^k LA^{1-\rho} R(\lambda, A).$$

Hence

$$R(\lambda, A + B) = R(\lambda, A) + A^\rho R(\lambda, A) M(\lambda) A^{1-\rho} R(\lambda, A) \qquad (13.1)$$

where $M(\lambda) := \sum_{k=0}^{\infty} (LAR(\lambda, A))^k L$ is R-bounded with bound $\leq C_1(1 - C_1 R_0)^{-1}$ by assumption.

The following lemma allows to compare fractional powers of A and $A + B$.

13.6 Lemma. *Assume that $A \in \mathcal{S}(X)$ is an R-sectorial operator having a bounded $H^\infty(\Sigma_\sigma)$-calculus. Let $B : D(A) \to X$ be a linear operator such that $A + B \in \mathcal{S}(X)$ is sectorial and satisfies, for some $\delta \in (0, 1)$,*

$$R(t, A + B) = R(t, A) + A^\delta R(t, A) M(t) A^{1-\delta} R(t, A), \quad t < 0, \qquad (13.2)$$

where $\{M(t) : t < 0\}$ is R-bounded in $B(X)$. Then

$$D((A+B)^\theta) \subset D(A^\theta) \text{ and } \|A^\theta x\| \leq C\|(A+B)^\theta x\|, \ x \in D((A+B)^\theta), \ (13.3)$$

for all $0 < \theta < 1 - \delta$ and

$$R(A^\theta) \subset R((A + B)^\theta)$$
$$\|(A + B)^{-\theta} x\| \leq C\|A^{-\theta} x\| \quad x \in R(A^\theta), 0 < \theta < \delta. \qquad (13.4)$$

Proof. We use the formula (cf. Theorem 15.18)

$$(A + B)^{-\theta} x = \frac{\sin(\pi\theta)}{\pi} \int_{-\infty}^{0} (-t)^{-\theta} R(t, A + B) x \, dt, \quad x \in R(A + B).$$

For $0 < \theta < 1 - \delta$ and $x^* \in D(A^*), x \in R(A + B)$ we get with (13.1)

$$\langle (A + B)^{-\theta} x, (A^*)^\theta x^* \rangle = \frac{\sin(\pi\theta)}{\pi} \int_{-\infty}^{0} (-t)^{-\theta} \langle A^\theta R(t, A) x, x^* \rangle dt$$

$$+ \frac{\sin(\pi\theta)}{\pi} \int_{-\infty}^{0} \langle M(t)[(-t)^\delta A^{1-\delta} R(t, A)] x, [(-t)^{1-\delta-\theta} (A^*)^{\theta+\delta} R(t, A)^*] x^* \rangle \frac{dt}{(-t)}$$

$$= \langle x, x^* \rangle + \frac{\sin(\pi\theta)}{\pi} \int_{0}^{\infty} \langle M(-t) \psi_{1-\delta}(t^{-1} A) x, \psi_{\delta+\theta}(t^{-1} A)^* x^* \rangle \frac{dt}{t}$$

where $\psi_\alpha(\lambda) = \lambda^\alpha (1 + \lambda)^{-1}$. With Lemma 12.6 we can estimate the last integral by

$$R\Big(\{M(t) : t < 0\}\Big) \, \sup_{t>0} \left\| \sum_k r_k \psi_{1-\delta}(t^{-1}2^{-k}A)x \right\|_{L_2(X)}$$

$$\sup_{t>0} \left\| \sum_k r_k \psi_{\theta+\delta}(t^{-1}2^{-k}A)^*x^* \right\|_{L_2(X^*)}.$$

Since $\theta + \delta < 1$ we can use boundedness of the H^∞-calculus and R-boundedness of $\{M(t)\}$ to obtain

$$\langle (A+B)^{-\theta}x, (A^*)^\theta x^* \rangle \le C\|x\| \cdot \|x^*\|.$$

Since X is reflexive we obtain $(A+B)^{-\theta}x \in D(A^\theta)$ and $\|A^\theta(A+B)^{-\theta}x\| \le C\|x\|$. Hence we have shown (13.3).

In a similar way we get for $0 < \theta < \delta$, $x \in R(A)$, $x^* \in D((A+B)^*)$

$$\langle A^\theta x, ((A+B)^*)^{-\theta}x^* \rangle$$
$$= \langle x, x^* \rangle + \frac{\sin(\pi\theta)}{\pi} \int_0^\infty \langle M(-t)\psi_{1-\delta+\theta}(t^{-1}A)x, \psi_\delta(t^{-1}A)^*x^* \rangle \frac{dt}{t}$$
$$\le C\|x\| \cdot \|x^*\|,$$

which means that (13.4) holds.   □

By (13.1) we may apply Lemma 13.6 for ρ in place of δ. Since $\rho \in (0, \delta)$ is arbitrary, we obtain

$$D((A+B)^\theta) \subset D(A^\theta)$$
$$\|A^\theta x\| \le C_\theta \|(A+B)^\theta x\|, \quad x \in D((A+B)^\theta), \, 0 < \theta < 1. \tag{13.5}$$

$$R(A^\theta) \subset R((A+B)^\theta)$$
$$\|(A+B)^{-\theta}x\| \le C_\theta \|A^{-\theta}x\| \quad x \in R(A^\theta), 0 < \theta < \delta.$$

Next we show that

$$\|A^{-\theta}x\| \le \|(A+B)^{-\theta}x\| \quad \text{for } x \in R(A^\theta) \text{ and } 0 < \theta < \delta. \tag{13.6}$$

Indeed, for $\theta \in (0, \delta)$, the assumption and (13.5) with $1 - \theta$ in place of θ yield

$$\|A^{-\theta}(A+B)^\theta x\| = \|[A^{-\theta}(A+B)A^{\theta-1}][A^{1-\theta}(A+B)^{\theta-1}]x\| \le C'_\theta \|x\|.$$

Hence we have proved

$$\|A^{-\theta}x\| \sim \|(A+B)^{-\theta}x\| \quad \text{for } \theta \in (0, \delta)$$

which means that the assumptions of Theorem 13.2 are satisfied for $u = 1$ and any $v \in (0, \delta)$.   □

13.7 Remark. Let $A \in \mathcal{S}(X)$ be a sectorial operator having a bounded H^∞-calculus. Suppose that $B : D(A) \to X$ is a linear operator such that $A + B \in \mathcal{S}(X)$ is R-sectorial. Then, by the resolvent equation,

$$R(\lambda, A + B) = R(\lambda, A) + R(\lambda, A)BR(\lambda, A + B)$$
$$= R(\lambda, A) + R(\lambda, A)[B + BR(\lambda, A + B)B]R(\lambda, A).$$

If, in addition, $L := A^{-\rho}BA^{\rho-1}$ is bounded, we obtain

$$R(\lambda, A + B) = R(\lambda, A) + A^\rho R(\lambda, A)[L + LA^{\rho-1}R(\lambda, A + B)A^\rho L]A^{1-\rho}R(\lambda, A).$$

Now the assumption of Lemma 13.6 boils down to the property that the corresponding version of $A + B$ is R–sectorial in the space $\dot{X}_{-\rho}$. In Theorem 13.5 this is assured by smallness of the norm of L and R-sectoriality of A.

Now we relax the assumptions of Theorem 13.5 by complex interpolation.

13.8 Corollary. *Let* $A \in \mathcal{S}(X)$ *be an* R-*sectorial operator having a bounded* $H^\infty(\Sigma_\sigma)$-*calculus in* X. *Let* $\delta \in (0, 1)$ *and suppose that* B *is a linear operator in* X *satisfying* $D(B) \supset D(A)$, $R(B) \subset R(A^\delta)$ *and*

$$\|Bx\| \leq C_0 \|Ax\|, \quad x \in D(A),$$
$$\|A^{-\delta}Bx\| \leq C_1 \|A^{1-\delta}x\|, \quad x \in D(A),$$

where $C_0 > 0$ *is sufficiently small and* $C_1 > 0$ *is arbitrary. Then* $A + B$ *has a bounded* $H^\infty(\Sigma_\nu)$-*calculus in* X *for any* $\nu > \sigma$.

Proof. We refer to Section 15 for the scale $\dot{X}_\alpha = \dot{X}_{\alpha, A}$. The first assumption on B means that B acts as a bounded operator $\dot{X}_1 \to X = \dot{X}_0$ with norm $\leq C_0$. The second assumption on B means that B acts as a bounded operator $\dot{X}_{1-\delta} \to \dot{X}_{-\delta}$ with norm $\leq C_1$. By complex interpolation we obtain that B acts as a bounded operator $[\dot{X}_1, \dot{X}_{1-\delta}]_\theta \to [\dot{X}_0, \dot{X}_{-\delta}]_\theta$ with norm $\leq C_0^{1-\theta}C_1^\theta$.

Since A has a bounded H^∞-calculus, A has bounded imaginary powers, and by Theorem 15.28 we have $[\dot{X}_1, \dot{X}_{1-\delta}]_\theta = \dot{X}_{1-\theta\delta}$ and $[\dot{X}_0, \dot{X}_{-\delta}]_\theta = \dot{X}_{-\theta\delta}$ with equivalent norms where equivalence constants may be chosen independent of $\theta \in (0, 1)$. Thus we obtain that, for all $\theta \in (0, 1)$, B acts as a bounded operator $\dot{X}_{1-\theta\delta} \to \dot{X}_{-\theta\delta}$ with norm $\leq cC_0^{1-\theta}C_1^\theta$. If C_0 is small then, for $\theta \in (0, \theta_0)$ this is $< 1/R_0$ where R_0 is as in Theorem 13.5. Thus we may apply Theorem 13.5 with θ_0 in place of δ. $\qquad\square$

The following gives another version of an additional assumption for perturbation of the H^∞-calculus where δ is replaced by $-\delta$.

13.9 Corollary. *Let* $A \in \mathcal{S}(X)$ *be an* R-*sectorial operator with* $0 \in \rho(A)$ *having a bounded* $H^\infty(\Sigma_\sigma)$-*calculus in* X. *Let* $\delta \in (0, 1)$ *and suppose that* B *is a linear operator in* X *satisfying* $D(B) \supset D(A)$, $B(D(A^{1+\delta}) \subset D(A^\delta)$ *and*

$$\|Bx\| \le C_0 \|Ax\|, \quad x \in D(A),$$
$$\|A^\delta Bx\| \le C_1 \|A^{1+\delta}x\|, \quad x \in D(A),$$

where $C_0 > 0$ is sufficiently small and $C_1 > 0$ is arbitrary. Then $A + B$ has a bounded $H^\infty(\Sigma_\nu)$-calculus in X for any $\theta > \nu$.

Proof. Let $\nu > \sigma$. Applying Corollary 13.8 in $X_{\delta,A}$ (cf. 15.21) we obtain a bounded $H^\infty(\Sigma_\nu)$-calculus for $A_\delta + B$ in $X_{\delta,A}$ and in $X_{1+\delta,A}$ (since $D(A_\delta + B) = D(A_\delta) = X_{1+\delta,A}$). By Theorem 15.28 we have $X_{1,A} = [X_{\delta,A}, X_{1+\delta}]_{1-\delta}$. By complex interpolation we thus obtain a bounded $H^\infty(\Sigma_\nu)$-calculus for the part of $A+B$ in $X_{1,A} = X_{1,A+B}$. By similarity, $A+B$ has an $H^\infty(\Sigma_\nu)$-calculus in X. $\qquad\square$

Next we prove an analog of Corollary 6.7.

13.10 Corollary. *Let $A \in \mathcal{S}(X)$ be an R-sectorial operator having an $H^\infty(\Sigma_\sigma)$-calculus in X. Let $\delta \in (0,1)$ and suppose that B is a linear operator in X satisfying $D(B) \supset D(A)$, and*

$$\|Bx\| \le a\|Ax\| + b\|x\|, \quad x \in D(A),$$
$$\|(1+A)^{-\delta}Bx\| \le C\|(1+A)^{1-\delta}x\|, \quad x \in D(A),$$

where a is sufficiently small and $b, C > 0$ are arbitrary. Then $\nu + A + B$ has a bounded $H^\infty(\Sigma_\theta)$-calculus in X for $\nu \ge 0$ sufficiently large and $\theta > \sigma$.

Proof. The first assumption on B yields for any $\nu > 0$ and $x \in D(A)$

$$\|B(\nu + A)^{-1}x\| \le a\|A(\nu + A)^{-1}x\| + b\|(\nu + A)^{-1}x\|$$
$$\le \left(a + \frac{b+1}{\nu}\right) \sup_{\lambda > 0} \|A(\lambda + A)^{-1}\| \cdot \|x\|.$$

If a is sufficiently small then, by choosing ν sufficiently large, the first assumption of Corollary 13.8 is satisfied for $\nu + A$ in place of A. But also the second assumption is satisfied since $\|(1+A)^{-\delta} \cdot \| \sim \|(\nu + A)^{-\delta} \cdot \|$ and $\|(1+A)^{1-\delta} \cdot \| \sim \|(\nu + A)^{1-\delta} \cdot \|$. Hence the assertion follows by an application of Corollary 13.8. $\qquad\square$

Finally, we apply the results presented in this section to some of the differential operators we met before.

Applications to differential operators

We start with **constant coefficient operators** on $\mathbb{R}^n$. For the notion of an (M, ω_0)–elliptic operator we refer back to Section 6. Here, we confine ourselves to scalar equations, i.e. to the case $N = 1$.

13.11 Proposition. *Let A be an (M, ω_0)-elliptic operator of order $2m$ with constant coefficients. Then, for $1 < p < \infty$, the realization A_p of A in $L_p(\mathbb{R}^n)$ has an $H^\infty(\Sigma_\nu)$-calculus for any $\nu > \omega_0$.*

Proof. By 10.2 b) the Laplace–operator $-\Delta$ has a bounded $H^\infty(\Sigma_\nu)$-calculus in $L_p(\mathbb{R}^n)$ for any $\nu > 0$. As in the second step of the proof of Theorem 11.14 (using 15.11) also $(-\Delta)^m$ has a bounded H^∞-calculus in $L_p(\mathbb{R}^n)$. We apply Theorem 13.2 for $u = v = 1$ with $(-\Delta)^m$ in place of A and A_p in place of B. By Theorem 6.2 the operator A_p is R-sectorial. All we have to check now is that $D(A_p) = D((-\Delta)^m)$ in L_p and $D(A_p^*) = D((-\Delta)^m)$ in $L_{p'}$ with equivalent *homogeneous* norms (cf. Lemma 13.3). But the operators $A_p^{\pm 1}(-\Delta)^{\mp m} = (-\Delta)^{\mp m}A_p^{\pm 1}$ are bounded in $L_p(\mathbb{R}^n)$, since their symbols are positively homogeneous of degree zero and C^∞ on $\mathbb{R}^n \setminus \{0\}$, and we can resort to Mihlin's theorem. $\qquad\square$

Next we consider elliptic operators in non-divergence form with **Hölder continuous coefficients** in the principle part. Recall that we studied operators with bounded uniformly continuous coefficients in Section 6. We need additional regularity of the coefficients for application of Theorem 13.5. Following the steps of Section 6 we start with small perturbations.

13.12 Proposition. *Let A_0 be a homogeneous (M, ω_0)-elliptic operator of order $2m$ with constant coefficients and $\gamma \in (0,1)$. Then there is a constant $\eta > 0$ such that, for any $B = \sum_{|\alpha|=2m} b_\alpha$ with $b_\alpha \in C^\gamma$ and $\sum_{|\alpha|=2m} \|b_\alpha\|_\infty < \eta$ the realization $1 + (A+B)_p$ of $1 + A + B$ in $L_p(\mathbb{R}^n)$ has a bounded H^∞-calculus.*

Proof. We have $D(1 + A_0) = W_p^{2m}(\mathbb{R}^n) = H_p^{2m}(\mathbb{R}^n)$ and $D((1 + A_0)^\delta) = H_p^{2m\delta}(\mathbb{R}^n)$ with equivalent norms. We apply the following fact (see, e.g., [213, 2.8.2. Cor.]):

$$\|mf\|_{H_p^s} \leq C\|m\|_{C^\gamma}\|f\|_{H_p^s}, \quad f \in H_p^s(\mathbb{R}^n), m \in C^\gamma(\mathbb{R}^n)$$

for all $|s| < \gamma$. Now the assertion follows from Corollary 13.8. $\qquad\square$

In order to allow for general Hölder continuous coefficients in the principle part we employ the localization procedure from Section 6.

13.13 Theorem. *Let $M > 0$, $\omega_0 \in [0, \pi)$, and $p \in (1, \infty)$. Let A be the L_p-realization of an (M, ω_0)-elliptic operator $\mathcal{A}(D) = \sum_{|\alpha|\leq 2m} a_\alpha D^\alpha$ of order $2m$ satisfying*

$$a_\alpha \in C^\gamma(\mathbb{R}^n, \mathbb{C}), \qquad |\alpha| = 2m, \tag{13.7}$$

for some $\gamma \in (0,1)$ and

$$a_\alpha \in L_\infty(\mathbb{R}^n, \mathbb{C}), \qquad |\alpha| \leq 2m - 1. \tag{13.8}$$

Then, for any $\theta \in (\omega_0, \pi)$, there exists a $\nu \geq 0$ such that $\nu + A$ has a bounded $H^\infty(\Sigma_\theta)$-calculus.

Proof. We localize as in Section 6 (cf. 6.6 and the paragraph preceding it) and use the same notations. The coefficients of the localized operators A_l now have to be Hölder continuous in addition to being an L_∞–small perturbation of constant coeficients. We therefore modify their definition as follows. For $l \in \Gamma$ and $|\alpha| = 2m$ we let

$$a_\alpha^l(x) := a_\alpha(l) + \psi_l(x)(a_\alpha(x) - a_\alpha(l)), x \in \mathbb{R}^n.$$

We define operators A_l with domain $W_p^{2m}(\mathbb{R}^n, \mathbb{C}^N)$ by

$$A_l u(x) := \sum_{|\alpha|=2m} a_\alpha^l(x) D^\alpha u(x).$$

Writing $A = A_\pi + A_{\mathrm{low}}$ where $A_\pi = \sum_{|\alpha|=2m} a_\alpha$ denotes the principle part of A, and A_{low} denotes the part containing the lower order terms of A, we have that $A_\pi u = A_l u$ if the support of $u \in W_p^{2m}(\mathbb{R}^n, \mathbb{C}^N)$ is contained in supp φ_l (recall that supp $\psi_l \subset Q_l$ and $\psi_l = 1$ on supp φ_l). The coefficients a_α^l are small in L_∞–norm and belong to $C^\gamma(\mathbb{R}^n)$. By Proposition 13.12 the operator $1 + A_l$ has a bounded H^∞–calculus with a constant independent of $l \in \Gamma$.

We define the operator $\mathbb{A}$ in $\mathbb{X}_p$ with domain $\mathbb{X}_p^{2m}$ as before (of course, we have to set $N = 1$ here since we are dealing with scalar equations). Then the operator $1 + \mathbb{A}$ has a bounded H^∞-calculus. Since the operators $\mathbb{B}$ and $\mathbb{D}$ are perturbations of lower order (cf. Section 6, before Corollary 6.7), we apply Proposition 13.1 and obtain bounded H^∞-calculi for suitable translates of $\mathbb{A}+\mathbb{B}$ and $\mathbb{A}+\mathbb{D}$. The representaion of resolvents of A in terms of resolvents of $\mathbb{A}+\mathbb{B}$ and $\mathbb{A}+\mathbb{D}$ then yields a bounded H^∞-calculus for a suitable translate of A. $\quad\square$

Now we study **operators in divergence-form** with **Hölder continuous coefficients** where we restrict ourselves to operators of second order without lower order terms, mainly for simplicity of notation. We assume that A_2 is a sectorial operator in $L_2(\mathbb{R}^n)$ given by the form

$$\mathfrak{a}(u, v) := \int_{\mathbb{R}^n} \sum_{j,k} a_{jk}(x) \partial_j \overline{v(x)} \partial_k u(x) \, dx,$$

where $a_{jk} \in C^\gamma(\mathbb{R}^n, \mathbb{C})$ for some $\gamma \in (0,1)$ and we assume ellipiticity in the form

$$\frac{1}{M}|\xi|^2 \leq \mathrm{Re}\left(\sum_{j,k} a_{jk}(x)\xi_j\xi_k\right) \leq M|\xi|^2, \quad x \in \mathbb{R}^n, \xi \in \mathbb{C}^n \qquad (13.9)$$

for a constant $M \geq 1$. The regularity assumptions are much stronger than in Section 8 where we only assumed $a_{jk} \in L_\infty(\mathbb{R}^n, \mathbb{C})$. On the other, we shall not only obtain a bounded H^∞-calculus in L_p but also in H_p^1 and H_p^{-1}. Moreover, the assumption allows to study the operator in the whole scale of spaces L_p, $1 < p < \infty$.

13.14 Theorem. *Let the assumptions above hold. Then A has a bounded H^∞-calculus in all spaces $L_p(\mathbb{R}^n, \mathbb{C})$, $1 < p < \infty$. Moreover, A has a bounded H^∞-calculus in all spaces $H_p^{-1}(\mathbb{R}^n, \mathbb{C})$ and $H_p^1(\mathbb{R}^n, \mathbb{C})$, $1 < p < \infty$.*

Proof. Again, we employ the classical three step method. But now we consider the operator as an operator in $H_p^{-1}(\mathbb{R}^n, \mathbb{C})$ with domain $H_p^1(\mathbb{R}^n, \mathbb{C})$, i.e., we write $A = -\sum_{j,k} \partial_j(a_{jk}\partial_k)$.

Step I: If the a_{jk} are constants then A has a bounded H^∞-calculus in $L_p(\mathbb{R}^n, \mathbb{C})$ by Proposition 13.11. Moreover, we obtain by Mihlin's theorem that $(1 + A)^{1/2} : L_p(\mathbb{R}^n) \to H_p^{-1}(\mathbb{R}^n)$ defines an isomorphism. Consequently, A has a bounded H^∞-calculus in $H_p^{-1}(\mathbb{R}^n)$. Moreover, letting $X = H_p^{-1}(\mathbb{R}^n)$ we have for the scale X_α of inhomogeneous spaces (cf. Section 15) that $X_\alpha = H_p^{-1+2\alpha}(\mathbb{R}^n)$.

Step II: Now let the coefficients be of the form $a_{jk}(x) = a_{jk}^0 + b_{jk}(x)$ where $\|b_{jk}\|_\infty \le \eta$ for small η and $b_{jk} \in C^\gamma(\mathbb{R}^n)$. Then we factorize the perturbation $B := -\sum_{j,k} \partial_j(b_{jk}(x)\partial_k)$ of the constant coefficient operator $A_0 = -\sum_{j,k} \partial_j(a_{jk}^0\partial_k) : H_p^1(\mathbb{R}^n) \to H_p^{-1}(\mathbb{R}^n)$ as follows. We have $\partial_k : H_p^1(\mathbb{R}^n) \to L_p(\mathbb{R}^n)$, $b_{jk}(x)\cdot : L_p(\mathbb{R}^n) \to L_p(\mathbb{R}^n)$ with norm $\le \eta$, and $\partial_j : L_p(\mathbb{R}^n) \to H_p^{-1}(\mathbb{R}^n)$. Hence $B : H_p^1(\mathbb{R}^n) \to H_p^{-1}(\mathbb{R}^n)$ has small norm, which means that B satisfies the first assumption in Corollary 13.8. But we may also consider $b_{jk}(x)\cdot : H_p^{-s}(\mathbb{R}^n) \to H_p^{-s}(\mathbb{R}^n)$ for $s \in (0, \gamma)$, and $\partial_k : H_p^{1-s}(\mathbb{R}^n) \to H_p^{-s}(\mathbb{R}^n)$, $\partial_j : H_p^{-s}(\mathbb{R}^n) \to H_p^{-1-s}(\mathbb{R}^n)$. Hence B satisfies also the second assumption in Corollary 13.8, and we obtain that $1 + A = 1 + A_0 + B$ has a bounded H^∞-calculus.

Step III: In the general case we employ a modification of the localization scheme in Section 6. We fix $r > 0$, choose a function φ such that (6.12) holds where the φ_l are defined as before (cf. the paragraph before 6.6). We use notations $\Gamma = r\mathbb{Z}^n$, $Q_l := l + (-r, r)^n$, and ψ_l as before. Recall that ψ_l has support in Q_l and $\psi_l = 1$ on $\operatorname{supp}\varphi_l$. For any $l \in \Gamma$ and $j, k \in \{1, \ldots, n\}$ we define coefficients $a_{jk}^l : \mathbb{R}^n \to \mathbb{C}$ by

$$a_{jk}^l(x) := a_{jk}(l) + \psi_l(x)(a_{jk}(x) - a_{jk}(l))$$

We define operators A_l with domain $W_p^{2m}(\mathbb{R}^n, \mathbb{C}^N)$ by

$$A_l u(x) := -\sum_{j,k} \partial_j(a_{jk}^l \partial_k u).$$

Then we have $Au = A_l u$ if the support of $u \in W_p^{2m}(\mathbb{R}^n, \mathbb{C}^N)$ is contained in $\operatorname{supp}\varphi_l$. Moreover, we may apply the result of Step II to each operator A_l.

For $|s| \le 1$, we now define

$$\mathbb{X}_p^s := l_p(\Gamma, H_p^s(\mathbb{R}^n)).$$

We define $J : H_p^{-1}(\mathbb{R}^n) \to \mathbb{X}_p^{-1}$ by $Jf := (\varphi_l f)$ and we define $P : \mathbb{X}_p^{-1} \to L_p(\mathbb{R}^n)$ by $P(f_l) := \sum_l \varphi_l f_l$. Then J and P are contiuous by arguments similar to those in 6.6. As before, we define the operator $\mathbb{A}$ in $\mathbb{X}_p^{-1}$ by $\mathbb{A}(u_l) := (A_l u_l)$. By Step II the operator $\mathbb{A}$ has an H^∞-calculus in $\mathbb{X}_p^{-1}$. We now proceed as in 6.6 to relate the operators A in $H_p^{-1}(\mathbb{R}^n)$ and $\mathbb{A}$ in $\mathbb{X}_p^{-1}$.

For any $l \in \Gamma$ we have

$$\varphi_l A u = A\varphi_l u + [\varphi_l A - A\varphi_l]u = A_l \varphi_l u + C_l u. \tag{13.10}$$

We take a closer look on the commutator C_l and find

$$C_l u = \sum_{j,k}\left[(\partial_k \varphi_l)\partial_j(a_{jk}u) + (\partial_j\partial_k\varphi_l)a_{jk}u + (\partial_j\varphi_l)a_{jk}\partial_k u\right]$$

$$=: \left(\sum_{j,k} C_{jkl}^{(1)} + C_{jkl}^{(2)} + C_{jkl}^{(3)}\right)u.$$

For $\nu = 1,2,3$ we let $C_l^{(\nu)} := \sum_{j,k} C_{jkl}^{(\nu)}$. By $\partial_k : H_p^1(\mathbb{R}^n) \to L_p(\mathbb{R}^n)$, $a_{jk}\cdot : L_p(\mathbb{R}^n) \to L_p(\mathbb{R}^n)$ and $\partial_j : L_p(\mathbb{R}^n) \to H_p^{-1}(\mathbb{R}^n)$ we obtain

$$C_l^{(1)} : L_p(\mathbb{R}^n) \to H_p^{-1}(\mathbb{R}^n), \quad C_l^{(2)} : L_p(\mathbb{R}^n) \to L_p(\mathbb{R}^n),$$

$$C_l^{(3)} : H_p^1(\mathbb{R}^n) \to L_p(\mathbb{R}^n)$$

as bounded operators. But $a_{jk} \in C^\gamma(\mathbb{R}^n)$ implies that a_{jk} acts as a pointwise multiplier in $H_p^{-s}(\mathbb{R}^n)$ for $s \in (0,\gamma)$. Hence we have for $s \in (0,\gamma)$ that $C_l^{(3)}$ acts as a bounded operator $H_p^{1-s}(\mathbb{R}^n) \to H_p^{-s}(\mathbb{R}^n)$, and finally, that $C_l := C_l^{(1)} + C_l^{(2)} + C_l^{(3)}$ acts as a bounded operator $H_p^{1-s}(\mathbb{R}^n) \to H_p^{-1}(\mathbb{R}^n)$.

Coming back to (13.10) we have

$$\varphi_l A u = A_l \varphi_l u + \sum_{m \bowtie l} C_l \varphi_m \varphi_m u$$

which means that $JA = (\mathbb{A} + \mathbb{B})J$ where $\mathbb{B}(u_l) := (\sum_{m\bowtie l} C_l\varphi_m u_m)_l$ acts boundedly $\mathbb{X}_p^{1-s} \to \mathbb{X}_p^{-1}$. Since $D(\mathbb{A}) = \mathbb{X}_p^1$ and $1 + \mathbb{A}$ has an H^∞-calculus we have $\mathbb{X}_p^{1-s} = D(\mathbb{A}^{1-s/2})$, and we can apply Proposition 13.1 to obtain a bounded H^∞-calculus for a suitable translate of $\mathbb{A} + \mathbb{B}$.

In a similar way we obtain

$$A\sum_l \varphi_l u_l = \sum_l (\varphi_l A_l u_l) + \sum_m C_m u_m = P\mathbb{A}(u_l) + \sum_l \varphi_l\left(\sum_{m\bowtie l}\varphi_l C_m u_m\right)$$

which means $AP = P(\mathbb{A} + \mathbb{D})$ where $\mathbb{D}(u_l) := (\sum_{m\bowtie l}\varphi_l C_m u_m)_l$ acts boundedly $\mathbb{X}_p^{1-s} \to \mathbb{X}_p^{-1}$. We apply Proposition 13.1 and obtain a bounded H^∞-calculus for a suitable translate of $\mathbb{A} + \mathbb{D}$. As in Section 6 we now obtain

$$(\lambda - A)^{-1} = P(\lambda - (\mathbb{A} + \mathbb{B}))^{-1}J = P(\lambda - (\mathbb{A} + \mathbb{D}))^{-1}J.$$

for all $\lambda \in \rho(\mathbb{A}+\mathbb{B})\cap\rho(\mathbb{A}+\mathbb{D})$. By definition of the functional calculus we hence obtain a bounded H^∞-calculus in $X = H_p^{-1}(\mathbb{R}^n)$ for $\nu+A$ and a suitable $\nu \geq 0$.

Since $X_1 = D(A) = H_p^1(\mathbb{R}^n)$ with equivalent norms, and $\nu + 1 + A : X_1 \to X$ is an isomorphism, we also have a bounded H^∞-calculus for the part of $\nu + A$ in $H_p^1(\mathbb{R}^n)$. Finally, we obtain a bounded H^∞-calculus for the part of $\nu + A$ in $L_p(\mathbb{R}^n) = [H_p^{-1}(\mathbb{R}^n), H_p^1(\mathbb{R}^n)]_{1/2}$ by complex interpolation. $\qquad\square$

Finally, we illustrate our results by an application to a simple boundary value problem.

13.15 Illustration. Suppose that $\Omega \subset \mathbb{R}^{n+1}$ is a bounded open subset with boundary of class C^2 and let $1 < p < \infty$. First we consider the $L_p(\Omega)$-realization A_p of an (M, ω_0)-elliptic operator A with constant coefficents and Dirichlet boundary conditions in $L_p(\Omega)$. We have $D(A_p) = \{u \in W_p^2(\Omega) : u|_{\partial\Omega} = 0\}$ and we know from 7.18 that A_p is R-sectorial, $\omega_R(A_p) \leq \omega_0$. Notice that $0 \in \rho(A)$ here and that $(A_p)^* = A_{p'}$ for real duality by Green's formula. The Dirichlet Laplacian $-\Delta_{D,p}$ has a bounded H^∞-calculus in $L_p(\Omega)$ (since, e.g., $\Delta_{D,p}$ generates a positive contraction semigroup in all $L_q(\Omega)$, $1 \leq q \leq \infty$, cf. also 8.2), also $(-\Delta_{D,p})^* = -\Delta_{D,p'}$. Hence we may apply Theorem 13.2 with $u = v = 1$, and obtain a bounded H^∞-calculus for A_p and $\omega_{H^\infty}(A_p) = \omega_0$.

By complex interpolation we obtain (cf. [213, Ch.3]) that, for $X = L_p(\Omega)$ and $A = A_p$, $\dot{X}_{\theta,A} = X_{\theta,A} = H_p^{2\theta}(\Omega)$ if $0 \leq 2\theta < 1/p$. Since the corresponding result also holds for the dual operator, we obtain $\dot{X}_{-\theta,A} = X_{-\theta,A} = H_p^{-2\theta}(\Omega)$ if $0 \leq \theta < 1 - 1/p$. Since functions in C^γ act as bounded pointwise multipliers on $H_p^{2\theta}(\Omega)$ for $|\theta| < \gamma/2$ (cf. [213, 3.3.2]), we may argue as before to obtain an analog of Proposition 13.12:

Assume that A_p is the $L_p(\Omega)$-realization of a purely second order (M, ω_0)-elliptic operator with Dirichlet boundary conditions. Let the second order coeffcients of A belong to C^γ for some $\gamma \in (0,1)$ and assume that they are an L_∞-small perturbation of an operator with constant coefficients. Then A_p has a bounded $H^\infty(\Sigma_\nu)$-calculus for any $\nu > \omega_0$.

By employing the localization procedure, we may reduce the case of a general (M, ω_0)-elliptic second order operator whose second order coefficients are Hölder continuous to the case we went to study. Thus we obtain:

Assume that A_p is the $L_p(\Omega)$-realization of a second order (M, ω_0)-elliptic operator with Dirichlet boundary conditions. Let the second order coeffcients of A belong to C^γ for some $\gamma \in (0,1)$. Then A_p has a bounded $H^\infty(\Sigma_\nu)$-calculus for any $\nu > \omega_0$.

Notes on Section 13:

N 13.1 is from [10].

N 13.2 - N 13.4 are simplified versions of results from [131]. In particular, in [131] assumptions on the Banach space X are removed and the assumptions on the operators are less restrictive.

N 13.5: The result is from [131] where also more general perturbations in the scales $\dot{X}_\alpha$ and X_α are considered and assumptions on the Banach space are removed. It is known ([177]) that A–smallness of B alone is not sufficient to obtain a bounded H^∞-calculus for $A + B$. Thus some kind of an additional assumption is needed. There are other additional assumptions in the literature, e.g., in [195] (the perturbation result there is formulated and proved for BIP, but it may be shown to hold for the H^∞-calculus as well) and in [65], which was shown independently of our results (we refer to the note on Corollary 13.9).

N 13.6 - N 13.7: More details can be found in [131]. There, we also prove a Banach space analog of a Hilbert space result in [225], [22], that characterizes coincidence of the domains of certain fractional powers of operators A and $A + B$.

N 13.8 - N 13.10 are again from [131] where these results are proved under weaker assumptions. Corollary 13.9 overlaps with the perturbation result for the H^∞-calculus in [65].

N 13.11: One may show by a direct application of Mihlin's Theorem that A_p has a bounded H^∞-calculus of optimal angle in $L_p(\mathbb{R}^n)$, $1 < p < \infty$ (cf. also [10], [66]).

N 13.12 - N 13.13: The assertions were first proved in [10] where the regularity assumptions are a little less restrictive. The assertion of Theorem 13.13 was proved in [79] under the assumption $a_\alpha \in BUC$, $|\alpha| = 2m$, in place of (13.7). For $m = 1$, this condition is further relaxed in [85] where it is only assumed that $a_\alpha \in VMO$, $|\alpha| = 2$. The proof in [79] uses techniques related to the $T(1)$-Theorem whereas the proof in [85] makes extensive use of wavelet techniques.

N 13.14: We do not know a reference for the result in $H_p^{-1}(\mathbb{R}^n)$ and $H_p^1(\mathbb{R}^n)$. We want to mention that the main result in [178] studies resolvent estimates in $H_p^{-m}(\mathbb{R}^n)$ for elliptic divergence form operators of order $2m$ whose coefficients are in BUC. The result on boundedness of the H^∞-calculus in $L_p(\mathbb{R}^n)$ in Theorem 13.14, however, may also be proved by Theorem 14.4 (or even the results in [119], [52]) since the semigroup operators satisfy Gaussian kernel bounds (cf. [23]).

N 13.15 is just a very simple case of a boundary value problem. Boundedness of the H^∞-calculus for general boundary value problems with constant coefficients on $\mathbb{R}_+^{n+1}$ has been established in [74]. Boundedness of the H^∞-calculus for general boundary value problems in a infinite-dimensional state space has been established in [65] under the assumption that the highest order coefficients are Hölder continuous.

14 H^∞-calculus for Divergence Operators

In this section, we are coming back to the class of operators we studied in Section 8. We are going to relax the regularity assumptions that we imposed in Theorem 13.14. In order to deal with **operators in divergence form** with **bounded measurable coefficients** we present the following abstract theorem, which may be viewed as a version of Theorem 8.6 for boundedness of the H^∞-calculus.

14.1 Theorem. *Let (Ω, d, μ) be a space of homogeneous type, and let $-A$ be the generator of a bounded analytic semigroup in $L_2(\Omega)$. Assume that A has a bounded H^∞-calculus. Let $1 \leq p < 2 < q \leq \infty$. Assume that the following weighted norm bound holds*

$$\|1_{B(x,t^{1/m})} e^{-tA} 1_{B(y,t^{1/m})}\|_{p \to q} \leq |B(x, t^{1/m})|^{-(1/p-1/q)} g\big(\frac{d(x,y)}{t^{1/m}}\big),$$

for all $t > 0$, $x, y \in \Omega$, where $g : [0, \infty) \to [0, \infty)$ is a non-increasing function that decays faster than any polynomial. Then, for all $s \in (p, q)$, the semigroup (e^{-tA}) induces a consistent and bounded analytic semigroup $(T_s(t))$ in $L_s(\Omega)$. If $-A_s$ denotes the generator of $(T_s(t))$ in $L_s(\Omega)$ then A_s has a bounded H^∞-calculus in $L_s(\Omega)$ for $s \in (p, q)$ and $\omega_{H^\infty}(A_s) = \omega_{H^\infty}(A) = \omega(A)$.

We shall give most parts of the proof to illustrate the ideas behind this result. We shall have to compare quantities $N_{p,r} f$ and $M_p f$ for different radii or at different points. The following lemma serves as a preparation.

14.2 Lemma. *Let $p \in [1, \infty]$ and Ω be a space of dimension D :*

$$|B(x, \lambda r)| \leq C_0 \lambda^D |B(x, r)| \text{ for all } x \in \Omega, \lambda \geq 1, r > 0.$$

(a) We have for all $R \geq r > 0$, $x \in \Omega$, and $f \in L_p(\Omega)$:

$$N_{p,r} f(x) \leq C_0^{1/p} \big(\frac{R}{r}\big)^{D/p} N_{p,R} f(x).$$

(b) We have for all $r > 0$, $x \in \Omega$, $y \in B(x, r)$, and $f \in L_p(\Omega)$:

$$N_{p,r} f(y) \leq C_0^{1/p} \big(\frac{r+d(x,y)}{r-d(x,y)}\big)^{D/p} M_p f(x).$$

(c) We have for all $r > 0$, $x \in \Omega$, $y \in B(x, r)$, and $f \in L_p(\Omega)$:

$$N_{p,r} f(y) \leq (C_0 3^D)^{1/p} N_{p,2r} f(x).$$

Proof. (a) is obvious:

$$N_{p,r} f(x) \leq |B(x, r)|^{-1/p} \|1_{B(x,R)} f\|_p \leq C_0^{1/p} \big(\frac{R}{r}\big)^{D/p} N_{p,R} f(x).$$

(b) Using the evident inclusions

$$B(x, r - d(x, y)) \subset B(y, r) \subset B(x, r + d(x, y))$$

we can estimate as follows :

$$N_{p,r}f(y) \leq |B(x, r - d(x,y))|^{-1/p}\|1_{B(y,r)}f\|_p$$

$$\leq C_0^{1/p}\left(\frac{r+d(x,y)}{r-d(x,y)}\right)^{D/p}|B(x, r + d(x,y))|^{-1/p}\|1_{B(y,r)}f\|_p$$

$$\leq C_0^{1/p}\left(\frac{r+d(x,y)}{r-d(x,y)}\right)^{D/p}M_pf(x).$$

(c) Since $B(y,r) \subset B(x,2r) \subset B(y,3r)$ we have $|B(x,2r)| \leq C_0 3^D|B(y,r)|$ and thus

$$N_{p,r}f(y) \leq (C_0 3^D)^{1/p}|B(x,2r)|^{-1/p}\|1_{B(y,r)}f\|_p \leq (C_0 3^D)^{1/p}N_{p,2r}f(x) \quad \square$$

The main idea of the proof of Theorem 14.1 is now the following:

- Show a weak (p,p)-bound for the H^∞-calculus of A,
- Get L_s-boundedness of the H^∞-calculus for $s \in (p,2)$ by interpolation,
- Apply the arguments in the dual situation: A^* has a bounded H^∞-calculus in L_2 and satisfies the same assumptions. We obtain a weak (q',q')-bound for the H^∞-calculus of A^*. By interpolation the H^∞-calculus for A^* is L_s-bounded for all $s \in (q',2)$. By redualization this means that A has a bounded H^∞-calculus in L_s for $s \in (2,q)$.

Hence the proof of weak (p,p)-boundedness is the key step. It uses techniques for singular integral operators, although the operators under consideration need not be integral operators. In particular, we shall use a Calderon-Zygmund decomposition but for L_p-functions.

The following is a straight forward generalization of the usual Calderon-Zygmund decomposition of L_1-functions. One has to reinspect the proof and we give full details here for convenience. For the precise formulation we recall some notation. Following [205, Ch. I] we denote by $\widetilde{M}$ the uncentered maximal operator given by

$$\widetilde{M}g(x) := \sup\{|B|^{-1}\int_B |g|\, d\mu : B \text{ ball }, x \in B\}.$$

By [205, Ch. I] the operator $\widetilde{M}$ is weak $(1,1)$-bounded and bounded in $L_s(\Omega)$ for all $1 < s \leq \infty$ if (Ω, d, μ) satisfies the doubling property.

For any $f \in L_{p,\text{loc}}(\Omega)$ and $\alpha > 0$ we denote by $E_\alpha f$ the set

$$E_\alpha f := \{x \in \Omega : \widetilde{M}(|f|^p)(x) > \alpha^p\}.$$

The restriction on α in Theorem 14.3 is a technical assumption. We will see in the proof of Theorem 14.1 that, if this assumption does not hold, then we do not need to decompose the function f.

14.3 Theorem. *Let Ω be a space of homogeneous type and $p \in [1,\infty)$. There are constants $M, c > 0$ only depending on the space such that, whenever $f \in L_p(\Omega)$ and $\alpha > 0$ with $(E_\alpha f)^c \neq \emptyset$, there are a function g and sequences (b_j) of functions and (B_j^*) of balls satisfying*

(i) $f = g + \sum_j b_j$,

(ii) $\|g\|_\infty \le c\alpha$,

(iii) $\mathrm{supp}\,(b_j) \subset B_j^*$ and $\#\{j : x \in B_j^*\} \le M$ for all $x \in \Omega$,

(iv) $\|b_j\|_p \le c\alpha |B_j^*|^{1/p}$,

(v) $\left(\sum_j |B_j^*|\right)^{1/p} \le (c/\alpha)\|f\|_p$.

Proof. We follow [205, I 4.1] in arguments and notation. Since $E_\alpha f$ has non-empty complement we apply [205, I 3.2], and find sequences (B_j) of balls and (Q_j) of mutually disjoint "cubes" (which may recursively be obtained from the balls by $Q_k := B_k^* \cap (\bigcup_{j<k} Q_j)^c \cap (\bigcup_{j>k} B_j)^c$) such that

$$B_j \subset Q_j \subset B_j^* \subset E_\alpha, \quad \bigcup_j B_j^* = E_\alpha, \quad B_j^{**} \not\subset E_\alpha.$$

If $B_j = B(x_j, r_j)$ then $B_j^* = B(x_j, c^* r_j)$ and $B_j^{**} = B(x_j, c^{**} r_j)$ where c^*, c^{**} are the constants from [205, I 3.2]. Now define g and the sequence (b_j) by

$$g(x) := \begin{cases} f(x) & , \quad x \notin E_\alpha \\ (|Q_j|^{-1} \int_{Q_j} |f(y)|^p\, dy)^{1/p} , & x \in Q_j \end{cases}$$

$$b_j(x) := 1_{Q_j}(x)\left[f(x) - \left(|Q_j|^{-1} \int_{Q_j} |f(y)|^p\, dy\right)^{1/p}\right]. \tag{14.1}$$

Then (i) holds. For each j we have $B_j^{**} \cap E_\alpha \ne \emptyset$, hence, by the definition of $\tilde{M}$,

$$|B_j^{**}|^{-1} \int_{B_j^{**}} |f(y)|^p\, dy \le \alpha^p$$

which leads to

$$\int_{Q_j} |f(y)|^p\, dy \le \int_{B_j^{**}} |f(y)|^p\, dy \le \alpha^p |B_j^{**}| \le c_1 \alpha^p |B_j| \le c_1 \alpha^p |Q_j|$$

by the doubling property. Hence $|g(x)| \le c_2^{1/p}\alpha$ for $x \in Q_j$ and clearly $|g(x)| \le \alpha$ for $x \notin E_\alpha$. Since [205, I 3.2] gives also the bounded intersection property for the B_j^* we have proved (ii) and (iii). To prove (iv) notice that $(a+b)^p \le 2^{p-1}(a^p + b^p)$ for $a, b \ge 0$, hence

$$\|b_j\|_p^p \le 2^p \int_{Q_j} |f(x)|^p\, dx \le 2^p \alpha^p c_2 |B_j^*|.$$

Now (v) follows by

$$\sum_j |B_j^*| \le c_3 \sum_j |B_j| \le c_3 \sum_j |Q_j| = c_3 |E_\alpha| \le c_4 \alpha^{-p} \|\,|f|^p\,\|_1 = \frac{c_4}{\alpha^p}\|f\|_p^p$$

where we used weak $(1,1)$-boundedness of $\tilde{M}$. □

Remark. Recall from the proof that the b_j have disjoint supports, hence

$$\|\sum_j b_j\|_p^p = \sum_j \|b_j\|_p^p \le c^p \alpha^p \sum_j |B_j^*| \le c^{2p}\|f\|_p^p$$

by (iv) and (v), which by (i) implies the following additional property:

$$(vi) \qquad \|g\|_p \le (c^2+1)\|f\|_p.$$

Outline of the proof of Theorem 14.1. We take functions $\psi \in H_0^\infty(\Sigma_\nu)$ for $\nu > \omega(A)$ such that $\|\psi\|_{H^\infty(\Sigma_\nu)} \le 1$, an let $S := \psi(A)$. Then $S \in B(L_2(\Omega))$ and $\|S\|_{2\to 2} \le C_\nu$ by assumption. We aim at showing a weak (p,p)-estimate with constant only depending on C_ν.

The following is the core of the argument, by which we mean the part of the proof which does not use any specific features of the H^∞-calculus. Everything boils down to a remaining maximal estimate (14.5). This estimate shall not be proved here. We refer to [34].

The core of the argument: We assume that S is a bounded linear operator in $L_2(\Omega)$. We assume that (T_t) is a family of linear operators satisfying the weighted norm bounds in Theorem 14.1 in place of (e^{-tA}). We aim at showing a weak (p,p)-bound for S.

We find constants $M, c > 0$ only depending on the space according to Theorem 14.3, and let $f \in L_p(\Omega)$, $\alpha > 0$. We recall the notation $E_\alpha f$ from above. In the following we use the symbol $\preceq$ to indicate domination up to constants only depending on the space, $\|S\|_{2\to 2}$, and the family (T_t), but independent of f and α.

If $E_\alpha f = \Omega$ then

$$|\{x \in \Omega : |Sf(x)| > \alpha\}| \le |E_\alpha f| \preceq \alpha^{-p}\|f\|_p^p$$

by weak $(1,1)$-boundedness of the uncentered maximal operator $\widetilde{M}$.

If $E_\alpha f \ne \Omega$ we consider the Calderon-Zygmund decomposition $f = g + \sum b_j$ at height α according to Theorem 14.3. We write $B_j = B(x_j, r_j)$ instead of B_j^* and let $t_j := (2r_j)^m$. For any $n \in \mathbb{N}$ we let

$$U_t^{(n)} := -\sum_{k=1}^n \binom{n}{k}(-1)^k T_{kt}, \quad t > 0.$$

We shall use the operators $U_t^{(n)}$ as regularizations. In the simplest case $n = 1$ we have $U_t^{(1)} = T_t$ which corresponds to the situation in [78], [82]. Observe that, letting $T_0 := I$, we have

$$I - U_t^{(n)} = \sum_{k=0}^n \binom{n}{k}(-1)^k T_{kt} =: D^n T(t).$$

Here D^n denotes a difference operator applied to the function $t \mapsto T_t$. Observe that, if f is a sufficiently smooth function, then $D^n f(t)/t^n \to (-1)^n f^{(n)}(0)$ as $t \to 0$. Hence we view $(U_t^{(n)})_{t>0}$ as higher order approximations of the identity.

We decompose $\sum_j b_j = h_1 + h_2$, where

$$h_1 = \sum_j U_{t_j}^{(n)} b_j \quad \text{and} \quad h_2 = \sum_j (I - U_{t_j}^{(n)}) b_j.$$

Then

$$|\{x \in \Omega : |Sf(x)| > \alpha\}|$$

$$\leq |\{x \in \Omega : |Sg(x)| > \alpha/3\}| + \sum_{k=1}^{2} |\{x \in \Omega : |Sh_k(x)| > \alpha/3\}|$$

and we shall estimate the three terms separately where we write α instead of $\alpha/3$. We start with the first term. Observe that $g \in L_p \cap L_\infty \subset L_2$. Hence the assumption $S \in B(L_2(\Omega))$ as well as the properties (ii) and (vi) of Theorem 14.3 imply

$$|\{x \in \Omega : |Sg(x)| > \alpha\}| \leq \alpha^{-2} \|Sg\|_2^2 \preceq \alpha^{-2} \|g\|_2^2 \leq \|g\|_\infty^{2-p} \|g\|_p^p \preceq \alpha^{-p} \|f\|_p^p.$$

We turn to the second term, i.e. the term involving h_1. We have

$$|\{x \in \Omega : |Sh_1(x)| > \alpha\}| \leq \alpha^{-2} \|Sh_1\|_2^2 \preceq \alpha^{-2} \|\sum_j U_{t_j}^{(n)} b_j\|_2^2,$$

where we used $S \in B(L_2(\Omega))$ again. We shall show

$$\|\sum_j U_{t_j}^{(n)} b_j\|_2 \preceq \alpha \|\sum_j 1_{B_j}\|_2. \tag{14.2}$$

By property (iii) of Theorem 14.3 this implies

$$\|\sum_j U_{t_j}^{(n)} b_j\|_2 \preceq \alpha (\sum_j |B_j|)^{1/2} \preceq \alpha^{1-p/2} \|f\|_p^{p/2}$$

which leads to the desired bound for the second term :

$$|\{x \in \Omega : |Sh_1(x)| > \alpha\}| \preceq \alpha^{-p} \|f\|_p^p.$$

For the proof of (14.2) we take $\phi \in L_2$. For any j we have by property (iv)

$$|\langle \phi, U_{t_j}^{(n)} b_j \rangle| = |\langle (U_{t_j}^{(n)})^* \phi, b_j \rangle| \leq \|1_{B_j} (U_{t_j}^{(n)})^* \phi\|_{p'} \|b_j\|_p$$

$$\preceq \alpha |B_j| N_{p',r_j} \left((U_{t_j}^{(n)})^* \phi \right)(x_j)$$

$$\preceq \alpha \int_{B_j} N_{p',t_j^{1/m}} \left((U_{t_j}^{(n)})^* \phi \right).$$

Here we used Lemma 14.2 (c) in the last step, recall $2r_j = t_j^{1/m}$. For any $k \in \{1, \ldots, n\}$, $x \in \Omega$, $t > 0$, we have by Lemma 14.2 (a) and Proposition 8.9

$$N_{p', t^{1/m}}((T_{kt})^* \phi)(x) \leq C_0 C_1 k^{D/(mp')} N_{p', (kt)^{1/m}}((T_{kt})^* \phi)(x)$$
$$\leq C_0 C_1 n^{D/(mp')} (M_{q'} \phi)(x)$$

Hence we obtain by definition of the $U_t^{(n)}$ a constant $C > 0$ such that

$$N_{p', t^{1/m}}\left((U_t^{(n)})^* \phi\right)(x) \leq C(M_{q'} \phi)(x), \quad t > 0.$$

Since $M_{q'}$ is L_2-bounded (here we use the hypothesis $2 < q$!) we thus end up with

$$|\langle \phi, \sum_j U_{t_j}^{(n)} b_j \rangle| \preceq \alpha \int (M_{q'} \phi) \sum_j 1_{B_j} \leq \alpha \|M_{q'} \phi\|_2 \|\sum_j 1_{B_j}\|_2$$
$$\preceq \alpha \|\phi\|_2 \|\sum_j 1_{B_j}\|_2.$$

Since $\phi \in L_2$ was arbitrary, (14.2) is proved.

We turn to the third term, i.e. the term which involves h_2. Denoting $B_j' := B(x_j, 20r_j)$ we have

$$|\{x \in \Omega : |Sh_2(x)| > \alpha\}| \leq \sum_j |B_j'| + |\{x \in \Omega \setminus \bigcup_j B_j' : |Sh_2(x)| > \alpha\}|. \quad (14.3)$$

For the first term on the right hand side of (14.3), (v) together with the volume doubling property yields

$$\sum_j |B_j'| \preceq \sum_j |B_j| \preceq \alpha^{-p} \|f\|_p^p.$$

The second term on the right hand side of (14.3) is estimated as follows by using (iii):

$$|\{x \in \Omega \setminus \bigcup_j B_j' : |Sh_2(x)| > \alpha\}| \leq \alpha^{-p} \|\sum_j S(I - U_{t_j}^{(n)}) b_j\|_{p, \Omega \setminus \bigcup B_j'}^p$$
$$= \alpha^{-p} \|\sum_j 1_{(B_j')^c} S(I - U_{t_j}^{(n)}) 1_{B_j} b_j\|_{p, \Omega \setminus \bigcup B_j'}^p$$
$$\leq \alpha^{-p} \|\sum_j G_j b_j\|_p^p$$

where $G_j := 1_{(B_j')^c} S(I - U_{t_j}^{(n)}) 1_{B_j}$. If we have established

$$\|\sum_j G_j b_j\|_p \preceq \alpha \|\sum_j 1_{B_j}\|_p \quad (14.4)$$

then we can argue as in the estimation of the second term above (observe that 2 is now replaced by p) to obtain the desired bound

$$|\{x \in \Omega \setminus \bigcup_j B'_j : |Sh_2(x)| > \alpha\}| \preceq \alpha^{-p}\|f\|_p^p.$$

But the proof of (14.4) can follow the lines of the proof of (14.2) (with 2 replaced by p) once we have shown the maximal estimate

$$N_{p',t^{1/m}}((G_j)^*\phi)(x) \le C(M_{q'}\phi)(x). \tag{14.5}$$

Knowing (14.5) we argue as in the proof of (14.2): For $\phi \in L_{p'}$ we have

$$|\langle\phi, G_j b_j\rangle| = |\langle G_j^*\phi, b_j\rangle| \le \|1_{B_j}G_j^*\phi\|_{p'}\|b_j\|_p \preceq \alpha|B_j|N_{p',r_j}(G_j^*\phi)(x_j)$$

$$\preceq \alpha \int_{B_j} N_{p',2r_j}(G_j^*\phi) \preceq \alpha \int_{B_j} M_{q'}\phi$$

where we used (14.5) in the last step. Now we may finish as before since, by $p < q$, the maximal function $M_{q'}$ is bounded in $L_{p'}$. We obtain

$$|\langle\phi, \sum_j G_j b_j\rangle| \preceq \alpha \int (M_{q'}\phi)\sum_j 1_{B_j} \le \alpha\|M_{q'}\phi\|_{p'}\|\sum_j 1_{B_j}\|_p$$

$$\preceq \alpha\|\phi\|_{p'}\|\sum_j 1_{B_j}\|_p.$$

which implies (14.4).

We conclude that S is of weak type (p,p) where the constant only depends on the space, the family (T_t), the integer n, and the norm $\|S\|_{2\to2}$ or – recalling $S = \psi(A)$ – on the constant C_ν of the $H^\infty(\Sigma_\nu)$-calculus of A in $L_2(\Omega)$.

Up to now, the special situation of Theorem 14.1 did not enter.

It is hence not surprising that the specific features of a bounded H^∞-calculus have to be used in the proof of (14.5), which takes the form

$$N_{p',t^{1/m}}\big(1_{B(z,t^{1/m}/2)}(\psi(A)(I - U_t^{(n)}))^*1_{B(z,10t^{1/m})^c}\phi\big)(x) \le C(M_{q'}\phi)(x) \tag{14.6}$$

where $x \in B(z,t^{1/m}/2)$, and $C > 0$ may only depend on C_ν but not on ψ. Note that we use the notation z here instead of the x_j and recall $2r_j = t_j^{1/m}$. For the proof of (14.6) we refer to [34] and [143]. □

Applying Theorem 14.1 to the class of operators we studied in Section 8 we obtain the following improvement of Theorem 8.24, which is also a partial improvement of Theorem 13.14.

14.4 Theorem. *For an operator A of the form (8.5) and p as in 8.21 we have for all $s \in (p,p')$ that A_s has a bounded H^∞-calculus and $\omega_{H^\infty}(A_s) = \omega(A_2)$.*

Proof. We only have to recall Propositions 8.21 and 8.22 and Remark 8.23 from Section 8 and the fact $\omega_{H^\infty}(A_2) = \omega(A_2)$ since L_2 is a Hilbert space and A_2 is associated to a closed sectorial form (cf. Section 11). □

Notes on Section 14:

N 14.1: The main idea of the proof of Theorem 14.1 is as in [78]. A well-known condition for singular operators to be weak $(1,1)$-bounded is the famous Hörmander condition. L_p-boundedness is then obtained by interpolation. The Hörmander condition was used in [81] to prove L_p-boundedness of the H^∞-calculus for elliptic operators whose highest order coefficients are Hölder continuous. The result in [78] uses kernel bounds of Poisson type and allows to deal with measurable coefficients provided these kernel bounds hold for the semigroup. The ideas used in [78] go back to [115]. They were put into a more general context in [82] where a *weakened Hörmander condition* is introduced. The weak type $(1,1)$ criterion from [82] was also applied to L_p-boundedness of Riesz transforms.

In spite of the fact that we are not necessarily dealing with integral operators a modification of the techniques allows to use similar arguments. The weak type $(1,1)$ criterion from [82] is generalized to a weak type (p,p)-criterion in [34]. The proof of Theorem 14.1 given here makes it transparent which condition has to be assumed in order to obtain a weak type (p,p)-boundedness of the operator S. It is the estimate (14.5) which serves as a replacement of the weakened Hörmander condition from [82]. In fact, in [34] the same arguments were used to prove a weak type (p,p) criterion for non-integral operators by adding the remaining maximal estimate (14.5) on the operators G_j as an assumption. We refer to the discussion in [34].

The result from [34] has been applied to Riesz transforms in [35] thus generalizing previous results in [51]. Independently, a similar result for second order operators has been obtained in [122]. We also refer to very recent work by P. Auscher ([18]).

In the proof of (14.5) above one has to extrapolate weighted norm estimates from real to complex times. This has been done in different settings, e.g., in [61], [78], [170]. For the setting we have here we refer to [143] and [36]. Notice, however, that, for the operators we deal with in Theorem 14.4, we already derived weighted norm estimates on sectors in Section 8. Hence we could avoid the extrapolation argument in this case. In any case, the first step in a proof of (14.6) in the given situation would be a representation of the operator $\psi(A)(I - U_t^{(n)})$ by means of the H^∞-calculus.

N 14.2 - N 14.3 are from [34]), cf. also [143].

N 14.4 is contained in [143] (cf. also [34]). Theorem 14.1 can be applied to other classes of elliptic operators including Schrödinger operators with singular potentials and second order operators with singular lower order terms, see [34]. Weighted norm bounds for the latter class of operators are proved, e.g., in [170].

15 Appendix: Fractional Powers of Sectorial Operators

In this appendix we give a concise introduction to fractional powers of sectorial operators and provide in several subsections the background we need including algebraic properties, integral representations and extrapolation scales. We shall define fractional powers via an extension of the H^∞-calculus. In particular, we make an attempt to avoid the assumption $0 \in \rho(A)$, which is made in many presentations of this material.

A. Restrictions and Duality

But first we study sectorial operators and their duals more closely. In Chapter II we required in addition to sectoriality that the operator is injective and has dense range. Recall that the notation $\mathcal{S}(X)$ stands for the space of injective sectorial operators in a Banach space X that have dense range. Here we start with a result on a decomposition of the underlying Banach space that, given a sectorial operator, allows to resort to a subspace where these additional properties hold. The situation here is even more general since we only assume (15.1) for the operator A and do not require that A is densely defined (which is part of the definition of *sectorial*). We recall that, for a linear operator A in a Banach space X and a subspace Y of X, the *part A_Y of A in Y* is the restriction of A to

$$D(A_Y) := \{y \in Y \cap D(A) : Ay \in Y\}.$$

We also recall the following lemma which we prove for convenience.

15.1 Lemma. *Let X be a Banach space and $Y \subset X$ be a Banach space for a stronger norm. Let A be a linear operator in X such that Y is invariant under resolvents of A. Then $\rho(A) \subset \rho(A_Y)$ and $R(\lambda, A_Y) = R(\lambda, A)|_Y$ for $\lambda \in \rho(A)$.*

Proof. If $\lambda \in \rho(A)$ then $\lambda - A_Y$ is injective as a restriction of the injective operator $\lambda - A$. Let $z \in Y$ and define $y := R(\lambda, A)z$. Then $y \in D(A)$, and $y \in Y$ by assumption. Moreover, $(\lambda - A)y = (\lambda - A)R(\lambda, A)y = z$, and $Ay = \lambda y - z \in Y$. Hence $y \in D(A_Y)$ and $(\lambda - A_Y)y = z$. This proves surjectivity of $\lambda - A_Y$ and $R(\lambda, A_Y) = R(\lambda, A)|_Y$. $\qquad\square$

15.2 Proposition. *Let A be a linear operator in X and $\omega \in (0, \pi)$ such that $\sigma(A) \subset \overline{\Sigma_\omega}$ and*

$$\sup\{\|\lambda R(\lambda, A)\| : |\arg \lambda| \in [\sigma, \pi]\} < \infty, \quad \sigma > \omega. \tag{15.1}$$

Then

a)
$$N(A) \oplus \overline{R(A)} = \left\{ x \in X : (\frac{1}{n}R(-\frac{1}{n}, A)x) \; converges \right\} =: X_0,$$

$$\overline{D(A)} = \left\{ x \in X : \lim_{n \to \infty} -nR(-n, A)x = x \right\} =: X_\infty,$$

$$\overline{R(A)} = \left\{ x \in X : \lim_{n \to \infty} \frac{1}{n}R(-\frac{1}{n}, A)x = 0 \right\} =: X_{00}.$$

b) *All spaces X_0, X_∞, and X_{00} are closed subspaces of X, the part A_∞ of A in X_∞ is sectorial of type ω, and the part $\tilde{A}$ of A in $\tilde{X} := X_\infty \cap X_{00}$ is sectorial of type ω, injective, and has dense range, i.e., $\tilde{A} \in \mathcal{S}(\tilde{X})$*

c) *If X is reflexive then $X = X_0 = X_\infty$.*

Proof. We first remark that (15.1) implies by standard arguments that all spaces X_0, X_∞, X_{00} are closed subspaces of X.

The inclusion $X_\infty \subset \overline{D(A)}$ is clear. For $x \in D(A)$ we have $-nR(-n, A)x - x = R(-n, A)Ax$, hence $x \in X_\infty$ by (15.1), and $\overline{D(A)} \subset X_\infty$ by closedness of X_∞.

If $x = Ay \in R(A)$ then, by (15.1), $-\frac{1}{n}R(-\frac{1}{n}, A)x = -\frac{1}{n} \cdot AR(-\frac{1}{n}, A)y \to 0$ as $n \to \infty$, hence $R(A) \subset X_{00}$, and $\overline{R(A)} \subset X_{00}$ by closedness of X_{00}. On the other hand, for $x \in X_{00}$ we have $x = \lambda R(\lambda, A)x - AR(\lambda, A)x$, which implies $x = -\lim_{n \to \infty} AR(-\frac{1}{n}, A)x \in \overline{R(A)}$.

We turn to the decomposition of X_0, and define $P : X_0 \to X$ by $Px := \lim_{n \to \infty} -\frac{1}{n}R(-\frac{1}{n}, A)x$. P is linear and bounded by (15.1). We claim that $R(P) = N(A) = N(I - P)$. Since $N(I - P) \subset R(P)$ is clear, we only have to show $N(A) \subset N(I - P)$ and $R(P) \subset N(A)$.

If $Ax = 0$ then $\lambda R(\lambda, A)x = x$ and $Px = x$, i.e., $x \in N(I - P)$. If $y = Px = \lim_{n \to \infty} -\frac{1}{n}R(-\frac{1}{n}, A)x$ then, by (15.1), $\lim_{n \to \infty} -\frac{1}{n} \cdot AR(-\frac{1}{n}, A)x = 0$, and $Ay = 0$ by closedness of A. Hence $R(P) = N(A) = N(I - P)$, and we have shown that $P : X_0 \to X_0$ is a projection and that $X_0 = N(A) \oplus N(P) = N(A) \oplus X_{00}$ as asserted.

Since X_∞ and $\tilde{X}$ are invariant under resolvents of A, it is clear that (15.1) holds if A is replaced by A_∞, by $\tilde{A}$, or by the part A_{00} of A in X_{00}. Defining in an obvious way spaces $(X_{00})_{00}$ and $(X_\infty)_\infty$ we obtain $(X_{00})_{00} = X_{00}$ and $(X_\infty)_\infty = X_\infty$, which means that A_{00} has dense range and A_∞ is densely defined. Moreover, we also obtain $(X_{00})_0 = X_{00} = \overline{R(A_{00})}$, which yields injectivity of A_{00}.

As a restriction of the injective operator A_{00}, the operator $\tilde{A}$ is injective. By $\tilde{X} \subset X_\infty$ we obtain $(\tilde{X})_\infty = \tilde{X}$, and $\tilde{A}$ is densely defined. By $\tilde{X} \subset X_{00}$ we obtain $(\tilde{X})_{00} = \tilde{X}$, and $\tilde{A}$ has dense range.

Now we assume that X is reflexive. We show $X = X_0$ and let $x \in X$. Then the bounded sequence $(-\frac{1}{n}R(-\frac{1}{n}, A)x)$ has a weakly convergent subsequence $(-\frac{1}{n_k}R(-\frac{1}{n_k}, A)x)$ with weak limit $y \in X$. Hence

$$-AR(-\frac{1}{n_k}, A)x = x + \frac{1}{n_k}R(-\frac{1}{n_k}, A)x \to x - y \quad \text{weakly},$$

and Mazur's theorem implies $x - y \in \overline{R(A)}$. For any $m \in \mathbb{N}$ we have

$$-AR(-\frac{1}{m}, A)(-\frac{1}{n_k}R(-\frac{1}{n_k}, A)x) = [I + \frac{1}{m}R(-\frac{1}{m}, A)](-\frac{1}{n_k}R(-\frac{1}{n_k}, A)x)$$
$$\rightarrow y + \frac{1}{m}R(-\frac{1}{m}, A)y \quad \text{weakly.}$$

Rewriting the left hand side we obtain

$$-AR(-\frac{1}{m}, A)(-\frac{1}{n_k}R(-\frac{1}{n_k}, A)x) = (-\frac{1}{n_k}R(-\frac{1}{n_k}, A))(-AR(-\frac{1}{m}, A)x$$
$$\rightarrow 0 \quad \text{in } \|\cdot\|$$

by $-AR(-\frac{1}{m}, A)x \in R(A) \subset X_{00}$. Hence $y = -\frac{1}{m}R(-\frac{1}{m}, A)y$ for all $m \in \mathbb{N}$, $Py = y$, and $y \in N(I-P) = N(A)$. Finally $x = y+x-y \in N(A)+\overline{R(A)} = X_0$.

The proof of $X = X_\infty$ is similar. For $x \in X$ we find a weakly convergent subsequence $(-n_k R(-n_k, A)x)$ of the bounded sequence $(-nR(-n, A)x)$ and a weak limit $z \in X$. Then $z \in \overline{D(A)}$ by Mazur's theorem, and $R(-1, A)(-n_k R(-n_k, A)x) \rightarrow R(-1, A)z$ weakly. On the other hand,

$$R(-1, A)(-n_k R(-n_k, A)x) = -n_k R(-n_k, A)R(-1, A)x$$
$$\rightarrow R(-1, A)x \quad \text{in } \|\cdot\|,$$

since $R(-1, A)x \in D(A) \subset X_\infty$. We conclude $R(-1, A)x = R(-1, A)z$ and finally $x = z \in \overline{D(A)} = X_\infty$. □

Remark. In particular, if X is reflexive then $X = \overline{R(A)} \oplus N(A)$, and A has dense range in X if and only if N is injective. Moreover, A takes the form $\begin{pmatrix} A_{00} & 0 \\ 0 & 0 \end{pmatrix}$ with respect to the decomposition $X = X_{00} \oplus N(A)$. The standard example for a sectorial operator A with $N(A) \neq \{0\}$ is the Neumann-Laplacian $-\Delta_{N,p}$ in $L_p(\Omega)$, $1 < p < \infty$, where $\Omega \subset \mathbb{R}^n$ is a bounded smooth domain. Here $N(-\Delta_{N,p}) = \text{span}\{1_\Omega\}$. Since $(-\Delta_{N,p})^* = -\Delta_{N,p'}$, we have for $f \in L_p(\Omega)$,

$$f \in \overline{R(-\Delta_{N,p})} \Leftrightarrow f \in N(-\Delta_{N,p'})^\perp \Leftrightarrow \int_\Omega f \, dx = 0.$$

Hence one considers the part of $-\Delta_{N,p}$ in

$$L_{p,0}(\Omega) := \{f \in L_p(\Omega) : \int_\Omega f \, dx = 0\}$$

This is done, e.g., in [204].

Next we study the dual situation. Let $A \in \mathcal{S}(X)$ be a sectorial operator of type ω. For the dual operator A^* we have $\rho(A^*) = \rho(A)$, and for all $z \in \rho(A)$ we have $R(z, A^*) = R(z, A)^*$. Hence (15.1) is satisfied for A^* in place of A.

But if X is non–reflexive then A^* need not be densely defined. If $\omega < \pi/2$ in this case, then $-A^*$ cannot be the generator of a bounded analytic semigroup.

Usually this problem is circumvented by considering the part $A^\odot$ of A^* in the sun-dual $X^\odot$ of X, which is defined to be the closure of $D(A^*)$ in the norm of X^*, i.e., $X^\odot = (X^*)_\infty$ in the notation of Proposition 15.2. Thus, if X is reflexive then $X^\odot = X^*$. In the general case, $X^\odot$ is still norming for X, i.e.,

$$\|x\|_\odot := \sup\{\,|\langle x, x^\odot\rangle| : x^\odot \in X^\odot, \|x^\odot\| \le 1\,\}$$

defines a norm on X which is equivalent to the original norm ([182, Thm. 1.3.5]).

The operator $A^\odot$ is densely defined in $X^\odot$ but need not have dense range (consider, e.g., $A(x_n) = (x_n/n)$ in l^1). Proposition 15.2 tells us that we have to resort to an even smaller subspace of X^*, namely $\widetilde{(X^*)}$. As we show below, this space is still norming for X.

15.3 Definition. *Let X be a Banach space and $A \in \mathcal{S}(X)$. The moon-dual $X^\sharp$ of X with respect to A is defined as $\overline{D(A^*)} \cap \overline{R(A^*)}$ where the closure is taken in the norm of X^*. The moon-dual operator $A^\sharp$ of A is the part of A^* in $X^\sharp$, i.e.,*

$$A^\sharp x^* = A^* x^* \text{ for } x^* \in D(A^\sharp) = \{\,x^* \in D(A^*) \cap \overline{R(A^*)} : A^* x^* \in \overline{D(A^*)}\,\}.$$

Remark. By Proposition 15.2 we know that $A^\sharp \in \mathcal{S}(X^\sharp)$ is sectorial. As we know from Section 9, this implies that $D(A^\sharp) \cap R(A^\sharp)$ is dense in $X^\sharp$. Hence we also have $X^\sharp = \overline{D(A^*)} \cap \overline{R(A^*)}$.

15.4 Proposition. *The embedding $\iota : X \to (X^\sharp)^*$, given by $(\iota x)(x^\sharp) := x^\sharp(x)$, is injective and maps X onto a closed subspace of $(X^\sharp)^*$.*

The assertion of this proposition means in other words that

$$\|x\|_\sharp := \sup\{\,|\langle x, x^\sharp\rangle| : x^\sharp \in X^\sharp, \|x^\sharp\| \le 1\,\}$$

defines a norm on X which is equivalent to the original norm, or that $X^\sharp$ is still norming for X.

Proof. We clearly have $\|x\|_\sharp \le \|x\|$ for all $x \in X$. For the lower bound we take an arbitrary $x^* \in X^*$ with $\|x^*\| \le 1$ and let

$$x_n^\sharp := \frac{1}{n} R(-\frac{1}{n}, A^*) x^* - n R(-n, A^*) x^*, \quad n \in \mathbb{N}.$$

Then $x_n^\sharp \in D(A^*)$, and by

$$x_n^\sharp = (\frac{1}{n} + A^* - A^*) R(-\frac{1}{n}, A^*) x^* + (-n - A^* + A^*) R(-n, A^*) x^*$$

$$= -x^* - A^* R(-\frac{1}{n}, A^*) x^* + x^* + A^* R(-n, A^*) x^*$$

$$= -A^* R(-\frac{1}{n}, A^*) x^* + A^* R(-n, A^*) x^*$$

we also have $x_n^\sharp \in R(A^*)$, hence $x_n^\sharp \in X^\sharp$. By 15.2 we have

$$\frac{1}{n} R(-\frac{1}{n}, A)x \to 0 \ (n \to \infty) \quad \text{for all } x \in \overline{R(A)} = X,$$

and

$$-nR(-n, A)x \to x \ (n \to \infty) \quad \text{for all } x \in \overline{D(A)} = X.$$

Thus we have for any $x \in X$

$$\langle x, x_n^\sharp \rangle = \langle \frac{1}{n} R(-\frac{1}{n}, A)x - nR(-n, A)x, x^* \rangle \to \langle x, x^* \rangle \ (n \to \infty).$$

By sectoriality of A we have $\|tR(-t, A)\| \le C$ for all $t > 0$. Consequently $\|x_n^\sharp\| \le 2C$, and we obtain

$$\|x\|_\sharp \ge \lim_{n \to \infty} |\langle x, x_n^\sharp/2C \rangle| = \frac{|\langle x, x^* \rangle|}{2C} \quad \text{for all } x \in X.$$

Taking the sup over all x^* with $\|x^*\| \le 1$ this yields $2C\|x\|_\sharp \ge \|x\|$.   □

The following proposition gathers properties of $A^\sharp$.

15.5 Proposition. *Let $A \in \mathcal{S}(X)$ be sectorial of type ω. Then:*

(a) *We have $\rho(A) = \rho(A^*) = \rho(A^\sharp)$ and for $z \in \rho(A^\sharp)$ the operator $R(z, A^\sharp)$ is the restriction of $R(z, A^*) = R(z, A)^*$ to $X^\sharp$.*

(b) *We have $A^\sharp \in \mathcal{S}(X^\sharp)$ and $A^\sharp$ is sectorial of type ω.*

(c) *If $\omega < \pi/2$ then the bounded analytic semigroup $(T_t^\sharp)$, which is generated by $-A^\sharp$, is the restriction of (T_t^*) to $X^\sharp$, where (T_t) denotes the bounded analytic semigroup generated by $-A$ in X.*

(d) *If A has a bounded $H^\infty(\Sigma_\theta)$–calculus, the same holds for $A^\sharp$.*

Proof. (a) It is well known that $\rho(A) = \rho(A^*)$ and $R(z, A^*) = R(z, A)^*$. Since $X^\sharp$ is invariant under resolvents of A^* we have $\rho(A) = \rho(A^*) \subset \rho(A^\sharp)$ and $R(\lambda, A^\sharp) = R(\lambda, A^*)|_{X^\sharp}$.

It rests to show that $\rho(A^\sharp) \subset \rho(A)$. So let $z \in \rho(A^\sharp)$. By Proposition 15.4 there exists an $\alpha > 0$ such that, for any $x \in X$, there is an $x^\sharp \in X^\sharp$ with $\|x^\sharp\| \le 1$ and $|\langle x, x^\sharp \rangle| \ge \alpha\|x\|$. For $x \in D(A)$ and the corresponding $x^\sharp$ we have, writing $\beta := \|R(z, A^\sharp)\|^{-1}$,

$$\|(z - A)x\| \ge \beta|\langle (z - A)x, R(z, A^\sharp)x^\sharp \rangle|$$
$$= \beta|\langle x, (z - A^\sharp)R(z, A^\sharp)x^\sharp \rangle| = \beta|\langle x, x^\sharp \rangle| \ge \alpha\beta\|x\|.$$

This proves injectivity of $z - A$ and closedness of $R(z - A)$.

It rests to show that $z - A$ has dense range. This is clear for $z = 0$, so let $z \ne 0$. If the range of $z - A$ is not dense there is an $x^* \in X^*$ with $x^* \ne 0$ and $\langle (z - A)x, x^* \rangle = 0$ for all $x \in D(A)$. Then $x^* \in D(A^*)$ and $(z - A)^*x^* = 0$, hence $A^*x^* = zx^* \in D(A^*)$ and $x^* = A^*x^*/z \in R(A^*)$. Thus we have $x^* \in D(A^\sharp)$ and $(z - A^\sharp)x^* = 0$. By $z \in \rho(A^\sharp)$ we obtain $x^* = 0$ which contradicts the choice of x^*.

(b) This follows from Proposition 15.2.

(c) By (b) the operator $-A^\sharp$ generates an analytic semigroup $(T_t^\sharp)$. Since $R(z, A^\sharp)$ is the restriction of $R(z, A)^*$ to $X^\sharp$ we obtain by the assertion by representing semigroups in terms of resolvents as in the proof of Theorem 2.20.

(d) By definition of the functinal calculus we have $\psi(A^\sharp) = \psi(A)^*|_{X^\sharp}$ for $\psi \in H_0^\infty(\Sigma_\theta)$. By $\|\psi(A)^*\| = \|\psi(A)\|$ this implies the assertion.   □

B. The Extended H^∞-calculus

From now on $A \in \mathcal{S}(X)$ is always a sectorial operator. So far we cannot form operators $A^n, A^{-n}, A^{1/2}$, or more generally A^α for $\alpha \in \mathbb{C}$ by means of our functional calculus, since these operators and also the functions $f(\lambda) = \lambda^\alpha$ for $\lambda \in \Sigma_\sigma$ are unbounded in general. Therefore we will extend the functional calculus of Section 9 to polynominally bounded functions. This will give us a framework to discuss fractional powers A^α in an efficient way.

15.6 Notation. Let $\alpha \geq 0$ and $0 < \sigma < \pi$. By $\mathcal{H}_\alpha(\Sigma_\sigma)$ we denote the space of all analytic functions $f : \Sigma_\sigma \to \mathbb{C}$ for which $|\lambda^\alpha f(\lambda)|$ is bounded for $|\lambda| \leq 1$ and $|\lambda^{-\alpha} f(\lambda)|$ is bounded for $|\lambda| \geq 1$. This is equivalent to

$$\|f\|_{\mathcal{H}_\alpha(\Sigma_\sigma)} = \sup\{|\varrho(\lambda)|^\alpha |f(\lambda)| : \lambda \in \Sigma_\sigma\} < \infty,$$

where $\varrho(\lambda) = \lambda(1 + \lambda)^{-2}$.

Recall from Theorem 9.2 that $\varrho(A) = A(I + A)^{-2}$ and therefore

$$\varrho(A)^{-1} = 2 + A + A^{-1}, \qquad D(\varrho(A)^{-k}) = R(\varrho(A)^k) = D(A^k) \cap R(A^k)$$

for $k \in \mathbb{N}$.

Now, if $f \in \mathcal{H}_\alpha(\Sigma_\sigma)$ and $k \in \mathbb{N}$, $k > \alpha$, we can write

$$f(\lambda) = \varrho(\lambda)^{-k}(\varrho^k f)(\lambda).$$

The second factor is in $H_0^\infty(\Sigma_\sigma)$ and we can apply the functional calculus of Section 9 to obtain the bounded operator $(\varrho^k f)(A)$. This motivates the following definition.

15.7 Definition. Let $A \in \mathcal{S}(X)$ be sectorial and $\pi > \sigma > \omega(A)$. For $f \in \mathcal{H}_\alpha(\Sigma_\sigma)$ and $k \in \mathbb{N}$, $k > \alpha$ we put

$$f(A) = \varrho(A)^{-k}(\varrho^k f)(A)$$
$$D(f(A)) = \{x \in X : (\varrho^k f)(A)x \in D(\varrho(A)^{-k})\}.$$

Of course, we have to show that $f(A)$ is well–defined. Recall the approximate identity $\varrho_n(\lambda) = \frac{n}{n+\lambda} - \frac{1}{1+n\lambda}$ from Proposition 9.4. We have that $\varrho_n(A) = n(n+A)^{-1} - \frac{1}{n}(\frac{1}{n} + A)^{-1}$, $R(\varrho_n(A)^k) = D(A^k) \cap R(A^k)$, and $\varrho_n(A)^k x \to x$ for $n \to \infty$ and all $k \in \mathbb{N}$.

15.8 Theorem. Let $A \in \mathcal{S}(X)$ be sectorial, $\pi > \sigma > \omega(A)$, $f \in \mathcal{H}_\alpha(\Sigma_\sigma)$ and $k \in \mathbb{N}$, $k > \alpha$. Then

a) $f(A)$ is a well–defined closed operator on X.

b) If $x \in D(A^k) \cap R(A^k)$, then $x \in D(f(A))$ and $f(A)x = (\varrho^k f)(A)\varrho(A)^{-k}x$.

c) $D(A^k) \cap R(A^k)$ is a core for $f(A)$.

d) $D(f(A)) = \{x \in X : \lim_{n \to \infty} (\varrho_n^k f)(A)x$ exists in $X\}$ and for $x \in D(f(A))$, we have

$$f(A)x = \lim_{n \to \infty} (\varrho_n^k f)(A)x.$$

Proof. a) First we check that our definition does not depend on the choice of k. If $l > k$ then by the multiplicativity of the functional calculus

$$\varrho(A)^{-l}(\varrho^l f)(A) = \varrho(A)^{-l}\varrho(A)^{l-k}(\varrho^k f)(A) = \varrho(A)^{-k}(\varrho^k f)(A).$$

To show the closedness of $f(A)$ choose $x_n \in D(f(A))$ so that $x_n \to x$ and $f(A)x_n \to y$ in X. Then $(\varrho^k f)(A)x_n \to (\varrho^k f)(A)x$ since $(\varrho^k f)(A)$ is bounded. By the closedness of $\varrho(A)^{-k}$ and $\varrho(A)^{-k}[(\varrho^k f)(A)x_n] = f(A)x_n \to y$ we have

$$(\varrho^k f)(A)x \in D(\varrho(A)^{-k}) \text{ and } f(A)x = \varrho(A)^{-k}[(\varrho^k f)(A)x] = y.$$

b) For $x \in D(A^k) \cap R(A^k)$ and $y = \varrho(A)^{-k}x$, we have

$$(\varrho^k f)(A)x = (\varrho^k f)(A)\varrho(A)^k y = \varrho(A)^k(\varrho^k f)(A)y \in D(\varrho(A)^{-k}).$$

So $x \in D(f(A))$ and $f(A)x = (\varrho^k f)(A)y$.

c) If $x \in D(f(A))$, then $x_n = \varrho_n^k(A)x \in D(A^k) \cap R(A^k)$ and $x_n \to x$ for $n \to \infty$. Since the functional calculus is multiplicative we also obtain $f(A)x_n = \varrho(A)^{-k}[(\varrho^k f)\varrho_n^k](A)x = \varrho_n^k(A)f(A)x \to f(A)x$ for $n \to \infty$.

d) If $x \in D(f(A))$ then $(\varrho^k f)(A)x \in D(\varrho(A)^{-k})$ and by Theorem 9.2 and Example 9.8 c)

$$(\varrho_n^k f)(A)x = (\varrho_n^k \varrho^{-k})(A)(\varrho^k f)(A)x = \varrho_n^k(A)(\varrho(A)^{-k}[(\varrho^k f)(A)x])$$
$$= \varrho_n^k(A)(f(A)x) \to f(A)x$$

for $n \to \infty$. This also shows the inclusion "$\subset$" of our last claim.

Conversely, assume that $\lim_{n \to \infty} (\varrho_n^k f)(A)x = z \in X$ and define $y_n = \varrho_n^k(A)(\varrho^k f)(A)x \in D(A^k) \cap R(A^k) = D(\varrho(A)^{-k})$. Then $y_n \to (\varrho^k f)(A)x$ and

$$\varrho(A)^{-k}y_n = (\varrho^{-k}\varrho_n^k)(A)(\varrho^k f)(A)x = (\varrho_n^k f)(A)x \to z.$$

By the closedness of $\varrho(A)^{-k}$ we have $(\varrho^k f)(A)x \in D(\varrho(A)^{-k})$ and therefore $x \in D(f(A))$. $\square$

The extended calculus inherits some of the properties of the Dunford calculus, but one has to be careful with domains. This can be a little tedious at times.

15.9 Proposition. *Let $A \in \mathcal{S}(X)$ be a sectorial operator, $\pi > \sigma > \omega(A)$ and $\alpha, \beta \geq 0$.*

a) *If $f, g \in \mathcal{H}_\alpha(\Sigma_\sigma)$, $u, v \in \mathbb{C}$ and $k > \alpha$, then $D(f(A)) \cap D(g(A)) \subset D((uf + vg)(A))$ and*

$$(uf + vg)(A)(x) = uf(A)x + vg(A)x, \quad x \in D(f(A)) \cap D(g(A)).$$

b) *If $f \in \mathcal{H}_\alpha(\Sigma_\sigma)$, $g \in \mathcal{H}_\beta(\Sigma_\sigma)$ and $k = l + m$, where $l, m \in \mathbb{N}$ with $l > \alpha$, $m > \beta$, then*

$$D(A^k) \cap R(A^k) \subset D(f(A)g(A)) = D((fg)(A)) \cap D(g(A))$$

and
$$(fg)(A)x = f(A)[g(A)x] \qquad for\ x \in D(f(A)g(A)).$$

c) *If $f \in \mathcal{H}_\alpha(\Sigma_\sigma)$, and if $f^{-1} \in \mathcal{H}_\beta(\Sigma_\sigma)$ exists, then $f(A)^{-1} = f^{-1}(A)$.*

d) *If $f_n \in \mathcal{H}_\alpha(\Sigma_\sigma)$ with $\|f_n\|_{\mathcal{H}_\alpha(\Sigma_\sigma)} \leq C$ and $f_n(\lambda) \to f(\lambda)$ for $\lambda \in \Sigma_\sigma$, then for all $x \in D(A^k) \cap R(A^k)$ with $k > \alpha$*

$$\lim_{n \to \infty} f_n(A)x = f(A)x.$$

Proof. a) For $x \in D(f(A)) \cap D(g(A))$ we have by Theorem 15.8 and Theorem 9.2 that the limit

$$\lim_{n \to \infty} [\varrho_n^k(uf + vg)](A)x = \lim_{n \to \infty} [u(\varrho_n^k f)(A)x + v(\varrho_n^k g)(A)x]$$
$$= uf(A)x + vg(A)x$$

exists and must equal $(uf + vg)(A)x$.

b) Let $x \in D(A^k) \cap R(A^k)$ and $y = \varrho(A)^{-k}x$. If $l > \alpha$, $m > \beta$ with $k = l + m$ then

$$(\varrho^l f)(A)g(A)x = (\varrho^l f)(A)[\varrho(A)^{-m}(\varrho^m g)(A)]\varrho(A)^k y$$
$$= \varrho(A)^l[(\varrho^l f)(A)(\varrho^m g)(A)y]$$

belongs to $D(\varrho(A)^{-l})$ and the first inclusion follows. If $x \in D(g(A))$ with $g(A)x \in D(f(A))$, then

$$(\varrho^k fg)(A)x = (\varrho^l f)(A)(\varrho^m g)(A)x = (\varrho^l f)(A)\varrho(A)^m g(A)x$$
$$= \varrho(A)^m(\varrho^l f)(A)(g(A)x) = \varrho(A)^m \varrho(A)^l f(A)(g(A)x).$$

Therefore $x \in D((fg)(A))$ and $(fg)(A)x = f(A)g(A)x$. To prove the remaining inclusion let $x \in D(g(A)) \cap D((fg)(A))$. We have to show $g(A)x \in D(f(A))$ and $f(A)g(A)x = (fg)(A)(x)$, i.e.,

$$(\varrho^k f)(A)g(A)x \in D(\varrho(A)^{-k}) \text{ and}$$
$$\varrho(A)^{-k}(\varrho^k f)(A)g(A)x = \varrho(A)^{-k}(\rho^k fg)(A)x. \tag{15.2}$$

By the inclusion just shown we have $D(g(A)) = D(g(A)) \cap D((\varrho^k f)(A)) \subset D((\varrho^k fg)(A)) = X$ and $(\varrho^k f)(A)g(A)x = (\varrho^k fg)(A)x \in D(\varrho(A)^{-k})$. This implies (15.2).

c) Let $k = l + m$, where $l, m \in \mathbb{N}$ with $l > \alpha$, $m > \beta$. By b) we have $f(A)f^{-1}(A)x = x$ for x in the core $D(A^k) \cap R(A^k)$ of $f(A)$ and $f^{-1}(A)$. Given $x \in D(f(A))$, we choose $x_n \in D(A^k) \cap R(A^k)$ with $x_n \to x$ and $f(A)x_n \to f(A)x$. Since $f^{-1}(A)f(A)x_n \to x$ and $f^{-1}(A)$ is closed we get that $f(A)x \in D(f^{-1}(A))$ and $f^{-1}(A)(f(A)x) = x$. Hence $R(f(A)) \subset D(f^{-1}(A))$ and $D(f(A)) \subset R(f^{-1}(A))$. Exchanging the role of $f(A)$ and $f^{-1}(A)$ gives the reverse inclusions.

d) Let $x \in D(A^k) \cap R(A^k)$ and $y = \varrho(A)^{-k}x$. Then

$$f(A)x - f_n(A)x = ([f\varrho^\alpha - f_n\varrho^\alpha]\varrho^{k-\alpha})(A)y \to 0$$

by Theorem 9.2 since $f\varrho^\alpha - f_n\varrho^\alpha$ is uniformly bounded and goes to zero pointwise. $\qquad\square$

15.10 Remark. In general one cannot expect equality of domains in part a) and b) of the theorem. More can be said if $g \in H_A^\infty(\Sigma_\sigma)$, i.e., if $g(A)$ is bounded:

a) $D(f(A)) = D(f(A) + g(A))$;

b) $D(f(A)g(A)) = D(fg(A))$.

Proof. a) Since $D(g(A)) = X$ the inclusion "$\subset$" follows from 15.9 a) and so does the reverse inclusion

$$D(f(A) + g(A)) \cap D(-g(A)) \subset D([f(A) + g(A)] - g(A)).$$

b) is clear from 15.9 b). $\qquad\square$

Next we give a partial result for the composition of functions.

15.11 Proposition. *Let $g \in \mathcal{H}_\beta(\Sigma_\sigma)$ for $\sigma > \omega(A)$, so that $g(A) \in \mathcal{S}(X)$ is a sectorial operator. Assume that $g(\Sigma_\omega) \subset \Sigma_{\omega'}$ for some ω, ω' with $\omega(A) < \omega < \sigma$ and $\omega(g(A)) < \omega'$. If $f \in \mathcal{H}_\alpha(\Sigma_{\sigma'})$ for $\sigma' > \omega'$ such that $f \circ g \in \mathcal{H}_\gamma(\Sigma_\omega)$ for some $\gamma > 0$ then*

$$f(g(A)) = (f \circ g)(A).$$

Proof. Choose $k > \alpha, \beta, \gamma$. Also fix $\nu \in (\omega(A), \omega)$ and $\nu' \in (\omega', \sigma')$. For $x \in D(A^k) \cap R(A^k)$ with $x = \varrho(A)^k y$ we have for all $n \in \mathbb{N}$

$$(\varrho_n^k f)(g(A))x = \frac{1}{2\pi i} \int_{\partial\Sigma_{\nu'}} \varrho_n^k(\lambda)f(\lambda)R(\lambda, g(A))(\varrho(A)^k y)d\lambda = (*).$$

Since $(\lambda - g(A))^{-1} = (\lambda - g(\cdot))^{-1}(A)$ by Proposition 15.9 c) we get

$$(*) = \left(\frac{1}{2\pi i}\right)^2 \int_{\partial \Sigma_{\nu'}} \varrho_n^k(\lambda) f(\lambda) \left(\int_{\partial \Sigma_\nu} \frac{\varrho(\mu)^k}{\lambda - g(\mu)} R(\mu, A) y \, d\mu\right) d\lambda$$

$$= \left(\frac{1}{2\pi i}\right)^2 \int_{\partial \Sigma_\nu} \left(\int_{\partial \Sigma_{\nu'}} \frac{\varrho_n^k(\lambda)}{\lambda - g(\mu)} f(\lambda) d\lambda\right) \varrho(\mu)^k R(\mu, A) y \, d\mu$$

by Fubini's theorem. Now we use Cauchy's formula:

$$(*) = \frac{1}{2\pi i} \int_{\partial \Sigma_{\nu'}} \varrho_n^k(g(\mu)) f(g(\mu)) \varrho(\mu)^k R(\mu, A) y \, d\mu$$

$$= ((\varrho_n^k \circ g)(f \circ g) \varrho^k)(A) y \to ((f \circ g) \varrho^k)(A) y$$

by Theorem 9.2. Therefore, by Theorem 15.8, $x \in D(f(g(A))$ and

$$f(g(A))x = (\varrho^k(f \circ g))(A) y = (f \circ g)(A) x$$

for $x \in D(A^k) \cap R(A^k)$. Since the closed operators $f(g(A))$ and $(f \circ g)(A)$ agree on a common core they are equal. □

15.12 Example. For $f_n(z) = z^n$, n an integer, we have $f_n(A) = A^n$, since $(\varrho^{n+1} f_n)(\lambda) = \lambda^{2n+1}(1 + \lambda)^{-2n-2}$ and therefore, by Example 9.8 c),

$$(\varrho^{n+1} f_n)(A) = A^n \varrho(A)^{n+1}.$$

We also see that the functional calculus gives the right result for all rational functions.

C. Fractional Powers

Now it is natural to define the fractional powers $A^z, z \in \mathbb{C}$, by applying the extended functional calculus to the function $f_z(\lambda) = \lambda^z = e^{z \log \lambda}$, where we use the branch of the logarithm analytic in $\mathbb{C} \setminus \mathbb{R}_-$. Since for $\lambda = re^{i\varphi} \in \Sigma_\sigma$

$$|f_z(\lambda)| = |r^z| |e^{i\varphi z}| = r^{\text{Re } z} e^{-\varphi \text{Im } z} \le |\lambda|^{\text{Re } z} e^{\sigma |\text{Im } z|},$$

f_z belongs to $\mathcal{H}_{|\text{Re } z|}(\Sigma_\sigma)$ for all $\sigma < \pi$. For a sectorial operator $A \in \mathcal{S}(X)$ we define

$$A^z := f_z(A), \quad z \in \mathbb{C}.$$

The properties of the extended functional calculus give us already a working knowledge about fractional powers.

15.13 Remark. Let $A \in \mathcal{S}(X)$ be sectorial. The fractional powers $A^z = f_z(A), z \in \mathbb{C}$, are closed, densely defined operators which satisfy the following rules:

a) Let $k = l + m$ with $l > |\operatorname{Re} z_1|$, $m > |\operatorname{Re} z_2|$. Then $A^{z_1+z_2}x = A^{z_1}(A^{z_2}x)$ for $x \in D(A^{z_1}A^{z_2}) = D(A^{z_1+z_2}) \cap D(A^{z_2})$, in particular for $x \in D(A^k) \cap R(A^k)$,

b) $(A^z)^{-1} = A^{-z}$.

c) If $c \in \mathbb{C} \backslash \{0\}$ with $|\arg c| < \pi - \omega(A)$, then cA is sectorial and $(cA)^z = c^z A^z$.

Proof. For a), b) see Proposition 15.9.
c) Since $(\mu - cA)^{-1} = c^{-1}(c^{-1}\mu - A)^{-1}$, the condition $|\arg(c)| < \pi - \omega(A)$ guarantees that cA is sectorial with $\omega(cA) \leq \omega(A) + |\arg c|$. Choose $\omega > \omega(A)$, so that $\omega_1 = \omega + |\arg c| < \pi$. Then, for $x \in D(A^k) \cap R(A^k)$ with $k > |\operatorname{Re} z|$,

$$(\varrho_n^k f_z)(cA)x = \frac{1}{2\pi i} \int_{\partial \Sigma_{\omega_1}} \varrho_n^k(\lambda) \lambda^z R(\lambda, cA)x \, d\lambda.$$

By Cauchy's theorem we can deform the path in the integral above to $\Gamma = c \cdot \partial \Sigma_\omega$ and obtain

$$(\varrho_n^k f_z)(cA)x = \frac{1}{2\pi i} \int_\Gamma \varrho_n^k(\lambda) \lambda^z c^{-1} R(c^{-1}\lambda, A)x \, d\lambda$$

$$= c^z \frac{1}{2\pi i} \int_{\partial \Sigma_\omega} \varrho_n^k(c\lambda) \lambda^z R(\lambda, A)x \, d\lambda$$

$$= c^z (\varrho_n^k(c \cdot) f_z)(A)x \to c^z A^z x,$$

as $n \to \infty$ by Proposition 15.9. So Theorem 15.8 yields that $(cA)^z = c^z A^z x$. Since $(cA)^z$ and $c^z A^z$ are closed and agree on a common core, they are equal.

$\square$

To discuss further properties of fractional powers we need norm estimates. Since A and A^{-1} are unbounded, it is not clear how to compare $\|A^\alpha x\|$ and $\|A^\beta x\|$ directly. But $\alpha \mapsto \|A^\alpha x\|$ satisfies a log–convexity property which is very useful.

15.14 Theorem. *Let $A \in \mathcal{S}(X)$ be a sectorial operator. For $\alpha, \beta, \gamma \in \mathbb{C}$ assume that $\operatorname{Re}\alpha < \operatorname{Re}\gamma < \operatorname{Re}\beta$ with $\operatorname{Re}\gamma = (1 - \theta)\operatorname{Re}\alpha + \theta\operatorname{Re}\beta$. If $x \in D(A^\alpha) \cap D(A^\beta)$ then $x \in D(A^\gamma)$ and*

a) $\|A^\gamma x\| \leq C(\epsilon^\theta \|A^\alpha x\| + \epsilon^{\theta-1}\|A^\beta x\|)$ for all $\epsilon > 0$,

b) $\|A^\gamma x\| \leq C\|A^\alpha x\|^{1-\theta}\|A^\beta x\|^\theta$.

where C depends on A, $\gamma - \alpha$ and $\beta - \gamma$.

Proof. Put $\alpha_0 = \operatorname{Re}\gamma - \operatorname{Re}\alpha$, $\beta_0 = \operatorname{Re}\beta - \operatorname{Re}\gamma$ and $\theta = \frac{\alpha_0}{\alpha_0 + \beta_0}$. Choose $m > \alpha_0 + 1, \beta_0 + 1$. We will use the auxiliary function

$$\psi(\lambda) = c\lambda^m (1 + \lambda)^{-2m}$$

where c is chosen so that $\int_0^\infty \psi(t) \frac{dt}{t} = 1$. Hence c only depends on α_0 and β_0. ψ is in $H_0^\infty(\Sigma_\sigma)$ for $\sigma > \omega(A)$ and by Lemma 9.13 we get for $x \in D(A^k) \cap R(A^k)$ with k large enough

$$A^\gamma x = \int_0^\infty \psi(t^{-1}A)(A^\gamma x)\frac{dt}{t}.$$

Now using $A^{z_1}A^{z_2}x = A^{z_1+z_2}x, (tA)^z x = t^z A^z x$, we get for all $\delta > 0$:

$$A^\gamma x = \int_0^\delta t^{\gamma-\alpha-1}[(t^{-1}A)^{\gamma-\alpha}\psi(t^{-1}A)](A^\alpha x)dt$$

$$+ \int_\delta^\infty t^{\gamma-\beta-1}[(t^{-1}A)^{\gamma-\beta}\psi(t^{-1}A)](A^\beta x)dt$$

$$= \int_0^\delta t^{\gamma-\alpha-1}[\psi_0(t^{-1}A)](A^\alpha x)dt$$

$$+ \int_\delta^\infty t^{\gamma-\beta-1}[\psi_\infty(t^{-1}A)](A^\beta x)dt$$

where $\psi_0(\lambda) = c\lambda^{-\alpha+\gamma+m}(1+\lambda)^{-2m}$ and $\psi_\infty(\lambda) = c\lambda^{\gamma-\beta+m}(1+\lambda)^{-2m}$. Hence

$$\|A^\gamma x\| \leq \frac{1}{\alpha_0}\delta^{\alpha_0}\sup_{t>0}\|\psi_0(t^{-1}A)\| \cdot \|A^\alpha x\|$$

$$+ \frac{1}{\beta_0}\delta^{-\beta_0}\sup_{t>0}\|\psi_\infty(t^{-1}A)\| \cdot \|A^\beta x\|.$$

Since ψ_0 and ψ_∞ are in $H_0^\infty(\Sigma_\sigma)$, we can estimate $\sup_{t>0}\|\psi_0(t^{-1}A)\|$ by (cf. Lemma 9.13 b))

$$c_1 \int_{\partial\Sigma_\sigma} |\psi_0(\lambda)|\frac{d|\lambda|}{|\lambda|} \leq 2c_1 c e^{\sigma|\text{Im}(\gamma-\alpha)|}\int_0^\infty \frac{r^{m+\alpha_0-1}}{|1+e^{i\sigma}r|^{2m}}dr$$

where c_1 only depends on A, and similarly for ψ_∞. Hence there is a constant c depending only on A, α_0, β_0 and $\text{Im}(\gamma-\alpha), \text{Im}(\beta-\gamma)$, so that

$$\|A^\gamma x\| \leq C(\delta^{\alpha_0}\|A^\alpha x\| + \delta^{-\beta_0}\|A^\beta x\|)$$

or for $\delta = \epsilon^{1/(\alpha_0+\beta_0)}$

$$\|A^\gamma x\| \leq C(\epsilon^\theta\|A^\alpha x\| + \epsilon^{\theta-1}\|A^\beta x\|).$$

If we choose $\delta = (\|A^\beta x\|/\|A^\alpha x\|)^{\frac{1}{\alpha_0+\beta_0}}$ (it follows from Remark 15.13 b) that $\|A^\alpha x\| \neq 0$) then

$$\|A^\gamma x\| \leq 2C\|A^\alpha x\|^{1-\theta}\|A^\beta x\|^\theta.$$

Finally, for arbitrary $x \in D(A^\alpha) \cap D(A^\beta)$ we can find $x_n \in D(A^k) \cap R(A^k)$ so that $x_n \to x, A^\alpha x_n \to A^\alpha x$ and $A^\beta x_n \to A^\beta x$. Our inequality (applied to $x_n - x_m$) shows that $A^\gamma x_n$ is also a Cauchy sequence. Since A^γ is closed it follows that $x \in D(A^\gamma)$.   □

Now we collect the basic operator theoretical properties of fractional powers. The next two theorems make A^z look like fractional powers, talk like fractional powers and walk like fractional powers.

15.15 Theorem. *Let $A \in \mathcal{S}(X)$ be sectorial. Then the fractional powers $A^z, z \in \mathbb{C}$, are closed and injective operators with dense domains and ranges. Furthermore,*

a) *for* $\operatorname{Re} z_1 < \operatorname{Re} z_2 < 0 < \operatorname{Re} z_3 < \operatorname{Re} z_4$ *we have*

$$D(A^{z_1}) \subset D(A^{z_2}), \quad D(A^{z_3}) \supset D(A^{z_4})$$
$$R(A^{z_1}) \subset R(A^{z_2}), \quad R(A^{z_3}) \supset R(A^{z_4}).$$

b) *Let* $\operatorname{Re} z_1 < \operatorname{Re} z_2$ *and* $x \in D(A^{z_1}) \cap D(A^{z_2})$ *and consider the strip* $S = \{\lambda : \operatorname{Re} z_1 < \operatorname{Re} \lambda < \operatorname{Re} z_2\}$. *Then* $z \mapsto A^z x$ *is analytic in* S.

c) $A^{z_1+z_2} = A^{z_1} A^{z_2}$ *for* $z_1, z_2 \in \mathbb{C}$ *with* $\operatorname{Re} z_1 \cdot \operatorname{Re} z_2 > 0$.

d) $(A^z)^{-1} = A^{-z}$ *for* $z \in \mathbb{C}$.

e) *If* $\mu \in \mathbb{C} \setminus \{0\}$ *with* $\arg(\mu) < \pi - \omega(A)$, *then* $(\mu A)^\alpha = \mu^\alpha A^\alpha$.

Proof. We know already that A^z is closed and densely defined. That A^z is injective and has dense range follows from d). But d) is a special case of Proposition 15.9 c).

a) If $\operatorname{Re} z_1 < \operatorname{Re} z_2 < 0$ then for $\alpha = z_1, \gamma = z_2$ and $\beta = 0$ Theorem 15.14 gives $D(A^{z_1}) \cap X \subset D(A^{z_2})$ and by d) $R(A^{-z_1}) \subset R(A^{-z_2})$. If $0 < \operatorname{Re} z_3 < \operatorname{Re} z_4$ we get $D(A^{z_4}) \cap X \subset D(A^{z_3})$.

b) If $x = \varrho^k(A)y$ with $y \in X$, and $k > |\operatorname{Re} z_1|, |\operatorname{Re} z_2|$, then for $\omega(A) < \omega < \pi$ (by Theorem 15.8)

$$A^z x = (\varrho^k f_z)(A)y = \frac{1}{2\pi i} \int_{\partial \Sigma_\omega} \lambda^z \varrho(\lambda)^k R(\lambda, A)y \, d\lambda$$

which is indeed analytic in S. If $x \in D(A^{z_1}) \cap D(A^{z_2})$ then $x \in D(A^z)$ for $z \in S$ and $A^z x = \lim_{n \to \infty} A^z(\varrho_n^k(x))$ is locally bounded (by Theorems 15.8 and 15.14)) and analytic in S as the pointwise limit of analytic functions.

c) follows from Remark 15.13 a) and part a) above.

e) was already proved in Remark 15.13 c). $\square$

Of particular interest are powers A^α with $\alpha \in \mathbb{R}$. In this case more can be said:

15.16 Theorem. *Let* $A \in \mathcal{S}(X)$ *be sectorial. Consider* $\alpha \in \mathbb{R}$ *with* $|\alpha| < \frac{\pi}{\omega(A)}$. *Then* A^α *is again a sectorial operator in* $\mathcal{S}(X)$ *with* $\omega(A^\alpha) = |\alpha|\omega(A)$. *Furthermore, for every* $z \in \mathbb{C}$ *we have*

$$(A^\alpha)^z = A^{\alpha z}.$$

We claim that $\mu(\mu - A^\alpha)^{-1}$ is bounded uniformly for $\arg \mu > \alpha\omega > \alpha\omega(A)$, if $\omega(A) < \omega < \min(\pi, \frac{\pi}{\alpha})$. Indeed, the functions $h_\mu(\lambda) = \mu(\mu - \lambda^\alpha)^{-1}$ belong to $H^\infty(\Sigma_\omega)$ and $\|h_\mu\|_{H^\infty(\Sigma_\omega)}$ is uniformly bounded. So if A has a bounded

$H^\infty(\Sigma_\omega)$–calculus our first claim follows. But in general we have to show that h_μ belongs to $H_A^\infty(\Sigma_\omega)$ for every sectorial operator $A \in \mathcal{S}(X)$ with $\omega(A) < \omega$.

However, this will follow from Lemma 9.12 and the next lemma, where we use the assumption of Theorem 15.16 for $\alpha > 0$ and the notation

$$\psi_\theta(\lambda) = \frac{e^{i\theta}\lambda + \lambda^\alpha}{(e^{i\theta} - \lambda^\alpha)(1 - \lambda)}.$$

Note that $\psi_\theta \in H_0^\infty(\Sigma_\sigma)$ for all $\sigma < |\theta|/\alpha$.

15.17 Lemma. *For all $\mu \notin \Sigma_{\alpha\omega(A)}$ with $\arg\mu = \theta$ we have $\mu \in \rho(A^\alpha)$ and*

$$\mu R(\mu, A^\alpha) = -|\mu|^{1/\alpha}R(-|\mu|^{1/\alpha}, A) + \psi_\theta(|\mu|^{-1/\alpha}A).$$

Proof. A straight forward calculation shows

$$h_\mu(\lambda) = \frac{\mu}{\mu - \lambda^\alpha} - \frac{|\mu|^{1/\alpha}}{|\mu|^{1/\alpha} + \lambda} = \frac{\mu\lambda + |\mu|^{1/\alpha}\lambda^\alpha}{(\mu - \lambda^\alpha)(|\mu|^{1/\alpha} + \lambda)} = \psi_\theta(|\mu|^{-1/\alpha}\lambda)$$

since $\mu|\mu|^{-1} = e^{i\theta}$. Hence h_μ is the sum of two functions in $H_A^\infty(\Sigma_\omega)$ with $\omega \in (\omega(A), \frac{|\theta|}{\alpha})$ and therefore belongs to the same class. Proposition 15.9 c) informs us that $(\frac{1}{\mu - \lambda^\alpha})(A)$ is indeed the inverse of $(\mu - \lambda^\alpha)(A) = \mu - A^\alpha$, so that $\mu \in \rho(A^\alpha)$ and

$$\mu R(\mu, A^\alpha) - (-|\mu|^{1/\alpha}R(-|\mu|^{1/\alpha}, A) = \psi_\theta(|\mu|^{-1/\alpha}A) \qquad \square$$

$\square$

Proof (of Theorem 15.16). Since $A^{-\alpha} = (A^\alpha)^{-1}$ by Theorem 15.15 we can restrict ourselves to $\alpha > 0$. The lemma shows that for $\theta > \alpha\omega(A)$ we have $\mu \in \rho(A)$ if $|\arg\mu| \geq \theta$. Furthermore, by Lemma 9.13, we have for $\theta' \in (\omega(A), \theta/\alpha)$

$$\sup\{\|\psi_\theta(tA)\| : t > 0, |\theta| \geq \theta_0\} \leq \frac{1}{2\pi}\int_{\partial\Sigma_{\theta'}} |\psi_\theta(\lambda)|, \frac{|d\lambda|}{|\lambda|}.$$

So the sectoriality of A implies the sectoriality of A^α with $\omega(A^\alpha) \leq \alpha\omega(A)$. Interchanging A and $A^{1/\alpha}$ gives equality. The last claim follows from Proposition 15.11 $\qquad \square$

D. Representation Formulas

So far we were concerned with the formal "algebraic" properties of fractional powers. In order to estimate the norm of $A^z x$ for some $x \in D(A^z)$ we also need integral representations for $A^z x$. As a matter of fact, usually the fractional powers are defined in such a way. It is clear, that the most natural representation

$$A^z x = \frac{1}{2\pi i} \int_{\partial \Sigma_\omega} \lambda^z R(\lambda, A) x \, d\lambda \qquad (15.3)$$

will only make sense in special cases (even as an improper integral at 0 and ∞). We do have that the formula holds in an approximative sense

$$A^z x = \lim_{n \to \infty} \frac{1}{2\pi i} \int_{\partial \Sigma_\omega} \varrho_n(\lambda)^k \lambda^z R(\lambda, A) x \, d\lambda \qquad (15.4)$$

where $k > |\mathrm{Re}\,\alpha|$, by Theorem 15.8. We will now derive variants of (15.3) which take into account the singularities at 0 and ∞ and therefore exist as Bochner integrals.

We will start from the well known formula

$$\int_0^\infty t^z (1+t)^{-1} \, dt = -\frac{\sin(\pi z)}{\pi}, \qquad -1 < z < 0. \qquad (15.5)$$

To derive this identity from Cauchy's fomula we parametrize the integral

$$1 = \frac{1}{2\pi i} \int_{\partial \Sigma_\omega} \frac{\lambda^z}{\lambda - 1} \, d\lambda$$

$$= \frac{-e^{i\omega} e^{i\omega z}}{2\pi i} \int_0^\infty \frac{t^z}{e^{i\omega} t - 1} \, dt + \frac{e^{-i\omega} e^{-i\omega z}}{2\pi i} \int_0^\infty \frac{t^z}{e^{-i\omega} t - 1} \, dt$$

and obtain (15.5) for $\omega \to \pi$. Traditionally, one uses similar manipulations with contour integrals of operator-valued functions to prove representation theorems for fractional powers. Instead, we will perform the necessary calculation in the space of analytic functions before we use the functional calculus to obtain operator equations.

15.18 Theorem. *Let $A \in \mathcal{S}(X)$ be sectorial on a Banach space X.*

a) *For $-1 < \mathrm{Re}\,z < 0$ and $x \in R(A)$ we have*

$$A^z x = -\frac{\sin(\pi z)}{\pi} \int_0^\infty t^z (t + A)^{-1} x \, dt.$$

b) *For $0 < \mathrm{Re}\,z < 1$ and $x \in D(A)$ we have*

$$A^z x = \frac{\sin(\pi z)}{\pi} \int_0^\infty t^{z-1} (t + A)^{-1} A x \, dt.$$

Proof. a) By analytic continuation (15.5) also holds for $-1 < \mathrm{Re}\,z < 0$. We substitute t by $t\lambda^{-1}$ for $\lambda > 0$ and obtain

$$\lambda^z = -\frac{\sin(\pi z)}{\pi} \int_0^\infty t^z (t + \lambda)^{-1} \, dt. \qquad (15.6)$$

For a fixed z with $-1 < \mathrm{Re}\,z < 0$ both sides are analytic functions in λ on Σ_σ for all $\sigma < \pi$. If we multiply both sides with $\varphi(\lambda) = (1 + \lambda^{-1})^{-1}$ then

$$\lambda^z \varphi(\lambda) = \frac{\lambda^z}{(1+\lambda^{-1})} = \int_0^\infty f(t,\lambda)\, dt$$

with $f(t,\lambda) = t^z(t+\lambda)^{-1}\varphi(\lambda)$. The functions $\lambda^z\varphi(\lambda)$ and $f(\cdot,\lambda)$ belong to $H_A^\infty(\Sigma_\sigma)$ and by Lemma 9.12 and the multiplicativity of the functional calculus we obtain for all $x \in X$

$$A^z[\varphi(A)x] = \int t^z(t+A)^{-1}\varphi(A)x\, dt.$$

Since $R(\varphi(A)) = D(A^{-1}) = R(A)$ we have shown a).

b) If $0 < \operatorname{Re} z < 1$ then (15.6) holds for $z - 1$:

$$\lambda^{z-1} = -\frac{\sin(\pi(z-1))}{\pi}\int_0^\infty t^{z-1}(t+\lambda)^{-1}\, dt$$

or

$$\lambda^z = \frac{\sin(\pi z)}{\pi}\int_0^\infty t^{z-1}(t+\lambda)^{-1}\lambda\, dt.$$

Again, we have an analytic function in λ on Σ_σ for $\sigma < \pi$ on both sides. We multiply by $\varphi(\lambda) = (1+\lambda)^{-1}$ and obtain as in the proof of a) with Lemma 9.12

$$A^z\varphi(A)x = \frac{\sin(\pi z)}{\pi}\int_0^\infty t^{z-1}(t+A)^{-1}A\varphi(A)x\, dt$$

for $x \in X$. Note that $R(\varphi(A)) = D(A)$ to complete the proof. □

In a similar way we can derive a more general representation theorem.

15.19 Theorem. *Let $A \in \mathcal{S}(X)$ be sectorial. Then A^z is a densely defined closed operator on X. For $\omega > \omega(A)$, $k > |\operatorname{Re} z|$ and all $x \in D(f(A))$ we have*

$$A^z x = \lim_{n\to\infty}\frac{1}{2\pi i}\int_{\partial\Sigma_\omega}\varrho_n^k(\lambda)\lambda^z R(\lambda,A)x\, d\lambda. \tag{15.7}$$

If $-m - 1 < \operatorname{Re} z < l$, where $m, l \in \mathbb{N}_0$ and z is not an integer, and $x \in D(A^l) \cap R(A^{m+1})$ then

$$A^z x = -\frac{\sin(\pi z)}{\pi}\Bigg[(-1)^m\int_0^1 t^{z+m}(t+A)^{-1}A^{-m}x\, dt$$

$$+ (-1)^l\int_1^\infty t^{z-l}(t+A)^{-1}A^l x\, dt + \sum_{j=-m}^{l-1}(-1)^{j+1}\frac{1}{z-j}A^j x\Bigg]$$

Proof. The first formula follows directly from Theorem 15.8.
For the second formula we need the identities (where $m, l \in \mathbb{N}$)

$$(t+\lambda)^{-1} - (-1)^m t^m(t+\lambda)^{-1}\lambda^{-m}$$

$$= (t+\lambda)^{-1}[1 - (-t/\lambda)^m] = \frac{(1+t/\lambda)}{t+\lambda}\sum_{j=0}^{m-1}(-t/\lambda)^j = \sum_{j=1}^m(-1)^{j+1}t^{j-1}\lambda^{-j}$$

and similarly

$$(t+\lambda)^{-1} - (-1)^l t^{-l}(t+\lambda)^{-1}\lambda^l = \sum_{j=0}^{l-1}(-1)^j t^{-j-1}\lambda^j.$$

Starting from the integral (15.5) again we obtain for $-1 < \operatorname{Re} z < 0$ and $\lambda \in \mathbb{C} \setminus \mathbb{R}_-$

$$\lambda^z = -\frac{\sin(\pi z)}{\pi}\int_0^\infty t^z(t+\lambda)^{-1}\,dt$$

$$= -\frac{\sin(\pi z)}{\pi}\int_0^1 t^z\Big[(-1)^m t^m(t+\lambda)^{-1}\lambda^{-m} + \sum_{j=1}^m (-1)^{j+1}t^{j-1}\lambda^{-j}\Big]\,dt$$

$$-\frac{\sin(\pi z)}{\pi}\int_1^\infty t^z\Big[(-1)^l t^{-l}(t+\lambda)^{-1}\lambda^l + \sum_{j=0}^{l-1}(-1)^j t^{-j-1}\lambda^j\Big]\,dt$$

$$= -\frac{\sin(\pi z)}{\pi}\Big[(-1)^m \int_0^1 t^{z+m}\lambda(t+\lambda)^{-1}\lambda^{-m-1}\,dt$$

$$+(-1)^l \int_1^\infty t^{z-l}(t+\lambda)^{-1}\lambda^l\,dt$$

$$+ \sum_{j=-m}^{l-1}(-1)^{j+1}\frac{1}{z-j}\lambda^j\Big]$$

since $\int_0^1 t^{z+j-1}\,dt = \frac{1}{z+j}$ and $\int_1^\infty t^{z-j-1}\,dt = -\frac{1}{z-j}$.

This formula certainly holds for $-1 < \operatorname{Re} z < 0$, but by analytic continuation it also holds in the domain $\{z : -m-1 < \operatorname{Re} z < l, z \notin \mathbb{Z}\}$. Fix now a z in this domain. Then we may consider the right and the left hand side as a holomorphic function on Σ_σ for all $0 < \sigma < \pi$. Put $\varphi(\lambda) = (1+\lambda^{-1})^{-m-1}(1+\lambda)^{-l} = \lambda^{m+1}(1+\lambda)^{-m-1}(1+\lambda)^{-l}$. If we multiply both sides of the equation with $\varphi(A)$ we obtain with Lemma 9.12

$$A^z\varphi(A)x = -\frac{\sin(\pi z)}{\pi}\Big[(-1)^m \int_0^1 t^{z+m}A(t+A)^{-1}A^{-m-1}\varphi(A)x\,dt$$

$$+(-1)^l \int_1^\infty t^{z-l}(t+A)^{-1}A^l\varphi(A)x\,dt$$

$$+ \sum_{j=-m}^{l-1}(-1)^{j+1}\frac{1}{z-j}A^j\varphi(A)x\Big]$$

for $x \in X$. Since $R(\varphi(A)) = R(A^{m+1}) \cap D(A^l)$ we are done. $\square$

15.20 Corollary. *Let $A \in \mathcal{S}(X)$ be sectorial with $\omega(A) < \frac{\pi}{2}$ and denote by T_t the semigroup generated by $-A$. Then for $0 < \operatorname{Re} z < 1$ we have*

$$A^{-z}x = \frac{1}{\Gamma(z)}\int_0^\infty t^{z-1}T_t x\,dt.$$

Proof. It is well known that $\Gamma(z) = \int_0^\infty s^{z-1}e^{-s}ds$. The substitution $s = \lambda t$ gives

$$\lambda^{-z} = \Gamma(z)^{-1} \int_0^\infty t^{z-1}e^{-\lambda t}\,dt.$$

Now we multiply both sides with $(1 + \lambda^{-1})^{-1}$ and obtain

$$\lambda^{-z}(1 + \lambda^{-1})^{-1} = \Gamma(z)^{-1} \int_0^1 t^{z-1}f(t,\lambda)\,dt + \Gamma(z)^{-1} \int_1^\infty t^{z-2}g(t,\lambda)\,dt$$

with $f(t,\lambda) = (1 + \lambda^{-1})^{-1}e^{-\lambda t}$ and $g(t,\lambda) = \frac{t\lambda}{1+\lambda}e^{-\lambda t}$. For $\lambda \in \Sigma_\sigma$ with $\sigma < \pi/2$ the assumptions of Lemma 9.12 are fulfilled and this yields

$$A^{-z}R(-1, A^{-1}) = \Gamma(z)^{-1} \int_0^\infty t^{z-1}T_tR(-1, A^{-1})\,dt.$$

Since $R(R(-1, A^{-1})) = D(A^{-1}) = R(A)$ the claim follows. □

E. Interpolation and Extrapolation Scales

We define two scales of interpolation and extrapolation spaces, both related to the fractional powers of a sectorial operator $A \in \mathcal{S}(X)$ in a Banach space X.

15.21 Definition. *For a sectorial operator $A \in \mathcal{S}(X)$ on X and $\alpha \in \mathbb{R}$ we define $\dot{X}_\alpha$ as the completion of $D(A^\alpha)$ with respect to the norm $\|x\|_{\dot{X}_\alpha} = \|A^\alpha x\|$, i.e.,*

$$(\dot{X}_\alpha, \|\cdot\|_{\dot{X}_\alpha}) := (D(A^\alpha), \|A^\alpha \cdot\|_X)^\sim.$$

In particular we have $\dot{X}_0 = X$. In addition to the scale $(\dot{X}_\alpha)$ we define another scale (X_α) by letting

$$(X_\alpha, \|\cdot\|_{X_\alpha}) := \begin{cases} (D(A^\alpha), \|(1 + A)^\alpha \cdot\|_X)\,, & \alpha \geq 0 \\ (X, \|(1 + A)^\alpha \cdot\|_X)^\sim\,, & \alpha < 0 \end{cases}.$$

The next lemma implies that, in fact, we can replace $(1 + A)^\alpha$ here by $(\lambda + A)^\alpha$ for any $-\lambda \in \varrho(A)$ and obtain an equivalent norm on $D(A^\alpha)$. In particular, $\dot{X}_\alpha = X_\alpha$ with equivalent norms for all $\alpha \in \mathbb{R}$ if $0 \in \rho(A)$. But for $0 \in \sigma(A)$ the spaces are different (apart from the obvious equality $\dot{X}_0 = X = X_0$).

15.22 Lemma. *Let $A \in \mathcal{S}(X)$ be a sectorial operator. Assume $\alpha > 0$ and $\epsilon > 0$. Then $D(A^\alpha) = D((\epsilon + A)^\alpha)$ and the norms $x \mapsto \|x\| + \|A^\alpha x\|$ and $x \mapsto \|(\epsilon + A)^\alpha\|$ are equivalent on $D(A^\alpha)$. Moreover, $(\epsilon + A)^{-\alpha} = ((\epsilon + A)^\alpha)^{-1} = ((\epsilon + A)^{-1})^\alpha$.*

Proof. The last assertion follows from 15.11 and all assertions are clear for $\alpha \in \mathbb{N}$. We now start with the case $\alpha \in (0,1)$ and show that $(\epsilon + A)^\alpha - A^\alpha \in B(X)$. This implies $D(A^\alpha) = D((\epsilon + A)^\alpha)$. Since $(\epsilon + A)^{-\alpha} \in B(X)$ it also implies equivalence of the norms by the open mapping theorem.

We let $\psi(\lambda) := (1 + \lambda)^\alpha - \lambda^\alpha - (1 + \lambda)^{-1}$. A moment's thought shows $\psi \in H_0^\infty(\Sigma_\sigma)$ for any $\sigma \in (0, \pi)$. Hence also $\psi_\epsilon(\lambda) := \epsilon^\alpha \psi(\lambda/\epsilon) = (\epsilon + \lambda)^\alpha - \lambda^\alpha - \epsilon^{\alpha+1}(\epsilon + \lambda)^{-1}$ belongs to H_0^∞. This implies that $\lambda \mapsto (\lambda + \epsilon)^\alpha - \lambda^\alpha$ belongs to H_A^∞ and thus $(\epsilon + A)^\alpha - A^\alpha \in B(X)$.

If $\beta = \alpha + n$ with $n \in \mathbb{N}$ and $\alpha \in (0,1)$ then, by Theorem 15.15 b), $D((\epsilon + A)^\beta) \subset D((\epsilon + A)^n)$. We observe $T_\epsilon x \in D(A^n)$ for $x \in D(A^n)$ since the operator T_ϵ defined above commutes with A. Thus we obtain, again by 15.15 b),

$$D((\epsilon + A)^\beta) = D((\epsilon + A)^n(\epsilon + A)^\alpha)$$
$$= \{x \in D((\epsilon + A)^n) : (\epsilon + A)^\alpha \in D((\epsilon + A)^n)\}$$
$$= \{x \in D(A^n) : (\epsilon + A)^\alpha \in D(A^n)\} = \{x \in D(A^n) : A^\alpha x \in D(A^n)\}$$
$$= D(A^n A^\alpha) = D(A^\beta).$$

Equivalence of norms now follows easily. □

The scale (X_α) is the usual Sobolev tower as defined, e. g. in [88]. For the Laplace operator $A = -\Delta$ on $X = L_p(\mathbb{R}^n)$ where $1 < p < \infty$ we have $\dot{X}_\alpha = \dot{H}_p^{2\alpha}(\mathbb{R}^n)$, a homogeneous or Riesz potential space, and $X_\alpha = H_p^{2\alpha}(\mathbb{R}^n)$, an inhomogeneous or Bessel potential space. If we want to emphasize the operator A, whose fractional domains we consider, we write $\dot{X}_{\alpha,A}$ and $X_{\alpha,A}$.

15.23 Proposition. *Let $A \in \mathcal{S}(X)$ be a sectorial operator in X. Then we have:*

a) *The operator $A^\alpha : D(A^\alpha) \to R(A^\alpha)$ extends to an isomorphism $\widetilde{A^\alpha} : \dot{X}_\alpha \to X$ whose inverse $(\widetilde{A^\alpha})^{-1}$ is an extension of the operator $A^{-\alpha} : R(A^\alpha) \to D(A^\alpha)$.*

b) *For $\alpha > 0$ the operator $(1 + A)^\alpha : D(A^\alpha) \to X$ is an isometry $X_\alpha \to X$ with inverse $(1 + A)^{-\alpha}$, and the operator $(1 + A)^{-\alpha} : X \to X$ extends to an isometry $J_{-\alpha} : X_{-\alpha} \to X$ whose inverse $(J_{-\alpha})^{-1}$ is an extension of the operator $(1 + A)^\alpha : D(A^\alpha) \to X$.*

c) *If X is reflexive, there are natural isomorphisms $(\dot{X}_{\alpha,A})^* = (X^*)_{-\alpha,A^*}$ and $(X_{\alpha,A})^* = (X^*)_{-\alpha,A^*}$. Moreover, $\dot{X}_{-\alpha,A} = ((X^*)_{\alpha,A^*})^*$ and $X_{-\alpha,A} = ((X^*)_{\alpha,A^*})^*$ with respect to the duality $\langle X, X^* \rangle$.*

Proof. a) and b) are clear.

c) Suppose $\phi : D(A^\alpha) \to \mathbb{C}$ is linear. Then ϕ is continuous for $\|A^\alpha \cdot\|_X$ if and only if $\langle \phi, A^{-\alpha} \cdot \rangle : D(A^{-\alpha}) \to \mathbb{C}$ extends to an element of X^*. But for $\phi \in X^*$ this is equivalent to ϕ belonging to the domain of $(A^{-\alpha})^* = (A^*)^{-\alpha}$. For the norms we obtain

$$\|\phi\|_{(\dot{X}_\alpha)^*} = \sup\left\{|\langle \phi, A^{-\alpha}x\rangle| : x \in D(A^{-\alpha}), \|x\|_X \leq 1\right\}$$
$$= \sup\left\{|\langle (A^*)^{-\alpha}\phi, x\rangle| : x \in D(A^{-\alpha}), \|x\|_X \leq 1\right\}$$
$$= \|(A^*)^{-\alpha}\phi\|_{X^*}.$$

This proves the assertion. □

Observe that one has to distinguish the extension $\widetilde{(A^\alpha)}^{-1}$ of the operator $A^{-\alpha} : R(A^\alpha) \rightarrow D(A^\alpha)$, which acts as an isometry $X \rightarrow \dot{X}_\alpha$ from the extension $\widetilde{A^{-\alpha}}$, which acts as an isometry $\dot{X}_{-\alpha} \rightarrow X$.

15.24 Proposition. *For $\alpha \in \mathbb{R}$ we define versions of A in each of the spaces $\dot{X}_\alpha$ by letting $\dot{A}_\alpha := \widetilde{(A^\alpha)}^{-1}A\widetilde{A^\alpha}$. In the inhomogeneous scale (X_α) we let, for $\alpha > 0$, A_α be the part of A in X_α and define the operator $A_{-\alpha} := (J_{-\alpha})^{-1}AJ_{-\alpha}$ in $X_{-\alpha}$. Then we have:*

a) *The operators $\dot{A}_\alpha$ in $\dot{X}_\alpha$ and A_α in X_α are similar to the operator A in X. In particular, $\dot{A}_\alpha \in S(\dot{X}_\alpha)$ and $A_\alpha \in S(X_\alpha)$ are sectorial.*

b) *Both operators $\dot{A}_\alpha$ and A are extensions of the operator $A^{-\alpha}AA^\alpha$. They coincide at least on*

$$D(A^{-\alpha}AA^\alpha) = \begin{cases} D(A^{1+\alpha}) & , \quad \alpha > 0 \\ D(A^\alpha) \cap D(A) , & \quad \alpha < 0 \end{cases}$$

Proof. a) is clear. b) follows from Remark 15.13. □

Concerning inclusions in the scales, we have $X_\beta \subset X_\alpha$ for $\alpha < \beta$ in a natural way, but for $0 \in \sigma(A)$ the spaces $\dot{X}_\alpha$ and $\dot{X}_\beta$ are not comparable (cf. the remark before Theorem 15.14). In order to study relation between these spaces we let $(Y^{(-m)}, \|\cdot\|_{(-m)}) := (X, \|(A(1+A)^{-2})^m \cdot \|_X)^\sim$ for $m \in \mathbb{N}$. It turns out that we may realize the completions $\dot{X}_\alpha$, $|\alpha| \leq m$, as subspaces of $Y^{(-m)}$.

15.25 Proposition. *Let $m \in \mathbb{N}$. Then each space $\dot{X}_\alpha$, $|\alpha| \leq m$, may be realized as a subspace of $Y^{(-m)}$. More precisely, writing*

$$Y^{(-m)} = \Big\{ \{(x_n) + (y_n) : y_n \in X, \|y_n\|_{(-m)} \rightarrow 0\} :$$
$$(x_n) \text{ is } \|\cdot\|_{(-m)}\text{-Cauchy sequence in } X \Big\}$$

we have

$$\dot{X}_\alpha = \Big\{ \{(x_n) + (y_n) : y_n \in X, \|y_n\|_{(-m)} \rightarrow 0\} :$$
$$(x_n) \text{ is } \|A^\alpha \cdot \|_X\text{-Cauchy sequence in } D(A^\alpha) \Big\}.$$

for $|\alpha| \leq m$.

Proof. We have to show that the inclusion $D(A^\alpha) \to X$ induces an injective embedding $\dot{X}_\alpha \hookrightarrow Y^{(-m)}$. To this end let (x_n) be a $\|A^\alpha \cdot\|$–Cauchy sequence in $D(A^\alpha)$ such that $\|x_n\|_{(-m)} \to 0$. We have to show $\|A^\alpha x_n\| \to 0$. Now $(A^\alpha x_n)$ is a Cauchy sequence in X and has a limit $x \in X$. By $|\alpha| \leq m$ the operator $A^{m-\alpha}(1+A)^{-2m}$ is bounded in X, and we obtain

$$A^m(1+A)^{-2m}x_n = A^{m-\alpha}(1+A)^{-2m}(A^\alpha x_n) \to A^{m-\alpha}(1+A)^{-2m}x.$$

On the other hand, we have by assumption $A^m(1+A)^{-2m}x_n \to 0$. Hence injectivity of $A^{m-\alpha}(1+A)^{-2m}$ yields $x = 0$. $\qquad\square$

Now we are able to study relations between the spaces $\dot{X}_\alpha$.

15.26 Proposition. a) *For any $m \in \mathbb{N}$ we have $Y^{(-m)} = \dot{X}_m + \dot{X}_{-m}$ and $\dot{X}_m \cap \dot{X}_{-m} = D(A^m) \cap R(A^m)$.*

b) *For $\alpha, \beta \geq 0$ we have $\dot{X}_\alpha \cap \dot{X}_{-\beta} = D(A^\alpha) \cap R(A^\beta)$.*

c) *For any $\alpha < \gamma < \beta$ we have $\dot{X}_\alpha \cap \dot{X}_\beta \hookrightarrow \dot{X}_\gamma$.*

d) *For $\alpha > 0$ we have $\dot{X}_\alpha \cap X = X_\alpha$ with equivalent norms.*

Proof. a) We have $\dot{X}_{\pm m} \hookrightarrow Y^{(-m)}$ by boundedness of $A^{m \mp m}(1+A)^{-2m}$. We now define coefficients a_j by

$$(1+z)^{4m-1} - (1+z^{2m})(1+z)^{2m-1} = \sum_{j=1}^{4m-2} a_j z^j, \qquad z \in \mathbb{C}.$$

Then, for $x \in X$,

$$x = \left[(1+A)^{-2m} + \sum_{j=1}^{2m-1} a_j A^j (1+A)^{-4m+1}\right]x$$

$$+ \left[A^{2m}(1+A)^{-2m} + \sum_{j=2m}^{4m-2} a_j A^j (1+A)^{-4m+1}\right]x$$

$$=: y + z$$

where $y \in D(A^m)$ and $z \in D(A^{-m})$. By boundedness of $A^j(1+A)^{-2m+1}$ for $j = 1, \ldots, 2m-1$ and boundedness of $A^{j-2m}(1+A)^{-2m+1}$ for $j = 2m, \ldots, 4m-2$ we obtain

$$\|A^m y\| + \|A^{-m} z\| \leq C\|A^m(1+A)^{-2m}x\| = C\|x\|_{(-m)}$$

for a constant C independent of $x \in X$. The second assertion is a special case of b).

b) By construction it is clear that $D(A^\alpha) \cap R(A^\beta) \subset \dot{X}_\alpha \cap \dot{X}_{-\beta}$. We now use the representation from Proposition 15.25 and fix $m \geq \max(\alpha, \beta)$. Let (x_n) be a sequence in $D(A^\alpha) \cap R(A^\beta)$ which is $\|A^\alpha \cdot \|$–Cauchy and $\|A^{-\beta} \cdot \|$–Cauchy.

Then $(A^\alpha x_n)$ and $(A^{-\beta} x_n)$ converge in X to $y \in X$ and $z \in X$, respectively. Since $A^{-\beta} x_n \in D(A^\alpha A^\beta) = D(A^{\alpha+\beta})$ and $A^\alpha x_n = A^{\alpha+\beta}(A^{-\beta} x_n)$, $n \in \mathbb{N}$, we obtain by the closedness of $A^{\alpha+\beta}$ that $z \in D(A^{\alpha+\beta})$ and $A^{\alpha+\beta} z = y$. Hence $x_0 := A^\beta z = A^{-\alpha} y$ belongs to $D(A^\alpha) \cap R(A^\beta)$ and $x_n \to x_0$ for $\|A^\alpha \cdot \|$ and $\|A^{-\beta} \cdot \|$.

c) The inclusion follows from Theorem 15.14.

d) We use b) with $\beta = 0$. Then

$$\dot{X}_\alpha \cap X = D(A^\alpha) \cap X = D(A^\alpha) = X_\alpha.$$

By boundedness of $A^{-\alpha}(1 + A)^\alpha$ and the open mapping theorem, the norms are equivalent.   □

Sometimes the following result on "shifting" in the scale $\dot{X}_\alpha$ is useful.

15.27 Proposition. *Let A be a sectorial operator in X and $\alpha \in \mathbb{R}$. Define $Y := X_{\alpha,A}$ and $B := A_\alpha$. Then we have, in a canonical way, $Y_{\beta,B} = X_{\alpha+\beta,A}$ and $B_\beta = A_{\alpha+\beta}$ for any $\beta \in \mathbb{R}$.*

Proof. We may assume $\alpha\beta \neq 0$. First we notice

$$B^\beta = (A_\alpha)^\beta = (\widetilde{A^\alpha})^{-1} A^\beta \widetilde{A^\alpha}.$$

Using Remark 15.13 we see that B^β and A^β coincide on

$$D := D(A^{-\alpha} A^\beta A^\alpha) = \begin{cases} D(A^{\alpha+\beta}) & , \quad \alpha\beta > 0 \\ D(A^\alpha) \cap D(A^\beta) , & \alpha\beta < 0 \end{cases}$$

which is a core for B^β in Y and for A^β and $A^{\alpha+\beta}$ in X. Thus for $y \in D$ we have $B^\beta y = A^{-\alpha} A^\beta A^\alpha y \in D(A^\alpha)$ and

$$\|B^\beta y\|_Y = \|A^\alpha B^\beta y\|_X = \|A^\beta A^\alpha y\|_X = \|A^{\alpha+\beta} y\|_X$$

which completes the proof.   □

We recall that a sectorial operator $A \in \mathcal{S}(X)$ in a Banach space X is said to have **bounded imaginary powers** (or BIP for short) if $A^{it} \in B(X)$, $t \in \mathbb{R}$, and there are constants $M \geq 1$, $\gamma \geq 0$ such that

$$\|A^{it}\| \leq M e^{\gamma|t|}, \qquad t \in \mathbb{R}. \tag{15.8}$$

The infimum of all such γ is denoted $\omega_{BIP}(A)$. If A has BIP then, by Remark 15.13, $D(A^\alpha) = D(A^{Re\,\alpha})$ and $\|A^\alpha \cdot \| \sim \|A^{Re\,\alpha} \cdot \|$ for all $\alpha \in \mathbb{C}$.

The following theorem is obtained by a reproduction of the proof of [212, Thm.1.15.2]. We give details and recall to this end the definition of the complex interpolation spaces. Let (Y_0, Y_1) be an interpolation couple of Banach spaces. We denote by S the strip $\{z \in \mathbb{C} : \mathrm{Re}\, z \in (0,1)\}$ and by $\mathcal{F}(Y_0, Y_1)$ the space of all functions $f \in C_b(\overline{S}, Y_0 + Y_1) \cap H(S, Y_0 + Y_1)$ such that, for $j = 0, 1$, the function $t \mapsto F(j + it)$ belongs to $C(\mathbb{R}, Y_j)$ and satisfies $\lim_{|t| \to \infty} \|f(j + it)\|_{Y_j} = 0$. The space $\mathcal{F}(Y_0, Y_1)$ is a Banach space for the norm

$$\|f\|_{\mathcal{F}(Y_0, Y_1)} := \max_{j=0,1} \sup_t \|f(j + it)\|_{X_j}.$$

Then, for $\theta \in (0, 1)$, the complex interpolation space $[Y_0, Y_1]_\theta$ is given by

$$[Y_0, Y_1]_\theta := \{y \in Y_0 + Y_1 : \exists f \in \mathcal{F}(Y_0, Y_1) \text{ with } f(\theta) = y\}.$$

It is a Banach space for

$$\|y\|_{[Y_0, Y_1]_\theta} := \inf\{\|f\|_{\mathcal{F}(Y_0, Y_1)} : f \in \mathcal{F}(Y_0, Y_1), f(\theta) = y\}.$$

For more details we refer to [212, 1.9] or [29, Ch.4]. Observe that $(\dot{X}_\alpha, \dot{X}_\beta)$ is an interpolation couple by Proposition 15.25.

15.28 Theorem. *Suppose that $A \in \mathcal{S}(X)$ is a sectorial operator in a Banach space X and that A has BIP as in (15.8). Then we have, for all real $\beta > \alpha$ and $\theta \in (0, 1)$,*

$$[\dot{X}_\alpha, \dot{X}_\beta]_\theta = \dot{X}_{(1-\theta)\alpha + \theta\beta}$$

and

$$C^{-1}\|x\|_{[\dot{X}_\alpha, \dot{X}_\beta]_\theta} \leq \|x\|_{\dot{X}_{(1-\theta)\alpha + \theta\beta}} \leq C\|x\|_{[\dot{X}_\alpha, \dot{X}_\beta]_\theta}$$

where C only depends on M, γ from (15.8) and on $\beta - \alpha$.

Proof. By Proposition 15.27 we may assume $\alpha = 0$, $\beta > 0$. We now reproduce the proof of [212, Thm.1.15.2] where we want to draw attention to the fact that $0 \in \rho(A)$ is part of the assumption there and that we allow for $0 \in \sigma(A)$.

We choose $m \in \mathbb{N}$ such that $m > \beta$, and let $x \in D(A^m)$. Then

$$\|x\|_{[X, \dot{X}_\beta]_\theta} \leq \|z \mapsto e^{(z-\theta)^2} A^{-(z-\theta)\beta} x\|_{\mathcal{F}(X, \dot{X}_\beta)}$$

$$= \max_{j=0,1} \sup_t \|e^{(j+it-\theta)^2} A^{-it\beta} A^{\theta\beta} x\|$$

$$\leq M \sup_t e^{(j-\theta)^2 - t^2 + \beta\gamma|t|} \|A^{\theta\beta} x\|$$

$$\leq Mee^{(\beta\gamma)^2/4} \|A^{\theta\beta} x\|.$$

On the other hand, by [212, 1.9] the set

$$\mathcal{L} := \text{span} \{e^{\delta z^2 + \lambda z} y : \delta > 0, \lambda \in \mathbb{R}, y \in D(A^m)\}$$

is dense in $\mathcal{F}(X, \dot{X}_\beta)$. We let

$$L(\theta, x) := \{g \in \mathcal{L} : g(\theta) = x\},$$

and obtain, for $x \in D(A^m)$,

$$\|x\|_{[X,\dot{X}_\beta]_\theta} = \inf_{g \in L(\theta,x)} \|g\|_{\mathcal{F}(X,\dot{X}_\beta)} = \inf_{g \in L(\theta,x)} \max_{j=0,1} \sup_t \|A^{j\beta}g(j+it)\|.$$

Now, for $g \in L(\theta,x)$ and $j = 0,1$,

$$\sup_t \|A^{j\beta}g(j+it)\| \geq \frac{1}{M} \sup_t e^{-\gamma\beta|t|} \|A^{\beta(j+it)}g(j+it)\|$$

$$\geq (Mee^{(\beta\gamma)^2/4})^{-1} \sup_t \|e^{(j+it-\theta)^2} A^{\beta(j+it)}g(j+it)\|.$$

Hence

$$\inf_{g \in L(\theta,x)} \|g\|_{\mathcal{F}(X,\dot{X}_\beta)} \geq (Mee^{(\beta\gamma)^2/4})^{-1} \inf_{g \in L(\theta,x)} \|z \mapsto e^{(z-\theta)^2} A^{\beta z}g(z)\|_{\mathcal{F}(X,X)}$$

$$\geq (Mee^{(\beta\gamma)^2/4})^{-1} \|A^{\theta\beta}x\|_{[X,X]_\theta}$$

$$= (Mee^{(\beta\gamma)^2/4})^{-1} \|A^{\theta\beta}x\|_X.$$

The theorem is proved by denseness of $D(A^m)$ in both spaces $\dot{X}_{\theta\beta}$ and $[X,\dot{X}_\beta]_\theta$.

The constant in the general case is $C := Mee^{(\gamma(\beta-\alpha))^2/4}$. □

Notes on Section 15:

N 15.1 is standard. Observe that we may use the lemma to obtain $\rho(A) = \rho(A_Y)$ if $D(A^k) \subset Y$ for some $k \in \mathbb{N}$. Also this application is standard.

N 15.2: This is [56, Thm 3.8]. We follow the proof given there.

N 15.3 - N 15.5 are taken from the dissertation of A.M. Fröhlich ([98]).

N 15.6 - N 15.20: We give a revised and expanded version of Section 10 from [147]. We introduce the extended functional calculus and fractional powers in the style of A. McIntosh ([176], [56]). The composition result 15.11 appeared first in [147]. A more general version was later given in [112]. The proof of 15.16 is from [22]. The results starting with 15.14 are from [138], see also [26], but the proof of 15.14 is from [147] and the proofs of 15.18 - 15.20 may be new. For other presentations based on the H^∞-calculus, see [3], [66], [72], [112]. For further information, in particular on the classical approach to fractional powers via integral representations, see also the monographs [43], [174].

N 15.21 - N 15.27: Interpolation and extrapolation spaces are widely used, see, e.g., [5], [172], [88]. For the inhomogeneous scale (X_α) we refer to, e.g., [88]. The homogeneous scale $(\dot{X}_\alpha)$ is introduced and used in [131] and [136]. The proof of 15.22 is from [112]. For $m = 1$ the space $Y_{(-m)}$ appears in [136] and 15.27 is from [131].

N 15.28: As already mentioned, the case $0 \in \rho(A)$ is covered by [212, Thm.1.15.3]. A converse statement is known to hold in Hilbert space (see [22]), but for general Banach spaces this seems to be open. We remark that in [131] an interpolation method is introduced such that coincidence of interpolation spaces with domains of fractional powers characterizes boundedness of the H^∞-calculus for A.

References

1. M. A. Akcoglu and L. Sucheston, *Dilations of positive contractions on L_p-space*, Can. J. Math. Bull. **20** (1977), 285–292.
2. D. Albrecht, *Functional calculi of commuting unbounded operators*, Ph.D. Thesis Monash University, Australia 1994.
3. D. Albrecht, X. Duong, and A. McIntosh, *Operator theory and harmonic analysis*, in: Instructional Workshop on Analysis
4. D. Albrecht, E. Franks, and A. McIntosh, *Holomorphic functional calculi and sums of commuting operators*, Bull. Austral. Math. Soc. **58** (1998), no. 2, 291–305.
5. H. Amann, *Linear and quasilinear parabolic problems. Vol. I*, Birkhäuser Boston Inc., Boston, MA, 1995, Abstract linear theory.
6. H. Amann, *Operator-valued Fourier multipliers, vector-valued Besov spaces, and applications*, Math. Nachr. **186** (1997), 5–56.
7. H. Amann, *Elliptic operators with infinite-dimensional state spaces*, J. Evol. Equ. **1** (2001), no. 2, 143–188.
8. H. Amann, *Linear parabolic equations with singular potentials*, J. Evol. Equations **3** (2003), 395-406.
9. H. Amann, *Maximal regularity for nonautonomous evolution equations*, preprint.
10. H. Amann, M. Hieber, and G. Simonett, *Bounded H_∞-calculus for elliptic operators*, Differential Integral Equations **7** (1994), no. 3-4, 613–653.
11. W. Arendt, C. J. K. Batty, and S. Bu, *Fourier multipliers for Hölder continuous functions and maximal regularity*, Studia Mathematica, no. 160, 2004, pp. 23-51.
12. W. Arendt, C. J. K. Batty, M. Hieber, and F. Neubrander, *Vector-valued Laplace transforms and Cauchy problems*, Birkhäuser Verlag, Basel, 2001.
13. W. Arendt and S. Bu, *Tools for maximal regularity*, Math. Proc. Cambridge Philos. Soc., no. 134 (2003), pp. 317-336.
14. W. Arendt and S. Bu, *The operator-valued Marcinkiewicz multiplier theorem and maximal regularity*, Math. Z., no. 240 (2002), pp. 311-343.
15. W. Arendt and S. Bu, *Operator-valued Fourier multipliers on periodic Besov spaces and applications*, Ulmer Seminare, no. Heft 6, 2001, pp. 26-41.
16. W. Arendt, S. Bu, and M. Haase, *Functional calculus, variational methods and Liapunov's theorem*, Arch. Math. (Basel) **77** (2001), no. 1, 65–75, Festschrift: Erich Lamprecht.
17. W. Arendt and A. F. M. ter Elst, *Gaussian estimates for second order elliptic operators with boundary conditions*, J. Operator Theory **38** (1997), no. 1, 87–130.
18. P. Auscher, *On necessary and sufficient conditions for L^p-estimates of Riesz transforms associated to elliptic operators on $\mathbb{R}^n$ and related estimates*, Preprint 2004.
19. P. Auscher, S. Hofmann, M. Lacey, J. Lewis, A. McIntosh, and P. Tchamitchian, *The solution of Kato's conjectures*, C. R. Acad. Sci. Paris Sér. I Math. **332** (2001), no. 7, 601–606.
20. P. Auscher, S. Hofmann, J. Lewis, and P. Tchamitchian, *Extrapolation of Carleson measures and the analyticity of Kato's square-root operators*, Acta Math. **187** (2001), no. 2, 161–190.

21. P. Auscher, S. Hofmann, A. McIntosh, and P. Tchamitchian, *The Kato square root problem for higher order elliptic operators and systems on* $\mathbb{R}^n$, J. Evol. Equ. **1** (2001), no. 4, 361–385, Dedicated to the memory of Tosio Kato.

22. P. Auscher, A. McIntosh, and A. Nahmod, *Holomorphic functional calculi of operators, quadratic estimates and interpolation*, Indiana Univ. Math. J. **46** (1997), no. 2, 375–403.

23. P. Auscher, A. McIntosh, and P. Tchamitchian, *Heat kernels of second order complex elliptic operators and applications*, J. Funct. Analysis **152** (1998), 22–73.

24. P. Auscher and M. Qafsaoui, *Equivalence between regularity theorems and heat kernel estimates for higher order elliptic operators and systems under divergence form*, J. Funct. Anal. **177** (2000), no. 2, 310–364.

25. P. Auscher and P. Tchamitchian, *Square root problem for divergence operators and related topics*, Astérisque no. 249, 1998.

26. A.V. Balakrishnan, *Fractional powers of closed operators and the semigroups generated by them*, Pac. J. Math. **10** (1960), 419–437.

27. A. Benedek, A.-P. Calderón, and R. Panzone, *Convolution operators on Banach space valued functions*, Proc. Nat. Acad. Sci. U.S.A. **48** (1962), 356–365.

28. A. Benedek, and R. Panzone, *The space L_p with mixed norm*, Duke Math. J. **28** (1961), pp. 301–324.

29. J. Bergh and J. Löfström, *Interpolation spaces. An introduction*, Springer-Verlag, Berlin, 1976, Grundlehren der Mathematischen Wissenschaften, No. 223.

30. E. Berkson and T. A. Gillespie, *Spectral decompositions and harmonic analysis on UMD spaces*, Studia Math. **112** (1994), no. 1, 13–49.

31. S. Blunck, *Maximal regularity of discrete and continuous time evolution equations*, Studia Math. **146** (2001), no. 2, 157–176.

32. S. Blunck, *Analyticity and discrete maximal regularity on L_p-spaces*, J. Funct. Anal. **183** (2001), no. 1, 211–230.

33. S. Blunck and P. C. Kunstmann, *Weighted norm estimates and maximal regularity*, Adv. Differential Equations **7** (2002), no. 12, 1513-1532.

34. S. Blunck and P. C. Kunstmann, *Calderon-Zygmund theory for non-integral operators and the H^∞ calculus*, Rev. Math. Iberoamericana, Vol. 19, No. 3 (2003), pp. 919–942.

35. S. Blunck and P. C. Kunstmann, *Weak type (p,p)-estimates for Riesz transforms*, to appear in Math. Z.

36. S. Blunck and P. C. Kunstmann, *Generalized Gaussian estimates and the Legendre transform*, to appear in J. Operator Theory.

37. J. Bourgain, *A Hausdorff-Young inequality for B-convex Banach spaces*, Pacific J. Math. **101** (1982), no. 2, 255–262.

38. J. Bourgain, *Some remarks on Banach spaces in which martingale difference sequences are unconditional*, Ark. Mat. **21** (1983), no. 2, 163–168.

39. J. Bourgain, *Vector-valued singular integrals and the H^1-BMO duality*, Probability theory and harmonic analysis (Cleveland, Ohio, 1983), Dekker, New York, 1986, pp. 1–19.

40. J. Bourgain, *Vector-valued Hausdorff-Young inequalities and applications*, Geometric aspects of functional analysis (1986/87), Springer, Berlin, 1988, pp. 239–249.

41. K. Boyadzhiev and R. deLaubenfels, *Semigroups and resolvents of bounded variation, imaginary powers, and H^∞-calculus*, Semigroup Forum **45** (1992), 372-384.

42. D. L. Burkholder, *Martingale transforms and the geometry of Banach spaces*, Probability in Banach spaces, III (Medford, Mass., 1980), Springer, Berlin, 1981, pp. 35–50.

43. P.L. Butzer and H. Berens, *Semi-groups of operators and approximation*, Berlin-Heidelberg-New York, Springer 1967.

44. P. Cannarsa and V. Vespri, *On maximal L^p regularity for the abstract Cauchy problem*, Boll. Un. Mat. Ital. B (6) **5** (1986), no. 1, 165–175.

45. P. Clément, J. Prüss, *Completely positive measures and Feller semigroups*. Math. Ann. 287, No.1, 73-105 (1990).

46. P. Clément, B. de Pagter, F. A. Sukochev, and H. Witvliet, *Schauder decomposition and multiplier theorems*, Studia Math. **138** (2000), no. 2, 135–163.

47. P. Clément and S. Li, *Abstract parabolic quasilinear equations and application to a groundwater flow problem*, Adv. Math. Sci. Appl. **3** (1993/94), no. Special Issue, 17–32.

48. P. Clément and J. Prüss, *An operator-valued transference principle and maximal regularity on vector-valued L_p-spaces*, Evolution equations and their applications in physical and life sciences (Bad Herrenalb, 1998), Dekker, New York, 2001, pp. 67–87.

49. P. Clément and J. Prüss, *Global existence for a semilinear parabolic Volterra equation*, Math. Z. **209** (1992), no. 1, 17–26.

50. R. Coifman and G. Weiss, *Transference Methods in Analysis*, CBMS Conf. series, Amer. Math. Soc., Providence **31** (1997).

51. T. Coulhon and X. T. Duong, *Riesz transforms for $1 \le p \le 2$*, Trans. Amer. Math. Soc. **351** (1999), no. 3, 1151–1169.

52. T. Coulhon and X. T. Duong, *Maximal regularity and kernel bounds: observations on a theorem by Hieber and Prüss*, Adv. Differential Equations **5** (2000), no. 1-3, 343–368.

53. T. Coulhon and X. T. Duong, *Riesz transforms for $p > 2$*, C. R. Acad. Sci. Paris Sér. I Math. **332** (2001), no. 11, 975–980.

54. T. Coulhon and D. Lamberton, *Régularité L^p pour les équations d'évolution*, Séminaire d'Analyse Fonctionelle 1984/1985, Univ. Paris VII, Paris, 1986, pp. 155–165.

55. T. Coulhon and D. Lamberton, *Quelques remarques sur la régularité L^p du semi-groupe de Stokes* Commun. Partial Differ. Equations **17** (1992), no. 1/2, 287-304.

56. M. Cowling, I. Doust, A. McIntosh, and A. Yagi, *Banach space operators with a bounded H^∞ functional calculus*, J. Austral. Math. Soc. Ser. A **60** (1996), no. 1, 51–89.

57. M. Cowling, *Harmonic analysis on semigroups*, Ann. of Math. (2) **117** (1983), no. 2, 267–283.

58. D. Daners, *Heat kernel estimates for operators with boundary conditions*, Math. Nachr. **217** (2000), 13-41.

59. G. Da Prato and P. Grisvard, *Sommes d'opérateurs linéaires et équations différentielles opérationnelles*, J. Math. Pures Appl. (9) **54** (1975), no. 3, 305–387.

60. E. B. Davies, *One parameter semigroups*, Academic Press (1980), London.

61. E. B. Davies, *Uniformly elliptic operators with measurable coefficients*, J. Funct. Anal. **132** (1995), no. 1, 141–169.

62. E. B. Davies, *Limits on L^p regularity of self-adjoint elliptic operators*, J. Differential Equations **135** (1997), no. 1, 83–102.

63. E. B. Davies, *Heat kernels and spectral theory*, Cambridge University Press, Cambridge, 1990.

64. Luciano de Simon, *Un'applicazione della teoria degli integrali singolari allo studio delle equazioni differenziali lineari astratte del primo ordine*, Rend. Sem. Mat. Univ. Padova **34** (1964), 205–223.

65. R. Denk, G. Dore, M. Hieber, J. Prüß, and A. Venni, *New thoughts on old ideas of Seeley*, Math. Ann. (to appear)

66. R. Denk, M. Hieber, and J. Prüß, *R-boundedness, Fourier multipliers and problems of elliptic and parabolic type*, Mem. Amer. Math. Soc., (to appear).

67. W. Desch, M. Hieber, and J. Prüß, *L^p-theory of the Stokes equation in a half space*, J. Evol. Equ. **1** (2001), no. 1, 115–142.

68. J. Diestel and J. J. Uhl, Jr., *Vector measures*, American Mathematical Society, Providence, R.I., 1977, With a foreword by B. J. Pettis, Mathematical Surveys, No. 15.

69. J. Diestel, H. Jarchow, and A. Tonge, *Absolutely summing operators*, Cambridge University Press, Cambridge, 1995.

70. G. Dore, *L^p regularity for abstract differential equations*, Functional analysis and related topics, 1991 (Kyoto), Springer, Berlin, 1993, pp. 25–38.

71. G. Dore, *Maximal regularity in L^p spaces for an abstract Cauchy problem*, Adv. Differential Equations **5** (2000), no. 1-3, 293–322.

72. G. Dore, *Fractional powers of closed operators*, preprint.

73. G. Dore and A. Venni, *On the closedness of the sum of two closed operators*, Math. Z. **196** (1987), no. 2, 189–201.

74. G. Dore and A. Venni, *H^∞ Functional calculus for an elliptic operator on a half-space with general boundary conditions*, Preprint 2001.

75. G. Dore and A. Venni, *H^∞ Functional calculus for sectorial and bisectorial operators*, Preprint 2001.

76. M. Duelli, *Diagonal operators and L_p-maximal regularity*, Ulmer Seminare in Funktionalanalysis und Differentialgleichungen 2000, 156–173.

77. N. Dunford and J. Schwartz, *Linear Operators I*, Wiley and Sous, (1957).

78. X. T. Duong and D. W. Robinson, *Semigroup kernels, Poisson bounds, and holomorphic functional calculus*, J. Funct. Anal. **142** (1996), no. 1, 89–128.

79. X. T. Duong and G. Simonett, *H_∞-calculus for elliptic operators with nonsmooth coefficients*, Differential Integral Equations **10** (1997), no. 2, 201–217.

80. X. T. Duong, *H_∞ functional calculus of second order elliptic partial differential operators on L^p spaces*, Miniconference on Operators in Analysis (Sydney, 1989), Austral. Nat. Univ., Canberra, 1990, pp. 91–102.

81. X. T. Duong and A. McIntosh, *Functional calculi of second-order elliptic partial differential operators with bounded measurable coefficients*, J. Geom. Anal. **6** (1996), no. 2, 181–205.

82. X. T. Duong and A. McIntosh, *Singular integral operators with non-smooth kernels on irregular domains*, Rev. Mat. Iberoamericana **15** (1999), no. 2, 233–265.

83. X. T. Duong and A. McIntosh, *Corrigenda: "Singular integral operators with non-smooth kernels on irregular domains"*, Rev. Mat. Iberoamericana **16** (2000), no. 1, 217.

84. X. T. Duong, H^∞-functional calculus of second order elliptic partial differential operators on L_p-spaces, Proc. Centre for Math. Analysis ANU, Canberra **24** (1998), 91–102.

85. X. T. Duong, Li Xi Yan, Bounded holomorphic functional calculus for non-divergence form operators, Differential Integral Equations **15** (2002), 709-730.

86. R. E. Edwards and G. I. Gaudry, Littlewood-Paley and Multiplier theory Springer (1977)

87. Yu. V. Egorov, M. A. Shubin, Foundations of the classical theory of partial differential equations, Transl. from the Russian by R. Cooke, 2nd printing of the 1st ed. 1992, Springer 1998.

88. K.-J. Engel and R. Nagel, One-parameter semigroups for linear evolution equations, Springer-Verlag, New York, 2000, With contributions by S. Brendle, M. Campiti, T. Hahn, G. Metafune, G. Nickel, D. Pallara, C. Perazzoli, A. Rhandi, S. Romanelli and R. Schnaubelt.

89. J. Escher, J. Prüß, and G. Simonett, Analytic solutions for the Stefan problem with Gibbs-Thomson correction, J. Reine Angew. Math., J. Reine Angew. Math **563** (2003), pp. 1–52.

90. G. Fendler, dilations of one-parameter semigroups of positive contractions on L_p-spaces, Can. J. Math. (1998), 731–748.

91. A. Favini, G. R. Goldstein, J .A. Goldstein, and S. Romanelli, Degenerate second order differential operators generating analytic semigroups in L^p and W_p^1, Math. Nachr. **238** (2002), 78-102.

92. A. Favini, G. R. Goldstein, J .A. Goldstein, and S. Romanelli, The heat equation with generalized Wentzell boundary conditions, J. Evolution Equations **2** (2002), 1-19.

93. C. Fefferman and E. M. Stein, Some maximal inequalities, Amer. J. Math. **93** (1971), 107–115.

94. E. Franks and A. McIntosh, Discrete quadratic estimates and holomorphic functional calculi in Banach spaces, Bull. Austral. Math. Soc. **58** (1998), no. 2, 271–290.

95. A. Fröhlich, Stokes- und Navier-Stokes-Gleichungen in gewichteten Funktionenräumen, Dissertation, TU Darmstadt 2001, Shaker Verlag.

96. A. Fröhlich, Maximal regularity for the non-stationary Stokes system in an aperture domain, J. Evol. Equ. **2** (2002), 471-493.

97. A. Fröhlich, The Stokes operator in weighted L^q-spaces I: Weighted estimates for the Stokes resolvent problem in a half space, J. Math. Fluid Mech. **5** (2003), 166-199.

98. A.M. Fröhlich, H^∞-Kalkül und Dilatationen, Dissertation, Univ. Karlsruhe 2003.

99. A.M. Fröhlich and L. Weis, H^∞-calculus of sectorial operators and dilations, Preprint.

100. J. García-Cuerva, K. S. Kazarian and V. I. Kolyada and J. L. Torrea, Vector-valued Hausdorff-Young inequality and applications, Russian Math. Surveys, **53**, (1998), pp. 435–513.

101. J. García-Cuerva and J. L. Rubio de Francia, Weighted norm inequalities and related topics, North-Holland Publishing Co., Amsterdam, 1985, Notas de Matemática , 104.

102. D. Gilbarg, and N. S. Trudinger, Elliptic partial differential equations of second order, Reprint of the 1998 ed., Classics in Mathematics, Springer, 2001.

103. M. Girardi and L. Weis, *Operator-valued Fourier multiplier theorems on Besov spaces*, Math. Nachr. **251** (2003), 34–51.

104. M. Girardi and L. Weis, *Operator-valued Fourier multiplier theorems on $L_p(X)$ and geometry of Banach spaces*, J. Funct. Anal. **204** (2003), 320–354.

105. M. Girardi and L. Weis, *Criteria for R-boundedness of operator families*, in Evolution Equations, edited by G.R. Goldstein, R. Nagel, and S. Romanelli, Lecture Notes in Pure and Applied Mathematics **234**, Marcel Dekker 2003, 203–221.

106. M. Girardi and L. Weis, *Vector-valued extensions of some classical theorems in harmonic analysis*, in Analysis and Applications - ISSAC 2001, edited by H.G.W Begehr, R.P. Gilbert, and M.W. Wong, Kluwer Academic Publishers 2003, International Society for Analysis, Applications and Computation **10**, 171–185.

107. M. Girardi and L. Weis, *Integral operators with operator-valued kernels*, J. Math. Anal. Appl. **290** (2004), 190–212.

108. G. R. Goldstein, J .A. Goldstein, and S. Romanelli, *A new approach to the analyticity of some classes of one-parameter semigroups in weighted L^p-spaces*, J. Math. Anal. Appl. **226** (1998), 393-413.

109. J. A. Goldstein, *Semigroups of linear operators and applications*, The Clarendon Press Oxford University Press, New York, 1985.

110. P. Grisvard, *Équations différentielles abstraites*, Ann. Sci. École Norm. Sup. (4) **2** (1969), 311–395.

111. M. Haase, *A characterization of group generators on Hilbert spaces and the H^∞-calculus*, Semigroup Forum **66** (2003), pp. 288–304

112. M. Haase, *Functional Calculus for Sectorial Operators and Similarity Methods*, Ph.D. thesis, Ulm 2003.

113. R. Haller, H. Heck, and M. Hieber, *Muckenhoupt weights and maximal L^p-regularity*, Arch. Math., (to appear).

114. R. Haller, H. Heck, and A. Noll, *Mikhlin's theorem for operator-valued multipliers in n-variables*, Math. Nachr. 244 (2002), pp. 110–130.

115. W. Hebisch, *A multiplier theorem for Schrödinger operators*, Colloq. Math. **60/61**, No.2 (1990), 659–664.

116. H. Heck, M. Hieber, *Maximal L^p-regularity for elliptic operators with VMO-coefficients*, J. Evol. Equ. 3, No.2, pp. 332-359 (2003).

117. M. Hieber, *Examples of pseudo-differential operators in L_p-spaces with unbounded imaginary powers*, Archiv Math. **66** (1996), 126–131.

118. M. Hieber and S. Monniaux, *Pseudo-differential operators and maximal regularity results for non-autonomous parabolic equations*, Proc. Amer. Math. Soc. **128** (2000), no. 4, 1047–1053.

119. M. Hieber and J. Prüß, *Heat kernels and maximal L^p-L^q estimates for parabolic evolution equations*, Comm. Partial Differential Equations **22** (1997), no. 9-10, 1647–1669.

120. M. Hieber and J. Prüß, *Functional calculi for linear operators in vector-valued L^p-spaces via the transference principle*, Adv. Differential Equations **3** (1998), no. 6, 847–872.

121. M. Hoffmann, N. Kalton, and T. Kucherenko, *R–bounded approximating sequences and applications to semigroups*, to appear in JMAA.

122. S. Hofmann and J.M. Martell, *Sharp L^p-bounds for Riesz transforms and square roots associated to second order elliptic operators*, Preprint 2002.

123. T. Hytönen, *Convolutions, multipliers, and maximal regularity on vector-valued Hardy spaces*, Helsinki University of Technology Institute of Mathematics Research Reports (Preprint).

124. T. Hytönen, *R-boundedness and multiplier theorems*, Master's thesis, Helsinki Univ. Techn. Inst. Math. Research Report C16, (2001).

125. T. Hytönen, *Fourier-embeddings and Mihlin-type multiplier theorems*, to appear in Math. Nachr.

126. T. Hytönen and L. Weis, *Singular convolution integrals with operator-valued kernels*, (submitted).

127. T. Hytönen and L. Weis, *Singular integrals in Besov spaces*, to appear in Math. Nachr.

128. M. Kac, *Statical independence in probability, analysis and number theory*, Amer. Math. Soc. (1959).

129. J.-P. Kahane, *Some random series of functions*, Cambridge University Press, (1985).

130. N.J. Kalton, *A remark on sectorial operators with an H^∞-calculus*, Trends in Banach spaces and Operator theory, 91-99 Contemp. Math. 321, AMS, Rhode Island (2003).

131. N. J. Kalton, P. C. Kunstmann, and Lutz Weis, *Comparison and perturbation theorems for the H^∞-calculus*, (in preparation).

132. N. J. Kalton and G. Lancien, *A solution to the problem of L_p-maximal regularity* Math. Z. **235** (2000), no. 3, 559–568.

133. N. J. Kalton and G. Lancien, *L_p-maximal regularity on Banach spaces with a Schauder basis*, Archiv der Mathematik **78** (2002), no. 397–408.

134. N. J. Kalton and L. Weis, *The H^∞-calculus and sums of closed operators*, Math. Ann. **321** (2001), no. 2, 319–345.

135. N. J. Kalton and L. Weis, *H^∞-functional calculus and square functions estimates* in preparation.

136. N.J. Kalton and L. Weis, in preparation.

137. T. Kato, *Perturbation theory for linear operators*, Springer, 1980.

138. H. Komatsu, *Fractional powers of operators*, Pac. J. Math. **19** (1966), 285–346.

139. U. Krengel, *Ergodic Theory*, De Guyter, (1985).

140. T. Kucherenko and L. Weis, *Operators with a bounded H^∞-calculus on L_1*, in preparation.

141. P. C. Kunstmann, *Maximal L_p-regularity for second order elliptic operators with uniformly continuous coefficients on domains*, Progress in Nonlinear Differential Equations and Their Applications, Vol. 55, 293-305, Birkhäuser 2003.

142. P. C. Kunstmann, *L_p-spectral properties of the Neumann Laplacian on horns, comets, and stars*, Math. Z. **242** (2002), 183–201.

143. P. C. Kunstmann, *L_p-spectral properties of elliptic differential operators*, Habilitationsschrift, Univ. Karlsruhe, 2002.

144. P. C. Kunstmann and Ž. Štrkalj, *H^∞-calculus for submarkovian generators*, Proc. Amer. Math. Soc. 131, 2081-2088 (2003).

145. P. C. Kunstmann and H. Vogt, *Weighted norm estimates and L_p-spectral independence of linear operators*, to appear in Colloq. Math.

146. P. C. Kunstmann and L. Weis, *Perturbation theorems for maximal L_p-regularity*, Ann. Sc. Norm. Sup. Pisa **XXX** (2001), 415–435.

147. P. C. Kunstmann and L. Weis, *Functional calculus and differential operators*, Internetseminar, Karlsruhe 2002.

148. S. Kwapień, *Isomorphic, characterization of inner product spaces by orthogonal series with vector-valued coefficients*, Studia Math. **44** (1972), no. 583–595.

149. S. Kwapień, *On Banach spaces containing c_0*, Studia Math. **52** (1974), 187–188,

150. O. A. Ladyzhenskaya, V. A. Solonnikov, and Ural'tseva, N.N. *Linear and quasilinear equations of parabolic type*, translated from the Russian by S. Smith, Translations of Mathematical Monographs 23, Providence, RI, American Mathematical Society. XI (1968).

151. D. Lamberton *Equations d'évolution linéaires associées à des semi-groupes de contractions dans les espaces L_p*, J. Funct. Anal. **72** (1987), 252–262.

152. F. Lancien, G. Lancien, and C. Le Merdy, *A joint functional calculus for sectorial operators with commuting resolvents*, Proc. London Math. Soc. (3) **77** (1998), no. 2, 387–414.

153. G. Lancien, *Counterexamples concerning sectorial operators*, Arch. Math. (Basel) **71** (1998), no. 5, 388–398.

154. G. Lancien and C. Le Merdy, *A generalized H^∞ functional calculus for operators on subspaces of L^p and application to maximal regularity*, Illinois J. Math. **42** (1998), no. 3, 470–480.

155. R. Latala and K. Oleszkiewicz, *On the best constant in the Khinchin-Kahane inequality*, Studia Mathematica **109** (1994), 101–104.

156. Y. Latushkin and F. Räbiger, *Fourier multipliers in stability and control theory*, (Preprint).

157. C. Le Merdy, *Counterexamples on L_p-maximal regularity*, Math. Z. **230** (1999), no. 1, 47–62.

158. C. Le Merdy, *H^∞-functional calculus and applications to maximal regularity*, Publ. Math. UFR Sci. Tech. Besançon. **16** (1998), 41-77.

159. C. Le Merdy, *Two results about H^∞ functional calculus on analytic UMD Banach spaces*, J. Austral. Math. Soc. 74 (2003), pp. 351–378.

160. C. Le Merdy, *The Weiss conjecture for bounded analytic semigroups*, J. London Math. Soc., **67**, No. 3 (2003), pp. 715–738.

161. C. Le Merdy, *On square functions associated to sectorial operators*, Bull. Soc. Math. France **132** (2004), 137–156.

162. C. Le Merdy, *The similarity problem for bounded analytic semigroups on Hilbert space*, Semigroup Forum **56** (1998), 205-224.

163. C. Le Merdy, *Similarities of ω-accretive operators*, Preprint.

164. J. Lindenstrauss and L. Tzafriri, *Classical Banach spaces*, Springer-Verlag, Berlin, 1973, Lecture Notes in Mathematics, Vol. 338.

165. J. Lindenstrauss and L. Tzafriri, *Classical Banach spaces. I*, Springer-Verlag, Berlin, 1977, Sequence spaces, Ergebnisse der Mathematik und ihrer Grenzgebiete, Vol. 92.

166. J. Lindenstrauss and L. Tzafriri, *Classical Banach spaces. II*, Springer-Verlag, Berlin, 1979, Function spaces.

167. V. A. Liskevich, *On C_0-semigroups generated by elliptic second order differential expressions on L^p-spaces*, Differential Integral Equations **9** (1996), 811-826.

168. V. A. Liskevich and M. A. Perel'muter, *Analyticity of sub-Markovian semigroups*, Proc. Amer. Math. Soc. **123** (1995), no. 4, 1097–1104.

169. V. A. Liskevich and Yu. A. Semenov, *Some problems on Markov semigroups*, Schrödinger operators, Markov semigroups, wavelet analysis, operator algebras, Akademie Verlag, Berlin, 1996, pp. 163–217.

170. V. A. Liskevich, Z. Sobol, and H. Vogt, *On L_p-theory of c_0-semigroups associated with second order elliptic operators II*, J. Funct. Anal., Vol. 193, No. 1 (2002), pp. 55–76.

171. V. A. Liskevich and H. Vogt, *On L^p-spectra and essential spectra of second-order elliptic operators*, Proc. London Math. Soc. (3) **80** (2000), no. 3, 590–610.

172. A. Lunardi, *Analytic semigroups and optimal regularity in parabolic problems*, Birkhäuser Verlag, Basel, 1995.

173. Z. M. Ma and M. Röckner, *Introduction to the theory of non symetric Dirichlet forms*, Springer (1992).

174. C. Martinez and M. Sanz, *The theory of fractional powers of operators*, North-Holland Mathematics Studies **187**, Amsterdam, Elsevier 2001.

175. T. R. McConnell, *On Fourier multiplier transformations of Banach-valued functions*, Trans. Amer. Math. Soc. **285** (1984), no. 2, 739–757.

176. A. McIntosh, *Operators which have an H_∞ functional calculus*, Miniconference on operator theory and partial differential equations (North Ryde, 1986), Austral. Nat. Univ., Canberra, 1986, pp. 210–231.

177. A. McIntosh and A. Yagi, *Operators of type ω without a bounded H^∞-functional calculus*, Proc. Cent. Math. Anal. Aust. Natl. Univ. **24** (1990), 159-172.

178. Y. Miyakazi, *The L^p resolvents of elliptic operators with unifomly continuous coefficients*, J. Differential Equations **188** (2003), 555–568.

179. S. Monniaux, *On uniqueness for the Navier-Stockes system in 3D-bounded Lipschitz domains*, J. Funct. Anal., Vol. 195, No. 1 (2002), pp. 1–11.

180. R. Nagel, (ed.), *One parameter semigroups of positive operators*, Lect. Notes in Math., **1184**, Springer, (1986).

181. B. Sz. Nagy, *On uniformly bounded linear trasformations in Hilbert spaces*, Acta. Univ. Szeged, Acta Sci Math. II: (1947),291–305.

182. J. van Neerven, *The adjoint of a semigroup of linear operators*, Lecture Notes in Mathematics **1529**, Springer 1992.

183. A. Noll and J. Saal, *H^∞-calculus for the Stokes operator on L_q-spaces*, Math. Z. 244, No. 3, pp. 651-688.

184. E.M. Ouhabaz, *L^∞-contractivity of semigroups generated by sectorial forms*, J. Lond. Math. Soc., II. Ser. **46** (1992), 529-542.

185. E.M. Ouhabaz, *Gaussian estimates and holomorphy of semigroups*, Proc. Am. Math. Soc. **123** (1995), 1465-1474.

186. A. Pazy, *Semigroups of linear operators and applications to partial differential equations*, Springer-Verlag, New York, 1983.

187. J. Peetre, *Sur la transformation de Fourier des fonctions à valeurs vectorielles*, Rend. Sem. Mat. Univ. Padova **42** (1969), 15–26.

188. G. Pisier, *Some results on Banach spaces without local unconditional structure*, Compositio Math. **37** (1978), no. 1, 3–19.

189. G. Pisier, *Factorization of linear operators and geometry of Banach spaces*, Published for the Conference Board of the Mathematical Sciences, Washington, DC, 1986.

190. G. Pisier, *The Volume of Convex Bodies and Banach Space Geometry*, Cambridge University Press (1989).

191. G. Pisier, Les Inégalités de Khinchine-Kahane, (d'après C. Borell) École Polyt. Palaiseau, *Sém. Géom. Espace de Banach* (1977/78), Exp.VII.

192. J. Prüß, *Maximal regularity for evolution equations in L_p-spaces*, Conferenze del Seminario di Matematica dell'Universita di Bari, No. **285**, 1–39, Aracne Roma 2002.

193. J. Prüß and R. Schnaubelt, *Solvability and maximal regularity of parabolic evolution equations with coefficients continuous in time*, J. Math. Anal. Appl. **256** (2001), no. 2, 405–430.

194. J. Prüß and H. Sohr, *On operators with bounded imaginary powers in Banach spaces*, Math. Z. **203** (1990), no. 3, 429–452.

195. J. Prüß and H. Sohr, *Imaginary powers of elliptic second order differential operators in L^p-spaces*, Hiroshima Math. J. **23** (1993), no. 1, 161–192.

196. J. L. Rubio de Francia, *Martingale and integral transforms of Banach space valued functions*, Lecture Notes in Math. 1221 (1986), pp. 195–222.

197. M. Reed and B. Simon, *Methods of modern mathematical physics II, Fourier analysis, Selfadjointness*, Academic Press (1975).

198. L. Saloff-Coste, *Aspects of Sobolev-type inequalities*, London Mathematical Society Lecture Note Series **289**, Cambridge University Press 2002.

199. G. Schreieck, *L_p-Eigenschaften der Wärmeleitungs-Halbgruppe mit singulärer Absorption*, doctoral dissertation, Aachen, 1996, Shaker Verlag.

200. G. Schreieck and J. Voigt, *Stability of the L_p-spectrum of Schrödinger operators with form-small negative part of the potential*, Functional analysis (Essen, 1991), Dekker, New York, 1994, pp. 95–105.

201. R. T. Seeley, *Complex powers of an elliptic operator*, Singular Integrals (Proc. Sympos. Pure Math., Chicago, Ill., 1966), Amer. Math. Soc., Providence, R.I., 1967, pp. 288–307.

202. Z. Sobol and H. Vogt, *On L_p-theory of c_0-semigroups associated with second order elliptic operators*, J. Funct. Anal., Vol. 193, No. 1, pp. 24–54.

203. P. E. Sobolevskiĭ, *Coerciveness inequalities for abstract parabolic equations*, Dokl. Akad. Nauk SSSR **157** (1964), 52–55.

204. H. Sohr and G. Thäter, *Imaginary powers of second order differential operators and L^q-Helmholtz decomposition in the infinite cylinder*, Math. Ann. **311** (1998), no. 3, 577–602.

205. E. M. Stein, *Harmonic analysis: real-variable methods, orthogonality, and oscillatory integrals*, Princeton University Press, Princeton, NJ, 1993, With the assistance of Timothy S. Murphy, Monographs in Harmonic Analysis, III.

206. E. M. Stein, *Singular Integrals and Differentiability Properties of Functions*, Princeton University Press (1970).

207. E. M. Stein, *Topics in harmonic analysis related to the Littlewood-Paley theory*, Princeton University Press and the University of Tokyo Press, 1970.

208. E. M. Stein and Guido Weiss, *Introduction to Fourier analysis on Euclidean spaces*, Princeton University Press, Princeton, N.J., 1971, Princeton Mathematical Series, No. 32.

209. Ž. Štrkalj, *R-Beschränktheit, Summensätze abgeschlossener Operatoren und operatorwertige Pseudodifferentialoperatoren*, doctoral dissertation, Universität Karlsruhe, Karlsruhe, Germany, 2000.

210. Ž. Štrkalj and Lutz Weis, *On operator-valued Fourier multiplier theorems*, to appear in Trans. Amer. Math. Soc.

211. H. Tanabe, *Equations of evolution*, Pitman (Advanced Publishing Program), Boston, Mass., 1979, Translated from the Japanese by N. Mugibayashi and H. Haneda.

212. H. Triebel, *Interpolation theory, function spaces, differential operators*, North-Holland Publishing Company 1978.

213. H. Triebel, *Theory of function spaces* Monographs in Mathematics, Vol. **78**, Birkhäuser 1983.

214. A. Venni: *A counterexample concerning imaginary power of linear operators* Lecture Notes in Mathematics 1540, Springer, Berlin (1993).

215. A. Venni, *Marcinkiewicz and Mihlin multiplier theorems, and R-boundedness*, in Iannelli, Mimmo (ed.) et al., Evolution equations: applications to physics, industry, life sciences and economics. Proceedings of the 7th international conference on evolution equations and their applications, EVEQ2000 conference, Levico Terme, Italy, October 30-November 4, 2000. Basel: Birkhäuser. Prog. Nonlinear Differ. Equ. Appl. 55, 403-413 (2003).

216. J. Voigt, *Abstract Stein Interpolation*, Mathematische Nachrichten, **157**, (1992), pp. 197–199.

217. J. Voigt, *The sector of holomorphy for symmetric sub-Markovian semigroups*, Functional analysis (Trier, 1994), de Gruyter, Berlin, 1996, pp. 449–453.

218. J. Voigt, *One-parameter semigroups acting simultaneously on different L_p-spaces*, Bull. Soc. R. Sci. Lige **61** (1992), 465-470.

219. L. Weis, *Stability theorems for semi-groups via multiplier theorems*, Differential equations, asymptotic analysis, and mathematical physics (Potsdam, 1996), Akademie Verlag, Berlin, 1997, pp. 407–411.

220. L. Weis, *A new approach to maximal L_p-regularity*, Evolution equations and their applications in physical and life sciences (Bad Herrenalb, 1998), Dekker, New York, 2001, pp. 195–214.

221. L. Weis, *Operator-valued Fourier multiplier theorems and maximal L_p-regularity*, Math. Ann. **319** (2001), no. 4, 735–758.

222. D. Werner, *Funktionalanalysis* (Functional analysis) 4. ed., Springer, 2002.

223. H. Witvliet, *Unconditional schauder decompositions and multiplier theorems*, doctoral dissertation, Technische Universiteit Delft, November 2000.

224. J. Wloka, *Partial differential equations*, Transl. from the German by C. B. and M. J. Thomas, Cambridge University Press, 1987.

225. A. Yagi, *Coincidence entre des espaces d'interpolation et des domaines de puissances fractionnaires d'opérateurs* C. R. Acad. Sci., Paris, Sér. I **299** (1984), 173-176.

226. F. Zimmermann, *On vector-valued Fourier multiplier theorems*, Studia Math. **93** (1989), no. 3, 201–222.

227. A. Zygmund, *Trigonometric Series*, Cambridge University Press (1935).

Optimal Control Problems and Riccati Equations for Systems with Unbounded Controls and Partially Analytic Generators-Applications to Boundary and Point Control Problems

Irena Lasiecka[1]

University of Virginia, Department of Mathematics, Kerchof Hall, P.O. Box 400137, Charlottesville, VA 22904-4137, USA.
il2v@weyl.math.virginia.edu

1 Introduction

The purpose of these notes is to present a relatively recent theory of optimal control problems and related Riccati Equations arising in the context of partially analytic dynamics with boundary and point controls. In the last 15 years or so there has been strong activity in the area of control theory with the so called "unbounded control actions" - see [6, 40, 14, 39, 24] and references therein. These include canonical examples of boundary and point controls which are motivated by numerous applications involving modern technology and, in particular, "smart materials". The results obtained depend heavily on the type of dynamics considered. The bottom line is that in the case of systems generated by *analytic semigroups* the theory is very regular with a lot of nice properties shared by solutions to Riccati equation. It is not so, however, in the case of general dynamics (i.e., generators of strongly continuous semigroups). Indeed, the unboundedness of the control actions propagates through the dynamics of control process having strong effect on lack of regularity of solutions to Riccati equations. In fact, in many cases one has to be very careful in defining solutions to Riccati equations, which may not exist in any reasonable classical sense (even as weak solutions - cf. [58, 61]). Moreover, classical results within "bounded control" theory such as boundedness of gain operators (critical property in applications) are no longer valid, see [39]. In fact, in some hyperbolic dynamics the gain operator may be even multivalued, see [58, 11]. This phenomena is reminiscent to the theory of HJB equations, where due to

[1] Research partially supported by the National Science Foundation under Grant DMS-0104305 and by the Army Research Office under Grant DAAD19-02-10179

limited regularity of value function the synthesis is often described by multi-valued operators. The consequence of this is that applicability of such results may be strongly limited. This mathematical phenomena, along with numerous examples arising in modern applications (more on this below), spurred interest in considering and studying other classes of dynamical systems which are not analytic but which posses certain special properties allowing for more regular Riccati Theory. In fact, this class of system is referred to as *systems with singular estimate*. The singular estimate property refers to the pair (A, B), where A is a generator of a given semigroup and B is a control operator. The property in question describes singular behaviour at the origin for the quantity $|e^{At}B|_{\mathcal{L}(U \to H)}$; $t > 0$. It is clear that in the case when B is a bounded operator, i.e., $B \in \mathcal{L}(U \to H)$, there is no singularity. On the other hand, analytic semigroups with relatively bounded operators B display singularity of the type $\frac{c}{t^\gamma}$ for some γ between zero and one. However, for more general unbounded control operators the above quantity may not be defined even away from the origin. So the class of the systems we are interested in displays a rather special behaviour. For this class of systems it has been shown that the Riccati theory and Optimal Control theory is sufficiently regular allowing for semiclassical solutions to nonlinear PDE's describing Riccati solutions. In addition, it turned out that this special class of systems is quite prevailing in applications. In fact, many (much more than originally expected) systems arising in control theory belong to the class of control systems with singular estimate. These typically involve coupled PDE systems where the interaction of two different types of dynamics (typically hyperbolic and parabolic) takes place.

Since the class of "singular estimate" control systems is defined through the pair: generator and control operator, one may also look at the modelling problem from the purely control point of view: find control operators which will induce a given dynamics to belong to this special singular estimate class. An obvious class of such systems are systems generated by analytic semigroups. But as stated before, we are interested in proper extension of the analytic class. In fact, we will see that even some of the hyperbolic systems may acquire desired properties.

It is the purpose of these notes to provide an exposition of theory relevant to this class of singular systems and to put this in the context of other optimal control theories available for PDE's. This will be the subject of Part I of these lectures. In the second Part II, our aim is to provide a number of concrete examples of control systems which not only illustrate applicability of the theory, but provide very motivation for this theory to exist.

These lectures focus on control problems with quadratic functional only within the context of linear dynamics. Extensions of the theory to H_∞ problems as well as to mini -max problems are given in [60]. Also, many nonlinear problems can be formulated within the context of models described. Some of these were already considered in the literature, however mostly from the stability point of view, see [34].

I. Abstract Theory

2 Mathematical Setting and Formulation of the Control Problem

In this section we consider an abstract formulation of optimal control problem which is governed by strongly continuous semigroups with unbounded control operators. The semigroup in question, along with control operators, will be assumed to satisfy the "singular" estimate.

2.1 Dynamics and Related Control Problem

Let H, U, Z be given separable Hilbert spaces. We consider the following abstract state equation:

$$\mathbf{y}_t(t) = A\mathbf{y}(t) + Bu(t) + F(t); \;\; on \; [\mathcal{D}(A^*)]'; \;\; \mathbf{y}(0) \in H \qquad (2.1)$$

subject to the following assumptions:

Hypothesis 2.1. • $A : \mathcal{D}(A) \subset H \to H$ *is a generator of a C_0 semigroup e^{At} on H with $\mathcal{D}(A) \subset H \subset [\mathcal{D}(A^*)]'$, where duality is considered with respect to the pivot space H.*
• *The linear operator $B : U \to [\mathcal{D}(A^*)]'$ satisfies the following condition:*

$$|R(\lambda, A)Bu|_H \le C_{Re\lambda}|u|_U; \;\; \lambda \in \rho(A) \qquad (2.2)$$

where $R(\lambda, A)$ denotes the resolvent operator for A and $\rho(A)$ denotes the resolvent set.
• *The operator $R : H \to Z$ is bounded.*
• *$F \in L_2(0, T; H)$ is a given element.*

With the the **dynamics** (2.1) we associate the following quadratic cost functional where $T > 0$ may be finite or infinite

$$J(u, \mathbf{y}) \equiv \int_0^T [|R\mathbf{y}|_Z^2 + |u|_U^2]dt. \qquad (2.3)$$

The corresponding Optimal Control Problem is the following:
Optimal Control Problem *For a fixed $F \in L_2(0, T; H)$, and fixed initial condition $\mathbf{y}(0) \in H$ minimize the functional $J(u, \mathbf{y})$ over all $u \in L_2(0, T; U)$ where $\mathbf{y} \in L_2(0, T; H)$ is a solution to (2.1).*

For *finite time horizon* problems $(T < \infty)$ a standard optimization argument (see [11, 6, 49]) provides existence and uniqueness of optimal solution $u^0 \in L_2(0, T; U)$; $\mathbf{y}^0 \in L_2((0, T); H)$ to the optimal control problem. For the *infinite time horizon* problems, the same conclusion follows provided that the so called FCC condition - *Finite Cost Condition* formulated in Hypothesis 3.4 - is satisfied. FCC condition is implied by appropriate stabilizability or controllability properties of the dynamics generated by A, B.

Our main aim is to derive the optimal synthesis for the control problem along with a characterization of optimal control via an appropriate Riccati Equation. As mentioned already in the introduction, Hypothesis 2.1 alone, will not suffice in order to obtain regular synthesis and well posed Riccati equations. While full characterization of optimal solution is obtainable under minimal assumption of Hypothesis 2.1, it is the wellposedness of Riccati Equation (RE) which requires additional hypotheses. In fact, without these it is known (see [58, 12, 61]) that classical wellposedness of RE may fail. The main technical issue is the wellposedness of the gain operator B^*P (which enters nonlinear term in the equation). It turns out, that this operator when defined on a natural domain $\mathcal{D}(A^*)$ may be multivalued, cf. [61, 58, 11].

In order to obtain satisfactory theory, additional assumptions are necessary. In what follows, the following **singular estimate** hypothesis imposed on a pair A, B will be assumed throughout.

2.2 Singular Estimate Assumptions

By Hypothesis 2.1, $R(\lambda_0, A)B \in \mathcal{L}(U, H)$ for, $\lambda_0 \in \rho(A)$, so $e^{At}B$ is well defined as an element of $[\mathcal{D}(A^*)]'$. More precisely, we always have

$$e^{At}B \in \mathcal{L}(U \to C([0, T], [\mathcal{D}(A^*)]')).$$

What we take as a standing assumption for this paper is the requirement that this operator is also defined pointwisely for $t > 0$ as an element of H. More specifically, we assume the following singular estimate:

Hypothesis 2.2 (Singular estimate). *There exists a constant* $0 \le \gamma < 1$ *such that*

$$|e^{At}Bu|_H \le \frac{C}{t^\gamma}|u|_U, \ 0 < t \le 1.$$

In what follows, in order to simplify the notation, we will assume that $0 \in \rho(A)$. The constant C will denote generic constant different in different occurrences.

Remark 2.3. These estimates are trivially satisfied if $B : U \to H$ is bounded. Also, for analytic semigroups e^{At} and relatively bounded control operators $B : U \to [\mathcal{D}(A^\gamma)]'$, (the latter condition being equivalent to $A^{-\gamma}B \in \mathcal{L}(U \to H)$), the above singular estimate follows from the following estimate available for analytic semigroups (see [55]) $|A^\gamma e^{At}|_{\mathcal{L}(H)} \le \frac{C}{t^\gamma}$, $0 < t \le 1$.

Our focus in these notes is on a class of *nonanalytic* semigroups e^{At} and unbounded operators B which still exhibit singular estimate. As we shall see later, there is a large class of dynamical systems which enjoy the above property. Thus, the present setup is a proper generalization of treatments in [39, 14] where Riccati theory was constructed for analytic semigroups.

Remark 2.4. The equivalent, dual version of singular estimate can be formulated as follows: There exists a constant $0 \leq \gamma < 1$ such that

$$|B^* e^{A^* t} y|_U \leq \frac{C}{t^\gamma} |y|_H, \ 0 < t \leq 1, y \in \mathcal{D}(A^*) \subset H$$

where $(B^* y, u)_U = (y, Bu)_{\mathcal{D}(A^*), [\mathcal{D}(A^*)]'}; y \in \mathcal{D}(B^*)$ and $\mathcal{D}(A^*) \subset \mathcal{D}(B^*)$.
   A convenient way of finding the adjoint B^* is as follows. By the Hypothesis 2.1 and assuming wlog $0 \in \rho(A)$, we can write $B = AG$ where $G \in \mathcal{L}(U, H)$. With this notation, according to the definition of B^* we can write

$$B^* v = G^* A^* v, \ v \in \mathcal{D}(A^*).$$

Taking advantage of additional regularity of G (if any) or equivalently of the boundedness of G^* on a space larger than H, say on some $H \subset \mathcal{D} \subset [\mathcal{D}(A)]'$ one may be able to extend the domain of B^* maximally. Indeed, assuming that $|G^* v|_U \leq C|v|_{\mathcal{D}}$, we have $\mathcal{D}(B^*) \supset \{v \in H : A^* v \in \mathcal{D}\}$ where A^* should be interpreted as a realization of A^* on $[\mathcal{D}(A)]'$ with the domain (of A^*) equal to H (typical examples involve spaces which are dual to suitable fractional powers of A).

   We note here that Hypothesis 2.2 does not necessarily imply that the control operator is admissible (in the terminology of system theory - see [57]). We recall that B is "admissible" iff the following inequality takes place for some $T > 0$

$$\int_0^T |B^* e^{A^* t} y|_U^2 dt \leq C_T |y|_H^2, \ y \in \mathcal{D}(A^*) \tag{2.4}$$

Remark 2.5. Standard semigroup argument shows that if (2.4) holds for some $T > 0$, it also holds for any $T > 0$.

   By duality argument (see [48]) one shows that the above condition (2.4) is equivalent to the continuity of the map $u \rightarrow \mathbf{y}(u)$ from $L_2(0, T; U) \rightarrow C([0, T]; H)$ where $\mathbf{y}(u)$ is a solution defined by the dynamics (2.1) with $\mathbf{y}(0) = 0$.

   Thus, validity of singular estimate in Hypothesis 2.2 does not imply continuity of the control to state map. In fact, in many situations of interest the control-to-state map denoted by L is not necessarily continuous when viewed as a map from $L_2(0, T; U) \rightarrow C([0, T]; H)$. The lack of this continuity is one of a major technical difficulties in the theory.
   Nevertheless, the validity of singular estimate provides, as we shall see below, full wellposedness of the Riccati theory. This includes critical statement that the gain operator $B^* P(t)$ is not only well defined on $\mathcal{D}(A^*)$, but in fact is a *bounded* operator $U \rightarrow H$.

2.3 Examples of Point and Boundary Control Models

In this section we provide several canonical examples of PDE dynamics with boundary or point control control which serve as an illustration of the model introduced in (2.1). The examples given in this section are particularly simple, well known in the literature (see [40, 39]) and in fact not that representative of models motivating the theory (these are given in Part II). The purpose of presenting these canonical examples here is to convey a clear idea, to less experienced readers, how to build semigroup control system from the underlying PDE. Later in sections 3.1-3.3 of Part II we shall be considering more complex control systems which are more representative of the theory presented in these notes.

Heat equation with boundary control

A canonical example of "unbounded" control is heat equation with control acting via Neumann or Dirichlet boundary conditions. We consider a bounded domain $\Omega \subset R^n$ with a boundary Γ. We shall assume that the domain is sufficiently regular - say satisfies Lipschitz cone condition. In $Q_T \equiv \Omega \times (0, T)$ we consider heat equation with Dirichlet boundary control:

$$y_t = \Delta y \ in \ Q_T$$
$$y(t = 0) = y_0 \in L_2(\Omega)$$
$$y = u \ on \ \Sigma_T \equiv \Gamma \times (0, T). \tag{2.5}$$

We wish to put this system into the semigroup framework. To accomplish this we introduce the following operators

- Positive, selfadjoint operator $A_D : \mathcal{D}(A_D) \subset L_2(\Omega) \to L_2(\Omega)$ defined by

$$A_D y = -\Delta y, \ y \in \mathcal{D}(A_D) = H^2(\Omega) \cap H_0^1(\Omega);$$

- Dirichlet map $D : L_2(\Gamma) \to L_2(\Omega)$ given by $Dg = v$ iff $\Delta v = 0 \ in \ \Omega$ and $v|_\Gamma = g, \ on \ \Gamma$.

The operator $-A_D$ is a generator of an analytic semigroup of contractions on $L_2(\Omega)$ (see [55]) and the following estimate holds

$$|A_D R(\lambda, -A_D)|_{\mathcal{L}(L_2(\Omega))} \le \frac{C}{|\lambda|}; \ Re \ \lambda > 0. \tag{2.6}$$

It is known from elliptic theory (see [51]) and characterization of fractional powers of operator A_D (see [26]) that

$$D \in \mathcal{L}(L_2(\Gamma) \to H^{1/2}(\Omega) \subset \mathcal{D}((A_D)^{1/4-\epsilon})), \forall \epsilon > 0. \tag{2.7}$$

It is easily seen that the original equation (2.5) corresponds to the following abstract equation

$$\mathbf{y}_t = -A_D \mathbf{y} + A_D Du; \quad on \ [\mathcal{D}(A_D)]' \ \mathbf{y}(0) = y_0 \qquad (2.8)$$

so we have in the notation of (2.1)

$$H = L_2(\Omega), \ U = L_2(\Gamma), \ A = -A_D, \ B = A_D D = -AD.$$

Indeed, this can be seen in variety of ways, cf. [39, 14]. We present here a rather short argument which is based on the following observation: let y be a sufficiently smooth solution of (2.5). Then

$$\Delta y = \Delta y - \Delta Du = \Delta(y - Du) = -A_D(y - Du)$$

where in the last step we have used the fact that $y - Du \in \mathcal{D}(A_D)$.

Thus equation (2.8), and consequently (2.1) is an abstract version of (2.5).

We verify next that our operator B complies with the original requirements in Hypothesis 2.1. Indeed, from the definition of B and regularity in (2.7) it follows at once that $-A^{-1}B = A_D^{-1}B = D \in \mathcal{L}(L_2(\Gamma)) \to L_2(\Omega))$. Moreover, from the estimate in (2.6) and (2.7) we also have

$$|R(\lambda, A)Bu|_H = |R(\lambda, -A_D)A_D Du|_{L_2(\Omega)} \le \frac{\tilde{c}|Du|_{L_2(\Omega)}}{|\lambda|} \le \frac{C|u|_{L_2(\Gamma)}}{|\lambda|}, \ Re \ \lambda > 0$$

so condition (2.2) in the Hypothesis 2.1 holds.

In order to determine B^* and its domain we write: with any $\epsilon > 0$

$$|(A^*v, A^{-1}Bu)_H| = |(A_D v, Du)_{L_2(\Omega)}| \le C|A_D^{3/4+\epsilon}v|_{L_2(\Omega)}|A^{1/4-\epsilon}Du|_{L_2(\Omega)}$$

$$\le C|A_D^{3/4+\epsilon}v|_{L_2(\Omega)}|u|_U$$

where we have used (2.7). Therefore, for any $\epsilon > 0$, $\mathcal{D}(B^*) \supset \mathcal{D}(A_D^{3/4+\epsilon})$ and $(B^*v, u)_U = (D^*A_D v, u)_{L_2(\Gamma)} = -(\Delta v, Du)_{L_2(\Omega)} = -(\frac{\partial}{\partial n}v, u)_{L_2(\Gamma)}, \ \forall v \in \mathcal{D}(A_D)$, where the last identity follows from Green's formula, cf. [40, 39]. Hence

$$B^*v = -\frac{\partial}{\partial n}v|_\Gamma, \ \forall v \in \mathcal{D}(A_D^{3/4+\epsilon}) = H^{3/2+2\epsilon}(\Omega) \cap H_0^1(\Omega).$$

Since $-A_D$ generates an analytic semigroup and by (2.7) we have

$$(-A)^{-3/4-\epsilon}B = A_D^{1/4-\epsilon}D \in \mathcal{L}(L_2(\Gamma) \to L_2(\Omega)),$$

the validity of singular estimate in Hypothesis 2.2 with $\gamma = 3/4 + \epsilon$ follows from the analytic estimate $|A^\gamma e^{At}|_{\mathcal{L}(H)} \le \frac{C}{t^\gamma}, \ 0 < t \le 1$.

Conclusion: The operators A, B satisfy Hypothesis 2.1, Hypothesis 2.2, however they do not satisfy (2.4).

Remark 2.6. • It is interesting to note that in this example the pair A_D, B do not satisfy admissibility condition (2.4). Indeed, (2.4) does not hold regardless the dimension of Ω - see [49].

• The same procedure as above applies to the Neumann boundary control. In this case the corresponding operators satisfy all the requirements in Hypothesis 2.1, singular estimate Hypothesis 2.2 with $\gamma = 1/4 + \epsilon$ and also condition (2.4), cf. [39].

Heat equation with point control

For this example we consider a bounded domain $\Omega \subset R^n, n \leq 3$, with a boundary Γ. Let x_0 be a specified point in the interior of Ω and let space of controls U coincide with R^1 (this is not essential, any other finite dimensional space can be easily accommodated). In $Q_T \equiv \Omega \times (0,T)$ we consider heat equation with point control which can be formally written as:

$$y_t = \Delta y + \delta(x_0)u \ in \ Q_T$$
$$y(t = 0) = y_0 \in L_2(\Omega)$$
$$y = 0 \ on \ \Sigma_T \equiv \Gamma \times (0,T). \tag{2.9}$$

So, "formally" the operator B coincides with $\delta(x_0)$. However the above form of the equation is only formal and the correct PDE representation should be given in variational form as

$$(y_t(t), \phi)_\Omega + (y(t), A_D\phi)_\Omega = \phi(x_0)u; \ \forall \phi \in \mathcal{D}(A_D)$$
$$y(t = 0) = y_0 \in L_2(\Omega) \tag{2.10}$$

where we have used the notation $(u, v)_\Omega \equiv \int_\Omega u(x)v(x)dx$.

Setting this problem within the semigroup framework with $H \equiv L_2(\Omega)$ leads to:

$$y_t = -A_Dy + Bu, \ on \ [\mathcal{D}(A_D)]'$$

where $(Bu, \phi)_H \equiv \phi(x_0)u, \forall \phi \in \mathcal{D}(A_D) = H^2(\Omega) \times H_0^1(\Omega), \ u \in R^1$

Setting $A = -A_D$ and writing $B = -A(A_D^{-1}B)$, we obtain $B = AG$ where the operator $G \equiv -(A_D^{-1}B) : R^1 \to L_2(\Omega)$ is defined by

$$(Gu, \psi)_H \equiv -(A_D^{-1}\psi(x_0))u, \ \forall \psi \in L_2(\Omega), \ u \in R^1$$

For $n \leq 3$, by the virtue of Sobolev's embeddings and identification of fractional powers of elliptic operators with appropriate Sobolev's spaces (see [26])

$$G \in \mathcal{L}(U \to \mathcal{D}(A_D^{1-n/4-\epsilon})) \tag{2.11}$$

So, in particular $G \in \mathcal{L}(U \to L_2(\Omega) = H)$ and the pair A, B in this example complies with the Hypothesis 2.1. Indeed, we have $A^{-1}B = G$ is bounded from $U \to H$. Since A generates an analytic semigroup of contractions with the estimate

$$|R(\lambda, A)A|_{\mathcal{L}(H)} \leq C; \forall Re\lambda > 0,$$

condition (2.2) is in place. Computing the adjoint $G^* \in \mathcal{L}(H \to U)$ leads to

$$G^*v = -(A_D^{-1}v)(x_0).$$

This allows to provide an explicit form for B^* which by the definition of B^* can be identified with the operator $G^*A^* = -G^*A_D$ maximally defined by the formula

$$B^*v = -G^*A_Dv = -(A_D^{-1}(A_Dv))(x_0) = v(x_0); v \in \mathcal{D}(B^*), \ \mathcal{D}(B^*) \supset \mathcal{D}(A_D).$$

From (2.11) we infer

$$G^* \in \mathcal{L}([\mathcal{D}(A_D^{1-n/4-\epsilon})]' \to U) \tag{2.12}$$

Hence

$$G^*A_D^{1-n/4-\epsilon} \in \mathcal{L}(L_2(\Omega) \to R^1) \Rightarrow B^*A_D^{-n/4-\epsilon} \in \mathcal{L}(H \to U) \tag{2.13}$$

and $\forall \epsilon > 0$

$$\mathcal{D}(B^*) \supset \mathcal{D}(A_D^{n/4+\epsilon}); \ B^*v = v(x_0); \ v \in \mathcal{D}(B^*).$$

Thus, in particular for $n = 1$, we have $\mathcal{D}(B^*) \supset H_0^{1/2+\epsilon}(\Omega)$, for $n = 2$ $\mathcal{D}(B^*) \supset H_0^{1+\epsilon}(\Omega)$, and for $n = 3$ $\mathcal{D}(B^*) \supset H^{3/2+\epsilon}(\Omega) \cap H_0^1(\Omega)$.

From (2.13) we also infer that

$$A_D^{-n/4-\epsilon}B \in \mathcal{L}(U \to H)$$

and by analyticity of e^{-A_Dt} we obtain

$$|e^{-A_Dt}B|_{\mathcal{L}(U \to H)} = |e^{-A_Dt}A_D^{n/4+\epsilon}A_D^{-n/4-\epsilon}B|_{\mathcal{L}(U \to H)} \leq C|e^{-A_Dt}A_D^{n/4+\epsilon}|_{\mathcal{L}(U \to h)}$$
$$\leq \frac{C}{t^{n/4+\epsilon}}$$

so the singular estimate in Hypothesis 2.2 is satisfied with $\gamma = n/4 + \epsilon$.

Conclusion: The pair A, B complies with the Hypothesis 2.1 and Hypothesis 2.2.

Remark 2.7. • For this example the regularity condition (2.2) is satisfied in the one dimensional case only.
• Similar analysis with similar results can be carried out for the Neumann or Robin boundary conditions imposed on the boundary.

Wave equation with boundary control.

We shall consider an example of wave operator with controls acting in Dirichlet boundary conditions. On a bounded domain $\Omega \subset R^n$ with a boundary boundary Γ we consider.

$$y_{tt} = \Delta y \ in \ Q_T \equiv \Omega \times (0,T)$$
$$y(t = 0) = y_0, y_t(t = 0) = y_1; (y_0, y_1) \in L_2(\Omega) \times H^{-1}(\Omega)$$
$$y = u \ on \ \Sigma_T \equiv \Gamma \times (0,T). \tag{2.14}$$

We wish to put this system into the semigroup framework.
We shall use the notation introduced in the previous section.
Define first $A : L_2(\Omega) \times [\mathcal{D}(A_D^{1/2})]' \to L_2(\Omega) \times [\mathcal{D}(A_D^{1/2})]' \equiv H$

$$A \equiv \begin{pmatrix} 0 & I \\ -A_D & 0 \end{pmatrix}$$

$$\mathcal{D}(A) = \mathcal{D}(A_D^{1/2}) \times L_2(\Omega) \sim H_0^1(\Omega) \times L_2(\Omega)$$

so that

$$[\mathcal{D}(A)]' = [\mathcal{D}(A_D^{1/2})]' \times [\mathcal{D}(A_D)]' \sim H^{-1}(\Omega) \times [\mathcal{D}(A_D)]'.$$

It is well known A generates strongly continuous semigroup of contractions on

$$H \equiv L_2(\Omega) \times [\mathcal{D}(A_D^{1/2})]' \sim L_2(\Omega) \times H^{-1}(\Omega).$$

With

$$G \equiv \begin{pmatrix} D \\ 0 \end{pmatrix}, \ U \equiv L_2(\Gamma)$$

we define $B : U \rightarrow [\mathcal{D}(A^*)]'$ as follows:

$$B \equiv AG.$$

Since $G = \begin{pmatrix} D \\ 0 \end{pmatrix} \in \mathcal{L}(L_2(\Gamma) \rightarrow H)$ and A is invertible, we clearly have $A^{-1}B = G \in \mathcal{L}(L_2(\Gamma) \rightarrow H)$ as desired for Hypothesis 2.1. With the above notation, the original wave equation can be represented as the first order abstract system in the variable $\mathbf{y} = (y, y_t) \in H$

$$\mathbf{y}_t = A\mathbf{y} + Bu; \ \mathbf{y}(0) = (y_0, y_1), \ \text{on } [\mathcal{D}(A)]'.$$

To find B^* we first notice that $G^*v = D^*v_1$, so with $v = (v_1, v_2) \in \mathcal{D}(A^*)$

$$B^*v = G^*A^*v = D^*[A^*v]_1 = -D^*v_2 = -D^*A_D A_D^{-1} v_2 = \frac{\partial}{\partial n} A_D^{-1} v_2 \quad (2.15)$$

which expression is well defined since $A_D^{-1}v_2 \in H^2(\Omega)$ for $v \in \mathcal{D}(A^*)$. Our next step is to extend the domain of B^*. In order to accomplish this we recall from (2.7) and duality $D^* \in \mathcal{L}([\mathcal{D}(A_D)^{1/4-\epsilon}]' \rightarrow L_2(\Gamma))$, where from (2.15)

$$\mathcal{D}(B^*) \supset \{v = (v_1, v_2); v_1 \in L_2(\Omega), v_2 \in [\mathcal{D}(A_D)^{1/4-\epsilon}]'\}.$$

It was shown in [38] that the operator pair A, B satisfies also regularity trace condition (2.4). This not straightforward regularity result is proved by PDE methods. As a consequence, the condition (2.2) is satisfied as well. Therefore, all requirements in Hypothesis 2.1 are fulfilled. However, singular estimate in Hypothesis 2.2 fails.

Conclusion: Operators A, B satisfy Hypothesis 2.1, condition (2.4) but not Hypothesis 2.2.

Remark 2.8. We note that wave equation with boundary controls in Neumann boundary conditions also can be set as an abstract equation on $H^1(\Omega) \times L_2(\Omega)$ space. In this case the B^* operator is defined as follows:

$$B^*v = v_2|_\Gamma, v_2 \in H^{1/2+\epsilon}(\Omega) \subset \mathcal{D}(B^*).$$

However, in this case the regularity condition (2.4) fails, unless the dimension of Ω is equal to one. Singular estimate does not hold either.

Wave equation with point control

We shall discuss wave equation subject to point control acting at the specified point $x_0 \in \Omega \subset R^3$ with scalar controls $u \in U = R^1$. Thus we consider

$$y_{tt} = \Delta y + \delta(x_0)u \ in \ Q_T = \Omega \times (0, T)$$
$$y(t = 0) = y_0, y_t(t = 0) = y_1; (y_0, y_1) \in L_2(\Omega) \times H^{-1}(\Omega)$$
$$y = 0 \ on \ \Sigma_T \equiv \Gamma \times (0, T). \qquad (2.16)$$

As before, we note that the above form of equation is only formal, and the PDE should be interpreted via duality as

$$(y_{tt}(t), \phi)_\Omega + (y(t), A_D\phi)_\Omega = \phi(x_0)u; \ \forall \phi \in \mathcal{D}(A_D)$$
$$y(t = 0) = y_0 \in L_2(\Omega), \ y_t(t = 0) = y_1 \in H^{-1}(\Omega). \qquad (2.17)$$

In order to put this problem in the semigroup framework, we define

$$H \equiv L_2(\Omega) \times [\mathcal{D}(A_D^{1/2})]'$$

and the corresponding wave generator is given by $A : L_2(\Omega) \times [\mathcal{D}(A_D^{1/2})]' \rightarrow L_2(\Omega) \times [\mathcal{D}(A_D^{1/2})]'$

$$A \equiv \begin{pmatrix} 0 & I \\ -A_D & 0 \end{pmatrix}$$

$$\mathcal{D}(A) = \mathcal{D}(A_D^{1/2}) \times L_2(\Omega) \sim H_0^1(\Omega) \times L_2(\Omega).$$

Hence

$$[\mathcal{D}(A)]' = [\mathcal{D}(A_D^{1/2})]' \times [\mathcal{D}(A_D)]' \sim H^{-1}(\Omega) \times [\mathcal{D}(A_D)]'.$$

We define next Green's map $G : U \rightarrow H$ by the formula:

$$(Gu, v)_H \equiv u \cdot (A_D^{-1}v_1)(x_0); \ \forall v = (v_1, v_2) \in H, \ \forall u \in R^1.$$

By Sobolev's embeddings, since $A_D^{-1}v_1 \in H^2(\Omega)$, $G \in \mathcal{L}(U \rightarrow H)$ and $G^*v = (A_D^{-1}v_1)(x_0)$. It is readily verified that the operator $B \equiv AG$ is the control operator corresponding to (2.17) i.e.,

$$(Bu, v)_H \equiv (Gu, A^*v)_H = (A_D^{-1}v_2)(x_0)u, \ \forall u \in R^1, \ \forall v \in \mathcal{D}(A)$$

and we obtain the abstract version of equation (2.17) written as

$$\mathbf{y}_t = A\mathbf{y} + Bu; \ \mathbf{y}(0) = (y_0, y_1) \in H.$$

Since G is bounded from $U \rightarrow H$ and A is invertible, we clearly have $A^{-1}B \in \mathcal{L}(U \rightarrow H)$. The operator B^*, unbounded from $H \rightarrow U$, is given by

$$B^*v = G^*A^*v = -G^*Av = -(A_D^{-1}v_2)(x_0)$$

originally defined on $\mathcal{D}(A)$. We are able to extend the domain of B^* by exploiting Sobolev's embeddings in three-dimensions. Indeed, by Sobolev's inequalities we have that $\forall \epsilon > 0$

$$G^* \in \mathcal{L}(L_2(\Omega) \times [\mathcal{D}(A_D^{1-3/4-\epsilon})]' \to U) \tag{2.18}$$

and therefore $\forall \epsilon > 0$

$$\mathcal{D}(B^*) \supset L_2(\Omega) \times \mathcal{D}(A_D^{3/4+\epsilon}); \ B^*v = (A_D^{-1}v_2)(x_0); \ v \in \mathcal{D}(B^*).$$

Moreover, by PDE methods one proves the following (non trivial) regularity result (see [50])

$$\int_0^T |(e^{At}v)_1(x_0)|^2 dt \le C[|v_2|_{L_2(\Omega)}^2 + |v_1|_{H_0^1(\Omega)}^2] \tag{2.19}$$

where by $(u)_1$ we denote the first coordinate of the vector u. Inequality displays an additional (extra) regularity of solutions to wave equation.

Recalling the definition of B^* inequality (2.19) is equivalent to the statement that

$$\int_0^T |B^* e^{A^*t}v|_U^2 dt \le C|v|_H^2, \ v \in \mathcal{D}(B^*) \tag{2.20}$$

which is regularity condition (2.4). This regularity condition, a-posteriori, implies that condition (2.2) is satisfied. Thus, we conclude

Conclusion: In this example the pair A, B complies with the Hypothesis 2.1 and regularity condition (2.2). However, the singular estimate in Hypothesis 2.2 is not satisfied.

Remark 2.9. • The analogous conclusions hold for dimensions $n = 1$ and $n = 2$ of Ω. However, in lower dimensions it is possible to choose higher regularity state space. In fact, in the one dimensional case one can work, cf. [39] vol. II Thm 9.8.2.2, in a classical finite energy space associated with the wave operator: $H^1(\Omega) \times L_2(\Omega)$. For the case of $n = 2$ one has $H = H^{1/2}(\Omega) \times H^{-1/2}(\Omega)$.
• One could consider higher singularities of δ - the so called dipoles. This however would require an appropriate calibration of the state space. For this particular example with $B \sim \delta'$ the regularity condition (2.2) is satisfied in the one dimensional case only.
• Similar analysis with similar results can be carried on for the Neumann or Robin boundary conditions imposed on the boundary.

Concluding remarks

Canonical examples presented in this section involve either heat or wave operator with point or boundary controls. These examples are too simple in order to be fully representative of the class of problems which is a focus of our treatment. Indeed, heat equation with either boundary or point control are examples of analytic semigroups with unbounded controls. Thus, the validity of singular estimate in Hypothesis 2.1 automatically follows from the said analyticity. On the other hand, wave equation examples do not satisfy singular estimate. So, per se, these particular examples are not the right objects for our analysis. However, the crux of the matter is that the *combination of these two dynamics: analytic and hyperbolic* when considered as a coupled interaction may satisfy singular estimate without the property of being an analytic semigroup. In fact, all the examples in Part II of these notes deal with such structures. There the main challenge and difficulty is to show that the singular estimate does hold indeed. This is achieved by appropriate propagation of regularizing effect of the dynamics related to "parabolic behaviour" onto the entire interaction which may have a dominant hyperbolic component.

3 Abstract Results for Control Problems with Singular Estimate

We shall distinguish two cases; *finite* and *infinite* horizon control problems. This means that we shall consider cases $T < \infty$ and $T = \infty$ separately. The finite horizon problem is associated with the Differential Riccati Equation (DRE) while the infinite horizon problem is characterized by Algebraic Riccati equation (ARE). We begin with the finite horizon case.

3.1 Finite Horizon Control Problem - Differential Riccati Equation (DRE)

The following theorem describes main abstract results for the finite horizon case:

Theorem 3.1 (DRE). *Consider the control problem governed by the dynamics described in (2.1), and the functional cost given in (2.3) with $T < \infty$. The control operator B and the generator A are subject to Hypothesis 2.2. Moreover, we assume that $F \in L_2(0,T;H)$. Then, for any initial condition $y_0 \in H$, there exists an unique optimal pair $(u^0, y^0) \in L_2(0,T;U \times H)$ with the following properties:*

(i) **regularity of the optimal pair**

$$u^0 \in C([0,T];U); \ \mathbf{y}^0 \in C([0,T];H).$$

(ii) **regularity of the gains and optimal synthesis** *There exist a selfadjoint positive operator $P(t) \in \mathcal{L}(H)$ with the property*

$$B^* P(\cdot) \in \mathcal{L}(H \to C([0,T];U)); \ P_t(\cdot) \in \mathcal{L}(\mathcal{D}(A) \to C([0,T];\mathcal{D}(A^*)'))$$

and an element $r \in C([0,T]; H)$ (depending on F) with the property

$$B^* r \in C([0,T]; U)$$

such that: $u^0(t) = -B^* P(t) \mathbf{y}^0(t) - B^* r(t); \ t \geq 0$.

(iii) **feedback evolution** *The operator $A_P(t) \equiv A - BB^* P(t) : \mathcal{D}(A_P(t)) \subset H \to D(A^*)'$ with $\mathcal{D}(A_{P(t)}) = \{y \in H; [I - A^{-1}BB^* P(t)] \in \mathcal{D}(A)\}$ generates a strongly continuous evolution on H.*

(iv) **Riccati equation** *The operator $P(t)$ is a unique (within the class of selfadjoint positive operators subject to the regularity in part (ii)) solution of the following operator Differential Riccati Equation (**DRE**):*

$$(P_t(t)x, y)_H = (A^* P(t)x, y)_H + (P(t)Ax, y)_H + (R^* Rx, y)_H - $$
$$(B^* P(t)x, B^* P(t)y)_U; \ for \ x, y \in \mathcal{D}(A); \ P(T) = 0. \ (3.1)$$

(v) **Equation for "r"** *With $A_P(t)$ defined in part (iii) the element $r(t)$ satisfies the differential equation*

$$r_t(t) = -A_P^*(t)r(t) - P(t)F; \ on \ [\mathcal{D}(A)]'; \ r(T) = 0. \qquad (3.2)$$

Remark 3.2. It should be noted that the boundedness of the gain operator $B^* P(t)$ is a very special feature which is not generally expected. It is a consequence of singular estimate assumption. Indeed, in general, one does not have (unless B is bounded) the boundedness of the gains, even for the simplest scalar hyperbolic equations, cf. [40].

Remark 3.3. In the case of analytic semigroups e^{At} stronger conclusions are valid, cf. [14, 39, 40]. Indeed, in the absence of the disturbance F the optimal solution $u^0(t), \mathbf{y}^0(t)$ is shown to be analytic in time with values in H for all $t \in (0, T)$. With the disturbance term present, the regularity of optimal pair is limited by that of the disturbance.

The evolution corresponding to $A_P(t)$ is always analytic on H. Moreover the Riccati operator has an additional regularizing effect

$$P(t) \in C([0,T]; \mathcal{D}(A^{1-\epsilon}), \forall \epsilon > 0.$$

As a consequence, one obtains the validity of DRE for all $x, y \in \mathcal{D}(A^\epsilon)$.

The result of Theorem 3.1 was first shown in [5] for a special case of structural acoustic interaction where elastic equations on the interface are modelled with Kelvin Voigt damping. Proof of Theorem 3.1 in the general case is given in [34] (see also [33]). A simpler and shorter version of this proof will appear in [46]. In the next section of these notes we shall provide proof of Theorem 3.1 by following the arguments of [46].

3.2 Infinite Horizon Control Problem - Algebraic Riccati Equations (ARE)

In order to obtain solvability of *Infinite Horizon Problem*, i.e., when $T = \infty$, one must assume that the *Finite Cost Condition* holds. *FCC* condition asserts an existence of at least one control u such that $J(u, y) < \infty$. Thus, in what follows we shall assume

Hypothesis 3.4. *[Finite cost condition (FCC) condition]* For any initial condition $\mathbf{y}_0 \in H$ there exists $u \in L_2(0, \infty; U)$ such that $J(u, y(u)) < \infty$, where $y(u)$ is the trajectory (2.1) corresponding to u and originating at $\mathbf{y}_0$.

The result stated below provides the wellposedness for Riccati Equations in the infinite horizon case.

Theorem 3.5 (ARE). *Consider the control problem governed by the dynamics described in (2.1), and the functional cost given in (2.3) with $T = \infty$. The control operator B and the generator A are subject to Hypothesis 2.2. Moreover, we assume that FCC condition in Hypothesis 3.4 is in place and $F \in L_2(0, \infty; H)$. Then, for any initial condition $\mathbf{y}_0 \in H$, there exists an unique optimal pair $(u^0, \mathbf{y}^0) \in L_2(0, \infty; U \times H)$ with the following properties:*

(i) **regularity of the optimal pair** *The following improved regularity of optimal solution holds:*

$$u^0 \in C([0, \infty); U); \quad \mathbf{y}^0 \in C([0, \infty); H).$$

(ii) **regularity of the gains and optimal synthesis** *There exist a selfadjoint positive operator $P \in \mathcal{L}(H)$ with the property*

$$B^* P \in \mathcal{L}(H \to U);$$

and an element $r \in C([0, \infty); H)$ (depending on F) with the property

$$B^* r \in C([0, \infty); U)$$

such that: $u^0(t) = -B^* P(t)\mathbf{y}^0(t) - B^* r(t); \ t \geq 0.$
(iii) **feedback semigroup** *The operator $A_P \equiv A - BB^* P : \mathcal{D}(A_P) \subset H \to \mathcal{D}(A^*)'$ generates a strongly continuous semigroup on H.*
(iv) **Algebraic Riccati Equation** *The operator P is a solution of the following operator Algebraic Riccati Equation (ARE) :*

$$(A^* Px, y)_H + (PAx, y)_H + (R^* Rx, y)_H - (B^* Px, B^* Py)_U \text{ for } x, y \in \mathcal{D}(A). \tag{3.3}$$

(v) **Equation for "r"** *With A_P defined in part (iii) the element $r(t)$ satisfies the differential equation*

$$r_t(t) = -A_P^* r(t) - PF; \text{ on } [\mathcal{D}(A)]'; \ \lim_{T \to \infty} r(T) = 0. \tag{3.4}$$

Remark 3.6. As usual, the uniqueness of ARE is guaranteed by imposing an appropriate "detectability condition". In this case the feedback semigroup $e^{A_P t}$ is exponentially stable on H. This, in turn, is guaranteed by the uniform stability of e^{At}. As we shall see later, the examples provided in this manuscript fall into this category.

Remark 3.7. As in the case of Differential Riccati Equation, the boundedness of the gain operator $B^* P$ is rather exceptional and results from the imposition of singular estimate. In general case one should not expect $B^* P$ to be bounded, but at most densely defined, cf. [25]. In addition, the quadratic term in (3.3) involves typically a suitable extension of $B^* P$ rather than $B^* P$ itself, cf. [12, 58].

Remark 3.8. As for the finite horizon problem, much stronger conclusions hold for the case of analytic semigroups e^{At}, cf. [40, 14, 39]. Indeed, Riccati operator has an additional regularity properties $P \in \mathcal{L}(H \to \mathcal{D}(A^{*(1-\epsilon)}))$ with any $\epsilon > 0$ and the semigroup $e^{A_P t}$ is analytic. Correspondingly, the ARE holds for all $x, y \in \mathcal{D}(A^\epsilon)$. Moreover, assuming that the forcing term F is analytic (differentiable), the same conclusion is implied for the optimal pair $u^0, \mathbf{y}^0$.

The proof of Theorem 3.5 was given in [34]. The streamlined and more general version of this proof is in [45]. Since the main strategy for the proof in the infinite horizon case is similar to that of finite horizon, we shall not repeat the arguments here and interested reader is referred to [45].

4 Finite Horizon Problem - Proofs

The strategy for our proof is based on a variational argument. By this we mean that we first "solve" control problem explicitly in terms of the data. We then extract a suitable representation for Riccati operator. Finally, we verify that this representation satisfies the DRE (Differential Riccati Equation). This last step is the most technical, as it requires number of regularity properties for various quantities (operators) involved. While in a classical - bounded control case - this is rather straightforward as the regularity needed follows basically from the C_0 semigroup theory, in the unbounded control case the situation is much more subtle. The unboundedness of controls makes certain critical quantities not well defined. Thus, additional regularity not implied by the dynamics needs to be established. In order to achieve the goal it is important to "trace" all the sources of additional regularity available for optimal solutions. These are provided by special properties of the underlying dynamics and also by the very nature of optimization process, which has some smoothing effect on solutions. This will become transparent through the course of the proof.

In these notes we concentrate on proofs for the finite horizon problem (i.e., $T < \infty$). The case of $T = \infty$, can be obtained via limiting procedure when $T \to \infty$, cf. [14]. For complete proof of the Algebraic Riccati Equation the reader is referred to [45].

4.1 Characterization of Optimal Solution

We already know that under sole Hypothesis 2.1 there exists a unique optimal solution. Our first goal is to provide explicit characterization of this solution.

To do this we introduce the so called solution operator, often also referred to as the "control-to-state" map.

$$(L_s u)(t) \equiv \int_s^t e^{A(t-z)} Bu(z) dz, \ 0 \le s \le t \le T. \tag{4.1}$$

Condition (2.2) in Hypothesis 2.1 is equivalent to the statement that the control-to-state operator L_s is topologically bounded L_2 in time. This is to say, for all $T < \infty$

$$L_s \in \mathcal{L}(L_2(s, T; U) \rightarrow L_2(s, T; H)). \tag{4.2}$$

One can easily verify - see [39] vol. II p. 773 - that the following operators are also bounded

$$[I + L_s^* R^* R L_s]^{-1} \in \mathcal{L}(L_2(s, T; U)); \ [I + L_s L_s^* R^* R]^{-1} \in \mathcal{L}(L_2(s, T; H)). \tag{4.3}$$

When $s = 0$ we shall write simply L. The same convention applies to other operators or quantities with the subscript s.

The effect of the deterministic noise is represented by the element

$$\mathcal{F}_s(t) \equiv \int_s^t e^{A(t-z)} F(z) ds$$

and from standard semigroup theory (see [55]) with $F \in L_1(s, T; H)$; $\mathcal{F}_s \in C([s, T]; H)$.

With the above notation, the solution of (2.1) due to control function u, initial condition $\mathbf{y}_0$ originating at $t = s$ and the disturbance F is represented via semigroup formula:

$$\mathbf{y}(t, s, \mathbf{y}_0, u, F) = e^{A(t-s)} \mathbf{y}_0 + (L_s u)(t) + \mathcal{F}_s(t) \tag{4.4}$$

where the notation $\mathbf{y}(t, s, \mathbf{y}_0, u, F)$ indicates the dependence on all parameters. In the case when the initial time for optimal trajectory x is equal 0 we shall use shorthand $\mathbf{y}^0(t) = \mathbf{y}^0(t, 0, \mathbf{y}_0, u^0, F)$.

In order to provide an expression for optimal synthesis it is convenient to introduce "evolution" operator $\Phi(t, s)$ defined by

$$\Phi(\cdot, s)x \equiv [I + L_s L_s^* R^* R]^{-1} e^{A(\cdot - s)} x \in L_2(s, T; H) \text{ for } x \in H. \tag{4.5}$$

From (4.3) it follows that

$$\Phi(\cdot, s) \in \mathcal{L}(H \rightarrow (L_2(s, T; H)); \ \forall s < T. \tag{4.6}$$

Clearly $\Phi(t,t) = I$. Moreover, this operator has "evolution" property, which in fact justifies the name "evolution operator", that is to say: for all $\tau > 0, x \in H$ the following relation holds

$$\Phi(t,s)\Phi(s,\tau)x = \Phi(t,\tau)x; \quad a.e.t > \tau, \tau < s < t. \tag{4.7}$$

There are several ways of proving evolution property. The origin of this property goes back to Bellman's optimality principle. However, classical methods of proof require some differentiability properties of this operator. Instead, by using explicit operator representation of the evolution given by (4.5) one can justify this property by operator calculus. This is based on the following algebraic identity (see [39] vol. I p. 32)

$$\{\mathcal{T}_s e^{A(\cdot - s)}[\mathcal{T}_\tau e^{(A\cdot - \tau)}x](s)\}(t) = [\mathcal{T}_\tau e^{(A\cdot - \tau)}x](t)$$

where we have used the notation $\mathcal{T}_t \equiv [I + L_t L_t^* R^* R]^{-1}$.

We note that the evolution property in (4.7) is not pointwise. However, later in the paper we shall prove that under Hypothesis 2.2 this property holds for every $\tau \le s \le t$.

The evolution operator allows us to define the "Riccati operator" $P(t)$ by the following formula (see [6, 40, 25, 39])

$$P(t) \equiv \int_t^T e^{A^*(s-t)} R^* R \Phi(s,t) ds. \tag{4.8}$$

From (4.6) we infer that

$$P(\cdot) \in \mathcal{L}(H \to C([0,T], H)). \tag{4.9}$$

Moreover, it is standard to show that $P(t)$ is selfadjoint and positive on H, cf. [25]. We define next the "adjoint state"

$$p(t) \equiv \int_t^T e^{A^*(z-t)} R^* R \mathbf{y}^0(z) dz \in C([0,T]; H). \tag{4.10}$$

Finally, we define the variable

$$r(t) \equiv p(t) - P(t)\mathbf{y}^0(t) \in L_2(0,T;H). \tag{4.11}$$

Since $B^* p = L^* R^* R y^0$, we have by (4.2) that

$$P(t)\mathbf{y}^0(t) + r(t) \in \mathcal{D}(B^*); \quad a.e \text{ in } t.$$

Here we note that at this point there is no claim that each element $P(t)\mathbf{y}^0(t)$ and $r(t)$ are in the domain of B^*. Again, on the strength of the singular estimate such claim will be made later.

By using the above notation and classical variational argument one obtains - see [39] vol. I sect. 6.2.3 - the explicit formulas for the optimal control. These are collected in the Lemma below:

Lemma 4.1. *With reference to the control problem stated in (2.3), subject to Hypothesis 2.1, with arbitrary initial condition* $\mathbf{y}_0 \in H$ *and* $F \in L_1(0, T; H)$, *there exists a unique optimal pair denoted by* $(u^0, \mathbf{y}^0)$ *with the following properties:*

(i) **optimal control**

$$u^0 = -[I + L^* R^* R L]^{-1} L^* R^* R [e^{A(\cdot)} \mathbf{y}_0 + \mathcal{F}] \in L_2(0, T; U).$$

(ii) **optimal trajectory**

$$\mathbf{y}^0 = [I + L L^* R^* R]^{-1} [e^{A(\cdot)} \mathbf{y}_0 + \mathcal{F}] \in L_2(0, T; H).$$

(iii) **optimal synthesis**

$$u^0(t) = -B^*[P(t)\mathbf{y}^0(t) + r(t)]; \quad a.e \text{ in } t \in (0, T),$$

where $P(t), r(t)$ *are given in (4.8), (4.11).*

Remark 4.2. In the case $T = \infty$ the formulas in Lemma 4.1 still hold under the additional assumption of exponential stability of e^{At}.

Remark 4.3. We note that the expression representing optimal control in (iii) of Lemma 4.1 can not be distributed. The reason is that at this point we do not know whether $r(t)$ or $P(t)$ are in the domain of B^*. This fact will be proved later by exploiting the strength of singular estimate.

We shall provide another representation for $r(t)$ which will be used later. To this end we introduce the notation: $\mathbf{y}_{F=0}^0(t, s, x)$ denotes the optimal trajectory at the time t corresponding to (2.1) with $F \equiv 0$ and some initial condition x originating at the time s. Explicitly

$$\mathbf{y}_{F=0}^0(\cdot, s, x) \equiv [I + L_s L_s^* R^* R]^{-1} e^{A(\cdot - s)} x = \Phi(\cdot, s)x, \qquad (4.12)$$

where $\mathbf{y}_F^0(t, s)$ denotes instead the optimal trajectory corresponding to (2.1) with the initial condition equal to zero, originating at the point s and disturbance equal to F. Explicitly

$$\mathbf{y}_F^0(\cdot, s) \equiv [I + L_s L_s^* R^* R]^{-1} \mathcal{F}_s.$$

Rewriting formula (4.11) with the above notation and using definition in (4.8) we obtain:

$$r(t) = \int_t^T e^{A^*(s-t)} R^* R[\mathbf{y}^0(s) - \Phi(s, t)\mathbf{y}^0(t)]ds$$

$$= \int_t^T e^{A^*(s-t)} R^* R[\mathbf{y}_{F=0}^0(s, 0, \mathbf{y}_0)) - \Phi(s, t)\mathbf{y}_{F=0}^0(t, 0, \mathbf{y}_0)]ds +$$

$$\int_t^T e^{A^*(s-t)} R^* R[\mathbf{y}_F^0(s, 0) - \Phi(s, t)\mathbf{y}_F^0(t, 0)]ds. \quad (4.13)$$

By (4.12) and evolution property in (4.7) we obtain

$$\Phi(s,t)\mathbf{y}^0_{F=0}(t,0,\mathbf{y}_0) = \Phi(s,t)\Phi(t,0)\mathbf{y}_0 = \Phi(s,0)\mathbf{y}_0 = \mathbf{y}^0_{F=0}(s,0,\mathbf{y}_0).$$

Thus, the first integral term on the RHS of (4.13) vanishes. The above leads to the following representation for $r(t)$.

$$
\begin{aligned}
r(t) &= \int_t^T e^{A^*(s-t)} R^* R[\mathbf{y}^0_F(s,0) - \Phi(s,t)\mathbf{y}^0_F(t,0)]ds \\
&= p_F(t) - P(t)\mathbf{y}^0_F(t,0)
\end{aligned}
\tag{4.14}
$$

where

$$p_F(t) \equiv \int_t^T e^{A^*(s-t)} R^* R\mathbf{y}^0_F(s,0)ds. \tag{4.15}$$

One of the main goals in optimal control theory is to provide an independent characterization of Riccati operator via Riccati Equation. And this is precisely the most delicate point where regularity of PDE dynamics has a very strong bearing on the well-posedness of Riccati Equations.

While full characterization of optimal solution is obtainable under minimal assumption, it is the wellposedness of Riccati Equation (RE) which requires additional hypotheses. In fact, without these additional assumptions, it is known (see [58, 12, 61]) that classical wellposedness of RE may fail. The main technical issue is the wellposedness of the gain operator $B^* P(t)$ (which enters the nonlinear term in the equation) on some dense domain in H.

In fact, subsequent sections will exploit the strength of the Hypothesis 2.2, which will then lead to more information on regularity and structure of Riccati operator P and the evolution Φ. Our ultimate goal is to provide rigorous derivation of Riccati Equation and the equation for $r(t)$ representing the effect of the disturbance. In order to achieve this we shall follow the following plan:

- **Step 1:** We first prove *continuity in time* of the optimal solution. This will allow to extend evolution property (4.7) to hold for all times $\tau \leq s \leq t$.
- **Step 2:** Continuity in time of evolution is critical to establish the boundedness of the operator $B^* P(t)$ and also of function $B^* r(t)$. The above properties allow to define the generator of the evolution $\Phi(t,s)$.
- **Step 3:** By using the boundedness of $B^* P(t)$ and the singular estimate Hypothesis 2.2 we will be able to show that the same singular estimate holds for the evolution $\Phi(t,s)$. This is critical for asserting differentiability of the evolution with respect to the second parameter.
- **Step 4:** We shall prove the differentiability of $\Phi(t,s)$ as a map from H into space of continuous functions with singularity at the origin.
- **Step 5:** Using the above differentiability, we derive differential form of Riccati equation.
- **Step 6:** We derive the equation satisfied by function $r(t)$.

4.2 Continuity in Time of Optimal Solutions

Our first step is to improve the L_2 in time regularity of u^0 and $\mathbf{y}^0$. We shall show that these elements are continuous in time. The above improvement is a critical step in the proof and it depends on validity of Hypothesis 2.2.

We begin with preliminary Proposition which displays reiterated regularity of operators L_s and L_s^* subject to Hypothesis 2.2.

Proposition 4.4 ([39, 46, 45]). *Assume Hypothesis 2.1 and Hypothesis 2.2. With reference to operators L_s and its adjoint introduced in (4.1), we have the following boundedness, continuity results, all uniformly, in the operator norm, with respect to the parameter s $s \leq 0 \leq T$, but with bounds which will generally depend on γ and T:*

(a) $L_s \in \mathcal{L}(L_2(s,T;U) \to L_2(s,T;H)) \cap \mathcal{L}(C([s,T];U) \to C([s,T];H))$
(b) $L_s^* \in \mathcal{L}(L_2(s,T;H) \to L_2(s,T;U)) \cup \mathcal{L}(C([s,T];H) \to C([s,T];U))$
(c) $[I_s + L_s^* R^* R L_s]^{-1} \in \mathcal{L}(L_2(s,T;U)) \cap \mathcal{L}(C([s,T];U))$
(d) $[I_s + L_s L_s^* R^* R]^{-1} \in \mathcal{L}(L_2(s,T;H)) \cap \mathcal{L}(C([s,T];H))$
(e) $L_s \in \mathcal{L}(L_2(s,T;U)) \to C([s,T];H))$, when $\gamma < 1/2$
(f) More generally $L_s \in \mathcal{L}(L_p(s,T;U)) \to L_{p+r}(s,T;H))$ where $p \geq 2$ and $r < \frac{4(1-\gamma))}{2\gamma-1}$ and $L_s^* \in \mathcal{L}(L_p(s,T;H)) \to L_{p+r}(s,T;U))$
(g) For $p > \frac{1}{1-\gamma}$, $L_s \in \mathcal{L}(L_p(s,T;U)) \to C([s,T];H))$
(h) Thus, a fortiori, there exist an integer n such that

$$[L_s^* R^* R L_s]^n \in \mathcal{L}(L_2(s,T;U)) \to C([s,T];H)).$$

By using the above Proposition along with Lemma 4.1 one concludes the following improved regularity result

Lemma 4.5. *Subject to Hypothesis 2.1 and Hypothesis 2.2 the following regularity holds (uniformly in s) with any $x \in H$ and $F \in L_1(s,T:H)$:*

$$u^0(\cdot, s, x) \in C([s,T];U); \quad \mathbf{y}^0(\cdot, s, x) \in C([s,T];H)$$

and

$$|\mathbf{y}^0(\cdot, s, x)|_{C([s,T];H)} + |u^0(\cdot, s, x)|_{L_2(s,T;H)} \leq C[|x|_H + |F|_{L_1(s,T;H)}]. \quad (4.16)$$

Remark 4.6. We note that the continuity in time of optimal solutions follows from the strength of Hypothesis 2.2. In fact, this occurs by the same mechanism exploited in the study of analytic semigroups, see [14, 39].

4.3 Regularity of Evolution $\Phi(t,s)$ and of Riccati Operator $P(t)$

One of the critical step in the derivation of Theorem 3.1 is the assessment of continuity (in time) of the evolution operator along with the boundedness of the so called gain operator $B^* P(t)$. These results are formulated below:

Lemma 4.7. *Under the Hypothesis 2.1 and Hypothesis 2.2 the following properties hold:*

(a) The operator $\Phi(\cdot, \cdot)$ satisfies the following regularity:

$$\Phi(\cdot, s) \in \mathcal{L}(H \to C([s, T]; H)), \quad uniformly \ in \ 0 \le s \le T$$
$$\Phi(t, \cdot) \in \mathcal{L}(H \to C([0, t]; H)), \quad uniformly \ in \ 0 \le t \le T. \quad (4.17)$$

Moreover, $\Phi(t, s)$ is an evolution operator, i.e.,
$$\Phi(t, t) = I, \ \Phi(t, s) = \Phi(t, \tau)\Phi(\tau, s) \ \forall \ 0 \le s \le \tau \le t \le T$$
(b) For each $t \ge 0$ $P(t)$ given by (4.8) satisfies $P(t) \in \mathcal{D}(B^*)$ and

$$B^* P \in \mathcal{L}(H \to C([0, T]; U)). \quad (4.18)$$

(c) $A_P(t) \equiv A - BB^* P(t) : H \to [\mathcal{D}(A^*)]'$ is a generator corresponding to the evolution $\Phi(t, s)$ and we have:

$$\frac{\partial}{\partial t}\Phi(t, s)x = A_P(t)\Phi(t, s)x; \ in \ [\mathcal{D}(A^*)]', s \le t, \quad (4.19)$$

$$\frac{\partial}{\partial t}\Phi(t, s) \in \mathcal{L}(H \to C([s, T]; [\mathcal{D}(A^*)]')). \quad (4.20)$$

(d) With $r(t)$ given by (4.11) we have

$$B^* r \in C([0, T]; H). \quad (4.21)$$

Proof:

- The proof of continuity of Φ in (4.17) follows from the definition of the evolution operator given by (4.5), a-priori regularity in (4.6) and "boot strap" argument based on repeated use of singular estimates in Proposition 4.4. Noting that all estimates are uniform in s, gives the first statement in (4.17). As for the second statement, continuity in $s \to \Phi(t, s)$, this is obtained by using continuity in the first variable $t \to \Phi(t, s)$ followed by standard argument (see for instance [39] Lemma I.4.6.2). Once the continuity in time of evolution is established, the derivation of pointwise evolution property is standard.
- For the regularity in (4.18) we first notice that by using singular estimate in Hypothesis 2.2 we infer that with any $x \in H$

$$e^{A^*(s-t)} R^* R\Phi(s, t)x \in \mathcal{D}(B^*) \ \forall t \ and \ a.e, s > t.$$

Moreover, using representation in (4.8) and regularity in (4.17)

$$|B^* P(t)x|_U \le C \int_t^T |B^* e^{A^*(s-t)} R^* R\Phi(s, t)x|_U ds$$

$$\le C_T \int_t^T \frac{1}{(s-t)^\gamma}|R^* R|_{\mathcal{L}(H)}|\Phi(s, t)x|_H ds$$

$$\le C_T \int_t^T \frac{1}{(s-t)^\gamma}|x|_H ds \le C_T(T-t)^{1-\gamma}|x|_H. \quad (4.22)$$

The final step for (4.18) is to establish the continuity in time of $|B^*P(t)|_{\mathcal{L}(H\to U)}$. This is done by exploiting the integral representation of the operator along with the singular estimate. This argument is standard and the same as in [39, 46].

- In order to prove that $A_P(t)$ is the generator of the evolution $\Phi(t, s)$ (in the dual sense as postulated by the Lemma) we consider the following integral equation: with $x \in H$

$$x(t, s) = e^{A(t-s)}x + \int_s^t e^{A(t-\tau)}BB^*P(\tau)x(\tau, s)d\tau. \qquad (4.23)$$

By applying contraction mapping principle, along with the singular estimate $|e^{At}Bu|_H \leq C\frac{1}{t^{\gamma}}|u|_U$ and regularity in (4.18) we obtain existence and uniqueness of solution $x(\cdot, s) \in C([s, T]; H)$ for each $x \in H$. On the other hand, by taking the inner product $(x(t, s), y)_H$ with $y \in \mathcal{D}(A^*)$ we obtain

$$\frac{d}{dt}(x(t, s), y)_H = (e^{A(t-s)}x, A^*y)_H$$

$$+ \left(\int_s^t e^{A(t-\tau)}BB^*P(\tau)x(\tau, s)d\tau, A^*y\right)_H$$

$$+ (A^{-1}BB^*P(t)x(t, s), A^*y)_H$$

$$= (x(t, s), A^*y)_H + (A^{-1}BB^*P(t)x(t, s), A^*y)_H$$

$$= (Ax(t, s), y)_{\mathcal{D}(A^*),[\mathcal{D}(A^*)]'}$$

$$+ (BB^*P(t)x(t, s), y)_{\mathcal{D}(A^*),[\mathcal{D}(A^*)]'}$$

$$= (A_P(t)x(t, s), y)_{\mathcal{D}(A^*),[\mathcal{D}(A^*)]'} \qquad (4.24)$$

which shows that $x(t, s)$ is an evolution associated with $A_P(t)$ and $\frac{d}{dt}$ $(x(t, s) \in C([s, T], [\mathcal{D}(A^*)]'))$. It is now straightforward to show that $\Phi(t, s)$ defined by (4.5) satisfies the integral equation (4.23), hence by the uniqueness of solutions to this equation we must have $\Phi(t, s)x = x(t, s)$. The proof of part (3) is completed.

- The regularity in (4.21) follows from representation in (4.11), regularity property (4.18), and Lemma 4.5 together with the statement that for all $y^0 \in H$

$$B^*p \in C([0, T]; U).$$

The validity of this last assertion is a consequence of singular estimate and Lemma 4.5 applied to

$$|B^*p(t)|_U = |\int_t^T B^*e^{A^*(s-t)}R^*R y^0(s)ds|_U$$

$$\leq C_T \int_t^T \frac{1}{(s-t)^{\gamma}}|R^*R|_{\mathcal{L}(H)}|y^0(s)|_H ds$$

$$\leq C_T(T-t)^{1-\gamma}|y_0|_H. \qquad (4.25)$$

$\square$

4.4 Singular Estimate and Left-differentiability for the Evolution Operator

Our next goal is to show that singular estimate is invariant under the optimal control process. This is to say, the estimate of Hypothesis 2.2 is also satisfied for the evolution operator $\Phi(t,s)$. This property is essential in derivation of the Riccati Equation.

Lemma 4.8. *Under the Hypothesis 2.1 and Hypothesis 2.2 the following estimate takes place:*

$$|\Phi(t,s)B|_{\mathcal{L}(U\to H)} \le C_T \frac{1}{(t-s)^\gamma}; \ 0 \le s \le t \le T.$$

Proof: We shall introduce the following spaces which measure singularity of functions. These spaces were originally introduced in [21] and later used in [14, 39]. For any Banach space X and a constant $0 \le r < \infty$ we introduce

$$C_r([0,T];X) \equiv \{f \in C((0,T];X); |f|_{C_r([0,T];X)} \equiv \sup_{0\le t\le T} t^r |f(t)|_X\}.$$

More generally we define

$$C_r([s,T];X) \equiv \{f \in C((s,T];X); |f|_{C_r([s,T];X)} \equiv \sup_{s\le t\le T} (t-s)^r |f(t)|_X\}.$$

It is convenient to express regularity of L operator in terms of the topology in C_r. The following properties are proved in [39] vol. I p. 46.

Proposition 4.9. *Under the Hypothesis 2.2 the following properties hold with any $0 < r < 1$.*

(a) $L \in \mathcal{L}(C_r([0,T];U) \to C_{r+\gamma-1}([0,T];H))$
(b) $L^* \in \mathcal{L}(C_r([0,T];H) \to C_{r+\gamma-1+\epsilon}([0,T];U)), \forall \epsilon > 0$
(c) *There exist m positive integer such that* $[L_s^* R^* R L_s]^m \in \mathcal{L}(C_r([s,T];U) \to C([0,T];U))$
(d) $[I + L_s^* R^* R L_s]^{-1} \in \mathcal{L}(C_r([s,T];U))$
(e) $[I + L_s L_s^* R^* R]^{-1} \in \mathcal{L}(C_r([s,T];H))$

Proof: The proof is essentially the same as in the case of analytic semigroups, cf. [14, 39]. For the orientation of the reader we shall sketch the main idea.

The first two statements in the Proposition describing regularity of L and L^* follow from singular estimate (Hypothesis 2.2) and Lemma 4.4. The third statement follows from a boot strap argument applied to (1) and (2) in Proposition 4.9. In order to establish (4) it suffices to write $[I + L_s^* R^* R L_s]^{-1} x = y \to y + L_s^* R^* R L_s y = x$, the expression valid a priori with $y \in C_\gamma([s,T];U)$. Applying in succession powers of the operator $L_s^* R^* R L_s$ and using property (3) above gives the desired result in (4). As for the last statement (5), this follows from the algebraic representation:

$$[I + L_s L_s^* R^* R]^{-1} = I - L_s [I + L_s^* R^* R L_s]^{-1} L_s^* R^* R$$

and properties (1)-(4). □

To continue with the proof of the Lemma, by virtue of Hypothesis 2.2 we have

$$e^{A(\cdot - s)} B \in \mathcal{L}(U, C_\gamma([s, T]; H)).$$

On the other hand, from (4.5) and Proposition 4.9

$$\Phi(t, s) B = [I + L_s L_s^* R^* R]^{-1} e^{A(\cdot - s)} B \in \mathcal{L}(U, C_\gamma([s, T]; H))$$

which gives the desired estimate in Lemma 4.8. □

Lemma 4.10. *Under the Hypothesis 2.1 and Hypothesis 2.2 we have the following differentiability (in the second variable) of evolution $\Phi(t, s)$ valid with any $x \in \mathcal{D}(A)$:*

$$\frac{\partial \Phi(t, s) x}{\partial s} = -\Phi(t, s)[A - BB^* P(s)] x = -\Phi(t, s) A_P(s) x \in C_\gamma([s, T]; H).$$
$$(4.26)$$

Moreover

$$\left| \frac{\partial \Phi(t, s) x}{\partial s} \right|_H \leq C_T [\|Ax\|_H + \frac{1}{(t - s)^\gamma} |x|_H].$$
$$(4.27)$$

Proof: From $\Phi(t, s) = [I + L_s L_s^* R^* R]^{-1} e^{A(\cdot - s)}$ we obtain

$$\Phi(t, s) x + \int_s^t e^{A(t - \tau)} BB^* \int_\tau^T e^{A^*(\sigma - \tau)} R^* R \Phi(\sigma, s) x \, d\sigma d\tau = e^{A(t - s)} x. \quad (4.28)$$

Taking $x \in \mathcal{D}(A)$ and differentiating (distributionally) both sides of (4.28) in the variable s yields

$$\frac{\partial \Phi(t, s) x}{\partial s} + \int_s^t e^{A(t - \tau)} BB^* \int_\tau^T e^{A^*(\sigma - \tau)} R^* R \frac{\partial \Phi(\sigma, s) x}{\partial s} d\sigma d\tau -$$
$$e^{A(t - s)} BB^* \int_s^T e^{A^*(\sigma - s)} R^* R \Phi(\sigma, s) x \, d\sigma = -e^{A(t - s)} Ax, \quad s < t \leq T \, (4.29)$$

or equivalently

$$[I_s + L_s L_s^* R^* R] \frac{\partial \Phi(\cdot, s) x}{\partial s} = e^{A(\cdot - s)} B[L_s^* R^* R \Phi(\cdot, s) x] - e^{A(\cdot - s)} Ax$$
$$= -e^{A(\cdot - s)} B u_{F=0}^0(s, s, x) - A e^{A(\cdot - s)} x, \quad (4.30)$$

where we recall the notation $u_{F=0}^0(t, s, x)$ to denote the optimal control corresponding to $F = 0$, initial data x originating at time $t_0 = s$.

For this undisturbed control process optimal trajectory coincides with the evolution $\Phi(t, s)$, so the formula (iii) in Lemma 4.1 yields

$$u^0_{F=0}(t, s, x) = -B^* P(t)\Phi(t, s)x. \tag{4.31}$$

According to (4.18) $B^* P(t)$ is a bounded operator we have

$$u^0_{F=0}(s, s, x) = -B^* P(s)x \in C([0, T], H), \forall x \in H$$

so $\forall x \in \mathcal{D}(A)$

$$- e^{A(\cdot - s)} B u^0_{F=0}(s, s, x) - A e^{A(\cdot - s)} x$$
$$= e^{A(\cdot - s)} B B^* P(s)x - A e^{A(\cdot - s)} x \in C_\gamma([s, T], H)$$

Combining the above identity with (4.30) and recalling property (5) in Proposition 4.9 gives

$$\frac{\partial \Phi(\cdot, s)x}{\partial s} = -[I_s + L_s L_s^* R^* R]^{-1} e^{A(\cdot - s)} [Ax - BB^* P(s)x]$$
$$= -\Phi(\cdot, s)[Ax - BB^* P(s)x] \tag{4.32}$$

which gives the first statement in the Lemma. The inequality in (4.27) follows now from (4.32) combined with regularity of $B^* P(t)$ and the singular estimate in Lemma 4.8. □

4.5 Differential Riccati Equation - DRE

By using critically the regularity of $B^* P$ and differentiability of the evolution in the second variable, we will be able to derive differential version of the Riccati Equation.

Lemma 4.11. *Assume Hypothesis 2.1 and Hypothesis 2.2. Then the operator $P(t)$ defined by (4.8) satisfies the following Differential Riccati Equation:* $\forall x, y \in \mathcal{D}(A)$

$$(P_t(t)x, y)_H + (Rx, Ry)_Z + (P(t)Ax, y)_H + (P(t)x, Ay)_H$$
$$= (B^* P(t)x, B^* P(t)y)_U. \tag{4.33}$$

Proof: We start with the definition of P given in (4.8):

$$(P(t)x, z)_H = \int_t^T (R\Phi(\tau, t)x, R e^{A(\tau - t)} z)_Z d\tau.$$

Taking $x, z \in \mathcal{D}(A)$ and differentiating in time the above expression with the use of Lemma 4.10 yield:

$$(P_t(t)x, z)_H = \int_t^T (R\frac{\partial \Phi(\tau, t)x}{\partial t}, Re^{A(\tau - t)}z)_H d\tau$$

$$- \int_t^T (R\Phi(\tau, t)x, Re^{A(\tau - t)}Az)_Z d\tau - (Rx, Rz)_Z \quad (4.34)$$

where differentiation is justified on the strength of singular estimates available for $\frac{\partial \Phi(\tau, t)x}{\partial t}$, which in turn leads to integrable terms under the integral sign. Applying now (4.26) along with formula (4.8) to the expression in (4.34) gives

$$(P_t(t)x, z)_H = \int_t^T (R\Phi(\tau, t)A_P x, Re^{A(\tau - t)}z)_Z d\tau - (P(t)x, Az)_Z$$

$$-(Rx, Rz)_Z = (P(t)A_P x, z)_H - (P(t)x, Az)_H - (Rx, Rz)_Z. \quad (4.35)$$

Regularity in (4.18) allows us to write

$$(P(t)A_P x, z)_H = (P(t)Ax - P(t)BB^* P(t)x, z)_H$$
$$= (P(t)Ax, z)_H - (B^* P(t)x, B^* P(t)z)_U$$

which expression, when inserted into (4.35), gives

$$(P_t(t)x, z)_H = -(P(t)x, Az)_H - (Rx, Rz)_Z - (P(t)Ax, z)_H$$
$$+ (B^* P(t)x, B^* P(t)z)_U \quad (4.36)$$

as desired. □

4.6 The Function $r(t)$

We define the adjoint to $A_P^*(t)$ by the following formula

$$(A_P^*(t)x, y)_H = (x, Ay - BB^* P(t)y)_{D(A^*),[D(A^*)]'}; \; \forall x \in D(A^*).$$

So $D(A^*) \subset D(A_P^*(t)), \forall t \geq 0$.

Lemma 4.12. *The operator $A_P^*(t)$ is a generator of strongly continuous evolution operator $Y(t, s)$ which has the following additional property:*

$$B^* Y(t, s) \in \mathcal{L}(H \to C_\gamma([s, T]; U)).$$

Proof: The proof is based on the analysis of the following integral equation

$$y(t, s) = e^{A^*(t-s)}x + \int_s^t e^{A^*(t-z)}P(z)BB^* y(z, s)dz \quad (4.37)$$

in the variable $y(t, s)$ where x is a given element in H. We introduce new variable

$$x(t,s) \equiv (t-s)^\gamma B^* y(t,s).$$

Then solving (4.37) is equivalent to

$$x(t,s) = (t-s)^\gamma B^* e^{A^*(t-s)} x$$
$$- (t-s)^\gamma \int_s^t B^* e^{A^*(t-z)} P(z) B(z-s)^{-\gamma} x(z,s) dz \quad (4.38)$$

On the strength of singular estimates in Hypothesis (2.2), the above equation can be solved in $C_\gamma([s,T];U))$. Indeed, this follows from contraction mapping principle and the fact that $(t-s)^\gamma \int_s^t \frac{1}{(t-z)^\gamma (z-s)^\gamma} dz$ can be made arbitrarily small when $s \to t$. Thus, for each $x \in H$ we obtain $x(t,s) \in C_\gamma([s,T];U)$. Going back to the equation for $y(t,s)$ we obtain

$$y(t,s) = e^{A^*(t-s)} x - \int_s^t e^{A^*(t-z)} P(z) B x(z,s)(z-s)^{-\gamma} dz$$

and by regularity in (4.18) this yields

$$y(\cdot,s) \in C([s,T];H), \ |y(\cdot,s)|_{C([s,T];H)} \le C|x|_H$$

as desired for the final conclusion. □

The function $r(t)$ - the effect of the disturbance in the system - is described by the following equation:

Lemma 4.13. *The function $r(t)$ is differentiable in $C([0,T];[\mathcal{D}(A)]')$ and satisfies the following equation.*

$$r_t(t) + A_P^*(t) r(t) = -P(t) F(t) \in [\mathcal{D}(A)]'; \ r(T) = 0. \quad (4.39)$$

Assuming for a moment validity of this Lemma 4.13, we use variation of parameter formula to write

$$r(t) = \int_t^T \Phi^*(\tau,t) P(\tau) F(\tau) d\tau$$

where we note that $\Phi^*(t,s)$ is the adjoint operator to $\Phi(t,s)$ (and not the adjoint evolution corresponding to $A_P^*(t)$ which was denoted by $Y(t,s)$). We recall from Lemma 4.8 that

$$B^* \Phi^*(\cdot,s) \in C_\gamma([s,T];U),$$

moreover $\Phi^*(t,s)$ satisfies

$$\frac{d}{dt} \Phi^*(s,t) = -A_P^*(t) \Phi^*(s,t) \text{ in } [\mathcal{D}(A)]' \text{ for } s > t.$$

By the virtue of singular estimate valid for the evolution - see Lemma 4.8 - we conclude that

Corollary 4.14. $r(t) \in \mathcal{D}(B^*)$, $\forall t \in [0, T] > 0$ *with the representation:*

$$B^* r(t) = \int_t^T B^* \Phi^*(\tau, 0) P(\tau) F(\tau) d\tau \in C([0, T]; H).$$

Thus it remains to prove Lemma 4.13.

Proof of Lemma 4.13: Step 1: We recall that $y_F^0(t)$ is the optimal trajectory corresponding to zero initial condition. Thus y_F^0 is driven only by the disturbance F.

Proposition 4.15. *The following relation takes place:*

$$\frac{\partial}{\partial s}[\Phi(t, s) y_F^0(s)] = [\Phi(t, s)B]B^* r(s) + \Phi(t, s) F(s) \in C_\gamma([s, T]; H). \quad (4.40)$$

Proof: From the definition of evolution, we recall

$$\Phi(t, s) y_F^0(s) = [I_s + L_s L_s^* R^* R]^{-1} e^{A(\cdot - s)} y_F^0(s).$$

Letting $X(t, s) \equiv \Phi(t, s) y_F^0(s)$ we obtain the following relation

$$X(t, s) + L_s L_s^* R^* R X(\cdot, s) = e^{A(t - s)} y_F^0(s). \quad (4.41)$$

We differentiate the above relation in s (the differentiation is justified for $t > s$ with the values in $[\mathcal{D}(A^*)]'$.

$$\frac{\partial}{\partial s} X(t, s) + L_s L_s^* R^* R \frac{\partial}{\partial s} X(t, s) = e^{A(t-s)} BB^* \int_s^T e^{A^*(z-s)} R^* R \Phi(z, s) y_F^0(s) dz$$

$$- A e^{A(t-s)} y_F^0(s) + e^{A(t-s)} \frac{d}{ds} y_F^0(s) = e^{A(t-s)} BB^* P(s) y_F^0(s)$$

$$- A e^{A(t-s)} y_F^0(s) + e^{A(t-s)} [A_P(s) y_F^0(s) - BB^* r(s) + F(s)]$$

$$= e^{A(t-s)} BB^* P(s) y_F^0(s) - A e^{A(t-s)} y_F^0(s) +$$

$$e^{A(t-s)} [A y_F^0(s) - BB^* P(s) y_F^0(s) - BB^* r(s) + F(s)] \quad (4.42)$$

where due to singular estimate and regularity of $B^* P(t)$ all the terms are well defined for $t > s$ with the values in $[\mathcal{D}(A^*)]'$. Noting cancellation of four terms we obtain:

$$\frac{\partial}{\partial s} X(t, s) + L_s L_s^* R^* R \frac{\partial}{\partial s} X(t, s)$$

$$= e^{A(t-s)} [-BB^* r(s) + F(s)] \in C_\gamma([s, T]; H) \quad (4.43)$$

On $C_\gamma([s, T]; H)$ we can invert $[I_s + L_s L_s^* R^* R]$ which leads to

$$X(t, s) = [I_s + L_s L_s^* R^* R]^{-1} e^{A(\cdot - s)} [-BB^* r(s) + F(s)] = \Phi(t, s)[-BB^* r(s) + F(s)]$$

as desired for the final conclusion. $\square$

Step 2: From the definition of $r(t)$ given by (4.14)

$$r(t) = p_F(t) - P(t)\mathbf{y}_F^0(t)$$

where consistently with our notation (4.15) $p_F(t)$ is the adjoint variable corresponding to zero initial condition. It is immediate to verify that p_F satisfies the following differential equation

$$\frac{d}{dt}p_F(t) = -A^* p_F(t) - R^* R\mathbf{y}_F^0(t), \ on \ [\mathcal{D}(A)]'. \tag{4.44}$$

By using the representation for $P(t)$ our starting formula is the following

$$r(t) = p_F(t) - \int_t^T e^{A^*(\tau - t)} R^* R\Phi(\tau, t) d\tau \mathbf{y}_F^0(t). \tag{4.45}$$

Using the differentiability property established in (4.40) we are in a position to differentiate in time (4.45). This procedure yields the following expression well defined on $[\mathcal{D}(A^*)]'$:

$$r_t(t) = \frac{d}{dt}p_F(t) + R^* R y_F^0(t)$$

$$+ A^* \int_t^T e^{A^*(\tau - t)} R^* R\Phi(\tau, t) d\tau y_F^0(t) - \int_t^T e^{A^*(\tau - t)} R^* R\frac{\partial}{\partial t}[\Phi(\tau, t) y_F^0(t)] d\tau$$

$$= -A^* p_F(t) - R^* R y_F^0(t) + R^* R y_F^0(t) + A^* P(t) y_F^0(t)$$

$$- \int_t^T e^{A^*(\tau - t)} R^* R\Phi(\tau, t)[BB^* r(t) + F(t)] d\tau \quad (4.46)$$

where in the last step we have used (4.44) and (4.40). Cancelling the terms and using the structure of Riccati operator along with the regularity (4.18), Lemma 4.8 and (4.21) one obtains:

$$r_t(t) = -A^*[p_F(t) - P(t)y_F^0(t)] - P(t)[BB^* r(t) + F(t)]$$

$$= -A^* r(t) - P(t)BB^* r(t) - P(t)F(t) \tag{4.47}$$

proving the differentiability $r_t \in C([0, T]; [\mathcal{D}(A)]')$ and the desired form of the equation for $r(t)$. □

As already mentioned before, the proofs of results dealing with infinite horizon are in [45]. Interested reader is referred to [45] for technical details. It suffices to say that the idea of the proof is very similar as in the finite horizon - starting with the formulas in Lemma 4.1 appropriately interpreted in the case when $T = \infty$. The above formulas allow to define Riccati operator in terms of the semigroup $\Phi(t)$ which is then verified to satisfy the Algebraic Riccati Equation. The crux of the matter are of course singular estimates - but here the idea is reminiscent to the one used in the finite horizon case.

II. Applications to Point and Boundary Control Problems

In Part II of these notes, we shall present several control systems which provide illustration - in fact motivation - for the abstract theory presented in Part I. We shall focus on concrete models of boundary/point control systems displaying singular estimate. Three applications will be considered: thermoelastic plates, composite plates and structural acoustic interactions. In all these examples we shall witness an interplay between parabolic and hyperbolic characteristics of the underlying dynamics.

The first example considered is a system of thermoelasticity whose dynamics consist of a natural coupling between parabolic (heat equation) and hyperbolic (plate equation) equations. This system is controlled via boundary of a two dimensional region with Dirichlet control actions. Other configurations of boundary control problems are also possible.

The second example involves composite materials, where the plate matrix consists of a composite (layered) material. As we shall see, this model also provides hyperbolic/parabolic configuration where parabolic heat equation is coupled with Kirchoff plate where the latter exemplifies hyperbolic behaviour. The composite structure is controlled from the boundary (an edge) of the plate.

Our third example, the most complex, involves structural acoustic interaction in a 3 dimensional acoustic chamber. This problem has been actually studied within the engineering community - see [23, 27, 13] and references therein - in the context of noise or pressure reduction in an acoustic environment (e.g. airplane). "Smart material" actuators and sensors, such as piezoelectric patches, magnetorheological fluids etc have been used as practical control devices which serve as examples of boundary or point controllers. As it turns out, the model fits exactly into the class of systems with singular estimate and consequently, all the theoretical findings, including the boundedness of the gain operator are applicable and have strong implications on design of robust control algorithms.

Based on these three canonical examples one may construct many other models by combining various configurations. For instance, one could take thermoelastic walls in structural acoustic interaction which will then consist of 3 equations coupled together: wave equation, plate equation and heat equation. It is clear that suitable theory can be developed for these models as well. In fact, stabilization issues with corresponding results describing decay rates for this model can be found in [36, 37]. Similarly, one may replace classical elastic wall by composite wall. This also will give rise to coupling of three different dynamics. Corresponding details of the theory will have to be worked out.

Sections which follow describe these three examples in detail. The main goal there is to illustrate how the results of abstract theory apply to these practical engineering applications.

Notation: In what follows we shall often use the following notation for Sobolev's norms.

$|u|_{s,\Omega} \equiv |u|_{H^s(\Omega)}$ for $s \geq 0$. For $s < 0$ we adopt standard convention $|u|_{s,\Omega} \equiv |u|_{[H_0^s(\Omega)]'}$. The same notation applies with Ω replaced by Γ etc.

5 Boundary Control Problems for Thermoelastic Plates

This section is devoted to an analysis of control problem governed by a thermoelastic system with boundary or point controls.

5.1 PDE Model

We consider the following dynamics described by thermoelastic plates with rotational inertia.

Let $\Omega \subset R^2$ be a smooth, bounded domain with a boundary Γ. In $\Omega \times (0,T)$ we consider vibrating plate subject to effects of thermoelasticity. The variable w denotes the vertical displacement of the plate, while θ denotes the temperature. We affect the dynamics of the plate by means of boundary control u acting on the boundary Γ. The corresponding coupled PDE system is the following

$$w_{tt} - \rho\Delta w_{tt} + \Delta^2 w = \Delta\theta$$
$$\theta_t - \Delta\theta + \Delta w_t = 0 \tag{5.1}$$

with the boundary conditions on $\Gamma \times (0,T)$

$$w = \Delta w = 0; \ \theta = u. \tag{5.2}$$

We consider this model on a state space given by

$$H \equiv H_w \times H_\theta, \ H_w \equiv (H^2(\Omega) \cap H_0^1(\Omega)) \times H_0^1(\Omega), \ H_\theta \equiv L_2(\Omega).$$

Thus, initial conditions are taken as $\mathbf{y}(0) \equiv (w(0), w_t(0), \theta(0)) \in H$. With equation (5.1) we associate the functional cost given by

$$J(u, w, \theta) \equiv \int_0^T \int_\Omega [|R_w(\begin{smallmatrix} w(t,x) \\ w_t(t,x) \end{smallmatrix})|^2 + |R_\theta \theta(t,x)|^2] dx dt + \int_0^T \int_\Gamma |u(t,x)|^2 dx dt \tag{5.3}$$

where $R_w \in \mathcal{L}(H^2(\Omega) \times H_0^1(\Omega) \to L_2(\Omega))$ and $R_\theta \in \mathcal{L}(L_2(\Omega) \to L_2(\Omega))$

Optimal control problem to be considered is the following:

Boundary Control Problem: Minimize $J(u, w, \theta)$ given by (5.3) subject to the dynamics described by (5.1) with (5.2).

We note that in the case when in eq (5.1) the parameter $\rho = 0$, the corresponding system is analytic (see [53, 42, 41]). Thus, in this case the well-posedness of standard Riccati theory follows from the "analytic LQR" theory, cf. [14, 39]. As mentioned before, our interest is in studying the nonanalytic case corresponding to $\rho > 0$. Indeed, if $\rho > 0$ the thermoelastic system is predominantly hyperbolic, i.e., it can be written as a compact perturbation of a group, cf. [44]. The main goal here is to show that the control problem defined above is regular in the sense of singular estimate introduced in Part I. In particular, that it satisfies Hypothesis 2.2, which, in turn, allows to deduce that the optimal control problem admits regular optimal synthesis including the wellposedness of Riccati equation in both finite and infinite horizon cases.

5.2 Semigroup Formulation

We find convenient to recast the PDE problem (5.1) in a semigroup framework. In order to achieve this we introduce several operators:

$$\mathcal{A} : L_2(\Omega) \to L_2(\Omega); \ \mathcal{A} = \Delta^2, \ \mathcal{D}(\mathcal{A}) = \{u \in H^4(\Omega) \cap H_0^1(\Omega); \ \Delta u = 0 \text{ on } \Gamma\}.$$

Operators A_D and D, introduced before in Section 2 are recalled below for reader's convenience

$$A_D : L_2(\Omega) \to L_2(\Omega); \ A_D \equiv -\Delta, \ \mathcal{D}(A_D) = H^2(\Omega) \cap H_0^1(\Omega),$$

$$D : L_2(\Gamma) \to L_2(\Omega); \ \Delta Dg = 0 \text{ in } \Omega, \ Dg = g \text{ on } \Gamma.$$

With the above notation we define

$$\mathcal{M} \equiv I + \rho A_D.$$

It is well known that $A_D, \mathcal{A}$ and $\mathcal{M}$ are selfadjoint, positive operators on $L_2(\Omega)$. Moreover, standard elliptic theory - see [51] - gives $D \in \mathcal{L}(L_2(\Gamma) \to H^{1/2}(\Omega))$. With the above notation our PDE system (5.1) can be rewritten as (cf. [44], see also section 2.3.1):

$$\mathcal{M} w_{tt} + \mathcal{A} w + A_D \theta = A_D D u$$
$$\theta_t + A_D \theta - A_D w_t = A_D D u. \tag{5.4}$$

We note that though control operator was acting only on heat equation in the PDE model. The same control affects both equations (heat and plate) when written abstractly. This is a new feature of the coupled system.

We introduce next an unbounded operator $A : H \to H$, and $B : L_2(\Gamma) \to [\mathcal{D}(A)]'$ given by

$$A \equiv \begin{pmatrix} 0 & I & 0 \\ -\mathcal{M}^{-1}\mathcal{A} & 0 & -\mathcal{M}^{-1}A_D \\ 0 & A_D & -A_D \end{pmatrix}, \tag{5.5}$$

$$Bu \equiv \begin{pmatrix} 0 \\ \mathcal{M}^{-1}A_D Du \\ A_D Du \end{pmatrix}. \tag{5.6}$$

It is known (see [53, 44]) that A is a generator of a C_0 semigroup on H. This semigroup contains a hyperbolic component, hence it is non-analytic, cf. [44].

With the above notation our plate control problem can be rewritten as a first order system:

$$\mathbf{y}_t = A\mathbf{y} + Bu \text{ on } [\mathcal{D}(A)]'; \ \mathbf{y} \equiv [w, w_t, \theta]; \ u \in L_2(0, T; U); \ U = L_2(\Gamma).$$

5.3 Main Result

Now we are in a position to state the main result of this section:

Theorem 5.1. *Control system described by (5.1) satisfies "singular estimate" of Hypothesis 2.2 with $\gamma = \frac{3}{4} + \epsilon$ on the space $H = H_w \times H_\theta$.*

The proof of Theorem 5.1 is relegated to the next subsection.

It has been shown recently (see [2, 3]) that thermoelastic semigroup is exponentially stable. This is to say $|e^{At}|_{\mathcal{L}(H)} \leq Ce^{-\omega t}$, $\omega > 0, t \geq 0$. Since exponentially stable systems automatically satisfy the *FCC* Condition in Hypothesis 3.4, we obtain the following Corollary:

Corollary 5.2. *All the statements of Theorem 3.1 and Theorem 3.5 apply to Boundary Control Problem with A, B specified in (5.5), (5.6).*

Remark 5.3. One can also consider different (than hinged) boundary conditions associated with (5.1), cf. [44]. These include clamped $\frac{\partial}{\partial n}w = 0, w = 0$ *on Γ* or free homogeneous boundary conditions given by

$$\Delta w + B_1 w = 0, \ \frac{\partial}{\partial n}\Delta w + B_2 w - \gamma\frac{\partial}{\partial n}w_{tt} = 0; \text{ on } \Gamma$$

where boundary operators B_1 - of the second order - and B_2 - of the third order - are standard boundary operators associated with moments and shears, cf. [31]. The analysis of singular estimates for these models requires additional technical details which need to be worked out.

Remark 5.4. One could also consider the following point control problem associated with this plate.

$$w_{tt} - \rho\Delta w_{tt} + \Delta^2 w = \Delta\theta \text{ in } \Omega$$
$$\theta_t - \Delta\theta + \Delta w_t = \delta_{x_0}u, \text{ in } \Omega,$$
$$w = \Delta w = \theta = 0 \text{ on } \Gamma \tag{5.7}$$

where x_0 is a designated point in Ω. The corresponding functional cost can be taken as

Point control problem: Minimize

$$J(w, \theta, u) = \int_0^T [\|R_w W(t)\|_{H_w}^2 + |R_\theta \theta(t)|_{L_2(\Omega)}^2 + |u(t)|^2]dt \qquad (5.8)$$

where $W \equiv (w, w_t)$ and $R_w \in \mathcal{L}(H_w)$, $R_\theta \in \mathcal{L}(L_2(\Omega))$ and u, w satisfying (5.7).

All the final statements of Theorem 5.1 and Theorems 3.1, 3.5 with $\gamma = 1/2 + \epsilon$ apply to this model as well. In fact, the arguments are simpler than in the case of boundary control considered in Theorem 5.1.

The remaining part of this section is devoted to the proof of Theorem 5.1.

5.4 Proof of Theorem 5.1:

We begin by verifying that our standing assumption $A^{-1}B \in \mathcal{L}(U \to H)$ is satisfied with $U \equiv L_2(\Gamma)$ and H defined above. Indeed, by direct computations and elliptic regularity of the Dirichlet problem:

$$A^{-1}B = \begin{pmatrix} 0 \\ 0 \\ -D \end{pmatrix} \in \mathcal{L}(L_2(\Gamma) \to H), \qquad (5.9)$$

thus the operators A, B qualify for application of abstract theory. In particular, the operator $e^{At}B$ is well defined from $H \to [\mathcal{D}(A)]'$. Indeed, it suffices to write $e^{At}B = Ae^{At}A^{-1}B \in \mathcal{L}(U \to [\mathcal{D}(A)]')$.

The key role in the proof is played by the singular estimate for the kernel $e^{At}B$, which establishes the meaning of this operator as acting $U \to H$.

Lemma 5.5 (Singular estimate). *With A, B given in (7.14), (7.16) the following estimate takes place:* $\forall \epsilon > 0$

$$|e^{At}B|_{\mathcal{L}(U \to H)} \leq \frac{C}{t^{3/4+\epsilon}}; \ 0 < t \leq 1.$$

Proof: Step 1: setting up integral equations: Denote $W \equiv [w, w_t]$. Then

$$\begin{pmatrix} W(t) \\ \theta(t) \end{pmatrix} = e^{At}Bu$$

is equivalent to:

$$\begin{cases} \mathcal{M}w_{tt} + \mathcal{A}w + A_D\theta = 0 \\ \theta_t + A_D\theta - A_D w_t = 0 \end{cases}, \qquad (5.10)$$

$$\begin{pmatrix} w(0) \\ w_t(0) \\ \theta(0) \end{pmatrix} = \begin{pmatrix} 0 \\ \mathcal{M}^{-1}A_D Du \\ A_D Du. \end{pmatrix}$$

Abstract equation can be rewritten equivalently as a perturbation of damped Kirchoff plate:

$$\begin{cases} \mathcal{M}w_{tt} + \mathcal{A}w + A_D w_t = \theta_t \\ \theta_t + A_D \theta - A_D w_t = 0 \end{cases}$$

We introduce another unbounded operator $A_1 : H_w \to H_w$ given by

$$A_1 \equiv \begin{pmatrix} 0 & I \\ -\mathcal{M}^{-1}\mathcal{A} & -\mathcal{M}^{-1}A_D \end{pmatrix}.$$

Operator A_1 is a standard generator of damped Kirchoff plate. Indeed, we have

$$W_t(t) = A_1 W(t) + \begin{pmatrix} 0 \\ \mathcal{M}^{-1}\theta_t(t) \end{pmatrix}.$$

Thus, by variation of parameters formula we obtain

$$W(t) = e^{A_1 t}W(0) + \int_0^t e^{A_1(t-s)} \begin{pmatrix} 0 \\ \mathcal{M}^{-1}\theta_t(s) \end{pmatrix} ds, \qquad (5.11)$$

$$\theta(t) = e^{-A_D t}\theta(0) + \int_0^t e^{-A_D(t-s)} A_D w_t(s) ds. \qquad (5.12)$$

These two integral equations are the main object of the analysis.

Step 2: analysis of W equation. Integrating equation (5.11) by parts yields

$$W(t) = e^{A_1 t}W(0) + e^{A_1(t-s)} \begin{pmatrix} 0 \\ \mathcal{M}^{-1}\theta(s) \end{pmatrix} |_0^t - \int_0^t e^{A_1(t-s)} A_1 \begin{pmatrix} 0 \\ \mathcal{M}^{-1}\theta(s) \end{pmatrix} ds. \qquad (5.13)$$

Hence

$$W(t) = e^{A_1 t}W(0) + \begin{pmatrix} 0 \\ \mathcal{M}^{-1}\theta(t) \end{pmatrix} - e^{A_1 t} \begin{pmatrix} 0 \\ \mathcal{M}^{-1}\theta(0) \end{pmatrix}$$
$$- \int_0^t e^{A_1(t-s)} \begin{pmatrix} \mathcal{M}^{-1}\theta(s) \\ -\mathcal{M}^{-1}A_D\mathcal{M}^{-1}\theta(s) \end{pmatrix} ds. \qquad (5.14)$$

By using the properties

- $\mathcal{M}^{-1} \in \mathcal{L}(L_2(\Omega) \to H^2(\Omega) \cap H_0^1(\Omega)) \cap \mathcal{L}(H^{-1}(\Omega) \to H_0^1(\Omega))$
- $\mathcal{M}^{-1}A_D\mathcal{M}^{-1} \in \mathcal{L}(H^{-1}(\Omega) \to H_0^1(\Omega))$

and noticing cancellation of terms in (5.14) we arrive at the estimate:

$$|W(t)|_{H_w} \le C[|\theta(t)|_{H^{-1}(\Omega)} + |\theta|_{L_1(0,t;L_2(\Omega))}]. \qquad (5.15)$$

Step 3: analysis of θ equation. Integrating by parts the second term in (5.12) yields

$$\theta(t) = e^{-A_D t}\theta(0) - e^{-A_D(t-s)}w_t(s)|_0^t - \int_0^t e^{-A_D(t-s)}w_{tt}(s)ds =$$

$$= e^{-A_D t}\theta(0) - w_t(t) + e^{-A_D t}w_t(0) - \int_0^t e^{-A_D(t-s)}w_{tt}(s)ds. \quad (5.16)$$

On the other hand by recalling original version of the plate equation in (5.10) we obtain:

$$w_{tt} = -\mathcal{M}^{-1}[\mathcal{A}w + A_D\theta].$$

Inserting the above relation into (5.16) and estimating terms yields

$$|\theta(t)|_{0,\Omega} \leq |e^{-A_D t}A_D Du|_{0,\Omega} + |e^{-A_D t}\mathcal{M}^{-1}A_D Du|_{0,\Omega} + |w_t(t)|_{0,\Omega}$$

$$+ |\int_0^t e^{-A_D(t-s)}\mathcal{M}^{-1}A_D\theta(s)ds|_{0,\Omega} + |\int_0^t e^{-A_D(t-s)}\mathcal{M}^{-1}\mathcal{A}w(s)ds|_{0,\Omega}$$

$$(5.17)$$

Since the semigroup $e^{-A_D t}$ is analytic and the Dirichlet map satisfies (see [39])

$$D \in \mathcal{L}(L_2(\Gamma) \to \mathcal{D}(A_D^{1/4-\epsilon}))$$

we infer

$$|e^{-A_D t}A_D D|_{\mathcal{L}(L_2(\Gamma)\to L_2(\Omega))} \leq \frac{C}{t^{3/4+\epsilon}}; \; 0 < t \leq 1. \quad (5.18)$$

Moreover the analyticity of $e^{-A_D t}$ along with the fact that $\mathcal{D}(A_D^\theta) \sim H_0^{2\theta}(\Omega)$, $0 \leq \theta \leq 1$, [26], also imply that $T_t(f) \equiv \int_0^t e^{-A_D(t-s)}f(s)ds$ satisfies (see [39, 14]):

$$T_t \in \mathcal{L}(L_\infty(0,t; H^{-1}(\Omega)) \to H_0^{1-\epsilon}(\Omega)); \; \forall t > 0. \quad (5.19)$$

Since $\mathcal{A} = A_D^2$

$$\mathcal{M}^{-1}\mathcal{A} \in \mathcal{L}(H_0^1(\Omega) \to H^{-1}(\Omega)). \quad (5.20)$$

Collecting (5.17), (5.18), (5.19), (5.20) we obtain:

$$|\theta(t)|_{0,\Omega} \leq C[|w_t(t)|_{0,\Omega} + \int_0^t |\theta(s)|_{0,\Omega}ds + |w|_{L_\infty(0,t; H_0^1(\Omega))} + \frac{|u|_U}{t^{3/4+\epsilon}}]; \; 0 < t \leq 1.$$

$$(5.21)$$

Gronwall's inequality applied to the above inequality gives

$$|\theta(t)|_{0,\Omega} \leq C[|w_t(t)|_{0,\Omega} + |w|_{L_\infty(0,t; H_0^1(\Omega))} + \frac{|u|_U}{t^{3/4+\epsilon}}], \; 0 < t \leq 1. \quad (5.22)$$

Step 4: decoupling W and θ given, respectively by (5.15) and (5.22)
   From (5.15)

$$|W(t)|_{H_w} \le C[|\theta(t)|_{-1,\Omega} + \int_0^t |w_t(s)|_{0,\Omega}ds +$$

$$t|w|_{L_\infty(0,t;H_0^1(\Omega))} + |u|_U \int_0^t \frac{C}{s^{3/4+\epsilon}}ds]$$

$$\le C sup_{0 \le s \le t}[|\theta(s)|_{-1,\Omega} + t|W(s)|_{H_w}] + Ct^{1/4-\epsilon}|u|_U \qquad (5.23)$$

and

$$(1 - Ct)|W|_{L_\infty((0,t),H_w)} \le C sup_{0 \le s \le t}|\theta(s)|_{-1,\Omega} + C|u|_U t^{1/4-\epsilon}. \qquad (5.24)$$

The H^{-1} norm of θ can be easily estimated from (5.12) as follows:

$$|\theta(t)|_{-1,\Omega} \le |A_D^{-1/2}e^{-A_D t}\theta(0)|_{0,\Omega} + |\int_0^t e^{-A_D(t-s)}A_D^{1/2}w_t(s)ds|_{0,\Omega}$$

$$\le C[\frac{|u|_U}{t^{1/4+\epsilon}} + t|w|_{L_\infty(0,t;H_0^1(\Omega))}] \, (5.25)$$

where in the last estimate we have used the basic semigroup estimate only.
Combining (5.24) and (5.25) allows us to decouple W variable

$$(1 - Ct)|W|_{L_\infty(0,t;H_w)} \le C|u|_U[\frac{1}{t^{1/4+\epsilon}} + |u|_U t^{1/4-\epsilon}]. \qquad (5.26)$$

Taking $t \le \frac{1}{2C}$ we obtain

$$|W(t)|_{H_w} \le C|u|_U \frac{1}{t^{1/4+\epsilon}} \qquad (5.27)$$

which then combined with (5.22) yields

$$|\theta(t)|_{0,\Omega} \le C\frac{|u|_U}{t^{3/4+\epsilon}}, \; 0 < t \le \frac{1}{2C}. \qquad (5.28)$$

The above argument can be now bootstrapped in order to obtain the estimate
for any $1/2C \le t \le T < \infty$. Prof of Lemma 5.5 is thus complete. $\quad\square$
   In order to assert validity of Theorem 5.1 it suffices to combine the result
of Lemma 5.5 with the fact that e^{At} is exponentially stable, cf. [2, 3]. $\quad\square$

6 Composite Beam Models with Boundary Control

In this subsection we focus on a model developed by [28] of a sandwich beam
consisting of thin, compliant middle layer and two identical stiff outer layers.
In this simplified model. The middle layer resists shear, but not bending,
and the thickness is assumed to be small enough, so that the mass may be
neglected or included in the outer layer. Damping is included in the middle
layer in such a way that shear motions are resisted by a force proportional
to the rate of a shear. The outer layers obey the usual assumptions of the
Euler-Bernoulli beam theory.

6.1 PDE Model and Main Results

In $\Omega \equiv (0,1)$ we consider $w(t,x)$ transversal displacement and $\xi(t,x)$ effective rotation of the angle with $x \in \Omega$ $0 \le t \le T$. Let $\rho > 0$ denotes a rotational moment of inertia, $d > 0$ is the stiffness of the beam, $\delta > 0$ is the stiffness of the middle layer, and β is the damping parameter for the middle layer.

In $\Omega \times (0,T)$ we consider the following coupled system developed in [28]:

$$w_{tt} - \rho w_{ttxx} - (d\xi_x)_{xx} = 0 \ in \ (0,T) \times \Omega$$

$$\frac{\beta}{3}(\xi_x + w_{tx}) - d(\xi_x + 1/4 w_{xx})_x + \frac{\delta}{3}(\xi + w_x) = 0 \ in \ (0,T] \times \Omega$$

$$w(t=0) = w_0, \ w_t(t=0) = w_1, \ \xi(t=0) = \xi_0 \ in \ \Omega \quad (6.1)$$

along with appropriate boundary conditions. This model is typically associated with either *clamped* or *hinged* boundary conditions imposed on $\Gamma \equiv \{0,1\}$.

$$w = 0, w_x = 0, \xi = 0 \ on \ \Sigma \equiv (0,T] \times \Gamma \ clamped, \quad (6.2)$$

$$w = 0, w_{xx} = 0, \xi_x = 0 \ on \ \Sigma \equiv (0,T] \times \Gamma \ hinged. \quad (6.3)$$

To focus attention, we shall consider boundary control problem with control acting via moments. This leads to:

$$w = 0, w_{xx} = 0, \xi_x = u \in L_2(0,T) \ on \ \Sigma. \quad (6.4)$$

Standard energy function associated with the model (see [28]) is the following

$$E(t) = \int_0^1 [w_t^2(t,x) + \rho w_{tx}^2(t,x) + d\xi_x^2(t,x) + \frac{d}{3}[\xi_x(t,x) + w_{xx}(t,x)]^2$$

$$+ \frac{4\delta}{9}[\xi(t,x) + w_x(t,x)]^2]dx \quad (6.5)$$

The above energy function motivates the following state space for the model with $\rho > 0$:

$$H \equiv H_w \times H_\xi, \ H_w \equiv H^2(\Omega) \cap H_0^1(\Omega) \times H_0^1(\Omega), \ H_\xi \equiv H^1(\Omega).$$

It can be easily verified (see [28]) that the topology determined by the energy function is equivalent to that of H defined above.

Thus, initial conditions $\mathbf{y}(0) \equiv (w(0), w_t(0), \xi(0)) \in H$ are taken in H.

With equation (6.1) and boundary conditions (6.4) we associate the functional cost given by

$$J(u, w, \xi) \equiv \int_0^T \int_\Omega [|R_w(\begin{matrix} w(t,x) \\ w_t(t,x) \end{matrix})|^2 + |R_\theta \theta(t,x)|^2 dxdt + \int_0^T |u(t)|^2 dt \quad (6.6)$$

where $R_w \in \mathcal{L}(H^2(\Omega) \times H_0^1(\Omega) \to L_2(\Omega))$ and $R_\theta \in \mathcal{L}(H^1(\Omega) \to L_2(\Omega))$

Optimal control problem to be considered is the following:

Boundary Control Problem: Minimize $J(u, w, \xi)$ given by (6.6) subject to the dynamics described by (6.1) with (6.4).

As in the case of thermal plates, (6.1) with the parameter $\rho = 0$, corresponds to an analytic semigroup, cf. [30]. Thus, in this case the wellposedness of standard Riccati theory follows from the "analytic LQR" theory, cf. [14, 39]. As mentioned before, our interest is in studying the nonanalytic case corresponding to $\rho > 0$. Indeed, if $\rho > 0$ the thermoelastic system is predominantly hyperbolic, i.e., it can be written as a compact perturbation of a group, cf. [30]. The main goal here is to show that the control problem defined above is regular in the sense described in Part I. In particular, we aim to show that the control system defined above satisfies Hypothesis 2.2, which, in turn, allows to deduce that the optimal control problem admits regular optimal synthesis including the classical wellposedness of Riccati equation in both finite and infinite horizon case.

Our main results are:

Theorem 6.1. *Control system described by (6.1) satisfies "singular estimate" of Hypothesis 2.2 with $\gamma = \frac{3}{4} + \epsilon$ on the space $H = H_w \times H_\xi$.*

The above result allows us to apply abstract results of Theorem 3.1, which implies the wellposedness of Finite Horizon optimal control synthesis and related Differential Riccati Equations. Since the uncontrolled semigroup is also exponentially stable - see [30] - the *FCC* condition is automatically satisfied. As the result, conditions of Theorem 3.5 are met as well. Therefore, we obtain the following final result

Corollary 6.2. *All the statements of Theorem 3.1 and Theorem 3.5 apply to Boundary Control Problem stated above.*

The remaining sections are devoted to the proof of Theorem 6.1.

6.2 Transformations of the Model

We shall see that this model has many features which are mathematically reminiscent of thermoelastic plates considered in the previous section. In order to exhibit this we shall perform several changes of variables.

First transformation

As in [30] we introduce

$$s(t, x) \equiv \xi(t, x) + w_x(t, x). \tag{6.7}$$

Physically, s is proportional to the shear in the middle layer. The transformed version of system (6.1) now becomes:

$$w_{tt} - \rho w_{ttxx} + dw_{xxxx} - ds_{xxx} = 0 \ in \ (0,T) \times \Omega$$
$$\beta s_t - 3d(s_{xx} - 3/4w_{xxx}) + \delta s = 0 \ in \ (0,T] \times \Omega$$
$$w(t=0) = w_0. \ w_t(t=0) = w_1, \ s(t=0) = s_0 = \xi_0 + w_x(0) \ in \ \Omega \quad (6.8)$$

along with boundary (uncontrolled) conditions

$$w = 0, w_x = 0, s = 0 \ on \ \Sigma \equiv (0,T] \times \Gamma \ \textbf{clamped}, \quad (6.9)$$

$$w = 0, w_{xx} = 0, s_x = 0 \ on \ \Sigma \equiv (0,T] \times \Gamma, \ \textbf{hinged}. \quad (6.10)$$

The controlled boundary conditions (6.4 become

$$w = 0, w_{xx} = 0, s_x = u \ on \ \Sigma. \quad (6.11)$$

Model (6.8) with boundary conditions (6.9) or (6.10) was successfully analyzed in [30] with major findings such as

- When $\rho = 0$ the corresponding system generates an analytic semigroup on H.
- When $\rho > 0$ the system has dominant hyperbolic component and it can be written as a compact perturbation of a group.
- The system is exponentially stable on H, regardless of value of ρ on H.

We notice that all these properties are very reminiscent of these displayed by thermoelastic plates, cf. [43]. In fact, this parallelism suggested possibility of another change of variables which closely resembles thermal plate equations (modulo lower order terms). It is this new transformation introduced in [59] which became very useful in proving validity of singular estimate for the controlled model. This transformation is described below.

Second change of variables

Following [59] we introduce the following change of variables

$$\theta \equiv s_x - 3/4w_{xx} \equiv \xi_x + 1/4w_{xx}.$$

Then the resulting model becomes

$$w_{tt} - \rho w_{ttxx} + \frac{1}{4}dw_{xxxx} - d\theta_{xx} = 0 \ in \ (0,T) \times \Omega$$

$$\beta\theta_t - 3d\theta_{xx} + \delta\theta + 3/4\beta w_{xtt} + \frac{3\delta}{4}w_{xx} = 0 \ in \ (0,T] \times \Omega$$

$$w(t=0) = w_0. \ w_t(t=0) = w_1, \ \theta(t=0) = \theta_0 \ in \ \Omega. \quad (6.12)$$

With this model hinged and clamped (uncontrolled) boundary conditions are translated into

$$w = 0, w_x = 0, \theta_x = 0 \ on \ \Sigma \equiv (0,T] \times \Gamma, \ \textbf{clamped}, \quad (6.13)$$

$$w = 0, w_{xx} = 0, \theta = 0 \ on \ \Sigma \equiv (0,T] \times \Gamma, \ \textbf{hinged}. \quad (6.14)$$

Thus, in the case of our control problem, we consider dynamics in (6.12) with the Dirichlet boundary control

$$w = 0, w_{xx} = 0, \theta = u \ on \ \Sigma \equiv (0,T] \times \Gamma. \quad (6.15)$$

6.3 Semigroup Formulation

We shall put PDE system (6.12) in a semigroup framework. To accomplish this we introduce several operators.

- $A_D = -\Delta, \mathcal{D}(A_D) = H^2(\Omega) \cap H_0^1(\Omega)$
- $\mathcal{M} \equiv I + \rho A_D$
- $\mathcal{A} \equiv A_D^2$
- $D : L_2(\Gamma) \to L_2(\Omega)$ given by $\Delta Du = 0$, in Ω, $Du|_\Gamma = u$, on Γ

and the spaces

$$H \equiv \mathcal{H}_w \times \mathcal{H}_\theta, \ \mathcal{H}_w \equiv \mathcal{D}(A_D) \times \mathcal{D}(\mathcal{M}^{1/2}), \ \mathcal{H}_\theta \equiv L_2(\Omega).$$

With the above notation we can rewrite the PDE given by (6.12), (6.15) as:

$$\mathcal{M}w_{tt} + \frac{d}{4}\mathcal{A}w + dA_D\theta = dA_D Du$$

$$\beta\theta_t + 3dA_D\theta + \delta\theta - \frac{3}{4}\beta A_D w_t - \frac{3\delta}{4}A_D w = A_D Du. \tag{6.16}$$

The above equation clearly ressembles thermoelastic model (5.4). We note that as in the thermoelastic case, the control function acts through both equations.

We introduce next $A : H \to H, \ B : L_2(\Gamma) \to \mathcal{D}(A)'$ given by

$$A \equiv \begin{pmatrix} 0 & I & 0 \\ -\frac{d}{4}\mathcal{M}^{-1}\mathcal{A} & 0 & -\mathcal{M}^{-1}A_D \\ \frac{3\delta}{4\beta}A_D & \frac{3}{4}A_D & -\frac{3d}{\beta}A_D - \frac{\delta}{\beta}I \end{pmatrix} \tag{6.17}$$

with the natural domain and

$$Bu \equiv \begin{pmatrix} 0 \\ \mathcal{M}^{-1}A_D Du \\ A_D Du \end{pmatrix} : U \to [\mathcal{D}(A)]' \tag{6.18}$$

so we obviously have $A^{-1}B \in \mathcal{L}(U \to H)$ as postulated in Hypothesis 2.1.

A useful observation is that the operator A in (6.17) is a bounded observation of the operator describing generator for thermoelastic system and given by (5.5). Indeed, the perturbation is of the form

$$\mathcal{P} \equiv \begin{pmatrix} 0 & 0 & 0 \\ 0 & 0 & 0 \\ \frac{3\delta}{4\beta}A_D & 0 & -\frac{\delta}{\beta}I \end{pmatrix} \tag{6.19}$$

so after accounting for the definition of the state space H we trivially obtain:

$$\mathcal{P} \in \mathcal{L}(H). \tag{6.20}$$

Therefore, by the same arguments as in the previous section and standard perturbation theorem (see [55]) we obtain that A is a generator of a strongly continuous semigroup on H, which however may not be contractive. It is also known (see [53, 44]) that A with $\rho = 0$ is a generator of an analytic semigroup on H. For $\rho > 0$ this semigroup contains a hyperbolic component, hence it is non-analytic, cf. [44].

With the above notation our composite beam problem can be rewritten as a first order system:

$$\mathbf{y}_t = A\mathbf{y} + Bu \text{ on } [\mathcal{D}(A)]'; \ \mathbf{y} \equiv [w, w_t, \theta] \in H; \ u \in L_2(0, T; U); \ U = L_2(\Gamma) = R^2.$$

6.4 Singular Estimate and Completion and Proper Proof of Theorem 6.1

We recall that singular estimate was already proved in Sect 4.3 Theorem 5.1 for the pair operator $A - \mathcal{P}, B$ where $A, \mathcal{P}, B$ are given by (6.17), (6.19), (6.18).

It is expected that validity of singular estimate should be invariant subject to bounded perturbation of the generator. In fact, this fact was proved in [59]

Lemma 6.3. *[59] Let A, B satisfy Hypothesis 2.1 and Hypothesis 2.2 with a given value $0 \leq \gamma < 1$. Let $\mathcal{P} \in \mathcal{L}(H)$. Then*

$$|e^{(A+\mathcal{P})t}Bu|_H \leq C\frac{1}{t^\gamma}|u|_U.$$

By applying the result of Lemma 6.3 and Theorem 5.1 in the context of operators A, B given by (6.17), (6.18) we obtain the final result stated in Theorem 6.1. $\square$

Remark 6.4. • One can consider control acting via clamped boundary conditions as well.
• Control problem with point control can be formulated and solved in a similar way as in the thermoelastic case.
• Though the sandwich model introduced above is one-dimensional, we purposely set up this problem in a more general abstract framework in order to allow for generalizations to higher dimensions as well.

Remark 6.5. Models of two dimensional composite plates are, as expected, more complicated, cf. [29]. However, the basic mathematical features of the problem are retained. This includes, in particular, strong analogy to thermoelastic plate models, cf. [29]. It is believed that the existent analysis of thermoelastic plates will prove helpful in establishing singular estimates for the higher dimensional composite plates.

7 Point and Boundary Control Problems in Acoustic-structure Interactions

In this section we shall study an optimal control problem governed by a model referred to as an abstract model of structural acoustic interactions. In these applications the goal of control is to reduce a noise entering an acoustic environment.

7.1 Description of the Model

The model under consideration consists of coupled wave equation with a dynamic plate equation interacting on an interface. This is a typical model arising in structural acoustic interactions, cf. [54, 13, 52].

Let Ω be a bounded domain in R^n, $n = 2, 3$, with a boundary Γ which consist of two parts Γ_0, Γ_1. We assume that $\Gamma = \overline{\Gamma_0} \cup \overline{\Gamma_1}$, Γ_0 is flat and $\Gamma_0 \cap \Gamma_1 = \emptyset$. We also assume that Ω is sufficiently smooth or convex.

In practical applications Ω will describe an acoustic chamber while Γ_i, $i = 0, 1$ denote the walls of the chamber. Γ_1 is referred to as "hard wall" and Γ_0 is flexible wall where the interaction with structural medium takes place. The acoustic medium in the chamber is described by the wave equation in the variable z (where the quantity $\rho_1 z_t$ is acoustic pressure)

$$z_{tt} = c^2 \Delta z - d_0(x) z_t + f \qquad \text{in} \quad \Omega \times (0, T)$$

$$\frac{\partial}{\partial n} z + d_1 z = 0 \qquad \text{on} \quad \Gamma_1 \times (0, T)$$

$$\frac{\partial}{\partial n} z + dD\, z_t = w_t \qquad \text{on} \quad \Gamma_0 \times (0, T).$$

Here c^2 is the speed of sound as usual, $f \in L^2((0, T) \times \Omega)$ is an external noise, $d_0(x) \geq 0$ in Ω, $d_1 > 0$. Coefficients $d, d_0 \geq 0$, represent potential boundary or interior damping. The operator $D : L_2(\Gamma_0) \to L_2(\Gamma_0)$ is positive and densely defined on $L_2(\Gamma_0)$ and subject to additional assumptions specified later. The operator D represents boundary damping on Γ_0 while d_0 represents internal damping.

The interaction between acoustic medium and structural medium is described by plate equation along with appropriate coupling (back pressure on the wall).

Let $\mathcal{A}$ denotes an elastic operator which is positive and selfadjoint on $L_2(\Gamma_0)$. We consider the following abstract model for the plate with structural damping (see [56])

$$w_{tt} + \mathcal{A}w + \rho \mathcal{A}^\alpha w_t + \rho_1 z_t|_{\Gamma_0} = 0 \; \rho > 0, \; \rho_1 > 0.$$

where $\rho_1 z_t$ denotes back pressure on the wall and $\rho \mathcal{A}^\alpha w_t$ denotes structural damping. It is known that for α between $1/2$ and 1 the above equation with

$\rho_1 = 0$ generates an analytic semigroup on $\mathcal{D}(\mathcal{A}^{1/2}) \times L_2(\Gamma_0)$, cf. [56, 18]. A canonical realization of $\mathcal{A}$ which is of interest to us is $\mathcal{A} = \Delta^2$ with associated (clamped, higed or free) boundary conditions.

The ultimate PDE model to be considered in this paper (wlog we take $\rho_1 = 1$) can be written as:

$$z_{tt} = c^2 \Delta z - d_0 z_t + f \qquad \text{in} \quad \Omega \times (0, T) \qquad (7.1a)$$

$$\frac{\partial}{\partial n} z + d_1 z = 0 \qquad \text{on} \quad \Gamma_1 \times (0, T) \qquad (7.1b)$$

$$\frac{\partial}{\partial n} z + dD z_t = w_t \qquad \text{on} \quad \Gamma_0 \times (0, T) \qquad (7.1c)$$

$$w_{tt} + \mathcal{A}w + \rho\mathcal{A}^\alpha w_t + z_t|_{\Gamma_0} = \mathcal{B}u \qquad \text{on} \quad \Gamma_0 \times (0, T). \qquad (7.1d)$$

Here the operator $\mathcal{B}$ is a control operator acting upon actuator $u(t) \in \mathcal{U}$, where $\mathcal{U}$ is a suitable control space. The control operator is typically unbounded such as they typically arise in the context of modelling "smart" materials.

Since for $\alpha \geq 1/2$ the uncoupled abstract plate equation is modelled by an analytic semigroup, the entire system (7.1) represents coupling between hyperbolic component (wave equation)) and analytic component (plate equation).

In concrete applications to structural acoustic, the last equation in (7.1) models plate equation with $\mathcal{A}$ being fourth order elliptic operator. The control operator $\mathcal{B}$ represents point controls realized via smart actuators such as piezoceramic or piezoelectric patches. In such instances $\mathcal{B}$ is just derivative of "delta" function supported either at some interior points of Γ (dim $\Gamma_0 = 1$) or on some closed curves in Γ_0 if dim $\Gamma_0 = 2$, cf. [23].

In this section we shall consider more general classes of operators $\mathcal{A}, \mathcal{B}$ which are defined by the following set of Hypotheses.

Hypothesis 7.1. $\mathcal{A} : \mathcal{D}(\mathcal{A}) \subset L_2(\Gamma_0) \to L_2(\Gamma_0)$ *is a positive, selfadjoint operator.*

Hypothesis 7.2. *There exists a positive constant r, $0 < r < 1/2$, such that*

$$\mathcal{A}^{-r}\mathcal{B} \in \mathcal{L}(\mathcal{U} \to L_2(\Gamma_0)); \quad \text{equivalently,} \quad \mathcal{B} \quad \text{continuous} : \mathcal{U} \to [\mathcal{D}(\mathcal{A}^r)]'; \qquad (7.2)$$

where $\mathcal{U}$ is another Hilbert space.

The following hypothesis is made regarding the boundary structural damping operator D.

Hypothesis 7.3. $D : L_2(\Gamma_0) \supset \mathcal{D}(D) \to L_2(\Gamma_0)$ *is a positive, selfadjoint operator, and there exists a constant r_0, $0 \leq r_0 \leq 1/4$, and positive constants δ_1, δ_2 such that*

$$\delta_1 |z|_{D(A^{r_0})} \le (Dz, z)_{L_2(\Gamma_0)} \le \delta_2 |z|_{D(A^{r_0})} \qquad \forall z \in D(A^{r_0}) \equiv \mathcal{D}(D^{1/2}).$$
$$(7.3)$$

Moreover, we assume that $H^1(\Gamma_0) \subseteq \mathcal{D}(A^{r_0})$.

Remark 7.4. If A models plate equations (case of interest to us), then with $r_0 = 1/4$ we have that $\mathcal{D}(A^{1/4})$ is topologically equivalent to $H^1(\Gamma_0)$ norm. In such case, the operator D corresponds to a *structural boundary damping* modelled by Laplace's Beltrami operator.

The controlled problem to be considered is the following:
Control Problem: Minimize

$$J(u, z, w) = \int_0^T [|\nabla z|_{0,\Omega}^2 + |z_t|_{0,\Omega}^2 + |A^{1/2}w|_{0,\Gamma_0}^2 + |w_t|_{0,\Gamma_0}^2 + |u(t)|_U^2] dt \quad (7.4)$$

subject to the dynamics in (7.1).

Remark 7.5. One could also consider other models for plates, including these where analyticity is generated by thermal effects, cf. [35]. Specific example can be given by considering the same system (7.1) with the fourth equation replaced by thermoelastic system in the variables w, θ defined on $\Gamma_0 \times (0, \infty)$.

$$w_{tt} + \Delta^2 w - \Delta\theta + \rho z_t = \mathcal{B}u$$
$$\theta_t - \Delta\theta + \Delta w_t = 0. \qquad (7.5)$$

With the above system we associate boundary conditions of either clamped $w = \frac{\partial}{\partial n} w = 0$, hinged $w = \Delta w = 0$ or free type, cf. [42]. The associated control problem can be formulated as follows:
Control Problem: Minimize

$$J(u, z, w) = \int_0^T [|\nabla z|_{0,\Omega}^2 + |z_t|_{0,\Omega}^2 + |\Delta w|_{0,\Gamma_0}^2 + |w_t|_{0,\Gamma_0}^2 + |\theta(t)|_{0,\Gamma_0}^2 + |u(t)|_U^2] dt$$
$$(7.6)$$

subject to equations (7.1 a-c) and (7.5).

Our goal is to show that structural acoustic interactions described above fit into the abstract framework described in Part I. As a consequence, the results of Theorem 3.1 and Theorem 3.5 will apply to these problems as well.

Remark 7.6. Structural acoustic control problems with active walls modelled by plate equations subject to Kelvin Voigt damping (i.e., $\alpha = 1$, $D = 0$ in (7.1) were considered in [8, 9] with numerous references therein. In fact, these papers were the first ones to bring to the attention of PDE control community mathematical models arising in structural acoustic problems. However, the focus of the analysis in [8] is on modelling, numerics and formal application of Riccati theory valid for *bounded control operators*. Although in [8] p. 128 and also [9] there is the statement that *"this theory can be extended to also include*

the case of unbounded B, i.e., $B \in \mathcal{L}(U, V^*)$, by invoking [7]", we wish to warn
the reader that the statement in such generality may be incorrect and, in fact,
contradicted by several examples where the gain operators are not defined at
all for such a large class of control operators, cf. [58, 61]. Such statement would
be, indeed, correct and fully justifiable , if the underlying semigroup were analytic, see [14, 39]. However, structural acoustic model is *not analytic*, so the
extension of Riccati theory to the case when $B \in \mathcal{L}(U \to V^*)$ is generally in-
valid. On the other hand, the emphasis, and the contribution of works such as
[8, 9] is on modelling, finite dimensional truncations, the resulting numerical
algorithms and engineering design of control systems, rather than on mathematical (PDE) analysis which would account for the effects of unboundedness
of control operators. That this analysis is critical, and not only for the sake
of mathematical rigour but above all for the validity of engineering design,
is now beyond any dispute. In fact, counterexamples with either blowing up
or not uniquely defined gain operators (see [58, 61, 11]) make a sufficiently
convincing point that a careful analysis of infinite-dimensional structure of
the problem is indeed critical. Even, in the cases where "numerics" provide
implementable results (as it is in the very special case of Kelvin Voigt damping with strong smoothing effect, cf. [8]), it is important to understand on
mathematical grounds why this is the case and what are the limitations of
the theory and, ultimately, of the design. And this precisely justifies the goal
of the analysis presented in these lectures. Such analysis, among other things,
provides justification and explanation for the results obtained either numer-
ically or experimentally in [8, 9] and references therein. In fact, we shall see
that the key to this understanding is precisely the validity of singular estimate
and the resulting boundedness of the gain operators. This, in turn, relies on a
rather deep PDE analysis including microlocal regularity of hyperbolic traces
and propagation of analyticity.

7.2 Semigroup Formulation

The following operators will be used in describing the PDE model given in
(7.1) - see [16].
Operators acting on Ω. (i) Let $A_N : L_2(\Omega) \supset \mathcal{D}(A_N) \to L_2(\Omega)$ be the
non-negative, selfadjoint operator defined by

$$A_N h = -c^2 \Delta h, \quad \mathcal{D}(A_N) = \left\{ h \in H^2(\Omega) : \; (\frac{\partial}{\partial n} h + d_1 h)|_\Gamma = 0 \right\}, \quad (7.7)$$

(ii) Let N be the Neumann map from $L_2(\Gamma_0)$ to $L_2(\Omega)$, defined by

$$\psi = Ng \iff \left\{ \Delta \psi = 0 \text{ in} \Omega ; \; \frac{\partial}{\partial n} \psi \Big|_{\Gamma_0} = g, \; (\frac{\partial}{\partial n} \psi + d_1 \psi) \Big|_{\Gamma_1} = 0 \right\}.$$

It is well known (see [51]) that N continuous : $L_2(\Gamma_0) \to H^{3/2}(\Omega) \subset$
$\mathcal{D}(A_N^{3/4-\epsilon})$, $\epsilon > 0$, Moreover, by Green's second theorem, the following trace

result holds true with any $h \in \mathcal{D}(A_N)$:

$$N^* A_N h = \begin{cases} h|_{\Gamma_0} & \text{on} \Gamma_0 \\ 0 & \text{on} \Gamma_1, \end{cases} \tag{7.8}$$

The validity of (7.8) is extended to all $h \in H^1(\Omega) \equiv \mathcal{D}(A_N^{1/2})$, since $N^* A_N$ is bounded on $\mathcal{D}(A_N^{1/2})$ and $\mathcal{D}(A_N^{1/2})$ is dense in $\mathcal{D}(A_N)$.

Second order abstract model. By using the Green operators introduced above, the coupled PDE problem (7.1) can be rewritten as the following abstract second order system - see [16]:

$$z_{tt} + A_N z + d A_N N D N^* A_N z_t + d_0 z_t - A_N N v_t = f \tag{7.9a}$$
$$v_{tt} + \mathcal{A} v + \rho \mathcal{A}^\alpha v_t + N^* A_N z_t = \mathcal{B} u, \tag{7.9b}$$

the first equation to be read on $[\mathcal{D}(A_N)]'$, the second one on $[\mathcal{D}(\mathcal{A})]'$.

Function spaces and operators. Let us define the following spaces

$$H_z \equiv \mathcal{D}(A_N^{1/2}) \times L_2(\Omega); \quad H_v \equiv \mathcal{D}(\mathcal{A}^{1/2}) \times L_2(\Gamma_0).$$

We note that since $d_1 > 0$, the topology of H_z is equivalent to that of $H^1(\Omega) \times L_2(\Omega)$.

On H_z we define (unbounded) operator $A_z : H_z \to H_z$ given by

$$A_z \equiv \begin{pmatrix} O & I \\ -A_N & -d A_N N D N^* A_N - d_0 \end{pmatrix} \tag{7.10}$$

with a natural maximal domain. Similarly, on H_v we define $A_v : H_v \to H_v$ by

$$A_v \equiv \begin{pmatrix} O & I \\ -\mathcal{A} & -\rho \mathcal{A}^\alpha \end{pmatrix}. \tag{7.11}$$

Coupling. Finally, we introduce the densely defined (unbounded, uncloseable) trace operator $C : H_z \supset D(C) \to H_v$ defined by

$$C \begin{bmatrix} z_1 \\ z_2 \end{bmatrix} := \begin{bmatrix} 0 \\ N^* A_N z_2 \end{bmatrix} = \begin{pmatrix} 0 & 0 \\ 0 & N^* A_N \end{pmatrix} \begin{bmatrix} z_1 \\ z_2 \end{bmatrix}, \tag{7.12}$$

with the domain

$$\mathcal{D}(C) = \{ [z_1, z_2] \in H_z : N^* A_N z_2 = z_2|_{\Gamma_0} \in L_2(\Gamma_0) \} \supset \mathcal{D}(A_N^{1/2}) \times \mathcal{D}(A_N^{1/4+\epsilon})$$

so that $\mathcal{D}(A_N^{1/2}) \times \mathcal{D}(A_N^{1/2}) \subset \mathcal{D}(C)$. Its adjoint $C^* : H_v \to \mathcal{D}(A_N^{1/2}) \times [\mathcal{D}(A_N^{1/4+\epsilon})]'$, in the sense of duality pairing $(Cy_1, y_2)_{H_v} = (y_1, C^* y_2)_{H_z}$; $y_1 \in \mathcal{D}(C), y_2 \in H_v$ is given by

$$C^* \begin{bmatrix} v_1 \\ v_2 \end{bmatrix} = \begin{bmatrix} 0 \\ A_N N v_2 \end{bmatrix} = \begin{pmatrix} 0 & 0 \\ 0 & A_N N \end{pmatrix} \begin{bmatrix} v_1 \\ v_2 \end{bmatrix}, \tag{7.13}$$

where $A_N N : L_2(\Gamma_0) \to [\mathcal{D}(A_N^{1/4+\epsilon})]'$.

First order abstract model. Finally, from (7.10), (7.11), (7.12), (7.13) we define the operator A acting on the space $H \equiv H_z \times H_v$

$$A := \begin{pmatrix} A_z & C^* \\ -C & A_v \end{pmatrix} : H \supset \mathcal{D}(A) \to H, \tag{7.14}$$

with the domain

$$\mathcal{D}(A) = \Big\{ [z_1, z_2, v_1, v_2] \in H : z_2 \in \mathcal{D}(A_N^{1/2}), \ v_2 \in \mathcal{D}(\mathcal{A}^{1/2}),$$
$$\mathcal{A}^{1-\alpha} v_1 + \rho v_2 \in \mathcal{D}(\mathcal{A}^\alpha), \ z_1 + dN DN^* A_N z_2 - N v_2 \in \mathcal{D}(A_N) \Big\}. \tag{7.15}$$

By applying Lumer Phillips Theorem (see [55]) it can be easily shown [16] that A generates a strongly continuous semigroup of contractions on H.

Control operator. Finally, we define the operator $B : U \to [\mathcal{D}(A^*)]'$, duality with respect to H, as a pivot space, as well as F by:

$$B \equiv [0, 0, 0, \mathcal{B}]^T; \ F \equiv [0, 0, 0, f]^T. \tag{7.16}$$

It is readily verified that

$$A^{-1}B = [0, 0, 0, -\mathcal{A}^{-1}\mathcal{B}]^T \in \mathcal{L}(U; H),$$

so that $B \in \mathcal{L}(U; [\mathcal{D}(A^*)]')$. Here we have used Hypothesis 7.2 to deduce that $A^{-1}B = \mathcal{A}^{r-1}\mathcal{A}^{-r}\mathcal{B}$ is bounded from U into $L_2(\Gamma_0)$, as $r < 1/2$.

Finally, returning to the second order abstract model (7.9), we see that these equations can be rewritten as the following first order abstract equation in the variable $\mathbf{y}(t) = [z(t), z_t(t), v(t), v_t(t)]$:

$$\mathbf{y}_t = A\mathbf{y} + Bu + F \quad \text{in} [\mathcal{D}(A^*)]', \ \mathbf{y}(0) = \mathbf{y}_0 \tag{7.17}$$

where A, B, F are defined in (7.14), (7.16), respectively.

7.3 Finite Horizon Control Problem

In order to apply abstract results of Part I, the key property to be verified is the singular estimate. As we shall see below, in the case of structural acoustic problem, this property is not always guaranteed by the analyticity of the semigroup generated by the plate equation. In fact, if the analyticity of the plate dynamics is not very strong (measured by the parameter α, $1/2 \le \alpha \le 1$), then the additional regularizing effect is needed in order to offset the unboundedness of B. It turns out that critical role in this is played by the boundary damping D. The above discussion motivates the following assumption.

Hypothesis 7.7. *We shall assume the following relation between the parameters representing damping in the system:*

(i) either $r_0 + \frac{\alpha}{2} \geq r$ and $d > 0$,
(ii) or else $\alpha - 2r \geq 1/6$ and $H^{1/3}(\Gamma_0) = \mathcal{D}(\mathcal{A}^{\frac{1}{12}})$

Remark 7.8. • In a canonical case of a plate equation, when $\mathcal{A}$ is the fourth order differential elliptic operator, the condition $H^{1/3}(\Gamma_0) = \mathcal{D}(\mathcal{A}^{\frac{1}{12}})$ (see [26]) is always satisfied. This follows from more general property $H^{4\theta}(\Gamma_0) = \mathcal{D}(\mathcal{A}^\theta); 0 \leq \theta < 1/8$.
• In the case of the fourth order elliptic operator $\mathcal{A}$ the first part of Hypothesis 7.7 is always fulfilled with $\alpha \geq 1/2$ and $r_0 = 1/4$. Thus any structurally damped plate ($\alpha \geq 1/2$) with boundary structural damping D represented by the Laplace's Beltrami operator ($r_0 = 1/4$) will always fulfill the requirements of the first part of the hypothesis.
• Regarding the second part of the hypothesis, this is always true for a strongly (Kelvin Voigt) damped plate equations when $\alpha = 1$. In this latter case we do not need any additional overdamping on the interface Γ_0.

The following singular estimate has been established in [16].

Theorem 7.9 (Singular estimate). *We assume Hypotheses 7.1-7.3. In addition we assume the first part of Hypothesis 7.7. Then control system described by (7.17) satisfies the "singular estimate".*

$$|e^{At}Bu|_H \leq \frac{C|u|_U}{t^\gamma}, \quad 0 < t \leq 1$$

with the value of γ given by

$$\gamma = \begin{cases} \frac{r}{\alpha}, & r \leq \frac{1}{2}\alpha \\ \frac{1/2 - \alpha + r}{1 - \alpha}, & r > \frac{1}{2}\alpha \end{cases}.$$

If the second part of Hypothesis 7.7 is in force, then singular estimate holds also with $d = 0$ in (7.1) i.e., $D \equiv 0$ and the value of γ is given by $\gamma = \frac{r}{\alpha} < 1/2$.

The proof of Theorem 7.9 given in [16] is technical and lengthy. It relies critically on two main ingredients: (i) characterization of fractional powers of elastic operators (see [19]), and (ii) sharp regularity of traces to wave equation with Neumann data (see [47]). By this we mean trace regularity which does not follow from usual trace theorems. In the special case when $\alpha = 1$ (Kelvin Voigt damping) this result was proved in [5].

Remark 7.10. We note that in the case when the operator B is unbounded (i.e., $r > 0$), the Hypothesis 7.7 forces certain amount of global damping in the system. The role of the damping is to offset the unboundedness resulting from the control operator. We have two sources of damping present in the model: structural damping yielding analyticity of the "plate" component (measured by the parameter $1/2 \leq \alpha \leq 1$) and boundary structural damping due to the presence of the operator D, measured by the parameter $0 \leq r_0 \leq 1/4$. The following interpretation of Hypothesis 7.7 can be given: the more analyticity in the system (i.e., the higher value of α), less boundary damping is needed (smaller value r_0) for controlling singularity at the origin. In the extreme case, when the plate equation has strong analyticity properties postulated in part (ii) of the Hypothesis 7.7, there is no need for boundary damping at all. In fact, the extreme case - $\alpha = 1$ - treated in [5] leads to singular estimate with $d = 0$ ie: $D = 0$ and $\gamma = 1/2 - \epsilon$. The result presented in the second part of the theorem extends the estimate in [2] to a larger range of parameters α and also provides more precise information on the singularity.

Remark 7.11. Singular estimate established in Theorem 7.9, when applied to the case of Kelvin Voigt damping ($\alpha = 1$), gives $|e^{At}Bu|_H \leq \frac{C|u|_U}{t^\gamma}$ where $\gamma < 1/2$. This implies, in particular, that the control-to-state map $u \to Lu$ is bounded from $L_2(0, T; U) \to C([0, T]; H)$, hence the control operator is admissible. We note that in [10], where the goal was to establish some sort of wellposedness of control-to-state map, the authors have shown that the control to state map L is bounded from $L_2(0, T; U) \to C([0, T], [\mathcal{D}(A)]')$. Thus, the result in [10] shows wellposedness in the sense of *distributions* and *not* in the sense of regular functions, as claimed above. The reason for this is that the analysis in [10] relies only on the commutativity of the generator with the semigroup and classical dual formulation of the variation of parameter formula. These arguments do not take any advantage of exceptional trace regularity for hyperbolic solutions and strong analyticity of the parabolic component - the two key ingredients behind singular estimate.

By applying abstract result in Theorem 3.1 along with singular estimate in Theorem 7.9 we infer the following final result

Theorem 7.12. *Under the hypotheses of Theorem 7.9 and with reference to finite horizon control problem consisting of (7.1) with functional cost (7.4), all the statements of abstract Theorem 3.1 apply with A, B, specified in (7.14) and (7.16).*

Remark 7.13. For the control problem governed by structural acoustic interaction with thermoelasticity , see Remark 7.5, the validity of singular estimate with $r_0 = 1/4$ and $\gamma = 2r$ was shown in [35]. Thus the same statement as in Theorem 7.12 is valid for this dynamics with the cost given in (7.6).

7.4 Infinite Horizon Control Problem

If the time T is infinite, the analysis of structural model is more complex. Indeed, we need to be concerned with the validity of *Finite Cost Condition*. This is typically guaranteed by some sort of stabilization result valid for the system under considerations. Unfortunately, in the case of structural acoustic problem, the coupled system is *stable* but not *uniformly stable*, cf. [4]. Thus, in order to enforce uniform stability, the corresponding model must be more complex. It is natural to impose some viscous damping in the interior of Ω. This corresponds taking $d_0 > 0$ in the first equation. In fact, this strategy works well, when there is no need for strong structural damping on the interface Γ_0. More precisely, the following result is known, cf. [1].

Theorem 7.14. *Consider (7.1) with $f = 0, u = 0, D \equiv 0$ (or $r_0 < 1/8$), and $d_0 > 0, d_1 > 0$. Then the corresponding system is exponentially stable on H That is to say, there exists $\omega > 0$ such that*

$$|e^{At}|_{\mathcal{L}(H)} \leq Ce^{-\omega t}; \ t \geq 0.$$

As we already know from the results of previous section, in order to assert singular estimate for the semigroup, depending on the value of α, we may need strong structural damping on the interface - i.e., the operator D must be unbounded. One may (naively) suppose that such damping should only enhance stability of the system. However, this is not the case as revealed in [16]. In spite of strong viscous damping in Ω with $d_0 > 0$, the system (7.1) is *not* uniformly stable, whenever $r_0 \geq 1/8$. Thus, there is a competition between regularity and stability, making the overall control problem much more subtle. A natural perception that "more damping" implies stronger decay rates is obviously false (in fact this is known among engineers as an overdamping phenomenon). In mathematical terms this is explained by noticing that the presence of strongly unbounded operator D introduces an element of continuous spectrum at the point 0. This is new phenomenon not present in structural acoustic models without the strong damping on the interface.

In view of the above, we are faced with the following dilemma. How to stabilize the system while preserving regularity guaranteed by the singular estimate?

The solution proposed below is based on the following idea: we counteract instability of the system by introducing an additional static feedback control. The role of the static damping is to remove the *continuous* spectrum from the spectrum of the generator. This leads to the following model:

$$z_{tt} = c^2 \Delta z - d_0(x)z_t + f \qquad \text{in} \quad \Omega \times (0,T) \qquad (7.18a)$$

$$\frac{\partial}{\partial n}z + d_1 z = 0 \qquad \text{on} \quad \Gamma_1 \times (0,T) \qquad (7.18b)$$

$$\frac{\partial}{\partial n}z + dD z_t + \beta Dz = w_t \qquad \text{on} \quad \Gamma_0 \times (0,T) \qquad (7.18c)$$

$$w_{tt} + \mathcal{A}w + \mathcal{A}^\alpha w_t + \partial_t z|_{\Gamma_0} = \mathcal{B}u \qquad \text{on} \quad \Gamma_0 \times (0,T). \qquad (7.18d)$$

The parameter $\beta \geq 0$ represents static damping on the interface Γ_0. If $\beta > 0$, it was shown in [16, 17] that the resulting system is exponentially stable also with a strong structural damping Dz_t. Precise formulation of this result is given below. Since we wish to consider cases when the damping d_0 is active only on a subportion of Ω we require the following geometric hypothesis.

Hypothesis 7.15. *We assume that either*

- $d_0(x) \geq d_0 > 0$ *in* Ω, *or*
- Ω *is convex and there exists* $x_0 \in R^n$ *such that* $(x - x_0) \cdot \nu \leq 0$ *on* Γ_1 *and* $d_0(x) \geq d_0 > 0$ *in* $\mathcal{U}(\Gamma_0) = \{x \in \Omega, dist(\Gamma_0, \Omega) \leq \delta\}$ *for some* $\delta > 0$.

Theorem 7.16 ([17]). *Consider (7.18) with* $f = 0, u = 0$ *and* $d > 0, d_1 > 0, \beta > 0$ *and* d_0 *subject to Hypothesis 7.15. Then the corresponding system is exponentially stable on* H. *That is to say: there exist positive constants* C, ω *such that*

$$|e^{At}|_{\mathcal{L}(H)} \leq Ce^{-\omega t}; \ t \geq 0.$$

Remark 7.17. If the parameter $r_0 < 1/4$, one can take $\beta = 0$.

Remark 7.18. The static damping βD can be replaced by a more general operator - say D_0 which shares the same estimates (see Hypothesis 7.3) as D.

The addition of static damping βDz has no effect on the validity of singular estimate. In fact, it was also shown in [16] that singular estimate of Theorem 7.9 still holds with $\beta > 0$. Thus, all the assumptions of abstract Theorem 3.5 are satisfied and we conclude with our final result:

Theorem 7.19. *With reference to system (7.18) subject to Hypothesis 7.1 - 7.15 and functional cost given by (7.4), all the statements of abstract Theorem 3.5 remain valid with* A, B *introduced in (7.14), (7.16) but with* A_z *replaced by*

$$A_z \equiv \begin{pmatrix} O & I \\ -A_N - \beta A_N NDN^* A_N & -dA_N NDN^* A_N - d_0 \end{pmatrix}. \qquad (7.19)$$

Remark 7.20. More detailed analysis of control problems arising in structural acoustic interactions, as well as structural-acoustic-thermal interactions can be found in [34].

Acknowledgment . The author would like to thank Delio Mugnolo, from the University of Tuebingen, for his careful reading of the manuscript and his valuable comments.

References

1. G. Avalos, The exponential stability of a coupled hyperbolic/parabolic system arising in structural a coustics, *Abstract and Applied Analysis*, vol 1, pp 203-219, 1996.
2. G. Avalos and I. Lasiecka, Exponential Stability of a Thermoelastic System without mechanical Dissipation, *Rend. ISTIT. Mat. Univ. Trieste*, vol XXVIII", pp 1-28, 1997.
3. G. Avalos and I. Lasiecka, Exponential Stability of a Thermoelastic Plates with free boundary conditions and without mechanical Dissipation, *SIAM J. Mathematical Analysis*, vol 29, pp 155-182, 1998.
4. G. Avalos and I. Lasiecka, The strong stability of a semigroup arising from a coupled hyperbolic/parabolic system, *Semigroup Forum*, vol 57, pp 278-292, 1998.
5. G. Avalos and I. Lasiecka, Differential Riccati equation for the active control of a problem in structural acoustics, *JOTA*, vol 91, pp 695-728, 1996.
6. A. V. Balakrishnan, *Applied Functional Analysis*, Springer-Verlag, 1975.
7. H.T. Banks and K. Ito, Approximations in LQR problems for infinite dimensional systems with unbounded input operators, Siam Conference on Control, San Francisco, 1989.
8. H.T. Banks and R.C. Smith. Feedback control of noise in a 2-D nonlinear structural acoustic model, *Discrete and Continuous Dynamical Systems*, vol 2, pp 119-149, 1995.
9. H.T. Banks and R. Smith. Active control of acoustic pressure fields using smart material technologies, in *Flow Control, IMA vol 68,* Proceedings (edited by M. Gunzburger) of IMA Conference, 1992. Published by Springer-Verlag.
10. H. T. Banks and R. Smith. Well-posedness of a model for structural acoustic coupling in a acvaity enclosed by a thin cylindrical shell, *Journal Mathematical Analysis and Applications* , vol 191, p 1-25, 1995.
11. V. Barbu, *Nonlinear Semigroups and Differential Equations in Banach Spaces*, Nordhof, 1976.
12. V. Barbu and I. Lasiecka and R. Triggiani, Extended Algebraic Riccati Equations arising in hyperbolic dynamics with unbounded controls, *Nonlinear Analysis*, vol 40, pp. 105-129, 2000.
13. J. Beale, Spectral properties of an acoustic boundary condition, *Indiana Univ. Math. J.*, vol 9, 895-917, 1976.
14. A. Bensoussan and G. Da Prato and M.C. Delfour and S.K. Mitter, *Representation and Control of Infinite Dimensional Systems*, Vol I,II, Birkhaeuser, Boston-Basel-Berlin, 1993.
15. H. Brezis, *Operateurs Maximaux Monotones*, North-Holland, Amsterdam 1973.
16. F. Bucci and I. Lasiecka and R. Triggiani. Singular estimate and uniform stability of coupled systems of hyperbolic/parabolic PDE's, *Abstract and Applied Analysis*, vol 7, pp 169-237, 2002.
17. F. Bucci and I. Lasiecka. Exponential decay rates for structural acoustic model with an overdamping on the interface and baoundary layer dissipation, *Applicable Analysis*, vol 81, pp 977-999, 2002.
18. S. Chen and R. Triggiani, Proof of extensions of two conjectures on structural damping for elastic systems, *Pacific J. of Mathematics*, vol 136, pp 15-55, 1989.

19. S. Chen and R. Triggiani, Characterization of Domains of Fractional Powers of Certain Operators Arising in Elastic Systems and Applications, *J. Differential Equations*, vol 64, pp 26-42, 1990.
20. C. Chicone and Y. Latushkin *Evolution Semigroups in Dynamical Systems*, AMS 1991.
21. G. Da Prato and A. Ichikawa, Riccati equations with unbounded coefficients, *Annali di mat. Pura et Applicata*, vol 140, pp 209-221, 1985.
22. G. Da Prato and I. Lasiecka and R. Triggiani, A Direct Study of Riccati Equations arising in Hyperbolic Boundary Co ntrol Problems, *Journal of Differential Equations*, pp 26-47, 1986.
23. E. K. Dimitriadis and C. R. Fuller and C. A. Rogers, Piezoelectric Actuators for Distributed Noise and Vibration Excitation of Thin Plates, *Journal of Vibration and Acoustics*, vol 13,pp 100-1017, 1991.
24. F. Flandoli, Algebraic Riccati equations arising in boundary control problems with distributed parameters, *Siam J. Control*, vol 25, pp 612-636, 1987.
25. F. Flandoli and I. Lasiecka and R. Triggiani, Algebraic Riccati Equations with non-smoothing observations arising in hyperbolic and Euler-Bernoulli boundary control problems, *Annali di Matematica Pura et. Applicata*, vol 153, pp 307-382, 1988.
26. P. Grisvard, Caractérisation de quelques espaces d'interpolation, *Archive Rational Mechanics and Analysis*, vol 25, pp 40-63, 1967.
27. K. Hoffman and N. Botkin, Homogenization of von Karman plates excited by piezoelectric patches, *ZAMM*, vol 19, pp 579-590, 2000.
28. S. Hansen, Modelling and analysis of multilayer laminated plates, *ESAIM Proceedings* vol 4, pp 117-135, 1998.
29. S. Hansen, Siam Conference, Philadelphia, 2002.
30. S. Hansen and I. Lasiecka, Analyticity, hyperbolicity and uniform stability of semigroups arising in models of composite beams, *Mathematical Models and Methods in Applied Sciences*, vol 10 pp 555-580. 2000.
31. J. Lagnese, *Boundary Stabilization of Thin Plates*, SIAM, 1989.
32. J. Lagnese and J.L. Lions, *Modelling, Analysis and Control of Thin Plates*, Masson, Paris, 1988.
33. I. Lasiecka, Mathematical control theory in structural acoustic problems, *Mathematical Models and Methods in Applied Sciences*, vol 8, pp 119-1153, 1998
34. I. Lasiecka, *Mathematical Control Theory of Coupled PDE Systems*, CMBS-NSF Lecture Notes, SIAM, vol 75, Philadelphia, 2002.
35. I. Lasiecka, Optimization problems for structural acoustic models with thermoelasticity and smart materials, *Disscussiones Mathematicae-Differential Inclusions Control and Optimization*, vol 20, pp 113-140, 2000.
36. I. Lasiecka and C. Lebiedzik Decay rates of interactive hyperbolic parabolic pde models with thermal effects on the interface, *Applied Math. Optimiz*, vol 42, pp 127-167, 2001.
37. I. Lasiecka and C. Lebiedzik, Boundary stabilizability of nonlinear acoustic models with thermal effects on the interface, *C. R. Acad. Sc. Paris*, vol 328 (2), pp 187-192, 2000.
38. I. Lasiecka and J.L. Lions and R. Triggiani, Nonhomogenuous boundary value problems for second order hyperbolic equations, *J. Math. Pure et Appli*, vol 65, pp 149-192, 1986.

39. I. Lasiecka and R. Triggiani, *Control Therory for Partial Differential Equations*, Vol I, Vol II, "Cambridge University Press", Encyklopedia of Mathematics and its Aplications, 2000.

40. I. Lasiecka and R. Triggiani, *Differential and Algebraic Riccati Equations with Applications to Boundary Point Control Problems, Continuous Theory and Approximation Theory*, Springer-Verlag, LNCIS vol 164, 1991.

41. I. Lasiecka and R. Triggiani, Two direct proofs on the analyticity of the S.C. semigroup arising in abstract thermo-elastic equations, *Advances in Differential Equations*, vol 3, pp 387-416, 1998.

42. I. Lasiecka and R. Triggiani, Analyticity of thermo-elastic semigroups with free B.C., *Annali di Scuola Normale Superiore*, vol XXVII, pp 457-482, 1998.

43. I.Lasiecka and R. Triggiani, Analyticity and lack thereof of thermoelastic semi-groups, *ESAIM*, vol 4, pp 199-222, 1998.

44. I.Lasiecka and R. Triggiani, Structural decomposition of thermoelastic semi-groups with rotational forces, *Semigroup Forum*, vol 60, pp 16-66, 2000.

45. I. Lasiecka and R. Triggiani, Optimal control and ARE under singular estimate for $e^{At}B$ in the absence of analyticity, *Differential Equations and Control Theory*, vol 225 LNPAM, Marcel Dekker, 2001.

46. I. Lasiecka and R. Triggiani, Optimal control and Differential Riccati equations under singular estimate for $e^{At}B$ in the absence of analyticity, *Advances in Dynamics and Control*, CRC - special volume in honour of A.V. Balakrishnan. To appear.

47. I. Lasiecka and R. Triggiani, Regularity Theory of hyperbolic equations with non-homogenous Neumann boundary conditions II: General Boundary data, *J. Differential Equations* , vol 94, pp 112-164, 1991.

48. I. Lasiecka and R. Triggiani, A lifting theorem for the time regularity of so-lutions to abstract equations with unbounded operators and applications to hyperbolic problems, *Proceedings of the AMS*, vol 104, pp 745-756, 1988.

49. J.L. Lions, *Controle Optimale des Systemes Gouvernes par des Equations aux Derivees Partielles*, Dunod, 1968.

50. J.L. Lions, *Contrôlabilité exacte, perturbations et stabilisation de systèmes distribués*, Masson, Paris, 1988.

51. J. L. Lions and E. Magenes, *Non-homogenous Boundary Value Problems and Applications*, Springer-Verlag, 1972.

52. W. Littman and B. Liu, On the spectral properties and stabilization of acoustic flow, *SIAM J. Applied Math.*, vol 1, pp 17-34, 1998.

53. Z. Liu and M. Renardy, A note on the equation of thermoelastic plate, *Applied Math. Letters*, vol 8, pp 1-6, 1995.

54. P.M. Morse and K.U. Ingard, *Theoretical Acoustics*, McGraw-Hill, New York, 1968.

55. A. Pazy *Semigroups of Linear Operators and Differential Equations*, Springer-Verlag, 1986.

56. D.L. Russell, Mathematical models for the elastic beam and their control-theoretic properties, *Semigroups Theory and Applications, Pitman Research Notes*, vol 152, pp 177-217, 1986.

57. D.L. Russell, Controllability and stabilizability theory for linear partial differential equations. Recent progress and open questions , *SIAM Review*, vol 20, pp 639-739, 1978.

58. R. Triggiani, The Algebraic Riccati Equations with Unbounded Coefficients; Hyperbolic Case Revisited, *Contemporary Mathematics: Optimization Methods in PDE's*, vol 209, pp 315-339, AMS, Providence, 1997.
59. R. Triggiani. The coupled PDE system of a composite sandwich beam revisited, *Discrete and Continuous Dynamical Syastems, Series B* - vol 3, pp 285-298, 2003.
60. R. Triggiani. Min-max game theory and optimal control with indefinite cost under a singular estimate for $e^{At}B$ in the absence of analyticity, *Evolutions Equations, Semigroups and Functional Analysis -in memory of B. Terreni*, Birkhauser, pp 353-380, 2002.
61. G. Weiss and H. Zwart, An example in LQ optimal control, *Systems and Control Letters*, vol 8, pp 339-349, 1998.

An Introduction to Parabolic Moving Boundary Problems

Alessandra Lunardi

Dipartimento di Matematica, Università di Parma, Via D'Azeglio 85/A, 43100 Parma, Italy.
`alessandra.lunardi@unipr.it`

1 Introduction

In these lectures I will give an overview on a class of free boundary evolution problems, such as the free boundary heat equation,

$$
\begin{cases}
u_t = \Delta u, & t > 0, \ x \in \Omega_t, \\[2mm]
u = 0, \ \dfrac{\partial u}{\partial \nu} = -1, & t > 0, \ x \in \partial\Omega_t,
\end{cases}
\tag{1.1}
$$

and its generalizations. The unknowns are the open sets $\Omega_t \subset \mathbb{R}^n$ for $t > 0$, and the real valued function u, defined for $t > 0$ and $x \in \Omega_t$. The initial set Ω_0 and the initial function $u(0, \cdot) = u_0 : \Omega_0 \mapsto \mathbb{R}$ are given data.

While this type of problems are well understood in space dimension $N = 1$ (see e.g. [27] and the review paper [24]) much less is known in the multidimensional case. For instance, uniqueness of the classical solution to the initial value problem for (1.1) is still an open problem.

I will describe some different approaches to (1.1), which lead to several properties of the solutions. Moreover I shall describe some special solutions, whose behavior is of great help in understanding the nature of the problem.

Many results for (1.1) can be extended to evolution problems of the type

$$
\begin{cases}
u_t(t, x) = \mathcal{L}u(t, x) + f(t, x, u(t, x), Du(t, x)), & t \geq 0, \ x \in \Omega_t, \\[2mm]
u(t, x) = g_0(t, x), & t \geq 0, \ x \in \partial\Omega_t, \\[2mm]
\dfrac{\partial u}{\partial \nu}(t, x) = g_1(t, x), & t \geq 0, \ x \in \partial\Omega_t, \\[2mm]
u(0, x) = u_0(x), & x \in \Omega_0.
\end{cases}
\tag{1.2}
$$

where the boundary data g_0, g_1 satisfy the transversality condition

$$\frac{\partial g_0}{\partial \nu}(0, x) \neq g_1(0, x), \quad x \in \partial \Omega_0. \tag{1.3}$$

Again, the unknowns are the open sets $\Omega_t \subset \mathbb{R}^N$ for $t > 0$, and the function u. The data are: the second order elliptic operator $\mathcal{L} = \sum_{i,j=1}^N a_{ij}(t, x)D_{ij}$, the functions $f : [0, T] \times \mathbb{R}^n \times \mathbb{R} \times \mathbb{R}^n \to \mathbb{R}$, $g_0, g_1 : [0, T] \times \mathbb{R}^n \mapsto \mathbb{R}$, the (possibly unbounded) initial set $\Omega_0 \subset \mathbb{R}^n$, and the initial function $u_0 : \overline{\Omega_0} \mapsto \mathbb{R}$.

However, all the substantial difficulties of (1.2) are already present in problem (1.1). Therefore, to simplify notation and to concentrate on the difficulties due to the structure of the problem, we shall consider mainly problem (1.1). It is motivated by models in combustion theory ([8, 26]) for equidiffusional premixed flames, where $u = \lambda(T^* - T)$, T is the temperature, T^* is the temperature of the flame, and $\lambda > 0$ is a normalization factor. It can be seen as the high activation energy limit of the regularizing problems $u_t = \Delta u - \beta_\epsilon(u)$ in $[0, T] \times \mathbb{R}^N$, where $T > 0$ is fixed and $\beta_\epsilon(s) = \beta_1(s/\epsilon)/\epsilon$ is supported in a small interval $[0, \epsilon]$. In [9], Caffarelli and Vazquez used this regularization to prove existence of global weak solutions to (1.1) for C^2 initial data (Ω_0, u_0), under suitable geometric assumptions on u_0. Such solutions may not be unique, and enjoy some regularity properties: the free boundary is locally Lipschitz continuous, and u is 1/2-Hölder continuous with respect to time, Lipschitz continuous with respect to the space variables. I shall describe the approach of Caffarelli and Vazquez in §4.

In §5 I shall describe a completely different approach, due to Baconneau and myself [2], which leads to existence, uniqueness, and time C^∞–regularity for $t > 0$ of local smooth (very smooth!) solutions (with u bounded, in the case of unbounded domains) to problem (1.1), and more generally to problem (1.2). It consists in transforming problem (1.1) into a fully nonlinear parabolic problem in the fixed domain Ω_0 for an auxiliary unknown w, for which the usual techniques of fully nonlinear parabolic problems in Hölder spaces (see e.g. [22, Ch. 8]) give a local existence and uniqueness result. Coming back to (1.1), for Ω_0 with $C^{3+\alpha+k}$ boundary and $u_0 \in C^{3+\alpha+k}(\overline{\Omega_0})$ we obtain a local solution with $\partial \Omega_t \in C^{2+\alpha+k}$ and $u(t, \cdot)$ in $C^{2+\alpha+k}(\overline{\Omega_t})$, $k = 0, 1$.

Both approaches lead to loss of regularity, in the sense that we are not able to show that the regularity of the initial datum is preserved throughout the evolution. Although this is natural for global weak solutions, as the first example in section 2 shows, it not wholly satisfactory for local classical solutions. One could expect that in a small time interval the solution remains at least as regular as the initial datum: this is what happens in other free boundary problems of parabolic type (see e.g. [11, 12] for boundary conditions of Stefan type), and in problem (1.2) for special initial data.

An interesting situation in which there is no loss of regularity, at least in small time intervals, is the case of initial data near special smooth solutions, such as stationary solutions, self-similar solutions, travelling waves. In particular, $C^{2+\alpha}$ initial data near any smooth stationary solution (Ω, U) of (1.2), with bounded Ω, were considered in the paper [3], where we studied stability of smooth stationary solutions, establishing a linearized stability principle for (1.2) in the time independent case. Since self-similar solutions to (1.1) become stationary solutions to a problem of the type (1.2) after a suitable change of coordinates, we can also consider initial data for (1.1) near self-similar solutions. In the unbounded domain case, $C^{2+\alpha}$ initial data near a planar travelling wave solution of (1.1) have been considered in [4]. The wave turns out to be orbitally stable, but the discussion is not trivial, because in dimension $N \geq 2$ this is a very critical case of stability.

Another example without loss of regularity was considered by Andreucci and Gianni in [1], where a two-phase version of (1.2) was studied in a strip for $C^{2+\alpha}$ initial data far from special solutions but satisfying suitable monotonicity conditions: u_0 is assumed to be strictly monotonic in the direction orthogonal to the strip. This allows to take u as new independent variable for small t, an old trick already used by Meirmanov in the Stefan problem and also in a two-dimensional version of problem (1.2), see [23].

Together with loss of regularity, the other big question about problem (1.2) is uniqueness of the classical solution. Indeed, the uniqueness results available up to now concern only particular situations, such as radially symmetric solutions of (1.1), studied in [13], and solutions in cylinders or strips, for initial data which are monotonic in the direction of the axis of the cylinder (see [23] in dimension $N = 2$ and [17] in any dimension), or in the direction orthogonal to the axis of the strip (see [1], for the two-phase case). The above mentioned papers [3, 4] give also uniqueness results in the parabolic Hölder space $C^{1+\alpha/2,2+\alpha}$, but only for solutions close to the special solutions considered. The paper [2] gives uniqueness of very regular solutions, in the space $C^{3/2+\alpha/2,3+\alpha}$.

Therefore, the problem of wellposedness for general initial data in dimension bigger than 1 still remains open, even in the simplest case (1.1).

2 The One Dimensional Case

Let us consider a one dimensional version of problem (1.1). To avoid technical difficulties we consider the simplest situation as far as local existence is concerned, i.e. the case where the initial set and the unknown sets Ω_t are left halflines,

$$
\begin{cases}
u_t(t,x) = u_{xx}(t,x), \quad t \geq 0, \ x < s(t), \\
u(t,s(t)) = 0, \ u_x(t,s(t)) = -1, \quad t \geq 0.
\end{cases}
\tag{2.1}
$$

The initial conditions are

$$
s(0) = s_0, \quad u(0,x) = u_0(x), \ x \leq s_0,
\tag{2.2}
$$

with u_0 smooth enough ($u_0 \in C^{2+\theta}((-\infty, s_0])$, $0 < \theta < 1$)), satisfying the compatibility conditions

$$(a)\ u_0(s_0) = 0, \quad (b)\ u_0'(s_0) = -1. \tag{2.3}$$

The trivial change of coordinates

$$\xi = x - s(t),$$

reduces (2.1) to a fixed boundary problem in $(-\infty, 0]$. Let $\widetilde{u}$ denote the function u in the new variables, i.e. $\widetilde{u}(t, \xi) = u(t, x)$. Then $\widetilde{u}$ has to satisfy

$$\begin{cases} \widetilde{u}_t(t, \xi) - \dot{s}(t)\widetilde{u}_\xi(t, \xi) = \widetilde{u}_{\xi\xi}(t, \xi), & t \geq 0, \ \xi \leq 0, \\ \widetilde{u}(t, 0) = 0, \ \widetilde{u}_\xi(t, 0) = -1, \ t \geq 0, \\ \widetilde{u}(0, \xi) = \widetilde{u}_0(\xi) = u_0(\xi + s_0), \ \xi \leq 0. \end{cases} \tag{2.4}$$

This system may be immediately decoupled using the boundary conditions: the first boundary condition $u(t, s(t)) \equiv 0$ gives $u_t(t, s(t)) + \dot{s}(t)u_\xi(t, s(t)) \equiv 0$, and since $u_\xi(t, s(t)) = -1 \neq 0$ we get

$$\dot{s}(t) = u_t(t, s(t)) = u_{\xi\xi}(t, s(t)) = \widetilde{u}_{\xi\xi}(t, 0). \tag{2.5}$$

Replacing in (2.4) we get

$$\widetilde{u}_t(t, \xi) = \widetilde{u}_{\xi\xi}(t, 0)\widetilde{u}_\xi(t, \xi) + \widetilde{u}_{\xi\xi}(t, \xi).$$

The final problem for $\widetilde{u}$ is therefore

$$\begin{cases} \widetilde{u}_t(t, \xi) = \widetilde{u}_{\xi\xi}(t, 0)\widetilde{u}_\xi(t, \xi) + \widetilde{u}_{\xi\xi}(t, \xi), & t \geq 0, \ \xi \leq 0, \\ \widetilde{u}_\xi(t, 0) = -1, \ t \geq 0, \\ \widetilde{u}(t, -\infty) = 0, \ t \geq 0, \\ \widetilde{u}(0, \xi) = \widetilde{u}_0(\xi) = u_0(\xi + s_0), \ \xi \leq 0. \end{cases} \tag{2.6}$$

I did not write the condition $\widetilde{u}(t, 0) = 0$ in (2.6) because it is contained in the other conditions. Indeed, writing the differential equation $\widetilde{u}_t(t, \xi) = \widetilde{u}_{\xi\xi}(t, 0)\widetilde{u}_\xi(t, \xi) + \widetilde{u}_{\xi\xi}(t, \xi)$ at $\xi = 0$ and replacing $\widetilde{u}_\xi(t, 0) = -1$ we get $\widetilde{u}_t(t, 0) = 0$ and hence $\widetilde{u}(t, 0) = \text{constant}$; the constant is 0 because the initial datum u_0 satisfies the compatibility condition $u_0(s_0) = 0$.

(2.6) is a nonlinear problem with a nonstandard nonlinearity, which depends on the spatial second order derivative of the solution, evaluated at $\xi = 0$. However, it can be solved, at least locally in time, with the usual techniques of nonlinear parabolic problems. Here we follow the abstract evolution equations technique, but other approaches work as well.

First, we transform easily problem (2.6) into a homogeneous at the boundary problem. This is made just to adapt our problem to the general theory of evolution equations in Banach spaces. It is sufficient to introduce the new unknown

$$w(t, \xi) = \widetilde{u}(t, \xi) + e^\xi, \ t \geq 0, \ \xi \leq 0.$$

Since $\widetilde{u}_t = w_t$, $\widetilde{u}_\xi = w_\xi - e^\xi$, $\widetilde{u}_{\xi\xi} = w_{\xi\xi} - e^\xi$, then w has to solve

$$\begin{cases} w_t(t,\xi) = (w_{\xi\xi}(t,0) - 1)(w_\xi(t,\xi) - e^\xi) + w_{\xi\xi}(t,\xi) - e^\xi, \ t \geq 0, \ \xi \leq 0, \\ w_\xi(t,0) = 0, \ t \geq 0, \\ w(0,\xi) = w_0(\xi) = u_0(\xi + s_0) + e^\xi, \ \xi \leq 0. \end{cases}$$

$$(2.7)$$

It is convenient to set problem (2.7) in the space

$$X = BUC((-\infty, 0])$$

of the uniformly continuous and bounded functions, endowed with the sup norm. The nonlinear function in the right hand side of the equation is well defined in the subspace

$$D = \{\varphi \in BUC^2((-\infty, 0]) : \ \varphi'(0) = 0\}.$$

D is endowed with the C^2 norm, and it is dense in X.

The function

$$F : D \mapsto X, \ \ F(\varphi)(\xi) = (\varphi''(0) - 1)(\varphi'(\xi) - e^\xi) + \varphi''(\xi) - e^\xi, \ \ \xi \leq 0$$

is smooth, and for each $w_0 \in D$ we have

$$F'(w_0)(\psi)(\xi) = (w_0''(0) - 1)\psi'(\xi) + \psi''(0)(w_0'(\xi) - e^\xi) + \psi''(\xi), \ \ \psi \in D, \ \xi \leq 0.$$

Therefore, $F'(w_0)$ is an elliptic operator $(A\psi = \psi'' + (w_0''(0) - 1)\psi')$ plus a perturbation $(B\psi = \psi''(0)(w_0'(\xi) - e^\xi))$ defined in D and not defined in any intermediate space between D and X. The operator $A : D(A) = D \mapsto$ generates an analytic semigroup in X, and its graph norm is equivalent to the norm of D. Moreover $B : D \mapsto X$ is a compact operator because its range is one dimensional. It follows (see e.g. [10]) that $A + B : D \mapsto X$ generates an analytic semigroup in X.

Now we may apply a local existence and uniqueness theorem for evolution equations in Banach spaces.

Theorem 2.1. *Let X, D be Banach spaces, with D continuously embedded in X. Let $\mathcal{O} \neq \emptyset$ be an open set in D, let $F : \mathcal{O} \mapsto X$ be a C^2 function. Assume that for each $w \in \mathcal{O}$ the operator $A = F'(w) : D(A) = D \mapsto X$ generates an analytic semigroup in X.*

Fix $\alpha \in (0, 1)$. For every $w_0 \in D$ such that $F(w_0) \in D_A(\alpha, \infty)$ there exist $\delta > 0$ and a unique solution $w \in C^\alpha([0, \delta]; D) \cap C^{1+\alpha}([0, \delta]; X)$ of problem

$$\begin{cases} w'(t) = F(w(t)), \ 0 \leq t \leq \delta, \\ w(0) = w_0, \end{cases}$$

$$(2.8)$$

Moreover $u'(t) \in D_A(\alpha, \infty)$ for each t, and $t \mapsto \|u'(t)\|_{D_A(\alpha, \infty)}$ is bounded in $[0, \delta]$.

We recall that $D_A(\alpha, \infty) = (X, D)_{\alpha, \infty}$ is the real interpolation space between X and D defined by

$$D_A(\alpha, \infty) = \{w \in X : [w]_\alpha = \sup_{0 < t < 1} t^{-\alpha} \| e^{tA} w - x \| < \infty\},$$

$$\|w\|_{D_A(\alpha, \infty)} = [w]_\alpha + \|w\|.$$

Here $\| \cdot \|$ denotes the norm in X.

The proof of theorem 2.1 relies on maximal regularity results in Hölder spaces for linear evolution equations in Banach spaces, due to [25].

Theorem 2.2. *Let $A : D(A) \mapsto X$ be the generator of an analytic semigroup e^{tA} in X. Fix $\alpha \in (0, 1)$ and $T > 0$. Let $f \in C^\alpha([0, \delta], X)$, with $0 < \delta \le T$, and $x \in D(A)$ be such that $Ax + f(0) \in D_A(\alpha, \infty)$.*

Then the linear problem

$$\begin{cases} u'(t) = Au(t) + f(t), & 0 \le t \le \delta, \\ u(0) = x, \end{cases} \tag{2.9}$$

has a unique solution $u \in C^{1+\alpha}([0, \delta], X) \cap C^\alpha([0, \delta], D(A))$. Moreover, u' is bounded with values in $D_A(\alpha, \infty)$, and there exists C such that

$$\|u\|_{C^{1+\alpha}([0,\delta];X)} + \|Au\|_{C^\alpha([0,\delta];X)} + \sup_{0 \le t \le \delta} \|u'(t)\|_{D_A(\alpha, \infty)}$$

$$\le C(\|f\|_{C^\alpha([0,\delta];X)} + \|x\|_{D(A)} + \|Ax + f(0)\|_{D_A(\alpha, \infty)}). \tag{2.10}$$

The constant C is independent of f, x, and δ.

Once theorem 2.2 is established, it is not hard to prove theorem 2.1 by a fixed point argument. It is convenient to look for the local solution of problem (2.8) as a fixed point of the operator Γ in the set $Y = \{w \in C^\alpha([0, \delta]; D) : w(0) = w_0, \|w - w_0\|_{C^\alpha([0,\delta];D)} \le r\}$ defined by $\Gamma w = u = $ the solution to problem

$$\begin{cases} u'(t) = Au(t) + F(w(t)), & 0 \le t \le \delta, \\ u(0) = w_0. \end{cases}$$

The operator Γ maps Y into itself and it is a $1/2$-contraction in Y provided δ and r are suitably chosen (roughly: r large and δ small). For the details see [21].

To apply theorem 2.1 we need to know $D_{f'(w_0)}(\alpha, \infty)$. It coincides with $D_\Delta(\alpha, \infty)$ where $\Delta : D \mapsto X$ is the realization of the second order derivative in X with Neumann boundary condition. For $\alpha < 1/2$ we have (see e.g. [22, thm. 3.1.30])

$$D_{f'(w_0)}(\alpha, \infty) = C^{2\alpha}((-\infty, 0]).$$

In our case $w_0 = u_0(\xi + s_0) + e^\xi \in C^{2+\theta}((-\infty, 0])$, and $w_0'(0) = 0$. Therefore $w_0 \in D$ and $f(w_0) \in D_{f'(w_0)}(\theta/2, \infty)$. We may apply theorem 2.1 with $\alpha = \theta/2$, which gives local existence and uniqueness of a solution w such that $t \mapsto w(t, \cdot) \in C^{1+\theta/2}([0, \delta]; BUC(-\infty, 0]) \cap C^{\theta/2}([0, \delta]; D)$, and $t \mapsto w_t(t, \cdot)$ is bounded with values in $C^{2+\theta}((-\infty, 0])$. This implies that w belongs to the parabolic Hölder space $C^{1+\theta/2, 2+\theta}([0, \delta] \times (-\infty, 0])$.

Now we come back to the original problem. Recall that $\dot{s} = \tilde{u}_{\xi\xi}(t, 0) = w_{\xi\xi}(t, 0) - 1$, so that

$$s(t) = \int_0^t w_{\xi\xi}(\sigma, 0)d\sigma - t + s_0,$$

$$u(t, \xi) = \tilde{u}(t, \xi - s(t)) = w(t, \xi - s(t)) - e^{\xi - s(t)}, \quad \xi \le s(t).$$

Since w satisfies (2.7), then the couple (s, u) satisfies (2.1), with the only exception that the condition $u(t, s(t)) = 0$ is replaced by $d/dt\, u(t, s(t)) = 0$, i.e. $u(t, s(t)) = $ constant. The constant to be 0 because of the compatibility condition $u_0(s_0) = 0$. So we get also $u(t, s(t)) = u(0, s_0) = 0$ for each $t \in [0, \delta]$.

We have proved the following result.

Proposition 2.3. *Let $s_0 \in \mathbb{R}$ and let $u_0 \in C^{2+\theta}((-\infty, s_0])$ satisfy the compatibility conditions (2.3). Then there exists $\delta > 0$ such that problem (2.1) has a unique solution (s, u) such that $s \in C^{1+\theta/2}([0, \delta])$ and $u \in C^{1+\theta/2, 2+\theta}(\overline{\Omega})$, where $\Omega = \{(t, x) : 0 < t < \delta,\ x < s(t)\}$.*

For another approach see [14].

Now I try to explain why the same approach does not work if the space dimension is bigger than 1. Let us represent $\mathbb{R}^N$ as $\mathbb{R} \times \mathbb{R}^{N-1}$, and let us write any element of $\mathbb{R}^N$ as (x, y), with $x \in \mathbb{R}$ and $y \in \mathbb{R}^{N-1}$. A natural generalization of problem (2.1) is to look for Ω_t of the type $\Omega_t = \{(x, y) \in \mathbb{R}^N : x < s(t, y)\}$ where s is an unknown function. Therefore, the free boundary at time t is the graph of the function $s(t, \cdot)$, the normal derivative is $-(D_x u, D_y u) = ((1 + |D_y s|^2)^{-1/2}, -D_y s(1 + |D_y s|^2)^{-1/2})$ and problem (1.1) is rewritten as

$$\begin{cases} u_t = \Delta u, & t > 0,\ x \le s(t, y),\ y \in \mathbb{R}^{N-1}, \\[2mm] u(t, s(t, y), y) = 0, & \\[2mm] u_x(t, s(t, y), y) = -\dfrac{1}{\sqrt{1 + |D_y s|^2}} & t > 0,\ y \in \mathbb{R}^{N-1}. \end{cases} \tag{2.11}$$

The initial conditions are now

$$s(0, y) = s_0(y),\ y \in \mathbb{R}^{N-1}, \quad u(0, x, y) = u_0(x, y),\ x \le s_0(y),\ y \in \mathbb{R}^{N-1}, \tag{2.12}$$

where s_0, u_0 are smooth enough and satisfy the compatibility conditions

$$(a)\ u_0(s_0(y),y)=0,\quad (b)\ D_x u_0(s_0(y),y)=-1/\sqrt{1+|D_y s_0(y)|^2}.\quad (2.13)$$

The change of coordinates which fixes the boundary is again

$$\xi = x - s(t,y),\quad (2.14)$$

and it reduces (2.11) to a fixed boundary problem in $(-\infty,0]\times\mathbb{R}^{N-1}$. Let $\tilde{u}$ denote the function u in the new variables, i.e. $\tilde{u}(t,\xi,y)=u(t,x,y)$. Then $\tilde{u}$ has to satisfy

$$\begin{cases} \tilde{u}_t - D_t s\tilde{u}_\xi = \Delta\tilde{u} - 2\langle D_y\tilde{u}_\xi, D_y s\rangle + \tilde{u}_{\xi\xi}|D_y s|^2 - \tilde{u}_\xi\Delta_y s,\ \xi\le 0,\ y\in\mathbb{R}^{N-1}, \\[2mm] \tilde{u}(t,0,y)=0,\ \tilde{u}_\xi(t,0,y) = -\dfrac{1}{\sqrt{1+|D_y s|^2}},\quad t\ge 0,\ y\in\mathbb{R}^{N-1}, \\[2mm] \tilde{u}(0,\xi,y)=\tilde{u}_0(\xi,y)=u_0(\xi+s_0(y),y),\ \ \xi\le 0,\ y\in\mathbb{R}^{N-1}. \end{cases}$$
$$(2.15)$$

The identity $u(t,s(t,y),y)\equiv 0$ may be still used to get an expression for s_t: differentiating with respect to time we get $u_t(t,s(t,y),y)+u_x(t,s(t,y),y)s_t(t,y)\equiv 0$, so that $s_t(t,y)=-u_t(t,s(t,y),y)/u_x(t,s(t,y),y)=-\Delta u(t,s(t,y),y)/u_x(t,s(t,y),y)$. Rewriting in terms of $\tilde{u}$ we get

$$s_t = -\frac{\Delta\tilde{u} - 2\langle D_y\tilde{u}_\xi, D_y s\rangle + \tilde{u}_{\xi\xi}|D_y s|^2 - \tilde{u}_\xi\Delta_y s}{\tilde{u}_\xi},$$

where all the derivatives of $\tilde{u}$ are evaluated at $(t,0,y)$. Recalling that $\tilde{u}_\xi(t,0,y)=-1/\sqrt{1+|D_y s(t,y)|^2}$, we get

$$s_t = \Delta_y s - 2\frac{\langle D_y^2 s Ds, Ds\rangle}{1+|D_y s|^2} + (\Delta\tilde{u}(t,0,y)+\tilde{u}_{\xi\xi}(t,0,y)|D_y s|^2)\sqrt{1+|D_y s|^2}.$$

This expression for s_t may be replaced in (2.15), to get a complicated system for $(\tilde{u},s)$,

$$\begin{cases} \tilde{u}_t = \tilde{u}_\xi\bigg(\Delta_y s - 2\dfrac{\langle D_y^2 s Ds, Ds\rangle}{1+|D_y s|^2} + \\[3mm] \qquad +(\Delta\tilde{u}(t,0,y)+\tilde{u}_{\xi\xi}(t,0,y)|D_y s|^2)\sqrt{1+|D_y s|^2}\bigg) + \\[3mm] \qquad +\Delta\tilde{u} - 2\langle D_y\tilde{u}_\xi D_y s\rangle + \tilde{u}_{\xi\xi}|D_y s|^2 - \tilde{u}_\xi \\[3mm] s_t = \Delta_y s - 2\dfrac{\langle D_y^2 s Ds, Ds\rangle}{1+|D_y s|^2} + \\[3mm] \qquad +(\Delta\tilde{u}(t,0,y)+\tilde{u}_{\xi\xi}(t,0,y)|D_y s|^2)\sqrt{1+|D_y s|^2} \\[3mm] \tilde{u}(t,0,y)=0,\quad \tilde{u}_\xi(t,0,y)=-\dfrac{1}{\sqrt{1+|D_y s|^2}} \\[3mm] \tilde{u}(0,\xi,y)=\tilde{u}_0(\xi,y) \end{cases}$$

with the conditions
$$t>0,\ \xi\le 0,\ y\in\mathbb{R}^{N-1},$$
$$t>0,$$
$$t>0,\ y\in\mathbb{R}^{N-1},$$
$$\xi\le 0,\ y\in\mathbb{R}^{N-1}.$$

Note that in dimension $N = 1$ all the derivatives with respect to y disappear, and the system reduces to (2.5)–(2.6). For $N > 1$ local solvability of the system, even for special initial data, is not clear.

3 Basic Examples

3.1 Self-similar Solutions in Bounded Domains

We give an example in which the domain $\Omega_t = B(0, r(t))$ shrinks to a single point in finite time T. We look for radial *self-similar* solutions of the form

$$u(t, x) = (T - t)^\alpha f(\eta), \; \eta = \frac{|x|}{(T - t)^\beta}, \; \Omega_t = \{|x| < r(T - t)^\beta\}, \; 0 \le t < T$$

(3.1)

with $f : [0, +\infty) \mapsto \mathbb{R}$, $r > 0$. Replacing in (1.1), we see that $\alpha = \beta = 1/2$, and that the function $g(x) = f(|x|)$ has to be an eigenfunction of an Ornstein-Uhlenbeck type operator in the ball $B(0, r)$,

$$\Delta g - \frac{1}{2}\langle x, Dg(x)\rangle + \frac{1}{2}g = 0,$$

with two boundary conditions, $g = 0$, $\partial g/\partial \nu = -1$. This amounts to say that

$$f''(\eta) + \frac{N - 1}{\eta}f'(\eta) + \frac{1}{2}f(\eta) = \frac{1}{2}\eta f'(\eta) \quad \text{for } 0 < \eta \le r;$$

$$f'(0) = f(r) = 0, \; f'(r) = -1.$$

(3.2)

Problem (3.2) is over determined. However, the following proposition holds [9].

Proposition 3.1. *There exist a unique $r > 0$ and a unique C^2 function f : $[0, r] \mapsto \mathbb{R}$ satisfying (3.2) and such that $f(\eta) > 0$ for $0 \le \eta < r$. Moreover f is analytic.*

To build the solution, first we find a solution g of the differential equation with the initial condition $g'(0) = 0$. All the solutions of the differential equation are linear combinations of two linearly independent solutions, one of which is singular at $\eta = 0$ and the other one may be written as a power series, $g(\eta) = c\sum_{n=0}^{\infty} a_n \eta^n$. Replacing in (3.2) we get

$$a_0 = 1,$$

$$a_{2n} = -\frac{(2n - 1)!!}{4^{n+1}(n + 1)!N(2 + N)(4 + N) \cdot \ldots \cdot (2n + N)}, \; a_{2n+1} = 0, \; n \in \mathbb{N}.$$

Since g is strictly decreasing for $\eta > 0$ and goes to $-\infty$ as $\eta \to +\infty$, there exists a unique $r > 0$ such that $g(r) = 0$. Now we choose $c = -1/g'(r)$. The function $f(\eta) = cg(\eta)$ satisfies all requirements of proposition 3.1.

The couple (Ω_t, u) defined in (3.1) with $\alpha = \beta = 1/2$, and f given by proposition 3.1, is a classical solution to (1.1) in the time interval $[0, T)$. Note that u is smooth for $t < T$; in particular Ω_0 is the ball centered at 0 with radius $r\sqrt{T}$ and $u(0, \cdot) \in C^{\infty}(\overline{\Omega}_0)$, but at $t = T$ the free boundary defined by $|x| = r\sqrt{T - t}$ is only 1/2-Hölder continuous with respect to time.

In [15] and in [13] it is shown that all radial solutions to (1.1) shrink to the origin in finite time and that they are asymptotically selfsimilar.

3.2 Travelling Waves in Unbounded Domains

Another important class of solutions are the *travelling waves*, with $u(t, x) = f(x + ct\mathbf{e})$, $\mathbf{e}$ being some unit vector, $c \neq 0$.

After a rotation we may assume that $\mathbf{e} = (1, 0, \ldots, 0)$. A *planar* travelling wave is a solution such that Ω_t is a halfspace delimited by the hyperplane $x_1 = -ct + c_0$, and $u(t, x) = f(x_1 + ct - c_0)$ where $f(\xi)$ is defined in the halfline $(-\infty, c_0]$ or $[c_0, +\infty)$. Since the problem is invariant with respect to the change of variables $x_1 \mapsto -x_1$, we may choose to consider only left halflines $(-\infty, c_0]$. Moreover, having fixed c and f, such solutions differ only by a translation in the x_1 variable. So we can look for solutions with $c_0 = 0$. Then we find $f(\xi) = -(e^{c\xi} - 1)/c$.

So we have an example of smooth solutions with existence in the large.

Stability of planar travelling waves is of physical interest. However, while stability under one-dimensional perturbations is relatively easy (see e.g. [14, 6]), stability for genuinely multidimensional perturbations comes out to be a rather complicated problem. See [4] for the heat equation, and [7, 5, 19, 20] for other free boundary problems of this type.

4 Weak Solutions

We describe here the approach of Caffarelli and Vazquez which provides weak solutions to (1.1), under suitable assumptions on the initial data. They look for an open set $\Omega \subset (0, T) \times \mathbb{R}^N$ with Lipschitz continuous lateral boundary Γ and for a continuous nonnegative function u, vanishing on Γ, such that for each test function $\phi \in C_0^{\infty}([0, T) \times \mathbb{R}^N)$ it holds

$$\int \int_{\Omega} u(\phi_t + \Delta\phi)dt\, dx + \int_{\Omega_0} u_o \phi\, dx = \int_{\Gamma} \phi \cos \alpha\, d\Sigma, \qquad (4.1)$$

where α is the angle between the exterior normal vector to Γ and the horizontal hyperplane $t = $ constant. Moreover, the free boundary should start at $\Gamma_0 = \partial\Omega_0$, i.e. the section $\Gamma_t = \{x : (t, x) \in \Gamma\}$ should converge to Γ_0 in a suitable sense as $t \to 0$.

If (Ω_t, u) is a regular solution to (1.1) in the interval $[0, T]$, setting $\Omega = \{(t, x) : 0 < t < T, \ x \in \Omega_t\}$, multiplying the equation by ϕ and integrating over Ω we obtain (4.1).

The assumptions on the initial data (Ω_0, u_0) are the following: Ω_0 is an open set with uniformly C^2 boundary Γ_0, $u_0 \in C^2(\overline{\Omega}_0)$ has positive values in Ω_0 and vanishes in Γ_0; moreover $|Du_0| \leq 1$ and there exists a number $\lambda > 0$ such that

$$\Delta u_0 + \lambda |Du_0| \leq 0.$$

The idea to construct such weak solutions is to follow backwards the procedure used by the physicists to obtain the free boundary problem (1.1) as a model in combustion theory. Problem (1.1) is approximated by a family of problems in $[0, T] \times \mathbb{R}^N$,

$$\begin{cases} D_t u_\varepsilon = \Delta u_\varepsilon - \beta_\varepsilon(u_\varepsilon), & 0 < t \leq T, \ x \in \mathbb{R}^N, \\ u_\varepsilon(0, x) = u_{\varepsilon 0}(x), & x \in \mathbb{R}^N, \end{cases} \tag{4.2}$$

where $\beta_\varepsilon(u) = \frac{1}{\varepsilon}\beta\left(\frac{u}{\varepsilon}\right)$, and $\beta : \mathbb{R} \mapsto \mathbb{R}$ is a nonnegative regular function supported in $[0, 1]$, increasing in $(0, 1/2)$ and decreasing in $(1/2, 1)$, with integral equal to $1/2$. $u_{\varepsilon 0}$ is suitable smooth approximation of the initial datum u_0 (more precisely, it is an approximation by convolution of the extension of u_0 which vanishes outside Ω_0).

The general theory of parabolic problems gives existence and uniqueness of a global bounded solution to (4.2).

Then the first step is to prove that the family $\{u_\varepsilon : \ \varepsilon > 0\}$ is bounded in the space $C^{1/2,1}([0, T] \times \mathbb{R}^N)$, and this is done using classical techniques of parabolic equations, under the only assumption that u_0 is a nonnegative function with bounded gradient. It follows that there is a sequence $\varepsilon_n \to 0$ such that u_{ε_n} converges uniformly on each compact subset of $[0, T] \times \mathbb{R}^N$ to a continuous function u, which in addition belongs to $C^{1/2,1}([0, T] \times \mathbb{R}^N)$.

The candidate to be a weak solution is the couple (Ω, u) where $\Omega = \{(t, x) \in (0, T) \times \mathbb{R}^N : u(t, x) > 0\}$. Note that if K is a compact set contained in Ω, then there is $C > 0$ such that for ε small enough we have $u_\varepsilon \geq C$; hence $\beta_\varepsilon(u_\varepsilon) = 0$ for ε small and u is a weak solution to the heat equation in K, so that it is smooth.

The second step is to pass to the limit in the obvious equality

$$\int_0^T \int_{\mathbb{R}^N} u_\varepsilon(\phi_t + \Delta\phi)dt\,dx + \int_{\mathbb{R}^N} u_{\varepsilon 0}\phi\,dx = \int_0^T \int_{\mathbb{R}^N} \beta_\varepsilon(u_\varepsilon)\phi\,dt\,dx \tag{4.3}$$

for every test function ϕ. The convergence of the left hand side is immediate. Concerning the right hand side, since $\beta_\varepsilon(u_\varepsilon)$ is bounded in L^1_{loc}, along a subsequence it converges to $\int_0^T \int_{\mathbb{R}^N} \phi\,d\mu$ where μ is some nonnegative measure. So we have

$$\int\int_\Omega u(\phi_t + \Delta\phi)dt\,dx + \int_{\Omega_0} u_0\phi\,dx = \int_0^T \int_{\mathbb{R}^N} \phi\,d\mu.$$

The measure μ has to be concentrated on the free boundary Γ. Indeed, it cannot be supported in Ω because there we have $u_t = \Delta u$. It cannot be supported in the interior of $N = \{(t, x) : u(t, x) = 0\}$ because if ϕ is a test function with support in N then (4.3) becomes

$$\int_0^T \int_{\mathbb{R}^N} u_\varepsilon (\phi_t + \Delta\phi) dt \, dx = \int_0^T \int_{\mathbb{R}^N} \beta_\varepsilon(u_\varepsilon) \phi \, dt \, dx$$

and the left hand side goes to 0 as $\varepsilon \to 0$. It cannot be supported in $\{(0, x) : x \in \mathbb{R}^N\}$ because of (4.3) again.

The third step is to prove that Γ is the graph of a Lipschitz continuous function $t = \theta(x)$, defined for $x \in \Omega_0$, and that $d\mu = \cos\alpha d\Sigma$ where $d\Sigma$ is the surface measure on Γ. This is not trivial at all, and requires all the further assumptions on (Ω_0, u_0) listed above.

The last step consists to prove that at each point $(x_0, t_0) \in \Gamma$ where the free boundary is a differentiable surface and the exterior normal is not directed along the time axis, the boundary condition $\frac{\partial u}{\partial \nu} = -1$ holds.

An interesting feature of this approach is a comparison principle between the self-similar solutions described in §3.1 and any weak solution obtained by this procedure. Let $(B(0, R), U)$ be the initial datum of any self-similar solution obtained in §3.1, and let (Ω_0, u_0) be the initial datum of a weak solution obtained by the procedure of this section, such that $\Omega_0 \subset B(0, R)$ and $u_0(x) \leq U(x)$ for each $x \in \Omega_0$. Then for each t the weak solution u stays below the self-similar solution (to be more precise, below the extension of the selfsimilar solution which vanishes outside the ball $B(0, r(t))$). This implies that Ω_t extinguishes in finite time.

5 The Fully Nonlinear Equations Approach

We assume that Ω_0 is an open set with uniformly $C^{3+\alpha}$ boundary, and that $u_0 \in C^{3+\alpha}(\overline{\Omega}_0)$. Since we look for a regular solution up to $t = 0$, we assume also that u_0 satisfies the compatibility conditions

$$u_0 = 0, \quad \frac{\partial u_0}{\partial \nu} = -1 \quad \text{at } \partial\Omega_0.$$

At the end of section 2 problem (1.1) was transformed to a fixed boundary problem in the special case where $\Omega_0 = \{(x, y) \in \mathbb{R} \times \mathbb{R}^{N-1} : x < s_0(y)\}$ is the left hand side of the graph of a regular function s_0, and we look for Ω_t of the same type, $\Omega_t = \{(x, y) \in \mathbb{R} \times \mathbb{R}^{N-1} : x < s(t, y)\}$. In that case, the change of coordinates which fixes the boundary is trivial, being just $(x, y) \mapsto (x - s(t, y), y)$. It transforms Ω_t into the halfspace $\Omega = \{(\xi, y) : \xi < 0, y \in \mathbb{R}^{N-1}\}$. Then $\partial\Omega_t$ may be seen as the range of a function defined in the boundary of Ω, $(0, y) \mapsto (s(t, y), y)$.

Of course this is not possible for general Ω_0, in particular when Ω_0 is bounded. But we can do something similar.

We look for $\partial\Omega_t$ as the range of a function $\xi \mapsto \xi + s(t, \xi)\nu(\xi)$, defined for $\xi \in \partial\Omega_0$, where $\nu(\xi)$ is the exterior unit normal vector to $\partial\Omega_0$ at ξ, and the unknown $s(t, \xi)$ is the signed distance of the point $x = \xi + s(t, \xi)\nu(\xi)$ from $\partial\Omega_t$.

By a natural change of coordinates we transform Ω_t into Ω_0, at least for small time. The change of coordinates is defined by $\xi \mapsto \xi + s(t, \xi)\nu(\xi)$ for $\xi \in \partial\Omega_0$, and it is extended smoothly to a $C^{2+\alpha}$ diffeomorphism $\xi \mapsto \xi + \Phi(t, \xi)$ in the whole of Ω_0.

More precisely, let $\mathcal{N}$ be a neighborhood of $\partial\Omega_0$ such that each $x \in \mathcal{N}$ may be written in a unique way as $x = \xi + s\nu(\xi)$ with $\xi \in \partial\Omega_0$ and $s \in \mathbb{R}$. For every $\xi \in \mathcal{N}$ we denote by ξ' the point of $\partial\Omega_0$ closest to ξ. For every $\xi \in \mathbb{R}^N$ we set

$$\widetilde{\nu}(\xi) = \begin{cases} \theta(\xi)\nu(\xi') & \text{if } \xi \in \mathcal{N}, \\ 0 & \text{otherwise,} \end{cases} \tag{5.1}$$

where θ is a C^∞ function with support contained in $\mathcal{N}$, such that $\theta \equiv 1$ in a neighborhood of $\partial\Omega_0$, $\theta \equiv 0$ outside a bigger neighborhood of $\partial\Omega_0$.

Then we set

$$\Phi(t, \xi) = \begin{cases} \theta(\xi)s(t, \xi')\nu(\xi') & \text{if } \xi \in \mathcal{N}, \\ 0 & \text{otherwise.} \end{cases} \tag{5.2}$$

For t small, the mapping

$$\xi \mapsto x(t, \xi) = \xi + \Phi(t, \xi), \quad \xi \in \mathbb{R}^N, \tag{5.3}$$

is a $C^{2+\alpha}$ diffeomorphism from $\mathbb{R}^N$ to itself, which maps Ω_0 onto Ω_t, and will be used to change coordinates. At $t = 0$, $\Phi(0, \cdot) \equiv 0$, so that the diffeomorphism is the identity. Moreover for any t, $\Phi(t, \cdot)$ vanishes in $\mathbb{R}^N \setminus \mathcal{N}$, so that the diffeomorphism differs from the identity only in a small neighborhood of $\partial\Omega_0$.

Note that if $\partial\Omega_0$ has a straight part Γ, up to translations and rotations the above change of coordinates coincides with (2.14) near Γ. Indeed, after a translation and a rotation, we may assume that $\Gamma = \{0\} \times U$, where U is an open set in $\mathbb{R}^{N-1}$, and that Ω_0 lies locally in the left hand side of Γ. The signed distance of $\xi = (\xi_1, \ldots, \xi_N)$ from Γ is just the first coordinate ξ_1, $\Phi(t, \xi) = s(t, \xi_2, \ldots, \xi_N)\mathbf{e}_1$, so that $\xi + \Phi(t, \xi) = (\xi_1 + s(t, \xi_2, \ldots, \xi_N), \xi_2, \ldots, \xi_N)$ and (5.3) coincides with (2.14) near Γ.

Denoting by $\widetilde{u}$ the unknown u expressed in the new coordinates (t, ξ), i.e.

$$\widetilde{u}(t, \xi) = u(t, \xi + \Phi(t, \xi)), \quad t \geq 0, \ \xi \in \overline{\Omega}_0, \tag{5.4}$$

we get a fixed boundary system for $(s, \widetilde{u})$, in the domain $[0, \delta] \times \overline{\Omega}_0$ with small δ,

$$\widetilde{u}_t - \langle D\widetilde{u}, (I + {}^tD\Phi)^{-1}\Phi_t \rangle = \mathcal{L}\widetilde{u}, \quad t \geq 0, \ \xi \in \overline{\Omega}_0, \tag{5.5}$$

where $[D\Phi]_{ij} = \frac{\partial\Phi_i}{\partial\xi_j}(t, \xi)$, the superscript t denotes the transposed matrix, and $\mathcal{L}$ is the new expression of the laplacian, i.e.

$$\mathcal{L} = \sum_{i,j,h=1}^{N} \frac{\partial \xi_j}{\partial x_i} \frac{\partial \xi_h}{\partial x_i} D_{jh} + \sum_{i,j=1}^{N} \frac{\partial^2 \xi_j}{\partial x_i^2} D_j.$$

Here we denote by $\xi = \xi(t, x)$ the inverse of (5.3).

The main step is now the introduction of a new unknown w defined by

$$\widetilde{u}(t, \xi) = u_0(\xi) + \langle Du_0(\xi), \Phi(t, \xi) \rangle + w(t, \xi). \tag{5.6}$$

At $t = 0$ we have $\widetilde{u}(0, \xi) = u_0(\xi)$, $\Phi(0, \cdot) \equiv 0$, so that

$$w(0, \xi) = 0, \quad \xi \in \Omega_0. \tag{5.7}$$

(5.6) allows to decouple the system using the boundary condition $u = 0$ at $\partial \Omega_t$, which gives

$$s(t, \xi) = w(t, \xi), \quad t \geq 0, \ \xi \in \partial \Omega_0, \tag{5.8}$$

so that

$$\Phi(t, \xi) = w(t, \xi') \widetilde{\nu}(\xi), \quad t \geq 0, \ \xi \in \overline{\Omega}_0. \tag{5.9}$$

Strictly speaking, the above expression is incorrect because ξ' is defined only for $\xi \in \mathcal{N}$. However, since $\widetilde{\nu}(\xi)$ vanishes outside $\mathcal{N}$, for any f defined in $\partial \Omega_0$ we keep the expression $f(\xi') \widetilde{\nu}(\xi)$ to denote the function which coincides with $f(\xi') \widetilde{\nu}(\xi)$ in $\mathcal{N}$ and with 0 outside $\mathcal{N}$.

Replacing (5.9) in (5.5) we get

$$w_t = \mathcal{F}_1(\xi, w, Dw, D^2 w) + \mathcal{F}_2(\xi, w, Dw) s_t, \quad t \geq 0, \ \xi \in \overline{\Omega}_0, \tag{5.10}$$

where $\mathcal{F}_1$, $\mathcal{F}_2$ are obtained respectively from

$$\mathcal{L}(u_0 + \langle Du_0, \Phi \rangle + w) \tag{5.11}$$

and from

$$-\langle Du_0 - (I + D\Phi)^{-1} D(u_0 + \langle Du_0, \Phi \rangle + w), \widetilde{\nu} \rangle \tag{5.12}$$

replacing $\Phi = w(t, \xi') \widetilde{\nu}(\xi)$.

Equation (5.10) still contains s_t; to eliminate it we use again the identity $s = w$ at the boundary which gives $s_t = w_t$. Replacing in (5.10) for $\xi \in \partial \Omega_0$ we get

$$s_t(1 - \mathcal{F}_2(\xi, w, Dw)) = \mathcal{F}_1(\xi, w, Dw, D^2 w), \quad t \geq 0, \ \xi \in \partial \Omega_0.$$

At $t = 0$ we have $w \equiv 0$, and $\mathcal{F}_2$ vanishes at $(\xi, 0, 0)$, so that, at least for t small, $\mathcal{F}_2(\cdot, w(t, \cdot), Dw(t, \cdot))$ is different from 1 and it is possible to get s_t as a function of w,

$$s_t = \mathcal{F}_3(\xi, w, Dw, D^2 w) = \frac{\mathcal{F}_1(\xi, w, Dw, D^2 w)}{1 - \mathcal{F}_2(\xi, w, Dw)}, \quad t \geq 0, \ \xi \in \partial \Omega_0, \tag{5.13}$$

which, replaced in (5.10), gives the final equation for w,

$$w_t = \mathcal{F}(w)(\xi), \quad t \geq 0, \ \xi \in \overline{\Omega}_0, \tag{5.14}$$

where

$$\mathcal{F}(w)(\xi) = \mathcal{F}_1(\xi, w, Dw, D^2 w) + \mathcal{F}_2(\xi, w, Dw)\mathcal{F}_3(\xi, w, Dw, D^2 w). \tag{5.15}$$

Although the explicit expression of $\mathcal{F}$ is rather complicated, we may note immediately that for ξ in $\Omega_0 \setminus \mathcal{N}$, that is far from the boundary, we have

$$\mathcal{F}(w)(\xi) = \Delta w + \Delta u_0.$$

This is because our change of coordinates reduces to the identity far from the boundary. Moreover, even near the boundary we have

$$\mathcal{F}(0)(\xi) = \Delta u_0(\xi). \tag{5.16}$$

For ξ near the boundary, the function $\mathcal{F}(v)$ is defined for $v \in C^2(\overline{\Omega}_0)$ with small C^1 norm. More precisely, it is defined for $v \in C^2(\overline{\Omega}_0)$ such that $1 - \mathcal{F}_2(\cdot, v(\cdot), Dv(\cdot)) \neq 0$. From formulas (5.8), (5.11), (5.12), (5.13) we see that $\mathcal{F}(v)(\xi)$ depends smoothly on v, Dv, $D^2 v$ and their traces at the boundary. Therefore the function $v \mapsto \mathcal{F}(v)$ is continuously differentiable from $\{v \in C^2(\overline{\Omega}_0) : \|v\|_{C^1} \leq r\}$ to $C(\overline{\Omega}_0)$, and from $\{v \in C^{2+\alpha}(\overline{\Omega}_0) : \|v\|_{C^1} \leq r\}$ to $C^\alpha(\overline{\Omega}_0)$.

In the next lemma we describe the linear part of $\mathcal{F}$ with respect to v near $v = 0$.

Lemma 5.1. $\mathcal{F}_v(0)$ *is the sum of the Laplacian plus a nonlocal differential operator of order* 1.

Proof — It is clear from the expression of $\mathcal{F}$ that the linear part of $v \mapsto \mathcal{F}(v)$ near $v = 0$ is a (possibly nonlocal) second order linear differential operator $\mathcal{A}$. To identify its principal part it is sufficient to consider $\mathcal{F}_1$. Indeed, the product $\mathcal{F}_2(\xi, v, Dv)\mathcal{F}_3(\xi, v, Dv, D^2 v)$ does not contribute to the principal part, because $\mathcal{F}_2$ does not depend on the second order derivatives of v, and it vanishes at $v = 0$.

Using (5.11) we see that only the term $\mathcal{L}(u_0 + \langle Du_0, \Phi \rangle + w)$ in $\mathcal{F}_1$ contributes to the principal part of $\mathcal{A}$. Its explicit expression is

$$\sum_{i,h,j=1}^{N} \frac{\partial \xi_j}{\partial x_i}(\xi + \Phi) \frac{\partial \xi_h}{\partial x_i}(\xi + \Phi) \, D_{jh}(u_0 + \langle Du_0, \Phi \rangle + v)$$

$$+ \sum_{i,j=1}^{N} \frac{\partial^2 \xi_j}{\partial x_i \partial x_i}(\xi + \Phi) \, D_j(u_0 + \langle Du_0, \Phi \rangle + v)$$

$$= \mathcal{L}_0(v, \Phi) + \mathcal{L}_1(v, \Phi),$$

where the matrix M with entries $m_{hk} = \partial \xi_h / \partial x_k$ is equal to $(I + D\Phi)^{-1}$. We recall that Φ depends on v through (5.8) and (5.2), i.e. $\Phi(\xi, v) = \mathcal{F}_0(\xi', v(\xi'))\tilde{\nu}(\xi)$ and $\Phi(\xi, 0) = 0$.

At $v = 0$, $\Phi = 0$, the linear part of $\mathcal{L}_0$ with respect to (v, Φ) is

$$\sum_{i=1}^{N} (D_{ii}v + D_{ii}\langle Du_0, \Phi \rangle) + \text{ lower order terms.}$$

From the developement

$$(I + D\Phi)^{-1} = I - D\Phi + (D\Phi)^2 (I + D\Phi)^{-1} \tag{5.17}$$

we get that the linear part of $\mathcal{L}_1$ with respect to (v, Φ) is

$$-\sum_{i,j=1}^{N} \frac{\partial^2 \Phi_j}{\partial x_i^2} D_j u_0 + \text{ lower order terms}$$

Therefore the second order derivatives of Φ cancel in the sum; what remains is

$$\mathcal{A} = \Delta + \text{ lower order terms,}$$

and the statement follows. □

Lemma 5.1 implies that equation (5.14) is parabolic near $t = 0$.

Equation (5.14) has to be supported with an initial and a boundary condition. The initial condition comes from the splitting (5.6),

$$w(0, \xi) = 0, \quad \xi \in \overline{\Omega}_0.$$

The boundary condition comes from $\partial u / \partial n = -1$. Since

$$n(t, \xi + \Phi) = \frac{(I + {}^t D\Phi)^{-1}\nu(\xi)}{|(I + {}^t D\Phi)^{-1}\nu(\xi)|}$$

and

$$Du(t, \xi + \Phi) = (I + {}^t D\Phi)^{-1} D\tilde{u}(t, \xi),$$

from $\partial u / \partial n = -1$ we get

$$\langle (I + {}^t D\Phi)^{-1}\nu, (I + {}^t D\Phi)^{-1} D(u_0 + \langle Du_0, \Phi \rangle + w \rangle + |(I + {}^t D\Phi)^{-1}\nu| = 0, \tag{5.18}$$

which gives a new boundary condition for w,

$$\mathcal{G}(\xi, w(t, \xi), Dw(t, \xi)) = 0, \quad t \geq 0, \; \xi \in \partial \Omega_0, \tag{5.19}$$

as soon as we replace $\Phi = w(t, \xi')\tilde{\nu}(\xi)$ in (5.18). The function $\mathcal{G}(\xi, u, p)$ and its first and second order derivatives with respect to u and p_i, $i = 1, \ldots, N$, are continuous in (ξ, u, p) and $C^{1/2+\alpha/2}$ in ξ. Moreover $\mathcal{G}(\xi, 0, 0) = 0$.

In the next lemma we identify the linear part of $\mathcal{G}$ with respect to (u, p) at $(\xi, 0, 0)$.

Lemma 5.2. *We have*

$$\mathcal{G}(\xi, w, Dw) = \mathcal{B}w - G(\xi, w, Dw)(\xi)$$

where $\mathcal{B}$ is the linear differential operator defined by

$$\mathcal{B}v = \frac{\partial v}{\partial \nu} + \frac{\partial^2 u_0}{\partial \nu^2} v, \tag{5.20}$$

and for every $\xi \in \partial \Omega_0$,

$$G(\xi, 0, 0) = G_u(\xi, 0, 0) = G_{p_i}(\xi, 0, 0), \quad i = 1, \ldots, N.$$

Proof — Taking (5.17) into account, we rewrite the first addendum of (5.18) as

$$\langle (I +{}^t D\Phi)^{-1}\nu, (I +{}^t D\Phi)^{-1} D(u_0 + \langle Du_0, \Phi \rangle + w) \rangle =$$
$$= \frac{\partial u_0}{\partial \nu} + \frac{\partial}{\partial \nu} \langle Du_0, \Phi \rangle + \frac{\partial w}{\partial \nu} - \langle ({}^t D\Phi)\nu, Du_0 \rangle -$$
$$- \langle \nu, ({}^t D\Phi)Du_0 \rangle + R(\xi, Dw, \Phi, D\Phi)$$

where $R(\xi, p, q, r)$ and its derivatives with respect to $p_i, q_i, r_{ij}, i, j = 1, \ldots, N$, vanish at $(\xi, 0, 0, 0)$.

In the above sum we have

$$\frac{\partial}{\partial \nu} \langle Du_0, \Phi \rangle = \frac{\partial^2 u_0}{\partial \nu^2} s, \quad ({}^t D\Phi)\nu = D^{tang} s, \quad \langle \nu, ({}^t D\Phi)Du_0 \rangle = 0,$$

where the last equality follows from $(D\Phi)\nu = 0$. Replacing $s = w$ at the boundary, we get

$$\langle (I +{}^t D\Phi)^{-1}\nu, (I +{}^t D\Phi)^{-1} D(u_0 + \langle Du_0, \Phi \rangle + w) \rangle$$

$$= \frac{\partial u_0}{\partial \nu} + \frac{\partial w}{\partial \nu} + \frac{\partial^2 u_0}{\partial \nu^2} w - \langle D^{tang} w, D^{tang} u_0 \rangle + R_1(\xi, w, Dw)$$

where $R_1(\xi, u, p)$ vanishes at $(\xi, 0, 0)$ as well as its derivatives with respect to $u, p_i, i = 1, \ldots, N$.

On the other hand, since $\langle ({}^t D\Phi)\nu, \nu \rangle = 0$, we have

$$|(I +{}^t D\Phi)^{-1}\nu| = 1 + R_3(\xi, \Phi, D\Phi)$$

where $R_3(\xi, p, q)$ and its derivatives with respect to $p_i, q_{ij}, i, j = 1, \ldots, N$, vanish at $(\xi, 0, 0)$. Replacing the expression (5.2) for Φ and taking into account that $s = w$ at $\partial \Omega_0$, we get

$$|(I +{}^t D\Phi)^{-1}\nu| = 1 + R_4(\xi, w, Dw)$$

where $R_4(\xi, u, p)$ and its derivatives with respect to (u, p_i) vanish at $(\xi, 0, 0)$. Summing up, the statement follows. □

Therefore, the final problem for the only unknown w may be rewritten as

$$\begin{cases} w_t = \mathcal{A}w + \mathcal{F}(w), & t \geq 0, \ \xi \in \overline{\Omega}_0, \\[2mm] \mathcal{B}w = \mathcal{G}(w), & t \geq 0, \ \xi \in \partial\Omega_0, \\[2mm] w(0, \cdot) = 0, & \xi \in \overline{\Omega}_0, \end{cases} \tag{5.21}$$

where $\mathcal{A}$, $\mathcal{B}$ are linear differential operators and $\mathcal{F}$, $\mathcal{G}$ are nonlinear functions. Problem (5.21) is fully nonlinear because $\mathcal{F}$ depends on w and on its space derivatives up to the second order, and $\mathcal{G}$ depends on w and on its first order space derivatives. Moreover, $\mathcal{F}$ is nonlocal in the second order derivatives of w. However, both $\mathcal{F}$ and $\mathcal{G}$ are smooth enough, and $\mathcal{F}_w$, $\mathcal{G}_w$ vanish at $t = 0$, $w = 0$. The second order linear operator $\mathcal{A}$ is the sum of Δ plus a nonlocal operator, acting only on w and on the first order space derivatives of w. This is crucial for our analysis: since the principal part of $\mathcal{A}$ coincides with the Laplacian, the linearized problem near $w_0 = 0$ is a good linear parabolic problem, for which optimal Hölder regularity results and estimates are available; we need such estimates to solve the nonlinear problem in a standard way, using the Contraction Theorem in a ball of the parabolic Hölder space $C^{1+\alpha/2,2+\alpha}([0,\delta] \times \overline{\Omega}_0)$, with $\delta > 0$ suitably small, or else in a ball of $C^{3/2+\alpha/2,3+\alpha}([0,\delta] \times \overline{\Omega}_0)$ if the data are more regular.

The above mentioned optimal Hölder regularity results and estimates are given by the well known Ladyzhenskaja – Solonnikov – Ural'ceva theorem (see e.g. [16, Thm. 5.3] or [22, Thm. 5.1.20]), which we write below.

Theorem 5.3. *Let $\alpha \in (0,1)$ and $T > 0$. Let Ω be an open set in $\mathbb{R}^n$ with uniformly $C^{2+\alpha}$ boundary, and let α_{ij}, β_i, $\gamma \in C^\alpha(\overline{\Omega})$ be such that*

$$\sum_{i,j=1}^{N} \alpha_{ij}(\xi)\eta_i\eta_j \geq m|\eta|^2, \ \xi \in \overline{\Omega},$$

for some $m > 0$. Moreover, let ϕ_i, $\psi \in C^{1/2+\alpha/2,1+\alpha}([0,T] \times \partial\Omega)$ be such that

$$\sum_{i=1}^{N} \phi_i(\xi)\nu_i(\xi) \neq 0, \ \xi \in \partial\Omega.$$

For every $v \in C^2(\overline{\Omega})$ set

$$(\mathcal{L}v)(\xi) = \sum_{i,j=1}^{N} \alpha_{ij}(\xi)D_{ij}v(\xi) + \sum_{i=1}^{N}(\beta_i(\xi)D_iv(\xi) + \gamma(\xi)v(\xi), \ \xi \in \overline{\Omega},$$

and

$$(\mathcal{B}v)(\xi) = \psi(\xi)v(\xi) + \sum_{i=1}^{N} \varphi_i(\xi)D_iv(\xi), \ \xi \in \partial\Omega.$$

Then for every $\delta \in (0, T]$ and for every $f_1 \in C^{\alpha/2,\alpha}([0,\delta] \times \overline{\Omega})$, $g_1 \in C^{1/2+\alpha/2,1+\alpha}([0,\delta] \times \partial\Omega)$ satisfying the compatibility condition

$$g_1(0, \cdot) = 0 \text{ in } \partial\Omega, \tag{5.22}$$

the problem

$$\begin{cases} v_t = \mathcal{L}v + f_1, \quad 0 \le t \le \delta, \ \xi \in \overline{\Omega}, \\ \\ \mathcal{B}v = g_1, \quad t \ge 0, \ \xi \in \partial\Omega, \\ \\ v(0, x) = 0, \quad \xi \in \overline{\Omega}, \end{cases} \tag{5.23}$$

has a unique solution $v \in C^{1+\alpha/2,2+\alpha}([0,\delta] \times \overline{\Omega})$. Moreover there exists $C > 0$ independent of δ such that

$$\|v\|_{C^{1+\alpha/2,2+\alpha}([0,\delta]\times\overline{\Omega})} \le C(\|f_1\|_{C^{\alpha/2,\alpha}([0,\delta]\times\overline{\Omega})} + \|g_1\|_{C^{1/2+\alpha/2,1+\alpha}([0,\delta]\times\partial\Omega)}). \tag{5.24}$$

Looking at the expression of $\mathcal{F}$, we see that it depends on third order derivatives of the initial datum u_0. Moreover both the boundary operator $\mathcal{B}$ and the nonlinear function $\mathcal{G}$ depend on second order derivatives of u_0.

Using theorem 5.3 it is possible to solve locally problem (5.21). To be more precise, first one extends the result of theorem 5.3 to the case where $\mathcal{L}$ is the Laplacian plus a nonlocal first order perturbation, and then one solves (5.21) by a fixed point theorem in the metric space $Y = \{u \in C^{1+\alpha/2,2+\alpha}([0,\delta] \times \overline{\Omega}_0) : u(0, \cdot) = 0, \|u\|_{C^{1+\alpha/2,2+\alpha}} \le R\}$.

Some comments about the application fixed point theorem are in order. For every $u \in Y$, the function $\mathcal{F}(u)$ is well defined if δ and R are suitably chosen. Note that $\mathcal{F}(u)$ depends on u_0 through its derivatives up to the third order. Since $u_0 \in C^{3+\alpha}(\overline{\Omega}_0)$, then $\mathcal{F}(u) \in C^{\alpha/2,\alpha}([0,\delta] \times \overline{\Omega}_0)$. Similarly, $\mathcal{G}(u)$ depends on u_0 through its derivatives up to the second order, so that $\mathcal{G}(u) \in C^{1/2+\alpha/2,1+\alpha}([0,\delta] \times \partial\overline{\Omega}_0)$; the coefficients of the boundary operator $\mathcal{B}$ depend on the second order derivatives of u_0 and hence they are in $C^{1/2+\alpha/2,1+\alpha}([0,\delta] \times \partial\overline{\Omega}_0)$. Therefore it is reasonable to look for a local solution to (5.21) as a fixed point of the operator Γ defined by $\Gamma u = w$, w being the solution of

$$\begin{cases} w_t = \mathcal{A}w + \mathcal{F}(u), \quad t \ge 0, \ \xi \in \overline{\Omega}_0, \\ \\ \mathcal{B}w = \mathcal{G}(u), \quad t \ge 0, \ \xi \in \partial\Omega_0, \\ \\ w(0, \cdot) = 0, \quad \xi \in \overline{\Omega}_0, \end{cases}$$

This procedure works, and it gives the following result.

Theorem 5.4. *There is $R_0 > 0$ such that for every $R \ge R_0$ and for every sufficiently small $\delta > 0$ problem (5.21) has a unique solution in the ball $B(0, R) \subset C^{1+\alpha/2,2+\alpha}([0,\delta] \times \overline{\Omega}_0)$.*

Once we have locally solved (5.21) we come back to the original problem (1.2) using (5.9) to define $\partial\Omega_t$. Note that s has the same regularity of w, i.e. it is in $C^{1+\alpha/2,2+\alpha}([0,\delta]\times\partial\Omega_0)$.

Then we define $\widetilde{u}$ through (5.6), where Φ is given by (5.2). Again, $\widetilde{u}$ has the same regularity of w. As a last step we define u through the change of coordinates, $u(t,x)=\widetilde{u}(t,\xi)$ where $x=\xi+\Phi(t,\xi)$.

It is clear now why we lose regularity: not only because of the change of coordinates, but mainly because w has the same space regularity of Du_0.

However, it is easy to see that uniqueness of w implies uniqueness of $(s,\widetilde{u})$, and hence uniqueness of the regular solution to the original problem (1.2). Moreover, using standard techniques of fully nonlinear problems, it is possible to prove that $t\mapsto w(t,\cdot)$ is smooth with values in $C^2(\overline{\Omega}_0)$ in the interval $(0,\delta)$ if the data are smooth. It follows that both the free boundary and u are smooth with respect to time for t in $(0,\delta)$.

By a covering argument we get eventually that the free boundary of any sufficiently smooth solution defined in an interval $[0,a]$ is C^∞ in time in $(0,a]$.

6 Special Geometries

6.1 Radially Symmetric Solutions

The paper [13] deals with a detailed study of radially symmetric solutions to (1.1), of the type $\Omega_t=B(0,s(t))$, $u(t,x)=v(t,|x|)$. The initial set is the ball $B_0=B(0,s(0))$, and the initial function u_0 has positive values in the interior of B_0. We may assume that u_0 is defined in the whole $\mathbb{R}^N$, and that it has negative values outside $B(0,s_0)$.

Problem (1.1) is transformed to a fixed boundary elliptic-parabolic one-dimensional problem, as follows.

Assume we have a radially symmetric solution in $[0,T]$, such that u has positive values in $B(0,s(t))$, and let $R>s(t)$ for $0\leq t\leq T$. We extend continuously u by a harmonic function outside $B(0,s(t))$. Then, still denoting by u the extension, u satisfies

$$(c(u)(t,x))_t=\Delta u(t,x),\ \ t\geq 0,\ |x|\neq s(t),$$

where

$$c(u)=\max\{0,u\}.$$

For $|x|>s(t)$, $u(t,\cdot)$ is harmonic, so that $(r^{N-1}u_r)_r=0$ and therefore $u_r=-(s(t)/r)^{N-1}$. Since $R>s(t)$ for each $t\in[0,T]$, this gives a Neumann boundary condition at $\partial B(0,R)$, i.e.

$$u_r(t,x)=-\left(\frac{s(t)}{R}\right)^{N-1},\ \ |x|=R.$$

On the other hand, since u vanishes at $|x| = s(t)$, from $u_r = -(s(t)/r)^{N-1}$ we get also (if $N > 2$)

$$u(t,x) = \int_{s(t)}^r \left(\frac{s(t)}{\rho}\right)^{N-1} d\rho = \frac{s(t)}{N-2}\left(\left(\frac{s(t)}{r}\right)^{N-2} - 1\right), \quad |x| = r > s(t).$$

This gives a Dirichlet boundary condition at $\partial B(0, R)$,

$$u(t,x) = \frac{s(t)}{N-2}\left(\left(\frac{s(t)}{R}\right)^{N-2} - 1\right), \quad |x| = R.$$

Eliminating $s(t)$ we get a nonlinear boundary condition for u,

$$u(t,R) = -\frac{R}{N-2}((-u_r(t,R))^{1/(N-1)} + u_r(t,R)) = G(u_r(t,R)).$$

To write it in the usual form $u_r = F(u)$ we note that G is not invertible, it has a minimum at $a_N = -(N-1)^{(N-1)/(2-N)} \in (-1,0)$, with $G(a_N) = -b_N R$. Since we are interested in the free boundary conditions $(u, u_r) = (0, -1)$ we take R close to s_0 and we want $(u(t,R), u_r(t,R))$ to be close to $(0, -1)$ for t small; so we invert G in $[-1, a_N]$ and we denote by $F : [-b_N R, 0] \mapsto [-1, a_N]$ its inverse. We extend F setting $F(u) = F(-b_N R) = a_N$ for $u < -b_N R$. Then our problem is rewritten as an elliptic-parabolic problem with nonlinear Neumann boundary condition,

$$\begin{cases} (c(u)_t) = \Delta u, \quad (t, x) \in Q_T = (0, T] \times B(0, R), \\ u_r(t, R) = F(u(t, R)), \quad 0 < t \leq T, \\ c(u(0, x)) = v_0(x), \quad |x| \leq R, \end{cases} \tag{6.1}$$

with $v_0 = c(u_0)$.

A function $u \in L^2(0, T; H^1(B(0, R)))$ is called a weak solution of (6.1) in $[0, T]$ if $c(u) \in C(Q_T)$ and

$$\int\int_{Q_T} (-\varphi_t c(u) + \langle Du, D\varphi \rangle) dx\, dt + \int_{B(0,R)} \varphi(T, x) c(u(T, x)) dx$$

$$= \int_{B(0,R)} \varphi(0, x) v_0(x) dx + \int_0^T \int_{\partial B(0,R)} \varphi(t, x) F(u(t, R)) dx\, dt,$$

for every $\varphi \in H^1(Q_T)$. The existence result of [13] is the following.

Theorem 6.1. Let $R > s_0 > 0$, and let v_0 be continuous and radially symmetric in $B(0, R)$, positive for $|x| < s_0$, zero for $|x| \geq s_0$. Then for R close to s_0 and T small there exists a unique radially symmetric solution of (6.1), $u \in L^\infty(0, T; H^1(B(0, R)))$ with $\Delta u \in L^2(Q_T)$. Moreover u satisfies the free boundary conditions almost everywhere.

In [13] other properties of the solution are shown. If the initial datum satisfies also the conditions of [9], then the solution of (6.1) given by theorem 6.1 coincides with the solution obtained by the approximation procedure of [9]. If the initial datum is C^2 and satisfies the natural compatibility conditions $u_0(x) = 0$, $u_0(x) = -1$ at $|x| = s_0$, then the solution is analytic for $t > 0$. Moreover, the behavior near the extinction time is studied.

6.2 Solutions Monotonic Along Some Direction

Another example without loss of regularity was considered in [1], where a two-phase version of (1.2) was studied in a strip for $C^{2+\alpha}$ initial data satisfying suitable monotonicity conditions: u_0 is assumed to be strictly monotonic in the direction orthogonal to the strip. This allows to take u as new independent variable for small t, an old trick already used by Meirmanov for the Stefan problem and also for a two-dimensional version of problem (1.2), see [23].

To fix the ideas, let us consider a modification of problem (1.1), as follows. Denote by (x, y) the coordinates in $\mathbb{R}^N$, with $x \in \mathbb{R}$, $y \in \mathbb{R}^{N-1}$. Fix $b \in \mathbb{R}$ and look for Ω_t of the type

$$\Omega_t = \{x \in \mathbb{R}^N : s(t, y) < x < b\}.$$

Here the free boundary Γ_t is not the whole boundary of Ω_t but it is only the graph of the unknown function $s(t, \cdot)$. On the fixed boundary $x = b$ the other unknown u is prescribed, say $u = c > 0$. The other conditions of (1.1) remain unchanged. Therefore we have the problem

$$\begin{cases} u_t = \Delta u, & t > 0, \ s(t, y) \leq x \leq b, \ y \in \mathbb{R}^{N-1}, \\ u = 0, \ \frac{\partial u}{\partial \nu} = -1, \ t > 0, \ x = s(t, y), \ y \in \mathbb{R}^{N-1}, \\ u = c, & t > 0, \ x = b, \end{cases} \tag{6.2}$$

supported by the initial conditions

$$s(0, y) = s_0(y), \ y \in \mathbb{R}^{N-1}; \ \ u(0, x, y) = u_0(x, y), \ y \in \mathbb{R}^{N-1}, \ s_0(y) \leq x \leq b. \tag{6.3}$$

Therefore, $\Omega_0 = \{(x, y) \in \mathbb{R}^N : s_0(y) < x < b\}$. The initial function s_0 has values in $[\varepsilon, b - \varepsilon]$ for some $\varepsilon > 0$, and belongs to $C^{2+\alpha}(\mathbb{R}^{N-1})$ for some $\alpha \in (0, 1)$. The initial datum u_0 is in $C^{2+\alpha}(\overline{\Omega_0})$, and the natural compatibility conditions

$$u_0(s_0(y), y) = 0, \ u_0(b, y) = c, \ \frac{\partial u}{\partial \nu}(s_0(y), y) = -1 \tag{6.4}$$

are satisfied.

The main assumption is that u_0 is increasing in the variable x; more precisely

$$D_x u_0 \geq \varepsilon, \quad (x, y) \in \Omega_0. \tag{6.5}$$

Since we look for a regular solution, for small $t > 0$ the unknown $u(t, \cdot)$ has to be increasing with respect to x. Therefore, it can be taken as a new coordinate ξ. Setting for fixed t

$$\Phi(x, y) = (\xi, y) = (u(t, (x, y)), y),$$

the change of coordinates Φ transforms Ω_t into the strip $(0, c) \times \mathbb{R}^{N-1}$, independent of time. The inverse change of coordinates is denoted by $(x, y) = \Phi(t, \xi)$ where

$$(x, y) = \Phi^{-1}(\xi, y) = (f(t, (\xi, y)), y),$$

and the function f is our new unknown. Deducing a problem for f is easy: observing that

$$D_x u = \frac{1}{D_\xi f}, \quad D_{y_i} u = -\frac{D_{y_i} f}{D_\xi f}, \quad i = 1, \ldots, N-1, \quad D_t u = -\frac{D_t f}{D_\xi f},$$

$$\Delta u = -\frac{\Delta_y f}{D_\xi f} + 2\frac{\langle D_y f_\xi, D_y f \rangle}{(D_\xi f)^2} - (|D_y f|^2 + 1)\frac{D_{\xi\xi} f}{(D_\xi f)^3},$$

we get

$$\begin{cases} f_t = \Delta_y f + \dfrac{D_{\xi\xi} f}{(D_\xi f)^2}(|D_y f|^2 + 1) - 2\dfrac{\langle D_y f_\xi, D_y f \rangle}{D_\xi f}, \\[2mm] t \geq 0, \ \xi \in [0, c], \ y \in \mathbb{R}^{N-1}, \\[2mm] D_\xi f(t, 0, y) = \sqrt{1 + |D_y f|^2}, \ y \in \mathbb{R}^{N-1}, \\[2mm] y(t, c, y) = b, \ y \in \mathbb{R}^{N-1}, \\[2mm] f(0, \xi, y) = f_0(\xi, y), \ \xi \in [0, c], \ y \in \mathbb{R}^{N-1}. \end{cases} \tag{6.6}$$

This is a quasilinear parabolic initial-boundary value problem with smooth nonlinearities, which is shown to have a unique regular (local in time) solution by classical methods of parabolic equations. Indeed it is standard to set it as a fixed point problem in the parabolic Hölder space $C^{1+\alpha/2, 2+\alpha}([0, T] \times ([0, c] \times \mathbb{R}^{N-1}))$, with T small, using again theorem 5.3 as a main tool.

Once a local solution $y \in C^{1+\alpha/2, 2+\alpha}([0, T] \times ([0, c] \times \mathbb{R}^{N-1}))$ exists, it is easy to come back to the original problem (6.2) by the inverse change of coordinates Φ^{-1}.

6.3 Solutions Close to Self-similar Solutions, Travelling Waves etc.

If the initial datum (Ω_0, u_0) is close to the initial datum (Ω, U) of a smooth given solution, the method of section 5 gives existence and uniqueness of a local classical solution without loss of regularity.

First of all, the free boundary problem is transformed into a fixed boundary problem in Ω by the changement of coordinates defined in (5.3), with Ω_0 replaced by Ω. Therefore, the unknown s is now the signed distance from $\partial\Omega$. Then the splitting (5.6) is replaced by

$$\tilde{u}(t,\xi) = U(\xi) + \langle DU(\xi), \Phi(t,\xi)\rangle + w(t,\xi), \quad \xi \in \Omega. \tag{6.7}$$

This gives again $s(t,\xi) = w(t,\xi)$ for $\xi \in \partial\Omega$. The final problem for w has $w_0 = \tilde{u}_0 - U - \langle DU, \Phi(0,\xi)\rangle$ which in general is not zero, but it is small. Fully nonlinear parabolic problems of the type (5.21) may be studied as well with nonzero initial data.

To fix the ideas, we consider the case where (Ω_0, u_0) is close to the initial datum of the self-similar solution $(B(0, r\sqrt{T-t}), f(|x|/\sqrt{T-t}))$ introduced in §3. It is convenient to transform the problem to selfsimilar variables

$$\hat{x} = \frac{x}{(T-t)^{\frac{1}{2}}}, \hat{t} = -\log(T-t), \hat{u}(\hat{x},\hat{t}) = \frac{u(x,t)}{(T-t)^{\frac{1}{2}}}, \hat{\Omega}_{\hat{t}} = \{\hat{x} : x \in \Omega_t\}. \tag{6.8}$$

Omitting the hats, we arrive at

$$u_t = \Delta u - \frac{1}{2}\langle x, Du\rangle + \frac{1}{2}u, \quad x \in \Omega_t, \tag{6.9}$$

with boundary conditions

$$u = 0, \quad \frac{\partial u}{\partial n} = -1 \text{ on } \partial\Omega_t. \tag{6.10}$$

The selfsimilar solution (3.1) is transformed by (6.8) into a stationary solution

$$U(x) = f(|x|), \quad \Omega = B_r = \{x \in \mathbb{R}^n : |x| < r\}, \tag{6.11}$$

of (6.9), (6.10).

Stability of stationary solutions to free boundary problems in bounded domains has been studied in [3]. The change of coordinates $(t,x) \mapsto (t,\xi)$ which transforms the unknown domain Ω_t into the fixed ball Ω is described in sect. 5. Setting $\tilde{u}(t,\xi) = u(t,x)$ and defining w by the splitting (5.7) we arrive at a final equation for w,

$$\begin{cases} w_t = \Delta w - \frac{1}{2}\langle \xi, Dw\rangle + \frac{w}{2} + \phi(w, Dw, D^2w), \quad t \geq 0, \ \xi \in \overline{\Omega}, \\[2mm] \dfrac{\partial w}{\partial n} + \left(\dfrac{N-1}{r} - \dfrac{r}{2}\right)w = \psi(w, Dw), \quad t \geq 0, \ \xi \in \partial\Omega, \\[2mm] w(0,\xi) = w_0(\xi), \quad \xi \in \overline{\Omega}, \end{cases} \tag{6.12}$$

where ϕ and ψ are smooth and quadratic near $w = 0$.

This problem is similar to (5.21). The differences are due to the fact that we linearize around a stationary solution instead of around the initial datum.

It comes out that the differential equation for w is a bit simpler, and the initial datum is not zero.

However, the technique to find a local solution is the same. We give here the proof of a slightly more general theorem, taken from the paper [3]. It gives existence of the solution for arbitrarily long time intervals, provided the initial datum is small enough. This may be seen as continuous dependence of the solution on the initial datum at $w_0 = 0$.

Theorem 6.2. *Let Ω be a bounded open set in $\mathbb{R}^N$ with $C^{2+\alpha}$ boundary, $0 < \alpha < 1$.*

Let $\mathcal{L} = a_{ij}D_{ij} + b_iD_i + c$ be a uniformly elliptic operator with coefficients a_{ij}, b_i, c in $C^\alpha(\overline{\Omega})$ and let $\mathcal{B} = \beta_iD_i + \gamma$ be a nontangential operator with coefficients β_i, γ in $C^{1+\alpha}(\partial\Omega)$.

Let $F : B(0,R) \subset C^2(\overline{\Omega}) \to C(\overline{\Omega})$, $G : B(0,R) \subset C^1(\overline{\Omega}) \to C(\partial\Omega)$ be smooth functions with smooth restrictions to $B(0,R) \subset C^{2+\alpha}(\overline{\Omega})$ and to $B(0,R) \subset C^{2+\alpha}(\overline{\Omega})$ with values in $C^\alpha(\overline{\Omega})$ and in $C^{1+\alpha}(\partial\Omega)$, respectively. Assume moreover

$$F(0) = 0, \ F'(0) = 0; \ \ G(0) = 0, \ G'(0) = 0.$$

Let $w_0 \in C^{2+\alpha}(\overline{\Omega})$ satisfy the compatibility condition

$$\mathcal{B}w_0 = G(w_0(\cdot))(\xi), \ \ \xi \in \partial\Omega. \tag{6.13}$$

Then for every $T > 0$ there are r, $\rho > 0$ such that the problem

$$\begin{cases} w_t = \mathcal{L}w + F(w), \ t \geq 0, \ \xi \in \overline{\Omega}, \\[2mm] \mathcal{B}w = G(w), \ t \geq 0, \ \xi \in \partial\Omega, \\[2mm] w(0,\xi) = w_0(\xi), \ \ \xi \in \overline{\Omega}, \end{cases} \tag{6.14}$$

has a solution $w \in C^{1+\alpha/2,2+\alpha}([0,T] \times \overline{\Omega})$ if $\|w_0\|_{C^{2+\alpha}(\overline{\Omega})} \leq \rho$. Moreover w is the unique solution in $B(0,r) \subset C^{1+\alpha/2,2+\alpha}([0,T] \times \overline{\Omega})$.

Proof — Let $0 < r \leq R$, and set

$$K(r) = \sup\{\|F'(\varphi)\|_{L(C^{2+\alpha}(\overline{\Omega}),C^\alpha(\overline{\Omega}))} : \varphi \in B(0,r) \subset C^{2+\alpha}(\overline{\Omega})\},$$

$$H(r) = \sup\{\|G'(\varphi)\|_{L(C^{2+\alpha}(\overline{\Omega}),C^{1+\alpha}(\partial\Omega))} : \varphi \in B(0,r) \subset C^{2+\alpha}(\overline{\Omega})\}.$$

Since $F'(0) = 0$ and $G'(0) = 0$, $K(r)$ and $H(r)$ go to 0 as $r \to 0$. Let $L > 0$ be such that, for all φ, $\psi \in B(0,r) \subset C^2(\overline{\Omega})$ with small r,

$$\|F'(\varphi) - F'(\psi)\|_{L(C^2(\overline{\Omega}),C(\overline{\Omega}))} \leq L\|\varphi - \psi\|_{C^2(\overline{\Omega})},$$

$$\|G'(\varphi) - G'(\psi)\|_{L(C^1(\overline{\Omega}),C(\partial\Omega))} \leq L\|\varphi - \psi\|_{C^1(\overline{\Omega})}.$$

For every $0 \leq s \leq t \leq T$ and for every $w \in B(0,r) \subset C^{1+\alpha/2,2+\alpha}([0,T] \times \overline{\Omega})$ with r so small that $K(r), H(r) < \infty$, we have

$$\|F(w(\cdot,t))\|_{C^{\alpha}(\overline{\Omega})} \leq K(r)\|w(\cdot,t)\|_{C^{2+\alpha}(\overline{\Omega})}, \quad \|F(w(\cdot,t)) - F(w(\cdot,s))\|_{C(\overline{\Omega})} \leq$$

$$Lr\|w(\cdot,t) - w(\cdot,s)\|_{C^2(\overline{\Omega})} \leq Lr|t-s|^{\alpha/2}\|w\|_{C^{1+\alpha/2,2+\alpha}([0,T]\times\overline{\Omega})},$$

and similarly

$$\|G(w(\cdot,t))\|_{C^{1+\alpha}(\partial\Omega)} \leq H(r)\|w(\cdot,t)\|_{C^{2+\alpha}(\overline{\Omega})} \leq H(r)\|w\|_{C^{1+\alpha/2,2+\alpha}([0,T]\times\overline{\Omega})},$$

$$\|G(w(\cdot,t)) - G(w(\cdot,s))\|_{C(\partial\Omega)} \leq$$

$$Lr\|w(\cdot,t) - w(\cdot,s)\|_{C^1(\overline{\Omega})} \leq Lr|t-s|^{1/2+\alpha/2}\|w\|_{C^{1+\alpha/2,2+\alpha}([0,T]\times\overline{\Omega})}.$$

Therefore, $(\xi,t) \to F(w(\cdot,t))(\xi)$ belongs to $C^{\alpha/2,\alpha}([0,T] \times \overline{\Omega})$, $(\xi,t) \to G(w(\cdot,t))(\xi)$ belongs to $C^{1/2+\alpha/2,1+\alpha}([0,T] \times \partial\Omega)$ and

$$\|F(w)\|_{C^{\alpha,\alpha/2}([0,T]\times\overline{\Omega})} \leq (K(r)+Lr)\|w\|_{C^{1+\alpha/2,2+\alpha}([0,T]\times\overline{\Omega})},$$

$$\|G(w)\|_{C^{1/2+\alpha/2,1+\alpha}([0,T]\times\partial\Omega)} \leq C_{\alpha}(2H(r)+Lr)\|w\|_{C^{1+\alpha/2,2+\alpha}([0,T]\times\overline{\Omega})}.$$

So, if $\|w_0\|_{C^{2+\alpha}(\overline{\Omega})}$ is small enough, we define a nonlinear map

$$\text{dom}\,(\Gamma) := \{w \in B(0,r) \subset C^{1+\alpha/2,2+\alpha}([0,T]\times\overline{\Omega}) : w(\cdot,0) = w_0\}$$
$$\Gamma : \text{dom}\,(\Gamma) \to C^{1+\alpha/2,2+\alpha}([0,T]\times\overline{\Omega}),$$

by $\Gamma w = v$, where v is the solution of

$$\begin{cases} v_t(t,\xi) = \mathcal{L}v + F(w(t,\cdot))(\xi), \ 0 \leq t \leq T, \ \xi \in \overline{\Omega}, \\ \mathcal{B}v = G(w(t,\cdot))(\xi), \ 0 \leq t \leq T, \ \xi \in \partial\Omega, \\ v(0,\xi) = w_0(\xi), \ \xi \in \overline{\Omega}. \end{cases}$$

Actually, thanks to the compatibility condition $\mathcal{B}w_0 = G(w_0)$ and the regularity of $F(w)$ and $G(w)$, the range of Γ is contained in $C^{1+\alpha/2,2+\alpha}([0,T] \times \overline{\Omega})$. Moreover, by theorem 5.3 there is $C = C(T) > 0$, independent of r, such that

$$\|v\|_{C^{1+\alpha/2,2+\alpha}([0,T]\times\overline{\Omega})} \leq$$

$$C(\|w_0\|_{C^{2+\alpha}(\overline{\Omega})} + \|F(w)\|_{C^{\alpha/2,\alpha}([0,T]\times\overline{\Omega})} + \|G(w)\|_{C^{1/2+\alpha/2,1+\alpha}([0,T]\times\partial\Omega)}),$$

so that

$$\|\Gamma(w)\|_{C^{1+\alpha/2,2+\alpha}([0,T]\times\overline{\Omega})} \leq$$

$$C(\|w_0\|_{C^{2+\alpha}(\overline{\Omega})} + (K(r) + Lr + C_{\alpha}(2H(r)+Lr))\|w\|_{C^{1+\alpha/2,2+\alpha}([0,T]\times\overline{\Omega})}).$$

Therefore, if r is so small that

$$C(K(r) + Lr + C_\alpha(2H(r) + Lr)) \leq 1/2, \tag{6.15}$$

and w_0 is so small that

$$\|w_0\|_{C^{2+\alpha}(\overline{\Omega})} \leq Cr/2,$$

Γ maps the ball $B(0,r)$ into itself. Let us check that Γ is a $1/2$-contraction. Let w_1, $w_2 \in B(0,r)$, $w_i(\cdot, 0) = w_0$. Writing $w_i(\cdot, t) = w_i(t)$, $i = 1, 2$, we have

$$\|\Gamma w_1 - \Gamma w_2\|_{C^{1+\alpha/2, 2+\alpha}([0,T] \times \overline{\Omega})} \leq$$

$$\leq C(\|F(w_1) - F(w_2)\|_{C^{\alpha/2, \alpha}([0,T] \times \overline{\Omega})} +$$

$$+ \|G(w_1) - G(w_2)\|_{C^{1/2+\alpha/2, 1+\alpha}([0,T] \times \partial\Omega)}),$$

and, arguing as above, for $0 \leq t \leq T$,

$$\|F(w_1(t)) - F(w_2(t))\|_{C^\alpha(\overline{\Omega})} \leq$$

$$K(r)\|w_1(t) - w_2(t)\|_{C^{2+\alpha}(\overline{\Omega})} \leq K(r)\|w_1 - w_2\|_{C^{1+\alpha/2, 2+\alpha}([0,T] \times \overline{\Omega})},$$

$$\|G(w_1(t)) - G(w_2(t))\|_{C^{1+\alpha}(\partial\Omega)} \leq$$

$$H(r)\|w_1(t) - w_2(t)\|_{C^{2+\alpha}(\overline{\Omega})} \leq H(r)\|w_1 - w_2\|_{C^{1+\alpha/2, 2+\alpha}([0,T] \times \overline{\Omega})},$$

while for $0 \leq s \leq t \leq T$

$$\|F(w_1(t)) - F(w_2(t)) - F(w_1(s)) - F(w_2(s))\|_{C(\overline{\Omega})} =$$

$$= \left\| \int_0^1 F'(\sigma w_1(t) + (1-\sigma)w_2(t)(w_1(\cdot, t) - w_2(t)) - \right.$$

$$\left. - F'(\sigma w_1(s) + (1-\sigma)w_2(s))(w_1(s) - w_2(s))d\sigma \right\|_{C(\overline{\Omega})} \leq$$

$$\leq \int_0^1 \|(F'(\sigma w_1(t) + (1-\sigma)w_2(t) - F'(\sigma w_1(s) +$$

$$+ (1-\sigma)w_2(s))(w_1(t) - w_2(t))d\sigma\|_{C(\overline{\Omega})} +$$

$$+ \int_0^1 \|F'(\sigma w_1(s))(w_1(t) - w_2(t) - w_1(s) + w_2(s))\|_{C(\overline{\Omega})} \leq$$

$$\leq \frac{L}{2} \left(\|w_1(t) - w_1(s)\|_{C^2(\overline{\Omega})} + \right.$$

$$+ \|w_2(t) - w_2(s)\|_{C^2(\overline{\Omega})}) \|w_1(t) - w_2(t)\|_{C^2(\overline{\Omega})} +$$

$$+ Lr \|w_1(t) - w_2(t) - w_1(s) + w_2(s))\|_{C^2(\overline{\Omega})} \leq$$

$$\leq 2Lr(t-s)^{\alpha/2} \|w_1 - w_2\|_{C^{1+\alpha/2, 2+\alpha}([0,T] \times \overline{\Omega})},$$

and similarly

$$\|G(w_1(\cdot, t)) - G(w_2(\cdot, t)) - G(w_1(\cdot, s)) - G(w_2(\cdot, s))\|_{C(\partial\Omega)}$$

$$\leq 2Lr(t-s)^{1/2+\alpha/2} \|w_1 - w_2\|_{C^{1+\alpha/2, 2+\alpha}([0,T] \times \overline{\Omega})}.$$

Therefore,

$$\|\Gamma w_1 - \Gamma w_2\|_{C^{1+\alpha/2, 2+\alpha}([0,T] \times \overline{\Omega})}$$

$$\leq C(K(r) + Lr + C_\alpha(2H(r) + Lr)) \|w_1 - w_2\|_{C^{1+\alpha/2, 2+\alpha}([0,T] \times \overline{\Omega})}$$

$$\leq \frac{1}{2} \|w_1 - w_2\|_{C^{1+\alpha/2, 2+\alpha}([0,T] \times \overline{\Omega})},$$

the last inequality being a consequence of (6.15). The statement follows. □

Local existence and uniqueness for initial data close to travelling wave solutions may be studied similarly. See [4] for the free boundary heat equation, [7] for a two phase Burgers equation, [18] for a one phase - two phase system from combustion theory.

References

1. D. ANDREUCCI, R. GIANNI: *Classical solutions to a multidimensional free boundary problem arising in combustion theory*, Comm. P.D.E's **19** (1994), 803-826.
2. O. BACONNEAU, A. LUNARDI: *Smooth solutions to a class of free boundary parabolic problems*, preprint Dipart. Mat. Univ. Parma **253** (2001) (submitted).
3. C.M. BRAUNER, J. HULSHOF, A. LUNARDI: *A general approach to stability in free boundary problems*, J. Diff. Eqns. **164** (2000), 16-48.
4. C.M. BRAUNER, J. HULSHOF, A. LUNARDI: *A critical case of stability in a free boundary problem*, J. Evol. Eqns. **1** (2001), 85-113.
5. C.M. BRAUNER, A. LUNARDI: *Instabilities in a combustion model with free boundary in R^2*, Arch. Rat. Mech. Anal. **154** (2000), 157-182.
6. C.M. BRAUNER, A. LUNARDI, CL. SCHMIDT-LAINÉ: *Une nouvelle formulation de modèles de fronts en problèmes complètement non linéaires*, C.R.A.S. Paris **311** (1990), Série I, 597-602.

7. C.M. BRAUNER, A. LUNARDI, CL. SCHMIDT-LAINÉ: *Stability analysis in a mul-tidimensional interface problem*, Nonlinear Analysis T.M.A. **44** (2001), 263-280.

8. J.D. BUCKMASTER, G.S.S. LUDFORD: *Theory of Laminar Flames*, Cambridge University Press, Cambridge, 1982.

9. L.A. CAFFARELLI, J.L. VAZQUEZ: *A free boundary problem for the heat equation arising in flame propagation*, T.A.M.S. **347** (1995), 411-441.

10. W. DESCH, W. SCHAPPACHER: *Some Perturbation Results for Analytic Semi-groups*, Math. Ann. **281** (1988), 157-162.

11. J. ESCHER, G. SIMONETT: *Maximal regularity for a free boundary problem*, NoDEA **2** (1995), 463-510.

12. J. ESCHER, G. SIMONETT: *Classical solutions of multidimensional Hele-Shaw models*, SIAM J. Math. Anal. **28** (1997), 1028-1047.

13. V.A. GALAKTIONOV, J. HULSHOF, J.L. VAZQUEZ: *Extinction and focusing be-haviour of spherical and annular flames described by a free boundary problem*, J. Math. Pures Appl. **76** (1997), 563-608.

14. D. HILHORST, J. HULSHOF: *An elliptic-parabolic problem in combustion theory: convergence to travelling waves*, Nonlinear Analysis TMA **17** (1991), 519-546.

15. D. HILHORST, J. HULSHOF: *A free boundary focusing problem*, Proc. AMS **121**, 1193-1202 (1994).

16. O.A. LADYZHENSKAJA, V.A. SOLONNIKOV, N.N. URAL'CEVA: *Linear and quasi-linear equations of parabolic type*, Nauka, Moskow 1967 (Russian). English transl.: Transl. Math. Monographs, AMS, Providence, 1968.

17. C. LEDERMAN, J.L. VÁZQUEZ, N. WOLANSKI: *Uniqueness of solution to a free boundary problem from combustion*, TAMS **353** (2001), 655-692.

18. L. LORENZI: *A free boundary problem stemmed from Combustion Theory. Part I: Existence, uniqueness and regularity results*, preprint Dipart. Mat. Univ. Parma **245** (2000), to appear in J. Math. Anal. Appl.

19. L. LORENZI: *A free boundary problem stemmed from Combustion Theory. Part II: Stability, instability and bifurcation results.*, preprint Dipart. Mat. Univ. Parma **246** (2000), to appear in J. Math. Anal. Appl.

20. L. LORENZI, A. LUNARDI: *Stability in a two-dimensional free boundary com-bustion model*, preprint Dipart. Mat. Univ. Parma **271** (2001), to appear in Nonlinear Analysis TMA.

21. A. LUNARDI: *An introduction to geometric theory of fully nonlinear parabolic equations*, in "Qualitative aspects and applications of nonlinear evolution equa-tions", T.T. Li, P. de Mottoni Eds., World Scientific, Singapore (1991),107-131.

22. A. LUNARDI: *Analytic semigroup and optimal regularity in parabolic problems*, Birkhäuser Verlag, Basel, 1995.

23. A.M. MEIRMANOV: *On a problem with free boundary for parabolic equations*, Math. USSR Sbornik **43** (1982), 473-484.

24. M. PRIMICERIO: *Diffusion problems with a free boundary*, Boll. U.M.I. A (5) **18** (1981), 11-68.

25. E. SINESTRARI: *On the abstract Cauchy problem in spaces of continuous func-tions*, J. Math. Anal. Appl. **107** (1985), 16-66.

26. J. L. VAZQUEZ: *The Free Boundary Problem for the Heat Equation with fixed Gradient Condition*, Proc. Int. Conf. "Free Boundary Problems, Theory and Applications", Zakopane, Poland, Pitman Research Notes in Mathematics 363, pp. 277-302, 1995.

27. T.D. VENTSEL: *A free-boundary value problem for the heat equation*, Dokl. Akad. Nauk SSSR **131** (1960), 1000-1003; English transl. Soviet Math. Dokl. **1** (1960).

Asymptotic Behaviour of Parabolic Nonautonomous Evolution Equations

Roland Schnaubelt

FB Mathematik und Informatik, Martin-Luther-Universität, D-06099 Halle, Germany.
`schnaubelt@mathematik.uni-halle.de`

1 Introduction

The long term behaviour of autonomous linear Cauchy problems on a Banach space X has been studied systematically and quite successfully by means of spectral theory and transform methods, see e.g. [8], [21], [50]. These techniques fail (almost) completely in the nonautonomous case if one tries to generalize them directly, as we indicate in Example 3.2. In fact, already nonautonomous ordinary differential equations lead to considerable difficulties and the available results are mostly restricted to perturbation type arguments, cf. [15], [16]. For infinite dimensional X (e.g., in the case of partial differential equations), the situation is much worse and we are far from a comprehensive theory. However, if one takes the ODE case as a standard, then we have now achieved quite satisfying results for the exponential dichotomy of parabolic linear homogeneous equations

$$u'(t) = A(t)u(t), \quad t > s, \qquad u(s) = x. \tag{1.1}$$

Roughly speaking, the *evolution family* (or, propagator) $U(t, s)$, $t \geq s$, solving (1.1) has an *exponential dichotomy* if there is a U–invariant, time varying splitting $X = X_0(s) \oplus X_1(s)$ such that the solution $u(t) = U(t, s)x$ decays exponentially as $t \to \infty$ (as $t \to -\infty$) if $x \in X_0(s)$ (if $x \in X_1(s)$), see [12], [27], [42], [54], and Section 3.1. Clearly, exponential dichotomy is a fundamental qualitative property of the Cauchy problem (1.1). Its importance relies in particular on the robustness, i.e., exponential dichotomy persists under 'small' (non–)linear perturbations. We will establish robustness and several criteria for exponential dichotomy of (1.1) in terms of the operators $A(t)$ in our core Section 4. If (1.1) has an exponential dichotomy, then one can study qualitative properties of inhomogeneous and nonlinear equations related to (1.1) by fixed point methods in a similar spirit as for ordinary differential equations, see [27], [42], [54], and Sections 5 and 6.

Thus our approach follows to some extend the influential monograph [27] by D. Henry. But in the linear case our setting is considerably more general:

Henry treats operators of the form $A(t) = A + B(t)$, where A generates an analytic C_0–semigroup and $B(\cdot)$ is a time dependent lower order perturbation. We deal with parabolic problems in the framework of the existence theory established by P. Acquistapace and B. Terreni, which can be considered as the most general setting available by now, see [2] and also [1], [3], [7], [63], [64], [65]. In terms of partial differential equations, this means that the coefficients of the highest order terms and of the boundary conditions may depend on time. In addition, the domains $D(A(t))$ need not to be dense (in contrast to [27]) so that Dirichlet type boundary conditions in a sup norm context are also covered. The lack of density poses several difficulties, e.g., the evolution family $U(t, s)$ is not strongly continuous at $t = s$.

In Section 2 we recall the results by P. Acquistapace and B. Terreni concentrating on those facts needed later. We can not describe the rather complicated proofs, but we show a few additional regularity and approximation results which are essential for our main theorems. We also give a brief introduction to interpolation spaces in the present context. The existence results are applied to parabolic partial differential equations of second order in divergence and non–divergence form formulated on $X = L^p(\Omega)$ or $X = C(\overline{\Omega})$.

The main obstacle to verify that a given problem has an exponential dichotomy is the construction of the required splitting of X. There are several approaches to obtain the corresponding projections, typically based on abstract characterizations of exponential dichotomy of an evolution family $U(t, s)$, see [12], [27], [35], [54], [58]. We prefer a characterization employing the spectra of the associated *evolution semigroup*

$$(T_U(t)f)(s) = U(s, s - t)f(s - t), \quad t \geq 0, \ s \in \mathbb{R}, \ f \in E,$$

acting on $E = C_0(\mathbb{R}, X)$. This technique works in a very general setting, is quite flexible, and allows for rather transparent proofs. We establish the main features of the spectral theory of evolution semigroups in Section 3.2 following our exposition in [21, §VI.9] (see the comprehensive treatise [12] for more general results). However, since we work with non–densely defined $A(t)$, we have to extend these results to certain evolution families $U(t, s)$ not being strongly continuous at $t = s$, see Section 3.3.

In a series of papers, [55], [56], [57], [59], we established the exponential dichotomy of parabolic evolution equations in various situations. There we generalized theorems from [27] dealing with a special class of parabolic problems, but we still assumed that the domains are dense. This additional assumption has been removed in Section 4. Here we show robustness of dichotomy under small perturbations of the same order as $A(t)$. We also obtain natural conditions for the exponential dichotomy of asymptotically autonomous problems and of equations with slowly oscillating coefficients. The lack of density of $D(A(t))$ has forced us to change the arguments from our previous works at several points, besides other simplifications and improvements. The results are applied to parabolic partial differential equations where we show how our hypotheses translate into conditions on the coefficients of the given equation.

In Sections 5 and 6 the main results of Section 4 are used to study in-homogeneous and quasilinear problems, respectively. Here we establish the existence of convergent solutions to asymptotically autonomous equations (extending an old result of Tanabe, see [62], on inhomogeneous problems) and of almost periodic solutions to almost periodic inhomogeneous equations.

Let us explain our principal arguments in a simplified setting. Assume that the domains $D(A(t))$, $t \in \mathbb{R}$, are dense and have uniformly equivalent graph norms and set $Y = D(A(0))$. Let the operators $A(t)$ generate analytic semigroups $(e^{\tau A(t)})_{\tau \geq 0}$ on X having uniform type and assume that $t \mapsto A(t) \in \mathcal{L}(Y, X)$ is globally Hölder continuous. Then there exists an evolution family $U(t, s)$ solving (1.1) which satisfies standard parabolic regularity (see Section 2). Let A generate the analytic C_0–semigroup $S(\cdot)$ on X and set $q := \sup_{t \in \mathbb{R}} \|A(t) - A\|_{\mathcal{L}(Y,X)}$. We suppose that the spectrum of A does not intersect the imaginary axis. Then it is well known, see e.g. [21], [42], that $S(\cdot)$ has an exponential dichotomy. We want to show that $U(\cdot, \cdot)$ also has an exponential dichotomy, if q is small enough. In Lemma 3.3 and Theorem 3.6 we will see that the exponential dichotomy of an evolution family $V(\cdot, \cdot)$ is equivalent to the invertibility of $I - T_V(1)$, where $T_V(\cdot)$ is the corresponding evolution semigroup. Thus we have to check that $\|T_U(1) - T_S(1)\| = \sup_{s \in \mathbb{R}} \|U(s, s - 1) - S(1)\|$ is sufficiently small. To this end, we first observe that $\partial_s U(t, s)x = -U(t, s)A(s)x$ for $x \in Y$. Therefore,

$$U(t + 1, t) - e^{A(t)} = - \int_t^{t+1} \frac{\partial}{\partial s} U(t + 1, s) e^{(s-t)A(t)} \, ds$$
$$= \int_t^{t+1} U(t + 1, s)[A(s) - A(t)] e^{(s-t)A(t)} \, d\tau.$$

Using standard regularity properties of analytic semigroups, we thus obtain

$$\|U(t + 1, t) - e^{A(t)}\| \leq c \int_t^{t+1} (s - t)^{\alpha/2} \, q^{1/2} \, (s - t)^{-1} \, ds = c' \, q^{1/2}.$$

For a suitable path $\Gamma \subset \mathbb{C}$, usual semigroup theory further implies

$$e^{A(t)} - S(1) = \int_\Gamma e^\lambda R(\lambda, A(t))[A(t) - A] R(\lambda, A) \, d\lambda,$$

so that $\|e^{A(t)} - S(1)\| \leq cq$. Putting everything together, we deduce that $U(\cdot, \cdot)$ has an exponential dichotomy provided that q is small enough. If $A(t)$ and A are given by elliptic partial differential operators, this smallness condition holds if the coefficients are close to each other, cf. Example 4.2.

Assume now that $U(\cdot, \cdot)$ has an exponential dichotomy. Then there exists *Green's function* $\Gamma(t, s)$, $t, s \in \mathbb{R}$, (see Definition 3.1) which satisfies $\|\Gamma(t, s)\| \leq N e^{-\delta|t-s|}$ for some constants $N, \delta > 0$. Let $f \in L^\infty(\mathbb{R}, X)$ be Hölder continuous. One can write the unique bounded solution of the inhomogeneous problem $u'(t) = A(t)u(t) + f(t)$, $t \in \mathbb{R}$, as

$$u(t) = \int_{\mathbb{R}} \Gamma(t,s) f(s)\, ds = \int_{\mathbb{R}} \Gamma(t, t+s) f(t+s)\, ds.$$

Consequently, if $f(t) \to 0$ as $t \to \infty$, then also $u(t) \to 0$ due to the theorem of dominated convergence. Based on such results, one can then set up fixed point arguments to study the asymptotic behaviour of quasilinear equations, see Section 6.

In fact, below we will treat far more complicated situations and obtain more detailed information. Accordingly our arguments will be much more involved, see e.g. Proposition 2.6 and Theorems 4.1 and 5.9. Nevertheless the above sketch already describes important features of our approach.

It is supposed that the reader is familiar with basic spectral theory of semigroups and with regularity properties of analytic semigroups. Our arguments make use of standard results and methods of operator theory and the functional analytic treatment of (parabolic) partial differential equations (in the spirit of Henry's book, say). We strove for a rather self–contained and systematic presentation. Of course, these lecture notes do not exhaust the range of possible results and applications, but they should give a concise picture of the field. Thus we hope that our text can serve as an introduction to the subject.

NOTATION. We denote by $D(A)$, $A\sigma(A)$, $\sigma(A)$, $\rho(A)$, and A^* the domain, (approximate point) spectrum, resolvent set, and adjoint of a closed (densely defined) linear operator A. We set $R(\lambda, A) := (\lambda - A)^{-1}$. X^* is the dual of the Banach space X, $\mathcal{L}(X, Y)$ is the space of bounded linear operators, $\mathcal{L}(X) := \mathcal{L}(X, X)$, and a subscript 's' indicates that $\mathcal{L}(X, Y)$ is endowed with the strong operator topology. Spaces of functions $f : U \to X$ (with $U \subseteq \mathbb{R}^n$) are designated as usual and endowed with their standard norms, where the subscript 'c' means 'compact support', $C_0(U, X)$ is the closure of $C_c(U, X)$ in the space of bounded continuous functions $C_b(U, X)$, $C_b^\alpha(U, X)$ is the space of bounded, globally Hölder continuous functions, and BUC means 'bounded and uniformly continuous'. Further, $\mathbb{T} = \{\lambda \in \mathbb{C} : |\lambda| = 1\}$. By $c = c(\alpha, \beta, \cdots)$ we denote a generic constant only depending on the constants in the hypotheses involved and on the quantities $\alpha, \beta, \cdots$.

2 Parabolic Evolution Equations

In this section we review several results on the existence and regularity of solutions to the parabolic evolution equation

$$\frac{d}{dt} u(t) = A(t)u(t) + f(t), \quad t > s, \qquad u(s) = x, \qquad (2.1)$$

for linear operators $A(t)$, $t \in J$, on a Banach space X, where $J \in \{[a, \infty), \mathbb{R}\}$ is the underlying time interval, $s \in J$, $x \in X$, and $f \in C(J, X)$. We use the following notions of solutions.

Definition 2.1. A classical solution *of (2.1) is a function* $u \in C([s, \infty), X) \cap C^1((s, \infty), X)$ *such that* $u(t) \in D(A(t))$ *for* $t > s$ *and (2.1) holds. If addition-ally* $u \in C^1([s, \infty), X)$, $x \in D(A(s))$, *and* $A(t)u(t) \to A(s)x$ *as* $t \to s$, *then* u *is called a* strict solution.

There is no unified theory for nonautonomous linear evolution equations. We surveyed various well–posedness theorems and the corresponding meth-ods in [58], where one can find further references. Many of the fundamental contributions are well documented in the monographs [7], [42], [47], [62], [63]. In these lecture notes we restrict ourselves to a class of parabolic problems[1] introduced by P. Acquistapace and B. Terreni in 1987, [2]; see also [1], [3], [7], [63], [64], [65].

(AT1) $A(t)$, $t \in J$, are linear operators on a Banach space X and there are constants $K \geq 0$, $w \in \mathbb{R}$, and $\phi \in (\pi/2, \pi)$ such that $\lambda \in \rho(A(t))$ and $\|R(\lambda, A(t))\| \leq \frac{K}{1+|\lambda-w|}$ for $\lambda \in \Sigma(\phi, w)$ and $t \in J$.

(AT2) There are constants $L \geq 0$ and $\mu, \nu \in (0, 1]$ with $\mu + \nu > 1$ such that

$$|\lambda|^\nu \|A_w(t)R(\lambda, A_w(t))(A_w(t)^{-1} - A_w(s)^{-1})\| \leq L |t - s|^\mu$$

for $A_w(t) := A(t) - w$, $t, s \in J$, and $|\arg \lambda| \leq \phi$.

Here we set $\Sigma(\phi, w) = \{w\} \cup \{\lambda \in \mathbb{C} \setminus \{w\} : |\arg(\lambda - w)| \leq \phi\}$. Observe that the domains are not required to be dense in X. If (AT1) and (AT2) hold, then we say that (AT) is satisfied. Operators A fulfilling (AT1) are called *sectorial (of type (ϕ, K, w))* (we always require that $\phi > \pi/2$). The equation (2.1) is called 'parabolic' because of the sectoriality of the operators $A(t)$.

Roughly speaking, in (AT2) the required Hölder exponent of the resolvents depends on the change of the domains $D(A(t))$. In the extreme case that $D(A(t))$, $t \in J$, equal a fixed Banach space X_1 with uniformly equivalent graph norms, (AT2) with $\nu = 1$ follows from the Hölder continuity of $A(t) : J \to \mathcal{L}(X_1, X)$ with any exponent $\mu > 0$. More variable domains than in (AT) are allowed in a different approach going back to work by T. Kato and H.Tanabe, see [62, §5.3], where it is assumed that $A(\cdot)^{-1} \in C^{1+\alpha}(J, \mathcal{L}(X))$ (besides another hypothesis). The Kato–Tanabe assumptions are logically independent of (AT), but the Acquistapace–Terreni conditions seem to be favourable since only a Hölder estimate is needed, see [2, §7] for a discussion of these matters and further literature.

It is known that a sectorial operator A generates the analytic semigroup

$$e^{tA} = \frac{1}{2\pi i} \int_\Gamma e^{\lambda t} R(\lambda, A) \, d\lambda, \quad t > 0, \qquad e^{0A} = I, \tag{2.2}$$

and that its fractional powers are given by

---

[1] In fact, Acquistapace and Terreni used a somewhat more general version of con-dition (AT2).

$$(w - A)^{-\alpha} = \frac{1}{2\pi i} \int_\Gamma (w - \lambda)^{-\alpha} R(\lambda, A)\, d\lambda, \quad \alpha > 0, \tag{2.3}$$

where we may choose the path $\Gamma = \{\lambda : \arg(\lambda - w) = \pm\phi\}$ oriented counterclockwise, see [7], [21], [42], [47], [62]. Throughout we will make extensive use of standard regularity properties of analytic semigroups, often without mentioning it explicitly. Recall that $e^{tA}X \subseteq D(A)$ for $t > 0$, $t \mapsto e^{tA}$ is analytic on $(0, \infty)$ with derivative Ae^{tA}, $e^{tA}x \to x$ as $t \to 0$ if and only if $x \in \overline{D(A)} =: X_0^A$.

Given a sectorial operator A and $0 < \alpha < 1$, we define the Banach spaces

$$X_{\alpha,\infty}^A := \{x \in X : \|x\|_\alpha^A < \infty\} \quad \text{and} \quad X_\alpha^A := \overline{D(A)}^{\|\cdot\|_\alpha^A} \tag{2.4}$$

$$\text{with the norm} \quad \|x\|_\alpha^A := \sup\{\|r^\alpha AR(r, A)x\| : r > w\}.$$

These spaces coincide with the *real interpolation spaces* of exponent α and coefficient ∞ introduced by J.L. Lions and J. Peetre and with the *continuous interpolation spaces* due to G. Da Prato and P. Grisvard, respectively. We further set $\|x\|_1^A := \|(w - A)x\|$ and $\|x\|_0^A := \|x\|$. By X_1^A we denote the domain of the part A_0 of A in X_0^A. The spaces X_1^A and $D(A)$ are endowed with the norm $\|x\|_1^A$. For convenience, we also write $X_{0,\infty}^A := X$ and $X_{1,\infty}^A := D(A)$. One can verify the continuous embeddings

$$X_1^A \subseteq D(A) \hookrightarrow X_\beta^A \subseteq X_{\beta,\infty}^A \hookrightarrow D((w - A)^\alpha) \hookrightarrow X_\alpha^A \hookrightarrow X_0^A \subseteq X \tag{2.5}$$

for $0 < \alpha < \beta < 1$, where '$\subseteq$' means 'being a closed subspace' and the domain of the fractional power is equipped with the norm $\|(w - A)^\alpha x\|$. The norms of these embeddings depend only on α, β, and the type of A. Moreover, X_A^1 is dense in $D((w-A)^\alpha)$ and X_α^A for $\alpha \in [0, 1]$. The part A_α of A in X_α^A, $\alpha \in [0, 1]$, generates an analytic strongly continuous semigroup being the restriction of e^{tA} (we mostly use the same symbol). Moreover,

$$\|A^k e^{tA}x\|_\beta^A \leq C t^{\alpha-k-\beta} \|x\|_\alpha^A \tag{2.6}$$

for $k = 0, 1, 0 \leq \alpha \leq \beta \leq 1, 0 < t \leq 1, x \in X_{\alpha,\infty}^A$, and a constant C depending only on the type of A. We also need the moment inequality

$$\|(w - A)^\beta x\| \leq c(\alpha, \beta, \gamma) \|(w - A)^\alpha x\|^{\frac{\gamma-\beta}{\gamma-\alpha}} \|(w - A)^\gamma x\|^{\frac{\beta-\alpha}{\gamma-\alpha}}, \tag{2.7}$$

where $\alpha < \beta < \gamma$, $x \in D((w - A)^\gamma)$, and the constant depends only on the exponents and the type of A. These facts follow from, e.g., [21, §II.5] or [42, §1, 2.2], where the interpolation theory of semigroups is developed in detail and further references can be found. We now assume that (AT) holds. We set

$$X_{\alpha,\infty}^t := X_{\alpha,\infty}^{A(t)}, \quad X_\alpha^t := X_\alpha^{A(t)}, \quad \text{and} \quad \|\cdot\|_\alpha^t := \|\cdot\|_\alpha^{A(t)}$$

for $\alpha \in [0, 1]$ and $t \in J$. Observing that

$$R(\lambda, A(t)) - R(\lambda, A(s))$$
$$= A_w(t)R(\lambda, A(t))\left(A_w(s)^{-1} - A_w(t)^{-1}\right)A_w(s)R(\lambda, A(s)) \qquad (2.8)$$

for $\lambda \in \rho(A(t)) \cap \rho(A(s))$, one easily deduces from (2.2) and (2.3) that

$$\|e^{\tau A(t)} - e^{\tau A(s)}\| \le c(t_0)\,\tau^{\nu-1}\,|t - s|^\mu \qquad (2.9)$$
$$\|(w - A(t))^{\alpha-1} - (w - A(s))^{\alpha-1}\| \le c(\alpha)\,|t - s|^\mu \qquad (2.10)$$

for $t, s \in J$, $0 < \tau \le t_0$, $0 \le \alpha < \nu$, and constants depending on the constants in (AT). Similarly,

$$\|(w - A(t))^{\alpha-1} - (w - A(s))^{\alpha-1}\|$$
$$\le \frac{1}{2\pi}\int_\Gamma \frac{\|R(\lambda, A(t)) - R(\lambda, A(s))\|^{1-\kappa}}{|w - \lambda|^{1-\alpha}}\,\|R(\lambda, A(t)) - R(\lambda, A(s))\|^\kappa\,|d\lambda|$$
$$\le c(\kappa)\,|t - s|^{\kappa\mu} \qquad (2.11)$$

for $t, s \in J$, $\alpha \in [\nu, 1)$, and $0 < \kappa < \frac{1-\alpha}{1-\nu}$.

Let E be a space of functions $f : J \to X$ and $B(t)$, $t \in J$, be linear operators on X with domains $D(B(t))$. Then the multiplication operator $B(\cdot)$ on E is defined by

$$(B(\cdot)f)(t) := B(t)f(t), \quad t \in J,$$
$$D(B(\cdot)) := \{f \in E : f(t) \in D(B(t)) \text{ for } t \in J,\ B(\cdot)f \in E\}, \qquad (2.12)$$

As stated in the next theorem, the solution of the homogeneous problem (2.1) with $f = 0$ is given by an *evolution family*, that is, by bounded linear operators $U(t, s)$, $t \ge s$, $s \in J$, satisfying

$$U(t, s) = U(t, r)U(r, s) \quad \text{and} \quad U(s, s) = I$$

for $t \ge r \ge s$ and $s \in J$. The evolution family is called *strongly continuous* if the mapping $(t, s) \mapsto U(t, s)$ is strongly continuous on the set $D_J := \{t, s \in J : t \ge s\}$. We say that $U(\cdot, \cdot)$ is *strongly continuous for $t > s$* if this map is strongly continuous on $D_J^0 := \{t, s \in J : t > s\}$.

Theorem 2.2. *Assume that (AT) holds. Then there is an evolution family $U(\cdot, \cdot)$ on X with time interval J such that $D_J^0 \ni (t, s) \mapsto U(t, s) \in \mathcal{L}(X)$ is continuous, $U(t, s)X \subseteq D(A(t))$ for $t > s$, and*

$$\|U(t, s)x\|_\alpha^t \le C\,(t - s)^{\beta-\alpha}\,\|x\|_\beta^s \qquad (2.13)$$
$$\|U(t, s) - e^{(t-s)A(s)}\| \le C\,(t - s)^{\mu+\nu-1}, \qquad (2.14)$$
$$\|U(t, s)(w - A(s))^\theta y\| \le C\,(\mu - \theta)^{-1}\,(t - s)^{-\theta}\,\|y\|, \qquad (2.15)$$

for $0 < t - s \le t_0$, $s \in J$, $0 \le \beta \le \alpha \le 1$, $x \in X_{\beta,\infty}^s$, $0 \le \theta < \mu$, $y \in D((w - A(s))^\theta)$, and a constant C only depending on the constants in

(AT) and t_0. Further, $U(\cdot, s) \in C^1((s, \infty), \mathcal{L}(X))$, $\partial_t U(t, s) = A(t)U(t, s)$, and $\partial_s^+ U(t, s)x = -U(t, s)A(s)x$ for $x \in X_s^1$ and $t > s$. We have $A(t)^k U(t, s)x \to A(s)^k x$ as $t \to s$ if $x \in X_k^s$ and $k = 0, 1$. Finally, the function

$$D_J \ni (t, s) \mapsto (w - A(t))^\alpha U(t, s)(w - A(s))^{-\alpha} f(s) \in X$$

is continuous for $\alpha \in [0, 1]$ and $f \in \overline{D(A(\cdot))}$ (here $A(\cdot)$ is considered as an operator in $E = C_b(J, X)$ endowed with the sup–norm).

In particular, $u = U(\cdot, s)x$ is a classical solution of (2.1) with $f = 0$ if $x \in X_0^s$ and a strict one if $x \in X_1^s$. The evolution family $U(\cdot, \cdot)$ is strongly continuous if all domains $D(A(t))$ are dense in X. Estimate (2.13) implies that

$$\|U(t', s)x - U(t, s)x\| \le \int_t^{t'} \|A(\tau)U(\tau, s)x\| \, d\tau \le c \, |t' - t|^\alpha \, \|x\|_\alpha^s \quad (2.16)$$

for $t' \ge t \ge s$ and $\alpha \in (0, 1]$.

Most assertions of the above theorem are established in [1, Thm.2.3]. There the evolution family $U(\cdot, \cdot)$ is defined by the formula

$$U(t, s) = e^{(t-s)A(s)} + \int_s^t Z(r, s) \, dr, \qquad t \ge s, \quad (2.17)$$

for certain operators $Z(r, s)$ with $\|Z(r, s)\| \le c \, (r - s)^{\mu+\nu-2}$, see equations (2.6) and (2.7) and Lemma 2.2 in [1]. These results imply in particular (2.14). The estimate (2.13) for $\alpha = \beta \in \{0, 1\}$ is proved in [1, Thm.2.3], and the other cases follow by interpolation and reiteration (see e.g. [42, Chap.2]). It can be seen by a (tedious) inspection of the proofs given in [1] that the constant C only depends on the constants in (AT), and not on s; see also [23]. The estimate (2.15) is proved in [65, Thm.2.3]. There it is further shown that

$$\|(w - A(t))^\alpha U(t, s)(w - A(s))^{-\alpha} - e^{(t-s)A(s)}\| \le c(t_0) \, (t - s)^{\mu+\nu-1} \quad (2.18)$$

for $\alpha \in (0, 1]$, whereas for $\alpha = 0$ the inequality (2.18) coincides with (2.14). Moreover, $\|e^{tA(s)}x - x\| \le ct\|A(s)x\|$ for $x \in D(A(s))$. This implies the final continuity assertion in Theorem 2.2 as $(t, s) \to (r, r)$. The other case easily follows from (2.11) and the norm continuity of $(t, s) \mapsto A(t)U(t, s) = A(t)U(t, \tau)U(\tau, s)$ for $t > s$. Clearly, (2.18) yields

$$\|(w - A(t))^\alpha U(t, s)(w - A(s))^{-\alpha}\| \le c(t_0), \quad 0 \le t - s \le t_0, \ 0 \le \alpha \le 1. \quad (2.19)$$

The representation formula (2.17) for $U(t, s)$ is rather implicit. H. Amann and A. Yagi developed different approaches to Theorem 2.2 in [7, Chap.IV], [64], [65], working in slightly different settings. These authors construct $U(t, s)$ as solutions of certain integral equations. Yagi's method is also employed in the proof of Proposition 2.4 below. It uses the Yosida approximations $A_n(t) = nA(t)R(n, A(t))$, $n > w$, which also satisfy (AT) as stated in the next lemma. Its elementary (but tedious) proof is omitted, cf. [2, Lem.4.2] or [64, Prop.2.1].

Lemma 2.3. *Assume that (AT) holds. Fix $w' > w$ and $\phi' \in (\frac{\pi}{2}, \phi)$. Then there are constants $\overline{n} > w$, $L' \geq L$, and $K' \geq K$ (only depending on the constants in (AT), w', and ϕ') such that the operators $A_n(t)$, $t \in J$, satisfy (AT) with constants $K', \phi', w', L', \mu, \nu$ for all $n \geq \overline{n}$. Moreover, for $n \geq \overline{n}$ and $\lambda \in \Sigma(\phi', w')$, we have*

$$R(\lambda, A_n(t)) = \tfrac{1}{\lambda+n}(n - A(t))R(\tfrac{\lambda n}{\lambda+n}, A(t)) = \tfrac{n^2}{(\lambda+n)^2}R(\tfrac{\lambda n}{\lambda+n}, A(t)) + \tfrac{1}{\lambda+n}$$

$$(2.20)$$

$$R(\lambda, A_n(t)) - R(\lambda, A(t)) = \tfrac{1}{\lambda+n}A(t)R(\tfrac{\lambda n}{\lambda+n}, A(t))A(t)R(\lambda, A(t)). \qquad (2.21)$$

Thus the evolution families $U_n(\cdot, \cdot)$ generated by $A_n(\cdot)$ satisfy the same estimates as $U(\cdot, \cdot)$ with constants independent of $n \geq \overline{n}$. This fact is used later on without further notice.

The following refinements of the regularity statements of Theorem 2.2 are taken from [57], see also [42, Cor.6.1.8] for the case of constant domains.

Proposition 2.4. *Let (AT) hold and $0 < \alpha < \mu + \nu - 1$. Then*

$$\|A(t)U(t, s)x\|_{\alpha}^{t} \leq C\,(t - s)^{-1-\alpha}\,\|x\|$$

$$\|A(t)U(t, s) - A(r)U(r, s)\| \leq C\,(t - r)^{\alpha}\,(r - s)^{-1-\alpha}$$

for $0 < t - s \leq t_0$, $s < r < t$, $x \in X$, and a constant C depending only on α, t_0, and the constants in (AT).

Proof. By rescaling we may assume that $w < w' \leq 0$ in (AT) and Lemma 2.3. We first prove the Hölder continuity. We infer from $A(t)U(t, s) = \partial_t U(t, s)$ and (2.17) that

$$A(t)U(t, s) - A(r)U(r, s) = A(s)(e^{(t-s)A(s)} - e^{(r-s)A(s)}) + Z(t, s) - Z(r, s)$$

$$= (e^{(t-r)A(s)} - I)\,A(s)e^{(r-s)A(s)} + Z(t, s) - Z(r, s)$$

for $s < r < t$. So [42, Prop.2.2.4, 2.2.2] and [1, Lem.2.2] yield the second assertion. (It can be verified by an inspection of the proofs given there that C only depends on the mentioned quantities.)

Note that the fractional powers $(-A_n(t))^{-\beta}$, $\beta > 0$, are uniformly bounded in $n \in \mathbb{N}$ and $t \in J$ and converge in operator norm to $(-A(t))^{-\beta}$ as $n \to \infty$ due to (2.21). Fix $\alpha \in (0, \mu + \nu - 1)$ and set $\theta := \frac{1}{2}(\alpha + \nu + 1 - \mu) \in (1 - \mu, \nu)$. Then we have

$$U_n(t, s) = e^{(t-s)A_n(t)} + \int_s^t A_n(t)e^{(t-\tau)A_n(t)}\,(A_n(t)^{-1} - A_n(\tau)^{-1})$$

$$\cdot\, A_n(\tau)U_n(\tau, s)\,d\tau,$$

$$(-A_n(t))^{\alpha+1}U_n(t, s) = (-A_n(t))^{\alpha+1}e^{(t-s)A_n(t)} - \int_s^t (-A_n(t))^{2+\alpha-\theta}e^{(t-\tau)A_n(t)}$$

$$\cdot\, (-A_n(t))^{\theta}(A_n(t)^{-1} - A_n(\tau)^{-1})\,A_n(\tau)U_n(\tau, s)\,d\tau$$

for $t \geq s$, $s \in J$, and $n \in \mathbb{N}$. We further define

$$W_n(t,s) := (-A_n(t))^{\alpha+1} U_n(t,s) - (-A_n(t))^{\alpha+1} e^{(t-s)A_n(t)} \quad \text{and} \quad (2.22)$$

$$R_n(t,s) := \int_s^t (-A_n(t))^{2+\alpha} e^{(t-\tau)A_n(t)} \left(A_n(t)^{-1} - A_n(\tau)^{-1} \right)$$
$$\cdot A_n(\tau) e^{(\tau-s)A_n(\tau)} \, d\tau.$$

This yields

$$W_n(t,s) = -R_n(t,s) + \int_s^t \left[(-A_n(t))^{2+\alpha-\theta} e^{(t-\tau)A_n(t)} \right. \qquad (2.23)$$
$$\left. \cdot (-A_n(t))^{\theta} (A_n(t)^{-1} - A_n(\tau)^{-1}) (-A_n(\tau))^{-\alpha} \right] W_n(\tau,s) \, d\tau$$

By [47, Thm.2.6.13], (AT), and (2.5), the kernel $[\cdots]$ of this integral equation can be estimated by $c\,(t-\tau)^{\mu+\theta-\alpha-2}$ for $0 < t - \tau \leq t_0$. As in [64, p.144], one sees that $\|R_n(t,s)\| \leq c\,(t-s)^{\mu+\theta-\alpha-2}$ for $0 < t - s \leq t_0$. Here the constants c only depend on t_0, α, and the constants in (AT). Note that $\mu + \theta - \alpha - 2 > -1$. Thus the solution $W_n(t,s)$ can be estimated in the same way due to [7, Thm.II.3.2.2]. The kernel of (2.23) and the operators $R_n(t,s)$ converge strongly to the same expressions without the index n by [64, Prop.2.1]. Since (2.23) is solved by a Neumann series, see [7, §II.3.2], $W_n(t,s)$ also converges strongly as $n \to \infty$. Hence, $(-A_n(t))^{\alpha+1} U_n(t,s)$ has a strong limit $V(t,s)$ and

$$\|V(t,s)\| \leq c\,(t-s)^{-\alpha-1} \quad \text{for} \ \ 0 < t - s \leq t_0 \qquad (2.24)$$

by (2.22) and $-\alpha - 1 < \mu + \theta - \alpha - 2$, where $c = c(t_0, \alpha, \mu, \nu, \phi, K, L)$. Since $A_n(t) U_n(t,s)$ tends strongly to $A(t) U(t,s)$ as $n \to \infty$ by [64, Prop.3.1.], we arrive at $A(t) U(t,s) = -(-A(t))^{-\alpha} V(t,s)$ for $t > s$. Thus the first assertion follows from (2.24) and (2.5).

The next approximation result is quite useful and due to C.J.K. Batty and R. Chill, [11, Prop.4.4], who employed ideas from [2]. We give here a different, more elementary proof (taken from [43]) leading to a different rate of convergence. Observe that (2.2) and (2.21) imply

$$\|(e^{tA} - e^{tA_n})(w - A)^{-k}\| \leq C\,n^{-1} t^{k-1} \qquad (2.25)$$

for $k = 0, 1$, $t \in (0, t_0]$, $n > \bar{n}$, cf. [2, Lem.4.3].

Proposition 2.5. *Let (AT) hold and $s \in J$. Fix $0 < t_0 < t_1$. Then*

$$\|U(t,s) - U_n(t,s)\| \leq c(t_1, \theta)\, n^{-\theta}$$

for $0 < t_0 \leq t - s \leq t_1$, $n \geq n_0(t_0) := \max\{\bar{n}, t_0^{-2/\mu}\}$, and any $0 < \theta < \min\{\mu/2, 1 - \mu/2, \mu(\mu + \nu - 1)/2\}$. Moreover,

$$\|(U(t,s) - U_n(t,s)) R(w, A(s))\| \leq c(t_1, \alpha)\, n^{-\alpha}$$

for $0 \leq t - s \leq t_1$, $n \geq \bar{n}$, and $\alpha \in (0, \mu)$.

Proof. Let $0 < h < t_0$, $n \geq \bar{n}$, $s \in J$, and $0 < t_0 \leq t - s \leq t_1$. Then we have

$$
\begin{aligned}
U(t,s) - U_n(t,s) &= (U(t,s+h) - U_n(t,s+h))U(s+h,s) - U_n(t,s+h) \\
&\quad \cdot [U(s+h,s) - e^{hA(s)} + e^{hA(s)} - e^{hA_n(s)} + e^{hA_n(s)} - U_n(s+h,s)] \\
&=: S_1 - S_2.
\end{aligned} \tag{2.26}
$$

Due to (2.25) and (2.14), we obtain

$$
\|S_2\| \leq c(t_1)\,(h^{\mu+\nu-1} + (hn)^{-1}). \tag{2.27}
$$

The other term can be transformed into

$$
\begin{aligned}
S_1 &= \int_{s+h}^{t} U_n(t,\sigma)(A(\sigma) - A_n(\sigma))U(\sigma,s)\,d\sigma \\
&= \int_{s+h}^{t} U_n(t,\sigma)(w' - A_n(\sigma))^{\alpha}\,(w' - A_n(\sigma))^{1-\alpha} \\
&\quad \cdot [R(w',A(\sigma)) - R(w',A_n(\sigma))](w' - A(\sigma))U(\sigma,s)\,d\sigma.
\end{aligned}
$$

where $\alpha \in (0,\mu)$ and $w' > w$ is fixed. The moment inequality (2.7) yields

$$
\|(w' - A_n(\sigma))^{1-\alpha}\| \leq c\,\|w' - A_n(\sigma)\|^{1-\alpha} \leq cn^{1-\alpha} \tag{2.28}
$$

This estimate as well as (2.15), (2.13), and (2.21) lead to

$$
\|S_1\| \leq c(t_1,\alpha)\,h^{-1}n^{-\alpha}. \tag{2.29}
$$

Combining (2.26), (2.27), and (2.29), we deduce

$$
\|U(t,s) - U_n(t,s)\| \leq c(t_1,\alpha)\,((nh)^{-1} + h^{\mu+\nu-1} + n^{-\alpha}h^{-1}).
$$

The first assertion now follows if we set $h := n^{-\mu/2}$. The second one can be shown using the formula

$$
\begin{aligned}
(U(t,s) - U_n(t,s))R(w,A(s)) &= \int_{s}^{t} U_n(t,\sigma)(w' - A_n(\sigma))^{\alpha}(w' - A_n(\sigma))^{1-\alpha} \\
&\quad \cdot [R(w',A(\sigma)) - R(w',A_n(\sigma))]\,(w' - A(\sigma))U(\sigma,s)R(w,A(s))\,d\sigma,
\end{aligned}
$$

together with (2.15), (2.28), (2.21), and (2.13).

Batty and Chill used the above result to establish that $U(\cdot,\cdot)$ depends continuously on $A(\cdot)$ in [11, Thm.4.7]. A part of their theorem (and a slight variant of it) are stated and proved in the next proposition.

Proposition 2.6. *Let $A(t)$ and $B(t)$, $t \in J$, satisfy (AT) and generate the evolution families $U(\cdot,\cdot)$ and $V(\cdot,\cdot)$, respectively. Fix $0 < t_0 < t_1$ and let*

$$
q(t,s) = \sup_{s \leq \tau \leq t} \|R(w,A(\tau)) - R(w,B(\tau))\|.
$$

Then there are numbers $q_0(t_0), \beta > 0$ (only depending on the constants in (AT)) such that

$$\|U(t,s) - V(t,s)\| \le c(t_1)\, q(t,s)^\beta$$

for $0 < t_0 \le t - s \le t_1$ and $q(t,s) \le q_0(t_0)$. Moreover,

$$\|(U(t,s) - V(t,s))R(w, A(s))\| \le c(\alpha, t_1)\, q(t,s)^\alpha$$

for $0 \le t - s \le t_1$ and $\alpha \in (0, \mu)$.

Proof. In the following we use Lemma 2.3 with $w' = w + 1$.

(1) Let $0 < h < t_0 \le t - s \le t_1$, $n \ge \bar{n}$, and set $q := q(t,s)$. We write

$$
\begin{aligned}
U(t,s) - V(t,s) &= [U(t,s) - U_n(t,s)] + [U_n(t, s+h) - V_n(t, s+h)] \\
&\quad \cdot U_n(s+h, s) + V_n(t, s+h)\big[U_n(s+h, s) - e^{hA_n(s)} + e^{hA_n(s)} \\
&\quad - e^{hB_n(s)} + e^{hB_n(s)} - V_n(s+h, s)\big] + [V_n(t,s) - V(t,s)] \\
&=: S_1 + S_2 + S_3 + S_4.
\end{aligned}
$$

Proposition 2.5 yields

$$\|S_1\| \le c(t_1, \theta)\, n^{-\theta} \quad \text{and} \quad \|S_4\| \le c(t_1, \theta)\, n^{-\theta}$$

with the number $\theta > 0$ given there. Formulas (2.2), (2.20), and (2.8) give

$$
\begin{aligned}
e^{hA_n(s)} - e^{hB_n(s)} &= \frac{1}{2\pi i} \int_\Gamma e^{\lambda h} \frac{n^2}{(n+\lambda)^2} [R(\tfrac{n\lambda}{n+\lambda}, A(s)) - R(\tfrac{n\lambda}{n+\lambda}, B(s))]\, d\lambda, \\
&= \frac{1}{2\pi i} \int_\Gamma e^{\lambda h} \frac{n^2}{(n+\lambda)^2} A_w(s) R(\tfrac{n\lambda}{n+\lambda}, A(s)) \\
&\quad \cdot (B_w(s)^{-1} - A_w(s)^{-1})\, B_w(s) R(\tfrac{n\lambda}{n+\lambda}, B(s))\, d\lambda
\end{aligned}
$$

for a suitable path $\Gamma \subseteq \Sigma(\phi', w')$. Thus $\|e^{hA_n(s)} - e^{hB_n(s)}\| \le cqh^{-1}$. Together with (2.14) this shows

$$\|S_3\| \le c(t_1)\,(h^{\mu+\nu-1} + qh^{-1}).$$

The remaining term can be transformed into

$$
\begin{aligned}
S_2 = \int_{s+h}^t V_n(t,\sigma)(w' - B_n(\sigma))^\alpha (w' - B_n(\sigma))^{1-\alpha} \quad\quad (2.30) \\
\cdot ((A_n(\sigma) - w')^{-1} - (B_n(\sigma) - w')^{-1})(A_n(\sigma) - w')U_n(\sigma, s)\, d\sigma.
\end{aligned}
$$

where $\alpha \in (0, \mu)$. Due to (2.20) and (2.8) we have

$$
\begin{aligned}
\|R(w', A_n(\sigma)) &- R(w', B_n(\sigma))\| \\
&= \big\| \tfrac{n^2}{(n+w')^2} [R(\tfrac{nw'}{n+w'}, A(\sigma)) - R(\tfrac{nw'}{n+w'}, B(\sigma))]\big\| \le cq
\end{aligned}
$$

for $n > ww'/(w' - w) = w^2 + w$. So (2.15), (2.28), and (2.13) imply

$$\|S_2\| \leq c(t_1, \alpha) \, h^{-1} n^{1-\alpha} q$$

for sufficiently large n. Putting everything together, we arrive at

$$\|U(t, s) - V(t, s)\| \leq c(t_1, \theta, \alpha) \, (n^{-\theta} + h^{-1} n^{1-\alpha} q + h^{\mu+\nu-1} + q h^{-1}).$$

Now take, e.g., $n = q^{1/4(\alpha-1)}$ and $h = \sqrt{q}$ to deduce the first assertion.
 (2) For the second assertion we write

$$
\begin{aligned}
&(U(t, s) - V(t, s))R(w, A(s)) \\
&= [U(t, s) - U_n(t, s)]R(w, A(s)) + [U_n(t, s) - V_n(t, s)]R(w', A_n(s)) \\
&\quad \cdot (w' - A_n(s))R(w, A(s)) + [V_n(t, s) - V(t, s)] \, [R(w, A(s)) - R(w, B(s))] \\
&\quad + [V_n(t, s) - V(t, s)]R(w, B(s))
\end{aligned}
$$

Here we let $n = q^{-1}$ supposing that $q < \bar{n}$. (For $q \geq \bar{n}$, the result is trivially true.) So the claim follows from Proposition 2.5 and formula (2.30) with $h = 0$.

We come back to the inhomogeneous problem (2.1). If u is a classical solution with $x \in X_0^s$ and $f \in C(J, X)$, then u is given by

$$u(t) = U(t, s)x + \int_s^t U(t, \tau)f(\tau)\, d\tau, \qquad t \geq s, \tag{2.31}$$

due to [1, Prop.3.2, 5.1]. This equality defines a function $u : [s, \infty) \to X$ for all $x \in X$ and $f \in C(J, X)$ which is called the *mild solution* of (2.1). The regularity of the first summand in (2.31) was discussed above. Using (2.13) and (2.16), one easily sees that the integral term is uniformly bounded in $X_{\alpha,\infty}^t$ for $\alpha < 1$ and Hölder continuous of exponent $\beta < 1$ on $[s, s+t_0]$. The mild solution becomes a classical one if one takes more regular data. This fact is stated in the next theorem which recalls a part of Theorems 6.1–6.4 of [2] in a simplified way. The asserted uniformity of the constants can be verified inspecting the proofs in [2].

Theorem 2.7. *Assume that (AT) holds and let $s \in J$, $t_0 > 0$, $\alpha \in (0, \mu+\nu-1]$, $f \in C([s, s+t_0], X)$, and $x \in X_0^s$.*
 (a) If $f \in C^\alpha([s, s+t_0], X)$, then the mild solution is the unique classical solution of (2.1) and

$$\|u'\|_{C^\alpha([s+\varepsilon, s+t_0], X)} + \|A(\cdot)u\|_{C^\alpha([s+\varepsilon, s+t_0], X)} + \sup_{s+\varepsilon \leq t \leq s+t_0} \|u'(t)\|_\alpha^t \tag{2.32}$$

$$\leq c(\varepsilon, t_0) \, [\|x\| + \|A(s)x\| + \|f\|_{C^\alpha([s, s+t_0], X)}]$$

for $\varepsilon > 0$. The solution is strict if $x \in D(A(s))$ and $A(s)x + f(s) \in X_0^s$. One can take $\varepsilon = 0$ in (2.32) if and only if $A(s)x + f(s) \in X_{\alpha,\infty}^s$, and then

$$\|u'\|_{C^{\alpha}([s,s+t_0],X)} + \|A(\cdot)u\|_{C^{\alpha}([s,s+t_0],X)} + \sup_{s\leq t\leq s+t_0} \|u'(t)\|_{\alpha}^t \qquad (2.33)$$

$$\leq c(t_0)\left[\|x\| + \|A(s)x\| + \|A(s)x + f(s)\|_{\alpha}^s + \|f\|_{C^{\alpha}([s,s+t_0],X)}\right].$$

(b) If $\|f(t)\|_{\alpha}^t$ is uniformly bounded for $t \in [s, s+t_0]$, then the mild solution is the unique classical solution of (2.1) and

$$\|A(\cdot)u\|_{C^{\alpha}([s+\varepsilon,s+t_0],X)} + \sup_{s+\varepsilon\leq t\leq s+t_0} (\|u'(t)\|_{\alpha}^t + \|A(t)u(t)\|_{\alpha}^t) \qquad (2.34)$$

$$\leq c(\varepsilon,t_0)\left[\|x\| + \|A(s)x\| + \sup_{s\leq t\leq s+t_0} \|f(t)\|_{\alpha}^t\right]$$

for $\varepsilon > 0$. The solution is strict if $x \in D(A(s))$ and $A(s)x + f(s) \in X_0^s$. One can take $\varepsilon = 0$ in the above estimate if and only if $A(s)x \in X_{\alpha,\infty}^s$, and then

$$\|A(\cdot)u\|_{C^{\alpha}([s,s+t_0],X)} + \sup_{s\leq t\leq s+t_0} (\|u'(t)\|_{\alpha}^t + \|A(t)u(t)\|_{\alpha}^t) \qquad (2.35)$$

$$\leq c(t_0)\left[\|x\| + \|A(s)x\| + \|A(s)x\|_{\alpha}^s + \sup_{s\leq t\leq s+t_0} \|f(t)\|_{\alpha}^t\right].$$

The constants only depend on the constants in (AT) and t_0, ε.

Concluding this section, we present two different ways to verify that (AT) holds for parabolic partial differential equations of second order under suitable regularity and ellipticity assumptions. The following two examples are taken from [64] and [1], respectively; see also [5], [42], [45], [62], [63], [65]. More general problems could be treated analogously.

Example 2.8. Consider the initial–boundary value problem

$$D_t\, u(t,x) = A(t,x,D)u(t,x) + h(t,x), \quad t > s \geq s, \ x \in \Omega,$$
$$B(t,x,D)u(t,x) = 0, \quad t > s \geq 0, \ x \in \partial\Omega, \qquad (2.36)$$
$$u(s,\cdot) = u_0.$$

Here we suppose that $D_t = \frac{\partial}{\partial t}$, $D_k = \frac{\partial}{\partial x_k}$, $\Omega \subseteq \mathbb{R}^n$ is a bounded domain with boundary $\partial\Omega$ of class C^2 being locally on one side of Ω and outer unit normal vector $n(x)$, and that

$$A(t,x,D) = \sum_{k,l=1}^{n} D_k\, a_{kl}(t,x)\, D_l + a_0(t,x),$$

$$B(t,x,D) = \sum_{k,l=1}^{n} n_k(x)\, a_{kl}(t,x)\, D_l\,.$$

We assume that the coefficients satisfy

$$a_{kl} \in C_b^{\mu}(\mathbb{R}_+, C(\overline{\Omega})) \cap C_b(\mathbb{R}_+, C^1(\overline{\Omega})),$$
$$a_0 \in C_b^{\mu}(\mathbb{R}_+, L^n(\Omega)) \cap C_b(\mathbb{R}_+, C(\overline{\Omega})) \qquad (2.37)$$

for $k, l = 1, \cdots, n$, $t \geq 0$, and a constant $\frac{1}{2} < \mu \leq 1$ and that (a_{kl}) is symmetric, real, and uniformly elliptic, i.e.,

$$\sum_{k,l=1}^{n} a_{kl}(t, x)\, v_k\, v_l \geq \eta\, |v|^2 \tag{2.38}$$

for a constant $\eta > 0$ and each $x \in \overline{\Omega}$, $t \geq 0$, $v \in \mathbb{R}^n$.

On $X_p = L^p(\Omega)$, $1 < p < \infty$, and $X_\infty = C(\overline{\Omega})$ we introduce the realizations $A_p(t)$, $1 < p < \infty$, and $A_\infty(t)$ of the differential operator $A(t, x, D)$ with domains

$$D(A_p(t)) = \{f \in W^{2,p}(\Omega) : B(t, \cdot, D)f = 0 \text{ on } \partial\Omega\}, \tag{2.39}$$

$$D(A_\infty(t)) = \{f \in \bigcap_{q>1} W^{2,q}(\Omega) : A(t, \cdot, D)f \in C(\overline{\Omega}),\ B(t, \cdot, D)f = 0 \text{ on } \partial\Omega\},$$

where the boundary condition is understood in the sense of traces if necessary. It is well known that these operators are sectorial of uniform type (ϕ, K, w), see e.g. [19], [42], [62], [63]. Moreover, the adjoint $A_p(t)^*$ of $A_p(t)$, $1 < p < \infty$, is given by $A_{p'}(t)$ on $W^{2,p'}(\Omega)$, where $1/p + 1/p' = 1$, and the resolvents of $A_p(t)$ and $A_q(t)$ coincide on $L^q(\Omega)$ if $q \in (p, \infty]$ and both resolvents exist. Therefore we will mostly omit the subscript p.

We verify (AT2) first for $p \in (1, \infty)$. Take $t, s \geq 0$, $f \in L^p(\Omega)$, and $g \in D((w - A(t)^*)^\nu)$ for some $\nu \in (1 - \mu, 1/2)$. Set $u = (R(w, A(t)) - R(w, A(s)))f \in W^{2,p}(\Omega)$ and $v = (w - A(t)^*)^{\nu-1}g \in D(A_{p'}(t))$. Integrating by parts three times, we deduce

$$\langle (R(w, A(t)) - R(w, A(s)))f, (w - A(t)^*)^\nu g \rangle = \int_\Omega u\, (w - A(t, \cdot, D))v\, dx$$

$$= \int_\Omega v\, (w - A(t, \cdot, D))u\, dx + \int_{\partial\Omega} v\, B(t, \cdot, D)u\, dS$$

$$= \int_\Omega v\, (A(t, \cdot, D) - A(s, \cdot, D))R(w, A(s))f\, dx - \int_{\partial\Omega} v\, B(t, \cdot, D)R(w, A(s))f\, dS$$

$$= \sum_{kl} \int_\Omega (a_{kl}(s, x) - a_{kl}(t, x))\, (D_l R(w, A(s))f)(x)\, (D_k(-A_w(t)^*)^{\nu-1}g)(x)\, dx$$

$$+ \int_\Omega (a_0(t, x) - a_0(s, x))(R(w, A(s))f)(x)\, ((w - A(t)^*)^{\nu-1}g)(x)\, dx \tag{2.40}$$

Observe that the operators $R(w, A(s)) : L^p(\Omega) \to W^{k,q}(\Omega)$ and $(w - A(t)^*)^{\nu-1} : L^{p'}(\Omega) \to W^{k,r}(\Omega)$ are bounded for $q \in [p, \infty]$, $r \in [p', \infty]$, and $k = 0, 1$ such that $\frac{1}{q} > \frac{1}{p} + \frac{k-2}{n}$ and $\frac{1}{r} > \frac{1}{p'} + \frac{k-2+2\nu}{n}$ due to e.g. [27, Thm.1.6.1] and $\nu < 1/2$. Using Hölder's inequality in (2.40), we then estimate

$$|\langle (R(w, A(t)) - R(w, A(s)))f, (w - A(t)^*)^\nu g \rangle|$$

$$\leq c_p \max_{kl}\{[a_{kl}]_{\mu,\infty}, [a_0]_{\mu,n}\}\, \|f\|_p\, \|g\|_{p'}\, |t - s|^\mu$$

where $[u]_{\mu,q}$ is the Hölder constant of $u : \mathbb{R}_+ \to L^q(\Omega)$ for the Hölder exponent $\mu \in (0,1)$. This leads to

$$\|(w - A(t))^\nu (R(w, A(t)) - R(w, A(s)))\|_p$$
$$\leq c_p \max_{kl}\{[a_{kl}]_{\mu,\infty}, [a_0]_{\mu,n}\} |t - s|^\mu. \quad (2.41)$$

We have verified (AT) for $p < \infty$ because of (2.5). In the case $p = \infty$ we fix $\nu \in (1 - \mu, 1/2)$, $\theta \in (0, 1/2 - \nu)$, and $p > \frac{n}{2\theta}$. Using again [27, Thm.1.6.1], we see that $(w - A_p(t))^{-\theta} : L^p(\Omega) \to C(\overline{\Omega})$ is bounded. Thus (2.41) with ν replaced by $\nu + \theta$ yields

$$\|(w - A(t))^\nu (R(w, A(t)) - R(w, A(s)))f\|_\infty$$
$$\leq c \|(w - A_p(t))^{\nu+\theta}(R(w, A(t)) - R(w, A(s)))f\|_p$$
$$\leq c |t - s|^\mu \max_{kl}\{[a_{kl}]_{\mu,\infty}, [a_0]_{\mu,n}\}\|f\|_p$$
$$\leq c |t - s|^\mu \max_{kl}\{[a_{kl}]_{\mu,\infty}, [a_0]_{\mu,n}\}\|f\|_\infty \quad (2.42)$$

for $f \in C(\overline{\Omega})$. So we deduce (AT) from (2.5) also for $p = \infty$. The above results now allow to solve the problem (2.36) under appropriate assumptions on $u_0 \in X$ and $h : \mathbb{R}_+ \to X$.

Example 2.9. We study again (2.36), but now with the differential expressions

$$A(t, x, D) = \sum_{k,l} a_{kl}(t, x) D_k D_l + \sum_k a_k(t, x) D_k + a_0(t, x),$$
$$B(t, x, D) = \sum_k b_k(t, x) D_k + b_0(t, x). \quad (2.43)$$

It assumed that $\partial\Omega$ is the disjoint union of two closed (possibly empty) subsets Γ_0 and Γ_1. We require that $a_{kl} = a_{lk} \in BUC(\mathbb{R}_+ \times \overline{\Omega})$, b_k, and b_0 are real, (2.38) holds, $a_{kl}, a_k, a_0 \in C_b^\mu(\mathbb{R}_+, C(\overline{\Omega}))$ and $b_k, b_0 \in C_b^\mu(\mathbb{R}_+, C^1(\overline{\Omega}))$ for some $\mu \in (1/2, 1)$, $b_0 = 1$ and $b_k = 0$ on Γ_0, and $\sum_{k=1}^n b_k(t, x)n_k(x) \geq \beta > 0$ for $x \in \Gamma_1$, $t \geq 0$, and $k, l = 1, \cdots, n$.[2]

As in the previous example, see (2.39), we define the operators $A_p(t)$ on X_p, $1 \leq p \leq \infty$. Observe that the closure of $D(A_\infty(t))$ is the space of those continuous functions on $\overline{\Omega}$ vanishing on Γ_0. Again it is well known that these operators are sectorial of uniform type (ϕ, K, w), see e.g. [19], [42], [62], [63]. For $f \in L^p(\Omega)$, $t, s \geq 0$, $|\arg \lambda| \leq \phi$, we set

$$v = -R(w, A(s))f \quad \text{and} \quad u = R(\lambda + w, A(t))(\lambda + w - A(s))v.$$

Then $u - v = (A(t) - w)R(\lambda + w, A(t))(R(w, A(t)) - R(w, A(s)))f$ and

$$(\lambda + w)u - A(t, \cdot, D)u = \lambda v - f, \quad (w - A(s, \cdot, D))v = -f, \quad \text{on } \Omega,$$
$$B(t, \cdot, D)u = 0, \qquad\qquad B(s, \cdot, D)v = 0 \quad \text{on } \partial\Omega.$$

---

[2] If $\Gamma_1 = \emptyset$, it suffices that $\mu > 0$.

This shows that

$$(\lambda + w)(u - v) - A(t, \cdot, D)(u - v) = (A(t, \cdot, D) - A(s, \cdot, D))v, \qquad \text{on } \Omega,$$
$$B(t, \cdot, D)(u - v) = (B(s, \cdot, D) - B(t, \cdot, D))v, \qquad \text{on } \partial\Omega.$$

The Agmon–Douglas–Nirenberg estimate and our assumptions now imply

$$\|u - v\|_p \le \frac{c_p}{|\lambda + w|}\left(|\lambda + w|^{1/2}\|((B(s, \cdot, D) - B(t, \cdot, D))v\|_{L^p(\Omega)}\right. \tag{2.44}$$
$$+ \|(B(s, \cdot, D) - B(t, \cdot, D))v\|_{W^{1,p}(\Omega)} + \|(A(t, \cdot, D) - A(s, \cdot, D))v\|_p\bigg)$$
$$\le c_p |\lambda + w|^{-1/2} |t - s|^\mu \|f\|_{L^p(\Omega)}$$

see e.g. [19], [42], [63, Thm.5.5], and the references therein. Thus (AT2) holds with μ and $\nu = 1/2$ on $L^p(\Omega)$ for $p \in (1, \infty)$. One can deal with the case $X = C(\overline{\Omega})$ as in (2.42) obtaining $\nu \in (1 - \mu, 1/2)$.[3] Theorems 2.2 and 2.7 again allow to solve (2.36).

Remark 2.10. Inspecting the proofs given above and in, e.g., [63], it can be seen in both examples that all constants c and c_p and the resulting constants in (AT) only depend on Ω, the ellipticity constants $\eta, \beta > 0$, the modulus of continuity w.r.t. the space variables of and a_{kl}, $k, l = 1, \cdots, n$, and the norms of the coefficients in the spaces indicated above.[4] Unfortunately, it is hard to control this dependence explicitly. In Example 4.16 we will modify the approach of Example 2.8 for $p = 2$ in order to obtain an explicit bound for the constants in (AT) which is important for some of our applications.

3 Exponential Dichotomy

Having discussed the solvability of parabolic evolution equations, we turn our attention to the long term behaviour of the solutions. In this section we introduce and characterize exponential dichotomy which is a fundamental qualitative property of evolution families. We formulate most of our results for general evolution families since the proofs do not use properties peculiar to parabolic problems and in later sections we occasionally employ evolution families which do not quite satisfy the Aquistapace–Terreni conditions.

3.1 Basic Observations

Let $U(\cdot, \cdot)$ be an evolution family with time interval $J \in \{\mathbb{R}, [a, \infty)\}$ on a Banach space X. Its *(exponential) growth bound* is defined by

$$\omega(U) = \inf\{\gamma \in \mathbb{R} : \exists M \ge 1 \text{ s.t. } \|U(t, s)\| \le Me^{\gamma(t-s)} \text{ for } t \ge s, \ s \in J\}.$$

---

[3] A modified argument also gives $\nu = 1/2$, see [1, §6].
[4] In fact, it suffices to consider the norms of b_k, $k = 0, \cdots, n$, in $C_b^\mu(\mathbb{R}_+, C^1(\partial\Omega))$.

If $\omega(U) < \infty$, then $U(\cdot,\cdot)$ is called *exponentially bounded*; if $\omega(U) < 0$, then $U(\cdot,\cdot)$ is *exponentially stable*. For instance, the evolution families $U(t,s) = e^{\pm(t^2-s^2)}I$ have growth bound $\pm\infty$. In contrast to the semigroup case, the location of $\sigma(U(t,s))$, $t \geq s$, has no influence on the asymptotic behaviour, in general, see [58, §3]. However, the implication

$$\|U(t,s)\| \leq C,\ 0 \leq t - s \leq t_0 \implies \|U(t,s)\| \leq C\exp(\tfrac{\ln C}{t_0}(t-s)),\ t \geq s,$$

holds because of the estimate

$$\|U(t,s)\| = \|U(t,s+nt_0)\cdots U(s+t_0,s)\| \leq C^{n+1} \leq C\exp(\tfrac{\ln C}{t_0}(t-s))$$

for $t = s + nt_0 + \tau$ with $n \in \mathbb{N}_0$ and $\tau \in [0, t_0)$. In a similar way one sees that

$$\omega(U) < \infty,\ \|U(s+t_1,s)\| \leq C_1,\ s \in J \implies \|U(t,s)\| \leq C_2\exp(\tfrac{\ln C_1}{t_1}(t-s))$$

for $t \geq s$ and a suitable constant $C_2 \geq 1$. In particular, the evolution family obtained in Theorem 2.2 is exponentially bounded.

The following notion combines forward exponential stability on certain subspaces with backward exponential stability on their complements. Here and below we set $Q = I - P$ for a projection P.

Definition 3.1. *An evolution family $U(\cdot,\cdot)$ on a Banach space X (with time interval $J = \mathbb{R}$ or $[a,\infty)$) has an* exponential dichotomy *(or is called* hyperbolic*) if there are projections $P(t)$, $t \in J$, and constants $N, \delta > 0$ such that, for $t \geq s$, $s \in J$,*

(a) $U(t,s)P(s) = P(t)U(t,s)$,
(b) the restriction $U_Q(t,s) : Q(s)X \to Q(t)X$ of $U(t,s)$ is invertible (and we set $U_Q(s,t) := U_Q(t,s)^{-1}$),
(c) $\|U(t,s)P(s)\| \leq Ne^{-\delta(t-s)}$, and $\|U_Q(s,t)Q(t)\| \leq Ne^{-\delta(t-s)}$.

The operator family $\Gamma(t,s)$, $t,s \in J$, given by

$$\Gamma(t,s) = U(t,s)P(s),\quad t \geq s,\qquad \Gamma(t,s) = -U_Q(t,s)Q(s),\quad t < s,$$

is called Green's function *corresponding to $U(\cdot,\cdot)$ and $P(\cdot)$.*

Note that $N \geq 1$ and $U(\cdot,\cdot)$ is exponentially stable if $P(t) = I$. Among the vast literature on exponential dichotomy, we refer to [12], [15], [16], [17], [21], [27], [42], [54]. The importance of this concept relies in particular on its robustness under small perturbations and on its impact on inhomogeneous and nonlinear problems as we will see later.

Let $U(\cdot,\cdot)$ be hyperbolic on J with projections $P(t)$. In computations involving Green's function it is useful to observe that

$$U_Q(t,s)Q(s) = U_Q(t,r)U_Q(r,s)Q(s) \text{ for } t,r,s \in J.$$

If $U(\cdot,\cdot)$ is strongly continuous for $t \geq s$, then $P(\cdot)$ is strongly continuous since then the expressions

$$P(t)x - P(s)x = P(t)(x - U(t,s)x) + U(t,s)P(s)x - P(s)x, \quad t \geq s,$$
$$Q(t)x - Q(s)x = U_Q(t,s)Q(s)(U(s,t)x - x) \tag{3.1}$$
$$+ (U(t,a) - U(s,a))U_Q(a,s)Q(s)x, \quad a \leq t \leq s,$$

tend to 0 as $t \to s$. In the parabolic case, the dichotomy projections are in fact Hölder continuous as shown in Proposition 3.18 below. For $t \leq s$ and $t' \leq s'$, we have

$$U_Q(t',s')Q(s')x - U_Q(t,s)Q(s)x \tag{3.2}$$
$$= U_Q(t',s')Q(s')\,[Q(s') - Q(s)]x + U_Q(t',s')Q(s')[U(s,t) - U(s',t')]$$
$$\cdot U_Q(t,s)Q(s)x + [Q(t') - Q(t)]U_Q(t,s)Q(s)x,$$

so that $J^2 \ni (t,s) \mapsto U_Q(t,s)Q(s)$ is strongly continuous for $t \geq s$ if $U(\cdot,\cdot)$ has the same property.

The dichotomy projections are not uniquely determined if $J = [a,\infty)$, in general, see [15, p.16]. However, the estimate

$$N^{-1}e^{\delta(t-s)}\|Q(s)x\| \leq \|U(t,s)Q(s)x\| \leq \|U(t,s)x\| + Ne^{-\delta(t-s)}\|P(s)x\|$$

for $t \geq s$ implies that

$$P(s)X = \{x \in X : \lim_{t\to\infty} U(t,s)x = 0\} \tag{3.3}$$

for $s \in J$. In the case $J = \mathbb{R}$, it is proved in Corollary 3.14 that the projections are unique and that the projections do not depend on t if the evolution family $U(t,s) = T(t-s)$ is given by a semigroup. Definition 3.1 thus agrees with the usual one for semigroups.

Recall that a semigroup $T(\cdot)$ is hyperbolic if and only if the unit circle $\mathbb{T}$ belongs to $\rho(T(t_0))$ for some/all $t_0 > 0$. Then the dichotomy projection P coincides with the *spectral projection*

$$P = \frac{1}{2\pi i} \int_{\mathbb{T}} R(\lambda, T(t_0))\, d\lambda, \tag{3.4}$$

see e.g. [21, §V.1.c]. We denote by A_P the restriction of the generator A of $T(\cdot)$ to PX and by A_Q the restriction of A to QX. Observe that A_P and A_Q generate the restriction of $T(\cdot)$ to PX and QX, respectively. Similar results hold for *periodic* evolution families $U(\cdot,\cdot)$, i.e., $U(t+p,s+p) = U(t,s)$ for $t \geq s$, $s \in \mathbb{R}$, and some $p > 0$. In this case, one has the equalities

$$\sigma(U(s+p,s)) = \sigma(U(p,0)) \quad \text{and} \quad U(t,s) = U(t,t-\tau)U(s+p,s)^n$$

where $t = s+np+\tau$, $n \in \mathbb{N}_0$, and $\tau \in [0,p)$. A periodic evolution family $U(\cdot,\cdot)$ is hyperbolic if and only if $\sigma(U(p,0)) \cap \mathbb{T} = \emptyset$ because of these relations. The dichotomy projections are then given by

$$P(s) = \frac{1}{2\pi i} \int_{\mathbb{T}} R(\lambda, U(s+p,s))\, d\lambda$$

for $s \in \mathbb{R}$, cf. [16], [17], [27], [42].

In the autonomous case the exponential dichotomy of a semigroup $T(\cdot)$ generated by A always implies the spectral condition $\sigma(A) \cap i\mathbb{R} = \emptyset$. In particular, the exponential stability of $T(\cdot)$ yields $s(A) := \sup\{\operatorname{Re}\lambda : \lambda \in \sigma(A)\} < 0$. The converse implications fail for general C_0–semigroups, but can be verified if the spectral mapping theorem

$$\sigma(T(t)) \setminus \{0\} = e^{t\sigma(A)}, \qquad t \geq 0, \tag{3.5}$$

holds. Formula (3.5) is fulfilled by eventually norm continuous semigroups, and hence by analytic or eventually compact semigroups. We refer to [21] and the references therein for these and related results.

In many cases it is thus possible to characterize the exponential dichotomy of a semigroup by the spectrum of its generator. This is an important fact since in applications usually the generator A is the given object. Unfortunately, in the nonautonomous case there is no hope to relate the location of $\sigma(A(t))$ to the asymptotic behaviour of the evolution family $U(\cdot, \cdot)$ generated by $A(\cdot)$ as is shown by the following example, cf. [15, p.3]. We point out that it deals with periodic evolution families $U_k(\cdot, \cdot)$ on $X = \mathbb{C}^2$ satisfying

$$s(A_1(t)) = -1 < \omega(U_1) \quad \text{and} \quad s(A_2(t)) = 1 > \omega(U_2) \quad \text{for } t \in \mathbb{R},$$

whereas $s(A) \leq \omega(T)$ for each semigroup $T(\cdot)$ with generator A.

Example 3.2. Let $A_k(t) = D(-t) A_k D(t)$ for $t \in \mathbb{R}$, where

$$D(t) = \begin{pmatrix} \cos t & \sin t \\ -\sin t & \cos t \end{pmatrix}, \qquad A_1 = \begin{pmatrix} -1 & -5 \\ 0 & -1 \end{pmatrix}, \qquad A_2 = \begin{pmatrix} 1 & 0 \\ 0 & -1 \end{pmatrix}.$$

The operators $A_1(t)$ and $A_2(t)$, $t \in \mathbb{R}$, generate the evolution families

$$U_1(t, s) = D(-t) \, \exp\left[(t - s) \begin{pmatrix} -1 & -4 \\ -1 & -1 \end{pmatrix}\right] D(s) \quad \text{and}$$

$$U_2(t, s) = D(-t) \, \exp\left[(t - s) \begin{pmatrix} 1 & 1 \\ -1 & -1 \end{pmatrix}\right] D(s),$$

respectively. Thus, $\omega(U_1) = 1$ and $\omega(U_2) = 0 = \omega(U_2^{-1})$, but $\sigma(A_1(t)) = \{-1\}$ and $\sigma(A_2(t)) = \{-1, 1\}$ for $t \in \mathbb{R}$. In other words, the exponential stability of $(e^{\tau A_1(t)})_{\tau \geq 0}$ and the exponential dichotomy of $(e^{\tau A_2(t)})_{\tau \geq 0}$ (with constants independent of t) are lost when passing to the nonautonomous problem.

In Example 3.2 we also have $\|R(\lambda, A_k(t))\|_2 = \|R(\lambda, A_k)\|_2$ for $t \in \mathbb{R}$, $\lambda \in \rho(A_k)$, and $k = 1, 2$. This shows that we cannot expect to deduce asymptotic properties of an evolution family from estimates on the resolvent of $A(t)$ along vertical lines as it is possible for a semigroup on a Hilbert space by virtue of Gearhart's theorem, see e.g. [12, Thm.2.16]. Positivity in the sense of order theory does not help, either: In [55, §5] we constructed a positive evolution family $U(\cdot, \cdot)$ such that $\omega(U) = +\infty$ and $U(\cdot, \cdot)$ solves the Cauchy problem corresponding to generators $A(t)$, $t \geq 0$, of uniformly bounded, positive semigroups on an L^1–space with $s(A(t)) = -\infty$ except for a sequence t_n where $A(t_n) = 0$.

3.2 Characterizations of Exponential Dichotomy

The above mentioned examples indicate that it is quite difficult to establish the exponential dichotomy of a given problem. In fact, the available results for ordinary differential equations are usually restricted to perturbation type arguments, see [15], [16]. Several of these theorems are extended to parabolic evolution equations in the next section. Our approach relies on characterizations of exponential dichotomy discussed below.

Let $U(\cdot, \cdot)$ be an exponentially bounded, strongly continuous evolution family with time interval $J = \mathbb{R}$ acting on a Banach space X and let $E = C_0(\mathbb{R}, X)$ be the space of continuous functions $f : \mathbb{R} \to X$ vanishing at infinity endowed with the sup–norm $\| \cdot \|_\infty$. We then introduce

$$(T_U(t)f)(s) := U(s, s - t)f(s - t), \quad s \in \mathbb{R}, \ t \geq 0, \ f \in E. \qquad (3.6)$$

It is easy to see that this definition yields a C_0–semigroup $T_U(\cdot)$ on E with $\omega(T_U) = \omega(U)$, which is called the *evolution semigroup* corresponding to $U(\cdot, \cdot)$, cf. [21, Lem.VI.9.10]. We denote its generator by G_U and omit the subscript if the underlying evolution family $U(\cdot, \cdot)$ is clear from the context.

Evolution semigroups were invented by J.S. Howland in 1974 for applications in scattering theory. They have further been used to study perturbation theory and well–posedness of evolution equations. In the nineties their relationship with the asymptotic behaviour of evolution equations was realized by several authors. The recent monograph [12] by C. Chicone and Y. Latushkin is entirely devoted to these matters (with some emphasis on spectral theory and applications to dynamical systems). The bibliographical notes of [12] and the survey article [58] give a detailed account of the field and provide plenty of relevant references. In the following we thus restrict ourselves to a sample of citations directly connected with subjects discussed here.

Evolution semigroups can be defined on a large variety of function spaces. In the present lecture notes we only need the space $E = C_0(\mathbb{R}, X)$ (and some slight variants). Therefore our treatment can be based on the short exposition given in [21, §VI.9], but extending it in several respects. We start with a simple, but important observation.

Lemma 3.3. *Let $U(\cdot, \cdot)$ be a strongly continuous, exponentially bounded evolution family with $J = \mathbb{R}$ on X and $T(\cdot)$ be the corresponding evolution semigroup on $E = C_0(\mathbb{R}, X)$ with generator G. Then $\sigma(T(t))$ is rotationally invariant for $t > 0$ and $\sigma(G)$ is invariant under translations along the imaginary axis. Moreover, $\|R(\lambda, T(t))\| = \|R(|\lambda|, T(t))\|$ and $\|R(\lambda, G)\| = \|R(\mathrm{Re}\,\lambda, G)\|$ for λ in the respective resolvent sets.*

Proof. We define an isometric isomorphism on E by $M_\mu f(s) := e^{i\mu s} f(s)$ for $\mu \in \mathbb{R}$. Clearly, $M_\mu T(t) M_{-\mu} = e^{i\mu t} T(t)$ for $t \geq 0$ so that the assertions follow from standard spectral and semigroup theory, see e.g. [21, §II.2.a].

In order to relate the exponential dichotomy of the evolution family $U(\cdot, \cdot)$ and its evolution semigroup we need two preliminary results.

Lemma 3.4. *Let $U(\cdot, \cdot)$ be a strongly continuous, exponentially bounded evolution family with $J = \mathbb{R}$ on X and $T(\cdot)$ be the corresponding evolution semigroup on $E = C_0(\mathbb{R}, X)$. Assume that $T(\cdot)$ is hyperbolic with dichotomy projection $\mathcal{P}$. Then $\varphi \mathcal{P} f = \mathcal{P}(\varphi f)$ for $\varphi \in C_b(\mathbb{R})$ and $f \in E$.*

Proof. Let $\varphi \in C_b(\mathbb{R})$ and $f \in E$. Since $T(t)\varphi \mathcal{P} f = \varphi(\cdot - t)T(t)\mathcal{P}f$, the identity (3.3) yields $\varphi \mathcal{P} f \in \mathcal{P}E$. On the other hand, we have

$$\|\mathcal{P}(\varphi \mathcal{Q} f)\|_\infty = \|\mathcal{P}\varphi T(t)T_Q^{-1}(t)\mathcal{Q}f\|_\infty = \|T(t)\mathcal{P}\varphi(\cdot + t)T_Q^{-1}(t)\mathcal{Q}f\|_\infty$$
$$\leq N^2 e^{-2\delta t} \|\varphi\|_\infty \|f\|_\infty, \qquad t \geq 0,$$

and thus $\mathcal{P}(\varphi \mathcal{Q} f) = 0$. As a result, $\mathcal{P}(\varphi f) = \mathcal{P}(\varphi \mathcal{P} f) + \mathcal{P}(\varphi \mathcal{Q} f) = \varphi \mathcal{P} f$.

Proposition 3.5. *A bounded operator $\mathcal{M}$ on $E = C_0(\mathbb{R}, X)$ is of the form $(\mathcal{M}f)(s) = M(s)f(s)$ for $M(\cdot) \in C_b(\mathbb{R}, \mathcal{L}_s(X))$ if (and only if) $\mathcal{M}(\varphi f) = \varphi \mathcal{M}f$ for $f \in E$ and $\varphi \in C_c(\mathbb{R})$. Moreover, $\|\mathcal{M}\|_{\mathcal{L}(E)} = \sup_{t \in \mathbb{R}} \|M(t)\|_{\mathcal{L}(X)}$.*

Proof. For $\varepsilon > 0$ and $t \in \mathbb{R}$, we choose a continuous function $\varphi_\varepsilon : \mathbb{R} \to [0, 1]$ with $\varphi_\varepsilon(t) = 1$ and $\operatorname{supp} \varphi_\varepsilon \subseteq [t - \varepsilon, t + \varepsilon]$. Let $f \in E$. Our assumption yields

$$\|(\mathcal{M}f)(t)\| = \|(\mathcal{M}\varphi_\varepsilon f)(t)\| \leq \|\mathcal{M}\| \|\varphi_\varepsilon f\|_\infty \leq \|\mathcal{M}\| \sup_{|t-s| \leq \varepsilon} \|f(s)\|.$$

Therefore, $f(t) = 0$ implies $(\mathcal{M}f)(t) = 0$. So we can define linear operators $M(t)$ on X by setting $M(t)x := (\mathcal{M}f)(t)$ for some $f \in E$ with $f(t) = x$. Clearly, $\sup_{t \in \mathbb{R}} \|M(t)\| = \|\mathcal{M}\|$ and $\mathcal{M} = M(\cdot)$. For $x \in X$ and $t \in \mathbb{R}$, take $f \in E$ being equal to x in a neighbourhood V of t. Then $V \ni t \mapsto M(t)x = (\mathcal{M}f)(t)$ is continuous.

Theorem 3.6. *Let $U(\cdot, \cdot)$ be an exponentially bounded evolution family with time interval $J = \mathbb{R}$ on a Banach space X being strongly continuous for $t \geq s$ and $T(\cdot)$ be the corresponding evolution semigroup on $E = C_0(\mathbb{R}, X)$. Then $U(\cdot, \cdot)$ has an exponential dichotomy on X with projections $P(\cdot) \in C_b(\mathbb{R}, \mathcal{L}_s(X))$ if and only if $T(\cdot)$ is hyperbolic on E with projection $\mathcal{P}$. If this is the case, then the formulas*

$$P(\cdot) = \mathcal{P} = \frac{1}{2\pi i} \int_\mathbb{T} R(\lambda, T(t)) \, d\lambda \qquad and \qquad (3.7)$$

$$R(\lambda, T(t))f = \sum_{n=-\infty}^{\infty} \lambda^{-(n+1)} \Gamma(\cdot, \cdot - nt)f(\cdot - nt) \qquad (3.8)$$

hold for $t > 0$, $\lambda \in \mathbb{T}$, $f \in E$, and Green's function $\Gamma(\cdot, \cdot)$ of $U(\cdot, \cdot)$.

Proof. (1) Assume that $U(\cdot, \cdot)$ is hyperbolic with strongly continuous projections $P(\cdot)$. Then $\mathcal{P}f := P(\cdot)f$ defines a bounded projection $\mathcal{P}$ on E commuting with $T(t)$ for $t \geq 0$ due to Definition 3.1(a). Using Definition 3.1(b) and (3.2), we see that the operator $T_Q(t) : \mathcal{Q}E \to \mathcal{Q}E$ has the inverse given by

$$\left(T_Q^{-1}(t)\mathcal{Q}f\right)(s) = U_Q(s, s+t)Q(s+t)f(s+t). \tag{3.9}$$

Finally, Definition 3.1(c) implies $\|T(t)\mathcal{P}\|, \|T_Q^{-1}(t)\mathcal{Q}\| \leq Ne^{-\delta t}$ for $t \geq 0$. As a result, $T(\cdot)$ is hyperbolic.

(2) Assume that $T(\cdot)$ is hyperbolic with spectral projection $\mathcal{P}$. By Lemma 3.4 and Proposition 3.5, we have $\mathcal{P} = P(\cdot) \in C_b(\mathbb{R}, \mathcal{L}_s(X))$. The operators $P(t)$ are projections and satisfy Definition 3.1(a) since $\mathcal{P}$ is a projection commuting with $T(t)$. Because of $T(t)\mathcal{Q}E = \mathcal{Q}E$, we obtain

$$Q(s)X = \{(\mathcal{Q}f)(s) : f \in E\} = \{(T(t)\mathcal{Q}g)(s) : g \in E\} = U(s, s-t)Q(s-t)X$$

for $s \in \mathbb{R}$ and $t \geq 0$. Let $s \in \mathbb{R}$, $t \geq 0$, $x \in Q(s-t)X$, and $\varepsilon > 0$. Choose $\varphi_\varepsilon \in C_c(\mathbb{R})$ with $\varphi_\varepsilon(s-t) = x$ and $0 \leq \varphi_\varepsilon \leq 1$ such that $f := \varphi_\varepsilon Q(\cdot)x$ satisfies $\|T(t)f\|_\infty \leq \|T(t)f(s)\| + \varepsilon$. Since $f = \mathcal{Q}f$ and $T(\cdot)$ is hyperbolic, there are constants $N, \delta > 0$ such that

$$N^{-1}e^{\delta t}\|x\| \leq N^{-1}e^{\delta t}\|f\|_\infty \leq \|T(t)f\|_\infty \leq \|U(s, s-t)x\| + \varepsilon. \tag{3.10}$$

This yields condition (b) and the second estimate in (c) of Definition 3.1. For the other estimate we take $f = \varphi_\varepsilon x$, where $\varphi_\varepsilon \in C_c(\mathbb{R})$ with $\varphi_\varepsilon(s-t) = 1$ and $0 \leq \varphi_\varepsilon \leq 1$. Then

$$\|U(s, s-t)P(s-t)x\| \leq \|T(t)\mathcal{P}f\|_\infty \leq Ne^{-\delta t}\|f\|_\infty = Ne^{-\delta t}\|x\|.$$

(3) Formula (3.7) follows from (3.4), and (3.8) can easily be checked using (3.9).

In our applications it is necessary to control the dichotomy constants of $U(\cdot, \cdot)$ by given quantities. This is rather simple for the exponent δ, but quite complicated for the constant N. We establish two lemmas which allow to overcome this problem. The first one is taken from [43, Lem.2.2]. To simplify notation, we set

$$U(t, s) := 0 \qquad \text{for } t < s. \tag{3.11}$$

Lemma 3.7. *Let $U(\cdot, \cdot)$ be a strongly continuous evolution family with $J = \mathbb{R}$ on X such that $\|U(t, s)\| \leq Me^{\gamma(t-s)}$ and $\|R(1, T(1))\| \leq C$ for the corresponding evolution semigroup on $E = C_0(\mathbb{R}, X)$. Then $U(\cdot, \cdot)$ has an exponential dichotomy with exponent $\delta \in (0, \log(1 + \frac{1}{C}))$ and a constant $N \geq 1$ depending only on C, M, γ, and δ.*

Proof. Lemma 3.3 and standard spectral theory, [21, Prop.IV.1.3], imply

$$\|R(\lambda, T(1))\| \leq \tilde{C} := \frac{C}{1 - (e^\delta - 1)C} \tag{3.12}$$

for $|\lambda| = e^\alpha$ and $0 \le |\alpha| \le \delta < \log(1 + \frac{1}{C})$. By Theorem 3.6 and a simple rescaling argument, we obtain the exponential dichotomy of $U(\cdot, \cdot)$ for every exponent $0 < \delta < \log(1 + \frac{1}{C})$. Fix such a δ. If the result were false, then there would exist evolution families $U_n(\cdot, \cdot)$ on Banach spaces X_n satisfying the assumptions (and thus (3.12)), real numbers t_n and s_n, and elements $x_n \in X_n$ such that $\|x_n\| = 1$ and

$$e^{\delta|t_n - s_n|} \|\Gamma_n(t_n, s_n)x_n\| \longrightarrow \infty \qquad \text{as } n \to \infty, \tag{3.13}$$

where $\Gamma_n(\cdot, \cdot)$ is Green's function of $U_n(\cdot, \cdot)$. By Lemma 3.3 and (3.7), the projections $P_n(t)$ are uniformly bounded for $t \in \mathbb{R}$ and $n \in \mathbb{N}$. Hence the operators $\Gamma_n(t, s)$, $s \le t \le s + 1$, $n \in \mathbb{N}$, are also uniformly bounded. Thus we have either $t_n > s_n + 1$ or $t_n < s_n$ in (3.13).

In the first case, we write $t_n = s_n + k_n + \tau_n$ for $k_n \in \mathbb{N}$ and $\tau_n \in (0, 1]$. Otherwise, $t_n = s_n - k_n + \tau_n$ for $k_n \in \mathbb{N}$ and $\tau_n \in (0, 1]$. Take continuous functions φ_n with $0 \le \varphi_n \le 1$, $\operatorname{supp} \varphi_n \subset (s_n + \frac{\tau_n}{2}, s_n + \frac{3\tau_n}{2})$, and $\varphi_n(t_n \mp k_n) = 1$ (here $t_n - k_n$ is used in the first case, and $t_n + k_n$ in the second). Set $\lambda = e^{\mp \delta}$ and $f_n(s) = e^{\pm \delta(s - s_n)} \varphi_n(s) U_n(s, s_n)x_n$ for $s \in \mathbb{R}$. Using (3.8) and a rescaling, we obtain

$$
\begin{aligned}
[R(\lambda, T_{U_n}(1))f_n](t_n) &= \sum_{k=0}^{\infty} \lambda^{-(k+1)} U_n(t_n, t_n - k) P_n(t_n - k) f_n(t_n - k) \\
&\quad - \sum_{k=1}^{\infty} \lambda^{k-1} U_{n,Q}(t_n, t_n + k) Q_n(t_n + k) f(t_n + k) \\
&= \sum_{k=0}^{\infty} e^{\pm \delta(t_n - s_n + 1)} \varphi_n(t_n - k) U_n(t_n, s_n) P_n(s_n)x_n \\
&\quad - \sum_{k=1}^{\infty} e^{\pm \delta(t_n - s_n + 1)} \varphi_n(t_n + k) U_{n,Q}(t_n, s_n) Q_n(s_n)x_n.
\end{aligned}
$$

Here exactly one term does not vanish, namely $k = k_n$ in the first sum if $t_n > s_n$ and $k = k_n$ in the second sum if $t_n < s_n$, so that

$$[R(\lambda, T_{U_n}(1))f_n](t_n) = e^{\pm \delta} e^{\delta|t_n - s_n|} \Gamma_n(t_n, s_n)x_n \,.$$

Thus (3.12) yields

$$\|e^{-\delta} e^{\delta|t_n - s_n|} \Gamma_n(t_n, s_n)x_n\| \le \|R(\lambda, T_{U_n}(1))f_n\|_\infty \le \tilde{C} M e^{2(\gamma + \delta)}.$$

This estimate contradicts (3.13).

In a second step we involve the generator G of the evolution semigroup $T(\cdot)$ and establish the spectral mapping theorem for $T(\cdot)$ and G. This fact is somewhat astonishing since the standard results from semigroup theory can not be applied here, cf. [21, §IV.3]. Instead one has to construct approximative eigenfunctions of G directly using formula (3.6).

Proposition 3.8. *Let $U(\cdot,\cdot)$ be an exponentially bounded evolution family with $J = \mathbb{R}$ on X being strongly continuous for $t \geq s$. Then the corresponding evolution semigroup $T(\cdot)$ on $E = C_0(\mathbb{R}, X)$ with generator G satisfies $\sigma(T(t)) \setminus \{0\} = \exp(t\sigma(G))$ for $t \geq 0$.*

Proof. By standard spectral theory of semigroups, see [21, Thm.IV.3.6., IV.3.7], and a rescaling argument, it suffices to show that $0 \in \sigma(G)$ if $1 \in A\sigma(T(t_0))$ for some $t_0 > 0$. So assume that $1 \in A\sigma(T(t_0))$. For each $n \in \mathbb{N}$, there exists $f_n \in C_0(\mathbb{R}, X)$ such that $\|f_n\|_\infty = 1$ and $\|f_n - T(kt_0)f_n\|_\infty < \frac{1}{2}$ for all $k = 0, 1, \ldots, 2n$. Hence,

$$\frac{1}{2} < \sup_{s \in \mathbb{R}} \|U(s, s - kt_0)f_n(s - kt_0)\| \leq 2 \tag{3.14}$$

for $k = 0, 1, \ldots, 2n$. Take $s_n \in \mathbb{R}$ such that $\|U(s_n, s_n - nt_0)x_n\| \geq \frac{1}{2}$ where $x_n := f_n(s_n - nt_0)$. Let $I_n = (s_n - nt_0, s_n + nt_0)$. Choose $\alpha_n \in C^1(\mathbb{R})$ such that $\alpha_n(s_n) = 1$, $0 \leq \alpha_n \leq 1$, $\operatorname{supp}\alpha_n \subseteq I_n$, and $\|\alpha_n'\|_\infty \leq \frac{2}{nt_0}$. Define $g_n(s) = \alpha_n(s)U(s, s_n - nt_0)x_n$ for $n \in \mathbb{N}$ and $s \in \mathbb{R}$ (recall (3.11)). Then $g_n \in E$, $\|g_n\|_\infty \geq \|g_n(s_n)\| \geq \frac{1}{2}$, and

$$T(t)g_n(s) = \alpha_n(s-t)U(s, s-t)U(s-t, s_n-nt_0)x_n = \alpha_n(s-t)U(s, s_n-nt_0)x_n$$

for $s - t \geq s_n - nt_0$ and $T(t)g_n(s) = 0$ for $s - t < s_n - nt_0$. Therefore, $g_n \in D(G)$ and $Gg_n(s) = -\alpha_n'(s)U(s, s_n - nt_0)x_n$. Each $s \in I_n$ can be written as $s = s_n + (k + \sigma - n)t_0$ for $k \in \{0, 1, \ldots, 2n\}$ and $\sigma \in [0, 1)$. Using $M := \sup\{\|U(t,s)\| : 0 \leq t - s \leq t_0\} < \infty$ and (3.14), we estimate

$$\|Gg_n(s)\| \leq \frac{2M}{nt_0}\|U(s_n + (k-n)t_0, s_n - nt_0)\,x_n\|$$
$$= \frac{2M}{nt_0}\|U(r_n, r_n - kt_0)f_n(r_n - kt_0)\| \leq \frac{4M}{nt_0}$$

for $s \in I_n$ and $r_n := s_n + (k - n)t_0$. Consequently, $0 \in A\sigma(A)$.

We have thus established our main characterization theorem.

Theorem 3.9. *Let $U(\cdot,\cdot)$ be an exponentially bounded evolution family with time interval $J = \mathbb{R}$ on a Banach space X being strongly continuous for $t \geq s$ and $T(\cdot)$ be the corresponding evolution semigroup on $E = C_0(\mathbb{R}, X)$ with generator G. Then the following assertions are equivalent.*

(a) $U(\cdot,\cdot)$ is hyperbolic with projections $P(t)$.
(b) $T(\cdot)$ is hyperbolic with projection $\mathcal{P}$.
(c) $\rho(T(t)) \cap \mathbb{T} \neq \emptyset$ for one/all $t > 0$.
(d) $\rho(G) \cap i\mathbb{R} \neq \emptyset$.

In (a) the projections are automatically strongly continuous. If (a)–(d) are true, then (3.7) and (3.8) hold and we have, for $\lambda \in i\mathbb{R}$,

$$(R(\lambda, G)f)(t) = \int_{\mathbb{R}} e^{-\lambda(t-s)}\Gamma(t,s)f(s)\,ds, \qquad t \in \mathbb{R}, \ f \in E. \tag{3.15}$$

Proof. Due to Theorem 3.6, Lemma 3.3, Proposition 3.8, (3.1), and a rescaling argument, it remains to verify (3.15) for $\lambda = 0$. Set $u = \int_{\mathbb{R}} \Gamma(\cdot, s) f(s)\, ds$ for $f \in E$. Then $u \in E$ and

$$(T(h)u)(t) = \int_{-\infty}^{t-h} U(t,s)P(s)f(s)\, ds - \int_{t-h}^{\infty} U_Q(t,s)Q(s)f(s)\, ds,$$

$$\frac{1}{h}\left[(T(h)u)(t) - u(t)\right] = -\frac{1}{h}\int_{t-h}^{t} U(t,s)f(s)\, ds$$

for $h > 0$. Therefore $u \in D(G)$ and $Gu = -f$.

The next lemma (taken from [56]) allows to control the dichotomy constants by $\|G^{-1}\|$.

Lemma 3.10. *Let $U(\cdot, \cdot)$ be a strongly continuous evolution family with $J = \mathbb{R}$ on X such that $\|U(t,s)\| \leq M^{\gamma(t-s)}$ and $\|G^{-1}\| \leq 1/\eta$ for the generator of the corresponding evolution semigroup on $E = C_0(\mathbb{R}, X)$ and some $\eta > 0$. Then $U(\cdot, \cdot)$ has exponential dichotomy with exponent $\delta \in (0, \eta)$ and constant N depending only on η, M, γ, and δ.*

Proof. Since $(-\eta, \eta) \subseteq \rho(G)$, the evolution family $U(\cdot, \cdot)$ is hyperbolic with exponent $\delta \in (0, \eta)$ by Theorem 3.9 and a simple rescaling argument. Fix some $\delta \in (0, \eta)$. Suppose the dichotomy constant $N = N(\delta)$ were not uniform in η, M, γ. Then there would exist evolution families $U_n(\cdot, \cdot)$ on Banach spaces X_n fulfilling the assumptions, vectors $x_n \in X_n$ with $\|x_n\| = 1$, and numbers $t_n, s_n \in \mathbb{R}$ such that Green's function $\Gamma_n(\cdot, \cdot)$ of $U_n(\cdot, \cdot)$ would satisfy

$$e^{\delta|t_n - s_n|}\, \|\Gamma_n(t_n, s_n)x_n\| \longrightarrow \infty \qquad \text{as } n \to \infty. \tag{3.16}$$

Set $f_n(s) = \varphi_n(s)e^{\pm\delta(s-s_n)} U_n(s, s_n)x_n$ for $s \in \mathbb{R}$, where $\varphi_n \in C(\mathbb{R})$ has compact support in $(s_n, s_n + 2)$, $0 \leq \varphi_n \leq 1$, $\int \varphi_n(s)\, ds = 1$, and we take $+\delta$ if $t_n \geq s_n$ and $-\delta$ if $t_n < s_n$. Using (3.15) and a rescaling, we obtain

$$R(\mp\delta, G_{U_n})f_n(t_n) = \int_{\mathbb{R}} e^{\pm\delta(t_n - s)}\Gamma_n(t_n, s)f_n(s)\, ds$$

$$= e^{\delta|t_n - s_n|} \int_{s_n}^{s_n+2} \varphi_n(s)\, \Gamma_n(t_n, s)U_n(s, s_n)x_n\, ds.$$

The right hand side equals $e^{\delta|t_n - s_n|}\, \Gamma_n(t_n, s_n)x_n$ if $t_n \notin [s_n, s_n + 2]$ and

$$e^{\delta|t_n - s_n|}\left[U_n(t_n, s_n)P_n(s_n)x_n \int_{s_n}^{t_n} \varphi_n(s)ds - U_n(t_n, s_n)Q_n(s_n)x_n \int_{t_n}^{s_n+2} \varphi_n(s)\, ds\right]$$

$$= e^{\delta|t_n - s_n|}\, \Gamma_n(t_n, s_n)x_n - e^{\delta|t_n - s_n|}\, U_n(t_n, s_n)x_n \int_{t_n}^{s_n+2} \varphi_n(s)\, ds$$

if $t_n \in [s_n, s_n + 2]$. In both cases, (3.16) implies that $\|R(\mp\delta, G_{U_n})f(t_n)\| \to \infty$ as $n \to \infty$. This contradicts the estimate

$$\|R(\mp\delta, G_{U_n})f_n\|_\infty \le \|G_{U_n}^{-1}(\mp\delta G_{U_n}^{-1} - I)^{-1}\|\,\|f_n\|_\infty \le \frac{M\,e^{2(\delta+\gamma)}}{\eta - \delta}.$$

Theorem 3.6 is essentially due to R. Rau, [52]. Y. Latushkin and S. Mont-gomery–Smith showed Proposition 3.8 in [33]. Above we presented a somewhat simpler proof taken from [51]. Using different methods, Theorem 3.6 was also proved in [33] and [34]. Theorem 3.9 remains valid if $E = C_0(\mathbb{R}, X)$ is replaced by $L^p(\mathbb{R}, X)$, $1 \le p < \infty$, see [12], [33], [34], [51]. For an alternative approach using 'mild solutions' of the inhomogeneous Cauchy problem, we refer to [10], [12], [15, §3], [16, §IV.3], [35], [44], and the references therein.

3.3 Extensions and the Parabolic Case

Theorem 3.9 is not quite sufficient for the applications in Section 4. First, the evolution family in the parabolic case is strongly continuous only for $t > s$ if the domains of $A(t)$ are not dense in X. Second, in some proofs we 'glue together' evolution families $U(\cdot, \cdot)$ and $V(\cdot, \cdot)$ at a point $a \in \mathbb{R}$; i.e., we set

$$W(t, s) = \begin{cases} U(t, s), & t \ge s \ge a, \\ U(t, a)V(a, s), & t \ge a \ge s, \\ V(t, s), & a \ge t \ge s. \end{cases} \tag{3.17}$$

This clearly defines an evolution family which is strongly continuous for $t \ge s$ if the same holds for $U(\cdot, \cdot)$ and $V(\cdot, \cdot)$. However, if $U(\cdot, \cdot)$ is only strongly continuous for $t > s$, then $t \mapsto W(t, a - 1)$ becomes discontinuous at $t = a$. Fortunately, in our setting we still obtain one–sided strong continuity of $s \mapsto W(s, s - t)$, see e.g. the proof of Theorem 4.1, and this property suffices to save Theorem 3.6.

We start with the second problem. Let $U(\cdot, \cdot)$ be an exponentially bounded evolution family with $J = \mathbb{R}$ such that $\mathbb{R} \ni s \mapsto U(s, s - t)$ is strongly continuous for each $t \ge 0$. Then (3.6) still defines an exponentially bounded semigroup on $E = C_0(\mathbb{R}, X)$. An inspection of the proof of Theorem 3.6 and of the auxiliary results leading to it shows that the arguments given there remain valid in the present setting. If we additionally assume that $t \mapsto U(t, s)$ is strongly continuous for $t > s$ and each $s \in \mathbb{R}$, then also Lemma 3.7 can be extended to this situation. Next, assume that $U(\cdot, \cdot)$ is an exponentially bounded evolution family with $J = \mathbb{R}$ such that $\mathbb{R} \ni s \mapsto U(s, s - t)$ is strongly continuous from the left for each $t \ge 0$. In this case (3.6) defines an exponentially bounded semigroup $T_U(\cdot) = T(\cdot)$ on the space

$$E_l = E_l(X) := \{f : \mathbb{R} \to X : f \text{ is bounded and left continuous}\}$$

endowed with the supremum norm. Observe that E_l is a Banach space and that the assertions of Lemmas 3.3 and 3.4 concerning $T(t)$ still hold. The other arguments need some modifications. In Proposition 3.5 one only obtains that $\mathcal{M} = M(\cdot)$ is strongly continuous from the left if $\mathcal{M}(\varphi f) = \varphi \mathcal{M} f$ for $f \in E_l(X)$ and $\varphi \in E_l(\mathbb{C})$. In the proof we have to replace the function φ_ε by, e.g., $\mathbb{1}_{(t-\varepsilon,t]}$. Analogously, we can only require strong continuity from the left of the dichotomy projection $P(\cdot)$ of $U(\cdot,\cdot)$ in Theorem 3.6. In the proof of this theorem one further has to change the function φ_ε in (3.10) into $\mathbb{1}_{(s-t-\eta,s-t]}$, where $\eta = \eta(\varepsilon)$ is sufficiently small. Lemma 3.7 still holds if we also assume that $t \mapsto U(t,s)$ is strongly continuous from the left for $t > s$ and each $s \in \mathbb{R}$. Of course, analogous arguments work for 'right continuity' if we use the space

$$E_r := \{f : \mathbb{R} \to X : f \text{ is bounded and right continuous}\}.$$

Thus we have shown the following facts.

Theorem 3.11. *Let $U(\cdot,\cdot)$ be an exponentially bounded evolution family with time interval $J = \mathbb{R}$ on a Banach space X such that the map $\mathbb{R} \ni s \mapsto U(s, s-t)$ is strongly continuous (from the left, resp. right) for each $t \geq 0$. Let $T(\cdot)$ be the corresponding evolution semigroup on E (on E_l, resp. E_r). Then $U(\cdot,\cdot)$ has an exponential dichotomy on X with projections $P(t)$ being strongly continuous in $t \in \mathbb{R}$ (from the left, resp. right) if and only if $T(\cdot)$ is hyperbolic on E (on E_l, resp. E_r) with projection $\mathcal{P}$ if and only if $\rho(T(t)) \cap \mathbb{T} \neq \emptyset$ for some/all $t > 0$. If this is the case, then the formulas (3.7) and (3.8) hold. Let $\|R(1, T(1))\| \leq C$. Then the dichotomy exponent δ can be chosen from the interval $(0, \log(1 + \frac{1}{C}))$. If, in addition, $(s, \infty) \ni t \mapsto U(t, s)$ is strongly continuous (from the left, resp. right) for each $s \in \mathbb{R}$, then the dichotomy constant N only depends on δ, C, and the exponential estimate of $U(\cdot,\cdot)$.*

We state two more variants of the above theorem as remarks which can be proved by the same reasoning.

Remark 3.12. One can replace E by $\tilde{E} = C_b(\mathbb{R}, X)$ in Theorem 3.11 in the case that $\mathbb{R} \ni s \mapsto U(s, s-t)$ is strongly continuous for $t \geq 0$.

Remark 3.13. One implication in Theorem 3.11 holds without any continuity assumption: Let $U(\cdot,\cdot)$ be an exponentially bounded evolution family with $J = \mathbb{R}$ on X which has an exponential dichotomy with projections $P(t)$. As in (3.6) we define the evolution semigroup $T(\cdot)$ on the space $B(\mathbb{R}, X)$ of bounded functions $f : \mathbb{R} \to X$ endowed with the sup–norm. Then $T(\cdot)$ is hyperbolic and (3.7) and (3.8) hold for $f \in B(\mathbb{R}, X)$.

The latter remark implies the uniqueness of the dichotomy projections $P(\cdot)$ if $J = \mathbb{R}$. For an invertible evolution family this fact can be proved directly, cf. [16, p.164].

Corollary 3.14. *Let $U(\cdot,\cdot)$ be an hyperbolic evolution family with $J = \mathbb{R}$. Then the dichotomy projections $P(s)$, $s \in \mathbb{R}$, of $U(\cdot,\cdot)$ are uniquely determined. If $U(s + t, s) = U(t)$ for $s \in \mathbb{R}$, $t \geq 0$, and an exponentially bounded semigroup $U(\cdot)$, then the projections $P(s)$, $s \in \mathbb{R}$, equal the spectral projection P of $U(\cdot)$.*

Proof. The first claim follows from formula (3.7) which holds due to Remark 3.13. If $U(s + t, s) = U(t)$, then we have $\sigma(U(t)) \cap \mathbb{T} = \emptyset$ by (the proof of) [12, Lem.2.38]. Thus $U(\cdot)$ has a spectral projection P, and $P \equiv P(t)$ due to the first assertion.

Theorem 3.11 yields a fundamental robustness result. The first part is well known for strongly continuous evolution families, see e.g. [12, Thm.5.23], [13, Thm.4.3], [27, Thm.7.6.10]. Equation (3.19) is taken from [57, Prop.2.3] and the last assertion from [43, Prop.2.1]. In particular, (3.19) implies that $V(\cdot, \cdot)$ inherits the exponential stability of $U(\cdot, \cdot)$.

Theorem 3.15. *Let $U(\cdot, \cdot)$ and $V(\cdot, \cdot)$ be exponentially bounded evolution families with $J = \mathbb{R}$ such that $\mathbb{R} \ni s \mapsto U(s+t, s)$ and $\mathbb{R} \ni s \mapsto V(s+t, s)$ are strongly continuous (from the left, resp. right) for each $t \geq 0$. Assume that $U(\cdot, \cdot)$ has an exponential dichotomy with projections $P_U(s)$ being strongly continuous (from the left, resp. right) and constants $N, \delta > 0$ and that*

$$q(\tau) := \sup_{s \in \mathbb{R}} \|U(s + \tau, s) - V(s + \tau, s)\| \leq \frac{(1 - e^{-\delta \tau})^2}{8N^2} \qquad (3.18)$$

for some $\tau > 0$. Then $V(\cdot, \cdot)$ has an exponential dichotomy with exponent $0 < \delta' < -\frac{1}{\tau} \log(2q(\tau)N + e^{-\delta \tau})$ and projections $P_V(s)$ which are strongly continuous (from the left, resp. right) and satisfy

$$\dim P_V(s)X = \dim P_U(s)X \quad \text{and} \quad \dim \ker P_V(s) = \dim \ker P_U(s) \quad (3.19)$$

for $s \in \mathbb{R}$. If also $(s, \infty) \ni t \mapsto V(t, s)$ is strongly continuous (from the left, resp. right) for each $s \in \mathbb{R}$, then the dichotomy constant N' of $V(\cdot, \cdot)$ depends only on N, δ, δ', and the exponential estimate of $V(\cdot, \cdot)$.

Proof. For simplicity we consider $\tau = 1$, the proof for arbitrary $\tau > 0$ is the same. We set $q := q(1)$. Formula (3.8) allows us to estimate

$$\|R(\lambda, T_U(1))\| \leq \frac{2N}{1 - e^{-\delta}} \qquad (3.20)$$

for $\lambda \in \mathbb{T}$ and the corresponding evolution semigroup $T_U(\cdot)$ on E (on E_l, resp. E_r). Therefore condition (3.18) implies that $\lambda - T_V(1)$ has the inverse

$$R(\lambda, T_V(1)) = R(\lambda, T_U(1)) \sum_{n=0}^{\infty} [(T_V(1) - T_U(1))R(\lambda, T_U(1))]^n \quad \text{and}$$

$$\|R(\lambda, T_V(1))\| \leq \frac{2N}{(1 - e^{-\delta})}\left(1 - \frac{2Nq}{(1 - e^{-\delta})}\right)^{-1} \leq \frac{8N}{3(1 - e^{-\delta})}. \qquad (3.21)$$

Thus $V(\cdot, \cdot)$ is hyperbolic by Theorem 3.11. Considering $e^{\pm \mu(t-s)}U(t, s)$ and $e^{\pm \mu(t-s)}V(t, s)$ for $\mu \in (0, \delta)$, we see that $I - e^{\pm \mu}T_V(1)$ is invertible if

$$e^\mu \frac{2qN}{1 - e^{(\mu-\delta)}} < 1.$$

This yields the asserted estimate for the exponent δ' because of Theorem 3.11. We further deduce from (3.7) that

$$P_V(\cdot) - P_U(\cdot) = \frac{1}{2\pi i} \int_{\mathbb{T}} [R(\lambda, T_V(1)) - R(\lambda, T_U(1))]\, d\lambda$$

$$= \frac{1}{2\pi i} \int_{\mathbb{T}} R(\lambda, T_V(1))\, [T_V(1) - T_U(1)]\, R(\lambda, T_U(1))\, d\lambda.$$

Together with (3.18), (3.20), (3.21), this identity yields

$$\|P_V(s) - P_U(s)\| \le q \frac{16\, N^2}{3(1 - e^{-\delta})^2} \le \frac{2}{3} \tag{3.22}$$

for $s \in \mathbb{R}$. Assertion (3.19) now follows from [32, p.298]. The last claim is a consequence of Theorem 3.11 and (3.21).

If $U(\cdot, \cdot)$ and $V(\cdot, \cdot)$ are strongly continuous, then the above result can be extended to time intervals $J = [a, \infty)$ by a suitable extension as in (3.17). This is formulated in a different setting in Theorem 4.1, where we also give further references and discuss the smallness assumption (3.18).

In a second step we want to extend the full Theorem 3.9 to evolution families being strongly continuous for $t > s$. As the following arguments indicate, probably this objective cannot be achieved without further assumptions. In view of the scope of the present lecture notes, we assume that the Acquistapace–Terreni conditions (AT) from Section 2 hold with $J = \mathbb{R}$. Thus we have operators $A(t)$, $t \in \mathbb{R}$, generating an exponentially bounded evolution family $U(\cdot, \cdot)$ on X being strongly continuous for $t > s$. It is easy to see that the multiplication operator $A(\cdot)$ (see (2.12)) generates the analytic semigroup $e^{tA(\cdot)}$ on $E = C_0(\mathbb{R}, X)$. On E we also define the first derivative $\frac{d}{dt} f = f'$ with domain $C_0^1(\mathbb{R}, X) = \{f \in C^1(\mathbb{R}, X) : f, f' \in C_0(\mathbb{R}, X)\}$.

By means of formula (3.6) we define an exponentially bounded evolution semigroup $T(\cdot) = T_U(\cdot)$ on $E = C_0(\mathbb{R}, X)$ being strongly continuous on $(0, \infty)$. Hence, we can apply Theorem 3.11 to this semigroup. We further introduce the space of strong continuity of $T(\cdot)$:

$$E_0 = \{f \in E : T(t)f \to f \text{ as } t \to 0\}.$$

Clearly, E_0 is closed subspace of E, $T(t)E_0 \subseteq E_0$, and the restriction $T_0(t) : E_0 \to E_0$ of $T(t)$ yields a C_0-semigroup. We denote its generator by G_0. Let $f \in E$. Observe that

$$T(t)f(s) - f(s) = U(s, s-t)[f(s-t) - f(s)] + [U(s, s-t) - e^{tA(s-t)}]f(s)$$
$$+ [e^{tA(s-t)} - e^{tA(s)}]f(s) + [e^{tA(s)} - I]f(s)$$

for $t \geq 0$ and $s \in \mathbb{R}$. Using (2.14) and (2.9), we see that

$$f \in E_0 \iff \|e^{tA(\cdot)}f - f\|_\infty \to 0 \text{ as } t \to 0 \iff f \in \overline{D(A(\cdot))}.$$

In other words, $T(t)$ and $e^{tA(\cdot)}$ possess the same space of strong continuity $E_0 = \overline{D(A(\cdot))}$. We also have $T(t)E \subseteq D(A(\cdot)) \subseteq E_0$ for $t > 0$ by Theorem 2.2. Hence, $T(t)$ induces the 0 operator on the quotient E/E_0 and $T_0(t)$ is not surjective for $t > 0$ (unless $D(A(s)) = X$ for all s, but in this case $T_0(t) = T(t)$). So we deduce from [21, Prop.IV.2.15] that $\sigma(T_0(t)) = \sigma(T(t))$ for $t \geq 0$.

In order to associate an operator G with $T(\cdot)$, we take the Laplace transform of $T(\cdot)$. This yields the bounded operator R_λ on E given by

$$(R_\lambda f)(t) := \left(\int_0^\infty e^{-\lambda \tau} T(\tau) f \, d\tau \right)(t) \tag{3.23}$$

$$= \int_{-\infty}^t e^{-\lambda(t-s)} U(t,s) f(s) \, ds \tag{3.24}$$

for $t \in \mathbb{R}$, $f \in E$, and $\operatorname{Re}\lambda > \omega(U)$. As in [8, Thm.3.1.7] one sees that R_λ fulfils the resolvent equation. In view of (3.24), the function $u = R_\lambda f$ satisfies

$$u(t) = e^{-\lambda(t-s)} U(t,s) u(s) + \int_s^t e^{-\lambda(t-\tau)} U(t,\tau) f(\tau) \, d\tau, \quad t \geq s. \tag{3.25}$$

Conversely, if (3.25) holds for $u, f \in E$, then one obtains $u = R_\lambda f$ by letting $s \to -\infty$ in (3.25). If $R_\lambda f = 0$ for some $\operatorname{Re}\lambda > \omega(U)$, then $R_\mu f = 0$ for each $\operatorname{Re}\mu > \omega(U)$ by the resolvent equation, and thus $T(t)f = 0$ for $t > 0$ due to the uniqueness of the Laplace transform and the continuity of $T(\cdot)f$. This means that $U(s+t,s)f(s) = 0$ for $t > 0$ and $s \in \mathbb{R}$. Estimate (2.14) then implies that $e^{tA(s)}f(s) \to 0$ as $t \to 0$, so that $f = 0$ because of [42, Prop.2.1.4]. Since the operators R_λ are injective, there exists a closed operator G with $D(G) = R_\lambda E$ such that $R_\lambda = R(\lambda, G)$ for $\operatorname{Re}\lambda > \omega(U)$, see the proof of e.g. [21, Prop.III.4.6].

For $u \in D(A(\cdot)) \cap C_0^1(\mathbb{R}, X)$ and $\operatorname{Re}\lambda > \omega(U)$, we set $f := \lambda u + u' - A(\cdot)u \in E$. Then u is a strict solution of (2.1) for the operators $A(t) - \lambda$, the initial value $u(s)$, and the inhomogeneity f. Thus u satisfies (3.25) by Theorem 2.7, and hence $u = R_\lambda f$. This means that G extends the operator $-\frac{d}{dt} + A(\cdot)$ defined on $D(A(\cdot)) \cap C_0^1(\mathbb{R}, X)$. Moreover, (3.25) (with $s = t-1$), (2.13), and (2.32) (with $\varepsilon = 1/2$) imply that $u \in R(\lambda, G)[E \cap C_b^\alpha(\mathbb{R}, X)]$ is differentiable, $u(t) \in D(A(t))$, and $u', A(\cdot)u \in C_b^\alpha(\mathbb{R}, X)$ for $\alpha \in (0, \mu + \nu - 1)$. We deduce

$$R(\lambda, G)[E \cap C_b^\alpha(\mathbb{R}, X)] \subseteq D(A(\cdot)) \cap C_0^1(\mathbb{R}, X)$$

by an interpolation argument using [42, Prop.0.2.2, 1.2.19], cf. (5.13). Consequently, $D(A(\cdot)) \cap C_0^1(\mathbb{R}, X)$ is a core of G and, hence, $D(G) \subseteq E_0$. Since $R(\lambda, G)f = R(\lambda, G_0)f$ for $f \in E_0$ and $\operatorname{Re}\lambda > \omega(U)$ due to (3.23), the operator G_0 is the part of G in E_0. This fact and the inclusion $D(G) \subseteq E_0$ yield $\sigma(G_0) = \sigma(G)$ thanks to [21, Prop.IV.2.17]. We summarize our observations in the next proposition.

Proposition 3.16. *Assume that $A(t)$, $t \in \mathbb{R}$, satisfy the conditions (AT) from Section 2. Let $U(\cdot, \cdot)$ be the evolution family on X generated by $A(\cdot)$ and $T(\cdot)$ be the induced evolution semigroup on $E = C_0(\mathbb{R}, X)$. Then the operator $-\frac{d}{dt} + A(\cdot)$ defined on $D(A(\cdot)) \cap C_0^1(\mathbb{R}, X)$ has a closure G in E whose resolvent is given the Laplace transform of $T(\cdot)$. The restriction $T_0(\cdot)$ of $T(\cdot)$ to $E_0 = \overline{D(A(\cdot))}$ is a C_0–semigroup generated by the part G_0 of G in E_0. Moreover, $\sigma(G) = \sigma(G_0)$ and $\sigma(T(t)) = \sigma(T_0(t))$ for $t \geq 0$.*

The desired extension of Theorem 3.9 now follows from the previous results.

Theorem 3.17. *In the situation of Proposition 3.16 the following assertions are equivalent.*

(a) $U(\cdot, \cdot)$ is hyperbolic on X with projections $P(t)$.
(b) $T(\cdot)$ is hyperbolic on E with projection $\mathcal{P}$.
(c) $T_0(\cdot)$ is hyperbolic on E_0.
(d) $\rho(T(t)) \cap \mathbb{T} \neq \emptyset$ for one/all $t > 0$.
(e) $\rho(T_0(t)) \cap \mathbb{T} \neq \emptyset$ for one/all $t > 0$.
(f) $\rho(G) \cap i\mathbb{R} \neq \emptyset$.
(g) $\rho(G_0) \cap i\mathbb{R} \neq \emptyset$.

If this is the case, then formulas (3.7), (3.8), and (3.15) hold for $T(\cdot)$ and G and, after restriction to E_0, also for $T_0(\cdot)$ and G_0. The dichotomy constants of $U(\cdot, \cdot)$ only depend on its exponential estimate and either $\|R(1, T(1))\|$ or $\|R(1, T_0(1))\|$ or $\|G^{-1}\|$ or $\|G_0^{-1}\|$.

Proof. By Theorem 3.11 assertions (a), (b), and (d) are equivalent and the concluding claims involving $T(\cdot)$ hold. Observe that Lemma 3.3 is valid for $T_0(\cdot)$. Then Proposition 3.16 yields the equivalence of (c), (d), and (e) as well as of (f) and (g). The proof of Proposition 3.8 can directly be extended to the semigroup $T_0(\cdot)$ so that also (e) and (g) are equivalent. Let $u = \int_\mathbb{R} \Gamma(\cdot, \tau) f(\tau) \, d\tau$ for $f \in E$. Then one has $u \in E$ and

$$u(t) = U(t, s)u(s) + \int_s^t U(t, \tau) f(\tau) \, d\tau, \quad t \geq s.$$

As above we see that $u \in D(G)$ and $Gu = -f$ if $f \in C_b^\alpha(\mathbb{R}, X) \cap E$. By approximation and rescaling we deduce (3.15) for all $f \in E$. One easily checks that $R(\lambda, T(t)) f = R(\lambda, T_0(t)) f$ and $R(\mu, G) f = R(\mu, G_0) f$ for $\lambda \in \mathbb{T}$, $\mu \in i\mathbb{R}$, and $f \in E_0$. Thus, (3.7), (3.8), and (3.15) are true for $T_0(t)$ and G_0. Finally, the proofs of Lemmas 3.7 and 3.10 work also for $T_0(\cdot)$, G_0, and G.

A related characterization of exponential dichotomy in terms of bounded strong solutions to the inhomogeneous problem was shown in [44, Thm.3.2].

Concluding this section we establish additional regularity properties of Green's function in the parabolic case. Refinements of the first part of the next proposition are (in principle) well known, compare [27, Lem.7.6.2] and [42, §6.3]. Parts (b) and (c), however, seem to be new for non–periodic parabolic problems. Their proofs are based on Theorem 3.17.

Proposition 3.18. *Assume that $A(t)$, $t \in J$, satisfy (AT) and that the generated evolution family $U(\cdot, \cdot)$ has an exponential dichotomy with constants $N, \delta > 0$ and projections $P(t)$ on $J \in \{\mathbb{R}, [a, \infty)\}$.*
(a) Let $s \geq t \geq a + \eta > a$ if $J = [a, \infty)$ and $s \geq t$ if $J = \mathbb{R}$. Then $Q(t)X \subseteq D(A(t))$, $P(t)D(A(t)) \subseteq D(A(t))$,

$$\|A(t)Q(t)\| \leq c, \ \|P(t)\|_{\mathcal{L}(D(A(t)))} \leq c, \ \|A(t)U_Q(t,s)Q(s)\| \leq ce^{\delta(t-s)}. \quad (3.26)$$

Also, $U_Q(\cdot, s)Q(s) \in C^1((a, \infty), \mathcal{L}(X))$ and $\frac{d}{dt}U_Q(t,s)Q(s) = A(t)U_Q(t,s)Q(s)$ (where $a := -\infty$ if $J = \mathbb{R}$). Let $t > s$. Then

$$\|A(t)U(t,s)P(s)\| \leq c \max\{1, (t-s)^{-1}\}e^{-\delta(t-s)}. \quad (3.27)$$

(b) If $J = \mathbb{R}$, then $P(\cdot) \in C_b^\kappa(\mathbb{R}, \mathcal{L}(X))$ for $\kappa = \mu + \nu - 1 > 0$.
(c) If $J = [a, \infty)$, then $P(\cdot) \in C_b^\kappa([a + \eta, \infty), \mathcal{L}(X))$ for each $\eta > 0$.
The constants c in (3.26) if $J = \mathbb{R}$ and (3.27) and the Hölder constants in (b) only depend on N, δ, and the constants in (AT). The constants in (3.26) if $J = [a, \infty)$ and (c) depend additionally on $\eta > 0$.

Proof. (a) We deduce (3.26) from $Q(t) = U(t,t')U_Q(t',t)Q(t)$ and 2.13, where $t' := \max\{t - 1, a\}$. For $t, s > a$ we take $r \in (a, t)$ and write $U_Q(t,s)Q(s) = U(t,r)U_Q(r,s)Q(s)$. This yields the asserted differential equation. Estimate (3.27) is clear for $s < t \leq s+1$ and follows from $A(t)U(t,s)P(s) = A(t)U(t,t-1)U(t-1,s)P(s)$ for $t \geq s + 1$.

(b) Let $\tau \in \mathbb{R}$. The evolution family $U_\tau(t,s) = U(t + \tau, s + \tau)$, $t \geq s$, (generated by $A(\cdot + \tau)$) has an exponential dichotomy with constants N, δ and projections $P(\cdot + \tau)$. Fix $t > 0$, $s \in \mathbb{R}$, and $x \in X$. Take $\varphi \in C(\mathbb{R})$ with $\varphi(s) = 1$ and $\operatorname{supp} \varphi \subseteq (s - t/2, s + t/2)$. Set $f = \varphi(\cdot)x$. Using (3.7) and (3.8) in Remark 3.13, we obtain

$$
\begin{aligned}
&P(s + \tau)x - P(s)x \\
&= (P(\cdot + \tau)f)(s) - (P(\cdot)f)(s) \\
&= \frac{1}{2\pi i} \int_{\mathbb{T}} \Big(R(\lambda, T_{U_\tau}(t))[T_{U_\tau}(t) - T_U(t)]R(\lambda, T_U(t))f \Big)(s) \, d\lambda \\
&= \frac{1}{2\pi i} \int_{\mathbb{T}} \sum_{k,l \in \mathbb{Z}} \lambda^{-k-l-2} \Gamma(s+\tau, s+\tau - kt) \big[U(s + \tau - kt, s + \tau - (k+1)t) \\
&\qquad\qquad - U(s - kt, s - (k+1)t) \big] \Gamma(s - (k+1)t, s - (k+1+l)t) \\
&\qquad\qquad \cdot \varphi(s - (k+1+l)t)x \, d\lambda.
\end{aligned}
$$

Observing that $\varphi(s - (k+1+l)t) = 1$ for $l = -k - 1$ and 0 otherwise, we arrive at

$$P(s+\tau) - P(s) = \sum_{k=-\infty}^{\infty} \Gamma(s+\tau, s+\tau - kt) \quad (3.28)$$

$$\cdot [U(s + \tau - kt, s + \tau - (k+1)t) - U(s - kt, s - (k+1)t)] \Gamma(s - (k+1)t, s).$$

Taking $t = 1$, this formula yields

$$\|P(s+\tau) - P(s)\| \leq \frac{2N^2 e^{-\delta}}{1 - e^{-2\delta}} \sup_{r \in \mathbb{R}} \|U(r+\tau, r+\tau-1) - U(r, r-1)\|. \quad (3.29)$$

If $|\tau| \leq 1/2$, we deduce from (2.13) and [1, Thm.2.3]

$$\|U(r+\tau, r+\tau-1) - U(r, r-1)\|$$
$$\leq \int_r^{r+\tau} \|A(\sigma)U(\sigma, r-1+\tau)\| \, d\sigma + \|U(r, r-1+\tau) - U(r, r-1)\|$$
$$\leq c(|\tau| + |\tau|^{\mu+\nu-1}). \quad (3.30)$$

Assertion (b) is a consequence of (3.29) and (3.30).

(c) We define $R = Q(a) - P(a)$ and

$$V(t, s) = \begin{cases} U(t, s), & t \geq s \geq a, \\ U(t, a)e^{\delta(a-s)R}, & t \geq a \geq s, \\ e^{\delta(t-s)R}, & a \geq t \geq s. \end{cases}$$

Since $e^{t\delta R} = e^{-t\delta}P(a) + e^{t\delta}Q(a)$, the evolution family $V(\cdot, \cdot)$ is exponentially dichotomic with constants N, δ and projections $P_V(t) = P(t)$ for $t \geq a$ and $P_V(t) = P(a)$ for $t \leq a$.

Let $s \geq a + \eta$ for some $\eta \in (0, 1/2]$. If $s - a \in [\eta, 3/2)$, we set $t := 2(s-a) \in [2\eta, 3]$. If $s - a = j + 1/2 + \tau$ for some $j \in \mathbb{N}$ and $\tau \in [0, 1)$, we set $t := 1 + \tau(j+1/2)^{-1} \in [1, 5/3]$. As a result, we have fixed $j \in \mathbb{N}_0$ and $t \in [2\eta, 3]$ such that $s - a = jt + t/2$. Take $|\tau| \leq t/4$. Replacing $U(\cdot, \cdot)$ by $V(\cdot, \cdot)$ in (3.28) (which is possible by Remark 3.13), we arrive at

$$\|P(s+\tau) - P(s)\| \leq \frac{2N^2 e^{-\delta t}}{1 - e^{-2\delta t}} \sup_{k \in \mathbb{Z}} \|V(s_k + \tau, s_k + \tau - t) - V(s_k, s_k - t)\|,$$

where $s_k = s - kt$. If $k \geq j+1$, then $s_k \leq a - t/2$ and thus

$$V(s_k + \tau, s_k + \tau - t) - V(s_k, s_k - t) = e^{-\delta t R} - e^{-\delta t R} = 0.$$

In the case $k \leq j-1$, we obtain $s_k - t \geq a + t/2$ and

$$V(s_k + \tau, s_k + \tau - t) - V(s_k, s_k - t) = U(s_k + \tau, s_k + \tau - t) - U(s_k, s_k - t).$$

As in (3.30) we estimate

$$\|V(s_k + \tau, s_k + \tau - t) - V(s_k, s_k - t)\| \leq c|\tau|^{\mu+\nu-1}, \qquad k \leq j-1. \quad (3.31)$$

For $k = j$, we have $s_k = a + t/2$ and

$$\|V(s_k + \tau, s_k + \tau - t) - V(s_k, s_k - t)\|$$
$$\leq \|(U(a + \tau + t/2, a) - U(a + t/2, a))e^{(\frac{t}{2} - \tau)\delta R}\|$$
$$+ \|U(a + t/2, a)(e^{(\frac{t}{2} - \tau)\delta R} - e^{\frac{t}{2}\delta R})\|$$
$$\leq c|\tau|$$

as in (3.30). Here the constants may depend on η. Assertion (c) follows by combining the above estimates.

Remark 3.19. In assertions (b) and (c) the projections are Lipschitz continuous in t if $A(t)$ is densely defined and their adjoints $A^*(t)$ also satisfy (AT). This follows if one uses in (3.30) and (3.31) the estimate $\|\frac{d}{ds}U(t,s)\| \leq c(t-s)^{-1}$ from [3, Thm.6.4].

4 Exponential Dichotomy of Parabolic Evolution Equations

Employing the theory presented in the previous two sections, we now derive sufficient conditions for exponential dichotomy of $U(\cdot,\cdot)$ in terms of the given operators $A(t)$. Our treatment is inspired by the monographs [15] by W.A. Coppel on ordinary differential equations and [27] by D. Henry on the case that $A(t) = A + B(t)$, where A is sectorial and densely defined and $B(\cdot) \in C_b^\alpha(J, D((w-A)^\beta))$ for some $\alpha, \beta \in (0,1)$.

4.1 Robustness

Assume that $A(\cdot)$ and $B(\cdot)$ generate the evolution families $U(\cdot,\cdot)$ and $V(\cdot,\cdot)$, respectively, and that $U(\cdot,\cdot)$ is hyperbolic. It is one of the main features of exponential dichotomy that $V(\cdot,\cdot)$ is also hyperbolic if $B(t)$ is 'close' to $A(t)$. The trivial example $A(t) = -\delta I$ shows that one really needs a smallness condition. One can formulate such results on the level of the evolution families as we did in Theorem 3.15. But, of course, this should only be a preliminary step in order to derive conditions on $B(t)$ itself. This was done in, e.g., [12], [13], [15], [16] for bounded perturbations, in [14], [27], [55] for unbounded perturbations of 'lower order', and in [37], [48] for the case of constant domains $D(A(t)) = D(B(t)) = Y$. (Here [13], [14], [48] are formulated in a different context and in [37] further conditions are required.) In the next theorem we allow for all parabolic problems in the framework of Section 2 and only assume that the resolvents of $A(t)$ and $B(t)$ are close in operator norm. In [59] our result was partially extended to the case of partial functional differential equations, see also [36] for ordinary functional differential equations.

Theorem 4.1. *Let $A(t)$ and $B(t)$ satisfy (AT) from Section 2 for $t \in J$, $J \in \{[a,\infty), \mathbb{R}\}$. Let $U(\cdot,\cdot)$ and $V(\cdot,\cdot)$ be the generated evolution families. Assume that $U(\cdot,\cdot)$ has an exponential dichotomy on J with constants $N, \delta > 0$ and projections $P_U(\cdot)$. Then there is a number $\tilde{q}$ depending only on N, δ, and the constants in (AT) such that if*

$$q := \sup_{t \in J} \|R(w, A(t)) - R(w, B(t))\| \leq \tilde{q}, \tag{4.1}$$

then $V(\cdot,\cdot)$ has an exponential dichotomy on $\mathbb{R}$ (if $J = \mathbb{R}$), resp. on $[a+\eta,\infty)$ for every $\eta > 0$ (if $J = [a,\infty)$), with Hölder continuous projections $P_V(s)$ and constants only depending on N, δ, and the constants in (AT) (and η if $J = [a,\infty)$). Moreover,

$$\dim P_V(s)X = \dim P_U(s)X \quad and \quad \dim \ker P_V(s) = \dim \ker P_U(s) \quad (4.2)$$

for $s \in \mathbb{R}$, resp. $s \in [a+\eta,\infty)$.

Proof. Recall that in the parabolic case the dichotomy projections are automatically Hölder continuous due to Proposition 3.18.

At first, let $J = [a,\infty)$. In order to apply Theorem 3.15 we extend $U(\cdot,\cdot)$ and $V(\cdot,\cdot)$ to the time interval $J = \mathbb{R}$ by setting

$$\tilde{U}(t,s) = \begin{cases} U(t,s), & t \geq s \geq b, \\ U(t,b)e^{(b-s)R}, & t \geq b \geq s, \\ e^{(t-s)R}, & b \geq t \geq s, \end{cases} \quad \tilde{V}(t,s) = \begin{cases} V(t,s), & t \geq s \geq b, \\ V(t,b)e^{(b-s)R}, & t \geq b \geq s, \\ e^{(t-s)R}, & b \geq t \geq s, \end{cases}$$

where $R = \delta Q(b) - dP(b)$, $d \geq \delta$, and $b = a+\eta$ for a fixed $\eta > 0$. Observe that $\mathbb{R} \ni s \mapsto W(s, s-t)$ and $(\sigma,\infty) \ni \tau \mapsto W(\tau,\sigma)$ are continuous from the left for each $t \geq 0$ and $\sigma \in \mathbb{R}$ and $W = \tilde{U}$ and $W = \tilde{V}$. Since $e^{tR} = e^{-td}P(a) + e^{t\delta}Q(a)$, the evolution family $\tilde{U}(\cdot,\cdot)$ is exponentially dichotomic with constants N, δ and projections $\tilde{P}(t) = P_U(t)$ for $t \geq b$ and $\tilde{P}(t) = P_U(b)$ for $t \leq b$. Moreover, both evolution families possess exponential bounds not depending on d. Set $\bar{q} = (1 - e^{-\delta})^2/(8N^2)$. For $b - 1 < s < b - 1/2$, we have

$$\tilde{U}(s+1, s) - \tilde{V}(s+1, s) = (U(s+1, b) - V(s+1, b))\,(P(b)e^{(s-b)d} + Q(b)e^{(b-s)\delta}).$$

We fix $d \geq \delta$ such that

$$\|(U(s+1, b) - V(s+1, b))P(b)e^{(s-b)d}\| \leq 2NM\,e^{-d/2} \leq \tfrac{\bar{q}}{2},$$

where $\|U(t,r)\|, \|V(t,r)\| \leq M$ for $0 \leq t - r \leq 1/2$. Propositions 2.6 and 3.18 further show that

$$\|(U(s+1, b) - V(s+1, b))Q(b)e^{(b-s)\delta}\|$$
$$\leq c\,\|(U(s+1, b) - V(s+1, b))R(w, A(b))\| \leq c\,q^\alpha$$

where $\alpha \in (0, \mu)$ is fixed and c depends only on N, δ, η, and the constants in (AT). Proposition 2.6 also yields

$$\|\tilde{U}(s+1, s) - \tilde{V}(s+1, s)\| \leq N(1 + e^\delta)\,\|(U(s+1, b) - V(s+1, b))\| \leq c\,q^\beta$$

for $b - 1/2 \leq s < b$ and $q \leq q_0(1/2)$, where $\beta > 0$ and $q_0(1/2) > 0$ are given by Proposition 2.6. Similarly,

$$\|\tilde{U}(s+1, s) - \tilde{V}(s+1, s)\| = \|U(s+1, s) - V(s+1, s)\| \leq c\,q^\beta$$

for $s \geq b$. Further $\tilde{U}(s+1, s) = \tilde{V}(s+1, s)$ for $s \leq b - 1$. So we find a number $\tilde{q} > 0$ such that $q \leq \tilde{q}$ implies $\|\tilde{U}(s+1, s) - \tilde{V}(s+1, s)\| \leq \overline{q}$ for each $s \in \mathbb{R}$. The assertions for $J = [a, \infty)$ now follow from Theorem 3.15.

In the case $J = \mathbb{R}$ one can directly estimate $U(s+1, s) - V(s+1, s)$ using Proposition 2.6 and then apply Theorem 3.15.

In the above arguments we used Theorem 3.15, where $T_U(1) - T_V(1)$ was estimated. One can also try to involve the difference $G_U - G_V$ of the corresponding generators, cf. [55]. It seems that this approach does not work in the same generality as the above theorem, but it yields better values of $\tilde{q}$. For instance, if $A(t)$ and $B(t)$, $t \in \mathbb{R}$, are bounded, then the smallness condition becomes $\|A(\cdot) - B(\cdot)\|_\infty < \frac{\delta}{2N}$, see [55, §4]. Refinements of this assumption are close to be optimal, see [46]. The next example indicates how the estimate (4.1) can be translated into conditions on the coefficients of a partial differential equation.

Example 4.2. Let the operators $A(t)$, $t \geq 0$, be given on $X = L^p(\Omega)$, $1 < p < \infty$, or $X = C(\overline{\Omega})$ as in Example 2.9. We assume that there are time-independent coefficients $\tilde{a}_{kl}, \tilde{a}_k, \tilde{a}_0 \in C(\overline{\Omega})$, and $\tilde{b}_k, \tilde{b}_0 \in C^1(\overline{\Omega})$, $k, l = 1, \cdots, n$, satisfying the assumptions of Example 2.9 and define the operator A on X as in (2.43) and (2.39) for these coefficients. Arguing as in (2.44), one sees that

$$\sup_{t \geq 0} \|R(w, A(t)) - R(w, A)\| \leq c \sup_{t \geq 0} \max_{kl} \left\{ \|a_{kl}(t, \cdot) - \tilde{a}_{kl}(\cdot)\|_{C(\overline{\Omega})}, \right.$$

$$\left. \|a_k(t, \cdot) - \tilde{a}_k(\cdot)\|_{C(\overline{\Omega})}, \|b_k(t, \cdot) - \tilde{b}_k(\cdot)\|_{C^1(\partial\Omega)} \right\}.$$

We also assume that $\sigma(A) \cap i\mathbb{R} = \emptyset$. Due to Theorem 4.3, the evolution family generated by $A(\cdot)$ has an exponential dichotomy if the quantity on the right hand side of the above estimate is sufficiently small, cf. Remark 2.10.

4.2 Asymptotically Autonomous Equations

We suppose that (AT) holds and that the resolvents $R(w, A(t))$ converge in operator norm as $t \to \infty$ to the resolvent $R(w, A)$ of a sectorial operator with $\sigma(A) \cap i\mathbb{R} = \emptyset$. It is reasonable to expect that then the evolution family $U(\cdot, \cdot)$ generated by $A(\cdot)$ has an exponential dichotomy, at least on an interval $[a, \infty) \subseteq \mathbb{R}_+$. In the case of constant domains (cf. Remark 4.5) and exponential stability, this was in fact shown by H. Tanabe already in 1961, see [62, §5.6]. Tanabe's result was extended by D. Guidetti, [26], to the case of Kato–Tanabe conditions. In [57], we have treated the case of exponential dichotomy under Acquistapace–Terreni conditions assuming the density of the domains and requiring convergence of $R(w, A(\cdot))$ in a slightly stronger norm. C.J.K. Batty and R. Chill, [11], then generalized this result in several directions: They also considered Kato–Tanabe conditions and asymptotically periodic problems, they managed to deal with resolvents converging in $\mathcal{L}(X)$, and they gave sufficient conditions for the almost periodicity of $U(s + \cdot, s)$. In the next theorem the case of non–dense domains is covered for the first time. In [59] we have partially extended our result to the case of partial functional differential equations, see also [36].

Theorem 4.3. *Let $A(t)$, $t \geq 0$, satisfy (AT) from Section 2 and let A satisfy (AT1). Assume that $\sigma(A) \cap i\mathbb{R} = \emptyset$ and $R(w, A(t)) \rightarrow R(w, A)$ in $\mathcal{L}(X)$ as $t \rightarrow \infty$. Then the evolution family $U(\cdot, \cdot)$ generated by $A(\cdot)$ has an exponential dichotomy on an interval $[a, \infty) \subseteq \mathbb{R}_+$ with projections $P(s)$ being globally Hölder continuous for $s \geq a$. Moreover, $P(s) \rightarrow P$ strongly and $U(s + \tau, s) \rightarrow e^{\tau A}$ in $\mathcal{L}(X)$ for $\tau \geq 0$ as $s \rightarrow \infty$, $\dim P(t)X = \dim PX$, and $\dim \ker P(t) = \dim \ker P$ for $t \geq a$, where P is the dichotomy projection of $e^{\tau A}$. The dichotomy constants of $U(\cdot, \cdot)$ only depend on the dichotomy constants of A, the type of A, and the constants in (AT). We can take the same number a for all $A(t)$ and A subject to the same constants and satisfying $\sup_{t \geq s} \|R(w, A(t)) - R(w, A)\| \leq q(s)$, $s \geq 0$, for a fixed function $q(s)$ converging to 0 as $s \rightarrow \infty$.*

Proof. Recall that e^{tA} has an exponential dichotomy with exponent $\delta > 0$ due to, e.g., [21, Thm.IV.3.12, V.1.17] or [42, §2.3]. For a given $a \geq 0$, we set

$$
A_a(t) = \begin{cases} A(t), & t \geq a, \\ A(a), & t < a, \end{cases} \qquad U_a(t, s) = \begin{cases} U(t, s), & t \geq s \geq a, \\ U(t, a)e^{(a-s)A(a)}, & t \geq a \geq s, \\ e^{(t-s)A(a)}, & a \geq t \geq s. \end{cases}
$$

Observe that $A_a(t)$, $t \in \mathbb{R}$, satisfy (AT) and generate $U_a(\cdot, \cdot)$. Moreover,

$$
\sup_{\tau \in \mathbb{R}} \|R(w, A_a(\tau)) - R(w, A)\| \leq q(a)
$$

for the function q from the statement. Proposition 2.6 thus yields

$$
\|U_a(s + t, s) - e^{tA}\| \leq c(t) \, q(a)^\beta
$$

for $s \in \mathbb{R}$, $t > 0$, $q(a) \leq q_0(t)$, and constants $\beta > 0$, $q_0(t) > 0$, $c(t) > 0$. Thanks to Theorem 3.15 the evolution family $U_a(\cdot, \cdot)$ has an exponential dichotomy and its projections have the same rank as those of e^{tA} if we choose a large enough. The projections $P_a(\cdot)$ are Hölder continuous by Proposition 3.18. Restricting to the time interval $[a, \infty)$, we establish the theorem except for asserted convergence. The convergence of $U(s + t, s)$ follows directly from Proposition 2.6, whereas the strong convergence of $P(\cdot)$ is a consequence of the next lemma.[5]

Lemma 4.4. *Let $U(\cdot, \cdot)$ be an exponentially bounded evolution family on X with $J = \mathbb{R}$ such that $s \mapsto U(s, s - t)$ is strongly continuous for $t \geq 0$ and let $S(\cdot)$ be a semigroup on X. Assume that $U(\cdot, \cdot)$ and $S(\cdot)$ have exponential dichotomies with projections $P(t)$ and P, respectively, and that $U(s + t, s) \rightarrow S(t)$ strongly for $t \geq 0$ as $s \rightarrow \infty$. Then $P(s) \rightarrow P$ strongly as $s \rightarrow \infty$.*

---

[5] In [57, Thm.3.3] we directly prove that $P(t)$ converges if the domains are dense.

Proof. Let $\tilde{E} := \{f \in C(\mathbb{R}, X) : f(t) \to 0 \text{ as } t \to -\infty, \ f(t) \to f_\infty \text{ as } t \to \infty\}$ be endowed with the sup–norm. Clearly,

$$(\tilde{T}_U(t)f)(s) := U(s, s - t)f(s - t), \quad t \geq 0, \ s \in \mathbb{R}, \ f \in \tilde{E}, \qquad (4.3)$$

defines an exponentially bounded semigroup $\tilde{T}_U(\cdot)$ on $\tilde{E}$. Its restriction to $E = C_0(\mathbb{R}, X)$ is the evolution semigroup $T_U(\cdot)$. The spaces $\tilde{E}$ and $E \times X =: \hat{E}$ are isomorphic via

$$\Phi : \tilde{F} \to \hat{F}, \ f \mapsto (f - Mf_\infty, f_\infty),$$

where $Mx := \varphi(\cdot)x$ for a fixed function $\varphi \in C(\mathbb{R})$ with support in $\mathbb{R}_+$ and $\lim_{t\to\infty} \varphi(t) = 1$. We consider the induced semigroup $\hat{T}_U(\cdot)$ on $\hat{E}$ given by

$$\hat{T}_U(t) := \Phi \tilde{T}_U(t)\Phi^{-1} = \begin{pmatrix} T_U(t) & \tilde{T}_U(t)M - MS(t) \\ 0 & S(t) \end{pmatrix}, \quad t \geq 0.$$

Due to the assumptions and Theorem 3.11, $\lambda - T_U(1)$ and $\lambda - S(1)$ are invertible for $\lambda \in \mathbb{T}$; hence $\lambda \in \rho(\hat{T}_U(1))$. This means that $\hat{T}_U(\cdot)$ and $\tilde{T}_U(\cdot)$ are hyperbolic with spectral projection $\hat{P}$ and $\tilde{P}$, respectively. On the other hand, formula (4.3) defines a semigroup $T_U^b(\cdot)$ on $C_b(\mathbb{R}, X)$ which is hyperbolic with projection

$$P(\cdot) = \mathcal{P}^b = \frac{1}{2\pi i} \int_\mathbb{T} R(\lambda, T_U^b(1)) \, d\lambda$$

by Remark 3.12 and the exponential dichotomy of $U(\cdot, \cdot)$. Since $R(\lambda, T_U^b(1))f = R(\lambda, \tilde{T}_U(1))f$ for $f \in \tilde{E}$, we obtain $\tilde{P}f = \mathcal{P}^b f = P(\cdot)f$ for $f \in \tilde{E}$. Therefore $P(t)$ converges strongly to a projection P' as $t \to \infty$. Finally,

$$\hat{P} = \frac{1}{2\pi i} \int_\mathbb{T} R(\lambda, \hat{T}_U(1)) \, d\lambda = \begin{pmatrix} P(\cdot) & * \\ 0 & P \end{pmatrix}$$

$$= \frac{1}{2\pi i} \int_\mathbb{T} \Phi R(\lambda, \tilde{T}_U(1))\Phi^{-1} \, d\lambda = \Phi P(\cdot)\Phi^{-1} = \begin{pmatrix} P(\cdot) & * \\ 0 & P' \end{pmatrix}$$

so that $P = P'$.

Remark 4.5. Assume that the operators $A(t)$ satisfy (AT1), that $D(A(t)) = D(A(0)) =: Y$, $t \geq 0$, with uniformly equivalent graph norms, and that $A(\cdot) : \mathbb{R}_+ \to \mathcal{L}(Y, X)$ is globally Hölder continuous of exponent $\mu > 0$. Let $A(t)$ converge in $\mathcal{L}(Y, X)$ to a closed operator A with domain Y as $t \to \infty$. If $\sigma(A) \cap i\mathbb{R} = \emptyset$, then the assumptions of Theorem 4.3 hold. Moreover, if $w = 0$ then $s(A) < 0$. We omit the straightforward proof.

The following simple example shows that strong convergence of the resolvents would not suffice in Theorem 4.3.

Example 4.6. Let $X = \ell^2$, $A = -I$, and

$$A(t)(x_k) = -(x_1, \cdots, x_{n-1}, (t - n + 1)x_n, 0, \cdots) \quad \text{for} \ n - 1 \leq t \leq n.$$

Then, $A(t)x \to -x$ for $x = (x_k) \in X$ as $t \to \infty$ and $s(A) = -1$. But $U(t, 0)x = x$ if $0 \leq t \leq n$ and $x_k = 0$ for $k = 1, \cdots, n$, so that $\|U(n, 0)\| = 1$.

Theorem 4.3 only gives an exponential dichotomy on an interval $[a, \infty)$ where a may be rather large. If $s(A) < 0$, then $U(\cdot, \cdot)$ is exponentially stable on $[a, \infty)$, and hence on $\mathbb{R}_+$. In the case of a non trivial exponential dichotomy, one cannot extend the exponential dichotomy to the left, in general, as the next example shows (in which (AT) is not quite fulfilled).

Example 4.7. On $X = L^1(\mathbb{R})$ we consider the semigroups $T_1(\cdot)$ generated by the second derivative and $T_2(t)f = e^t f|\mathbb{R}_+ + e^{-t}f|\mathbb{R}_-$, and define the evolution family

$$U(t, s) = \begin{cases} T_1(t - s), & s \le t \le 1, \\ T_2(t - 1)T_1(1 - s), & s \le 1 \le t, \\ T_2(t - s), & 1 \le s \le t. \end{cases}$$

Clearly, $U(\cdot, \cdot)$ has exponential dichotomy on $[1, \infty)$ with $Q(t)f = f|\mathbb{R}_+$ for $t \ge 1$. However, $U(1, 0)f$ is a smooth function so that condition (b) of Definition 3.1 cannot be satisfied on $J = [0, \infty)$.

The next result gives sufficient conditions to extend exponential dichotomy to larger time intervals. It is essentially due to [36, §2].

Proposition 4.8. *Let $U(\cdot, \cdot)$ be an exponentially bounded evolution family on the time interval $[a, \infty)$ having an exponential dichotomy on $[b, \infty)$ for $b > a$ with projections $P(t)$ and constants $N, \delta > 0$. If $\dim Q(t)X < \infty$ for some/all $t \ge b$ and $U(b, a)^* x^* \ne 0$ for every $x^* \in Q(b)^* X^* \setminus \{0\}$, then $U(\cdot, \cdot)$ has an exponential dichotomy on $[a, \infty)$ with exponent δ and projections $\tilde{P}(t)$, $t \ge a$. Moreover, for $t \ge b$,*

$$\dim \ker \tilde{P}(t) = \dim \ker P(t), \qquad \tilde{P}(t)X = P(t)X, \tag{4.4}$$

$$\|\tilde{P}(t) - P(t)\| \le c\, e^{-2\delta(t-b)}. \tag{4.5}$$

If $U(\cdot, \cdot)$ is strongly continuous on $[a, \infty)$, then $\tilde{P}(t)$ and $U_{\tilde{Q}}(t, s)\tilde{Q}(s)$ are strongly continuous for $t, s \ge a$.

Proof. (1) We introduce the closed subspaces $X_-(t) := \{x \in X : U(b, t)x \in P(b)X\}$ for $t \in [a, b]$. We first want to verify

$$\operatorname{codim} X_-(a) = \operatorname{codim} P(b)X < \infty. \tag{4.6}$$

Observe that $\operatorname{codim} P(b)X = \dim Q(t)X$ for all $t \ge b$. We define $X_+^*(b) := Q(b)^* X^*$ and $X_+^*(a) := U(b, a)^* X_+^*(b)$. The assumptions imply that $X_+^*(a)$ and $X_+^*(b)$ are finite dimensional and that $U(b, a)^* : X_+^*(b) \to X_+^*(a)$ is an isomorphism. Using [30, Lem.III.1.40], we obtain

$$[P(b)X]^\perp := \{x^* \in X^* : \langle x, x^* \rangle = 0 \quad \forall x \in P(b)X\} \tag{4.7}$$
$$= \{x^* \in X^* : \langle y, P(b)^* x^* \rangle = 0 \ \forall y \in X\} = \ker P(b)^* = X_+^*(b),$$
$$\operatorname{codim} P(b)X = \dim [P(b)X]^\perp = \dim X_+^*(b) = \dim X_+^*(a). \tag{4.8}$$

Let $x \in X_-(a)$ and $x^* \in X_+^*(b)$. By (4.7) we have $0 = \langle U(b,a)x, x^* \rangle = \langle x, U(b,a)^* x^* \rangle$, which gives $x \in {}^\perp[X_+^*(a)] := \{y \in X : \langle y, y^* \rangle \quad \forall y^* \in X_+^*(a)\}$. Reversing this argument, we deduce $X_-(a) = {}^\perp[X_+^*(a)]$. Since $\dim X_+^*(a) < \infty$, we have $X_+^*(a) = [{}^\perp[X_+^*(a)]]^\perp$ by [53, Thm.4.7], so that

$$\dim X_+^*(a) = \dim [{}^\perp[X_+^*(a)]]^\perp = \dim X_-(a)^\perp = \operatorname{codim} X_-(a)$$

by [30, Lem.III.1.40] again. Combined with (4.8), this equality shows (4.6).

(2) Due to (4.6), there exists a finite dimensional complement $X_+(a)$ of $X_-(a)$ in X. We further define

$$X_+(t) := U(t,a)X_+(a) \quad \text{for} \ \ t \geq a \quad \text{and} \quad X_-(t) := P(t)X \quad \text{for} \ \ t \geq b.$$

Observe that $X_\pm(t)$, $t \geq a$, are closed subspaces. We want to show that $X_-(t) \oplus X_+(t) = X$ for $t \geq a$. Let $x \in X_-(t) \cap X_+(t)$. Then $x = U(t,a)y$ for some $y \in X_+(a)$. If $t \leq b$, we have $U(b,a)y = U(b,t)x \in P(b)X$, hence $y \in X_-(a)$. This yields $y = 0$ and $x = 0$. If $t > b$, we have $x \in P(t)X$. Thus $U(s,b)U(b,a)y = U(s,t)x \to 0$ as $s \to \infty$, and so $U(b,a)y \in P(b)X$ by (3.3). Again we obtain $y = x = 0$. This argument also shows that $U(t,a)y \neq 0$ for $y \in X_+(a) \setminus \{0\}$ and $t \geq a$. Therefore $U(t,s) : X_+(s) \to X_+(t)$ is an isomorphism for $t \geq s \geq a$. From this fact and (4.6) we deduce

$$\dim X_+(t) = \dim X_+(a) = \operatorname{codim} P(b)X = \operatorname{codim} P(t)X \qquad (4.9)$$

for $t \geq b$. Consequently, $X_-(t) \oplus X_+(t) = X$ in this case. If $t \in [a,b]$ and $x \in X$, then $U(b,t)x = y_+ + y_-$ with $y_\pm \in X_\pm(b)$. There exists $x_+ \in X_+(t)$ with $U(b,t)x_+ = y_+$. Hence, $U(b,t)(x - x_+) = y_- \in X_-(b)$ which means that $x - x_+ \in X_-(t)$. So we arrive at $X_-(t) \oplus X_+(t) = X$ for all $t \geq a$.

(3) We denote by $\tilde{P}(t)$ the projection onto $X_-(t)$ with kernel $X_+(t)$ for $t \geq a$. Assertion (4.4) holds due to (4.9). It is clear that $U(t,s)\tilde{P}(s)X \subseteq \tilde{P}(t)X$ and $U(t,s) : \tilde{Q}(s)X \to \tilde{Q}(t)X$ is an isomorphism for $t \geq s \geq a$. As a consequence, $U(t,s)$ and $\tilde{P}(t)$ satisfy (a) and (b) of Definition 3.1 for $t \geq s \geq a$. For $a \leq t \leq s \leq b$ we have

$$\|U_{\tilde{Q}}(t,s)\tilde{Q}(s)\| \leq \|U(t,a)U_{\tilde{Q}}(a,b)\tilde{Q}(b)U(b,s)\| \leq M^2 \|U_{\tilde{Q}}(a,b)\tilde{Q}(b)\|.$$

where $\|U(t,s)\| \leq M$ for $a \leq s \leq t \leq b$. So we obtain the boundedness of $\tilde{P}(t)$ and of Green's function of $U(\cdot,\cdot)$ on $[a,b]$. Further, $\tilde{P}(t) = P(t)\tilde{P}(t)$ and $P(t) = \tilde{P}(t)P(t)$ for $t \geq b$ since $P(t)$ and $\tilde{P}(t)$ have the same range. These facts lead to

$$\tilde{P}(t) = \tilde{P}(t)P(t) + \tilde{P}(t)Q(t) = P(t) + \tilde{P}(t)U(t,b)U_Q(b,t)Q(t)$$
$$= P(t) + U(t,b)P(b)\tilde{P}(b)U_Q(b,t)Q(t),$$
$$\|\tilde{P}(t) - P(t)\| \leq N^2 \|\tilde{P}(b)\| \, e^{-2\delta(t-b)}$$

for $t \geq b$. Thus (4.5) holds and $\tilde{P}(t)$ is uniformly bounded on $[a, \infty)$. The first estimate in Definition 3.1(c) now follows from $U(t,s)\tilde{P}(s) = U(t,s)P(s)\tilde{P}(s)$; the second one from

$$\tilde{Q}(t) = \tilde{Q}(t)Q(t) + \tilde{Q}(t)P(t) = \tilde{Q}(t)Q(t)$$
$$= \tilde{Q}(t)U(t,s)U_Q(s,t)Q(t) = U(t,s)\tilde{Q}(s)U_Q(s,t)Q(t),$$
$$U_{\tilde{Q}}(s,t)\tilde{Q}(t) = \tilde{Q}(s)U_Q(s,t)Q(t)$$

for $t \geq s \geq b$. If $U(\cdot,\cdot)$ is strongly continuous, then $\tilde{P}(t)$ and $U_{\tilde{Q}}(t,s)\tilde{Q}(s)$ are strongly continuous for $t, s \geq a$ due to (3.1) and (3.2).

In the context of Theorem 4.3, the above assumptions are satisfied if, in addition, the resolvent of A is compact and the adjoint system possesses 'backward uniqueness' (at least) on the unstable subspaces, as we demonstrate in the next example.

Example 4.9. Let the operators $A(t)$ be given on $X = L^p(\Omega)$, $1 < p < \infty$, or $X = C(\overline{\Omega})$ as in Example 2.8. Assume there are real $\tilde{a}_{kl} \in C^1(\overline{\Omega})$, $k,l = 1, \cdots, n$, and $\tilde{a}_0 \in C(\overline{\Omega})$ such that $(\tilde{a}_{kl})$ is symmetric, satisfies (2.38), and

$$a_{kl}(t,\cdot) \to \tilde{a}_{kl}(\cdot) \quad \text{in} \quad L^n(\Omega) \quad \text{and} \quad a_0(t,\cdot) \to \tilde{a}_0(\cdot) \quad \text{in} \quad L^{n/2}(\Omega)$$

as $t \to \infty$, where $n/2$ is replaced by 1 if $n = 1$. Define the operator A on X as in (2.39) for $\tilde{a}_{kl}, \tilde{a}_0$. Arguing as in (2.40) with $\nu = 0$, one sees that $R(w, A(t)) \to R(w, A)$ in $\mathcal{L}(X)$. Thus, due to Theorem 4.3, the evolution family generated by $A(\cdot)$ has exponential dichotomy on an interval $[a, \infty)$ provided that $\sigma(A) \cap i\mathbb{R} = \emptyset$.

We further want to show that here the exponential dichotomy of $U(\cdot,\cdot)$ can be extended to $\mathbb{R}_+$. To that purpose we restrict ourselves to $X = L^p(\Omega)$, $1 < p < \infty$, and assume additionally that $a_{kl}, a_0 \in C^1(\mathbb{R}_+, L^\infty(\Omega))$. Since A has compact resolvent, the range of $Q(t)$ is finite dimensional by Theorem 4.3. Observe that the adjoints $U(t,s)^*$ on X^* have properties analogous to those stated Theorem 2.2 since the operators $A(t)^*$, $t \geq 0$, also satisfy (AT). Thus $v(s) = U(a,s)^*y^*$ solves the adjoint backward problem

$$v'(s) = A(s)^*v(s), \quad s < a, \qquad v(a) = y^*, \tag{4.10}$$

on $X^* = L^{p'}(\Omega)$, $1/p + 1/p' = 1$. Due to [33, Thm.1.1], $v(0) = 0$ implies $y^* = 0$ if $y^* \in D(A_2(a))$.[6] Here we have $y^* = Q(a)^*x^*$ for some $x^* \in X^*$. By duality the formula $A(a)^*U(a+1,a)^*U_Q(a,a+1)^*Q(a)^* = A(a)^*Q(a)^*$ holds, so that $Q(a)^*X^* \subseteq D(A(a)^*) = D(A_{p'}(a))$. Thus the assumptions of Proposition 4.8 are satisfied if $p' \geq 2$, i.e., if $p \leq 2$. In the case $p > 2$, we use that for each $a' > a$ there is $z^* \in X^*$ such that $Q(a)^*x^* = U(a',a)^*Q(a')^*z^*$. For $a < a'' < a'$, we have $Q(a)^*x^* = U(a'',a)^*U(a',a'')^*Q(a')^*z^*$. Observe that $U(a',a'')^*Q(a')^*z^* \in D(A(a'')^*) \subseteq W^{2,p'}(\Omega) \hookrightarrow L^r(\Omega)$ with $r = np'/(n-2p')$ for $n > 2p'$ and r is arbitrarily large if $n \leq 2p'$, by Sobolev's embedding theorem. Moreover, the backward evolution families on different spaces $L^{p'}(\Omega)$ coincide on the intersection of these spaces by uniqueness of the problem

---

[6] In fact, it suffices that $y^* \in W^{1,2}(\Omega)$, cf. [33].

(4.10). Hence, $Q(a)^*x^* \in D(A_r(a))$. This argument can be iterated until we reach an exponent $\tilde{r} \geq 2$. Therefore Proposition 4.8 can be applied for every $p \in (1, \infty)$ and $U(\cdot, \cdot)$ has an exponential dichotomy on $\mathbb{R}_+$.

4.3 Slowly Oscillating Coefficients

We have seen in Example 3.2 that the hyperbolicity of the semigroups $(e^{\tau A(t)})_{\tau \geq 0}$ (with common constants $N, \delta > 0$) does not imply the exponential dichotomy of the evolution family $U(\cdot, \cdot)$ generated by $A(\cdot)$. However, one can expect that $U(\cdot, \cdot)$ has an exponential dichotomy if in addition $A(\cdot)$ does not 'oscillate too much'. Such results were established by W.A. Coppel in [15, Prop.6.1] for matrices $A(t)$ and by A.G. Baskakov in [9] for bounded operators $A(t)$. Both authors required that the Lipschitz constant of $A(\cdot)$ is small compared to $N, \delta, \|A(\cdot)\|_\infty$. M.P. Lizana used Coppel's result to prove an analogous theorem for ordinary functional differential equations, [38, Thm.4]. For parabolic problems one knows estimates of the type $\|U(t,s)\| \leq c_1 e^{(w+c_2 L^\kappa)(t-s)}$, where w and L are given by (AT) and $c_k, \kappa > 0$ are constants, see Theorems II.5.1.1 and IV.2.3.2 in [7], [23, Thm.2.3], or [27, Thm.7.4.2]. In particular, $U(\cdot, \cdot)$ is exponentially stable if $w < 0$ and the Hölder constant L is small. In our main Theorem 4.13 we show that $U(\cdot, \cdot)$ also inherits the exponential dichotomy (with the same ranks) of $A(t)$ if L is sufficiently small. Here we extend [55, Thm.3.7] and [56, Thm.5], where the case of (constant) dense domains $D(A(t))$ was treated. Our approach is based on Theorem 3.17, more precisely, we show that the 'generator' G of the evolution semigroup $T(\cdot)$ on E is invertible (recall that $T(\cdot)$ is only strongly continuous on $(0, \infty)$ if the domains are not dense). The rank of the dichotomy projections is computed by means of (3.7) and (3.8) and a continuity argument.

We start with several preparations. In order to simplify the proof of Theorem 4.13, we choose a slightly different Hölder condition in (AT) and require

(AT') $A(t)$, $t \in J$, $J \in \{[a, \infty), \mathbb{R}\}$, satisfy (AT1), have uniformly bounded inverses, and

$$\|r^\nu A(t)R(r, A(t))\left(A(t)^{-1} - A(s)^{-1}\right)\| \leq \ell\,|t-s|^\mu$$

for $t, s \in J$, $r > w$, and constants $\ell \geq 0$ and $\mu, \nu \in (0, 1]$ with $\mu + \nu > 1$.

If $J = [a, \infty)$, we set $A(t) := A(a)$ for $t \leq a$. This definition preserves (AT'). It is straightforward to verify that (AT') is stronger than (AT) and that (AT) implies (AT') (possibly after replacing w by $w' > w$) if the operators $A(t)$ have uniformly bounded inverses.

We introduce the completion X_{-1}^t of $X_0^t = \overline{D(A(t))}$ with respect to the norm $\|x\|_{-1}^t := \|A^{-1}(t)x\|$, see e.g. [7, Chap.V], [21, §II.5].[7] Observe that $A(t)$ has a unique continuous extension $A_{-1}(t) : X_0^t \to X_{-1}^t$. Since one can also

---

[7] One usually takes the norm $\|R(w, A(t))x\|$ which is equivalent to $\|x\|_{-1}^t$ (uniformly in $t \in \mathbb{R}$) by (AT').

extend $R(\lambda, A(t))$ to an operator in $\mathcal{L}(X^t_{-1}, X^t_0)$, $A_{-1}(t)$ is sectorial in X^t_{-1} of the same type. Now we can proceed as in Section 2. We define

$$X^t_{\alpha-1,\infty} := (X^t_{-1})^{A_{-1}(t)}_{\alpha,\infty} \quad \text{and} \quad X^t_{\alpha-1} := (X^t_{-1})^{A_{-1}(t)}_\alpha,$$

$$\text{with the norm} \quad \|x\|^t_{\alpha-1} := \sup_{r>w} \|r^\alpha R(r, A_{-1}(t))x\|$$

for $0 < \alpha < 1$ and $t \in \mathbb{R}$. The domain $D(A(t))$ is dense in $X^t_{\alpha-1}$ for all $\alpha \in [0,1]$ and the continuous embeddings

$$X^t_0 \subseteq X \hookrightarrow X^t_{\beta-1} \hookrightarrow X^t_{\beta-1,\infty} \hookrightarrow X^t_{\alpha-1} \hookrightarrow X^t_{-1} \tag{4.11}$$

hold for $0 < \alpha < \beta < 1$. The norms of the embeddings depend only on the type of $A(t)$ and the norm of its inverse. The operator $A(t)$ can be extended to an isometric isomorphism $A_{\alpha-1}(t) : X^t_\alpha \to X^t_{\alpha-1}$ for $0 \le \alpha < 1$ which coincides with the part of $A_{-1}(t)$ in $X^t_{\alpha-1}$ (and analogously for $X^t_{\alpha,\infty}$). The semigroup $e^{\tau A(t)}$ extends to an analytic C_0–semigroup on $X^t_{\alpha-1}$ generated by $A_{\alpha-1}(t)$, $0 \le \alpha < 1$. In addition, (2.6) yields

$$\|A(t)^k e^{sA_{-1}(t)}x\|_X \le C_\alpha\, s^{\alpha-k-1} \|x\|^t_{\alpha-1}, \quad \text{where } C_\alpha \le C, \tag{4.12}$$

for $k = 0, 1$, $0 < \alpha < 1$, $x \in X^t_{\alpha-1,\infty}$, $0 < s \le 1$, $t \in \mathbb{R}$, and constants C_α and C depending only on the type of $A(t)$ and $\|A^{-1}(t)\|$.

Analogously we treat the multiplication operator $A(\cdot)$ on $E = C_0(\mathbb{R}, X)$ as defined in (2.12). There exist the corresponding inter- and extrapolation spaces $E_\alpha := E^{A(\cdot)}_\alpha$, $-1 \le \alpha \le 1$, and $E_{\alpha,\infty} := E^{A(\cdot)}_{\alpha,\infty}$, $-1 < \alpha < 1$, where $E_0 = \overline{D(A(\cdot))}$. The extrapolated operator $A(\cdot)_{-1} : E_0 \to E_{-1}$ is given by the multiplication operator $A_{-1}(\cdot)$ and $f \in E_{-1}$ can be identified with an element of $\prod_{t\in\mathbb{R}} X^t_{-1}$, cf. [25, Thm.4.7]. As above we see that $E_{\alpha-1}$ is the closure of $D(A(\cdot))$ in E_{-1} with respect to the norm

$$\|f\|_{\alpha-1} := \sup_{r>w} \sup_{s\in\mathbb{R}} \|r^\alpha R(r, A_{-1}(s))f(s)\|,$$

$A_{\alpha-1}(\cdot) : E_\alpha \to E_{\alpha-1}$ is an isometric isomorphism for $0 \le \alpha < 1$, and

$$\begin{aligned} D(A(\cdot)) \hookrightarrow E_\beta \subseteq E_{\beta,\infty} \hookrightarrow D((w - A(\cdot))^\alpha) \hookrightarrow E_\alpha \hookrightarrow E_0 \subseteq E \\ E \hookrightarrow E_{\beta-1} \subseteq E_{\beta-1,\infty} \hookrightarrow E_{\alpha-1} \hookrightarrow E_{-1} \end{aligned} \tag{4.13}$$

for $0 < \alpha < \beta < 1$ (the domain of the fractional power is endowed with the norm $\|(w - A(\cdot))^\alpha f\|_\infty$).

Due to (2.15), (2.5), the reiteration theorem [42, Thm.1.2.15], and $\omega(U) < \infty$, we can extend $U(t, s)$ to an operator $\overline{U}(t, s) : X^s_{\alpha-1} \to X$ with norm

$$\|\overline{U}(t, s)\| \le c(\alpha)\, ((t - s)^{\alpha-1} \vee 1)\, e^{\gamma(t-s)} \tag{4.14}$$

for $1 - \mu < \alpha < 1$ and constants $c(\alpha) \ge 0$ and $\gamma \in \mathbb{R}$ depending on the constants in (AT).

Let $T(\cdot)$ be the evolution semigroup on $E = C_0(\mathbb{R}, X)$ induced by $U(\cdot, \cdot)$ with 'generator' $G = \overline{A(\cdot) - d/dt}$, compare Proposition 3.16. Due to (4.14) and (3.24) we can extend $R(\lambda, G)$ to a bounded operator $\overline{\mathbb{K}}_\lambda : E_{\alpha-1} \to E$, $1 - \mu < \alpha < 1$, $\lambda > \gamma$, given by

$$\left(\overline{\mathbb{K}}_\lambda f\right)(t) = \int_{-\infty}^t e^{-\lambda(t-s)} \overline{U}(t, s) f(s)\, ds, \quad t \in \mathbb{R},\ f \in E_{\alpha-1}.$$

The following 'integration by parts' formula is crucial to prove Theorem 4.13.

Lemma 4.10. *Let (AT) hold, $\lambda > \gamma$, and $f \in C_0^1(\mathbb{R}, X) \cap D((w - A(\cdot))^\alpha)$ for some $\alpha \in (1 - \mu, \nu)$. Then $R(\lambda, G)f' = f - \lambda R(\lambda, G)f + \overline{\mathbb{K}}_\lambda A_{-1}(\cdot)f$.*

Proof. By rescaling we may assume that $\lambda = 0 > \gamma$ and $w = 0$. Let $f \in D(A(\cdot))$, $\alpha \in (1 - \mu, \nu)$, $s \in \mathbb{R}$, and $0 < h \le 1$. We write

$$D_{h,f}(s) := (-A(s+h))^{\alpha-1} \tfrac{1}{h} [U(s+h, s) - I] f(s) + (-A(s))^\alpha f(s)$$

$$= \frac{1}{h} \int_0^h \left[(-A(s+h))^{\alpha-1} - (-A(s+\tau))^{\alpha-1}\right] A(s+\tau) U(s+\tau, s) f(s)\, d\tau$$

$$- \frac{1}{h} \int_0^h \left[(-A(s+\tau))^\alpha\, U(s+\tau, s)(-A(s))^{-\alpha} - I\right] (-A(s))^\alpha f(s)\, d\tau.$$

Due to (2.10), (2.13), and (2.5) the norm of the first integral can be estimated by

$$\frac{c}{h} \int_0^h (h - \tau)^\mu\, \tau^{\alpha-1}\, d\tau\, \|(-A(\cdot))^\alpha f\|_\infty = c\, h^{\alpha+\mu-1}\, \|(-A(\cdot))^\alpha f\|_\infty$$

for constants c independent of h and s. One verifies that the second integral also tends to 0 as $h \searrow 0$ uniformly in s using (2.19) and $(-A(\cdot))^\alpha f \in C_0(\mathbb{R}, X)$ for large s and the strong continuity of $(s, \tau) \mapsto (-A(s + \tau))^\alpha U(s + \tau), s)(-A(s))^{-\alpha}$ (proved in Theorem 2.2) for s in compact sets. As a result,

$$\lim_{h \to 0} \|D_{h,f}\|_\infty = 0 \tag{4.15}$$

for $f \in D(A(\cdot))$. In the same way one sees that $\|D_{h,f}\|_\infty \le c\, \|(-A(\cdot))^\alpha f\|_\infty$ so that (4.15) holds for $f \in D((-A(\cdot))^\alpha)$ by approximation. For $f \in C_0^1(\mathbb{R}, X) \cap D((-A(\cdot))^\alpha)$, we further compute

$$-\int_{-\infty}^t U(t, s) f'(s)\, ds = \lim_{h \to 0} \int_{-\infty}^{t-h} U(t, s+h) \frac{1}{h}(f(s+h) - f(s))\, ds$$

$$= \lim_{h \to 0} \left(\frac{1}{h} \int_{t-h}^t U(t, s) f(s)\, ds + \frac{1}{h} \int_{-\infty}^{t-h} U(t, s+h)[U(s+h, s) - I] f(s)\, ds\right)$$

$$= f(t) + \lim_{h \to 0} \frac{1}{h} \int_{-\infty}^{t-h} U(t, s+h)[U(s+h, s) - I] f(s)\, ds, \tag{4.16}$$

where we used $f \in E_0$ in the last line. To determine the remaining limit, we note that the operator $U(t,s)(-A(s))^{1-\alpha}$, $t > s$, defined on $D((-A(s))^{1-\alpha})$ has a unique bounded extension $V(t,s) : X_0^s \to X_0^t$ satisfying

$$\|V(t,s)\| \le c(\alpha)\,((t-s)^{\alpha-1} \vee 1)\,e^{\gamma(t-s)} \qquad (4.17)$$

for $t > s$ and a constant $c(\alpha)$ by (2.15) and $\omega(U) < \infty$. Approximating f in $D((-A(\cdot))^{\alpha})$ by $f_n \in D(A(\cdot))$, one sees that $\overline{U}(t,s)A_{-1}(s)f(s) = -V(t,s)(-A(s))^{\alpha}f(s)$. Hence,

$$\frac{1}{h} \int_{-\infty}^{t-h} U(t,s+h)[U(s+h,s)-I]f(s)\,ds - \int_{-\infty}^{t} \overline{U}(t,s)A_{-1}(s)f(s)\,ds$$

$$= \int_{-\infty}^{t-h} V(t,s+h)\{(-A(s+h))^{\alpha-1}\tfrac{1}{h}[U(s+h,s)-I]f(s)+(-A(s))^{\alpha}f(s)\}ds$$

$$- \int_{-\infty}^{t-h} V(t,s+h)\,[(-A(s))^{\alpha}f(s)-(-A(s+h))^{\alpha}f(s+h)]\,ds. \qquad (4.18)$$

Due to (4.17), (4.15), and the uniform continuity of $(-A(\cdot))^{\alpha}f(\cdot)$, the right hand side of (4.18) tends to 0 in X as $h \searrow 0$, and the assertion follows from (4.16).

Our second assumption reads as follows.

(ED)   Each semigroup $(e^{\tau A(t)})_{\tau \ge 0}$, $t \in J$, has an exponential dichotomy with constants $N, \delta > 0$ and projection P_t.

Notice that the extension $A(t) := A(a)$ for $t < a$ preserves (ED). The next observation says that (ED) could also be formulated in terms of resolvents. We use the variant stated in (ED) to simplify the presentation.

Lemma 4.11. *Let A be a sectorial operator of type (ϕ, K, w).*

(a) If $[-\delta, \delta] + i\mathbb{R} \subseteq \rho(A)$ and $\|R(\lambda, A)\| \le r$ for $\lambda \in \pm\delta + i\mathbb{R}$ and constants $\delta, r > 0$, then $(e^{tA})_{t \ge 0}$ is hyperbolic with constants δ and $N = N(\phi, K, w, \delta, r)$ and with the projection on the unstable subspace

$$Q = \frac{1}{2\pi i} \int_{\Gamma} R(\lambda, A)\,d\lambda, \qquad (4.19)$$

where Γ is a suitable path around the spectral set $\{\lambda \in \sigma(A) : \operatorname{Re}\lambda \ge \delta\}$.
(b) If $(e^{tA})_{t \ge 0}$ is hyperbolic with constants $N, \delta > 0$, then we have $[-\delta', \delta'] + i\mathbb{R} \subseteq \rho(A)$ and $\|R(\lambda, A)\| \le \frac{2N}{\delta-\delta'}$ for $\lambda \in \pm\delta + i\mathbb{R}$ and each $\delta' \in [0, \delta)$,

Proof. (a) The exponential dichotomy of $(e^{tA})_{t \ge 0}$ and the formula (4.19) is shown e.g. in [42, Prop.2.3.3]. Clearly, the norms of Q and $P = I - Q$ can be estimated by a constant c_1 depending on ϕ, K, w, δ, r. Recall that there is a constant c_2 depending only on the type of A such that $\|e^{tA}\| \le c_2$ for $0 \le t \le (w+\delta)^{-1}$, see e.g. [21, p.98], [42, p.36]. Further,

$$e^{t(A+\delta)} P = \frac{1}{2\pi i} \int_{\Gamma'} e^{\lambda t} R(\lambda, A + \delta) P \, d\lambda,$$

where the path Γ' consists of the straight line between $\pm i(w + \delta)\tan(\pi - \phi)$ and the rays of angle $\pm\phi$ starting at $\pm i(w + \delta)\tan(\pi - \phi)$. A straightforward computation then yields

$$\|e^{t(A+\delta)} P\| \le c_3 \, c_1 \qquad \text{for} \ \ t \ge (w + \delta)^{-1}$$

and a constant $c_3 = c_3(\phi, K, w, \delta, r)$. Thus, $\|e^{tA}P\| \le N e^{-\delta t}$ for $t \ge 0$ and $N = N(\phi, K, w, \delta, r)$. In the same way one derives the exponential estimate for $e^{-tA_Q}Q$.

(b) By the spectral mapping theorem [42, Cor.2.3.7] and rescaling, the strip $[-\delta', \delta'] + i\mathbb{R}$ belongs to $\rho(A)$ for $0 \le \delta' < \delta$. For $|\operatorname{Re}\lambda| \le \delta'$ we have

$$R(\lambda, A) = \int_0^\infty e^{-\lambda t} e^{tA} P \, d\lambda - \int_0^\infty e^{\lambda t} e^{-tA_Q} Q \, d\lambda$$

which implies (b).

Notice that Lemma 4.11 and (ED) yield that $\|A(t)^{-1}\| \le \frac{2N}{\delta}$ for $t \in \mathbb{R}$. Hypothesis (ED) allows us to define

$$\Gamma_t(s) := \begin{cases} e^{sA_P(t)} P_t, & s \ge 0, \ t \in \mathbb{R}, \\ -e^{sA_Q(t)} Q_t, & s < 0, \ t \in \mathbb{R}. \end{cases}$$

Lemma 4.12. *If (AT') and (ED) hold, then the following assertions are true.*

(a) $t \mapsto P_t \in \mathcal{L}(X)$ *is globally Hölder continuous for* $t \in \mathbb{R}$ *and* $(t, s) \mapsto \Gamma_t(s) \in \mathcal{L}(X)$ *is locally Hölder continuous for* $t \in \mathbb{R}$ *and* $s \in \mathbb{R} \setminus \{0\}$.

(b) P_t *and* $\Gamma_t(s)$ *leave* X_α^t *and* $X_{\alpha,\infty}^t$ *invariant and have unique bounded extensions to* $X_{\alpha-1}^t$ *and* $X_{\alpha-1,\infty}^t$ *for* $0 < \alpha \le 1$ *and* $t, s \in \mathbb{R}$ *(the extensions are denoted by the same symbol). Moreover,*

$$\|A(t)^k \, \Gamma_t(s)x\| \le \begin{cases} NC_\alpha \, s^{\alpha-1-k} \, \|x\|_{\alpha-1}^t, & 0 < s \le 1, \\ NC_\alpha \, e^{-\delta(s-1)} \, \|x\|_{\alpha-1}^t, & s \ge 1, \\ NC_\alpha \, e^{\delta(s-1)} \, \|x\|_{\alpha-1}^t, & s < 0, \end{cases}$$

for $k = 0, 1$, $\alpha \in (0, 1)$, $t \in \mathbb{R}$, *and* $x \in X_{\alpha-1,\infty}^t$, *where* $C_\alpha \le C$ *are given by (4.12).*

(c) *Set* $\psi(s) := \sup_{t \in \mathbb{R}} \|A(t)\Gamma_t(s)\|_{\mathcal{L}(X_{\nu-1,\infty}^t, X)}$ *for* $s \ne 0$. *Then the upper integral* $q := \bar{\int}_\mathbb{R} \psi(s)|s|^\mu \, ds$ *is finite; more precisely,*

$$q \le NC_\nu \left(\int_0^1 s^{\nu+\mu-2} \, ds + \int_0^\infty \left((s+1)^\mu + e^{-\delta} s^\mu \right) e^{-\delta s} \, ds \right).$$

Proof. (a) By (2.9) the map $t \mapsto e^{sA(t)}$ is Hölder continuous uniformly for $s \in [s_0, s_1] \subseteq (0, \infty)$ Moreover, $s \mapsto e^{sA(t)}$ is locally Lipschitz continuous for $s > 0$ uniformly in $t \in \mathbb{R}$ since the semigroup is analytic. Equation (2.8) shows that $\|R(\lambda, A(t)) - R(\lambda, A(\tau))\| \leq c \, |t - \tau|^\mu$ for a constant c, $t, \tau \in \mathbb{R}$, and λ contained in the path Γ used in (4.19). The global Hölder continuity of $\mathbb{R} \ni t \mapsto Q_t \in \mathcal{L}(X)$ and $\mathbb{R} \ni t \mapsto P_t \in \mathcal{L}(X)$ now follow from (4.19). Finally,

$$e^{-sA_Q(t)}Q_t - e^{-\sigma A_Q(\tau)}Q_\tau = e^{-sA_Q(t)}Q_t (Q_t - Q_\tau) + (Q_t - Q_\tau) e^{-\sigma A_Q(\tau)}Q_\tau$$
$$+ e^{-sA_Q(t)}Q_t (e^{\sigma A(\tau)} - e^{sA(t)})e^{-\sigma A_Q(\tau)}Q_\tau$$

for $t, \tau \in \mathbb{R}$ and $s, \sigma > 0$. Hence, (a) is proved.

(b) The first sentence is an easy consequence of (4.19) and the definition of the inter- and extrapolation spaces. To establish the asserted estimates, we write

$$A(t)^k \, \Gamma_t(s) = \begin{cases} P_t A(t)^k \, e^{sA(t)}, & 0 < s \leq 1, \\ e^{(s-1)A(t)} P_t A(t)^k \, e^{A(t)}, & s \geq 1, \\ -e^{(s-1)A_Q(t)}Q_t A(t)^k \, e^{A(t)}, & s < 0, \end{cases}$$

for $t \in \mathbb{R}$ and use (4.12). Assertion (c) follows from (b).

We now come to the main result in this paragraph.

Theorem 4.13. *Assume that (AT') and (ED) hold and let q be given by Lemma 4.12. If $q\ell < 1$, then the evolution family $U(\cdot, \cdot)$ generated by $A(\cdot)$ has an exponential dichotomy with exponent $\delta' \in (0, (1 - q\ell)/\kappa)$ and constant N'. Here κ depends on N, δ, and the constants in (AT') and N' depends on N, δ, q, δ', and the constants in (AT'). Moreover, the Hölder continuous dichotomy projections $P(\cdot)$ of $U(\cdot, \cdot)$ satisfy*

$$\dim P(t)X = \dim P_s X \quad and \quad \dim \ker P(t) = \dim \ker P_s, \quad t, s \in J. \quad (4.20)$$

Proof. If $A(\cdot)$ is given on $J = [a, \infty)$, we extend it to $\mathbb{R}$ by $A(t) := A(a)$ for $t < a$. Then the extension satisfies (AT') and (ED) with the same constants (but observe that q was defined in Lemma 4.12 for the extended family). By restriction the assertions then follow from the case $J = \mathbb{R}$ treated below.

(1) Assume that $Gf = 0$ for some $f \in D(G)$. Then there are $f_n \in D(A(\cdot)) \cap C_0^1(\mathbb{R}, X)$ such that $f_n \to f$ and $Gf_n = A(\cdot)f_n - f_n' \to 0$ in E as $n \to \infty$ (cf. Proposition 3.16). By [1, Prop.3.2,5.1] we have

$$f_n(t) = U(t, s)f_n(s) - \int_s^t U(t, \tau)(Gf_n)(\tau) \, d\tau, \quad t \geq s.$$

Hence, $f(t) = U(t, s)f(s)$ for $t \geq s$ so that $f'(t) = A(t)f(t)$ exists by Theorem 2.2. Taking $s = t - 1$ we see that $f \in D(A(\cdot)) \cap C_0^1(\mathbb{R}, X)$. For $g \in E$, the function

$$(Lg)(t) = \int_{-\infty}^\infty \Gamma_t(t - s)g(s) \, ds, \quad t \in \mathbb{R},$$

belongs to E by Lemma 4.12. Using [42, Prop.2.1.4(c)], we obtain $(Lf)(t) \in D(A(t))$ and

$$(Lf')(t) = \lim_{h \to 0} \tfrac{1}{h} L(f(\cdot + h) - f)(t)$$
$$= \lim_{h \to 0} \left(\frac{1}{h}(e^{hA(t)} - I)(Lf)(t) + \frac{1}{h} \int_t^{t+h} e^{(t+h-s)A(t)} f(s)\, ds \right)$$
$$= f(t) + A(t)(Lf)(t), \qquad t \in \mathbb{R},$$

since the second summand in the middle line converges due to $f(t) \in \overline{D(A(t))}$. We also define

$$(Vf)(t) := \int_{\mathbb{R}} \Gamma_t(t - s)(A(s) - A_{-1}(t))f(s)\, ds$$
$$= \int_{\mathbb{R}} A(t)\Gamma_t(t - s)(A(t)^{-1} - A(s)^{-1})A(s)f(s)\, ds$$

for $t \in \mathbb{R}$, where the integral in the first line is understood in the topology of X_{-1}^t. Combining these formulas, we derive

$$0 = LGf = -Lf' + LA(\cdot)f = Vf - f, \quad \text{i.e.,} \quad Vf = f.$$

Employing (AT'), $\|A(t)^2 \Gamma_t(\tau)\|_{\mathcal{L}(X_{\nu,\infty}^t, X)} = \|A(t)\Gamma_t(\tau)\|_{\mathcal{L}(X_{\nu-1,\infty}^t, X)}$, and Lemma 4.12(c), we further deduce that

$$\|A(t)f(t)\| = \|A(t)(Vf)(t)\| \le \int_{\mathbb{R}} \|A(t)^2 \Gamma_t(\tau)\|_{\mathcal{L}(X_{\nu,\infty}^t, X)} \, \ell \, |\tau|^\mu \, d\tau \, \|A(\cdot)f\|_\infty$$
$$\le q\ell \, \|A(\cdot)f\|_\infty < \|A(\cdot)f\|_\infty .$$

Since each $A(t)$, $t \in \mathbb{R}$, is invertible, this estimate shows that $f = 0$. Hence G is injective.

(2) For $f \in E$ and $t \in \mathbb{R}$ we define

$$(Rf)(t) = \int_{-\infty}^{\infty} \Gamma_s(t - s)f(s)\, ds.$$

By Lemma 4.12, R is a bounded operator on E which can be extended to an operator $\tilde{R} : E_{\alpha-1} \to E$ for $\alpha \in (0, 1)$ with norm

$$\|\tilde{R}\|_{\mathcal{L}(E_{\alpha-1}, E)} \le NC_\alpha \left(\frac{1}{\alpha} + \frac{1 + e^{-\delta}}{\delta} \right) =: \rho_\alpha. \tag{4.21}$$

For $f \in D(A(\cdot))$, one easily sees that $Rf \in C_0^1(\mathbb{R}, X)$ and $\frac{d}{dt}Rf = f + RA(\cdot)f$. Let $\alpha \in (1 - \mu, \nu]$. Then (AT') and Lemma 4.12 imply

$$\sup_{r > w} \left\| r^\alpha A(t)R(r, A(t)) \int_{\mathbb{R}} (A(s)^{-1} - A(t)^{-1})\Gamma_s(t - s)A(s)f(s)\, ds \right\|$$
$$\le c(\alpha) \int_{\mathbb{R}} \ell \, |t - s|^\mu \, \|A(s)\Gamma_s(t - s)f(s)\| \, ds \le c'(\alpha) \, \|f\|_{\alpha-1}$$

for $f \in D(A(\cdot))$ and constants $c(\alpha)$ and $c'(\alpha)$ with $c'(\nu) = q\ell$. Therefore $Rf \in E_{\nu,\infty}$ and the function

$$(Sf)(t) := A_{-1}(t) \int_{\mathbb{R}} (A(s)^{-1} - A(t)^{-1})A(s)\Gamma_s(t-s)f(s)\,ds, \quad t \in \mathbb{R},$$

belongs to $E_{\nu-1,\infty}$. Moreover, S can be extended to a bounded operator from $E_{\alpha-1}$ to $E_{\alpha-1,\infty}$ having the same representation and then restricted to an operator $\tilde{S} : E_{\nu-1,\infty} \to E_{\nu-1,\infty}$ with norm less than $q\ell$. We set $\tilde{G}h := -h' + A_{-1}(\cdot)h \in E_{-1}$ for $h \in C_0^1(\mathbb{R}, X) \cap E_0$. For $f \in D(A(\cdot))$ the above observations yield

$$\tilde{G}Rf = -f - RA(\cdot)f + A_{-1}(\cdot)Rf = (S - I)f. \tag{4.22}$$

For a given $g \in E$, there exists $f := (\tilde{S}-I)^{-1}g \in E_{\nu-1,\infty}$. Fix $\alpha \in (1-\mu, \nu)$. There are $f_n \in D(A(\cdot))$ converging to f in $\|\cdot\|_{\alpha-1}$. Equation (4.22) now gives

$$\tilde{G}Rf_n = (S - I)f_n \longrightarrow g \quad \text{in } \|\cdot\|_{\alpha-1} \text{ as } n \to \infty.$$

Since $Rf_n \in E_{\nu,\infty} \hookrightarrow D((w - A(\cdot))^\alpha)$, Lemma 4.10 shows that

$$\overline{\mathbb{K}}_\lambda \tilde{G}Rf_n = \overline{\mathbb{K}}_\lambda A_{-1}(\cdot)Rf_n - R(\lambda, G)\tfrac{d}{dt}Rf_n = \lambda R(\lambda, G)Rf_n - Rf_n.$$

Letting $n \to \infty$ and using the continuity of $\overline{\mathbb{K}}_\lambda, \tilde{R} : E_{\alpha-1} \to E$, we deduce

$$R(\lambda, G)g = \overline{\mathbb{K}}_\lambda g = \lambda R(\lambda, G)\tilde{R}f - \tilde{R}f.$$

Therefore, $\tilde{R}f \in D(G)$ and $G\tilde{R}f = g$. This means that G is surjective. Together with step (1) we have proved the invertibility of G.

The above argument also shows that $G^{-1}g = \tilde{R}(\tilde{S} - I)^{-1}g$ for $g \in E$, and thus $\|G^{-1}\| \leq \frac{\sigma\rho}{1-q\ell}$ where $\rho := \rho_\nu$ is given by (4.21) and $\sigma := \sup_{r>w, t \in \mathbb{R}} \|r^\nu R(r, A(t))\|$. Hence Theorem 3.17 and Proposition 3.18 establish the theorem except for (4.20).

(3) Recall that the dimensions of the kernel and the range of two projections P and $\hat{P}$ coincide if $\|P - \hat{P}\| < 1$, see [32, p.298]. So Lemma 4.12(a) shows that the dimensions of $\ker P_t$ and $P_t X$ do not depend on t.

We set $A_\varepsilon(t) := A(\varepsilon t)$ for $0 \leq \varepsilon \leq 1$ and $t \in \mathbb{R}$. Observe that $A_\varepsilon(\cdot)$ satisfies (AT') and (ED) with the same constants and the same q. Hence the corresponding evolution families $U_\varepsilon(\cdot, \cdot)$ fulfill the same exponential estimate and are hyperbolic with common constants $\delta', N' > 0$ and projections $P_\varepsilon(t)$ due to steps (1) and (2). Note that $P_1(t) = P(t)$ and $P_0(t) = P_0$ for $t \in \mathbb{R}$. Applying [32, p.298] once more, it remains to show that $\varepsilon \mapsto P_\varepsilon(t) \in \mathcal{L}(X)$ is continuous for each $t \in \mathbb{R}$. Since (AT) holds, we have

$$\|R(w, A_\varepsilon(t)) - R(w, A_\eta(t))\| \leq c|t|^\mu |\varepsilon - \eta|^\mu$$

for $t \in \mathbb{R}$ and $\varepsilon, \eta \in [0,1]$. Thus Proposition 2.6 shows that

$$\|U_\varepsilon(s, s-1) - U_\eta(s, s-1)\| \leq \tilde{c}(1+r)^{\mu\beta} |\varepsilon - \eta|^{\mu\beta} \tag{4.23}$$

for $|s| \le r$ and some $\beta > 0$ provided that $(1 + r)|\varepsilon - \eta|$ is sufficiently small, where the constant $\tilde{c}$ does not depend on r, ε, η.

Let $T_\varepsilon(\cdot) = T_{U_\varepsilon}(\cdot)$. Formula (3.8) yields $\|R(\lambda, T_\varepsilon(1))\| \le 2N'(1 - e^{-\delta'}) =: d$ for each $|\lambda| = 1$ and $0 \le \varepsilon \le 1$. Moreover, $M := \sup_{s, \varepsilon} \|U_\varepsilon(s, s - 1)\| < \infty$. Given $t \in \mathbb{R}$ and $x \in X$, we set $f = \varphi(\cdot)x$ for a function $\varphi \in C(\mathbb{R})$ with support in $(t - \frac{1}{2}, t + \frac{1}{2})$, $\varphi(t) = 1$, and $0 \le \varphi \le 1$. Let $0 \le \varepsilon, \eta \le 1$. Then (3.7), (3.8), and (4.23) imply

$$\|P_\eta(t)x - P_\varepsilon(t)x\| \le \|P_\eta(\cdot)f - P_\varepsilon(\cdot)f\|_\infty$$

$$\le d \sup_{|\lambda| = 1} \|(T_\eta(1) - T_\varepsilon(1))R(\lambda, T_\varepsilon(1))f\|_\infty$$

$$\le d \max\{ \sup_{|s| \le r, |\lambda| = 1} \|U_\varepsilon(s, s - 1) - U_\eta(s, s - 1)\| \, \|R(\lambda, T_\varepsilon(1))f\|_\infty,$$

$$\sup_{|s| \ge r, |\lambda| = 1} 2M \|R(\lambda, T_\varepsilon(1))f(s - 1)\|\}$$

$$\le \max\{d^2 \tilde{c}(1 + r)^{\mu\beta} |\varepsilon - \eta|^{\mu\beta}, 2dMN' \sup_{|s| \ge r} \sum_{n \in \mathbb{Z}} e^{-\delta'|n|} \varphi(s - 1 - n)\} \|x\|$$

$$\le \max\{d^2 \tilde{c}(1 + r)^{\mu\beta} |\varepsilon - \eta|^{\mu\beta}, 2dMN' e^{-\delta' N(r)}\} \|x\|$$

for each $r > \max\{t + \frac{1}{2}, -t - \frac{3}{2}\}$ and sufficiently small $|\varepsilon - \eta|$ (depending on r), where $N(r)$ is the minimum of the integer parts of $r - t - \frac{1}{2}$ and $r + t + \frac{3}{2}$. Therefore $P_\eta(t) \to P_\varepsilon(t)$ in $\mathcal{L}(X)$ as $\eta \to \varepsilon$. As observed above, (4.20) follows from this fact.

In the above proof we have used the operators $A(\varepsilon t)$ and noted that they fulfill (AT') and (ED) with the same constants and the same q. More precisely, in (AT') the Hölder constant ℓ can be replaced by $\varepsilon^\mu \ell$. This observation leads to the next corollary.

Corollary 4.14. *Assume that the operators $A(t)$ satisfy (AT') and (ED). Let $0 < \varepsilon < \varepsilon_0 := (q\ell)^{-1/\mu}$. Then the evolution family $U_\varepsilon(\cdot, \cdot)$ generated by $A(\varepsilon t)$ is hyperbolic with constants not depending on $\varepsilon \in (0, \varepsilon_1]$ if $\varepsilon_1 < \varepsilon_0$. Moreover, its dichotomy projections $P_\varepsilon(t)$ satisfy*

$$\dim P_\varepsilon(t)X = \dim P_s X \quad and \quad \dim \ker P_\varepsilon(t) = \dim \ker P_s, \quad t, s \in J. \quad (4.24)$$

The operators $\frac{1}{\varepsilon}A(t)$, $t \in J$, generate the evolution family $\hat{U}_\varepsilon(t, s) = U_\varepsilon(\frac{t}{\varepsilon}, \frac{s}{\varepsilon})$, $t \ge s$, where $U_\varepsilon(\cdot, \cdot)$ is generated by $A(\varepsilon t)$. So we obtain the following result on singular perturbation, cf. [32, §IV.1]. Similar theorems are contained in [35, §10.7] for bounded $A(t)$, in [38, Thm.6] for delay equations, and in [27, p.215] for a special class of exponentially stable parabolic equations. Notice that the dichotomy exponent blows up as $\varepsilon \searrow 0$.

Corollary 4.15. *Assume that $A(\cdot)$ satisfies (AT') and (ED). Let $0 < \varepsilon < \varepsilon_0 := (q\ell)^{-1/\mu}$. Then the solutions of*

$$\varepsilon u'(t) = A(t)u(t), \quad t \ge s, \quad u(s) = x,$$

have exponential dichotomy with constants δ'/ε and N', where δ' and N' do not depend on $\varepsilon \in (0, \varepsilon_1]$ for $\varepsilon_1 < \varepsilon_0$. Moreover, (4.24) holds.

Example 4.16. We consider the situation of Example 2.8 assuming additionally that Ω is connected, $p = 2$, $a_0 = 0$, but requiring instead of (2.37) only that $a_{kl} \in C_b^\mu(\mathbb{R}_+, L^\infty(\Omega))$ for some $\mu > \frac{1}{2}$. Our aim is to calculate the constants q and ℓ in Theorem 4.13 explicitly (which is hardly possible in the framework of Example 2.8 and 2.9). We combine the approach of [64] with Hilbert space techniques. On $X = L^2(\Omega)$ we define the operator $A(t)$ by the closed symmetric quadratic form

$$a_t(f, g) = -\sum_{k,l=1}^n \int_\Omega a_{kl}(t, x)\, D_k f(x)\, \overline{D_l g(x)}\, dx$$

for $f, g \in D(a_t) = W^{1,2}(\Omega)$, cf. [30, Chap.6], [62, §2.2, 3.6]. If we can verify (AT) for $A(\cdot)$, then we obtain functions $u \in C^1((s, \infty), L^2(\Omega)) \cap C([s, \infty), L^2(\Omega))$ satisfying (2.36) in a weak sense w.r.t. the space variables (see Example 2.8 concerning better regularity).

Since the domain is connected, 0 is a simple eigenvalue of the Neumann Laplacian Δ_N on Ω. By the Rayleigh–Ritz formula, see e.g. [18, §4.5], 0 is also a simple eigenvalue of $A(t)$ and the first non–zero eigenvalue of $A(t)$ is smaller than $\eta \lambda_1$, where λ_1 is the first nonzero eigenvalue of Δ_N and η is the ellipticity constant from (2.38). Set

$$\delta := -\tfrac{1}{2}\, \eta\, \lambda_1. \tag{4.25}$$

Then $\tilde{A}(t) := A(t) + \delta$ has an exponential dichotomy with constants δ and $N = 1$ and orthogonal projection P_t, where $\dim Q_t X = 1$. Moreover $\tilde{A}(t)$ is invertible and sectorial of type (ϕ, K, w) for all $\phi \in (\frac{\pi}{2}, \pi)$, $w > \delta$, and a constant $K = K(w, \phi)$. We fix some $w > \delta$. Using the orthogonality of P_t and the functional calculus, see e.g. [18, Thm.2.5.3], one sees that

$$\|\tilde{A}(t)e^{\tau \tilde{A}(t)}\| \le \frac{1}{e\tau} \quad \text{and} \quad \|\tilde{A}(t)^2 e^{\tau \tilde{A}(t)}\| = \|\tilde{A}(t)e^{\tau \tilde{A}(t)}\|_{\mathcal{L}(X_{-1}^t, X)} \le \frac{4}{e^2\tau^2}$$

for $t \ge 0$ and $0 < \tau \le \frac{2}{\delta}$, where X_{-1}^t is the extrapolation space for $\tilde{A}(t)$. By real interpolation, see e.g. [42, Prop.1.2.6], this implies

$$\|\tilde{A}(t)e^{\tau \tilde{A}(t)} f\| \le 2e^{-3/2}\, \tau^{-3/2}\, \|\|f\|\|_{-\frac{1}{2}}^t$$

for $0 < \tau \le \frac{2}{\delta}$ and $f \in X_{-\frac{1}{2}, \infty}^t$, where

$$\|\|f\|\|_{-\frac{1}{2}}^t := \sup_{s>0} s^{-\frac{1}{2}}\, K(s, f),$$

$$K(s, f) := \inf\{\|u\|_{-1}^t + s\, \|v\| : f = u + v,\ u \in X_{-1}^t,\ v \in X\}.$$

For $s < \frac{1}{w}$, set $u := -\tilde{A}_{-1}(t)R(\frac{1}{s}, \tilde{A}_{-1}(t))f$ and $v := \frac{1}{s}R(\frac{1}{s}, \tilde{A}_{-1}(t))f$. Then

$$\sup_{0<s<1/w} s^{-\frac{1}{2}} K(s,f) \leq \sup_{0<s<1/w} 2s^{-\frac{1}{2}} \|R(\tfrac{1}{s}, \tilde{A}_{-1}(t))f\| = 2\,\|f\|_{-\frac{1}{2}}^t.$$

For $s \geq \frac{1}{w}$ we choose $u := f$ and $v := 0$ and obtain

$$\sup_{s\geq 1/w} s^{-\frac{1}{2}} K(s,f) \leq \sup_{s\geq 1/w} s^{-\frac{1}{2}} \|\tilde{A}_{-1}(t)^{-1}f\| = w^{\frac{1}{2}} \|\tilde{A}_{-1}(t)^{-1}f\|$$

$$= \|(w - \tilde{A}(t))\tilde{A}(t)^{-1}\| \; w^{\frac{1}{2}} \,\|R(w, \tilde{A}_{-1}(t))f\|$$

$$\leq (1 + \tfrac{w}{\delta}) \,\|f\|_{-\frac{1}{2}}^t$$

since $\|\tilde{A}(t)^{-1}\| = \frac{1}{\delta}$. As a result,

$$\|\tilde{A}(t)e^{\tau \tilde{A}(t)}\|_{\mathcal{L}(X_{-\frac{1}{2},\infty}^t, X)} \leq 2e^{-3/2}\left(1 + \tfrac{w}{\delta}\right)\tau^{-3/2} =: c_1(w)\tau^{-3/2} \qquad (4.26)$$

for $0 < \tau \leq \frac{2}{\delta}$. This allows to estimate q from Lemma 4.12 for $\tilde{A}(\cdot)$ by

$$q \leq c_1 \int_0^{2/\delta} s^{\mu-3/2}\,ds + c_1\left(\frac{\delta}{2}\right)^{\frac{3}{2}} \int_{2/\delta}^{\infty} e^{-\delta(s-2/\delta)} s^\mu\,ds \qquad (4.27)$$

$$+ c_1\left(\frac{\delta}{2}\right)^{\frac{3}{2}} \int_0^{\infty} e^{-\delta(s+2/\delta)} s^\mu\,ds$$

$$= c_1\left[\frac{1}{\mu - 1/2}\left(\frac{2}{\delta}\right)^{\mu-\frac{1}{2}} + \left(\frac{\delta}{2}\right)^{\frac{3}{2}} \int_0^{\infty} e^{-\delta s}\left((s+\tfrac{2}{\delta})^\mu + e^{-2}s^\mu\right)ds\right] =: c_2(w).$$

For instance, $c_2(w) = c_1(w)\left[2^{\frac{3}{2}} + 2^{-\frac{1}{2}} + (1 + e^{-2})\,2^{-\frac{3}{2}}\right]\delta^{-\frac{1}{2}}$ if $\mu = 1$.

Let $[a]_\mu$ be the maximum of the Hölder constants of a_{kl}, $k,l = 1,\cdots,n$, $\varepsilon := w - \delta > 0$, $f \in L^2(\Omega)$, and $g \in D((\varepsilon - A(t))^{\frac{1}{2}})$. Then (2.40) implies

$$|((\varepsilon - A(t))^{-\frac{1}{2}}[R(\varepsilon, A(t)) - R(\varepsilon, A(s))]f, g)|$$

$$\leq n\,[a]_\mu\,|t - s|^\mu\,\|\nabla R(\varepsilon, A(s))f\|_2\,\|\nabla(\varepsilon - A(t))^{-\frac{1}{2}}g\|_2.$$

Further, (2.38) yields

$$\|\nabla R(\varepsilon, A(s))f\|_2^2 = \sum_{k=1}^{n} \int_\Omega |D_k R(\varepsilon, A(s))f(x)|^2\,dx$$

$$\leq \frac{1}{\eta} \sum_{k,l=1}^{n} \int_\Omega a_{kl}(s,x)\,D_k R(\varepsilon, A(s))f(x)\,\overline{D_l R(\varepsilon, A(s))f(x)}\,dx$$

$$= \frac{1}{\eta}\left(-A(s)R(\varepsilon, A(s))f, R(\varepsilon, A(s))f\right)$$

$$\leq \frac{1}{\eta}\|f\|^2(1 + \varepsilon\|R(\varepsilon, A(s))\|)\,\|R(\varepsilon, A(s))\| = \frac{2}{(w-\delta)\eta}\,\|f\|^2.$$

In the same way we obtain

$$\|\nabla(\varepsilon - A(t))^{-\frac{1}{2}} g\|_2^2 \leq \frac{1}{\eta} \left(-A(t)(\varepsilon - A(t))^{-\frac{1}{2}} g, (\varepsilon - A(t))^{-\frac{1}{2}} g\right)$$
$$= \frac{1}{\eta} \left(-A(t) R(\varepsilon, A(t)) g, g\right) \leq \frac{2}{\eta} \|g\|^2.$$

So we have shown that

$$\|(\varepsilon - A(t))^{\frac{1}{2}} [R(\varepsilon, A(t)) - R(\varepsilon, A(s))]\|_{\mathcal{L}(X)} \leq [a]_\mu \frac{2n}{\eta(w-\delta)^{\frac{1}{2}}} |t-s|^\mu. \quad (4.28)$$

It remains to relate (4.28) with the Hölder estimate for $\tilde{A}(t)$ in (AT'). To do this, we first write for $r > w$

$$r^{\frac{1}{2}} \tilde{A}(t) R(r, \tilde{A}(t)) (\varepsilon - A(t))^{-\frac{1}{2}}$$
$$= \frac{1}{\pi} r^{\frac{1}{2}} \tilde{A}(t) R(r, \tilde{A}(t)) \int_0^\infty (rs)^{-\frac{1}{2}} R(rs + \varepsilon, A(t)) \, r \, ds$$
$$= \frac{1}{\pi} \left(\int_0^1 s^{-\frac{1}{2}} r R(r - \delta, A(t)) \tilde{A}(t) R(rs + \varepsilon + \delta, \tilde{A}(t)) \, ds \right.$$
$$\left. + \int_1^\infty s^{-\frac{3}{2}} rs R(rs + \varepsilon, A(t)) \tilde{A}(t) R(r, \tilde{A}(t)) \, ds \right)$$

using a standard formula for fractional powers, see e.g. [47, (2.6.4)]. So we can estimate

$$\|r^{\frac{1}{2}} \tilde{A}(t) R(r, \tilde{A}(t)) (\varepsilon - A(t))^{-\frac{1}{2}}\|$$
$$\leq \frac{1}{\pi} \left(\int_0^1 s^{-\frac{1}{2}} \frac{r}{r-\delta} \left(\frac{rs + \varepsilon + \delta}{rs + \varepsilon} + 1 \right) ds + \int_1^\infty s^{-\frac{3}{2}} \frac{rs}{rs + \varepsilon} \left(\frac{r}{r-\delta} + 1 \right) ds \right)$$
$$\leq \frac{2}{\pi} \left(1 + \frac{w}{w-\delta} \right)^2 =: c_3(w). \quad (4.29)$$

Finally, we compute

$$\tilde{A}(s)^{-1} - \tilde{A}(t)^{-1}$$
$$= (\varepsilon - A(t))(\delta + A(t))^{-1} [R(\varepsilon, A(t)) - R(\varepsilon, A(s))](\varepsilon - A(s))(\delta + A(s))^{-1}$$
$$= (w(A(t) + \delta)^{-1} - I) [R(\varepsilon, A(t)) - R(\varepsilon, A(s))] (w(A(s) + \delta)^{-1} - I).$$

Putting all this together, we conclude that

$$\|\tilde{A}(t)^{-1} - \tilde{A}(s)^{-1}\|_{\mathcal{L}(X, X_{\frac{1}{2}, \infty}^t)} \leq c_3 \frac{2n(1 + \frac{w}{\delta})^2}{\eta(w-\delta)^{1/2}} [a]_\mu |t-s|^\mu$$
$$=: c_4(w) [a]_\mu |t-s|^\mu. \quad (4.30)$$

Thus (AT') holds for $\tilde{A}(\cdot)$ with $\ell := c_4 [a]_\mu$. As a consequence of Theorem 4.13, the operators $\tilde{A}(t)$ generate an evolution family having exponential dichotomy with exponent $\delta' > 0$ and $\dim Q(t) = 1$ provided that

$$[a]_\mu < (c_2 c_4)^{-1}, \quad (4.31)$$

where c_2 and c_4 depend on $\mu, w, \eta, \lambda_1, n$ as described in (4.25), (4.26), (4.27), (4.29), and (4.30). Consequently, there are subspaces $X_1(s)$ of codimension 1 such that the solutions of (2.36) starting in $X_1(s)$ tend to 0 exponentially fast. On the other hand, (2.36) has the constant solution $u(t, x) = c$ since $a_0 = 0$. Estimate (4.31) can always be achieved if we replace $a_{kl}(t, x)$ by $a_{kl}(\varepsilon t, x)$ for a small $\varepsilon > 0$.

5 Inhomogeneous Problems

In this section we investigate the inhomogeneous problems

$$u'(t) = A(t)u(t) + f(t), \quad t \in \mathbb{R}, \tag{5.1}$$
$$u'(t) = A(t)u(t) + f(t), \quad t > a, \qquad u(a) = x, \tag{5.2}$$

assuming that $f \in C_b(J, X)$, $x \in X$, and $A(t)$, $t \in J$, $J \in \{\mathbb{R}, [a, \infty)\}$, satisfy (AT) and that the evolution family $U(\cdot, \cdot)$ generated by $A(\cdot)$ has an exponential dichotomy with projections $P(t)$ and Green's function $\Gamma(\cdot, \cdot)$. We want to show that u inherits the convergence and almost periodicity of f. We have seen in Section 2 that the classical solution of (5.2) with $x \in \overline{D(A(a))}$ is given by the mild solution

$$u(t) = U(t, a)x + \int_a^t U(t, \tau)f(\tau)\, d\tau, \qquad t \geq a. \tag{5.3}$$

Conversely, the mild solution is a classical or strict one if the data are sufficiently regular due to Theorem 2.7. Writing $f(\tau) = P(\tau)f(\tau) + Q(\tau)f(\tau)$, we deduce from (5.3) that

$$u(t) = U(t, a)\left(x + \int_a^\infty U_Q(a, \tau)Q(\tau)f(\tau)\, d\tau\right) + \int_a^\infty \Gamma(t, \tau)f(\tau)d\tau \tag{5.4}$$

for $t \geq a$. Since $U(\cdot, \cdot)$ is exponentially dichotomic, the function u is bounded if and only if the term in brackets belongs to $P(a)X$ if and only if

$$Q(a)x = -\int_a^\infty U_Q(a, \tau)Q(\tau)f(\tau)\, d\tau. \tag{5.5}$$

This condition automatically holds if $U(\cdot, \cdot)$ is exponentially stable. In the general case it says that the unstable part of the initial value of a bounded solution is determined by the inhomogeneity (and by U_Q). If (5.5) is valid, then the mild solution is given by

$$u(t) = U(t, a)P(a)x + \int_a^\infty \Gamma(t, \tau)f(\tau)\, d\tau =: v_1(t) + v_2(t), \qquad t \geq a. \tag{5.6}$$

In the case that $J = \mathbb{R}$ we can apply the above arguments for $x = u(s)$ and all $s = a \in \mathbb{R}$. So we call a function $u \in C(\mathbb{R}, X)$ satisfying

$$u(t) = U(t, s)u(s) + \int_s^t U(t, \tau)f(\tau)\, d\tau \quad \text{for all } t \geq s \qquad (5.7)$$

the *mild solution* of (5.1). Such a function is continuously differentiable, belongs to $D(A(t))$, and solves (5.1) if f is locally Hölder continuous or $\|f(t)\|_\alpha^t$ is locally bounded by virtue of Theorem 2.7. It is easy to verify that

$$u(t) = \int_{\mathbb{R}} \Gamma(t, \tau)f(\tau)\, d\tau, \qquad t \in \mathbb{R}, \qquad (5.8)$$

is a bounded mild solution. Conversely, letting $a \to -\infty$ in (5.6) with $x = u(a)$, we see each bounded mild solution is given by (5.8).

5.1 Asymptotically Autonomous Problems

We first study (5.2) in the framework of Paragraph 4.2 where we require that $f \in C([a, \infty), X)$ converges to f_∞ in X as $t \to \infty$. Supposing the assumptions of Theorem 4.3, the evolution family $U(\cdot, \cdot)$ has an exponential dichotomy on an interval $[a, \infty)$.[8] Our next result generalizes a theorem due to H.Tanabe from 1961, see [62, Thm.5.6.1], where the situation of Remark 4.5 with $w = 0$ was studied. In [26, Thm.1.3] Tanabe's result was extended to the case of Kato–Tanabe conditions for f converging in a weaker sense, but still for the case of exponential stability. In [57, Thm.4.1] we proved our theorem for dense domains $D(A(t))$. It was partly extended to partial functional differential equations in [59]. Recall that the condition (5.5) is equivalent to the boundedness of the mild solution of (5.2).

Theorem 5.1. *Assume that the operators $A(t)$, $t \geq 0$, satisfy (AT) from Section 2, the operator A satisfies (AT1), $R(w, A(t)) \to R(w, A)$ in $\mathcal{L}(X)$ as $t \to \infty$, and $\sigma(A) \cap i\mathbb{R} = \emptyset$. Fix the number $a \geq 0$ as obtained in Theorem 4.3. Suppose that $f \in C([a, \infty), X)$ tends to f_∞ in X as $t \to \infty$ and that $x \in X$ satisfies (5.5). Then the mild solution u of (5.2) is given by (5.6) and converges to $u_\infty := -A^{-1}f_\infty$ as $t \to \infty$. If, in addition, either*

(a) $f \in C_b^\alpha([a, \infty), X)$ for some $\alpha > 0$ or
(b) $\sup_{t \geq a} \|f(t)\|_\beta^t < \infty$ for some $\beta > 0$,

then $u \in C^1((a, \infty), X)$, $u(t) \in D(A(t))$ for $t > a$, and (5.2) holds. Moreover, $u'(t) \to 0$ and $A(t)u(t) \to Au_\infty$ as $t \to \infty$. Finally, u is a classical solution if $x \in \overline{D(A(a))}$.

Proof. (1) Let u be the mild solution given by (5.3) or (5.6). Theorems 2.2 and 2.7 show that $u \in C^1((a, \infty), X)$, $u(t) \in D(A(t))$ for $t > a$, and (5.2) holds if (a) or (b) are true. Moreover, $u(t) \to x$ as $t \to a$ if $x \in \overline{D(A(a))}$.

We use the functions v_k defined in (5.6), where we extend f by 0 to $\mathbb{R}$. Thus v_2 is given by (5.8) for $t \geq a$. Clearly, $v_1(t) \to 0$ and

---

[8] We recall that Proposition 4.8 provides conditions allowing to take $a = 0$.

$$\|v_1'(t)\| = \|A(t)v_1(t)\| \le \|A(t)U(t, t-1)\| \, \|U(t-1, a)P(a)x\| \longrightarrow 0$$

as $t \to \infty$ due to (2.13). In Theorem 4.3 we proved that $U(s+t, s) \to e^{tA}$ and $P(s) \to P$ strongly as $s \to \infty$, where P is the dichotomy projection of e^{tA}. We further have

$$v_2(t) = \int_0^\infty U(t, t-\tau)P(t-\tau)f(t-\tau)\, d\tau - \int_0^\infty U_Q(t, t+\tau)Q(t+\tau)f(t+\tau)\, d\tau$$

for $t \ge a$. The first integrand converges to $e^{\tau A}Pf_\infty$ as $t \to \infty$. Concerning the second integrand, we observe that

$$\begin{aligned}
U_Q(t, t+\tau)&Q(t+\tau)f(t+\tau) - e^{-\tau A_Q}Qf_\infty \\
&= U_Q(t, t+\tau)Q(t+\tau)\left[Q(t+\tau)f(t+\tau) - Qf_\infty\right] + U_Q(t, t+\tau)Q(t+\tau) \\
&\quad \cdot \left[e^{\tau A} - U(t+\tau, t)\right]e^{-\tau A_Q}Qf_\infty + \left[Q(t+\tau) - Q\right]e^{-\tau A_Q}Qf_\infty
\end{aligned}$$

tends to 0 as $t \to \infty$. Hence,

$$\begin{aligned}
\lim_{t\to\infty} u(t) = \lim_{t\to\infty} v_2(t) &= \int_0^\infty e^{\tau A_P}Pf_\infty\, d\tau - \int_0^\infty e^{-\tau A_Q}Qf_\infty\, d\tau \\
&= -A_P^{-1}f_\infty + (-A_Q^{-1})f_\infty = -A^{-1}f_\infty \qquad (5.9)
\end{aligned}$$

by the theorem of dominated convergence and standard semigroup theory. The remaining assertions follow from the claim

$$\lim_{t\to\infty} v_2'(t) = 0. \qquad (5.10)$$

To this establish this fact, we will use the formula

$$v_2(t) = U(t, r)v_2(r) + \int_r^t U(t, \tau)f(\tau)\, d\tau, \quad t \ge r \ge a, \qquad (5.11)$$

which is a consequence of (5.3).

(2) Assume that (a) holds. We first want to show that v_2' is uniformly Hölder continuous on $[a+1, \infty)$. In view of (5.11) and Theorems 2.2 and 2.7, v_2' is given by

$$v_2'(t) = A(t)U(t, r)v_2(r) + A(t)\int_r^t U(t, \tau)f(\tau)\, d\tau + f(t) \qquad (5.12)$$

for $t > r \ge a$. For $t \ge s \ge r + \frac{1}{2}$, Proposition 2.4 shows that

$$\begin{aligned}
\|A(t)U(t, r)v_2(r) - A(s)U(s, r)v_2(r)\| &\le c\,(t-s)^\alpha\,(s-r)^{-1-\alpha}\,\|v_2\|_\infty \\
&\le c\,(t-s)^\alpha\,\|f\|_\infty
\end{aligned}$$

for constants c independent of t, s, r satisfying $r+1 \ge t \ge s \ge r + \frac{1}{2}$. Further, (2.32) with $\varepsilon = 1/2$ yields

$$\left\| A(t) \int_r^t U(t,\tau)f(\tau)d\tau - A(s) \int_r^s U(s,\tau)f(\tau)d\tau \right\| \le c\,(t-s)^\alpha \, \|f\|_{C^\alpha}$$

if $r + 1 \ge t \ge s \ge r + \frac{1}{2}$, where c does not depend on t, s, r. As a result,

$$\|v_2'(t) - v_2'(s)\| \le c\,(t-s)^\alpha \, \|f\|_{C^\alpha}$$

where c is independent of $t \ge s$ provided we take $t, s \in [r + \frac{1}{2}, r + 1]$. This can be achieved for $0 \le t - s \le 1/2$ by choosing $r = t - 1 \ge a$. Since $v_2'(t)$ is bounded for $t \ge a + 1$ due to (5.12) (for $r = t - 1$), (2.13), and (2.32), we obtain that $v_2' \in C_b^\alpha([a + 1, \infty), X)$.

Set $\tilde{v}_2(t) := v_2(t) - u_\infty$. The interpolation result [42, Prop.1.2.19] and [42, Prop.0.2.2] then imply

$$\sup_{n \le t \le n+1} \|v_2'(t)\| \le \|\tilde{v}_2\|_{C^\gamma([n,n+1],X)} \le c\,\|\tilde{v}_2\|_{C([n,n+1],X)}^{1-\theta} \, \|\tilde{v}_2\|_{C^{1+\alpha}([n,n+1],X)}^{\theta}$$

$$\le c \sup_{n \le t \le n+1} \|v_2(t) - u_\infty\|^{1-\theta} \tag{5.13}$$

for some $\gamma \in (1, 1 + \alpha)$, $\theta := \frac{\gamma}{1+\alpha}$, and constants independent of n. Thus (5.10) is true.

(3) Now assume that (b) holds. To verify again (5.10), we want to employ

$$\sup_{t \ge a+1} \|v_2'(t)\|_\beta^t < \infty. \tag{5.14}$$

This fact is an immediate consequence of (5.12) (for $r = t-1$), Proposition 2.4 and (2.34) with $\varepsilon = 1/2$. On the other hand, (5.2) implies

$$R(w, A(t))v_2'(t) = -v_2(t) + wR(w, A(t))v_2(t) + R(w, A(t))f(t).$$

The right hand side converges to 0 as $t \to \infty$ because of (5.9) and $R(w, A(t)) \to R(w, A)$. This means that

$$\lim_{t \to \infty} R(w, A(t))v_2'(t) = 0. \tag{5.15}$$

Let $\gamma \in (0, \beta)$ and $\theta = \frac{1}{1+\gamma}$. Using (2.7), (2.5), and (5.14), we derive

$$\|v_2'(t)\| \le c\,\|(w - A(t))^{-1}v_2'(t)\|^{1-\theta} \, \|(w - A(t))^\gamma v_2'(t)\|^\theta$$
$$\le c\,\|R(w, A(t))v_2'(t)\|^{1-\theta}$$

for constants c independent of t. Now (5.15) yields (5.10).

Example 5.2. One can directly apply the above theorem to the situation of Example 4.9.

5.2 Almost Periodic Equations

We assume in this paragraph that the resolvents $R(w, A(\cdot))$ and the inhomogeneity f are almost periodic (in $\mathcal{L}(X)$ and X, respectively) in the sense of the following definition due to H. Bohr, compare [8], [35], [50].

Definition 5.3. *Let Y be a Banach space. A continuous function $g : \mathbb{R} \to Y$ is* almost periodic *if for every $\epsilon > 0$ there exist a set $P(\varepsilon) \subseteq \mathbb{R}$ and a number $\ell(\varepsilon) > 0$ such that each interval $(a, a + \ell(\varepsilon))$, $a \in \mathbb{R}$, contains an* almost period $\tau = \tau_\varepsilon \in P(\varepsilon)$ *and the estimate $\|g(t + \tau) - g(t)\| \le \varepsilon$ holds for all $t \in \mathbb{R}$ and $\tau \in P(\varepsilon)$. The space of almost periodic functions is denoted by $AP(\mathbb{R}, Y)$.*

We recall that $AP(\mathbb{R}, Y)$ is a closed subspace of the space of bounded and uniformly continuous functions, see [35, Chap.1]. In addition to (AT) we make the following assumptions.

(AP) $R(w, A(\cdot)) \in AP(\mathbb{R}, \mathcal{L}(X))$ with pseudo periods $\tau = \tau_\varepsilon$ belonging to sets $P(\varepsilon, A)$.

(H)   The evolution family $U(\cdot, \cdot)$ generated by $A(\cdot)$ is hyperbolic with projections $P(t)$ and constants $N, \delta > 0$.

The almost periodicity of inhomogeneous problems has been studied by many authors in the autonomous and the periodic case, see [8], [35], [50], and the references given there and in [43], [58]. Equations with almost periodic $A(\cdot)$ are treated in, e.g., [15] and [22] for $X = \mathbb{C}^n$ and in [27] for a certain class of parabolic problems, see also [11], [35]. For general evolution families $U(\cdot, \cdot)$ (but subject to an extra condition not assumed here), it is shown in [28] that $U(\cdot, \cdot)$ has an exponential dichotomy *with* an almost periodic Green's function if and only if there is a unique almost periodic mild solution u of (5.1) for each almost periodic f, see also [15, Prop.8.3]. Our main Theorem 5.9 extends [15, Prop.8.4], [22, Thm.7.7], and [27, p.240], and complements [28, Thm.5.4] in the case of parabolic evolution equations. Theorem 5.9 is taken from [43], where we also treat problems on the time interval $\mathbb{R}_+$.

Below we first show the almost periodicity of Green's function $\Gamma(\cdot, \cdot)$ of $U(\cdot, \cdot)$ and then deduce from (5.8) that u is almost periodic. Our strategy is similar to Henry's approach in [27, §7.6] who derived the almost periodicity of $\Gamma(\cdot, \cdot)$ from a formula for $\Gamma(t + \tau, s + \tau) - \Gamma(t, s)$ (compare (5.16)). In the context of [27], this equation allows to verify almost periodicity by straightforward estimates in operator norm if τ is a pseudo period of $A(\cdot)$. However, in the present more general situation one obtains such a formula only on a subspace of X (see [43, Cor.4.3]) so that we cannot proceed in this way.

The Yosida approximations $A_n(t) = nA(t)R(n, A(t))$, $n \ge w$, allow us to overcome this difficulty. By Lemma 2.3 there is a number $\bar{n}$ such that $A_n(\cdot)$ satisfies (AT) with constants K', L', ϕ', and $w' := w + 1$ for $n \ge \bar{n}$, and thus generates an evolution family $U_n(\cdot, \cdot)$. Proposition 2.5, Theorem 3.15, and (3.22) imply that $A_n(\cdot)$ also fulfills (H) for large n and that the projections converge.

Corollary 5.4. *Let (AT) and (H) hold. Then there is a number $n_1 \geq n_0(1) \geq \bar{n}$ such that $U_n(\cdot,\cdot)$ has an exponential dichotomy for $n \geq n_1$ with constants $\delta' \in (0,\delta)$ and $N' = N'(\delta')$ independent of n. Moreover, the dichotomy projections $P_n(t)$ of $U_n(\cdot,\cdot)$ satisfy $\|P_n(t) - P(t)\| \leq c\,n^{-\theta}$ for $t \in \mathbb{R}$, where $\theta \in (0,1)$ and $n_0(1)$ are given by Proposition 2.5.*

The almost periodicity of the resolvents is inherited by the Yosida approximations, too.

Lemma 5.5. *If (AT) and (AP) hold, then $R(w', A_n(\cdot)) \in AP(\mathbb{R}, \mathcal{L}(X))$ for $n \geq \bar{n}$, with pseudo periods belonging to $P(\epsilon/\kappa, A)$, where $\kappa := (1 + (1+w)K)^2$.*

Proof. Let $\tau > 0$ and $t \in \mathbb{R}$. Equations (2.20) and (2.8) yield

$$R(w', A_n(t+\tau)) - R(w', A_n(t))$$
$$= \tfrac{n^2}{(w'+n)^2} [R(\tfrac{w'n}{w'+n}, A(t+\tau)) - R(\tfrac{w'n}{w'+n}, A(t)],$$
$$\|R(w', A_n(t+\tau)) - R(w', A_n(t))\|$$
$$\leq (1 + (1+w)K)^2 \|R(w, A(t+\tau)) - R(w, A(t))\|.$$

The assertion thus follows from (AP). $\qquad\blacksquare$

We next establish the almost periodicity of Green's function Γ_n of $U_n(\cdot,\cdot)$.

Lemma 5.6. *Assume that (AT), (AP), and (H) hold. Let $n \geq n_1$, $\eta > 0$, and $\tau \in P(\eta/\kappa, A)$, where n_1 and κ were given in Corollary 5.4 and Lemma 5.5. Then*

$$\|\Gamma_n(t+\tau, s+\tau) - \Gamma_n(t,s)\| \leq c\,\eta\,n^2\,e^{-\frac{\delta}{2}|t-s|} \quad \text{for } t, s \in \mathbb{R}.$$

Proof. The operators $\Gamma_n(t,s)$ exist by Corollary 5.4. It is easy to see that

$$g_n(\sigma) := \frac{d}{d\sigma}\Big(\Gamma_n(t,\sigma)\Gamma_n(\sigma+\tau, s+\tau)\Big)$$
$$= \Gamma_n(t,\sigma)(A_n(\sigma+\tau) - A_n(\sigma))\Gamma_n(\sigma+\tau, s+\tau)$$
$$= \Gamma_n(t,\sigma)(A_n(\sigma) - w')((A_n(\sigma) - w')^{-1} - (A_n(\sigma+\tau) - w')^{-1})\cdot$$
$$\cdot (A_n(\sigma+\tau) - w')\Gamma_n(\sigma+\tau, s+\tau),$$

for $\sigma \neq t, s$ and $n \geq n_1$. Hence,

$$\int_{\mathbb{R}} g_n(\sigma)\,d\sigma = \begin{cases} \int_{-\infty}^{s} g_n(\sigma)d\sigma + \int_{s}^{t} g_n(\sigma)d\sigma + \int_{t}^{\infty} g_n(\sigma)d\sigma, & t \geq s, \\ \int_{-\infty}^{t} g_n(\sigma)d\sigma + \int_{t}^{s} g_n(\sigma)d\sigma + \int_{s}^{\infty} g_n(\sigma)d\sigma, & t < s, \end{cases}$$

$$= \begin{cases} -\Gamma_n(t,s)Q_n(s+\tau) + P_n(t)\Gamma_n(t+\tau, s+\tau) - \Gamma_n(t,s)P_n(s+\tau) \\ \quad + Q_n(t)\Gamma_n(t+\tau, s+\tau), \quad t \geq s, \\ P_n(t)\Gamma_n(t+\tau, s+\tau) - \Gamma_n(t,s)Q_n(s+\tau) + Q_n(t)\Gamma_n(t+\tau, s+\tau) \\ \quad - \Gamma_n(t,s)P_n(s+\tau), \quad t < s, \end{cases}$$

$$= \Gamma_n(t+\tau, s+\tau) - \Gamma_n(t,s).$$

We have shown that

$$\Gamma_n(t+\tau, s+\tau) - \Gamma_n(t,s) = \int_{\mathbb{R}} \Gamma_n(t,\sigma)(A_n(\sigma) - w')\, [R(w', A_n(\sigma+\tau))$$
$$- R(w', A_n(\sigma))]\,(A_n(\sigma+\tau) - w')\Gamma_n(\sigma+\tau, s+\tau)\, d\sigma \quad (5.16)$$

for $s, t \in \mathbb{R}$ and $n \geq n_1$. Lemma 5.5 and Corollary 5.4 now yield

$$\|\Gamma_n(t+\tau, s+\tau) - \Gamma_n(t,s)\| \leq c\eta n^2 \int_{\mathbb{R}} e^{-\frac{3\delta}{4}|t-\sigma|}\, e^{-\frac{3\delta}{4}|\sigma-s|}\, d\sigma$$

for $\tau \in P(\eta/\kappa, A)$, which implies the asserted estimate.

Lemma 5.7. *Assume that (AT) and (H) hold. Fix $0 < t_0 < t_1$ and let $\theta > 0$, $n_0(t_0)$, and n_1 be given by Proposition 2.5 and Corollary 5.4. Then $\|\Gamma(t,s) - \Gamma_n(t,s)\| \leq c(t_1)\, n^{-\theta}$ holds for $t_0 \leq |t-s| \leq t_1$ and $n \geq \max\{n_0(t_0), n_1\}$.*

Proof. If $t_0 \leq t - s \leq t_1$, we write

$$\Gamma_n(t,s) - \Gamma(t,s) = (U_n(t,s) - U(t,s))P_n(s) + U(t,s)(P_n(s) - P(s)).$$

For $-t_1 \leq t - s \leq -t_0$, we have

$$\Gamma(t,s) - \Gamma_n(t,s) = U_{n,Q}(t,s)Q_n(s) - U_Q(t,s)Q(s)$$
$$= U_Q(t,s)Q(s)(Q_n(s) - Q(s)) + (Q_n(t) - Q(t))U_{n,Q}(t,s)Q_n(s)$$
$$- U_Q(t,s)Q(s)(U_n(s,t) - U(s,t))U_{n,Q}(t,s)Q_n(s).$$

In both cases the claim is a consequence of Proposition 2.5 and Corollary 5.4.

Proposition 5.8. *Assume that (AT), (AP), and (H) hold. Let $\varepsilon > 0$, $h > 0$, and $|t - s| \geq h$. Then*

$$\|\Gamma(t+\tau, s+\tau) - \Gamma(t,s)\| \leq \varepsilon\, e^{-\frac{\delta}{2}|t-s|}$$

holds for $\tau \in P(\eta/\kappa, A)$, where $\kappa = (1 + (1+w)K)^2$ and $\eta = \eta(\varepsilon, h) \to 0$ as $\varepsilon \to 0$ and h is fixed.

Proof. Let $\varepsilon > 0$ and $h > 0$ be fixed. Then there is a $t_\varepsilon > h$ such that

$$\|\Gamma(t+\tau, s+\tau) - \Gamma(t,s)\| \leq \epsilon\, e^{-\frac{\delta}{2}|t-s|}$$

for $|t - s| \geq t_\varepsilon$. Let $h \leq |t-s| \leq t_\epsilon$, $\eta > 0$, and $\tau \in P(\eta/\kappa, A)$. Lemma 5.6 and 5.7 yield

$$\|\Gamma(t+\tau, s+\tau) - \Gamma(t,s)\| \leq (c(t_\epsilon)e^{\frac{\delta}{2}t_\epsilon}\, n^{-\theta} + c\eta n^2)\, e^{-\frac{\delta}{2}|t-s|}$$

for $n \geq \max\{n_0(h), n_1\}$. We now choose first a large n and then a small $\eta > 0$ (depending on ε and h) in order to obtain the assertion.

Theorem 5.9. *Assume that (AT), (AP), and (H) hold. Then $r \mapsto \Gamma(t+r, s+r)$ belongs to $AP(\mathbb{R}, \mathcal{L}(X))$ for $t, s \in \mathbb{R}$, where we may take the same pseudo periods for t and s with $|t - s| \geq h > 0$. If $f \in AP(\mathbb{R}, X)$, then the unique bounded mild solution $u = \int_{\mathbb{R}} \Gamma(\cdot, s) f(s) \, ds$ of (5.1) is almost periodic.*

Proof. In Lemma 5.6 we have seen that $P_n(\cdot) \in AP(\mathbb{R}, \mathcal{L}(X))$. Corollary 5.4 then shows that $P(\cdot) \in AP(\mathbb{R}, \mathcal{L}(X))$. Thus the first assertion follows from Proposition 5.8. Further, for $\tau, h > 0$, $t \in \mathbb{R}$, we can write

$$
u(t + \tau) - u(t) = \int_{\mathbb{R}} \Gamma(t + \tau, s + \tau) f(s + \tau) \, ds - \int_{\mathbb{R}} \Gamma(t, s) f(s) \, ds
$$

$$
= \int_{\mathbb{R}} \Gamma(t + \tau, s + \tau)(f(s + \tau) - f(s)) \, ds
$$

$$
+ \int_{|t-s| \geq h} (\Gamma(t + \tau, s + \tau) - \Gamma(t, s)) f(s) \, ds
$$

$$
+ \int_{|t-s| \leq h} (\Gamma(t + \tau, s + \tau) - \Gamma(t, s)) f(s) \, ds.
$$

For $\bar{\varepsilon} > 0$ let $\eta = \eta(\bar{\varepsilon}, h)$ be given by Proposition 5.8. Let $P(\epsilon, A, f)$ be the set of pseudo periods for the almost periodic function $t \mapsto (f(t), R(\omega, A(t)))$, cf. [35, p.6]. Taking $\tau \in P(\eta/\kappa, A, f)$, we deduce from Proposition 5.8 and (AP) that

$$
\|u(t + \tau) - u(t)\| \leq \tfrac{2N}{\delta \kappa} \eta(\bar{\varepsilon}, h) + (\tfrac{4}{\delta} \bar{\varepsilon} + 4Nh) \|f\|_\infty.
$$

for $t \in \mathbb{R}$. Given an $\varepsilon > 0$, we can take first a small $h > 0$ and then a small $\bar{\varepsilon} > 0$ such that $\|u(t + \tau) - u(t)\| \leq \varepsilon$ for $t \in \mathbb{R}$ and $\tau \in P(\eta/\kappa, A, f) =: P(\varepsilon)$.

Remark 5.10. For $g \in AP(\mathbb{R}, Y)$ and $\lambda \in \mathbb{R}$ the means

$$
\lim_{t \to \infty} \frac{1}{2t} \int_{-t}^{t} e^{-i\lambda s} g(s) \, ds
$$

exist and they are different from zero for at most countable many λ, which are called the *frequencies* of g, see e.g. [8, §4.5], [35, §2.3]. The *module* of g is the smallest additive subgroup of $\mathbb{R}$ containing all frequencies of g. By [35, p.44] (see also [22, Thm.4.5]) and the proof of Theorem 5.9 the module of the solution u to (5.1) is contained in the joint module of f and $R(w, A(\cdot))$ (which is the smallest additive subgroup of $\mathbb{R}$ containing the frequencies of the function $t \mapsto (f(t), R(\omega, A(t)))$). Similarly, the modules of $\Gamma(t + \cdot, s + \cdot)$, $t \neq s$, are contained in those of $R(w, A(\cdot))$.

Example 5.11. In the situation of Example 2.8 we replace the time interval $\mathbb{R}_+$ by $\mathbb{R}$ and assume additionally that $p \in (1, \infty)$, $a_{kl} \in AP(\mathbb{R}, L^n(\Omega))$ and $a_0 \in AP(\mathbb{R}, L^{n/2}(\Omega))$ for $k, l = 1, \cdots, n$. (We replace $n/2$ by 1 if $n = 1$.) Using (2.40) with $\nu = 0$, one can then verify that $R(w, A(\cdot))$ is almost periodic. As a result, the mild solution of (5.1) is almost periodic if $f \in AP(\mathbb{R}, X)$ and the evolution family generated by $A(t)$ has an exponential dichotomy.

6 Convergent Solutions for a Quasilinear Equation

In this section we want to illustrate how the linear theory developed so far can be used to investigate the long term behaviour of nonlinear equations. Here we concentrate on the quasilinear autonomous problem

$$u'(t) = A(u(t))u(t) + f(u(t)), \quad t > a, \qquad u(a) = x. \tag{6.1}$$

One way to deal with such problems is to choose a suitable function space F, fix some $u \in F$ in the arguments of the nonlinear operators, and solve the corresponding inhomogeneous, non–autonomous, linear problem

$$v'(t) = A(u(t))v(t) + f(u(t)), \quad t > a, \qquad v(a) = x. \tag{6.2}$$

This procedure should yield a mapping $u \mapsto v =: \Phi(u)$ whose fixed points $\Phi(u) = u$ correspond to solutions of (6.1). Indeed, in many cases one can employ this approach to obtain local (in time) solutions of the quasilinear problem. T. Kato developed and applied this method in particular for hyperbolic problems, see e.g. [29], [31]. Local and global solvability of parabolic equations was studied by several authors within this framework in differing situations, see e.g. [4], [5], [6], [39], [40], [61], [65]. In the same context principles of linearized stability of equilibria were established in [20], [23], [24], [41], [49], and center manifolds were constructed in [60]. We also refer to the monographs [27] for semilinear and [42] for fully nonlinear problems.

We want to find convergent solutions of (6.1) assuming in particular that $\sigma(A(y)) \cap i\mathbb{R} = \emptyset$. Our arguments are based on Theorems 4.3 and 5.1. For technical reasons we have to restrict ourselves to the case of constant domains $D(A(y))$, cf. Remark 6.4. Thus we are led to the following hypotheses, where $X_1 \hookrightarrow X$ are Banach spaces, $X_\beta := (X, X_1)_{\beta,\infty}$ is the real interpolation space with exponent $\beta \in (0, 1)$ and coefficient ∞ (see e.g. [42, Chap.1]), $\|\cdot\|$ is the norm on X, and $\|\cdot\|_\beta$ is the norm on X_β for $0 < \beta \leq 1$.

(A)   There are $R > 0$ and $\alpha \in (0, 1)$ such that for $y \in X_\alpha$ with $\|y\|_\alpha \leq R$ the operators $A(y)$ are sectorial of type (ϕ, K, w), $D(A(y)) = X_1$ with uniformly equivalent graph norms, and $\|A(y) - A(z)\|_{\mathcal{L}(X_1, X)} \leq L^A \|y - z\|_\alpha$ for $\|y\|_\alpha, \|z\|_\alpha \leq R$.

(S) $i\mathbb{R} \subseteq \rho(A(y))$ and $\|R(i\tau, A(y))\| \leq \tilde{K}$ for $\|y\|_\alpha \leq R$, $\tau \in \mathbb{R}$, and a constant $\tilde{K} > 0$.

(f) $f(0) = 0$ and $\|f(y) - f(z)\| \leq L(r) \|y - z\|_\alpha$ for $\|y\|_\alpha, \|z\|_\alpha \leq r \leq R$, where $L(r) \to 0$ as $r \to 0$.

Observe that (f) implies that $\|f(y)\| \leq rL(r)$ for $\|y\|_\alpha \leq r \leq R$. We fix $\beta \in (\alpha, 1)$, denote by κ the norm of the embedding $X_\beta \hookrightarrow X_\alpha$ and set $R' := R/\kappa$. We introduce the space

$$F = \{u : \mathbb{R}_+ \to X \text{ such that } \|u(t)\|_\beta \leq R', \ \|u(t) - u_\infty\|_\alpha \leq (1 + t)^{-1},$$
$$\|u(t) - u(s)\| \leq M_2 |t - s|^\beta \text{ for } t, s \geq 0 \text{ and some } u_\infty \in X_\alpha\}$$

for a fixed number $M_2 > 0$. It is clear that $A_u(t) := A(u(t))$ are sectorial of uniform type for $u \in F$ and $t \geq 0$. Assumption (A) and the reiteration theorem [42, Thm.1.2.15] yield

$$\|A_u(t) - A_u(s)\|_{\mathcal{L}(X_1,X)} \leq L^A \|u(t) - u(s)\|_\alpha \leq L^A M_2 (R')^{\alpha/\beta} |t - s|^{\beta-\alpha}$$

for $t, s \geq 0$. Thus $A_u(t)$, $t \geq 0$, satisfy (AT) with $\nu = 1$, $\mu = \beta - \alpha$, and $L = cL^A M_2(R')^{\alpha/\beta}$. We denote by $U_u(\cdot, \cdot)$ the evolution family generated by $A_u(\cdot)$. Moreover,

$$\|R(w, A_u(t)) - R(w, A(u_\infty))\| \leq c \|A(u(t)) - A(u_\infty)\|_{\mathcal{L}(X_1,X)} \leq cL^A(1 + t)^{-1}$$

for $t \geq 0$. By (S) and Lemma 4.11, the semigroups generated by $A(y)$, $\|y\|_\alpha \leq R$, have an exponential dichotomy with common constants $\tilde{N}, \tilde{\delta} > 0$. Theorem 4.3 now shows that the evolution families $U_u(\cdot, \cdot)$, $u \in F$, have exponential dichotomies with uniform constants $N \geq 1$ and $\delta \in (0, \tilde{\delta})$ on an interval $[a, \infty)$ for some $a \geq 0$. We fix this a and some $\delta' \in (0, \delta)$, choose $M_1 > 0$ such that $M_1(1 + t) \leq e^{\delta' t}$ for $t \geq 0$, and define

$$F_r = \{u : [a, \infty) \to X \text{ such that } \|u(t)\|_\beta \leq r, \ \|u(t) - u_\infty\|_\alpha \leq M_1 e^{-\delta' t},$$
$$\|u(t) - u(s)\| \leq M_2 |t - s|^\beta \text{ for } t, s \geq a \text{ and some } u_\infty \in X_\alpha\}$$

for $r \leq R'$. Functions in F_r are extended to $\mathbb{R}$ by setting $u(t) = u(a)$ for $t \leq a$ and thus considered as elements of F. The norm

$$\|\|f\|\|_\beta = \sup_{t \geq a} \|u(t)\|_\beta$$

induces a complete metric on F_r. For a given $x \in X_\beta$ we define

$$(\Phi_x u)(t) = U_u(t, a)P_u(a)x + \int_a^\infty \Gamma_u(t, \tau) f(u(\tau)) \, d\tau$$

for $t \geq a$ and $u \in F_{R'}$. Here $P_u(t)$ and $\Gamma_u(\cdot, \cdot)$ denote the projections and Green's function of $U_u(\cdot, \cdot)$. The following two lemmas provide the estimates establishing that Φ is a strict contraction on F_r for small $r > 0$ and $\|x\|_\beta$.

Lemma 6.1. Let (A) and (S) hold. Fix β, a, and δ' as above. Let $u, v \in F_{R'}$ and $t, s \geq a$. Then there is a constant c independent of u, v, t, s such that

$$\|\Gamma_u(t, s) - \Gamma_v(t, s)\|_{\mathcal{L}(X_\beta)} \leq c e^{-\delta'|t-s|} \sup_{\tau \geq a} \|u(\tau) - v(\tau)\|_\alpha.$$

Proof. Recall that we have extended $u(t)$ constantly to $\mathbb{R}$. As seen in the proof of Theorem 4.3, the operators $A_u(t)$, $t \in \mathbb{R}$, generate an evolution family $U_u(\cdot, \cdot)$ having an exponential dichotomy on $J = \mathbb{R}$, with Green's function $\Gamma_u(t, s)$. Moreover, the corresponding projections $P_u(t)$ are Hölder continuous in t. Take $u, v \in F_{R'}$. Employing [42, Prop.6.2.6], the formula

$U_{Q,u}(t,\tau)Q_u(\tau) = U_{Q,u}(t,t')Q_u(t')U_u(t',\tau)$ for $t < \tau < t'$, and Proposition 3.18, we conclude that

$$\tfrac{\partial}{\partial\tau}\,\Gamma_u(t,\tau)\Gamma_v(\tau,s)x = \Gamma_u(t,\tau)\left[A_v(\tau) - A_u(\tau)\right]\Gamma_v(\tau,s)x$$

for $\tau \neq t, s$. Because of (2.13) and Proposition 3.18, the right hand side of this equality is integrable in $\tau \in \mathbb{R}$ if $x \in X_\beta$. Moreover, $\tau \mapsto \Gamma_u(t,\tau)\Gamma_v(\tau,s)x$ has one-sided limits as $\tau \to t, s$ due to (3.2), $x \in X_\beta \subseteq \overline{X_1}$, and the continuity of $P_u(\cdot)$. Thus we obtain as in Lemma 5.7 that

$$\Gamma_v(t,s)x - \Gamma_u(t,s)x = \int_{\mathbb{R}} \Gamma_u(t,\tau)\left[A_v(\tau) - A_u(\tau)\right]\Gamma_v(\tau,s)x\,d\tau \qquad (6.3)$$

for $t, s \in \mathbb{R}$ and $x \in X_\beta$. Using Proposition 3.18, (2.13), and (A), we estimate

$$\|\Gamma_u(t,s)x - \Gamma_v(t,s)x\|_\beta$$
$$\leq c\int_{\mathbb{R}} \lceil t-\tau\rceil^{-\beta}e^{-\delta|t-\tau|}\,\|u(\tau) - v(\tau)\|_\alpha \,\lceil\tau - s\rceil^{\beta-1}e^{-\delta|\tau - s|}\|x\|_\beta\,d\tau$$
$$\leq c\,e^{-\delta'|t-s|}\|x\|_\beta \sup_{\tau\geq a}\|u(\tau) - v(\tau)\|_\alpha$$

where $\lceil t\rceil := t$ if $0 < t \leq 1$ and $\lceil t\rceil := 1$ if $t \leq 0$ or $t \geq 1$.

Lemma 6.2. *Let (A), (S), and (f) hold. Fix β, a, and δ' as above. Take $t \geq a$ and $u, v \in F_r$ with $r \leq R'$. Then there are constants c independent of u, v, t, r such that*

$$\left\|\int_a^\infty \left(\Gamma_u(t,\tau)f(u(\tau)) - \Gamma_v(t,\tau)f(v(\tau))\right)d\tau\right\|_\beta \leq cL(r)\sup_{\tau\in\mathbb{R}}\|u(\tau) - v(\tau)\|_\alpha$$

$$\left\|\int_a^\infty \left(\Gamma_u(t,\tau)f(u(\tau)) - \Gamma_v(t,\tau)f(v(\tau))\right)d\tau\right\|_\beta$$
$$\leq cL(r)e^{-\delta't}\left(r + \sup_{\tau\geq a}e^{\delta'\tau}\|u(\tau) - v(\tau)\|_\alpha\right).$$

Proof. We only show the second assertion since the first one can be treated similarly. Let $t \geq a$ and $u, v \in F_r$. We write

$$\int_a^\infty e^{\delta't}\left(\Gamma_u(t,\tau)f(u(\tau)) - \Gamma_v(t,\tau)f(v(\tau))\right)d\tau$$
$$= \int_a^\infty e^{\delta'(t-\tau)}\Gamma_u(t,\tau)\,e^{\delta'\tau}(f(u(\tau)) - f(v(\tau)))\,d\tau$$
$$+ \int_a^\infty e^{\delta't}\left(\Gamma_u(t,\tau) - \Gamma_v(t,\tau)\right)f(v(\tau))\,d\tau =: I_1 + I_2.$$

We estimate $\|I_1\|_\beta \leq cL(r)\sup_{\tau\geq a}e^{\delta'\tau}\|u(\tau) - v(\tau)\|_\alpha$ using Proposition 3.18, (2.13), and (f). In order to deal with I_2, we employ as in Lemma 5.6 the Yosida approximations $A_n(u(t))$ of $A(u(t))$ for $t \in \mathbb{R}$ and $n \geq \bar{n}$. As in Corollary 5.4

and Lemma 5.7, we see that the evolution families $U_{u,n}(\cdot,\cdot)$ generated by $A_n(u(t))$ have exponential dichotomies with uniform constants for sufficiently large n and that the corresponding Green's functions $\Gamma_{u,n}(t,s)$ converge in operator norm to $\Gamma_u(t,s)$ as $n \to \infty$, for $t,s \in \mathbb{R}$. Arguing as in (5.16), we deduce

$$
\begin{aligned}
I_2 &= \lim_{n\to\infty} \int_a^\infty e^{\delta' t}\left(\Gamma_{u,n}(t,\tau) - \Gamma_{v,n}(t,\tau)\right)f(v(\tau))\,d\tau \\
&= -\lim_{n\to\infty}\left(\int_a^\infty \int_a^\infty e^{\delta'(t-\sigma)}\Gamma_{u,n}(t,\sigma)\,e^{\delta'\sigma}[A_n(v(\sigma)) - A_n(u(\sigma))]\,\Gamma_{v,n}(\sigma,\tau)\right.\\
&\qquad\left. \cdot f(v(\tau))\,d\sigma\,d\tau + \int_a^\infty e^{\delta' t}\,\Gamma_{u,n}(t,a)\Gamma_{v,n}(a,\tau)f(v(\tau))\,d\tau\right)\\
&=: -\lim_{n\to\infty}\left(I_{21}^n + I_{22}^n\right).
\end{aligned}
$$

It is easy to see that $\|I_{22}^n\|_\beta \le crL(r)$ due to Proposition 3.18, (2.13), and (f). Interchanging the order of integration in I_{21}^n yields

$$
\begin{aligned}
I_{21}^n &= \int_a^\infty e^{\delta'(t-\sigma)}\Gamma_{u,n}(t,\sigma)\,e^{\delta'\sigma}\,[A_n(v(\sigma)) - A_n(u(\sigma))]\\
&\qquad \cdot \int_a^\infty \Gamma_{v,n}(\sigma,\tau)f(v(\tau))\,d\tau\,d\sigma,\\
&= \int_a^\infty e^{\delta'(t-\sigma)}\Gamma_{u,n}(t,\sigma)\,nR(n, A(u(\sigma)))\,e^{\delta'\sigma}\,[A(v(\sigma)) - A(u(\sigma))]\\
&\qquad \cdot R(w, A(v(\sigma)))n(w - A(v(\sigma)))R(n, A(v(\sigma)))\int_a^\infty \Gamma_{v,n}(\sigma,\tau)f(v(\tau))\,d\tau\,d\sigma,\\
\|I_{21}^n\|_\beta &\le c\sup_{\sigma \ge a}\left\{e^{\delta'\sigma}\|u(\sigma) - v(\sigma)\|_\alpha\left(\left\|\int_a^\infty \Gamma_{v,n}(\sigma,\tau)f(v(\tau))\,d\tau\right\|\right.\right.\\
&\qquad\left.\left. + \left\|A_n(v(\sigma))\int_a^\infty \Gamma_{v,n}(\sigma,\tau)f(v(\tau))\,d\tau\right\|\right)\right\} \qquad (6.4)
\end{aligned}
$$

by Proposition 3.18 and (A). The integral in (6.4) is equal to

$$
\begin{aligned}
&P_{v,n}(\sigma)\int_{(\sigma-1)\vee a}^\sigma U_{v,n}(\sigma,\tau)f(v(\tau))\,d\tau\\
&+ U_{v,n}(\sigma,\sigma-1)\int_a^{(\sigma-1)\vee a} U_{v,n}(\sigma-1,\tau)P_{v,n}(\tau)f(v(\tau))\,d\tau\\
&- U_{v,n}(\sigma,\sigma-1)\int_\sigma^\infty U_{v,n,Q}(\sigma-1,\tau)Q_{v,n}(\tau)f(v(\tau))\,d\tau.
\end{aligned}
$$

Using Proposition 3.18, [42, Lem.6.2.1] or [65, Thm.2.4], and (2.13), we conclude that

$$
\|I_{21}^n\|_\beta \le c\,\|f(v)\|_{C_b^{\beta-\alpha}([a,\infty),X)}\sup_{\tau \ge a} e^{\delta'\tau}\|u(\tau) - v(\tau)\|_\alpha.
$$

Finally, assumption (f) and the reiteration theorem, [42, Thm.1.2.15], give

$$\|f(v(\tau)) - f(v(\tau'))\| \le L(r) \|v(\tau) - v(\tau')\|_\alpha$$
$$\le L(r) \|v(\tau) - v(\tau')\|_\beta^{\alpha/\beta} \|v(\tau) - v(\tau')\|^{1-\alpha/\beta}$$
$$\le c\, r^{\alpha/\beta} L(r) |\tau - \tau'|^{\beta-\alpha} \tag{6.5}$$

for $\tau, \tau' \ge a$. The second assertion is verified combining the above estimates.

In the next theorem we establish the existence and uniqueness of convergent solutions to (6.1). Recall from Paragraph 5.1 that we can only prescribe the stable part of the initial value of a bounded mild solution to (6.2).

Theorem 6.3. *Assume that (A), (S), and (f) hold. Fix $\beta \in (\alpha, 1)$ and the initial time $a > 0$ as obtained in Theorem 4.3. Then there are numbers $r_0, \rho_0 > 0$ such that for $x \in X_\beta$ with $\|x\|_\beta \le \rho_0$ there is a unique $u \in F_{r_0} \cap C^1((a, \infty), X) \cap C((a, \infty), X_1)$ solving*

$$u'(t) = A(u(t))u(t) + f(u(t)), \qquad t > a, \tag{6.6}$$

such that $P_u(a)u(a) = P_u(a)x$. Moreover,

$$u(a) = P_u(a)x - \int_a^\infty U_{u,Q}(0, \tau) Q_u(\tau) f(u(\tau))\, d\tau, \tag{6.7}$$

$\|u(t) - u_\infty\|_\alpha \le M_1 e^{-\delta' t}$ for $t \ge a$, the above chosen $\delta', M_1 > 0$, and $u_\infty \in X_1$ satisfying $A(u_\infty)u_\infty = -f(u_\infty)$, and $u'(t) \to 0$ and $A(u(t))u(t) \to A(u_\infty)u_\infty$ in X as $t \to \infty$.

Proof. To construct a solution, we employ the spaces F_r and the operator Φ_x introduced above. Let $u \in F_r$ and $\|x\|_\beta \le \rho$. Then

$$\|(\Phi_x u)(t)\|_\beta \le c\|x\|_\beta + c\int_a^\infty \lceil t - \tau \rceil^{-\beta} e^{-\delta|t-\tau|} \|f(u(\tau))\|\, d\tau$$
$$\le c\,(\rho + rL(r)) \tag{6.8}$$

for $t \ge a$ by Proposition 3.18, (2.13), and (f). For $t' \ge t \ge a$ we have

$$(\Phi_x u)(t') - (\Phi_x u)(t) = (U_u(t', t) - I)(\Phi u)(t) + \int_t^{t'} U_u(t, \tau) f(u(\tau))\, d\tau.$$

Thus (2.16) and (6.8) yield

$$\|(\Phi_x u)(t') - (\Phi_x u)(t)\| \le c\,(\rho + rL(r)) |t' - t|^\beta. \tag{6.9}$$

For the function $v \equiv u_\infty$, we calculate

$$e^{\delta' t} \Phi_x u(t) - e^{\delta' t} R(0, A(u_\infty)) f(u_\infty)$$
$$= e^{\delta' t} U_u(t, a) P_u(a) x + e^{\delta' t} \int_a^\infty (\Gamma_u(t, \tau) f(u(\tau)) - \Gamma_{u_\infty}(t - \tau) f(u_\infty)) d\tau$$
$$- e^{\delta' t} e^{(t-a) A(u_\infty)} P_{u_\infty} \int_{-\infty}^a e^{(a-\tau) A(u_\infty)} P_{u_\infty} f(u_\infty) \, d\tau.$$

Due to Proposition 3.18, (2.13), Lemma 6.2, and (f) this identity leads to the estimate

$$e^{\delta' t} \| \Phi_x u(t) - R(0, A(u_\infty)) f(u_\infty) \|_\alpha$$
$$\leq c \|x\|_\alpha + cL(r)(r + \sup_{\tau \geq a} e^{\delta' \tau} \|u(\tau) - u_\infty\|_\alpha) + crL(r)$$
$$\leq c (\rho + rL(r) + L(r) M_1). \tag{6.10}$$

The above constants do not depend on u, v, t, t', r, ρ. In view of (6.8), (6.9), and (6.10), we may choose $r_1 > 0$ and $\rho_1 > 0$ such that Φ_x maps F_r into F_r if $r \leq r_1$ and $\|x\|_\beta \leq \rho_1$. Using Lemmas 6.1 and 6.2 and (2.5), we find $r_0 \in (0, r_1]$ and $\rho_0 \in (0, \rho_1]$ such that $\Phi_x : F_r \to F_r$ is a strict contraction for $\| \cdot \|_\beta$ if $r \leq r_0$ and $\|x\|_\beta \leq \rho_0$. The resulting unique fix point $\Phi_x u = u \in F_r$ has the asserted properties due to Theorem 5.1 and (6.5).

Remark 6.4. We have not striven for optimal regularity results in the above theorem. In principle, our approach can be extended to time varying domains in the framework of the Acquistapace–Terreni conditions, cf. [23], [24], [65]. However, since the operator Φ changes the value of the function at $t = a$, some difficulties with compatibility conditions arise which lead to unsatisfactory restrictions on x. For this reason and to obtain a simpler exposition, we have chosen the present setting.

Example 6.5. We consider the quasilinear initial–boundary value problem

$$D_t u(t, x) = A(x, u(t, x), \nabla u(t, x), D) u(t, x) + g(x, u(t, x), \nabla u(t, x)), \tag{6.11}$$
$$t > a, \ x \in \Omega,$$
$$u(t, x) = 0, \quad t > a, \ x \in \partial\Omega,$$
$$u(a, \cdot) = u_0,$$

on a bounded domain $\Omega \subseteq \mathbb{R}^n$ with boundary $\partial\Omega$ of class C^2, where

$$A(x, \sigma, \tau, D) = \sum_{kl} a_{kl}(x, \sigma, \tau) D_k D_l + \sum_k a_k(x, \sigma, \tau) D_k + a_0(x, \sigma, \tau).$$

It is assumed that the coefficients $a_{kl} = a_{lk}, a_k, a_0, g$ belong to $C(\overline{\Omega} \times [\tilde{R}, \tilde{R}] \times \overline{B}(0, \tilde{R}))$, are real–valued, $g(x, 0, 0) = 0$, and

$$\sum_{k,l=1}^n a_{kl}(x, \sigma, \tau) v_k \, v_l \geq \eta \, |v|^2$$

for $k, l = 1, \cdots, n$, constants $\tilde{R}, \eta > 0$, and $x \in \overline{\Omega}$, $\sigma \in [\tilde{R}, \tilde{R}]$, $\tau \in \overline{B}(0, \tilde{R})$, $v \in \mathbb{R}^n$. Here $\overline{B}(0, \tilde{R})$ is the closed ball in $\mathbb{R}^n$ with radius $\tilde{R}$ and center 0. We further require that

$$|a_\gamma(x, \sigma, \tau) - a_\gamma(x, \sigma', \tau')| \leq \tilde{L}(|\sigma - \sigma'| + |\tau - \tau'|)$$
$$|g(x, \sigma, \tau) - g(x, \sigma', \tau')| \leq \tilde{L}(r)(|\sigma - \sigma'| + |\tau - \tau'|)$$

for $x \in \overline{\Omega}$ and $|\sigma|, |\sigma'|, |\tau|, |\tau'| \leq r \leq \tilde{R}$, where $\tilde{L}(r) \to 0$ as $r \to 0$ and $a_\gamma \in \{a_{kl}, a_k, a_0 : k, l = 1, \cdots, n\}$. Fix $p > n$ and $\alpha \in (1/2 + n/(2p), 1)$. We set $X = L^p(\Omega)$ and $X_1 = W^{2,p}(\Omega) \cap W_0^{1,p}(\Omega)$. Then $X_\alpha \hookrightarrow C^{1+\varepsilon}(\overline{\Omega})$ for some $\varepsilon \in (0, 1)$ due to e.g. [27, Thm.1.6.1] and (2.5), where we denote the norm of this embedding by $d > 0$. Thus we can define

$$A(\psi)\varphi = A(\cdot, \psi(\cdot), \nabla\psi(\cdot))\varphi \text{ with } D(A(\psi)) = X_1, \quad f(\psi) = g(\cdot, \psi(\cdot), \nabla\psi(\cdot)),$$

for $\psi \in X_\alpha$ with $\|\psi\|_\alpha \leq R := d\tilde{R}$. It is then straightforward to check that (A) and (f) hold using standard elliptic regularity, cf. [42], [62], [63]. If also (S) is valid, we can fix an initial time a in (6.11) as obtained in Theorem 4.3. Theorem 6.3 then shows the existence of converging solutions to (6.11) for suitable initial values u_0.

References

1. P. Acquistapace, *Evolution operators and strong solutions of abstract linear parabolic equations,* Differential Integral Equations **1** (1988), 433–457.
2. P. Acquistapace, B. Terreni, *A unified approach to abstract linear nonautonomous parabolic equations,* Rend. Sem. Mat. Univ. Padova **78** (1987), 47–107.
3. P. Acquistapace, B. Terreni, *Regularity properties of the evolution operator for abstract linear parabolic equations,* Differential Integral Equations, 5 (1992), 1151–1184.
4. H. Amann, *Dynamic theory of quasilinear parabolic equations – I. Abstract evolution equations,* Nonlinear Anal. **12** (1988), 895–919.
5. H. Amann, *Dynamic theory of quasilinear parabolic equations – II. Reaction-diffusion systems,* Differential Integral Equations **3** (1990), 13–75.
6. H. Amann, *Dynamic theory of quasilinear parabolic equations – III. Global Existence,* Math. Z. **202** (1989), 219–250.
7. H. Amann, *Linear and Quasilinear Parabolic Problems. Volume 1: Abstract Linear Theory,* Birkhäuser, 1995.
8. W. Arendt, C.J.K. Batty, M. Hieber, F. Neubrander, *Vector Valued Laplace Transforms and Cauchy Problems,* Birkhäuser, 2001.
9. A.G. Baskakov, *Some conditions for invertibility of linear differential and difference operators,* Russian Acad. Sci. Dokl. Math. **48** (1994), 498–501.
10. A.G. Baskakov, *Semigroups of difference operators in spectral analysis of linear differential operators,* Funct. Anal. Appl. **30** (1996), 149–157.
11. C.J.K. Batty, R. Chill, *Approximation and asymptotic behaviour of evolution families,* Differential Integral Equations **15** (2002), 477–512.

12. C. Chicone, Y. Latushkin, *Evolution Semigroups in Dynamical Systems and Differential Equations*, Amer. Math. Soc., 1999.

13. S.-N. Chow, H. Leiva, *Existence and roughness of the exponential dichotomy for skew product semiflow in Banach space*, J. Differential Equations **120** (1995), 429–477.

14. S.-N. Chow, H. Leiva, *Unbounded perturbations of the exponential dichotomy for evolution equations*, J. Differential Equations **129** (1996), 509–531.

15. W.A. Coppel, *Dichotomies in Stability Theory*, Springer–Verlag, 1978.

16. J.L. Daleckij, M.G. Krein, *Stability of Solutions of Differential Equations in Banach Spaces*, Amer. Math. Soc., 1974.

17. D. Daners, P. Koch Medina, *Abstract Evolution Equations, Periodic Problems and Applications*, Longman, 1992.

18. E.B. Davies, *Spectral Theory and Differential Operators*, Cambridge University Press, 1995.

19. R. Denk, M. Hieber, J. Prüss, *R-Boundedness, Fourier Multipliers and Problems of Elliptic and Parabolic Type*, to appear in Memoirs Amer. Math. Soc..

20. A.-K. Drangeid, *The principle of linearized stability for quasilinear parabolic evolution equations*, Nonlinear Anal. **13** (1989), 1091–1113.

21. K.J. Engel, R. Nagel, *One–Parameter Semigroups for Linear Evolution Equations*, Springer–Verlag, 2000.

22. A.M. Fink, *Almost Periodic Differential Equations*, Springer–Verlag, 1974.

23. K. Furuya, A. Yagi, *Linearized stability for abstract quasilinear evolution equations of parabolic type*, Funkcial. Ekvac. **37** (1994), 483–504.

24. K. Furuya, A. Yagi, *Linearized stability for abstract quasilinear evolution equations of parabolic type II, time non homogeneous case*, Adv. Math. Sci. Appl. **3** (1993/94), 285–300.

25. T. Graser, *Operator multipliers, strongly continuous semigroups, and abstract extrapolation spaces*, Semigroup Forum **55** (1997), 68–79.

26. D. Guidetti, *On the asymptotic behavior of solutions of linear nonautonomous parabolic equations*, Boll. Un. Mat. Ital. B (7) **1** (1987), 1055–1076.

27. D. Henry, *Geometric Theory of Semilinear Parabolic Equations*, Springer–Verlag, 1981.

28. W. Hutter, F. Räbiger, *Spectral mapping theorems for evolution semigroups on spaces of almost periodic functions*, preprint.

29. T. Kato, *Quasilinear equations of evolution, with applications to partial differential equations*, in: N.V. Everitt (Ed.), "Spectral Theory and Differential Equations" (Proceedings Dundee 1974), Springer–Verlag, 1975, pp. 25–70.

30. T. Kato, *Perturbation Theory for Linear Operators*, Springer–Verlag, 1980.

31. T. Kato, *Abstract evolution equations, linear and quasi–linear, revisited*, in: H. Komatsu (Ed.), "Functional Analysis and Related Topics" (Proc. Kyoto 1991), Springer–Verlag, 1993, pp. 103–125.

32. S.G. Krein, *Linear Differential Equations in Banach Spaces*, Amer. Math. Soc., 1971.

33. Y. Latushkin, S. Montgomery–Smith, *Evolutionary semigroups and Lyapunov theorems in Banach spaces*, J. Funct. Anal. **127** (1995), 173–197.

34. Y. Latushkin, T. Randolph, *Dichotomy of differential equations on Banach spaces and an algebra of weighted translation operators*, Integral Equations Operator Theory **23** (1995), 472–500.

35. B.M. Levitan, V.V. Zhikov, *Almost Periodic Functions and Differential Equations*, Cambridge University Press, 1982.

36. X.B. Lin, *Exponential dichotomies and homoclinic orbits in functional differential equations,* J. Differential Equations **63** (1986), 227–254.

37. X.B. Lin, *Exponential dichotomies in intermediate spaces with applications to a diffusively perturbed predator–prey model,* J. Differential Equations **108** (1994), 36–63.

38. M.P. Lizana, *Exponential dichotomy of singularly perturbed linear functional differential equations with small delays,* Appl. Anal. **47** (1992), 213–225.

39. A. Lunardi, *Abstract quasilinear parabolic equations,* Math. Ann. **267** (1984), 395–415.

40. A. Lunardi, *Global solutions of abstract quasilinear parabolic equations,* J. Differential Equations **58** (1985), 228–242.

41. A. Lunardi, *Asymptotic exponential stability in quasilinear parabolic equations,* Nonlinear Anal. **9** (1985), 563–586.

42. A. Lunardi, *Analytic Semigroups and Optimal Regularity in Parabolic Problems,* Birkhäuser, 1995.

43. L. Maniar, R. Schnaubelt, *Almost periodicity of inhomogeneous parabolic evolution equations,* to appear in: G. Ruiz Goldstein, R. Nagel, S. Romanelli (Eds.): "Recent contributions to evolution equations", Dekker.

44. A. Marchesi, *Exponential dichotomy and strong solutions for abstract parabolic non–autonomous equations,* Commun. Appl. Anal. **6** (2002), 147–162. (1997).

45. S. Monniaux, J. Prüss, *A theorem of Dore–Venni type for non–commuting operators,* Trans. Amer. Math. Soc. **349** (1997), 4787–4814.

46. D.G. Obert, *Examples and bounds in the general case of exponential dichotomy roughness,* J. Math. Anal. Appl. **174** (1993), 231–241.

47. A. Pazy, *Semigroups of Linear Operators and Applications to Partial Differential Equations,* Springer–Verlag, 1983.

48. V.A. Pliss, G.R. Sell, *Robustness of exponential dichotomies in infinite-dimensional dynamical systems,* J. Dynam. Differential Equations **11** (1999), 471–514.

49. M. Potier–Ferry, *The linearization principle for the stability of solutions of quasilinear parabolic equations. I.,* Arch. Rational Mech. Anal. **77** (1981), 301–320.

50. J. Prüss, *Evolutionary Integral Equations and Applications,* Birkhäuser, 1993.

51. F. Räbiger, R. Schnaubelt, *The spectral mapping theorem for evolution semigroups on spaces of vector–valued functions,* Semigroup Forum **52** (1996), 225–239.

52. R. Rau, *Hyperbolic evolution semigroups on vector valued function spaces,* Semigroup Forum **48** (1994), 107–118.

53. W. Rudin, *Functional Analysis,* McGraw–Hill, 1989.

54. R.J. Sacker, G.R. Sell, *Dichotomies for linear evolutionary equations in Banach spaces,* J. Differential Equations **113** (1994), 17–67.

55. R. Schnaubelt, *Sufficient conditions for exponential stability and dichotomy of evolution equations,* Forum Math. **11** (1999), 543–566.

56. R. Schnaubelt, *A sufficient condition for exponential dichotomy of parabolic evolution equations,* in: G. Lumer, L. Weis (Eds.), "Evolution Equations and their Applications in Physical and Life Sciences" (Proceedings Bad Herrenalb, 1998), Marcel Dekker, 2000, pp. 149–158.

57. R. Schnaubelt, *Asymptotically autonomous parabolic evolution equations,* J. Evol. Equ. **1** (2001), 19–37.

58. R. Schnaubelt, *Well–posedness and asymptotic behaviour of non–autonomous linear evolution equations,* in: Progress in Nonlinear Differential Equations and their Applications **50**, Birkhäuser, 2002, pp. 311–338.

59. R. Schnaubelt, *Parabolic evolution equations with asymptotically autonomous delay,* Report No.2, Fachbereich Mathematik und Informatik, Universität Halle, 2001.

60. G. Simonett, *Center manifolds for quasilinear reaction–diffusion systems,* Differential Integral Equations **8** (1995), 753–796.

61. P.E. Sobolevskii, *Equations of parabolic type in a Banach space,* Amer. Math. Soc. Transl. **49** (1966), 1-62. (Russian 1961)

62. H. Tanabe, *Equations of Evolution,* Pitman, 1979.

63. H. Tanabe, *Functional Analytic Methods for Partial Differential Equations,* Marcel Dekker, 1997.

64. A. Yagi, *Parabolic equations in which the coefficients are generators of infinitely differentiable semigroups II,* Funkcial. Ekvac. **33** (1990), 139–150.

65. A. Yagi, *Abstract quasilinear evolution equations of parabolic type in Banach spaces,* Boll. Un. Mat. Ital. B (7) **5** (1991), 351–368.

Printing and Binding: Strauss GmbH, Mörlenbach

4. Manuscripts should in general be submitted in English. Final manuscripts should contain at least 100 pages of mathematical text and should always include
   - a general table of contents;
   - an informative introduction, with adequate motivation and perhaps some historical remarks: it should be accessible to a reader not intimately familiar with the topic treated;
   - a global subject index: as a rule this is genuinely helpful for the reader.

5. Lecture Notes volumes are, as a rule, printed digitally from the authors' files. We strongly recommend that all contributions in a volume be written in the same LaTeX version, preferably LaTeX2e. To ensure best results, authors are asked to use the LaTeX2e style files available from Springer's web-pages at

   www.springeronline.com

   [on this page, click on <Mathematics>, then on <For Authors> and look for <Macro Packages for books>]. Macros in LaTeX2.09 and TeX are available on request from: lnm@springer.de. Careful preparation of the manuscripts will help keep production time short besides ensuring satisfactory appearance of the finished book in print and online. After acceptance of the manuscript authors will be asked to prepare the final LaTeX source files (and also the corresponding dvi-, pdf- or zipped ps-file) together with the final printout made from these files. The LaTeX source files are essential for producing the full-text online version of the book. For the existing online volumes of LNM see: http://www.springerlink.com .
   The actual production of a Lecture Notes volume takes approximately 8 weeks.

6. Volume editors receive a total of 50 free copies of their volume to be shared with the authors, but no royalties. They and the authors are entitled to a discount of 33.3 % on the price of Springer books purchased for their personal use, if ordering directly from Springer.

7. Commitment to publish is made by letter of intent rather than by signing a formal contract. Springer-Verlag secures the copyright for each volume. Authors are free to reuse material contained in their LNM volumes in later publications: A brief written (or e-mail) request for formal permission is sufficient.

Addresses:

Professor J.-M. Morel, CMLA,
École Normale Supérieure de Cachan,
61 Avenue du Président Wilson, 94235 Cachan Cedex, France
E-mail: Jean-Michel.Morel@cmla.ens-cachan.fr

Professor F. Takens, Mathematisch Instituut,
Rijksuniversiteit Groningen, Postbus 800,
9700 AV Groningen, The Netherlands
E-mail: F.Takens@math.rug.nl

Professor B. Teissier, Université Paris 7
Institut Mathématique de Jussieu, UMR 7586 du CNRS
Équipe "Géométrie et Dynamique", 175 rue du Chevaleret
75013 Paris, France
E-mail: teissier@math.jussieu.fr

Springer-Verlag, Mathematics Editorial, Tiergartenstr. 17,
69121 Heidelberg, Germany,
Tel.: +49 (6221) 487-8410
Fax: +49 (6221) 487-8355
E-mail: lnm@springer.de